XXI International Conference on
DIFFERENTIAL GEOMETRIC METHODS IN THEORETICAL PHYSICS

Proceedings Supplements
International Journal of Modern Physics A (1993)

Published volumes:

Physics and Experiments with Linear Colliders
Saariselkä, Finland, September 9–14, 1991 (ed. R Orava, P Eerola & M Nordberg)

XVth International Conference on High Energy Accelerators
Hamburg, Germany, July 20–24, 1992 (ed. J. Rossbach)

Forthcoming volumes:

International Workshop on Theoretical Physics
String Quantum Gravity and Physics at the Planck Energy Scale
Erice, Sicily, June 21–28, 1992 (ed. N Sanchez)

Batavia Meeting
1992 Meeting of the Division of Particles and Fields
of the American Physical Society
Batavia, Illinois, USA, October 13–17, 1992 (ed. R Raja & J Yoh)

Trieste Workshop on the Search for New Elementary Particles:
Status and Prospects
Trieste, Italy, May 20–22, 1992 (ed. G. Herten, L. Beers & M. L. Perl)

Subscription Information

For 1993 there will be five volumes (ten issues). The total subscription price is US$580. The special price of US$470 is available to subscribers of the International Journal of Modern Physics A, D and E. For postage and handling add US$80. Subscription orders may be sent to any of the following World Scientific offices:

USA: World Scientific Publishing Co. Inc., Suite 1B,
1060 Main Street, River Edge, NJ 07661, USA

UK: World Scientific Publishing Co., 73 Lynton Mead,
Totteridge, London, N20 8DH, England

Singapore: World Scientific Publishing Co. Pte. Ltd.
Farrer Road, P O Box 128, Singapore 9128

Int. J. Mod. Phys. A (Proc. Suppl.) 3A (1993)

Proceedings of the

XXI INTERNATIONAL CONFERENCE ON

DIFFERENTIAL GEOMETRIC METHODS IN THEORETICAL PHYSICS

Tianjin, China *5 – 9 June 1992*

Editors

C. N. YANG
State University of New York
Stony Brook, USA

M. L. GE
X. W. ZHOU
Nankai Institute of Mathematics
Tianjin, China

World Scientific
Singapore • New Jersey • London • Hong Kong

Published by

World Scientific Publishing Co. Pte. Ltd.

5 Toh Tuck Link, Singapore 596224

USA office: 27 Warren Street, Suite 401-402, Hackensack, NJ 07601

UK office: 57 Shelton Street, Covent Garden, London WC2H 9HE

British Library Cataloguing-in-Publication Data
A catalogue record for this book is available from the British Library.

XXI International Conference on
DIFFERENTIAL GEOMETRIC METHODS IN THEORETICAL PHYSICS

ISBN-13 978-981-02-1210-0
ISBN-10 981-02-1210-0

21st International Conference on
Differential Geometric Methods in Theoretical Physics
June 5–9, 1992

organized by
Nankai Institute of Mathematics
Nankai University, Tianjin, People's Republic of China

co-organized by
National Natural Science Foundation of China

International Advisory Committee:
S. S. Chern (Berkeley), K. Bleuler (Bonn), W. Nahm (Bonn)

Local Organizing Committee:
M. L. Ge (Secretary)
G. Z. He (Secretary)
B. Y. Hou, G. D. Hu
C. N. Yang

Programme Coordinators
Y. M. Cho, X. W. Zhou

The organizers acknowledge the generous support of
Municipal Government of Tianjin
Wei Lun Foundation Ltd., Hong Kong
K. C. Wong Education Foundation Ltd., Hong Kong
Prof. K. Bleuler, Bonn
World Scientific Publishing Co. Pte. Ltd., Singapore
Southeast Asia Theoretical Physics Association, Singapore
ICTP, Trieste

Professor Bleuler was born on Sept. 23rd, 1912, in Herzogenbuchsee, Switzerland, to a family of economically successful engineers. In 1931 he began to study engineering, but soon switched to physics, where he learned about Riemannian geometry and Einstein's gravitational theory. He was strongly influenced by Pauli and by the mathematician Heinz Hopf. In 1942 he received his Ph.D. in mathematics under Polya, and subsequently worked in Geneva and Zürich, where he obtained a titular professorship in 1945.

In 1950, he and S. Gupta published what became known as the Gupta-Bleuler formalism of quantum electrodynamics. In 1957 he became ordinary professor at the university of Neuchâtel where he introduced Pauli to the writer

Dürrenmatt. Their intensive discussions resulted in Dürrenmatt's piece "the physicists", which included a character based on Bleuler. In 1959 he received offers from Freiburg and Bonn, and chose the latter. In the same year he married Tinette Specogna. He later founded the Institute for Theoretical Nuclear Physics in Bonn.

In discussions with physicists and mathematicians, and in particular with R. Nevanlinna, he had the idea of creating this series of conferences, which started in 1971 in an almost private setting next to his home and continued every year without interruption. He died on January 1st, 1992, leaving two children and three grandchildren.

W. Nahm

Preface

The XXI International Conference on Differential Geometric Methods in Theoretical Physics (XXI DGMTP) was organized by Nankai Institute of Mathematics and co-organized by National Science Foundation of China from 5 to 9 June, 1992 at Nankai University, Tianjin, China.

It was Prof. K. Bleuler, a pioneer of DGMTP, and Prof. W. Nahm who first suggested that this conference be held at Nankai. Unfortunately Prof. K. Bleuler had passed away in the meantime and could not join us at the Conference. His name would be long remembered by us all who work in this field. The main goal of this conference is to gather together physicists and mathematicians actively working in mathematical physics to promote interactions among the various intensively studied areas of this field, in particular Quantum Groups, Algebraic Methods, Knot Theory and Quantum Topology including Chern–Simons, Differential Geometry, and Differential Equations in Mathematics, and Classical and Quantum Integrable Systems, Yang–Baxter Equations, Non-commutative Geometry, Statistical Mechanics, CFT, and Unifications in Physics.

The XXI DGMTP brought together more than 220 active researchers in these areas, more than half of them from outside China. We were particularly happy that Prof. S. S. Chern joined the Conference and presented a most interesting talk not included in the proceedings. Prof. C. N. Yang personally participated in the planning and organization of this Conference in its various aspects. We were most happy to see that his theories have given rise to the key topics of this Conference.

The scientific program, coordinated by Prof. Y. M. Cho and Prof. X. W. Zhou, consisted of plenary talks, parallel talks and poster presentations. Actually, some of the talks and presentations, had to be scheduled in parallel and poster sessions for technical reasons. Many have contributed to make this conference a success. We are particularly grateful to the speakers at the various sessions, not only for their reports of new discoveries, but also for their kind cooperation in many ways. Many thanks are also due to the organizers of the parallel sessions for a most excellent job.

Last but not least, we would like to thank Nankai University, particularly the staff and graduate students of Nankai Institute of Mathematics, for their support and assistance throughout the conference.

None of this would have happened without the generous financial support of our sponsors: NSF of China, Municipal Government of Tianjin, Wei Lun Foundation, Ltd., K. C. Wong Education Foundation Ltd., Southeast Asia Theoretical Physics Association, ICTP (Italy), and World Scientific Publishing, to whom we acknowledge our most sincere thanks.

Mo-Lin Ge

Contents

Parallel Sessions I: Geometry of Quantum Group and Applications

Parallel Sessions 2: Quantum Group and Knot Theory

Parallel Sessions 4: Fractional Statistics and Lattice Models

Parallel Sessions 6: Differential Equations and Differential Geometry

Plenary Sessions

The Pattern Calculus for q-Tensor Operators in the q-Group $U_q(u(n))$*†

L. C. BIEDENHARN
Department of Physics, Duke University
P.O. Box 90305, Durham NC 27708-0305

M. D. GOULD
Department of Applied Mathematics, University of Queensland
St. Lucia, Queenland, Australia 4072

Abstract

We have recently constructed an explicit evaluation for all matrix elements of all unit norm q-tensor-operators belonging to $U_q(u(n))$, having extremal operator shift patterns. These complicated results will be shown to be easily comprehensible in terms of a *diagrammatic calculus of patterns*.

1. Introduction

Quantum groups are of considerable current interest, in both mathematics and physics, and have been the focus of much recent research.[1,2,3] Although quantum groups are not groups, they share many group-like features. For generic values of q—the deformation parameter—the representation theory for compact quantum groups is surprisingly close to that of standard compact Lie groups, and it is of interest to see how far this similarity extends for more complicated structures, such as tensor operator algebras. We shall demonstrate that a pattern calculus

* Invited paper presented at the XXIst International Conference on Differential Geometric Methods in Theoretical Physics, Nankai, China 5-9 June 1992

† Supported in part by the National Science Foundation, Grant No. PHY-9008007

for tensor operators in the $U_q(u(n))$ quantum group exists and is very similar in *form*—but not detail—to the pattern calculus known for tensor operators on Lie group representations.

The pattern calculus is actually:

(a) a diagrammatic procedure whereby the diagram correlates (or more precisely, implies) the explicit (algebraic) matrix element associated (by the pattern calculus rules) to the diagram; and

(b) a procedure for the construction of all tensor operators. The elementary tensor operators are given directly by the pattern calculus, and by extending the calculus so that patterns act on patterns, one obtains an algebraic basis for the construction of general tensor operators.

It is the merit of the pattern calculus that it 'explains' and makes accessible *structurally* the otherwise difficult-to-comprehend explicit matrix elements (which can be of arbitrarily large complexity and length).

The pattern calculus was first constructed (for $U(n)$) by Biedenharn and Louck[4], exploiting the Jordan-Schwinger map[5]. These results were later verified using an algebraic version of the Borel-Weil construction by LeBlanc and Hecht[6] and, using a more conceptual approach, by Gould[7,8].

The approach of Gould[7] evaluated the required tensor operator matrix elements explicitly, by using a projection operator realization. To understand the concepts underlying this method by an elementary example, consider the total angular momentum operator $\vec{J}$ as the sum of two independent angular momenta $\vec{L}$ and $\frac{\vec{\sigma}}{2}$, that is, $\vec{J} = \frac{\vec{\sigma}}{2} + \vec{L}$. Then $2\vec{\sigma}\cdot\vec{L} = \vec{J}^2 - \vec{L}^2 - \frac{3}{4}$ is an invariant operator and this operator may be used to construct: (a) the Casimir operator: $\vec{L}^2 = \mathrm{tr}((\vec{\sigma}\cdot\vec{L})^2)$ and (b) the projection operators: $P_\pm : (\mathbf{J} \to j = \ell \pm \frac{1}{2}) : \ P_+ = \frac{\sigma\cdot L+\ell+1}{2\ell+1}, \ P_- = \frac{\ell-\sigma\cdot L}{2\ell+1}$. Matrix elements of $P_\pm$ then yield the desired spin-$\frac{1}{2}$ tensor operator results.

The extension to an arbitrary Lie group considers the algebra of tensor products, $\mathcal{A} \equiv U(g) \otimes U(g)$, where $g \in$ Lie (G) and $U(g)$ is the universal enveloping algebra of Lie (G), and we use the diagonal co-product: $\partial(g) = \mathbf{1} \otimes g + g \otimes \mathbf{1}$. [Co-products are usually denoted by the symbol Δ but for tensor operator theory this symbol has a different significance.] The analog of spin-$\frac{1}{2}$ is now an irrep λ of G, with the associated module V_λ, and Lie algebra generators $\pi_\lambda(g)$ playing the rôle of $\frac{\vec{\sigma}}{2}$. Generating elements of the algebra $\mathcal{A}_\lambda \equiv$ End $(V_\lambda) \otimes U(g)$ are then given by: $\partial_\lambda(g) \equiv (\pi_\lambda \otimes \mathbf{1})(\partial(g))$.

The Casimir element, C, of G is: $C = \sum_r x_r x^r$, where $\{x_r\}$ is a basis for Lie G and $\{x^r\}$ is the dual basis defined by the Killing form. The analog to $\vec{\sigma}\cdot\vec{L}$ is then: $A_\lambda = \frac{1}{2}(\pi_\lambda(C) \otimes \mathbf{1} + \pi_\lambda(\mathbf{1}) \otimes C - \partial_\lambda(C))$. It is easily seen that A_λ lies in the center of End $(V_\lambda) \otimes U(g)$, that is: $[\partial_\lambda(g),\ A_\lambda] = 0$.

The invariants of the Lie group G, that is, the center of $U(g)$, is generated as an algebra by the elements $z_m = \mathrm{tr}_\lambda((A_\lambda)^m)$. Similarly, projection operators for irreducible modules of $\mathcal{A}_\lambda$ can be constructed as polynomials in A_λ. These projection operators can be given explicitly when one has determined the eigenvalue spectrum of A_λ.

Since this construction has been formulated purely algebraically, the extension to quantum groups is immediate: one simply uses the algebra (and co-algebra) structure relevant to the quantum group, and the quantum group analog of the trace operation. As shown in detail in ref. (9), this general procedure suffices to determine explicitly the desired quantum group matrix elements of quantum group tensor operators. The *interpretation* of these results in terms of a calculus of patterns is the goal of the present paper.

2. Tensor Operator Concepts

A given symmetry group has the implication for physics not only of distinguishing possible physical states by symmetry based quantum numbers but also of classifying (and partially determining) physical transitions between these states. This latter structure is abstracted as the *algebra of tensor operators* of the symmetry group.

Consider a compact Lie group G to be a physical symmetry. This symmetry allows one to determine a Hilbert space basis for physical states using vectors carrying the irreps of G and in general additional parameters from configuration space. To discuss the problem of interactions, which induce transitions within and between irreps, it is useful to abstract from the physical Hilbert space a *model space*[10,11], $\mathbf{M}$, defined to be the direct sum of vectors carrying unitary irreps of the group G, each equivalence class of irreps occurring *once and only once*. The operators on $\mathbf{M}$ are defined to belong to the linear space $\mathbf{T}$ with the action: $\mathbf{T}$: $\mathbf{M} \rightarrow \mathbf{M}$. Using equivariance, the symmetry G can be exploited to give a partial classification of the operators belonging to $\mathbf{T}$. Let $g \in G$ and $\mathbf{M} \rightarrow \mathbf{U}(g)\mathbf{M}$. The equivariance condition is:

$$\mathbf{U}(g)\, \mathbf{T}_i\, \mathbf{U}^{-1}(g) \;=\; \sum_j \mathbf{T}_j\, D_{ji}(g), \tag{2.1}$$

where $\mathbf{D}(g)$ is a representation. It follows that a tensor operator $\{\mathbf{T}_i\}$—*a set of operators*—can be classified by an irrep label of G, with individual operators in the tensor operator set being labelled as vectors carrying the irrep.

The tensor operator classification is not categoric; e.g. multiplying a given irreducible tensor operator by an invariant operator leaves the classification unchanged. To eliminate this freedom one defines ***unit tensor operators*** whose norm is unity. For $G = U(n)$ with $n \geq 3$, *distinct* unit tensor operators with the same irrep labels $[m_{1n} \ldots m_{nn}]$ exist, but it has been shown[12] that a *canonical* (natural) labelling splitting this multiplicity exists for $n = 3$, and believed to exist[13] for all n.

DEFINITION: *Let* $\mathbf{T}$ *denote the vector space of operators mapping the model space* $\mathbf{M}$ *of the compact quantum group* G_q *into itself:* $\mathbf{M} \xrightarrow{\mathbf{T}} \mathbf{M}$. *An irreducible q-tensor operator is a set of operators,* $\{\mathbf{t}_{\Xi,\xi}\} \in \mathbf{T}$ *which carries a finite-dimensional irrep* Ξ, *with vectors* ξ, *of the quantum group* G_q. *That is:*

$$E_\alpha(\mathbf{t}_{\Xi,\xi}) = \sum_{\xi'} \langle \Xi, \xi' | E_\alpha | \Xi, \xi \rangle \mathbf{t}_{\Xi,\xi'}, \tag{2.2}$$

where E_α *is a generator of* G_q, $E_\alpha(\mathbf{t}_{\Xi,\xi})$ *denotes an action of* E_α *on* $\mathbf{T}$, *and* $\langle \cdots \rangle$ *denotes the matrices of the generators for the irrep* Ξ. *A q-tensor operator is accordingly a linear combination of irreducible q-tensor operators, with coefficients invariant under the q-group action.*

THEOREM[14]: *If* $\{\mathbf{t}_{\Xi,\xi}\}$ *is a q-tensor operator of the compact quantum group* G_q *such that the co-product of* G_q *is compatible with the action* $E_\alpha(\mathbf{t}_{\Xi,\xi})$, *then the matrix elements of* $\{\mathbf{t}_{\Xi,\xi}\}$ *in* $\mathbf{M}$ *are proportional to the q-WCG coefficients of* G_q *with the constant of proportionality an invariant.*

Tensor operators thus form an algebra with invariant operators as scalars. This algebra is generated by the *unit* tensor operators, that is, tensor operators that are normed to unity on all irreps not in the characteristic null space of the operator. ***This algebra of tensor operators contains as a sub-algebra the universal enveloping algebra.*** Expressed differently, the algebra of tensor operators is a generalization of the concept of a universal enveloping algebra.[11]

Unit q-tensor operators are characterized by two structures:

(a) the quantum group action, by which the unit tensor operators may be divided into irreducible sets carrying the labels of a quantum group irrep; this is the label $[M_{1n} M_{2n} \ldots M_{nn}]$ for the q-group $U_q(u(n))$. The operators belonging to a finite dimensional irrep may be further split by labels characterizing the various *components* of the operators, which carry representation labels in the canonical group-subgroup chain, $U_q(u(n)) \supset U_q(u(n-1)) \supset \ldots \supset U_q(u(1))$, written in the form of a *Gel'fand-Weyl pattern:*

$$(M) \equiv \begin{pmatrix} M_{1n} & & M_{2n} & \cdots & & M_{nn} \\ & M_{1,n-1} & & \cdots & M_{n-1,n-1} & \\ & & M_{23} & \cdots & M_{22} & \\ & & & M_{11} & & \end{pmatrix}, \tag{2.3}$$

which is a triangular pattern of *integers* obeying the *betweenness constraints*: $M_{ij} \geq M_{i,j-1} \geq M_{i+1,j}$. This pattern is a 'geometric' realization, via constraints, of the *Weyl branching theorem for* $U(n)$ which carries over to the quantum group $U_q(u(n))$ when q is generic, in consequence of the Lusztig-Rosso theorem[15,16].

(b) The Gel'fand-Weyl patterns in (a) do not fully characterize a unit tensor operator, and a further splitting is necessary. A suitable labelling (unique to within ± 1 phase conventions) is afforded by a second pattern, called an *operator pattern* (Γ), which is structurally similar to a Gel'fand-Weyl pattern in being integer-valued and obeying the betweenness constraints:

$$(\Gamma) \equiv \begin{pmatrix} M_{1n} & & M_{2n} & \cdots & & M_{nn} \\ & \Gamma_{1,n-1} & \cdots & \cdots & \Gamma_{n-1,n-1} & \\ & & \Gamma_{11} & & & \end{pmatrix} \tag{2.4}$$

For $U(3)$ and for $U_q(u(3))$ the operator pattern (Γ) has been proved to be *canonical*, that is, unique to within equivalence and free of any arbitrary choices. (It is conjectured[13] that this is true for all $U(n)$ and $U_q((u(n))$.)

The individual elements of an irreducible unit tensor operator set are labelled by the symbol: $\left\langle \begin{matrix} (\Gamma) \\ [M] \\ (M) \end{matrix} \right\rangle$. The angular brackets denote a unique unit tensor operator, labelled by the $U_q(u(n))$ irrep labels $[M]$—*common to both the operator pattern* (Γ) *and the Gel'fand-Weyl pattern* (M). For brevity the operator pattern Γ is inverted (this leaves the betweenness relations invariant). The significance of these labels can be seen from the properties:

(a) the *weights* $(W) = (W_1 \ldots W_n)$ associated by the equivariant quantum group action are given by

$$W_i = \sum_j (M_{j,i} - M_{j,i-1}), \quad M_{ij} \equiv 0,\ i > j. \tag{2.5}$$

(b) the *shifts* associated to the operator, that is, the action *by* the operator *on* the irrep labelled by $[m_{1n}, m_{2n}, \ldots m_{nn}]$ yields the irrep $[m_{1n} + \Delta_1, m_{2n} +$

$\Delta_2, \ldots, m_{nn} + \Delta_n]$ (or the null irrep) where the shifts $\{\Delta_i\} = (\Delta_1, \Delta_2, \ldots \Delta_n)$ are given by:

$$\Delta_i = \sum_j (\Gamma_{j,i} - \Gamma_{j,i-1}), \quad \Gamma_{ij} \equiv 0, \; i > j. \tag{2.6}$$

The sub-set of unit tensor operators belonging to the universal enveloping algebra is the sub-set characterized by $(\Delta) = (\dot{k})$, that is, constant shift k for all labels (M), which (for $U_q(su(n))$ model space) is equivalent to zero shift $(\Delta) = (\dot{0})$.

It is a consequence of the definition of a unit tensor operator that the matrix elements of the operator:

$$\left\langle \begin{matrix} (\Gamma) \\ [M] \\ (M) \end{matrix} \right\rangle \longrightarrow \left\langle \begin{matrix} [m^{\text{final}}] \\ (m^f) \end{matrix} \right| \left\langle \begin{matrix} (\Gamma) \\ [M] \\ (M) \end{matrix} \right\rangle \left| \begin{matrix} [m^{\text{initial}}] \\ (m^i) \end{matrix} \right\rangle \tag{2.7}$$

are *exactly* the same as the matrix elements of the matrix which brings the tensor product of the irreps $[M] \otimes [m^{\text{initial}}] \rightarrow [m^{\text{final}}]$ to fully reduced (block-diagonal) form, ordered by the standard ordering on irrep labels $[M]$. In other words the construction of explicit unit tensor operators is equivalent to the complete resolution of the Clebsch-Gordan (tensor) product, including the resolution of all multiplicities.

For brevity we will refer to the tensor operator in (2.7) as $\langle [M] \rangle$.

A particularly simple class of unit tensor operators is the class of such operators for which the shift labels (Δ) are a permutation of the operator irrep labels $[M]$. For this class, there is no multiplicity. This class includes all *elementary* unit tensor operators, where an elementary unit tensor operator (distinct from the unit operator) is defined to be an irreducible unit tensor operator with irrep labels of the form: $[M] = [\underbrace{1\ldots1}_{k}\underbrace{0\ldots0}_{n-k}], \quad (k = 1, 2, \ldots, n-1)$.

The operator $\langle [M] \rangle$, considered as an explicitly determined algebraic object, is enormously complicated. What is needed is an understanding of the *structure* implicit in such an object. For the quantum group $U_q(u(n))$, this desired structure is implied by the chain: $U_q(u(n)) \supset U_q(u(n-1)) \supset \ldots U_q(1)$. For this chain we have the result:

$$\left\langle \begin{matrix} [m^f] \\ (m^i) \end{matrix} \right| \left\langle \begin{matrix} (\Gamma) \\ [M] \\ (M) \end{matrix} \right\rangle \left| \begin{matrix} [m^i] \\ (m^i) \end{matrix} \right\rangle = \sum_{(\Gamma')} \left\langle \begin{matrix} [m^f]_n \\ [m^f]_{n-1} \end{matrix} \right| \left[\begin{matrix} (\Gamma) \\ [M] \\ (\Gamma') \end{matrix} \right] \left| \begin{matrix} [m^i]_n \\ [m^i]_{n-1} \end{matrix} \right\rangle$$

$$\times \left\langle \begin{matrix} [m^f]_{n-1} \\ (m^f) \end{matrix} \right| \left\langle \begin{matrix} (\Gamma') \\ (M)_{n-1} \end{matrix} \right\rangle \left| \begin{matrix} [m^i]_{n-1} \\ (m^i) \end{matrix} \right\rangle. \tag{2.8}$$

This result, stated in words, asserts that a matrix element for a unit tensor operator belonging to $U_q(u(n))$ can be expressed as a sum of products of two factors:

(a) a *reduced operator coefficient*:

$$\left\langle \begin{matrix} [m^f]_n \\ [m^f]_{n-1} \end{matrix} \middle| \begin{bmatrix} (\Gamma) \\ [M] \\ (\Gamma') \end{bmatrix} \middle| \begin{matrix} [m^i]_n \\ [m^i]_{n-1} \end{matrix} \right\rangle, \tag{2.9}$$

which, by construction, is an invariant under the $U_q(u(n-1))$ subgroup; and (b) an irreducible unit tensor operator matrix element:

$$\left\langle \begin{matrix} [m^f]_{n-1} \\ (m^f) \end{matrix} \middle| \left\langle \begin{matrix} (\Gamma') \\ (M)_{n-1} \end{matrix} \right\rangle \middle| \begin{matrix} [m^i]_{n-1} \\ (m^i) \end{matrix} \right\rangle, \tag{2.10}$$

which belongs to a unit tensor operator in the sub-q-group $U_q(u(n-1))$, where the irrep labels are contained in the patterns (Γ') and $(M)_{n-1}$.

Thus one has a recursive approach to the explicit determination of tensor operators in which one need only obtain generic matrix elements for the *reduced operators*, denoted by $\begin{bmatrix} (\Gamma) \\ [M] \\ (\Gamma') \end{bmatrix}$, abstracting the operator from the matrix elements above. (This abstraction is *unique* since the labels of the reduced operator determine completely the final state, given the initial state irrep data.) *Note that the reduced operators act in a $U_q(u(n))$: $U_q(u(n\text{-}1))$ subspace, labelled by a truncated (two-rowed) Gel'fand-Weyl pattern.*

3. The Pattern Calculus for the Quantum Group $\mathrm{U_q(u(n))}$

Matrix elements for reduced operators have been obtained explicitly using projection operator techniques in tensor product modules in ref. 9. In the present paper we will demonstrate that these matrix elements (for elementary operators) are characterized by a very simple structure which can be directly related to the shifts themselves. These relations constitute the *pattern calculus rules.*

The pattern calculus rules determine the explicit reduced matrix element in the form:

$$\left\langle \begin{matrix} [\mu+\Delta(\Gamma)] \\ [\mu_0+\Delta_0(\Gamma_0)] \end{matrix} \middle\| \begin{bmatrix} (\Gamma) \\ [\lambda] \\ [\lambda_0] \\ (\Gamma_0) \end{bmatrix} \middle\| \begin{matrix} [\mu] \\ [\mu_0] \end{matrix} \right\rangle = (\pm) q^{\alpha} \cdot \left| \frac{N}{D} \right|, \tag{3.1}$$

directly from the initial state date–$[\mu]$ and $[\mu_0]$–using the operator data $[\lambda], [\lambda_0], (\Gamma)$ and (Γ_0). The final state data are uniquely determined from this information.

The pattern calculus rules apply to extremal operator patterns, that is, operator patterns such that the shift Δ implies the pattern Γ. For such patterns, the shifts Δ are permutations of the maximal weight of the irrep characterizing the operator.

Now arrange the two shifts Δ and Δ_0 in the form of a two-rowed pattern, similar to a (truncated) Gel'fand-Weyl pattern, but *without the betweenness constraints.* We call this the *shift pattern.*

We can now state the pattern calculus rules for elementary tensor operators.

The Pattern Calculus Rules for Elementary Tensor Operators in $U_q(u(n))$:

Rule 1. Write out two rows of dots with n dots in row n and n-1 dots in row n-1, in the matter displayed below:

$$\begin{array}{ccccccccccccl} \bullet & & \bullet & & \bullet & & & & \bullet & & \bullet & & \bullet \quad \text{row } n \\ & \bullet & & \bullet & & \bullet & \cdots & & & \bullet & & \bullet & \quad \text{row } n-1 \end{array} .$$

Rule 2. In row n, assign to the ith dot the $U_q(u(n))$ partial hook: $p_{i,n} \equiv m_{i,n} + n - i$; do the same in row n-1 using the $U_q(u(n\text{-}1))$ partial hooks: $p_{i,n-1} = m_{i,n-1} + n - 1 - i$.

Rule 3. Draw arrows between dots as follows: Select a dot i in row n and a dot j in row $n-1$. If $\Delta_{i,n} > \Delta_{j,n-1}$, draw an arrow *from* dot i *to* dot j; if $\Delta_{i,n} < \Delta_{j,n}$, draw the arrow *from* j *to* dot i. Carry out this procedure for all dots in rows n and $n-1$. (If $\Delta_{in} = \Delta_{j,n-1}$ go to another pair.) This yields a *numerator arrow pattern* with arrows going *between* rows.

Carry out this same procedure for dots *within* row n and dots *within* row n-1. This yields a *denominator arrow pattern* with arrows going *within* rows.

Rule 4. Assign to each arrow the *q-integer factor*: $[p(\text{tail}) - p(\text{head}) + e(\text{tail})]$

where: $p(\text{tail/head})$ is the partial hook associated with the dot(tail/head), and

$$e(\text{tail}) \equiv \begin{cases} 1 & \text{if the tail of the arrow is on row } n-1 \\ 0 & \text{if the tail of the arrow is on row } n. \end{cases}$$

Rule 5. Write out the products:

$$N^2 = \text{product of all } q\text{-integer factors for numerator arrow pattern},$$
$$D^2 = \text{product of all } q\text{-integer factors for denominator arrow pattern}.$$

Rule 6. *Sign Convention*

The ($\pm$) sign for the matrix element is obtained in this way.

Take the shift Δ—which consists of k ones and (n-k) zeroes in some order—and assign the k integers $i_1 < i_2 < \ldots < i_k$ which denote the k places in the shift pattern Δ where the ones occur. (Example: $\Delta = 01100 \Rightarrow (i_1 i_2) = (23)$.) Similarly assign $j_1 < j_2 < \ldots j_k$ for the shift Δ_0. (Note that we have formally extended the shift Δ_0 to n places by embedding λ_0 in $U_q(u(n))$.)

The sign of the matrix element is then defined to be: $(-1)^{k(k-1)/2} \prod_{\ell,s=1}^{k} S(j_\ell - i_s)$, for $i_1 < i_2 < \ldots i_k$ and $j_1 < j_2 < \ldots < j_k$, where $S(j-i) \equiv \begin{cases} +1 & j \geq i \\ -1 & j < i \end{cases}$.

Rule 7. *q-phase factor* $\equiv q^\alpha$

There are two cases to consider:

(a) The shift pattern Δ and Δ_0 both have k ones, to which we assign (cf. rule 6) the sequences $(i_1, i_2 \ldots i_k)$ and $(j_1 j_2 \ldots j_k)$ respectively.

The q-phase factor is then given by:

$$q^\alpha = q^{\frac{1}{4}\sum_{s=1}^{k}(p_{i_s,n} - p_{j_s,n-1} - 1)}.$$

(b) The shift pattern Δ contains k ones, and the shift pattern Δ_0 (k-1) ones. For this case we label the (n-k) *zeroes* appearing in Δ and Δ_0 by labels $\bar{\imath}_1 < \bar{\imath}_2 < \ldots < \bar{\imath}_{n-k}$ for Δ, and by $\bar{\jmath}_1 < \bar{\jmath}_2 < \ldots < \bar{\jmath}_{n-k}$ for Δ_0 exactly as we labelled the ones in rule 6 by $(i_1 \ldots i_k)$ and $(j_1 \ldots j_k)$.

The q-phase factor is then given by:

$$q^\alpha = q^{\frac{1}{4}\sum_{s=1}^{n-k}(p_{\bar{\imath}_s,n} - p_{\bar{\jmath}_s,n-1})}.$$

Rule 8. The desired matrix element of the $U_q(u(n))$ elementary tensor operator $\langle[\underbrace{1\ldots1}_{k}\underbrace{0\ldots0}_{n-k}]\rangle$ is then given by:

$$\left\langle \begin{matrix} [m_{i,n} + \Delta_n(\Gamma)] \\ [m_{i,n-1} + \Delta_{n-1}(\Gamma)_0] \end{matrix} \middle| \left[\begin{matrix} (\Gamma) \\ [1_k 0_{n-k}] \\ (\Gamma)_0 \end{matrix} \right] \middle| \begin{matrix} [m_{i,n}] \\ [m_{i,n-1}] \end{matrix} \right\rangle = (\pm 1) \cdot (q^\alpha) \cdot \left| \left(\frac{N^2}{D^2} \right)^{\frac{1}{2}} \right|,$$

where each of the terms have been given by earlier rules.

These pattern calculus rules determine explicitly all reduced matrix elements of all elementary tensor operators acting on arbitrary weight vectors in the quantum group structure $U_q(u(n)) : U_q(u(n\text{-}1))$. In the limit $q \to 1$, they become exactly the pattern calculus rules[4] *for the family of unitary groups $U(n)$.*

Extension of these pattern calculus rules to q-tensor operators with extremal operator patterns is given in reference (9).

References

1. C. N. Yang and M. L. Ge (Editors), *Braid Group, Knot Theory and Statistical Mechanics*, Advanced Series in Mathematical Physics, Vol. 9, World Scientific, Singapore (1989).
2. T. Curtright, D. Fairlie and C. Zachos (Editors), *Quantum Groups*, Proceedings of the Argonne Workshop, April - May 1990, World Scientific, Singapore (1991).
3. H.-D. Doebner, J.-D. Hennig (Editors), *Quantum Groups, Lecture Notes in Physics,* **370**, Proceedings of 8th International Workshop on Mathematical Physics, July 1989, Springer-Verlag, Berlin (1990).
4. L. C. Biedenharn and J. D. Louck, *Commun. Math. Phys.* **8** (1968) 89.
5. L. C. Biedenharn and J. D. Louck, *Angular Momentum in Quantum Physics, Theory and Application*, Encyclopedia of Mathematics and its Applications, Vol. 8, Reading, Massachusetts: Addison-Wesley (1981). Reprinted by Cambridge University Press (1989).
6. R. LeBlanc and K. T. Hecht, *J. Phys. A: Math. Gen.* **20** (1987) 4613-4635.
7. M. D. Gould, *J. Math. Phys.* **27** (1986) 1944.
8. M. D. Gould, *J. Math. Phys.* **27** (1986) 1964.
9. M. D. Gould and L. C. Biedenharn, The Pattern Calculus for Tensor Operators in Quantum Groups, to be published in J. Math. Phys.
10. I. M. Gel'fand and V. Zelevinsky, *Societé Math. de France, Astérique, hors series* **117** (1985).
11. D. E. Flath and L.C. Biedenharn, *Can. J. Math.* **37** (1985) 710-729.
12. G. E. Baird and L. C. Biedenharn, *J. Math. Phys.* **5** (1964) 1730.
13. L. C. Biedenharn, A. Giovannini and J. D. Louck, *J. Math. Phys.* **8** (1967) 691.
14. L. C. Biedenharn and M. Tarlini. *Lett. in Math. Phys.* **20** (1990) 271-278.
15. G. Lusztig, *Adv. in Math.* **70** (1988) 237-249.
16. M. Rosso, *Commun. Math. Phys.* **117** (1988) 581-593.

An Integrable Approach to 4-d Quantum Field Theories: Yang-Mills, Yang-Baxter, and Local Quantum Group

LING-LIE CHAU* AND ITARU YAMANAKA*

*Physics Department, University of California, Davis, CA 95616

We have constructed a quantum field theory for the Self-Dual Yang-Mills system in four dimensions (4-d) in terms of the group-valued gauge-invariant local fields $\hat{J}$. We show that the commutation relations among the quantum fields $\hat{J}$ are best written in the form of quantum-exchange-algebras, of which the structure-matrix $\hat{R}$ explicitly satisfies Yang-Baxter equations such that products of $\hat{J}$ are well defined. The form of the $\hat{R}$ matrix also dictates many essential features of the fields $\hat{J}$; nontrivial critical exponents (Hausdroff or fractal dimensions) of products of $\hat{J}$ fields at short-distances, normal-order-products, and transformation properties. Most striking is the necessary call for a local-quantum-group (Hopf algebra) to which fields $\hat{J}$ form noncommutative-vector-spaces. We obtain the quantum-field Hamiltonian and equations of motion. They have interesting interaction forms and symmetry properties that are local $\{Sl(2)\} \times \{U_q[Sl(2)]\}$ in the specific case of $Sl(2)$-group-valued $\hat{J}$. We also identify the generators for these symmetries and further construct the quantum-field affine-algebra currents, Virasoro-algebra fields, and hierarchies of linear and nonlinear systems.

One of the most pressing current problems in particle physics is to formulate four-dimensional (4-d) quantum field theories nonperturbatively and to find nonperturbative quantum-field-theoretical solutions to the Yang-Mills equations for strong interactions as well as to quantum gravity. Choosing the 4-d Self-Dual Yang-Mills (SDYM) system as a concrete example, in a previous paper[1] we had given a canonical formulation for the SDYM system in terms of the group-valued gauge-invariant fields J.[2] In this paper, we give the results of quantizing the SDYM system.

As we know that except for Dirac's rule of quantization,[3] replacing Dirac-brackets by commutators, the route going from a given canonical formulation to a corresponding quantum field theory is not unique nor obvious. We consider the SDYM $\hat{\mathbf{J}}$-quantum-field-theory developed here[4] satisfactory because for $\hbar \to 0$ the $\hat{\mathbf{J}}$-quantum-field-theory reduces to the given canonical formulation of Ref. [1]. It also has the remarkable feature of reducing to the WZNW $\hat{\mathbf{g}}$-quantum-field-theory[5,6] when reduced to two-dimensions (2-d).

The nonperturbative results obtained here are based upon the integrable-system approach; it would be hard to obtain these results in the conventional way of formulating quantum field theories by quantizing the Lie-algebra-valued gauge-potential fields $\mathbf{A}_\mu$ as had been done for QED or QCD. We are hopeful that these results will give a start for calculating 4-d physical quantities nonperturbatively. These nonperturbative results are significant and will stay with full 4-d quantum field theories, since the SDYM $\hat{\mathbf{J}}$-quantum-field-theory is a prototype and the classical J-field-like formulations had already been done for the full 4-d $n \geq 3$ supersymmetric Yang-Mills theories[7] and the 4-d $n \geq 5$ conformal supergravity theories.[8][9].

The highlights of this paper are:

1. We formulate the SDYM $\hat{\mathbf{J}}$-quantum-field-theory in a 4-d space with light-cone and hyperbolic coordinates $y, \bar{y}, z, \bar{z}$. The theory is spanned by four-sets of equal-time relations, choosing each one of the four light-cone coordinates as time.
2. We have obtained quantum-exchange-algebras for the quantum fields $\hat{\mathbf{J}}$, which replace the conventional commutation-relations of quantum fields and in the $\hbar \to 0$ limit become the Dirac-brackets of quadratic-algebras in the canonical formulation of Ref. [1].
3. We show that the structure-matrix $\hat{\mathbf{R}}$ satisfies explicitly the Yang-Baxter equations so that products of the noncommuting quantum-fields $\hat{\mathbf{J}}$ have well defined meaning. We

find that the $\hat{\mathbf{R}}$ matrix also dictates many essential properties of the $\hat{\mathbf{J}}$ fields, which are listed below.

4. We derive the non-trivial critical exponents (Hausdroff or fractal dimensions) of products of fields $\hat{\mathbf{J}}$ at short distances from the structure of $\hat{\mathbf{R}}$; thus normal-order-products of fields $\hat{\mathbf{J}}$ can be defined.

5. Dictated by the $\hat{\mathbf{R}}$ matrix and the nontrivial critical exponents, we find that the $\hat{\mathbf{J}}$ fields form noncommutative vector-spaces of a local-quantum-group (Hopf algebra) on the one side and commutative-vector-spaces of a regular local-group on the other side. In Minkowski $(+---)$-signatured space, they are $\{Sl(2,C)\}_{\mathcal{L}} \times \{U_q[Sl(2,C)]\}_{\mathcal{R}}$ for y as time; and $\mathcal{L}, \mathcal{R}$ interchanged for $\bar{y}$ as time, (where $\mathcal{L}$ stands for left and $\mathcal{R}$ for right). In $(+-+-)$-signatured space, the Lie-group involved reduces from $Sl(2,C)$ to $Sl(2)$. We also identify the generators for these transformations.

6. We construct the quantum-field Hamiltonian and equations of motion which have interesting interaction forms. They are invariant under the symmetry transformations described previously in 5.

7. We can also construct the local quantum-field affine-current-algebras, Virasora algebras, as well as the hierarchies of linear and non-linear systems corresponding to those given in the cannonical formulation of Ref. [1].

First, we describe the y-time case in which the local quantum-fields are group-valued local fields $\hat{\mathbf{J}}(y;\bar{y},k,\bar{k})$, where the $z, \bar{z}$ coordinates are discretized for the conventional regularization and the integers $k, \bar{k}$ are the discretized coordinates, respectively, of $z \equiv ka$, $\bar{z} \equiv \bar{k}a$, and a the lattice spacing; thus $\delta(z-z'), \delta(\bar{z}-\bar{z}') \rightarrow (1/a)\delta_{k,k'}, (1/a)\delta_{\bar{k},\bar{k}'}$ respectively, and $\int dz, \int d\bar{z} \rightarrow a\Sigma_k, a\Sigma_{\bar{k}}$, respectively. Due to the fact that the Dirac brackets of classical canonical fields form quadratic algebras,[1] the equal-time relations among the $\hat{\mathbf{J}}$ fields are best written in the form of quantum-exchange-algebras.

To acquaint the reader with the quantum-exchange-algebras, we first describe it for the Abelian case. [Besides being the simplest, the Abelian case is relevant for the $Gl(n)$-group-valued field which can be shown to be decomposable into a direct-product of an Abelian field $\hat{\mathbf{J}}^A$ and a $Sl(n)$-group-valued field $\hat{\mathbf{J}}$]. Quantizing the quadratic-algebra obtained in the canoncial formulation of Ref. [1], we obtain the fixed-y-time Abelian-fields-$\hat{\mathbf{J}}^A$ exchange-algebra, (From now on we drop the word "quantum" in front of exchange-algebra and reserve the term exchange-algebra for quantum fields only. To emphasize the quantum-field characteristics, we use a hat superscript " ^ " on those quantities.)

$$\hat{\mathbf{J}}^A(y;\bar{y},k,\bar{k})\hat{\mathbf{J}}^A(y;\bar{y}',k',\bar{k}') = \hat{\mathbf{J}}^A(y;\bar{y}',k',\bar{k}')\hat{\mathbf{J}}^A(y;\bar{y},k,\bar{k})\mathbf{R}^A(y;\bar{y},\bar{y}',k,k',\bar{k},\bar{k}'), \tag{1}$$

$$\mathbf{R}^A(y;\bar{y},\bar{y}',k,k',\bar{k},\bar{k}') = [e^{i\hbar/(4\alpha a^2)}]^{\epsilon(\bar{y}-\bar{y}')\delta_{kk'}\delta_{\bar{k}\bar{k}'}}, \tag{2}$$

where α is the dimensional constant appearing in front of the action and the Hamiltonian[1] (αa^2 is dimensionless) and $\epsilon(\bar{y}-\bar{y}') \equiv \text{sign}\ (\bar{y}-\bar{y}')$. We call $\mathbf{R}^A$ the structure-constant of the exchange-algebra and it is important to note that the dependence of $\mathbf{R}^A$ on $\bar{y},\bar{y}'$ is only on the difference of $(\bar{y}-\bar{y}')$ via $\epsilon(\bar{y}-\bar{y}')$. The Kronecker deltas $\delta_{kk'}, \delta_{\bar{k},\bar{k}'}$ imply that $\hat{\mathbf{J}}$ fields commute for $k \neq k'$, or $\bar{k} \neq \bar{k}'$. For $\hbar \to 0$, the exchange-algebra becomes a commutator, $[\hat{\mathbf{J}}^A(y;\bar{y},k,\bar{k}), \hat{\mathbf{J}}^A(y;\bar{y}',k',\bar{k}')\ ] \approx \hat{\mathbf{J}}^A(y;\bar{y}',k',\bar{k}')\hat{\mathbf{J}}^A(y;\bar{y},k,\bar{k})[i\hbar/(4\alpha a^2)]\epsilon(\bar{y}-\bar{y}')\delta_{kk'}\delta_{\bar{k}\bar{k}'}$. This is precisely the quadratic-algebra of the Dirac bracket we obtained in Ref. [1] for the $U(1)$ $\hat{\mathbf{J}}$-field according to the Dirac-quantization-rule[3]: commutators of fields divided by $\hbar$ corresponding to Dirac brackets, $[\ ,\ ]i/\hbar = \{\ \ ,\ \ \}_D$.

Here we can discuss the relevance of the Yang-Baxter equation. Due to the fact that the $\hat{\mathbf{J}}^A$'s do not commute, we need to make sure that products of the $\hat{\mathbf{J}}^A$'s have an unique and well-defined meaning irrespective to the history of how the product is obtained. For example, the product $\hat{\mathbf{J}}^A(\bar{y}_3)\hat{\mathbf{J}}^A(\bar{y}_2)\hat{\mathbf{J}}^A(\bar{y}_1)$ can be related to $\hat{\mathbf{J}}^A(\bar{y}_1)\hat{\mathbf{J}}^A(\bar{y}_2)\hat{\mathbf{J}}^A(\bar{y}_3)$ in two ways using associativity and the exchange-algebra. The first way is $\hat{\mathbf{J}}^A(\bar{y}_1)\hat{\mathbf{J}}^A(\bar{y}_2)\hat{\mathbf{J}}^A(\bar{y}_3) = [\hat{\mathbf{J}}^A(\bar{y}_1)\hat{\mathbf{J}}^A(\bar{y}_2)]\hat{\mathbf{J}}^A(\bar{y}_3) = \cdots = \hat{\mathbf{J}}^A(\bar{y}_3)\hat{\mathbf{J}}^A(\bar{y}_2)\hat{\mathbf{J}}^A(\bar{y}_1)[\mathbf{R}^A(\bar{y}_2-\bar{y}_3)\mathbf{R}^A(\bar{y}_1-\bar{y}_3)\mathbf{R}^A(\bar{y}_1-\bar{y}_2)]$; the second way is $\hat{\mathbf{J}}^A(\bar{y}_1)\hat{\mathbf{J}}^A(\bar{y}_2)\hat{\mathbf{J}}^A(\bar{y}_3) = \hat{\mathbf{J}}^A(\bar{y}_1)[\hat{\mathbf{J}}^A(\bar{y}_2)\hat{\mathbf{J}}^A(\bar{y}_3)] = \cdots = \hat{\mathbf{J}}^A(\bar{y}_3)\hat{\mathbf{J}}^A(\bar{y}_2)\hat{\mathbf{J}}^A(\bar{y}_1)$ $[\mathbf{R}^A(\bar{y}_1-\bar{y}_2)\ \mathbf{R}^A(\bar{y}_1-\bar{y}_3)\mathbf{R}^A(\bar{y}_2-\bar{y}_3)]$. The two ways should be equivalent; thus the two triple-products of the $\mathbf{R}^A$'s should be equal: $\mathbf{R}^A(\bar{y}_1-\bar{y}_2)\mathbf{R}^A(\bar{y}_1-\bar{y}_3)\mathbf{R}^A(\bar{y}_2-\bar{y}_3) = \mathbf{R}^A(\bar{y}_2-\bar{y}_3)\mathbf{R}^A(\bar{y}_1-\bar{y}_3)\mathbf{R}^A(\bar{y}_1-\bar{y}_2)$, which is the Yang-Baxter equation. (Notice that for simplicity of expression we suppress all the dependence of $\mathbf{R}^A$ on $y,k,\bar{k}$ and all the $\mathbf{R}^A$'s in the Yang-Baxter equation are at the same values of $y,k,\bar{k}$.) Using the explicit expression of $\mathbf{R}^A$ in Eq. (2), we see that the Yang-Baxter equation is satisfied; both sides are given by $[e^{i\hbar/(4\alpha a^2)}]^{\epsilon(\bar{y}_1-\bar{y}_2)+\epsilon(\bar{y}_1-\bar{y}_3)+\epsilon(\bar{y}_2-\bar{y}_3)}$. Keeping only up to the $\hbar$–term in $\mathbf{R}^A \approx 1 - i\hbar\mathbf{r}^A$, the Yang-Baxter equation reduces to its classical form $[\mathbf{r}^A(\bar{y}_1-\bar{y}_2), \mathbf{r}^A(\bar{y}_3-\bar{y}_1)] + [\mathbf{r}^A(\bar{y}_2-\bar{y}_3), \mathbf{r}^A(\bar{y}_1-\bar{y}_2)] + [\mathbf{r}^A(\bar{y}_3-\bar{y}_1), \mathbf{r}^A(\bar{y}_2-\bar{y}_3)] = 0$, which is satisfied trivally.

We see that the Yang-Baxter equation is a very general requirement for noncommutative quantities; like Jacobi identity for Lie algebra and Bianchi identity for geometry. It was originally observed in 2-d statistical-mechanics models[10,11]. Here we see the appearance of Yang-Baxter equation in the context of a 4-d quantum-field-theory.

The innocent looking $\epsilon(\bar{y}-\bar{y}')$ of the exchange algebra has the following significant implication. It tells us that the product of the two $\hat{\mathbf{J}}^A$ fields has a branch point at $\bar{y}=\bar{y}'$. Using the identity $\epsilon(\bar{y}-\bar{y}') = -[\ln(\bar{y}-\bar{y}'+i\epsilon)-\ln(\bar{y}'-\bar{y}+i\epsilon)]/\pi i$, we conclude that the product of two $\hat{\mathbf{J}}^A$ fields at short-distance, $(\bar{y}-\bar{y}')\to 0$, has the non-trivial behavior $\hat{\mathbf{J}}^A(\bar{y})\hat{\mathbf{J}}^A(\bar{y}') \propto [e^{i\hbar/(4\alpha a^2)}]^{\delta_{kk'}\delta_{\bar{k}\bar{k}'}(-)\ln(\bar{y}-\bar{y}'+i\epsilon)/\pi i} = (\bar{y}-\bar{y}')^{-\hbar/(4\pi\alpha a^2)\delta_{kk'}\delta_{\bar{k}\bar{k}'}}$. Identifying $[-\hbar/(4\pi\alpha a^2)]\delta_{kk'}\delta_{\bar{k}\bar{k}'}$ as the critical exponent (Hausdorff or fractal dimension) of $\hat{\mathbf{J}}^A(\bar{y})\hat{\mathbf{J}}^A(\bar{y}')$, $\bar{y}\to\bar{y}'$, we can define normal-order-product $:\hat{\mathbf{J}}^A(\bar{y})\hat{\mathbf{J}}^A(\bar{y}'):$ as

$$\hat{\mathbf{J}}^A(y;\bar{y},k,\bar{k})\hat{\mathbf{J}}^A(y;\bar{y}',k',\bar{k}') = (\bar{y}-\bar{y}')^{-[\hbar/(4\pi\alpha a^2)]\delta_{kk'}\delta_{\bar{k}\bar{k}'}} :\hat{\mathbf{J}}^A(y;\bar{y},k,\bar{k})\hat{\mathbf{J}}^A(y;\bar{y},k,\bar{k}): \quad . \tag{3}$$

This is a nonperturbative result for a 4-d quantum field theory. Next we shall discuss the non-Abelian cases.

To be specific for the non-Abelian cases, we shall discuss the $Sl(2)$ case explicitly. The generalization to cases of $Sl(n)$ is not difficult. The $Sl(2)$-group-valued-$\hat{\mathbf{J}}$ is a 2×2 matrix-field. Quantizing the quadratic-algebras of Ref. [1], we obtain the following exchange-algebras for the fields $\hat{\mathbf{J}}$,

$$\hat{\mathbf{J}}_I(y;\bar{y},k,\bar{k})\hat{\mathbf{J}}_{II}(y;\bar{y}',k',\bar{k}') = \mathbf{1}_{I,II}\hat{\mathbf{J}}_{II}(y;\bar{y}',k',\bar{k}')\hat{\mathbf{J}}_I(y;\bar{y},k,\bar{k})\hat{\mathbf{R}}_{I,II}(y;\bar{y},\bar{y}',k,k',\bar{k},\bar{k}'), \tag{4}$$

which has the identity $\mathbf{1}_{I,II}$ as the left-structure-matrix and $\hat{\mathbf{R}}_{I,II}$ as the right-structure-matrix. To avoid writing out matrix-element sub-indexes of $\hat{\mathbf{J}}$ we have adopted the tensor-product notation in Eq. (4) with the Roman subnumerals I, II denoting the two different $Sl(2)$-matrix-tensor-product spaces, *i.e.*, $\hat{\mathbf{J}}\otimes\mathbf{1}\equiv\hat{\mathbf{J}}_I, \mathbf{1}\otimes\hat{\mathbf{J}}\equiv\hat{\mathbf{J}}_{II}$. When needed, tensor-product spaces can be extended, *e.g.*, $\mathbf{1}\otimes\mathbf{1}\otimes\hat{\mathbf{J}}\equiv\hat{\mathbf{J}}_{III}$. The matrix $P_{I,II}$ performs the function of interchanging the location of a matrix in tensor spaces I and II, *i.e.*, $P_{I,II}\hat{\mathbf{J}}_I P_{I,II} = \hat{\mathbf{J}}_{II}$, $(P_{I,II})^2 = \mathbf{1}_{I,II}$ and for the specific case of $Sl(2)$, $P_{I,II} = (1/2)[1_I 1_{II} + (\sigma_3)_I(\sigma_3)_{II}] + (\sigma_+)_I(\sigma_-)_{II} + (\sigma_-)_I(\sigma_+)_{II}$.

In Ref.[4], we had given the explicit form of $\hat{\mathbf{R}}_{I,II}$, which is a complicated function of the zero-mode first-class-constrain fields. After an extensive search, we find the following a much more illuminating way of writing the $\hat{\mathbf{R}}$ matrix:

$$\begin{aligned}\hat{\mathbf{R}}_{I,II}(y;\bar{y}-\bar{y}',k,k',\bar{k},\bar{k}') = P_{I,II}\{[e^{-i\hbar/(2\alpha a^2)}]^{(-1/2)\epsilon(\bar{y}-\bar{y}')\delta_{kk'}\delta_{\bar{k}\bar{k}'}}\hat{\mathbf{T}}_{I,II}(y;k,\bar{k}) \\ -[e^{-i\hbar/(2\alpha a^2)}]^{(3/2)\epsilon(\bar{y}-\bar{y}')\delta_{kk'}\delta_{\bar{k}\bar{k}'}}\hat{\mathbf{S}}_{I,II}(y;k,\bar{k})\},\end{aligned} \tag{5}$$

where $\hat{\mathbf{T}}_{I,II}, \hat{\mathbf{S}}_{I,II}$ are projection operators which satisfy the projector-operator equations: $(\hat{\mathbf{T}}_{I,II})^2 = \hat{\mathbf{T}}_{I,II}; (\hat{\mathbf{S}}_{I,II})^2 = \hat{\mathbf{S}}_{I,II};\ \hat{\mathbf{T}}_{I,II}\hat{\mathbf{S}}_{I,II} = 0 = \hat{\mathbf{S}}_{I,II}\hat{\mathbf{T}}_{I,II};$ *and* $\hat{\mathbf{T}}_{I,II} + \hat{\mathbf{S}}_{I,II} = \mathbf{1}_{I,II}$. Similarly, we can write the identity left-structure-matrix of Eq.(4) as the sum of two projector operators $\mathbf{1}_{I,II} = s_{I,II}+t_{I,II}$. Later, we shall give more explicit forms of them when we discuss the vector-space interpretation of the fields $\hat{\mathbf{J}}$.

We now demonstrate that the form of $\hat{\mathbf{R}}_{I,II}$ expressed in Eq. (5) already indicates nontrivial critical exponents and dictates the way to define normal-order products. Multiplying $\hat{\mathbf{T}}_{I,II}, \hat{\mathbf{S}}_{I,II}$ from the right and $s_{I,II}, t_{I,II}$ from the left of Eq.(4), respectively, we obtain $t_{I,II}\hat{\mathbf{J}}_I(\bar{y})\hat{\mathbf{J}}_{II}(\bar{y}')\hat{\mathbf{T}}_{I,II} = t_{I,II}\hat{\mathbf{J}}_I(\bar{y}')\hat{\mathbf{J}}_{II}(\bar{y})\hat{\mathbf{T}}_{I,II}[e^{-i\hbar/(2\alpha a^2)}]^{(-1/2)\epsilon(\bar{y}-\bar{y}')\delta_{kk'}\delta_{\bar{k}\bar{k}'}}$, $t_{I,II}\hat{\mathbf{J}}_I(\bar{y})\hat{\mathbf{J}}_{II}(\bar{y}')$ $\hat{\mathbf{S}}_{I,II} = -t_{I,II}\hat{\mathbf{J}}_I(\bar{y}')\ \hat{\mathbf{J}}_{II}(\bar{y})\ \hat{\mathbf{S}}_{I,II}[e^{-i\hbar/(2\alpha a^2)}]^{(3/2)\epsilon(\bar{y}-\bar{y}')\delta_{kk'}\delta_{\bar{k}\bar{k}'}}$, and, except for a sign change, same equations for replacing $t_{I,II}$ by $s_{I,II}$. [The $(-)$ sign comes from the $(-)^{1/2+1/2-0}$ rule. In obtaining these results, we have moved $P_{I,II}$ of $\mathbf{R}_{I,II}$ next to $\mathbf{1}_{I,II}$ in Eq.(4) and used $t_{I,II}P_{I,II} = t_{I,II}$ and $s_{I,II}P_{II,II} = -s_{I,II}$.] As discussed in the Abelian case, these equations tell us that each of the $\hat{\mathbf{T}}_{I,II}$ and $\hat{\mathbf{S}}_{I,II}$ right-projected states of $\hat{\mathbf{J}}_I\hat{\mathbf{J}}_{II}$ has well defined critical exponent at short distance of $(\bar{y}-\bar{y}') \to 0$; *i.e.*, $\hat{\mathbf{J}}_I(\bar{y})\hat{\mathbf{J}}_{II}(\bar{y}')\hat{\mathbf{T}}_{I,II} \propto (\bar{y}-\bar{y}')^{(-1/2)[\hbar/(2\pi\alpha a^2)]\delta_{kk'}\delta_{\bar{k}\bar{k}'}}$ and $\hat{\mathbf{J}}_I(\bar{y})\hat{\mathbf{J}}_{II}(\bar{y}')\hat{\mathbf{S}}_{I,II} \propto (\bar{y}-\bar{y}')^{(3/2)[\hbar/(2\pi\alpha a^2)]\delta_{kk'}\delta_{\bar{k}\bar{k}'}}$. The normal-order-products of the $(\hat{\mathbf{J}}_I\hat{\mathbf{J}}_{II})$-product fields can be defined only for those projected from the right by $\mathbf{T}_{I,II}$, or $\mathbf{S}_{I,II}$:

$$\hat{\mathbf{J}}_I(\bar{y})\hat{\mathbf{J}}_{II}(\bar{y}')\hat{\mathbf{T}}_{I,II} = (\bar{y}-\bar{y}')^{\Delta_{(1/2)(1/2),1}[\hbar/(2\pi\alpha a^2)]\delta_{kk'}\delta_{\bar{k}\bar{k}'}}\{:\hat{\mathbf{J}}_I(\bar{y})\hat{\mathbf{J}}_{II}(\bar{y}')\hat{\mathbf{T}}_{I,II}:\}, \tag{6}$$

$$\hat{\mathbf{J}}_I(\bar{y})\hat{\mathbf{J}}_{II}(\bar{y}')\hat{\mathbf{S}}_{I,II} = (\bar{y}-\bar{y}')^{\Delta_{(1/2)(1/2),0}[\hbar/(2\pi\alpha a^2)]\delta_{kk'}\delta_{\bar{k}\bar{k}'}}\{:\hat{\mathbf{J}}_I(\bar{y})\hat{\mathbf{J}}_{II}(\bar{y}')\hat{\mathbf{S}}_{I,II}:\}, \tag{7}$$

where the common dependences of $\hat{\mathbf{J}}, \hat{\mathbf{T}}$ and $\hat{\mathbf{S}}$ on $y, k, \bar{k}$ are not written out explicitly. The Δ's are the critical exponents (Hausdroff or fractal dimensions) at short-distance $(\bar{y}-\bar{y}') \to 0$, $\Delta_{(1/2)(1/2),1} = 2{\cdot}(1/2)(1/2{+}1){-}1(1{+}1) = -1/2$, $\Delta_{(1/2)(1/2),0} = 2{\cdot}(1/2)(1/2{+}1){-}0(0{+}1) = 3/2$, which we have writen in terms of the quantum numbers of irreducible representations of the vector spaces that rows of $\hat{\mathbf{J}}$ and $(\hat{\mathbf{J}}_I\hat{\mathbf{J}}_{II})$ form; and can be obviously generalized for calculating critical exponents of higher products of $\hat{\mathbf{J}}$ fields and for $\hat{\mathbf{J}}$ belonging to groups other than $Sl(2)$.

Next, we discuss the transformation properties of $\hat{\mathbf{J}}$. From our canonical formulation[1], we can also obtain the commutation relations between $\hat{\mathbf{J}}$ and the affine-algebra currents $\hat{\mathcal{J}}_l$, $l = \pm, 3$,

$$[\hat{\mathcal{J}}_l(\bar{y}, \bar{k}), \hat{\mathbf{J}}(y; \bar{y}', k', \bar{k}')] = -(\hbar/2a)\sigma_l \hat{\mathbf{J}}(y; \bar{y}', k', \bar{k}')\delta(\bar{y} - \bar{y}')\delta_{\bar{k}\bar{k}'}, \tag{8}$$

where the subscript l specifies Lie-algebra components of $\hat{\mathcal{J}}, \hat{\mathcal{J}} = \hat{\mathcal{J}}_3\sigma_3 + 2\hat{\mathcal{J}}_+\sigma_+ + 2\hat{\mathcal{J}}_-\sigma_-$, which from our canonical formulation of Ref. [1] satisfy the affine-current-algebra, $[\hat{\mathcal{J}}_I(\bar{y}, \bar{k}), \hat{\mathcal{J}}_{II}(\bar{y}', \bar{k}')] = \hbar 2an\delta'(\bar{y} - \bar{y}')\delta_{\bar{k}\bar{k}'} + (\hbar/a)P_{I,II}(\hat{\mathcal{J}}_I - \hat{\mathcal{J}}_{II})\delta(\bar{y} - \bar{y}')\delta_{\bar{k},\bar{k}'}$, and the δ' term is the Schwinger-term with central-charge $\hbar 2an$ and n the lattice size in the z direction. Eq. (8) means that the affine currents $\hat{\mathcal{J}}_l$ are the generators for the $Sl(2)$-left-transformation of $\hat{\mathbf{J}}$. (Later, we shall explicitly construct $\hat{\mathcal{J}}_l$ in terms of $\hat{\mathbf{J}}$.) From Eq.(8) we can also obtain the transformations that $\hat{\mathcal{J}}_l$'s generate on the tensor-product $:\hat{\mathbf{J}}_I\hat{\mathbf{J}}_{II}:$,

$$[\hat{\mathcal{J}}_l(\bar{y}, \bar{k}), :\hat{\mathbf{J}}_I\hat{\mathbf{J}}_{II}:(y; \bar{y}', k', \bar{k}')] = -(\hbar/2a)(\sigma_{I,II})_l : \hat{\mathbf{J}}_I\hat{\mathbf{J}}_{II} : (y; \bar{y}', k', \bar{k}')\delta(\bar{y} - \bar{y}')\delta_{\bar{k}\bar{k}'}, \tag{9}$$

where $(\sigma_{I,II})_l \equiv (\sigma_l)_I + (\sigma_l)_{II}$. We neglect writing the $\hat{\mathbf{T}}_{I,II}, \hat{\mathbf{S}}_{I,II}$ factors in the normal order product, since the equation is true for both cases.

Eqs.(8) and (9) imply that each column matrix-elements of $\hat{\mathbf{J}}$ form a $|\frac{1}{2}\frac{1}{2}>$ doublet ket-vector-space for $\hat{\mathcal{J}}_l$, and $\hat{\mathbf{J}}_I\hat{\mathbf{J}}_{II}$ form a $|\frac{1}{2}\frac{1}{2}>|\frac{1}{2}\frac{1}{2}>, |\frac{1}{2}\frac{1}{2}>|\frac{1}{2}-\frac{1}{2}>, |\frac{1}{2}-\frac{1}{2}>|\frac{1}{2}\frac{1}{2}>, |\frac{1}{2},-\frac{1}{2}>|\frac{1}{2},-\frac{1}{2}>$ tensor-product ket-vector-space for $\hat{\mathcal{J}}_l$. We can decompose the tensor-product vector-space into irreducible vector-spaces for $\hat{\mathcal{J}}_l$: the triplet ket-vector-space formed by the ket-vectors $|\frac{1}{2}\frac{1}{2}>|\frac{1}{2}\frac{1}{2}>, \frac{1}{2}\{|\frac{1}{2}\frac{1}{2}>|\frac{1}{2}-\frac{1}{2}> + |\frac{1}{2}-\frac{1}{2}>|\frac{1}{2}\frac{1}{2}>\}$, and $|\frac{1}{2},-\frac{1}{2}>|\frac{1}{2},-\frac{1}{2}>$; and singlet ket-vector-space by the ket-vector $\frac{1}{2}\{|\frac{1}{2}\frac{1}{2}>|\frac{1}{2}-\frac{1}{2}> - |\frac{1}{2}-\frac{1}{2}>|\frac{1}{2}\frac{1}{2}>\}$. The projection-matrices on the left of $\hat{\mathbf{J}}_I\hat{\mathbf{J}}_{II}$ to form these ket-triplet and ket-singlet, respectively, are

$$t_{I,II} = \begin{pmatrix} 1 & \begin{matrix} 0 & 0 \end{matrix} & 0 \\ \begin{matrix} 0 \\ 0 \end{matrix} & \frac{1}{2}\begin{pmatrix} 1 & 1 \\ 1 & 1 \end{pmatrix} & \begin{matrix} 0 \\ 0 \end{matrix} \\ 0 & \begin{matrix} 0 & 0 \end{matrix} & 1 \end{pmatrix}, \quad s_{I,II} = \begin{pmatrix} 0 & \begin{matrix} 0 & 0 \end{matrix} & 0 \\ \begin{matrix} 0 \\ 0 \end{matrix} & \frac{1}{2}\begin{pmatrix} 1 & -1 \\ -1 & 1 \end{pmatrix} & \begin{matrix} 0 \\ 0 \end{matrix} \\ 0 & \begin{matrix} 0 & 0 \end{matrix} & 0 \end{pmatrix}. \tag{10}$$

They satisfy the projection-matrix relations: $(t_{I,II})^2 = t_{I,II}, (s_{I,II})^2 = s_{I,II}$ and $t_{I,II}s_{I,II} = 0 = s_{I,II}t_{I,II}$. Eq. (9) can now be written seperately for the left-triplet and left-singlet, respectively, as

$$[\hat{\mathcal{J}}_l(\bar{y}, \bar{k}), t_{I,II} : \hat{\mathbf{J}}_I\hat{\mathbf{J}}_{II} : (y; \bar{y}', k'\bar{k}')] = -(\hbar/2a)(\sigma_{I,II})_l \; t_{I,II} : \hat{\mathbf{J}}_I\hat{\mathbf{J}}_{II} : (y; \bar{y}'k', \bar{k}')\delta(\bar{y} - \bar{y}')\delta_{\bar{k}\bar{k}'},$$

$$(11)$$

$$[\hat{\mathcal{J}}_l(\bar{y},\bar{k}), s_{I,II} : \hat{\mathbf{J}}_I\hat{\mathbf{J}}_{II} : (y;\bar{y}',k'\bar{k}')] = 0, \tag{12}$$

since $(\sigma_{I,II})_l s_{I,II} = 0$. Note that $t_{I,II}, s_{I,II}$ can be constructed out of the Casimir matrices of $(\sigma_{I,II})_l$, $t_{I,II} = (1/4)\{(\sigma_{I,II})_3(\sigma_{I,II})_3/2 + (\sigma_{I,II})_+(\sigma_{I,II})_- + (\sigma_{I,II})_-(\sigma_{I,II})_+\}$ and $s_{I,II} = 1_{I,II} - t_{I,II}$.

The reason that we go into such detail of this quite obvious relation between transformations in tensor-product ket-vector-spaces and the original ket-vector-space is to prepare for the later discussions on quantum-group vector-spaces formed by each row matrix-elements of $\hat{\mathbf{J}}$, and $\hat{\mathbf{J}}_I\hat{\mathbf{J}}_{II}$. The simple and conventional transformation properties of the tensor-product ket-vector-spaces formed by each column matrix-elements $\hat{\mathbf{J}}_I\hat{\mathbf{J}}_{II}$ are consequences of the fact that the left-structure-matrix of the exchange-algebra is the simple identity $\mathbf{1}_{I,II}$, Eq. (4). The right-structure-matrix being nontrivial signals new transformation properties of the tensor-product bra-vector-spaces. We shall show that each row matrix-elements of $\hat{\mathbf{J}}$ and those of $(\hat{\mathbf{J}}_I\hat{\mathbf{J}}_{II})$ form non-commutative bra-vector-space and tensor-product bra-vector-spaces, respectively, of a quantum-group.

In the general frames of the zero-mode first-class-constraint fields,[1,4] $\hat{\mathbf{T}}_{I,II}$ and $\hat{\mathbf{S}}_{I,II}$ of $\hat{\mathbf{R}}_{I,II}$ in Eq. (5) are very complicated 4×4 matrix functions of the zero-mode first-class-constraint fields. It took us some effort to decode their full meaning, but once understood they became clear and beautiful. We found out that the dynamical-variable-dependent projection matrices $\hat{\mathbf{T}}_{I,II}$, $\hat{\mathbf{S}}_{I,II}$ are related to c-number projection matrices $\mathbf{t}^q_{I,II}, \mathbf{s}^q_{I,II}$ by a similarity transformation

$$\hat{\mathbf{T}}_{I,II} = \hat{U}_{I,II}(y;k,\bar{k})\ \mathbf{t}^q_{I,II}\ \hat{U}^{-1}_{I,II}(y;k,\bar{k}), \quad \hat{\mathbf{S}}_{I,II} = \hat{U}_{I,II}(y;k,\bar{k})\ \mathbf{s}^q_{I,II}\ \hat{U}^{-1}_{I,II}(y;k,\bar{k}), \tag{13}$$

$$\mathbf{t}^q_{I,II} = \begin{pmatrix} 1 & 0 & 0 & 0 \\ 0 & \multicolumn{2}{c}{} & 0 \\ 0 & d\begin{pmatrix} q & 1 \\ 1 & q^{-1}\end{pmatrix} & & 0 \\ 0 & 0 & 0 & 1 \end{pmatrix}, \quad \mathbf{s}^q_{I,II} = \mathbf{1}_{I,II} - \mathbf{t}^q_{I,II} = \begin{pmatrix} 0 & 0 & 0 & 0 \\ 0 & & & 0 \\ 0 & d\begin{pmatrix} q^{-1} & -1 \\ -1 & q\end{pmatrix} & & 0 \\ 0 & 0 & 0 & 0 \end{pmatrix}, \tag{14}$$

$q \equiv e^{-i\hbar/(2\alpha a^2)}$ and $d = 1/(q+q^{-1})$. The similarity-transformation matrices $\hat{U}_{I,II}(y;k,\bar{k})$ and its inverse $\hat{U}^{-1}_{I,II}(y;k,\bar{k})$ are complicated 4×4 matrices depending on the zero-mode first-class-constraint field. We shall publish their explicit forms elsewhere[4] since our following discussions do not depend on them.

Eq.(13) and Eq.(14) imply that the relevant triplet-states and singlet-state formed from the row-matrix-elements of $(\hat{\mathbf{J}}_I\hat{\mathbf{J}}_{II})$, that have definite critical exponents, are q-deformed: the bra-triplet-states are $\hat{U}_{I,II}$-transformed q-deformed-states $< \frac{1}{2}\frac{1}{2}| < \frac{1}{2}\frac{1}{2}|, (q+q^{-1})^{-1}\{q < \frac{1}{2}\frac{1}{2}| < \frac{1}{2} - \frac{1}{2}| + < \frac{1}{2} - \frac{1}{2}| < \frac{1}{2}\frac{1}{2}|\}$, and $< \frac{1}{2} - \frac{1}{2}| < \frac{1}{2} - \frac{1}{2}|$; and the bra-singlet-state is the $\hat{U}_{I,II}$-transformed q-deformed-state $(q+q^{-1})^{-1}\{q^{-1} < \frac{1}{2}\frac{1}{2}| < \frac{1}{2} - \frac{1}{2}| - < \frac{1}{2} - \frac{1}{2}| < \frac{1}{2}\frac{1}{2}|\}$. $\hat{\mathbf{T}}_{I,II}$ and $\hat{\mathbf{S}}_{I,II}$ are the right-projection operators which project the bra-tensor-product vector-spaces $(\hat{\mathbf{J}}_I\hat{\mathbf{J}}_{II})$ into the q-deformed bra-triplet and bra-singlet states, respectively. These are noncommutative irreducible tensor-product vector-spaces of the quantum-groups (Hopf algebras).

In order to guarantee the products of $\hat{\mathbf{J}}$ fields have a well-defined meaning, the non-trivial structure-matrix $\hat{\mathbf{R}}$ must satisfy the following two conditions (a) and (b):[10–14]

(a) the Yang-Baxter equation

$$\hat{\mathbf{R}}_{I,II}(\bar{y}_1-\bar{y}_2)\check{\mathbf{R}}_{I,III;II}(\bar{y}_1-\bar{y}_3)\hat{\mathbf{R}}_{II,III}(\bar{y}_2-\bar{y}_3) = \check{\mathbf{R}}_{II,III;I}(\bar{y}_2-\bar{y}_3)\hat{\mathbf{R}}_{I,III}(\bar{y}_1-\bar{y}_3)\check{\mathbf{R}}_{I,II;III}(\bar{y}_1-\bar{y}_2), \tag{15}$$

where the matrices $\check{\mathbf{R}}$'s are defined by $\hat{\mathbf{J}}_{III}\check{\mathbf{R}}_{I,II;III} = \hat{\mathbf{R}}_{I,II}\hat{\mathbf{J}}_{III}$, and others by permuting the Roman subnumerals accordingly [Note again that we have neglected writing out the same point of $y, k, \bar{k}$ that all the $\hat{\mathbf{R}}$'s are at]; their explicit expressions were given in Ref. [4]. As mentioned before in the Abelian case, this is to make the two ways of arriving at $\hat{\mathbf{J}}_{III}(\bar{y}_3)\hat{\mathbf{J}}_{II}(\bar{y}_2)\hat{\mathbf{J}}_I(\bar{y}_1)$ from $\hat{\mathbf{J}}_I(\bar{y}_1)\hat{\mathbf{J}}_{II}(\bar{y}_2)\hat{\mathbf{J}}_{III}(\bar{y}_3)$ the same. We can show[4] that the explicit expressions of our $\hat{\mathbf{R}}$ and $\check{\mathbf{R}}$ matrices satisfy the Yang-Baxter equation, Eq.(15). In the $\hbar \to 0$ limit, $\hat{\mathbf{R}} \approx 1 - i\hbar\hat{\mathbf{r}}$ and $\check{\mathbf{R}} - \hat{\mathbf{R}} \approx 1 - \hbar^2\check{\mathbf{r}}$, the Yang-Baxter equation Eq. (15) reduces to its classical form $[\hat{\mathbf{r}}_{I,II}, \hat{\mathbf{r}}_{III,I}] + [\hat{\mathbf{r}}_{II,III}, \hat{\mathbf{r}}_{I,II}] + [\hat{\mathbf{r}}_{III,I}, \hat{\mathbf{r}}_{II,III}] + \check{\mathbf{r}}_{I,II;III} + \check{\mathbf{r}}_{II,III;I} + \check{\mathbf{r}}_{III,I;II} = 0$, where the $\mathbf{r}$'s are the same points in $\bar{y}$ (as well as in $y, k, \bar{k}$) as their corresponding $\mathbf{R}$'s;

(b) q-deformation and quantum-group operations; this is to ensure consistency of the exchange-algebras for the vector-space as well as for the tensor-product $[\hat{\mathbf{J}}_I(\bar{y}_1)\,\hat{\mathbf{J}}_{II}(\bar{y}_2)]\hat{\mathbf{J}}_{III}(\bar{y}_3) \equiv \hat{\mathbf{J}}_{III}(\bar{y}_3)[\hat{\mathbf{J}}_I(\bar{y}_1)\hat{\mathbf{J}}_{II}(\bar{y}_2)]\hat{\mathbf{R}}_{I\otimes II,III} = \hat{\mathbf{J}}_I(\bar{y}_1)\hat{\mathbf{J}}_{II}(\bar{y}_3)\hat{\mathbf{J}}_{II}(\bar{y}_2)\hat{\mathbf{R}}_{II,III}(\bar{y}_2-\bar{y}_3) = \hat{\mathbf{J}}_{III}(\bar{y}_3)\hat{\mathbf{J}}_I(\bar{y}_1)\hat{\mathbf{J}}_{II}(\bar{y}_2)$ $\check{\mathbf{R}}_{I,III;II}(\bar{y}_1 - \bar{y}_3)\hat{\mathbf{R}}_{II,III}(\bar{y}_2 - \bar{y}_3)$, where $\hat{\mathbf{R}}_{I\otimes II,III}$ is the structure-matrix for the exchange-algebra of the tensor-product $[\hat{\mathbf{J}}_I\hat{\mathbf{J}}_{II}]$ and $\hat{\mathbf{J}}_{III}$. This condition requires

$$\hat{\mathbf{R}}_{I\otimes II,III}(\bar{y}_1 - \bar{y}_2) = \check{\mathbf{R}}_{I,III;II}(\bar{y}_1 - \bar{y}_3)\hat{\mathbf{R}}_{II,III}(\bar{y}_2 - \bar{y}_3). \tag{16}$$

The explicit form of $\hat{\mathbf{R}}_{I\otimes II,III}$ can be calculated. Before we give it, we need to write out

$\hat{U}_{I,II}$ more explicitly. In the $\hat{U}_{I,II}$-untransformed frame where $\hat{\mathbf{J}}$ becomes $\tilde{\mathbf{J}}$,

$$\hat{\mathbf{J}}\hat{\mathbf{u}} \equiv \tilde{\mathbf{J}} \quad and \quad \hat{U}_{I,II} \equiv \breve{\mathbf{u}}_{I;II}\hat{\mathbf{u}}_{II}, \tag{17}$$

where $\breve{\mathbf{u}}$ is related to $\hat{\mathbf{u}}$ by $\tilde{\mathbf{J}}_{II}\breve{\mathbf{u}}_{I;II} \equiv \hat{\mathbf{u}}_I\tilde{\mathbf{J}}_{II}$. Note that both $\hat{\mathbf{u}}$ and $\breve{\mathbf{u}}$ are functions of the zero-mode first-class-constraint fields which depend on y, k, and $\bar{k}$. The right-structure-matrix for the $\tilde{\mathbf{J}}$-field exchange-algebra is $\mathbf{R}^q_{I,II}$ and related to $\hat{\mathbf{R}}_{I,II}$ by $P_{I,II}\mathbf{R}^q_{I,II} = \hat{U}^{-1}_{I,II}P_{I,II}\hat{\mathbf{R}}_{I,II}\hat{U}_{I,II}$, *i.e.*, $\mathbf{R}^q_{I,II}$ is expressed by Eq.(5) of $\hat{\mathbf{R}}_{I,II}$ with $\hat{\mathbf{T}}_{I,II}$ and $\hat{\mathbf{S}}_{I,II}$ replaced by $\mathbf{t}^q{}_{I,II}$ and $\mathbf{s}^q{}_{I,II}$, respectively. In this $\tilde{\mathbf{J}}$-frame, $\mathbf{R}^q{}_{I\otimes II,III}$ is just a representation of Drinfeld's universal $\mathbf{R}$ matrix,[14] $\mathbf{R}^q_{I\otimes II,III} = q^{-\frac{1}{2}(\sigma^q{}_{I,II})_3(\sigma_3)_{III}} \exp_q\{-(q-q^{-1})q^{-(\sigma^q_{I,II})_3}[(\sigma^q_{I,II})_+][(\sigma_- q^{\sigma_3})_{III}]\}$ where $\exp_q\{x\} \equiv \Sigma^{\infty}_{n=0} q^{-\frac{1}{2}n(n-1)}\, x^n[\prod^n_{m=1}(q^m - q^{-m})/(q-q^{-1})]^{-1}$. Using the relation $\mathbf{R}^q_{I\otimes II,III} = \mathbf{R}^q_{I,III}\mathbf{R}^q_{II,III}$ and the relation between $\mathbf{R}^q$ to $\hat{\mathbf{R}}$, we can obtain $\hat{\mathbf{R}}_{I\otimes II,III}$.

It is interesting to note that condition (a) alone, satisfying the Yang-Baxter equation, does not dictate the q-deformation of the $\hat{\mathbf{R}}$ matrix, Eqs.(13) and (14). For example $P_{I,II}$ satisfies the Yang-Baxter equation but does not satisfy condition (b). Consistently, as $\hbar \to 0$, $\hat{\mathbf{R}}_{I,II} \to \mathbf{1}_{I,II}$;not $\to P_{I,II}$.

From these results we have demonstrated the need of the quantum-group in 4-d quantum-field-theory due to the physical requirement of defining irreducible representations for physical fields that have distinct critical-exponents. The quantum-group structures in physics were first observed in 2-d quantum-integrable-systems,[12,13] and Drinfeld[14] identified them to be the known Hopf algebras in mathematics. They are algebras that deal with transformation-rules on tensor-products of vector-spaces that have noncommutative elements. The noncommutative vector-space is formed here by the noncommutative row matrix-elements of the $\hat{\mathbf{J}}$ matrix-field and the noncommutative tensor-product-vector-spaces are formed by the row matrix-elements of $(\hat{\mathbf{J}}_I\hat{\mathbf{J}}_{II})$ and higher products of fields $\hat{\mathbf{J}}$. The quantum-group structures appear for the right-transformations since, in this y-time quantization, the right-structure-matrix $\hat{\mathbf{R}}$ in the exchange-algebra Eq. (4) is nontrivial.

In the process of developing these results, we have remained as critical as possible and made sure that there are no other natural alternatives; thus we have shown the necessity of Yang-Baxter equations and quantum-groups in 4-d quantum field theories. Quantizing the group-valued-fields $\mathbf{J}$, rather than conventional Lie-algebra-valued fields $\mathbf{A}_\mu$, is the main reason why we can see these beautiful structures which so far have been observed only in

2-d quantum-field-theories. The integrable-system approach makes it possible to quantize group-valued J-like fields of 4-d interacting theories.[9]

For simplicity, from now on we shall give equations for $\tilde{\mathbf{J}}$, which can be easily transformed into equations for $\hat{\mathbf{J}}$ using Eq.(17).

From the known quantum-group (Hopf algebra) structures[14], we can identify the quantum-group generators $\hat{\mathbf{e}}_l, (l = \pm, 3)$ for $\tilde{\mathbf{J}}$ and $:\tilde{\mathbf{J}}_I\tilde{\mathbf{J}}_{II}:$. Here we need to use the adjoint-operations of $\hat{\mathbf{e}}_l$ defined as $ad\hat{\mathbf{e}}_3 \circ A \equiv [\hat{\mathbf{e}}_3, A]_q \equiv [\hat{\mathbf{e}}_3, A]$, $ad\hat{\mathbf{e}}_- \circ A \equiv [\hat{\mathbf{e}}_-, A]_q \equiv q^{\hat{\mathbf{e}}_3/\hbar}[q^{-\hat{\mathbf{e}}_3/\hbar}\hat{\mathbf{e}}_-, A]$, $ad\hat{\mathbf{e}}_+ \circ A \equiv [\hat{\mathbf{e}}_+, A]_q \equiv [\hat{\mathbf{e}}_+ q^{-\hat{\mathbf{e}}_3/\hbar}, A]q^{\hat{\mathbf{e}}_3/\hbar}$, where [,] is the regular commutator bracket. We then obtain the following transformation relations,

$$ad\{\hat{\mathbf{e}}_l(y;k',\bar{k}')\} \circ \{: \tilde{\mathbf{J}}_I(y;\bar{y},k,\bar{k})\tilde{\mathbf{J}}_{II}(y;\bar{y},k,\bar{k})\mathbf{t}^q_{I,II} :\} \\ = \hbar\{: \tilde{\mathbf{J}}_I(y;\bar{y},k,\bar{k})\tilde{\mathbf{J}}_{II}(y;\bar{y},k,\bar{k})\mathbf{t}^q_{I,II} :\}(\sigma^q_{I,II})_l \; \delta_{k,k'}\delta_{\bar{k}\bar{k}'}, \tag{18}$$

$$ad\{\hat{\mathbf{e}}_l(y;k',\bar{k}')\} \circ \{: \tilde{\mathbf{J}}_I(y;\bar{y},k,\bar{k})\tilde{\mathbf{J}}_{II}(y;\bar{y},k,\bar{k})\mathbf{s}^q_{I,II} :\} = 0 \quad ; \quad \text{and} \tag{19}$$

$$ad\{\hat{\mathbf{e}}_l(y;k',\bar{k}')\} \circ \tilde{\mathbf{J}}(y;\bar{y},k,\bar{k}) = \hbar\tilde{\mathbf{J}}(y;\bar{y},k,\bar{k}) \; \sigma_l \; \delta_{kk'}\delta_{\bar{k}\bar{k}'}, \tag{20}$$

where $l = \pm, 3$; $(\sigma^q_{I,II})_+ \equiv (\sigma_+)_I(q^{-\sigma_3})_{II} + (\sigma_+)_{II}$, $(\sigma^q_{I,II})_- \equiv (\sigma_-)_I + (q^{\sigma_3})_I(\sigma_-)_{II}$, and $(\sigma^q_{I,II})_3 \equiv (\sigma_3)_I + (\sigma_3)_{II}$. The projection operators $\mathbf{t}^q_{I,II}, \mathbf{s}^q_{I,II}$ can be constructed in terms of the Casimir operator C of $(\sigma^q_{I,II})_l$, *i.e.*, $\mathbf{t}^q_{I,II} = (C - \rho_1)/(\rho_3 - \rho_1)$, $\mathbf{s}^q_{I,II} = \mathbf{1}_{I,II} - \mathbf{t}^q_{I,II}$, $C \equiv [q^{(\sigma^q_{I,II})_3+1} + q^{-(\sigma^q_{I,II})_3-1} - 2]/(q - q^{-1})^2 + [(\sigma^q_{I,II})_-][(\sigma^q_{I,II})_+]$, and $\rho_i = (q^i + q^{-i} - 2)/(q - q^{-1})^2$ with $i = 1$ and 3. The $\hat{\mathbf{e}}$'s satisfy the $U_q[sl(2)]$ quantum-group algebras: $[\frac{1}{\hbar}ad\hat{\mathbf{e}}_+, \frac{1}{\hbar}ad\hat{\mathbf{e}}_-] = (q^{\frac{1}{\hbar}ad\hat{\mathbf{e}}_3} - q^{-\frac{1}{\hbar}ad\hat{\mathbf{e}}_3})/(q - q^{-1})$, $[\frac{1}{\hbar}ad\hat{\mathbf{e}}_3, \frac{1}{\hbar}ad\hat{\mathbf{e}}_\pm] = \pm\frac{2}{\hbar}ad\hat{\mathbf{e}}_\pm$.

We shall see that the $\bar{k}$-summed transformations $ad\{\Sigma_{\bar{k}}\hat{\mathbf{e}}_l(y;k,\bar{k},)\}\circ \equiv ad\{\hat{\mathbf{E}}_l(y;k)\}\circ$ leave the interaction Hamiltonian $\hat{\mathbf{H}}_{\text{int}}$ and the equations of motion invariant. Note that $\hat{\mathbf{E}}_l(y;k)$'s are local quantum-group generators.

Here we see the non-trivial right-structure-matrix $\hat{\mathbf{R}}$ of the exchange-algebra dictates the short-distance nontrivial critical-exponents, which in turn requires the quantum-group $U_q[Sl(2)]$-right-transformations, [this actually breaks the original symmetry of the regular $Sl(2)$-right-transformations generated by the zero-mode first-class-constraint fields].[4]

In order to derive the quantum Hamiltonian and equation of motion we need to construct $(\tilde{\mathbf{J}})^{-1}$. We can show that the following defined field

$$(\tilde{\mathbf{J}})^{-1} \equiv \left[\begin{pmatrix} 0 & -1 \\ 1 & 0 \end{pmatrix} \tilde{\mathbf{J}} \begin{pmatrix} 0 & 1 \\ -q^{-1} & 0 \end{pmatrix}\right]^T, \tag{21}$$

with the superscript T meaning transpose, has the following properties; $(\tilde{\mathbf{J}}\tilde{\mathbf{J}}^{-1})$ commutes with $\tilde{\mathbf{J}}$, *i.e.*, $\tilde{\mathbf{J}}_I(\bar{y})[\tilde{\mathbf{J}}(\bar{y}')\tilde{\mathbf{J}}^{-1}(\bar{y}'')]_{II} = [\tilde{\mathbf{J}}(\bar{y}')\tilde{\mathbf{J}}^{-1}(\bar{y}'')]_{II}\tilde{\mathbf{J}}_I(\bar{y})$, for $\bar{y}$ not in between $\bar{y}'$ and y''; $(\hat{\mathbf{J}}\tilde{\mathbf{J}}^{-1})$ forms a singlet under both transformations Eq.(8) and Eq.(20) with the singlet-critical-dimension,[4] *i.e.*, $\tilde{\mathbf{J}}(\bar{y})[\tilde{\mathbf{J}}(\bar{y}')]^{-1} = (\bar{y}-\bar{y}')^{\Delta_{(1/2)(1/2),0}[\hbar/(2\pi\alpha a^2)]}[1_{2\times 2} + \mathcal{O}(\bar{y}-\bar{y}')]$; thus Eq. (21) defines the inverse of $\tilde{\mathbf{J}}$ under the procedure of normal-order-product, *i.e.*, $:\tilde{\mathbf{J}}\tilde{\mathbf{J}}^{-1}:(y) = 1_{2\times 2}$.

Now we give the quantum interaction Hamiltonian

$$\begin{aligned}\hat{\mathbf{H}}_{\rm int} = & -\alpha \int\int\int d\bar{y}dzd\bar{z}\Big\{Tr\{(\partial_{\bar{z}}\tilde{\mathbf{J}}^{-1})(\partial_z\tilde{\mathbf{J}})\} \\ & + \int_0^1 d\rho Tr\{(\tilde{\mathbf{J}}^{-1}\partial_\rho\tilde{\mathbf{J}})[(\partial_{\bar{z}}\tilde{\mathbf{J}}^{-1})(\partial_z\tilde{\mathbf{J}}) - (\partial_z\tilde{\mathbf{J}}^{-1})(\partial_{\bar{z}}\tilde{\mathbf{J}})]\}\Big\},\end{aligned} \tag{22}$$

from which we can derive the equation of motion, $\partial_y(:\tilde{\mathbf{J}}\partial_{\bar{y}}\tilde{\mathbf{J}}^{-1}:) = \frac{i}{\hbar}[\hat{\mathbf{H}}_{\rm int}, :\tilde{\mathbf{J}}\partial_y\tilde{\mathbf{J}}^{-1}:]$,

$$\partial_y(:\tilde{\mathbf{J}}\partial_{\bar{y}}\tilde{\mathbf{J}}^{-1}:) = -\partial_z(\tilde{\mathbf{J}}\partial_{\bar{z}}\tilde{\mathbf{J}}^{-1}). \tag{23}$$

In Eqs. (22) and (23) we use differentiation in ∂_z and $\partial_{\bar{z}}$ as a short-hand for the discretized differences, $\partial_z\hat{A}(z) \equiv [\hat{A}(k+1) - \hat{A}(k-1)]/2a$; the ρ direction is also similarly discretized. Using this way of taking differences, procedures of normal-order-product are avoided.

It is easy to show that $\hat{\mathbf{H}}_{\rm int}$ and the equation of motion are invariant under the transformations Eq. (8) and Eq. (20); thus the symmetries of the theory are local $\{Sl(2)\}_{\mathcal{L}} \times \{U_q[Sl(2)]\}_{\mathcal{R}}$, generated by $\hat{\mathcal{J}}_l(\bar{y},\bar{z})$ and $\hat{\mathbf{E}}_l(y;k)$, $(l=\pm,3)$, for the left- and right- transformations on $\tilde{\mathbf{J}}$, respectively. These symmetries correspond to the classical transformations[15] of $\mathbf{J} \to V(\bar{y},\bar{z})\mathbf{J}V(y,z)$.

Now the affine-algebra-current can also be constructed explicitly,

$$\hat{\mathcal{J}}(\bar{y},\bar{k}) \equiv 2\alpha a\Sigma_k\{:\hat{\mathbf{J}}(y;\bar{y},k,\bar{k})\partial_{\bar{y}}\hat{\mathbf{J}}^{-1}(y;\bar{y},k,\bar{k}):\} \tag{24}$$

which can be shown to be conserved $\partial_y\hat{\mathcal{J}}(\bar{y},\bar{k}) = 0$ by using the equation of motion; thus there is no dependence on y in $\hat{\mathcal{J}}(\bar{y},\bar{k})$. Note the generators $\hat{\mathcal{J}}$'s and $\hat{\mathbf{E}}_l$'s can be shown to commute, as they should since they generate separately the left and right transformations respectively on $\tilde{\mathbf{J}}$.

We can also construct the corresponding quantum Virasoro-algebra fields and hierarchies of nonlinear and linear systems given in our canonical formulation of Ref. [1] for quantum fields $\tilde{\mathbf{J}}$ in a rather straight forward way.

Besides these equal-y-time results discussed so far, there are the equal $\bar{y}$-, z-, and $\bar{z}$-times relations. For $\bar{y}$ as time, all the equations given above hold except with $y \leftrightarrow \bar{y}$. For z or $\bar{z}$ as time, all the equations given above hold except $z \leftrightarrow y$, $\bar{y} \leftrightarrow \bar{z}$; the left-structure-matrix of the exchange algebra is the nontrivial $\hat{\mathbf{R}}$ and the right-structure-matrix is the trivial identity $\mathbf{1}_{I,II}$ in Eq. (4); and symmetries are left ($\mathcal{L}$) right ($\mathcal{R}$) interchanged , *i.e.*$\{Sl(2)\}_{\mathcal{R}} \times \{U_q[Sl(2)]\}_{\mathcal{L}}$.

It is easy to see when reduced to two-dimensions this SDYM $\hat{\mathbf{J}}$-quantum-field-theory nicely reduces to the 2-d WZNW $\hat{\mathbf{g}}$-quantum-field-theory;[6] for example, when we take away the $z, \bar{z}$ dependences, Eqs. (22) and (23) reduce to $\hat{\mathbf{H}}_{\text{int}} = 0$ and $\partial_y : \tilde{\mathbf{J}}\partial_{\bar{y}}\tilde{\mathbf{J}} := 0$ which are the interaction-free WZNW equations. This reflects the fact that the classical SDYM systems can be reduced to many lower-dimensional integrable-systems.[16][9] Therefore, we expect that our $\hat{\mathbf{J}}$-quantum-field-theory will also have many interesting reduced quantum-field-systems. The SDYM system has remarkable properties. It is not a 2-d integrable-system nor a 2-d conformal field theory, thus not confined by the 2-d bondages, yet it has the resemblances of both. Even though these resemblances are not essential for us to develop this 4-d $\hat{\mathbf{J}}$-quantum-field-theory; nonetheless, they give us a feeling of having roots and continuity.

From this $\hat{\mathbf{J}}$-quantum-field-theory formulation, we finally begin to decipher the 4-d quantum-field-theoretical consequences implied by the classical integrable-system[9] and conformal characteristics[17] of the 4-d $\mathbf{J}$ fields. The interacting quantum-field Hamiltonian will tell us what blossoms in four-dimensions the theory will bring us for physics and mathematics.

Acknowledgement:

We would like to thank Professor S.S. Chern, Professor C.N. Yang, and the organizers of the conference for the interesting conference and for warm hospitality.

This work is partially supported by the U.S. Department of Energy.

References

1. L.L. Chau and I. Yamanaka, Phys. Rev. Lett. 68 (1992) 1807.

2. The classical field J of the SDYM system was introduced by C.N. Yang, Phys. Rev. Lett. 38 (1977) 1377; the classical J-field SDYM system has many nice integrable-system properties, see M.K. Prasad, A. Sinha, and L.L. Chau Wang, Phys. Rev. Lett. 43 (1979) 750; Phys. Rev. D 24 (1981) 1574; and L.L. Chau. Proceedings of "Nonlinear Phenomena," Oaxtepec, Mexico, Nov. 19 – Dec. 17, 1982, Lecture Notes in Physics#189, edited by K.B. Wolf. We have made the change (in Ref. [1] and here) to use light-cone and hyperbolic coordinates in $(+-+-)$-signatured 4-d space or $(+---)$-signatured 4-d space rather than a complex 4-d Euclidean space. Classical SDYM system had also been studied in the twistor formulation, see "Spinors and Space-time" by R. Penrose and W. Rindler, Cambridge; S. Ward, Phys. Lett. 61A (1977) 81; however, as far as we know, the twistor approach so far has only used the Lie-algebra-valued fields $\mathbf{A}_\mu$.

3. P.A.M. Dirac, "Lectures on Quantum Mechanics," and "Lectures on Quantum Field Theory", Yeshiva University, New York (1964).

4. This paper is an extended work of our previous preprint, L.L. Chau and I. Yamanaka, UCDPHYS-PUB-91-19, September 1991 (unpublished), where the exchange-algebras were given in the general frame of the zero-mode first-class-constraint field, $\hat{\mathbf{F}}(y;k,\bar{k}) = \sum_{l=\pm,3} \hat{\mathbf{f}}_l(y;k,\bar{k})\sigma_l$ and $\hat{\mathbf{R}}_{I,II}$ is a complicated 4×4 matrix depending on $\hat{\mathbf{f}}_l$. We eventually found the similarity transformation that bring $\hat{\mathbf{F}}$ to the diagonal form, $\hat{\mathbf{F}}^D(y;k,\bar{k}) = \hat{\mathbf{f}}_\hbar(y;k,\bar{k})\sigma_3$, $\hat{\mathbf{f}}_\hbar$ the Casimir operator $\hat{\mathbf{f}}_\hbar \equiv [\hbar^2 + \Sigma_l(\hat{\mathbf{f}}_l)^2]^{1/2}$; and the corresponding similarity transformations for $\hat{\mathbf{J}}, (\hat{\mathbf{J}}_I\hat{\mathbf{J}}_{II})$, and $\hat{\mathbf{R}}$ as used in the text. We shall present the detail discussions and explicit transformations in a future publication.

5. J. Wess and B. Zumino, Phys. Lett. 37B (1971) 95; Serge Novikov, Uspekhi Matematicheskikh Nauk, 5 (1982) 3; E. Witten, Commun. Math. Phys. 92 (1984) 455.

6. L.L. Chau and I. Yamanaka, "Quantization of the WZNW Model in the Light-Cone Coordinate: Quadratic Algebra and Hierarchy", UC Davis preprint UCDPHYS-PUB-91-5, Feb., 1991 (unpublished); and extended version May, 1992; the field of quantization is the group-valued g, in contrast to the Lie-algebra-valued fields $A_\mu = g\partial_\mu g^{-1}$ in Witten's paper of Ref.(5).

7. E. Witten, Phys. Lett. 77B (1978) 394; M. Sohnius, Nucl. Phys. B136 (1978) 461; For classical J-field formulation for the full supersymmetric Yang-Mills field, see L.L. Chau, in "Vertex Operators in Mathematics and Physics," Eds: J. Lepowsky, S. Madelstam,

and I.M. Singer, by Springer Verlag, 1983; I.V. Volovich, Phys. Lett. 129B (1983) 429 and Theor. Math. Phys. 57 (1983) 1269; C. Devchand, Nucl. Phys. B238 (1984) 333; L.L. Chau, M.-L. Ge and C.-S. Lim, Phys. Rev. 33 (1986) 1056;

8. L.L. Chau and C.S. Lim, Phys. Rev. Lett. 56 (1986) 294; Classical J-field-like formulation for the full conformal supergravity was done by L.L. Chau, Phys. Lett. 202 (1988) 238.

9. For an overall review see L.L. Chau, in Nankai Lectures on Math. Phys., "Integrable System," 1987, World Scientific.

10. C.N. Yang, Phys. Rev. Lett. 19 (1967) 1312.

11. R.J. Baxter, "Exactly Solved Models for Statistical Mechanics," Academic Press 1982.

12. L.D. Faddeev, in Proceedings Les Houches XXXIX (1982), Elsevier Science Publishers B.V., 1984; P. Kulish and N. Reshetikhin, Zap. Nauch. Semin. LOMI, 101 (1981) 101 (in Russian); For a recent review, see L. Faddeev, N. Reshetikhin and L. Takhtajan, in "Algebraic Analysis", Eds: M. Kashiwara and T. Kawai, Academic Press 1988.

13. M. Jimbo, Comm. Math. Phys. 102 (1986), 537; Lett. Math. Phys. 10 (1985) 63.

14. V. Drinfeld, Akad. Nauk SSSR, 283 (1985) 1060. V. Drinfeld, "Quantum groups", Proc. Int. Congr. Math. Berkeley, 1986, pp. 798.

15. M.K. Prasad, A. Sinha, and L.L. Chau Wang, Phys. Rev. D. 23 (1981) 2321. The $V(y,z)$, $\bar{V}(\bar{y},\bar{z})$ symmetry of the J-field SDYM equations was extensively discussed in this paper and was used to prove the reality of the multimonopole solutions.

16. L. Witten, Phys. Rev. D19 (1979) 718; R.S. Ward, Phil. Trans. Roy. Soc. Lond. A315 (1985) 451; L.J. Mason and G.A.J. Sparling, Phys. Lett. A137 (1989) 29; S. Chakravarty, M.J. Ablowitz and P.A. Clarkson, Phys. Rev. Lett. 65 (1990) 1085; L.-L. Chau and I. Yamanaka, UC Davis preprint (1991).

17. The conformal connections of classical J-field SDYM system are twofold; one is through its reduction to the 2-d g-field WZNW model, and the other is through the linear systems for the J-fields (which are generalizations of Cauchy-Riemann equations). For linear systems in J fields see L.L. Chau, Proceedings of the Guanzhou (Canton) conference on the Theoretical Particle Physics, Jan. 1980, Van Notstrand Reinhold and Science Press; K. Pohlmeyer, Comm. Math. Phys. 72 (1980) 37. The quantum field connections to the latter will be worked out in a future publication.

Gauge Theory of Diff S^1 and Affine Lie Groups

Y. M. Cho

Department of Physics
Seoul National University
Seoul 151-742, Korea

Abstract

We present an effective field theory of an internal string by constructing a gauge theory of the Virasoro-Kac-Moody symmetry associated with $P \otimes G$, where P is the Poincaré group and G is the grand unification group. The theory automatically breaks the Virasoro-Kac-Moody group down to $P \otimes H \otimes U(1)$, where H is a subgroup of G and $U(1)$ is the Cartan's subgroup of the Virasoro group. After the spontaneous symmetry breaking the mass spectrum of the theory is characterized by two completely different mass scales, the Planck scale which is responsible for the harmonic oscillator spectrum of the string and the elementary particle scale which is responsible for the fine-structure of the mass spectrum. The theory allows a straightforward supersymmetric generalization.

During the last decade the string theory [1,2] has become a leading candidate of the theory of everything, the ultimate theory which unifies all the interactions in nature, and has been studied extensively. In spite of the extensive studies, however, an effective field theory of string which could describe the dynamics of a propagating colored string in the 4-dimensional curved space-time, has not been available so far. The purpose of this talk is to review a recent development on the subject [3,4] and to present a gauge theory of the Virasoro-Kac-Moody symmetry associated with the Poincaré group P and the grand unification group G, which could provide a field-theoretic description of a colored string [5]. After an inevitable spontaneous symmetry breaking *the particle spectrum of the theory consists of an infinite tower of massive spin-two and spin-one particles and a finite number of the massless particles made of the graviton as well as the gauge fields of the unbroken subgroup* $H \otimes U(1)$, *where* H *is a subgroup of* G *and* $U(1)$ *is the Cartan's subgroup of the Virasoro group. In addition the theory contains light scalar particles, the dilaton and an adjoint multiplet of* H, *as the pseudo-Goldstone particles of the symmetry breaking.* These are precisely what one can expect for a string theory.

In the limit where the size of the string can be neglected, an effective field theory of string must satisfy the following minimal requirements. First of all such a theory must have the Virasoro invariance before an inevitable spontaneous symmetry breaking [3,4]. This is so because the Virasoro group is the fundamental symmetry of the string theory which assures the reparametrization invariance of the string. Secondly it must have the general invariance to accomodate the gravitational interaction. In fact it is more likely that the theory must have the Virasoro-Kac-Moody symmetry associated with the Poincaré group P, if it can succesfully describe the propagation of the string in a 4-dimensional curved space-time [5,6]. Finally it must have the Kac-Moody symmetry associated with the color $SU(3)$ to accomodate the strong interaction [7,8]. Indeed it must have the Kac-Moody symmetry associated with the grand unification group G to achieve the full unification of all the interactions. So what we need for an effective field theory of string (in the limit that the string can be approximated to a point) is a gauge theory of the Virasoro-Kac-Moody symmetry associated with the Poincaré group and the grand unification group.

To construct such a theory we need a general representation of the Virasoro-Kac-Moody symmetry associated with an arbitrary non-abelian group G. We first present a representation which we call *the product representation* [5,7]. The Virasoro-Kac-Moody group is the semi-direct product group of the Virasoro group and the Kac-Moody group whose algebra is given by

$$\begin{aligned}
[L_m, L_n] &= f_{mn}{}^k L_k = (m-n)L_{m+n} \\
[L_m, T_{an}] &= f_{m\ an}{}^{ck} T_{ck} = -nT_{am+n} \\
[T_{am}, T_{bn}] &= f_{am\ bn}{}^{ck} T_{ck} = f_{ab}{}^c T_{cm+n},
\end{aligned} \tag{1}$$

where k, m, n are the indices of the Virasoro group and a, b, c are the indices of the associated group G. Notice that here we do not consider the central extension because we identify the symmetry as a group, not as an algebra. The product representation is given by

$$\begin{aligned}
(L_m)_n{}^k &= [(\alpha+1)m + \beta - k]\delta_{m+n}{}^k \\
(T_{am})_{\beta n}{}^{\alpha k} &= (T_a)_\beta{}^\alpha (T_m)_n{}^k = (T_a)_\beta{}^\alpha \delta_{m+n}{}^k.
\end{aligned} \tag{2}$$

Notice that L_m forms an arbitrary (α, β)-representation of the Virasoro group [3,9] and T_{am} forms a representation of the Kac-Moody group made of two matrices T_a and T_m, where T_a is an arbitrary matrix representation of G which acts only on the indices of the associated group and T_m is an infinite-dimensional matrix which acts only on the indices of the Virasoro group. When T_a forms an adjoint representation, we obtain the following product representation ϕ^{ak}

$$\begin{aligned}
\delta_\xi \phi^{ck} &= i\xi^m (L_m)_n{}^k \phi^{cn} = i[(\alpha+1)m + \beta - k]\xi^m \phi^{ck-m} \\
\delta_{\vec{\alpha}} \phi^{ck} &= \alpha^{am} (T_{am})_{bn}{}^{ck} \phi^{bn} = f_{ab}{}^c \alpha^{am} \phi^{bk-m},
\end{aligned} \tag{3}$$

where ξ^m and α^{am} are the infinitesimal parameters of the Virasoro-Kac-Moody group.

With this preliminary we now show how to construct a gauge theory of the Virasoro-Kac-Moody symmetry associated with the Poincaré group and the grand unification group. The simplest way to construct such a theory is to start from the Einstein-Yang-Mills theory based on the principal fibre bundle $P(M_5, G)$, whose fibre is the grand unification group G and the base manifold is a 5-dimensional manifold M_5 which itself forms a fibre bundle $B(M_4, S^1)$ made of the closed string S^1 and the 4-dimensional space-time M_4. In a coordinate basis $(\partial_\mu \oplus \partial_5)$ the most general metric on M_5 can be written as [4,10]

$$\gamma_{AB} = \begin{pmatrix} \gamma_{\mu\nu} + e^2\kappa^2\phi^2 A_\mu A_\nu & e\kappa A_\mu \phi^2 \\ e\kappa\phi^2 A_\nu & \phi^2 \end{pmatrix}, \tag{4}$$

where e is the coupling constant of the Virasoro group and κ is a scale parameter which characterizes the size of the string. But in the block-diagonal basis $(\hat{\partial}_\mu \oplus \partial_5)$ where [4,10]

$$\begin{aligned} &\hat{\partial}_\mu = \partial_\mu - e\kappa A_\mu \partial_5 \\ &[\hat{\partial}_\mu, \hat{\partial}_\nu] = -e\kappa F_{\mu\nu}\partial_5 \ , \ [\hat{\partial}_\mu, \partial_5] = e(\partial_\theta A_\mu)\partial_5 \\ &F_{\mu\nu} = \partial_\mu A_\nu - \partial_\nu A_\mu - e(A_\mu \partial_\theta A_\nu - A_\nu \partial_\theta A_\mu), \end{aligned} \tag{5}$$

we have

$$\gamma_{AB} = \begin{pmatrix} \gamma_{\mu\nu} & 0 \\ 0 & \phi^2 \end{pmatrix}. \tag{6}$$

Notice that here we have identified x^5 with $\kappa\theta$, where θ is the dimensionless string coordinate of S^1.

Let the gauge potentials of G in the coordinate basis $(\partial_\mu \oplus \partial_5)$ be $A_\mu{}^a$ and ϕ^a. Clearly the field strengths of the potentials are given by

$$\begin{aligned} F_{\mu\nu}{}^c &= \partial_\mu A_\nu{}^c - \partial_\nu A_\mu{}^c + g f_{ab}{}^c A_\mu{}^a A_\nu{}^b \\ F_{\mu 5}{}^c &= \partial_\mu \phi^c + g f_{ab}{}^c A_\mu{}^a \phi^b - \frac{1}{\kappa}\partial_\theta A_\mu{}^c \\ &= \hat{D}_\mu \phi^c. \end{aligned} \tag{7}$$

In the basis $(\hat{\partial}_\mu \oplus \partial_5)$, however, we have [11]

$$\begin{aligned} \hat{F}_{\mu\nu}{}^c &= \hat{\partial}_\mu \hat{A}_\nu{}^c - \hat{\partial}_\nu \hat{A}_\mu{}^c + g f_{ab}{}^c \hat{A}_\mu{}^a \hat{A}_\nu{}^b + e\kappa F_{\mu\nu}\phi^c \\ \hat{F}_{\mu 5}{}^c &= \hat{\partial}_\mu \phi^c + g f_{ab}{}^c \hat{A}_\mu{}^a \phi^b - \frac{1}{\kappa}\partial_\theta \hat{A}_\mu{}^c - e(\partial_\theta A_\mu)\phi^c, \end{aligned} \tag{8}$$

where $\hat{A}_\mu{}^c = A_\mu{}^c - e\kappa A_\mu \phi^c$ is the gauge potential in the basis $\hat{\partial}_\mu$. Notice that

$$\begin{aligned} \hat{F}_{\mu\nu}{}^c &= F_{\mu\nu}{}^c + e\kappa A_\mu(\hat{D}_\nu \phi^c) - e\kappa A_\nu(\hat{D}_\mu \phi^c) \\ \hat{F}_{\mu 5}{}^c &= F_{\mu 5}{}^c. \end{aligned} \tag{9}$$

Obviously the Einstein-Yang-Mills theory on M_5 must be invariant under the following (infinitesimal) general coordinate transformation

$$
\begin{aligned}
&x^\mu \to x^{\mu'} = x^\mu + \xi^\mu(x,\theta)\\
&\theta \to \theta' = \theta + \xi(x,\theta).
\end{aligned} \tag{10}
$$

The coordinate transformation generated by ξ^μ describes the gauge transformation of the Kac-Moody symmetry associated with the Poincaré group (or more precisely the group of the 4-dimensional general coordinate transformation), and the one generated by ξ describes the gauge transformation of the Virasoro group [4,12]. Under the Virasoro transformation we have

$$
\begin{aligned}
&\delta_\xi \gamma_{\mu\nu} = -\xi\partial_\theta \gamma_{\mu\nu}\\
&\delta_\xi A_\mu = -\frac{1}{e}\partial_\mu \xi + (\partial_\theta\xi)A_\mu - \xi\partial_\theta A_\mu\\
&\delta_\xi \phi = -(\partial_\theta\xi)\phi - \xi\partial_\theta\phi\\
&\delta_\xi \hat{A}_\mu{}^c = -\xi\partial_\theta \hat{A}_\mu{}^c\\
&\delta_\xi \phi^c = -(\partial_\theta\xi)\phi^c - \xi\partial_\theta\phi^c,
\end{aligned} \tag{11}
$$

from which we obtain

$$
\begin{aligned}
&\delta_\xi F_{\mu\nu} = (\partial_\theta\xi)F_{\mu\nu} - \xi\partial_\theta F_{\mu\nu}\\
&\delta_\xi \hat{F}_{\mu\nu}{}^c = -\xi\partial_\theta \hat{F}_{\mu\nu}{}^c\\
&\delta_\xi(\hat{D}_\mu\phi^c) = -(\partial_\theta\xi)\hat{D}_\mu\phi^c - \xi\partial_\theta(\hat{D}_\mu\phi^c).
\end{aligned} \tag{12}
$$

Upon the Fourier decomposition (11) can be written as [4,7]

$$
\begin{aligned}
&\delta_\xi \gamma_{\mu\nu}{}^k = i(m-k)\xi^m \gamma_{\mu\nu}{}^{k-m}\\
&\delta_\xi A_\mu{}^k = -\frac{1}{e}\partial_\mu \xi^k + i(2m-k)\xi^m A_\mu{}^{k-m}\\
&\delta_\xi \phi^k = -ik\xi^m \phi^{k-m}\\
&\delta_\xi \hat{A}_\mu{}^{ck} = i(m-k)\xi^m \hat{A}_\mu{}^{ck-m}\\
&\delta_\xi \phi^{ck} = -ik\xi^m \phi^{ck-m}.
\end{aligned} \tag{13}
$$

The result shows that $\gamma_{\mu\nu}{}^k$, $\hat{A}_\mu{}^{ck}$, ϕ^k, ϕ^{ck} forms a $(0,0)$, $(0,0)$, $(-1,0)$, $(-1,0)$ representation of the Virasoro group, respectively. More importantly it shows that $A_\mu{}^k$ transforms exactly as the gauge potential of the Virasoro group. From (13) we have

$$
\begin{aligned}
&\delta_\xi F_{\mu\nu}{}^k = i(2m-k)\xi^m F_{\mu\nu}{}^{k-m}\\
&\delta_\xi \hat{F}_{\mu\nu}{}^{ck} = i(m-k)\xi^m \hat{F}_{\mu\nu}{}^{ck-m}\\
&\delta_\xi(\hat{D}_\mu\phi^{ck}) = -ik\xi^m(\hat{D}_\mu\phi^{ck-m}),
\end{aligned} \tag{14}
$$

which shows that $F_{\mu\nu}{}^k$, $\hat{F}_{\mu\nu}{}^{ck}$, $\hat{D}_\mu\phi^{ck}$ forms a $(1,0)$, $(0,0)$, $(-1,0)$ representation of the Virasoro group, respectively.

Under the infinitesimal gauge transformation of G generated by $\vec{\alpha}(x,\theta)$, we have

$$\begin{aligned}
&\delta_{\vec{\alpha}}\gamma_{\mu\nu} = 0, \quad \delta_{\vec{\alpha}}A_\mu = 0, \quad \delta_{\vec{\alpha}}\phi = 0 \\
&\delta_{\vec{\alpha}}\hat{A}_\mu{}^c = -\frac{1}{g}\partial_\mu\alpha^c + f_{ab}{}^c\alpha^a\hat{A}_\mu{}^b \\
&\delta_{\vec{\alpha}}\phi^c = -\frac{1}{g\kappa}\partial_\theta\alpha^c + f_{ab}{}^c\alpha^a\phi^b.
\end{aligned} \tag{15}$$

This can be written as

$$\begin{aligned}
&\delta_{\vec{\alpha}}\gamma_{\mu\nu}{}^k = 0, \quad \delta_{\vec{\alpha}}A_\mu{}^k = 0, \quad \delta_{\vec{\alpha}}\phi^k = 0 \\
&\delta_{\vec{\alpha}}\hat{A}_\mu{}^{ck} = -\frac{1}{g}\partial_\mu\alpha^{ck} + i\frac{e}{g}m\alpha^{cm}A_\mu{}^{k-m} + f_{ab}{}^c\alpha^{am}\hat{A}_\mu{}^{bk-m} \\
&\delta_{\vec{\alpha}}\phi^{ck} = -i\frac{k}{g\kappa}\alpha^{ck} + f_{ab}{}^c\alpha^{am}\phi^{bk-m},
\end{aligned} \tag{16}$$

from which we have

$$\begin{aligned}
&\delta_{\vec{\alpha}}F_{\mu\nu}{}^k = 0, \\
&\delta_{\vec{\alpha}}\hat{F}_{\mu\nu}{}^{ck} = f_{ab}{}^c\alpha^{am}\hat{F}_{\mu\nu}{}^{bk-m} \\
&\delta_{\vec{\alpha}}(\hat{D}_\mu\phi^{ck}) = f_{ab}{}^c\alpha^{am}(\hat{D}_\mu\phi^{bk-m}).
\end{aligned} \tag{17}$$

Obviously the gauge transformation corresponds to the Kac-Moody symmetry associated with G. But there are two points to be emphasized here. First (16) tells us that *the gauge transformation is non-linearly realized.* The scalar multiplet ϕ^{ck} transforms non-linearly, but nevertheless admits the unique covariant derivative which transforms linearly under the Kac-Moody transformation. The non-linear realization has a very important physical implication as we will see in the following. Secondly (16) tells us that, up to the normalization factor ie/g which can easily be absorbed to $A_\mu{}^k$, *$A_\mu{}^k$ and $\hat{A}_\mu{}^{ck}$ form the gauge potentials of the Virasoro-Kac-Moody symmetry associated with G.* The corresponding field strengths of the Virasoro-Kac-Moody group are given by $F_{\mu\nu}{}^k$ and $\tilde{F}_{\mu\nu}{}^{ck}$, where $\tilde{F}_{\mu\nu}{}^{ck}$ is defined by

$$\begin{aligned}
\tilde{F}_{\mu\nu}{}^{ck} &= \partial_\mu\hat{A}_\nu{}^{ck} - \partial_\nu\hat{A}_\mu{}^{ck} + g[i\frac{e}{g}(f_{m\ an}{}^{ck}A_\mu{}^m\hat{A}_\nu{}^{an} \\
&\quad + f_{am\ n}{}^{ck}\hat{A}_\mu{}^{am}A_\nu{}^n) + f_{am\ bn}{}^{ck}\hat{A}_\mu{}^{am}\hat{A}_\nu{}^{bn}] \\
&= \hat{F}_{\mu\nu}{}^{ck} - e\kappa F_{\mu\nu}{}^m\phi^{ck-m}.
\end{aligned} \tag{18}$$

Notice that the above definition of $\tilde{F}_{\mu\nu}{}^{ck}$ explicitly incorporates the semi-direct product structure of the Virasoro-Kac-Moody group. From (18) we have

$$\delta_{\vec{\alpha}}\tilde{F}_{\mu\nu}{}^{ck} = i\frac{e}{g}m\alpha^{cm}\tilde{F}_{\mu\nu}{}^{k-m} + f_{ab}{}^c\alpha^{am}F_{\mu\nu}{}^{bk-m}, \tag{19}$$

which indeed confirms the fact that, again up to the normalization factor ie/g, $F_{\mu\nu}{}^k$ and $\tilde{F}_{\mu\nu}{}^{ck}$ form the field strengths of the Virasoro-Kac-Moody group.

Now we are ready to write down the gauge theory of the Virasoro-Kac-Moody group. After the dimensional reduction of the 5-dimensional Einstein-Yang-Mills Lagrangian $\mathcal{L}_5$ down to M_4, we obtain

$$\mathcal{L} = \int \mathcal{L}_5 \frac{d\theta}{2\pi} = \mathcal{L}_1 + \mathcal{L}_2,$$

$$\begin{aligned} \mathcal{L}_1 &= \frac{(\sqrt{\gamma})^{-k}}{16\pi G} \Big[\phi^{k-l} R^l + \frac{e^2\kappa^2}{4} (\phi^3)^{k-m-n-p-q} \gamma^{\mu\nu m} \gamma^{\alpha\beta n} F_{\mu\alpha}{}^p F_{\nu\beta}{}^q \\ &\quad - \frac{pq}{4\kappa^2} (\phi^{-1})^{k-m-n-p-q} \gamma^{\mu\nu m} \gamma^{\alpha\beta n} (\gamma_{\mu\alpha}{}^p \gamma_{\nu\beta}{}^q - \gamma_{\mu\nu}{}^p \gamma_{\alpha\beta}{}^q) \Big] \\ \mathcal{L}_2 &= (\sqrt{\gamma})^{-k} \Big[-\frac{1}{4} \phi^{k-m-n-p-q} \gamma^{\mu\nu m} \gamma^{\alpha\beta n} \hat{F}_{\mu\alpha}{}^{ap} \hat{F}_{\nu\beta}{}^{aq} \\ &\quad - \frac{1}{2} (\phi^{-1})^{k-l-m-n} \gamma^{\mu\nu l} (\hat{D}_\mu \phi^{am})(\hat{D}_\nu \phi^{an}) \Big], \end{aligned} \tag{20}$$

where R^k is the Fourier components of the 4-dimensional scalar cuvature R in the basis $\hat{\partial}_\mu$. Notice that $\mathcal{L}_1$ describes a gauge theory of the Virasoro-Kac-Moody symmetry associated with P [4,12], and $\mathcal{L}_2$ describes a gauge theory of the Kac-Moody symmetry associated with G [7]. Remarkably *the Lagrangian does not, in fact can not, contain any Higgs-type potential for the scalar multiplets ϕ^k and* ϕ^{ak}. The gauge invariance simply forbids any polynomial potential for the scalar multiplets from the Lagrangian.

To discuss the physical content of the theory notice that $\mathcal{L}_5$ has the unique vacuum

$$< \gamma_{\mu\nu}{}^k >= \eta_{\mu\nu} \delta_0{}^k, \quad < \phi^k >= \delta_0{}^k, \quad < \phi^{ak} >= 0. \tag{21}$$

The vacuum breaks the Virasoro-Kac-Moody group down to $P \otimes G \otimes U(1)$, where $U(1)$ is the Cartan's subgroup of the Virasoro group. This must be clear from (13) and (16). Expanding the fields around the vacuum and putting

$$\gamma_{AB} \simeq \begin{pmatrix} \eta_{\mu\nu} + e\kappa h_{\mu\nu} & e\kappa A_\mu \\ e\kappa A_\nu & 1 + \frac{2}{\sqrt{3}} e\kappa\varphi \end{pmatrix} \tag{22}$$

we obtain, *with* $e^2\kappa^2 = 16\pi G$, the following [4]

$$\begin{aligned} \mathcal{L}_1 &= -\frac{1}{4} [(\partial_\mu \rho_{\alpha\beta}{}^{-n})(\partial_\mu \rho_{\alpha\beta}{}^n) - (\partial_\mu \rho_{\alpha\alpha}{}^{-n})(\partial_\mu \rho_{\beta\beta}{}^n) \\ &\quad + 2(\partial_\mu \rho_{\mu\nu}{}^{-n})(\partial_\nu \rho_{\alpha\alpha}{}^n) - 2(\partial_\mu \rho_{\mu\alpha}{}^{-n})(\partial_\nu \rho_{\nu\alpha}{}^n) \\ &\quad - \frac{n^2}{\kappa^2} (\rho_{\alpha\beta}{}^{-n} \rho_{\alpha\beta}{}^n - \rho_{\alpha\alpha}{}^{-n} \rho_{\beta\beta}{}^{-n})] \\ &\quad - \frac{1}{2} (\partial_\mu \varphi^0)^2 - \frac{1}{4} (\partial_\mu A_\nu{}^0 - \partial_\nu A_\mu{}^0)^2 + \textit{interactions}, \end{aligned} \tag{23}$$

where all the 4-dimensional contractions are made with the flat metric $\eta_{\mu\nu}$ and

$$\begin{aligned} \rho_{\mu\nu}{}^0 &= h_{\mu\nu}{}^0 + \frac{1}{\sqrt{3}} \eta_{\mu\nu} \varphi^0 \\ \rho_{\mu\nu}{}^n &= h_{\mu\nu}{}^n + i\frac{\kappa}{n} (\partial_\mu A_\nu{}^n + \partial_\nu A_\mu{}^n) - \frac{2}{\sqrt{3}} \frac{\kappa^2}{n^2} \partial_\mu \partial_\nu \varphi^n \quad (n \neq 0). \end{aligned} \tag{24}$$

So, after the symmetry breaking, $\mathcal{L}_1$ contains an infinite tower of massive spin-two particles $\rho_{\mu\nu}{}^n$ $(n \neq 0)$, the massless graviton $\rho_{\mu\nu}{}^0$, the massless gauge fields $A_\mu{}^0$ of $U(1)$, and the massless dilaton φ^0. Notice that *it is not* $h_{\mu\nu}{}^0$, *but* $\rho_{\mu\nu}{}^0$, *which describes the massless spin-two graviton* [4,13]. Now, expanding $\mathcal{L}_2$ around the vacuum (21) we obtain [7]

$$\begin{aligned} \mathcal{L}_2 &= -\frac{1}{4} H_{\mu\nu}{}^{a-n} H_{\mu\nu}{}^{an} - \frac{1}{2}(\frac{n}{\kappa})^2 C_\mu{}^{a-n} C_\mu{}^{an} \\ &\quad -\frac{1}{2}(D_\mu \phi^{a0})^2 + interactions, \end{aligned} \tag{25}$$

where $H_{\mu\nu}{}^{ak}$ is the field strength of $C_\mu{}^{ak}$ given by

$$\begin{aligned} C_\mu{}^{a0} &= A_\mu{}^{a0} \\ C_\mu{}^{an} &= A_\mu{}^{an} + i\frac{\kappa}{n}\partial_\mu \phi^{an} \quad (n \neq 0). \end{aligned} \tag{26}$$

So $\mathcal{L}_2$ contains an infinite tower of the massive spin-one fields $C_\mu{}^{an}$ $(n \neq 0)$, the massless gauge fields $C_\mu{}^{a0}$ of G, and the massless scalar multiplet ϕ^{a0} which forms an adjoint representation of G. This shows that *the mass spectrum of the Lagrangian* (20) *becomes that of a harmonic oscillator whose mass scale is determined by the Planck scale.* Furthermore all the massive modes are doubly degenerate. The degeneracy follows from the hermiticity of the Lagrangian (20), which guarantees the charge conjugation invariance of the theory under the unbroken subgroup $U(1)$ of the Virasoro group. These are precisely what one expects from a string theory.

Certainly the vacuum (21) is unique from the 5-dimensional point of view. After the dimensional reduction, however, there is nothing which can forbid ϕ^{ak} to acquire a non-vanishing expectation value. So the Lagrangian (20) in general has the following vacuum

$$< \gamma_{\mu\nu}{}^k >= \eta_{\mu\nu}\delta_0{}^k, \quad < \phi^k >= \delta_0{}^k, \quad < \phi^{ak} >= \mu^a \delta_0{}^k. \tag{27}$$

Remarkably this vacuum breaks the Kac-Moody group down to H, where H is the subgroup of G which forms the little group of μ^a. More significantly the symmetry breaking generates the following mass matrix for the gauge fields $A_\mu{}^{ak}$ [7]

$$M^2_{am,bn} = [(\frac{m}{\kappa})^2 \delta_{ab} - 2ig\frac{m}{\kappa} f_{ab}{}^c \mu^c + g^2 f_{ac}{}^e f_{bd}{}^e \mu^c \mu^d] \delta_{m+n}{}^0. \tag{28}$$

So *the mass spectrum of the vector fields is characterized by two completely different mass scales, the Planck scale which is responsible for the harmonic oscillator spectrum of the string and the elementary particle scale which is responsible for the fine-structure of the mass spectrum.* The mass spectrum of the spin-two particles remains unchanged, and is characterized by the Planck scale.

In summary we have succeeded to construct a prototype gauge theory of the Virasoro-Kac-Moody symmetry associated with the Poincaré group and the grand unification group, which is capable of describing the dynamics of a propagating colored string in the 4-dimensional curved space-time. We conclude with the following remarks:

1. A most remarkable characteristics of the theory is that *it describes a theory which is not based on the unitary representations of the underlying symmetry group*. Indeed none of the fields which appear in the Lagrangian (20) form a unitary representation under the Virasoro group [3,4]. Nevertheless the Hamiltonian of the theory becomes explicitly positive-definite as we have demonstrated. This is because after the symmetry breaking all the physical fields become explicitly unitary under the unbroken subgroup. This guarantees the unitary of the theory.

2. Another remarkable characteristics of the theory is that *it generates a geometric symmetry breaking* [14]. Furthermore *the geometric symmetry breaking makes the theory a non-linearly realized gauge theory*. In spite of the fact that the non-linear gauge invariance precludes any Higgs-type potential from the Lagrangian, the symmetry breaking must take place. The theory simply does not admit a vacuum which remains invariant under the gauge transformation. Notice that the non-linear realization described by (16) is of a novel type, totally different from the standard Callan-Coleman-Wess-Zumino non-linear realization [15].

3. The spontaneous symmetry breaking must leave the gauge fields of the unbroken subgroup massless. But the theory also has the massless scalar fields which can be interpreted as the Nambu-Goldstone fields of the symmetry breaking. The interpretation follows from the fact that the vacuum of the theory is degenerate, because there is no potential for the scalar fields. Notice, however, that after the symmetry breaking there is nothing which can guarantee the scalar fields to remain massless. So *the scalar fields become the pseudo-Goldstone fields of the symmetry breaking*.

4. Notice that *when the 5-dimensional metric* (4) *becomes flat, the Lagrangian* (20) *describes a gauge theory of the Kac-Moody group associated with G*, which is non-linearly realized. The gauge theory of the Kac-Moody group is very interesting in its own right, and certainly is expected to be renormalizable [7,8]. Assuming the renormalizability one may be able to calculate the masses of the pseudo-Goldstone particles through the quantum correction.

5. The theory may be capable of generating a second-step symmetry breaking through the higher-order quantum correction. Indeed it is quite possible that one may start from the vacuum (21) but end up with the vacuum (27) through the quantum correction which induces a Coleman-Weinberg type effective potential. According to this scenario the quantum correction can generate the elementary particle scale which provides the fine-structure to the mass spectrum.

6. As we have emphasized the physical metric of the unified theory which describes the massless graviton is $\rho_{\mu\nu}{}^0$ defined by (24). Futhermore the dilaton ϕ^0 becomes a pseudo-Goldstone particle which should acquire a small mass after the quantum correction. These facts should have important physical implications in the cosmology and the fifth force [13,16].

7. Obviously the theory allows a straightforward supersymmetric generalization. With the supersymmetric generalization the theory becomes an effective field theory

of an internal superstring.

A detailed discussion of the subject with a supersymmetric generalization will be published elsewhere [17].

Acknowledgements

The work is supported in part by the Ministry of Education and by the Korea Science and Engineering Foundation.

References

[1] Y. Nambu, Lecture at the Copenhagen Summer Symposium (1970); T. Goto, Prog. Theor. Phys. *46*, 1560 (1971).

[2] M. B. Green, J. H. Schwartz, and E. Witten, *Superstring Theory*, Cambridge University Press (Cambridge) 1987; M. Kaku, *Introduction to Superstrings*, Springer-Verlag (New York), 1988.

[3] Y. M. Cho, Phys. Rev. Lett. *67*, 3469 (1991). See also Y. M. Cho, in *Proceedings of XXV-th International Conference on High Energy Physics*, edited by K. K. Phua and Y. Yamaguchi, World Scientific (Singapore) 1991.

[4] Y. M. Cho and S. W. Zoh, Rapid Communication, Phys Rev. *D46* in press; Y. M. Cho and S. W. Zoh, Phys. Rev. *D46* in press.

[5] Y. M. Cho, SNUTP 91-57, submitted to Phys. Rev. Lett.

[6] I. Bars and S. Yanckielowicz, Phys. Rev. *D35*, 3878 (1987); Phys. Lett. *196B*, 329 (1987); I. Bars, Nucl. Phys. *B317*, 395 (1988); M. Kaku, Phys. Rev. *D38*, 3052 (1988).

[7] Y. M. Cho and S. W. Zoh, Phys. Lett. *B* in press.

[8] A. C. Cadavid and R. J. Finkelstein, Lett. Math. Phys. *16*, 279 (1988); *17*, 159 (1989).

[9] I. Kaplansky, Comm. Math. Phys. *86*, 49 (1982); B. L. Feigin and D. B. Fuks, Funct. Anal. Appl. *16*, 47 (1982).

[10] Y. M. Cho, J. Math. Phys. *16*, 2029 (1975); Y. M. Cho and P. G. O. Freund, Phys. Rev. *D12*, 1711 (1975); Y. M. Cho and P. S. Jang, Phys. Rev. *D12*, 3789 (1975).

[11] Y. M. Cho, Rapid Communication, Phys. Rev. *D35*, 2628 (1987); Phys. Lett. *B186*, 38 (1987); *B199*, 358 (1987); Y. M. Cho and D. S. Kimm, J. Math. Phys. *30*, 1570 (1989).

[12] L. Dolan and M. J. Duff, Phys. Rev. Lett. *52*, 14 (1984); C. S. Aulakh and D. Sahdev, Phys. Lett. *B164*, 293 (1985).

[13] Y. M. Cho, Phys. Rev. Lett. *68*, 3133 (1992).

[14] Y. M. Cho, Phys. Rev. Lett. *55*, 2932 (1985); Y. M. Cho and P. Y. Pac, Mod. Phys. Lett. *A3*, 1473 (1988).

[15] S. Coleman, J. Wess, and B. Zumino, Phys. Rev. *177*, 2239 (1969); C. G. Callan, S. Coleman, J. Wess, and B. Zumino, Phys. Rev. *177*, 2247 (1969); Y. M. Cho, Phys. Rev. *D18*, 2810 (1978).

[16] Y. M. Cho, Phys. Rev. *D41*, 2462 (1990); Y. M. Cho and D. H. Park, Gen. Rel. Gravit. *23*, 741 (1991). See also Y. M. Cho, in *Lectures in Mathematical Physics*, edited by Y. M. Cho, World Scientific (Singapore) 1989.

[17] Y. M. Cho, in *Proceedings of the V-th Asia Pacific Physics Conference*, edited by S. P. Chia and K. S. Low, World Scientific (Singapore) 1992.

NON-STANDARD SOLUTIONS OF YANG-BAXTER EQUATIONS AND THEIR PROPERTIES

Mo-Lin Ge* ** Yi-Wen Wang*
and
Kang Xue*#
*Theoretical Physics Section, Nankai Institute of Mathematics, Tianjin, 300071, P.R.China
#Dept. of Phys. Northeast Normal Univ. Changchun, 130024, P.R. China

ABSTRACT

The non-standard solutions of YBE associated with spin models and other simple Lie algebras are systematically discussed from the point of view of representation of Quantum Algebras. The relationship for the solutions, Alexander link polynomials and Birman-Wenzl algebraic structure are set up. The color-extended Birman-Wenzl algebra is established.

(I) Introduction

R-matrices related to the Quantum Algebras associated with simple Lie algebras can be expressed by $R=\Sigma e_a \otimes e^a$ where e_a is basis[1-5]. Correspondingly, $\check{R}=PR$ satisfies

$$\check{R}_{12}\check{R}_{23}\check{R}_{12}=\check{R}_{23}\check{R}_{12}\check{R}_{23} \tag{1.1}$$

Following the Drinfeld-Jimbo-Reshetikhin construction[2,4,5] representations of R can be given by $R=\Sigma\tau_i(q)P_i(q)$ where $\tau_i(q)$ are related to the Casimirs and $P_i(q)$ the q-deformed projectors. We call the R "standard" which takes r-matrix as the classical limit when q goes to 1. There exist non-standard R-matrices[6-11] which may differ from the standard ones in the following properties:

(a) By Faddeev-Reshetikhin-Takhtajan(F-R-T)[12] approach non-standard R can lead no longer to the standard quantum algebraic structures[6,7].
(b) Some of them are super-extension of standard ones, but some others are not.
(c) Non-standard R does not have the usual classical limit. Some of them cannot be diagonalizable, i.e. there are not the projectors. Hence the loop-extension of Jimbo meets difficulty in the case.
(d) For some $\check{R}$ it has $tr(h)=0$ where h is diagonal matrix in the Markov trace theories[13,14], so that it leads to Alexander type of link polynomials[15-17] rather than Jones-Kauffman type[18], as was first pointed out by H.C. Lee[15].
(e) So far most of the known non-standard R's obey the Quantum Double of Drinfeld[1,19], but there is not a general proof.

In this presentation we would like to discuss two important families of non-standard R and the related properties. One is for spin models. We shall show that the colored non-standard R associated with spin models can be interpreted in terms of the representation theory of Quantum Algebra $SL_q(2)$ at q a root of unity and the Alexander link polynomials can be constructed. The other is concerned with simple Lie algebras A_n, B_n, C_n and D_n. The Birman-Wenzl algebraic(BWA) behavior will be discussed to set up the Alexander link polynomials.

** Talk at XXIst DGMTP Conference, Tianjin, June 5-9, 1992.

(II) Non-standard R-Matrices Associated with Spin Models

(1) The "non-standard" family of R-matrices associated with the spin model was firstly discussed by H.C. Lee et al[6] and can simply be interpreted in terms of the representations of Quantum Algebra[19-21]. In refs[19-21] we presented a general formula of colored R-matrices on the basis of a direct application of Drinfeld's theory[1] at q a root of unity. The main calculations were made through the q-boson realization of Quantum Algebra for $SL_q(2)$[22,23]

It is known that for the q-commutators

$$[J_+,J_-]=[J_0], \quad [J_0,J_\pm]=\pm 2J_\pm, \quad [n]=(q^n-q^{-n})/(q-q^{-1}) \tag{2.1}$$

the mapping $SL_q(2)\to B_q$ defines a q-boson realization of $SL_q(2)$ with generators $J_\pm$ and J_0

$$\begin{aligned} &J_+\to B(J_+)=a^+\alpha(N), \quad J_-\to B(J_-)=a\beta(N), \quad J_0\to B(J_0)=2N-\lambda \\ &aa^+-q^{-1}a^+a=q^N, \quad [N,a^+]=a^+, \quad [N,a]=-a, \quad (N=a^+a) \end{aligned} \tag{2.2}$$

if $\alpha(N)$ and $\beta(N)$ satisfy

$$\alpha(N-1)\beta(N)=[\lambda+1-N] \qquad \lambda\in C.$$

After calculations we find that the formula of Drinfeld gives[1]

$$\begin{aligned} R(\lambda_1,\lambda_2)=q^{(2N-\lambda_1)\otimes(2N-\lambda_2)/2}\sum_{n=0}^{\infty}\frac{(1-q^{-2})^n}{[n]!}q^{\frac{n}{2}(n-1)} \\ \{q^{N-\lambda_1/2}a^+\alpha(N,\lambda_1)\otimes q^{-N+\lambda_2/2}a\beta(N,\lambda_2)\}^n \\ \alpha(N-1,\lambda_i)\beta(N,\lambda_i)=[\lambda_i+1-N] \qquad i=1,2 \end{aligned} \tag{2.3}$$

Eq(2.3) lead to the universal Ř-matrix form[19-21]

$$\begin{aligned} (\check{R}^{j_1j_2}(\lambda,\mu))^{ab}_{mi}=q^{2(j_1+b-\frac{\lambda}{2})(j_2+a-\frac{\mu}{2})}\{\delta^a_i\delta^b_m+\sum_{n=0}^{k}\frac{(1-q^{-2})^n}{[n]!}q^{-\frac{1}{2}n(n-1)+n(j_1-j_2+b-a-\frac{\lambda}{2}+\frac{\mu}{2})} \\ \prod_{L=0}^{n}\alpha_{j_1,m+L-1}(\lambda)\beta_{j_2,i-L+1}(\mu)[j_2+i-L+1]\delta^b_{m+n}\delta^a_{i-n}\} \end{aligned} \tag{2.4}$$

where $k=\min(2j_1,2j_2)$, $\alpha_{j,m}(\mu)=\alpha(j+m,\mu)$, $\beta_{j,m}(\mu)=\beta(j+m,\mu)$ and $q^{2k+1}=1$. It is noted that q is a cyclic parameter whereas λ and μ are continuous ones and are referred to colors. Eq(2.4) satisfies

$$\check{R}_{12}(\lambda,\mu)\check{R}_{23}(\lambda,\nu)\check{R}_{12}(\mu,\nu)=\check{R}_{23}(\mu,\nu)\check{R}_{12}(\lambda,\nu)\check{R}_{23}(\lambda,\mu) \tag{2.5}$$

All of the known colored or non-colored non-standard Ř-matrices associated with SU(2) are special case of Eq(2.4)[21].

Next we shall firstly calculate the inverse of $\check{R}^{j_1j_2}(\lambda,\mu)$ and construct the Markov trace. Secondly we shall prove that by making the colored symmetry breaking transformation Eq(2.4) leads to all the results of refs[6].

(2) Markov Trace Properties

Eq(2.4) can be recast into the form

$$(\check{R}^{j_1j_2}(\lambda,\mu))_{mi}^{ab}=q^{2(j_1+b-\frac{\lambda}{2})(j_2+a-\frac{\mu}{2})}\{\delta_i^a\delta_m^b+\sum_{n=1}^{k}A(n;a,b,m,i)\delta_{i-n}^a\delta_{m+n}^b\} \tag{2.6}$$

$$\text{where}\quad A(n;a,b,m,i)=\frac{(q-q^{-1})^n}{[n]!}q^{\frac{-n(n+1)}{2}+n(j_1-j_2-\frac{\lambda}{2}+\frac{\mu}{2}+b-a)}C(n;m,i)\ ,$$
$$C(n;m,i)=\prod_{L=1}^{n}\alpha_{j_1,m+L-1}(\lambda)\beta_{j_2,i-L+1}(\mu)[j_2+i-l+1] \tag{2.7}$$

To satisfy

$$\check{R}^{j_1j_2}(\lambda,\mu)\bar{\check{R}}^{j_1j_2}(\lambda,\mu)=I \tag{2.8}$$

we find that

$$(\bar{\check{R}}^{j_1j_2}(\lambda,\mu))_{cd}^{ab}=q^{-\Delta}\left\{\delta_d^a\delta_c^b+\sum_{n=1}^{k}\frac{(q^{-1}-q)^n}{[n]!}q^{\frac{n(n+1)}{2}+n(j_2-j_1+\frac{\lambda}{2}-\frac{\mu}{2}+b-a)}C(n;b,a)\delta_{d+n}^a\delta_{c-n}^b\right\} \tag{2.9}$$
$$\Delta=2(j_1+m-\frac{\lambda}{2})(j_2+i-\frac{\mu}{2})$$

where C(n;i,m) are given by eq(2.7) with m ⟷ i. In the derivation of eq(2.9) the relation

$$\prod_{m=0}^{n-1}(1+q^{-2m}Z)=\sum_{m=0}^{n}\prod_{i=1}^{m}\frac{(1-q^{n-i+1})}{(1-q^i)}q^{m(1-n)}Z^m$$
$$\text{when}\quad Z=-1\ \text{it is}\qquad \sum_{m=0}^{n}\frac{(-1)^m[n]!}{[m]![n-m]!}q^{m(1-n)}=0 \tag{2.10}$$

has been used. When $\lambda=\mu$ and $j_1=j_2$ eq(2.9) turns to be the known results given by ref[13,14].

In order to discuss Markov trace[13,14] and Alexander polynomials[15-17] only the case $\lambda=\mu$ should be taken into account and a diagonal matrix h (or h')should be found such that[16]

$$tr_2(\check{R}(I\otimes h))=\tau I\ ,\quad tr_2(\bar{\check{R}}(I\otimes h'))=\bar{\tau}I \tag{2.11}$$

where τ and $\bar{\tau}$ are scalars and tr_2 means that the trace is taken on the second space only. Through direct calculations we obtain

$$h=h'=diag(1,q^2,\cdots,q^{4j}),\quad trh=0,$$
$$\tau=q^{\frac{1}{2}\lambda^2-4j\lambda+1}\ ,\quad \bar{\tau}=q^{-\frac{1}{2}\lambda^2} \tag{2.12}$$

The general proof of satisfaction of Murakami's redundant conditions[16] for the enhanced Yang-Baxter operators associated with $\check{R}$ is difficult. The rigorous verification of the existence of Alexander type of link polynomials for j=1 had been given explicitly in ref[25]

For instance, when $j_1=j_2=1, \lambda=\mu$ we have

$$\check{R}^{11}=block\ diag(A_1,A_2,A_3,A_4,A_5)$$

$$A_1=1,\quad A_5=\omega t^4,\quad A_2=\begin{bmatrix}0 & t\\ t & 1-t^2\end{bmatrix},\quad A_4=\begin{bmatrix}0 & \omega^2t^3\\ \omega^2t^3 & -t^2(1-\omega t^2)\end{bmatrix}$$
$$A_3=\begin{bmatrix}0 & 0 & t^2\\ 0 & \omega t^2 & i\omega tz\\ t^2 & i\omega tz & (1-t^2)(1-\omega t^2)\end{bmatrix} \tag{2.13}$$
$$\omega=q^{-2},\quad \omega^3=1,\quad t=e^{\lambda},\quad z=\{(1-t^2)(1-\omega t^2)\}^{1/2}$$

Following F-R-T[12] it gives rise to the representation of $SL_\omega(2)$. It re-checks the general result eq(2.4). The basic properties for this solution are listed bellow.

(i) $\check{R}^{11}$ does not obey the Birman-Wenzl algebra though it possesses three distinct eigenvalues[25].

(ii) It can be Yang-Baxterized in terms of the prescription in ref[10].

(iii) For the case $h=diag(1,\omega^{-1},\omega)$ it turns to satisfy the redundancy condition of Murakami for the enhanced Yang-Baxter operators[16]. Actually for $r_i=I\times I\times\ldots\times I\times\check{R}^{11}\times I\times\ldots\times I$, we have 23 bases among which 21 bases contain r_2 or r_2^{-1} linearly. However calculation shows that the other two $r_2r_1^{-1}r_2$ and $r_2^{-1}r_1r_2^{-1}$, can be expressed by the former 21 bases linearly. Hence the Alexander link polynomials can be constructed based on Murakami's theorem[16].

(3) Colored Symmetry Breaking Transformation and The Consequences

In refs[23-24] Akutsu et al presented the solutions of $\check{R}$-matrix

$$G^{ab}_{cd}(\alpha,\beta;+)=\begin{bmatrix}a\\ d\end{bmatrix}_\omega\begin{pmatrix}c\\ b\end{pmatrix}_{\beta,\omega}\beta^d\omega^{bd}\alpha^{ub+vc}\beta^{-ud-va}f(\alpha,\beta,\omega)\,\frac{F(\alpha,a)F(\beta,b)}{F(\alpha,d)F(\beta,c)}e^{w(a+d-b-c)+p(ab-cd)} \tag{2.14}$$

$$\begin{bmatrix}m\\ n\end{bmatrix}_q=\begin{cases}\dfrac{(q;m)_q}{(q;m-n)_q(q;n)_q} & for\ \ m-n\geq 0\\ 0 & for\ \ m-n<0\end{cases},$$
$$\begin{pmatrix}m\\ n\end{pmatrix}_{z,q}=\frac{(z;m)_q}{(z;n)_q}\quad for\ \ m,n\geq 0\ ,\quad (z;n)_q=\begin{cases}\prod_{k=0}^{n-1}(1-zq^k) & for\ \ n>0\\ 0 & for\ \ n=0\end{cases} \tag{2.15}$$

We shall prove that eq(2.14) is the consequence of colored symmetry breaking transformation of eq(2.6) with $j_1=j_2$. The proof is based on the following statements:

If $\check{R}(\alpha,\beta)$ satisfies

$$\sum_{j,k,l} \check{R}^{ab}_{jk}(\alpha,\beta)\check{R}^{kn}_{lf}(\alpha,\gamma)\check{R}^{jl}_{cd}(\beta,\gamma)=\sum_{j,k,l}\check{R}^{bn}_{lj}(\beta,\gamma)\check{R}^{al}_{ck}(\alpha,\gamma)\check{R}^{kj}_{df}(\alpha,\beta) \tag{2.16}$$

than $$\bar{R}^{ab}_{cd}(\alpha,\beta)=E(\alpha,\beta)\frac{F(\alpha,a)F(\beta,b)}{F(\alpha,d)F(\beta,c)}(f(\alpha))^{\zeta(a,b,c,d)}(f(\beta))^{\xi(a,b,c,d)}\check{R}^{ab}_{cd}(\alpha,\beta) \tag{2.17}$$

satisfies the same equations for

$$\begin{aligned} &(a) \quad \zeta(a,b,c,d)=ub+vc, \qquad \xi(a,b,c,d)=-ud-vb \\ or \quad &(b) \quad \zeta(a,b,c,d)=w(a+d)(c-b), \quad \xi(a,b,c,d)=-w(a-d)(c+b) \end{aligned} \tag{2.18}$$

By making the transformation eq(2.17) then eq(2.14) can be simplified to

$$\bar{G}^{ab}_{cd}(\alpha,\beta;+)=\begin{bmatrix}a\\d\end{bmatrix}_{\omega}\begin{pmatrix}c\\d\end{pmatrix}_{\beta,\omega}\beta^{d}\omega^{bd} \tag{2.19}$$

and so does for its inverse

$$\bar{G}^{ab}_{cd}(\alpha,\beta;-)=\begin{bmatrix}b\\c\end{bmatrix}_{1/\omega}\begin{pmatrix}d\\a\end{pmatrix}_{1/\alpha,1/\omega}\alpha^{-c}\omega^{-ac} \tag{2.20}$$

Both of them satisfy eq(2.16). Eq(2.20) can be rewritten as

$$\begin{aligned} \bar{G}^{ab}_{cd}(\alpha,\beta;-)&=Z^{2c}q^{2ac}\prod_{j=1}^{n}\frac{1-q^{2c+2j}}{1-q^{2j}}\prod_{k=0}^{n-1}(1-Z^{2}q^{2a+2k})\big|_{n=b-c=d-a} \\ &=q^{2ab+n(b-a)-\mu(2b-n)-\frac{n(n+1)}{2}}\frac{(q-q^{-1})^{n}}{[n]!}\prod_{j=1}^{n}[c+j][\mu-a-j-1] \\ &\qquad where \quad \alpha^{-1}=Z^{2}, \quad \omega^{-1}=q^{-2\mu} \end{aligned} \tag{2.21}$$

Later we shall see that the variables q and μ used here are just the same ones appearing in eq(2.4). On the other hand eq(2.4) can be rewritten in the form

$$\begin{aligned} \check{R}^{ab}_{cd}(\mu,\lambda)&=q^{\rho}\frac{(q-q^{-1})^{n}}{[n]!}\prod_{j=1}^{n}\frac{[\alpha-j+1]}{\alpha(d-j,\mu)}\frac{\alpha(c+j-1,\lambda)}{[c+j]}[c+j][\mu-a-j+1] \\ \rho&=\frac{1}{2}\lambda\mu+\mu b-\lambda a+\frac{n}{2}(\mu-\lambda)-n\mu \end{aligned} \tag{2.22}$$

In comparison to eq(2.21) we obtain

$$\begin{aligned} \check{R}^{ab}_{cd}(\mu,\lambda)&=q^{\frac{1}{2}\lambda\mu}(q^{\frac{\mu}{2}})^{(b+c)}(q^{\frac{\lambda}{2}})^{-(a+d)}\frac{H(\mu,a)H(\lambda,b)}{H(\mu,d)H(\lambda,c)}G^{ab}_{cd}(\alpha,\beta;-) \\ where \quad H(\mu,a)&=\frac{\prod_{j=0}^{a-1}\alpha_{a-j}(\mu)}{[a]!}, \quad \alpha=q^{2\mu}, \quad and \quad \omega=q^{-2} \end{aligned} \tag{2.23}$$

Obviously the coefficients before $G_{cd}^{ab}(\alpha,\beta;-)$ in eq(2.23) are nothing but the transformations eq(2.17). We thus conclude that eq(2.14) and eq(2.4) are related each other by successive colored symmetry breaking transformations in the straight forward way. The relationship between the parameters is given by $\alpha=q^{2\mu}$ and $\omega=q^{-2}$ that are consistent with the previous calculations[25].

(III) Non-Standard $\check{R}$-Matrices and Birman-Wenzl Algebra

(1) The non-standard $\check{R}$-matrices for the fundamental representations of B_n, C_n and D_n and their quantum algebraic structures have been found in refs[9,26] based on the weight-conservation[27] and extended diagrammatic technique[28]. For instance, the general form of the $\check{R}$-matrices for C_n and D_n possess the form

$$\check{R}=\sum_{k\neq 0} u_k e_{kk}\otimes e_{kk}+w\sum_{\substack{k<m\\ k\neq -m}} e_{kk}\otimes e_{mm}+\sum_{k\neq\pm m} e_{km}\otimes e_{mk}+\sum_{k,m} a_{km}e_{k\,-m}\otimes e_{-k\,m} \tag{3.1}$$

$$\text{where}\quad w=q-q^{-1},\quad (e_{km})_{ab}=\delta_{ka}\delta_{mb}\quad\text{and}\quad u_k=q\ \text{or}\ -q^{-1},\quad u_{-k}=u_k$$

The a_{km} in eq(3.1) are given by

$$a_{km}=\begin{cases} u_k^{-1} & (k=m)\\ w[1-\epsilon u_m^{-1}(\prod_{j=1}^{m-1/2} u_{j-1/2}^{-2})u_{1/2}] & (k=-m<0)\\ -wu_m^{-1/2}u_k^{-1/2}(\prod_{j=1}^{m-1/2} u_{j-1/2}^{-2})u_k^{-1/2} & (0<k<m\ \text{or}\ k<m<0)\\ -\epsilon wu_m^{-1/2}(\prod_{j=|k|+1/2}^{m-1/2} u_{j-1/2}^{-1})(\prod_{i=1}^{|k|+1/2} u_{i-1/2}^{-2})u_k^{1/2}u_{1/2}^{\epsilon} & (k<0,\ m>0,\ k+m\neq 0) \end{cases} \tag{3.2}$$

$$\text{where}\quad \epsilon=-1\ \text{for}\ C_n^{(1)}\quad\text{and}\quad \epsilon=1\ \text{for}\ D_n^{(1)}$$

$$\text{When}\ \left(\prod_{j=1}^{n} u_{j-1/2}^{-2}\right)u_{1/2}^{-1}=\pm q\ \text{or}\ \mp q^{-1}\ \text{for}\ C_n,\ \left(\prod_{j=1}^{n} u_{j-1/2}^{-2}\right)u_{1/2}=\pm q\ \text{or}\ \pm q^{-1}\ \text{for}\ D_n\ \text{the}$$

eigenvalue equation of eq(3.1) is reduced to $(\check{R}-q)^n(\check{R}+q^{-1})^n=0$. However, it cannot always be reduced to $(\check{R}-q)(\check{R}+q^{-1})=0$. As an example for C_2 we have

$$(\check{R}-q)^2(\check{R}+q^{-1})=0 \tag{3.3}$$

where the multiplicity two is essential that indicates that the $\check{R}$ cannot be diagonalized. Their quantum algebras can be found in ref[26]. It is interesting to note that the serre relations for B_n become, under the representation, $(X_i^{\pm})^2=0$ $(i=1,\ldots,n-1)$ and $(X_n^{\pm})^3=0$ that means no longer super-extension.

(2) Alexander Link Polynomials

Since the general solutions (either standard or non-standard) of $\check{R}$-matrices associated with the fundamental representations of B_n,C_n and D_n obey BWA[29]. It is natural to set up their link polynomials. However for the type of eq(3.3) it can be shown that

$$\mathrm{tr(h)}=0 \tag{3.4}$$

where h stands for the diagonal matrix in Markov trace theory[13,14]. Eq(3.4) means that this type of $\check{R}$'s leads to Alexander-Conway-Lee link polynomials[6,16,17] rather than Jones-Kauffman type[18]. However the following theorem holds[16]:

If a braid group representation obeys BWA than

$$B_n=B_{n-1}+B_{n-1}b_{n-1}B_{n-1}+B_{n-1}b_{n-1}^{-1}B_{n-1} \tag{3.5}$$

where B_n is a set of braids generated by 1, b_1, ...,b_{n-1} and b_j is generated by braid between j-th and (j+1)-th strings.

Based on eq(3.5) we then can prove that either standard or non-standard solutions of $\check{R}$ for B_n, C_n and D_n satisfy the redundant condition of Murakami[16], namely, the Alexander link polynomials can be constructed[29].

(3) Colored BWA

We can solve colored BGR (see eq(2.5)) and verify the solutions associated with B_n, C_n and D_n from the extended colored BWA[30]. We introduce the operator $I_i(\alpha, \beta)$, $G_i(\alpha, \beta)$ and $E_i(\alpha, \beta)$ instead of I, G_i and E_i in ref[29], They satisfy the relations(here, G is the $\check{R}$).

$$\begin{aligned} G_i(\alpha,\beta)G_{i+1}(\alpha,\gamma)G_i(\beta,\gamma) &= G_{i+1}(\beta,\gamma)G_i(\alpha,\gamma)G_{i+1}(\alpha,\beta) \\ G_i(\alpha,\beta)G_j(\gamma,\mu) &= G_j(\gamma,\mu)G_i(\alpha,\beta), \quad |i-j|\geq 2 \end{aligned} \tag{3.6}$$

and

$$\begin{aligned} I_i(\alpha,\beta)I_{i+1}(\alpha,\gamma)I_i(\beta,\gamma) &= I_{i+1}(\beta,\gamma)I_i(\alpha,\gamma)I_{i+1}(\alpha,\beta) \\ I_i(\alpha,\beta)I_j(\gamma,\mu) &= I_j(\gamma,\mu)I_i(\alpha,\beta), \quad |i-j|\geq 2 \\ I_i(\alpha,\beta)I_i(\beta,\alpha) &= I_i(\alpha,\alpha) = I_i(\beta,\beta) = I \end{aligned} \tag{3.7}$$

where I is unit. We define the following algebraic relations

$$\begin{aligned} E_i(\alpha,\beta) &= m^{-1}(G_i(\alpha,\beta)+G_i^{-1}(\beta,\alpha))-I_i(\alpha,\beta) \\ E_i(\alpha,\beta)G_i(\beta,\alpha) &= G_i(\alpha,\beta)E_i(\beta,\alpha) = l^{-1}E_i(\alpha,\beta)I_i(\beta,\alpha) \\ &= l^{-1}I_i(\alpha,\beta)E_i(\beta,\alpha) \\ E_i(\alpha,\beta)E_{i+1}(\alpha,\gamma)E_i(\beta,\gamma) &= I_i(\alpha,\beta)I_{i+1}(\alpha,\gamma)E_i(\beta,\gamma) \\ &= E_i(\alpha,\beta)I_{i+1}(\alpha,\gamma)I_i(\beta,\gamma) = I_{i+1}(\beta,\gamma)E_i(\alpha,\gamma)I_{i+1}(\alpha,\beta) \\ E_{i+1}(\beta,\gamma)E_i(\alpha,\gamma)E_{i+1}(\alpha,\beta) &= I_{i+1}(\beta,\gamma)I_i(\alpha,\gamma)E_{i+1}(\alpha,\beta) \\ &= E_{i+1}(\beta,\gamma)I_i(\alpha,\gamma)I_{i+1}(\alpha,\beta) = I_i(\alpha,\beta)E_{i+1}(\alpha,\gamma)I_i(\beta,\gamma) \\ G_i(\alpha,\beta)G_{i+1}(\alpha,\gamma)G_i(\beta,\gamma) &= E_{i+1}(\beta,\gamma)G_i(\alpha,\gamma)G_{i+1}(\alpha,\beta) \\ &= E_{i+1}(\beta,\gamma)E_i(\alpha,\gamma)I_{i+1}(\alpha,\beta) = I_i(\alpha,\beta)E_{i+1}(\alpha,\gamma)E_i(\beta,\gamma) \\ E_i(\alpha,\beta)G_{i+1}(\alpha,\gamma)G_i(\beta,\gamma) &= G_{i+1}(\beta,\gamma)G_i(\alpha,\gamma)E_{i+1}(\alpha,\beta) \\ &= E_i(\alpha,\beta)E_{i+1}(\alpha,\gamma)I_i(\beta,\gamma) = I_{i+1}(\beta,\gamma)E_i(\alpha\gamma)E_{i+1}(\alpha,\beta) \\ G_i(\alpha,\beta)I_{i+1}(\alpha,\gamma)I_i(\beta,\gamma) &= I_{i+1}(\beta,\gamma)G_i(\alpha,\gamma)I_{i+1}(\alpha,\beta) = I_i(\alpha,\beta)I_{i+1}(\alpha,\gamma)G_i(\beta,\gamma) \\ G_{i+1}(\beta,\gamma)I_i(\alpha,\gamma)I_{i+1}(\alpha,\beta) &= I_i(\alpha,\beta)G_{i+1}(\alpha,\gamma)I_i(\beta,\gamma) = I_{i+1}(\beta,\gamma)I_i(\alpha,\gamma)G_{i+1}(\alpha,\beta) \end{aligned} \tag{3.8}$$

where m, l are arbitrary parameters, $G_i^{-1}(\beta,\alpha)$ is the inverse of $G_i(\beta,\alpha)$. In terms of the algebraic relations eqs(3.6)-(3.8), it is easy to derive other algebraic relations given by

$$\begin{aligned}
G_i(\alpha,\beta)I_i(\beta,\alpha) &= I_i(\alpha,\beta)G_i(\beta,\alpha) \\
E_i(\alpha,\beta)I_i(\beta,\alpha) &= I_i(\alpha,\beta)G_i(\beta,\alpha) \\
E_i(\alpha,\beta)E_i(\beta,\alpha) &= (m^{-1}(l+l^{-1})-1)E_i(\alpha,\beta)I_i(\beta,\alpha) \\
G_i(\alpha,\beta)G_i(\beta,\alpha) &= m(G_i(\alpha,\beta)+l^{-1}E_i(\alpha,\beta))I_i(\beta,\alpha)-I \\
G_i(\alpha,\beta)G_i(\beta,\alpha)G_i(\alpha,\beta) &= (m+l^{-1})G_i(\alpha,\beta)G_i(\beta,\alpha)I_i(\alpha,\beta) \\
&\quad -(l^{-1}m+1)G_i(\alpha,\beta)+l^{-1}I_i(\alpha,\beta) \\
G_i(\alpha,\beta)E_{i+1}(\alpha,\gamma)E_i(\beta,\gamma) &= G_{i+1}(\beta,\gamma)E_i(\alpha,\gamma)I_{i+1}(\alpha,\beta) = I_i(\alpha,\beta)G_{i+1}^{-1}(\gamma,\alpha)E_i(\beta,\gamma) \\
G_{i+1}(\beta,\gamma)E_i(\alpha,\gamma)E_{i+1}(\alpha,\beta) &= G_i^{-1}(\beta,\alpha)E_{i+1}(\alpha,\gamma)I_i(\beta,\gamma) = I_{i+1}(\beta,\gamma)G_i^{-1}(\gamma,\alpha)E_{i+1}(\alpha,\beta)
\end{aligned} \tag{3.9}$$

$$\begin{aligned}
E_i(\alpha,\beta)E_{i+1}(\alpha,\gamma)G_i(\beta,\gamma) &= E_i(\alpha,\beta)G_{i+1}^{-1}(\gamma,\alpha)I_i(\beta,\gamma) = I_{i+1}(\beta,\gamma)E_i(\alpha,\gamma)G_{i+1}^{-1}(\beta,\alpha) \\
E_{i+1}(\beta,\gamma)E_i(\alpha,\gamma)G_{i+1}(\alpha,\beta) &= E_{i+1}(\beta,\gamma)G_i^{-1}(\gamma,\alpha)I_{i+1}(\alpha,\beta) = I_i(\alpha,\beta)E_{i+1}(\alpha,\gamma)G_i^{-1}(\gamma,\beta) \\
E_i(\alpha,\beta)G_{i+1}(\alpha,\gamma)E_i(\beta,\gamma) &= l\,I_{i+1}(\beta,\gamma)E_i(\alpha,\gamma)I_{i+1}(\alpha,\beta) \\
E_{i+1}(\beta,\gamma)G_i(\alpha,\gamma)E_{i+1}(\alpha,\beta) &= I_i(\alpha,\beta)E_{i+1}(\alpha,\gamma)I_i(\beta,\gamma) \\
G_i(\alpha,\beta)E_{i+1}(\alpha,\gamma)G_i(\beta,\gamma) &= G_{i+1}^{-1}(\gamma,\beta)E_i(\alpha,\gamma)G_{i+1}^{-1}(\beta,\alpha) \\
G_{i+1}(\beta,\gamma)E_i(\alpha,\gamma)G_{i+1}(\alpha,\beta) &= G_i^{-1}(\beta,\alpha)E_{i+1}(\alpha,\gamma)G_i^{-1}(\gamma,\beta) \\
E_i(\alpha,\beta)I_{i+1}(\alpha,\gamma)E_i(\beta,\gamma) &= (m^{-1}(l+l^{-1})-1)E_i(\alpha,\beta)I_{i+1}(\alpha,\gamma)I_i(\beta,\gamma) \\
E_{i+1}(\beta,\gamma)I_i(\alpha,\gamma)E_{i+1}(\alpha,\beta) &= (m^{-1}(l+l^{-1})-1)I_{i+1}(\beta,\gamma)I_i(\alpha,\gamma)E_{i+1}(\alpha,\beta) \\
G_i(\alpha,\beta)I_{i+1}(\alpha,\gamma)G_i(\beta,\gamma) &= (m(G_i(\alpha,\beta)+l^{-1}E_i(\alpha,\gamma))+I_i(\alpha,\beta)I_{i+1}(\alpha,\gamma)I_i(\beta,\gamma) \\
G_{i+1}(\beta,\gamma)I_i(\alpha,\gamma)G_{i+1}(\alpha,\beta) &= (m(G_{i+1}(\beta,\gamma)+l^{-1}E_{i+1}(\beta,\gamma)) \\
&\quad +I_{i+1}(\beta,\gamma)I_i(\alpha,\gamma)I_{i+1}(\alpha,\beta)
\end{aligned} \tag{3.10}$$

Theorem[29,31]: For Colored BWA it holds

$$B_{n+1}=B_n+B_nA_nB_n \tag{3.11}$$

where $A_n=\{E_n(\alpha,\beta)I_n(\beta,\alpha),\ G_n(\alpha,\beta)I_n(\beta,\alpha)\}$.

The proof is based on the fact that any braid block can be expressed by(for simplicity the color indices have been omitted)

$$A=\sum_{\{a\}}\prod_{\{i\}}G_i^aE_j^bI_k^c(E_lI_l)^d(G_mI_m)^e \tag{3.12}$$

with {a}={a,b,c,d,e}, {i}={i,j,k,l,m} ({a} taking value 0 or 1). Eq(3.12) is easy to be verified by the definitions of $E_n(\alpha,\beta)$ and $I_n(\alpha,\beta)$. Furthermore, a detail analysis proves the following statements

For a colored BWA there exists

$$B'_{n+1}=B'_n+B'_nC_nB'_n \tag{3.13}$$

where B_{n+1}' stands for any braid block and $C_n=\{G_n(\alpha,\beta),\ E_n\alpha,\beta),\ I_n(\alpha,\beta),\ E_n(\alpha,\beta)I_n(\beta,\alpha),\ G_n(\alpha,\beta)I_n(\beta,\alpha)\}$. Eq(3.13) can be proved[31] by the induction on n and the definitions of the colored BWA. Since only this strings with the same color can be closed to form a knot or a link, so in eq(3.11) we take A_n in stead of C_n. Obviously, this theorem is the colored extension of eq(3.5).

(4) Colored BWA and Yang-Baxterization[30]

It is more complicated than the case without color[20]. In this case besides eqs(3.6) and (3.7) we assume that $\check{R}(\alpha,\beta)$ and $I(\alpha,\beta)$ satisfy the following relations

$$\begin{aligned} \check{R}_{12}(\alpha,\beta)I_{23}(\alpha,\gamma)I_{12}(\beta,\gamma) &= I_{12}(\alpha,\beta)I_{23}(\alpha,\gamma)\check{R}_{12}(\beta,\gamma) \\ I_{23}(\beta,\gamma)I_{12}(\alpha,\gamma)\check{R}_{23}(\alpha,\beta) &= \check{R}_{23}(\beta,\gamma)I_{12}(\alpha,\gamma)I_{23}(\alpha,\beta) \end{aligned} \tag{3.14}$$

then for three distinct eigenvalues

$$\check{R}(\alpha,\alpha)^2 = (\sum_{i=1}^{3}\rho_i)\check{R}(\alpha,\alpha)-(\sum_{i<j}^{3}\rho_i\rho_j)I+(\prod_{i=1}^{3}\rho_i)\check{R}^{-1}(\alpha,\alpha) \tag{3.15}$$

the colored BWA eqs(3.6)-(3.10) leads to the Yang-Baxterization.

$$\check{R}(\alpha,\beta;x) = Ax(x-1)\check{R}^{-1}(\beta,\alpha)+BxI(\alpha,\beta)+C(x-1)\check{R}(\alpha,\beta) \tag{3.16}$$

where $\check{R}(\alpha,\beta;x)$ satisfies YBE with colors. There are some examples which can be found in ref[30].

The authors would like to think X.F. Liu and C.P. Sun for discussions. We are indebted to Prof. C.N. Yang and Prof. L.H. Kauffman for enlightening discussions.

This work is in part supported by NSF of China.

References

1. V.G. Drinfeld, *Quantum Group*, Proc. ICM Berkeley, 1986 pp.798-820.
2. M. Jimbo, *Lett. Math. Phys.* 10 (1985) 267-302.
   Commun. Math. Phys. 102 (1986) 537.
3. M. Jimbo, *Quantum Group and Quantum Integrable Systems*, ed. M.L. Ge, Nankai Lectures on Mathematical Physics, World Scientific, 1991, pp.1-62.
4. N.Yu. Reshetikhin, *LOMI* E-4-87, E-17-87, 1988.
5. L.A. Takhtajan, *Introduction to Quantum Group and Integrable Massive Models of QFT*, eds. M.L. Ge and B.H. Zhao, Nankai Lectures on Mathematical Physics, World Scientific, 1990, pp.69-107.
6. H.C. Lee, M. Couture and N.C. Schmeing, *Physics Geometry and Topology*, ed. H.C.Lee, Plenum press, New York (1990), pp.573-582.
7. M.L. Ge and K. Xue, *J. Phys.* 24A (1991) 2679-2690.
8. M. Couture, Y. Cheng, M.L. Ge and K. Xue, *Inter. J. Mod. Phys.* 6A (1991) 559-576.
9. M.L. Ge and K. Xue, *J. Math. Phys.* 32 (1991) 1301.
10. M.L. Ge, Y.S. Wu and K. Xue, *Inter. J. Mod. Phys.* 6A (1991) 3735.

11. For A_n case:
C.L. Schultz, *Phys. Rev. Lett.* 46 (1981) 629.
J.H.H. Perk and C.L. Schultz, *Proceedings of RIMS Symposium on Non-Linear Integrable Systems*, eds. M. Jimbo and T. Miwa. World Scientific, Singapore, 1983, pp.137-150.
H.J.de Vega, *Nucl. Phys.* **B**240 (1984) 495.
O. Balelon, H.J.de Vega and C.M. Viallet, *Nucl. Phys.* **B**190 (1981) 542.
12. L.D. Faddeev, N.Yu. Reshetikhin and L.A. Takhtajan, *Braid Group, Knot Theory and Statistical Mechanics* eds. C.N. Yang and M.L. Ge, World Scientific, Singapore, 1989, pp.97-110.
13. Y. Akutsu, T. Deguchi and M. Wadati, *Braid Group, Knot Theory and Statistical Mechanics*, eds. C.N. Yang and M.L. Ge, World Scientific, Singapore, pp.151-200.
14. Y. Akutsu and M. Wadati, *J. Phys. Soc. Jpn.* 56 (1987) 839.
15. H.C. Lee, "*Twisted Quantum Groups of A_n and the Alexander-Conway Link Polynomials*", CRNL-TP-90-0220, 1990, Canada.
16. J. Murakami, *Alexander Polynomials of Colored Links,* Osaka preprint, 1990; *Lecture Notes in Math. 1510*, eds. P.P. Kulish, Springer-Verlag, 1991, pp.350-353.
17. L.H. Kauffman and H. Saleur,
Free Fermions and Alexander-Conway Polynomials, EFI-90-42, 1990;
QFT for the Multivariable Alexander Conway Polynomials YCTP-p20-91, 1991.
18. L.H. Kauffman, *Knots and Physics*, World Scientific, Singapore 1991.
F.Y. Wu, *Knot Theory and Statistical Mechanics*, Northeastern Univ. preprint, 1992.
19. M.L. Ge, X.F. Liu and C.P. Sun, *Phys. Lett.* 155**A** (1991) 137.
20. M.L. Ge, C.P. Sun and K. Xue, *Inter. J. Mod. Phys.* 7**A** (1992) 2763.
21. C.P. Sun and M.L. Ge, *Quantum Group and Quantum Integrable Systems*, ed. M.L. Ge World Scientific, Singapore 1992 pp.133-229.
22. L.C. Biedenharn, *J. Phys.* 22**A** (1989) L873.
A.J. Macfarlan, *J. Phys.* 22**A** (1989) 4581.
C.P. Sun and H.C. Fu, *J. Phys.* 22**A** (1989) L983.
P.P. Kulish and E. Damashinsky, *J. Phys.* 23**A** (1990) L415.
23. Y. Akutsu and T. Deguchi, *Phys. Rev. Lett.* 67 (1991) 777.
24. T. Deguchi and Y. Akutsu, *Knot Theory and Its Ramifications*, 1 (1992) 161-184.
25. M.L. Ge, C.P. Sun and Y.W. Wang, *Non-Birman-Wenzl Algebraic Properties and Redundancy of Exotic Enhanced YB Operator for Spin Model*, Nankai preprint, 1992.
26. M.L. Ge, G.C. Liu and K. Xue, *J. Phys.* A24 (1991) 2676.
27. M.L. Ge, C.P. Sun, L.Y. Wang and K. Xue, *J. Phys.* A23 (1990) L645.
28. M.L. Ge, Y.Q. Li and K. Xue, *J. Phys.* A23 (1990) 605-619.
29. Y. Cheng, M.L. Ge and K. Xue, *Commun. Math. Phys.* 136 (1991) 195; *Knot Theory and Its Ramifications*, 1 (1992) 31.
30. M.L. Ge and K. Xue, *New solutions of YBE without Additivity and Birman-Wenzl Algebra with Color*, Stony Brook preprint, ITP-SB-92-04, 1992.
31. M.L. Ge, Y.W. Wang and K. Xue, *Alexander Link Polynomials Associated with Colored BWA*, Nankai preprint, 1992.

THE QUANTUM GROUP STRUCTURE AND GEOMETRY OF (GENERALIZED) TWO-DIMENSIONAL GRAVITIES

JEAN-LOUP GERVAIS
Laboratoire de Physique Théorique de l'École Normale Supérieure[1],
24 rue Lhomond, 75231 Paris CEDEX 05, France.

ABSTRACT

Recent progress in understanding the quantum-group structure of the Liouville approach to 2D gravity are reviewed. The fusion and braiding matrices are derived in closed form, and shown to involve additional coupling constants, besides the quantum-group symbols. In the second part, one reviews the geometrical interpretation of the A_n-Toda theories, just put forward, that relate W geometries with the external geometry of particular (W) surfaces.

1 Introduction

The connection between conformally invariant field theories (CIFT) and integrable models was already present in the early treatments of Liouville theory[4, 5, 6, 18]. It has turned out to be more and more useful in recent years. A particular role is played in this connection by the conformal Toda theories that generate, upon quantization, a whole class of the most important CIFT's. Two-dimensional gravity in the conformal gauge is notoriously equivalent to the Liouville theory which is the conformal Toda theory associated with the Lie algebra A_1. A step to generalize this situation was made in [3] where it was shown that the Toda theory associated with any given simple Lie algebra gives two Noether realizations of the corresponding W-algebra. Thus, if the above A_1-Liouville scheme is repeated for the other Lie algebras, there should exist generalizations of two-dimensional gravity (called W-gravities) which are invariant by generalized diffeomorphisms, and coincide with the conformal Toda theories when a particular local coordinate frame is used.

These lecture notes cover recent developments.

In section 2, Liouville theory is discussed using the operator approach[6, 12, 14, 15, 17], where the quantum-group structure plays a fundamental role. After recalling some background material, one reviews recent results:
1) The complete derivation[20] of the holomorphic operator-algebra (braided category) where the role of the quantum group structure is fully established. The main

[1] *Unité Propre du Centre National de la Recherche Scientifique, associée à l'École Normale Supérieure et à l'Université de Paris-Sud.*

new result here is that the fusion and braiding are not completely given by the quantum-group 3-j or 6-j symbols, as is usually assumed: There are coupling constants, which are not trigonometrical functions, and are not determined solely by the quantum-group symmetry.
2) The derivation[17] of the gravity-matter coupling in the weak-coupling regime, where the present method is found to agree with matrix-model calculations. One crucial point here is to show that the continuation in the number of screening operators which is made in the Coulomb gas picture is equivalent to the symmetry between quantum-group spin J and $-J-1$ put forward in my previous study[13, 14] of the strong-coupling regime.

The remaining sections cover recent progress about the more general Toda theories, considered at the purely classical level. The basic point of section 3 is the recent proposal[1, 2] that one can regard the W-geometries as the extrinsic geometry of particular two dimensional surfaces (W-surface) embedded into target spaces that are higher dimensional Kähler manifolds (We will restrict ourselves to the simplest particular situation, i.e. our target space is CP^n which corresponds to the A_n-type W-geometry). Instead of introducing higher-spin gauge generators, our approach makes use of the *extrinsic* curvatures of the embedded surface at its regular points, and relates it with the Toda dynamics mentioned above. The main virtue of our approach is that it is very simple to begin with. A W surface is characterized by the specific *chiral structure* of its embedding which we call chiral for short.

Section 4 deals with the geometry of the Toda hierachy. The dynamical variables of this hierarchy are shown to give particular coordinates of the higher dimensional Kähler manifolds. They are delt with by means of the free-fermion formalism, which is shown to be deeply connected with the concept of analytic continuation and with its W-generalizations. An important point here is to show that this allows us to extend the W transformations as reparametrizations of the target space, so that they become linear.

Finally, in section 5 we first reformulate our approach in terms of the intrinsic geometry of the family of associated surfaces in the Grassmannians $G_{n+1,k+1}$, $k = 1, \cdots, n$. This is needed to study singular points and global aspects of W-surfaces following the general scheme of refs. [1, 2], where these surfaces were shown to be intanton solutions of the associated non-linear σ models. The aim is to establish the generalization of the Gauss-Bonnet theorem to the W-surfaces discussed above. The instanton-number associated with each mapping are the global invariant of W-geometry. They are connected with the singularity indices of the W-surface.

2 Two dimensional gravity

2.1 The basic chiral operator-algebras

Denote by C the central charge of gravity. The standard screening charges $-\alpha_\pm$ of the Liouville theory[6, 18] are such that

$$\alpha_\pm = \frac{1}{2}\Big(\sqrt{\frac{C-1}{3}} \pm \sqrt{\frac{C-25}{3}}\Big),$$

$$\alpha_\pm = \frac{Q}{2} \pm \alpha_0, \quad Q = \sqrt{\frac{C-1}{3}}, \quad \alpha_0 = \frac{1}{2}\sqrt{\frac{C-25}{3}} \tag{2.1}$$

Q, and α_0 are introduced so that they agree with the standard notations. Kac's formula may be written as

$$\Delta_{Kac}(J,\widehat{J};C) = -\frac{1}{2}\beta(J,\widehat{J};C)\big(\beta(J,\widehat{J};C)+Q\big), \quad \beta(J,\widehat{J};C) = J\alpha_- + \widehat{J}\alpha_+, \tag{2.2}$$

where $2J$ and $2\widehat{J}$ are positive integers. We shall deal with generic values of C in order to avoid the complications of quantum group representations. As a result, we have to deal with h and $\widehat{h}$ on the same basis. This is in fact crucial for dealing with the strong-coupling regime[18, 13, 14]. There are two quantum-group parameters $h = \pi(\alpha_-)^2/2$, with $q = e^{ih}$, and $\widehat{h} = \pi(\alpha_+)^2/2$, with $\widehat{q} = e^{i\widehat{h}}$. Thus the number of fields doubles with respect with the classical case just recalled. The holomorphic operator will have a hat, or not, if they are related with $\widehat{h}$, or h, respectively; and the antiholomorphic operators are distinguished by an additional bar. According to Eq.2.2, the most general Liouville field is to be written as $\exp(-(J\alpha_- + \widehat{J}\alpha_+)\Phi)$. These fields have decompositions onto holomorphic and antiholomorphic operators to which we shall come below.

First we concentrate on holomorphic fields which are only functions of $x = \sigma - i\tau$. Then the whole structure may be described on the unit circle $\tau = 0$. The holomorphic operators associated with h form a subfamily which we next describe following refs.[12, 14, 20]. There are two useful basis of operators. First, the vertex operators $V_m^{(J)}$, with $-J \le m \le J$, and $2J$ a positive integer. They diagonalize the monodromy matrix

$$V_m^{(J)}(\sigma+2\pi) = e^{2ihm\varpi}e^{2ihm^2}\,V_m^{(J)}(\sigma), \tag{2.3}$$

where ϖ is the Liouville quasi-momentum (zero mode) around the cylinder. The meaning of the index m is to specify the shift of ϖ:

$$V_m^{(J)}(\sigma)\,\varpi = (\varpi+2m)\,V_m^{(J)}(\sigma). \tag{2.4}$$

The Liouville momentum has been normalized so that this shift is integer. ϖ characterizes the Verma module. We shall work on the sphere, where the spectrum of ϖ is of the form $\varpi_J = \varpi_0 + 2J$. The momentum $\varpi_0 = 1 + 2\pi/h$ is the one of the $sl(2, C)$ invariant state. There is a Verma modules $\mathcal{H}_J$ for each ϖ_J. The Moore Seiberg (MS) chiral vertex-operators connect three specified Verma modules and are thus of the form $\phi^{J_2}_{J_3,J_1}$. The operators $V^{(J)}_m$, on the contrary, act in the direct sum $\mathcal{H} = \oplus_J \mathcal{H}_J$. It is quite obvious, according to Eq.2.4, that the two are related by the projection operator $\mathcal{P}_J$:

$$\mathcal{P}_J \mathcal{H} = \mathcal{H}_J, \quad \mathcal{P}_{J_3} V^{(J_2)}_m \equiv \phi^{J_2}_{J_3,J_3+m} \tag{2.5}$$

The V fields are such that $< \varpi_2 | V^{(J)}_m | \varpi_1 >$ is equal to one of $\varpi_1 = \varpi_3 + 2m$, and is equal to zero otherwize. This normalization is required by the symmetry between three legs (sphere with three punctures). The complete fusion and braiding algebras take the form:

$$\mathcal{P}_K \, V^{(J_1)}_{m_1} \, V^{(J_2)}_{m_2} = \sum_{J=|J_1-J_2|}^{J_1+J_2} F_{K+m_1,J}\begin{bmatrix} J_1 & J_2 \\ K & K+m_1+m_2 \end{bmatrix}$$

$$\sum_{\{\nu\}} \mathcal{P}_K \, V^{(J,\{\nu\})}_{m_1+m_2} < \varpi_J, \{\nu\} | V^{(J_1)}_{J_2-J} | \varpi_{J_2} > \tag{2.6}$$

where $\mathcal{P}_K$ is the projector onto $\mathcal{H}_K$, and,

$$\mathcal{P}_K \, V^{(J_1)}_{m_1} \, V^{(J_2)}_{m_2} = \sum_{n_2} B_{K+m_1,K+n_2}\begin{bmatrix} J_1 & J_2 \\ K & K+m_1+m_2 \end{bmatrix} \mathcal{P}_K \, V^{(J_2,)}_{n_2} V^{(J_1)}_{m_1+m_2-n_2} \tag{2.7}$$

For simplicity of notation we omitted the dependence upon the world-sheet variables, which is standard. Using Eq.2.5, one may verify that these expressions have the general MS form. In ref.[20], it was shown that

$$F_{J_{23},J_{12}}\begin{bmatrix} J_1 & J_2 \\ J_{123} & J_3 \end{bmatrix} = \frac{g^{J_{12}}_{J_1 J_2} \, g^{J_{123}}_{J_{12} J_3}}{g^{J_{23}}_{J_2 J_3} \, g^{J_{123}}_{J_1 J_{23}}} \left\{ \begin{matrix} J_1 & J_2 \\ J_3 & J_{123} \end{matrix} \middle| \begin{matrix} J_{12} \\ J_{23} \end{matrix} \right\} \tag{2.8}$$

where $\left\{ \begin{matrix} J_1 & J_2 \\ J_3 & J_{123} \end{matrix} \middle| \begin{matrix} J_{12} \\ J_{23} \end{matrix} \right\}$ denote the 6-j symbols of $U_q(sl(2))$, with $q = e^{ih}$. This term was of course expected, in view of the quantum-group structure previously exhibited, in particular, in refs.[12, 14]. However, there appear, in addition, coupling constants $g^{J_{12}}_{J_1 J_2}$, which are not trigonometrical functions of h. Their general expression is

$$g^{J}_{J_1 J_2} = \prod_{n=1}^{J_1+J_2-J} \sqrt{\frac{G(J_1 - n/2) G(J_2 - n/2) G(J + n/2)}{G((n-1)/2)}}, \tag{2.9}$$

where

$$G(J) \equiv \frac{\Gamma(1 + (1+2J)h/\pi)}{\Gamma(-(1+2J)h/\pi)}. \tag{2.10}$$

$\Gamma(x)$ is the usual (not q-deformed) Γ-function. The relation between braiding and fusion matrices is standard[19], and one may verify that the appearance of the coupling constant does not spoil the consistency conditions (pentagonal, and so on) satisfied by the 6-j symbols. In terms of the V fields the operator-algebra is entirely expressed in terms of quantum group invariant. Thus, it is clearly symmetric, but one does not see how the quantum group acts. This last feature is exhibited[11] by going to another basis of operators denoted $\xi_M^{(J)}$ (the quantum group basis) where M will really be the "third" component of the quantum-group spin. The change of basis is of the form[11, 12]

$$\mathcal{P}_K \xi_M^{(J)}(\sigma) := \sum_m C_{KK+2m}(J,M)\, \mathcal{P}_K V_m^{(J)}(\sigma). \tag{2.11}$$

where the coefficient $C_{KK+2m}(J,M)$ is proportional to a q-deformed hypergeometric function. The fusion properties become[20]

$$\xi_{M_1}^{(J_1)}\xi_{M_2}^{(J_2)} = \sum_{J=|J_1-J_2|}^{J_1+J_2} g_{J_1 J_2}^{J}(J_1,M_1;J_2,M_2|J) \sum_{\{\nu\}} \xi_{M_1+M_2}^{(J,\{\nu\})} < \varpi_J, \{\nu\}|V_{J_2-J}^{(J_1)}|\varpi_{J_2}>, \tag{2.12}$$

where $(J_1,M_1;J_2,M_2|J)$ denotes the q-deformed Clebsch-Gordan (CG) coefficients. The braiding relations are[2] as follows[12]

$$\xi_{M_1}^{(J_1)}\xi_{M_2}^{(J_2)} = \sum_{N_1 N_2} (J_1,J_2)_{M_1 M_2}^{N_2 N_1}\, \xi_{M_2}^{(J_2)}\xi_{M_1}^{(J_1)}. \tag{2.13}$$

The braiding matrix $(J_1,J_2)_{M_1 M_2}^{N_2 N_1}$ is given by the matrix element of the universal R-matrix of $U_q(sl(2))$:

$$(J_1,J_2)_{M_1 M_2}^{N_2 N_1} = \Big(<< J_1,M_1| \otimes << J_2,M_2|\Big)\, R\, \Big(|J_1,N_1>> \otimes |J_2,N_2>>\Big). \tag{2.14}$$

$|J,M>>$ are group theoretic states which span the representation of spin J of $U_q(sl(2))$, and R is the universal R-matrix:

$$R = e^{-2ihJ_3\otimes J_3} \sum_{n=0}^{\infty} \frac{(1-e^{2ih})^n\, e^{ihn(n-1)/2}}{\lfloor n \rfloor !} e^{-ihnJ_3}(J_+)^n \otimes e^{ihnJ_3}(J_-)^n, \tag{2.15}$$

$J_\pm$, and J_3 are the quantum-group generators, which satisfy

$$\big[J_+,J_-\big] = \lfloor 2J_3 \rfloor, \quad \big[J_3,J_\pm\big] = \pm J_\pm. \tag{2.16}$$

We define $\lfloor r \rfloor \equiv \sin(hr)/\sinh$ in general. The fusion and braiding of the ξ fields are covariant under the quantum group action

$$J_3\,\xi_M^{(J)} = M\xi_M^{(J)}, \quad J_\pm\,\xi_M^{(J)} = \sqrt{\lfloor J \mp M \rfloor \lfloor J \pm M + 1 \rfloor}\, \xi_{M\pm 1}^{(J)}. \tag{2.17}$$

Indeed, first in the the operator-product Eq.2.12, the term $< \varpi_J, \{\nu\}|V_{J_2-J}^{(J_1)}|\varpi_{J_2}>$ is a quantum-group invariant since it does not involves the indices M_1, or M_2. Thus, by the basic property of the Clebsch-Gordan coefficients, it follows that $\xi_{M_1}^{(J_1)}\xi_{M_2}^{(J_2)}$

[2]We only coonsider one of the two possible orderings on the unit circle

also gives a representation of the quantum group algebra 2.16 with the co-product generators

$$\Lambda_\pm = J_\pm \otimes e^{ihJ_3} + e^{-ihJ_3} \otimes J_\pm, \quad \Lambda_3 = J_3 \otimes 1 + 1 \otimes J_3. \tag{2.18}$$

The tensor product is defined so that

$$(A \otimes B)\big(\xi_{M_1}^{(J_1)}(\sigma)\, \xi_{M_2}^{(J_2)}(\sigma')\big) := (A\xi_{M_1}^{(J_1)}(\sigma))\,(B\xi_{M_2}^{(J_2)}(\sigma')), \tag{2.19}$$

Roughly speaking the operator-product correspond to "adding" the quantum group spins. Since the result of the fusion should not depend upon the ordering, the non-commuativity of the ξ fields as quantum operators should precisely cancell the lack of symmetry of the co-product, and Clebsch-Gordan coefficients. This is why the braiding relations are given by the unversal R-matrix. **Thus the mathematical deformation of $sl(2)$, is precisely governed by the truly quantum mechnical effects of the theory.** Second, since the universal R-matrix which gives the braiding commutes with the co-product, the braiding relation Eq.2.13 is also covariant. This quantum-group symmetry, does not prevent invariant coupling constants $g_{J_1J_2}^{J_3}$ to appear[3]. The value of these coupling constants makes the difference between Liouville theory and the $SU(2)$ WZNW model at this level.

An other important feature is that the fusion and braiding matrices do not depend upon the Verma module to which the fusion and braiding relations are applied. This is in contrast with the usual MS expressions. In the latter case, this means that the fusion and braiding matrices are not c-numbers since they depend upon the eigenvalue of the Liouville momentum. The corresponding solution of the Yang-Baxter equation thus takes a form that differs from the solutions obtained from the R-matrix of a quantum group. It is only when (and if) this dependence upon the Verma module may be removed that a direct connection with quantum-group representations is established. Note that, for A_n-W-algebras ($n \neq 2$), the same procedure works but gives[16] a deformation of $sl(n)$ that differs from the standard one.

2.2 The Liouville field, and matter-gravity couplings

We consider closed surfaces following ref.[17]. The above ξ fields together with their antiholomorphic counterparts $\bar{\xi}$ allow us to reconstruct the fields $\exp -J\alpha_-\Phi(\sigma,\tau)$. Imposing locality, and conservation of the winding number determines this operator completely. One finds

$$e^{-J\alpha_-\Phi(\sigma,\tau)} = \frac{1}{\sqrt{\varpi}} \sum_{M=-J}^{J} (-1)^{J-M}\, e^{ih(J-M)}\, \xi_M^{(J)}(x_+)\, \bar{\xi}_{-M}^{(J)}(x_-)\sqrt{\varpi} \tag{2.20}$$

[3] we simply see here an application of the well-known Wigner-Eckart theorem, which is so much used, for instance, in atomic physics

The quantum-group action on the $\overline{\xi}$ field is given by:

$$\overline{J}_3\overline{\xi}_M^{(J)} = M\overline{\xi}_M^{(J)}, \quad \overline{J}_\pm\overline{\xi}_M^{(J)} = \sqrt{\lfloor J \mp M\rfloor \lfloor J \pm M+1\rfloor}\,\overline{\xi}_{M\pm1}^{(J)}. \tag{2.21}$$

Thus if we define

$$\mathcal{J}_\pm = J_\pm e^{-ih\overline{J}_3} + e^{ihJ_3}\overline{J}_\pm, \quad \mathcal{J}_3 = J_3 + \overline{J}_3, \tag{2.22}$$

we obtain a representation of the quantum-group algebra Eq.2.17. Then one easily checks that

$$\mathcal{J}_\pm \exp(-J\alpha_-\Phi) = \mathcal{J}_3\exp(-J\alpha_-\Phi) = 0, \tag{2.23}$$

so that the quantized Liouville field is a quantum-group invariant. This is the quantized version of a well known property of Liouville's classical solution[17] . Thus the classical $sl(2)$ symmetry has been deformed by the quantization in the mathematically standard way. The most general Liouville field $\exp(-(J\alpha_- + \hat{J}\alpha_+)\Phi)$, is given by a similar expression in terms of the fields ξ, $\hat{\xi}$, and $\overline{\xi}$, and $\hat{\overline{\xi}}$ (see ref[17] for details).

Next we cansider the dressing of minimal models with central charge D by Liouville with central charge C. The balance of central charge requires

$$C + D = 26. \tag{2.24}$$

We shall be concerned with the case $D < 1$, where the Liouville theory is in its weakly coupled regime $C > 25$. The quantum structure of the Liouville theory just recalled is basically a consequence of operator differential equations which are equivalent to the Virasoro Ward-identities that describe the decoupling of null vectors. Thus the same operator algebra, with appropriate quantum deformation parameters also describes the matter with $D < 1$. We will thus have another copy of the quantum-group structure recalled above. It will be distinguished by primes. Thus we let

$$D = 1 + 6(\frac{h'}{\pi} + \frac{\pi}{h'} + 2) = 1 + 6(\frac{\hat{h}'}{\pi} + \frac{\pi}{\hat{h}'} + 2), \quad \text{with} \quad h'\hat{h}' = \pi^2, \tag{2.25}$$

The appropriate dressing by gravity is such that one is concerned with matrix elements of operators of the type

$$\mathcal{V}_{J',\hat{J}'}(\sigma,\tau) \equiv e^{((\hat{J}'+1)\alpha_- - J'\alpha_+)\Phi(\sigma,\tau)}\, e^{-(J'\alpha'_- + \hat{J}'\alpha'_+)X(\sigma,\tau)}, \tag{2.26}$$

where X is the matter field that is similar to the Liouville field. In ref.[17], the present approach was used to determine the three-point gravity-matter couplings:

$$\langle\prod_{\ell=1}^{3}\mathcal{V}_{J'_\ell,\hat{J}'_\ell}(z_\ell, z_\ell^*)\rangle = C_{1,2,3}\Big/\Big(\prod_{k,l}|z_k - z_l|^2\Big). \tag{2.27}$$

For positiv J', Eq.2.26 involve negative quantum-group spins. This difficulty is equivalent with the need for negative number of dressing operator that is encountered in the Coulomb gas approach. Remarkably, this problem is precisely solved by symmetry $J \to -J-1$. which was found in refs.[13, 14]. Finally the result is very simple

$$C_{1,2,3} = \prod_l \frac{B_{J_l', \hat{J}_l'}}{\left[\Gamma\left(1+2\hat{J}_l+(1+2J_l)h/\pi\right)\right]^2}. \tag{2.28}$$

It factorises, and thus gives back results of matrix models for ratios of correlators that do not depend upon the normalizations. The outcome of ref.[17] is that the symmetry between quantum-group spins J and $-J-1$ is the explanation of the continuation in the number of screening operators discovered by Goulian and Li. Moreover, and contrary to the previous discussions of this problem, the present approach clearly separates the emission operators for each leg. This clarifies the structure of the dressing by gravity. It is shown, in particular that the end points are not treated on the same footing as the mid point. Since the outcome is completely symmetric this suggests the possibility of a picture-changing mechanism.

Before leaving this topic we quote an older result[21] concerning open surfaces. The equivalent of Eq.2.20 is

$$e^{-\alpha_- J\Phi(\sigma)} = \left(\frac{h}{16\pi^3}\right)^J \sum_{M,N} < J, M|A|J, N > \xi_M^{(J)}(\sigma)\, \xi_N^{(J)}(2\pi-\sigma) \tag{2.29}$$

where

$$A = e^{-ihJ_3^2} \sum_{r,s=0}^{\infty} e^{ih(r+s)J_3} \frac{(J_+)^{r+s} e^{ihrs/2}}{\lfloor r \rfloor! \lfloor s \rfloor!}. \tag{2.30}$$

The matrix A is the universal reflection matrix associated with $U_q(sl(2))$, a concept that is gradually recognized as very important for integrable systems with boundaries.

3 GEOMETRY OF CHIRAL SURFACES

3.1 The background

The present discussion is concerned with the generalization of the situation. described in section 2 for 2D gravity. On the one hand, the above Liouville dynamics is a particular case of the Toda dynamics, which, as just recalled, exists for each simple Lie algebra. The Liouville case is associated with A_1. It was shown in ref.[3] that general Toda systems are related, through Noether theorem to the non-linear extensions of the Virasoro algebra called W-algebras. Thus going from Liouville to general Toda should correspond, in general cordinates to going from Einstein gravity to W gravity. As a first step towards formulating these fascinating theories, we next unravell the geometrical meaning of Toda systems. This is, at least a way to see the geometrical meaning of W algebras. We will show that it corresponds to the extrinsic geometry of embeddings of special (W) surfaces, a viewpoint which is natural since in general, we expect that conformal systems and their W generalizations are to be connected with string theories. The material covered untill the end is a summary of refs.[1, 2].

3.2 Chiral embeddings

The basic objects are two-dimensional surfaces embedded in Kähler manifolds. We shall only consider C^n here explicitly. The case of CP^n is treated by using homogeneous coordinates in C^{n+1}. We call z and $\bar{z}$ the surface parameters. One may think of z as an ordinary complex number, in which case the parametrization is Euclidian, and $\bar{z}$ is the complex conjugate of z; or take z, and $\bar{z}$ to be real, in which case $x_0 \equiv (z+\bar{z})/2$, and $x_1 \equiv (z-\bar{z})/2$ are coordinates of the Minkowsky type. The adjective chiral means function of a single variable z or $\bar{z}$ (if z is a complex variable this is equivalent to analytic or anti-analytic). We parametrize C^n by coordinates X^A, $\bar{X}^{\bar{A}}$, $1 \leq A, \bar{A} \leq n$. A chiral embedding is defined by equations of the form

$$X^A = f^A(z),\ A = 1, \cdots, n, \quad \bar{X}^{\bar{A}} = \bar{f}^{\bar{A}}(\bar{z}),\ \bar{A} = 1, \cdots, n. \tag{3.1}$$

where f and $\bar{f}$ are arbitrary f unctions. We call a W-surface the corresponding manifold Σ. We shall see that its extrinsic geometry is directly related to W transformations. It is convenient to introduce the matrix of inner products:

$$g_{i\bar{j}} \equiv \sum_A f^{(i)A}(z)\, \bar{f}^{(\bar{j})A}(\bar{z}), \quad 1 \leq i, \bar{j} \leq n. \tag{3.2}$$

We use ∂ and $\bar{\partial}$ as short hands for $\partial/\partial z$ and $\partial/\partial\bar{z}$ respectively. $f^{(i)A}$ stands for $(\partial)^i f^A$, and $\bar{f}^{(\bar{j})A}$ stands for $(\bar{\partial})^{\bar{j}} \bar{f}^A$. Later on we shall exhibit a particular parametrization of C^n, called W-parametrization, where the derivatives with i or $\bar{j}$ larger than one will be replaced by first order derivatives in other variables, so that the covariance properties of the present discussion will become more transparent. This section is concerned with generic regular points of Σ where the Taylor expansions of f^A and $\bar{f}^{\bar{A}}$ give linearly independent vectors. Then $f^{(a)}$, and $\bar{f}^{(a)}$, $a = 1, \cdots, n$, span the following

Definition 1 ***Moving frame in C^n.*** *Consider the vectors e_a, and $\bar{e}_a$, $a = 1, \cdots, n$, with components*

$$e_a^A = \frac{1}{\sqrt{\Delta_a \Delta_{a-1}}} \begin{vmatrix} g_{1\bar{1}} & \cdots & g_{a\bar{1}} \\ \vdots & & \vdots \\ g_{1\overline{a-1}} & \cdots & g_{a\overline{a-1}} \\ f^{(1)A} & \cdots & f^{(a)A} \end{vmatrix}, \qquad e_a^{\bar{A}} = 0, \tag{3.3}$$

$$\bar{e}_a^A = 0, \qquad \bar{e}_a^{\bar{A}} = \frac{1}{\sqrt{\Delta_a \Delta_{a-1}}} \begin{vmatrix} g_{\bar{1}1} & \cdots & g_{\bar{a}1} \\ \vdots & & \vdots \\ g_{\bar{1}a-1} & \cdots & g_{\bar{a}a-1} \\ \bar{f}^{(1)\bar{A}} & \cdots & \bar{f}^{(a)\bar{A}} \end{vmatrix}, \tag{3.4}$$

Δ_a is the determinant

$$\Delta_a \equiv \begin{vmatrix} g_{1\bar{1}} & \cdots & g_{a\bar{1}} \\ \vdots & & \vdots \\ g_{1\bar{a}} & \cdots & g_{a\bar{a}} \end{vmatrix}. \tag{3.5}$$

One may verify that the moving frame defined above is orthonormal, that is,

$$(e_a, e_b) = (\bar{e}_a, \bar{e}_b) = 0, \quad (e_a, \bar{e}_b) = \delta_{a,b}. \tag{3.6}$$

The vectors e_1 and $\bar{e}_1$ are tangents to the surface, while the other vectors are clearly normals. Thus the Gauss-Codazzi equations may be derived by studying their derivatives along the W-surface Σ. One derives equations of the form

$$\partial e_a = \sum_b \omega_{za}^{\ b} e_b, \quad \bar{\partial} e_a = \sum_b \omega_{\bar{z}a}^{\ b} e_b, \quad \bar{\partial} \bar{e}_a = \sum_b \bar{\omega}_{\bar{z}a}^{\ b} \bar{e}_b, \quad \partial \bar{e}_a = \sum_b \bar{\omega}_{za}^{\ b} \bar{e}_b. \tag{3.7}$$

which may be regarded as generalized Frenet-Serret formulae. Next we recall the

Definition 2 ***CP^n target space.*** *The complex projective space C^n is defined to be the quotient of the space C^{n+1} by the equivalence relation*

$$X \sim Y, \qquad if \quad X^A = Y^A \rho(Y), \quad and \quad \bar{X}^{\bar{A}} = \bar{Y}^{\bar{A}} \bar{\rho}(\bar{Y}). \tag{3.8}$$

where ρ and $\bar{\rho}$ are arbitrary functions of $n+1$ variables.

Thus, the modification to go from C^n to CP^n is to use $n+1$ coordinates, so that now A, and $\bar{A}$ run from 0 to n, and to write formulae that are covariant under rescaling. This is achieved by letting the indices of the matrix $g_{i\bar{j}}$ run from 0 to n in the CP^n formulae for the moving frame. for this one includes derivatives of order 0. Our basic result is the following:

Theorem 1 ***Gauss-Codazzi equations.***
Define Toda fields by

$$\phi_a = -\ln(\tau_a), a = 1, \cdots, n; \ \tau_a \equiv \begin{vmatrix} g_{00} & \cdots & g_{a0} \\ \vdots & & \vdots \\ g_{0a} & \cdots & g_{aa} \end{vmatrix}. \tag{3.9}$$

The integrability conditions of the Frenet-Serret equations for the embedding in CP^n coincide with the Toda equations associated with A_n:

$$\partial \bar{\partial} \phi_i + \exp\left(\sum_{j=1}^{n} K_{ij} \phi_j\right) = 0. \tag{3.10}$$

K is the Cartan matrix associated with A_n. The functions τ_a relevant for CP^n are similar to the Δ_a of Eq.3.5. except that they include the derivatives of zeroth order.

In conclusion: The A_n Toda dynamics describes the extrinsic geometry of W surfaces.

3.3 Some basic facts about A_n Toda theories

Its general solution is of the form

$$e^{-\phi_k} = \sum_{i_1<\cdots<i_k} \begin{vmatrix} \chi^{i_1} & \cdots & \chi^{i_k} \\ \chi^{(1)i_1} & \cdots & \chi^{(1)i_k} \\ \vdots & \cdots & \vdots \\ \chi^{(k-1)i_1} & \cdots & \chi^{(k-1)i_k} \end{vmatrix} \begin{vmatrix} \bar\chi^{i_1} & \cdots & \bar\chi^{i_k} \\ \bar\chi^{(1)i_1} & \cdots & \bar\chi^{(1)i_k} \\ \vdots & \cdots & \vdots \\ \bar\chi^{(k-1)i_1} & \cdots & \bar\chi^{(k-1)i_k} \end{vmatrix}. \tag{3.11}$$

where k runs from 1 to n. It is expressed in terms of n functions of z $\chi^1, \cdots, \chi^n$, and n functions of $\bar z$ $\bar\chi^1, \cdots, \bar\chi^n$. Upper indices in between parenthesis denotes derivatives. These n functions χ^k (resp. $\bar\chi^k$) are restricted to be solution of the same differential equation $\chi^{(n+1)k} - \sum_{\ell=0}^{n-1} U_{n+1-\ell}\chi^{(\ell)k} = 0$ (resp. $\bar\chi^{(n+1)k} - \sum_{\ell=0}^{n-1} \bar U_{n+1-\ell}\bar\chi^{(\ell)k} = 0$). The set of potentials $\{U_\ell, \ell = 2, \cdots n+1\}$ and $\{\bar U_\ell, \ell = 2, \cdots n+1\}$, each generate a realization of the A_n W algebra by Poisson brackets. These are non-linear generalizations of the Virasoro algebra (conformal transformations). The Toda dynamics is non-chiral, and this is why the W algebra appears twice (for the holomorphic and anti-holomorphic components).

It follows that from the above geometrical derivation of the Toda eqautions, we may discuss the geometrical meaning of the W transformations.

3.4 Connection with the WZNW models

It is known, in general, that there is one W algebra associated with each simple[4] Lie algebra $\mathcal{G}$. This appears in several ways. First, as we have already seen, there is a Toda theory, and, hence, two PB realizations of the associated W-algebra for any given $\mathcal{G}$. On the other hand, consider the affine Lie algebra $\tilde{\mathcal{G}}$ associated with $\mathcal{G}$. The associated non-chiral theory is the WZNW model whose quantum solutions are given by representations of $\tilde{\mathcal{G}}$. It is possible to derive the Toda theory from the WZNW model by conformal reduction[9]. Here we have the

Theorem 2 ***Gauss decomposition from moving frame.*** *The moving-frame equations may be written as*

$$e_a = \sum_{b\le a} C_{ab}(z,\bar z)\sqrt{\frac{\tau_a}{\tau_{a+1}}}\, f^{(b)}(z), \quad \textit{with}\ C_{aa} = 1, \tag{3.12}$$

$$\bar e_a = \sum_{b\le a} A_{ba}(z,\bar z)\sqrt{\frac{\tau_a}{\tau_{a+1}}}\, \bar f^{(b)}(\bar z), \quad \textit{with}\ A_{aa} = 1. \tag{3.13}$$

The matrix $\theta = g^{-1}$ s such that

[4]The non-simple Lie algebras are trivially reduced to the simple case by separating the invariant subalgebras

1) It has the Gauss decomposition

$$\theta = ABC, \tag{3.14}$$

where A, *and* C, *which appear in eqs.3.13, and 3.12 are triangular with diagonal elements equal to one, and*

$$B = \exp\left(\sum_i h_i \phi_{i+1}\right).$$

h_i *are the Cartan generators. 2) It is a solution of the conformally reduced WZNW model associated with* A_n.

4 Geometry of Toda hierarchy

The Toda equations are a subsystem of the Toda hierarchy[7]. (This is the non-chiral version of the fact that the Virasoro algebra is identical with the second Poisson bracket of KdV, and that W algebras are obtained from KP hierarchies and Gelfand Dicki brackets). Introduce the additional variables as coordinates in our geometrical embedding problem. This is best done using the free-fermion formalism. Let

$$\begin{aligned} [\psi_n, \psi_m]_+ &= \left[\psi_n^+, \psi_m^+\right]_+ = 0, \\ \left[\psi_n, \psi_m^+\right]_+ &= \delta_{n,m}, \qquad (\, n, m \;=\; 0,\; 1,\; \cdots) \end{aligned} \tag{4.1}$$

$$\psi_n|\emptyset\rangle = 0, \qquad \langle\emptyset|\psi_n^+ = 0 \qquad \forall n. \tag{4.2}$$

We use the semi-infinite indices $n = 0, 1, 2, \cdots, \infty$ for the fermion-operators. The vacuum states $|\emptyset\rangle$ and $\langle\emptyset|$ correspond to the no-particle states. The n-particle ground state is created from them in the standard way:

$$|n\rangle = \psi_{n-1}^+\psi_{n-2}^+\cdots\psi_0^+|\emptyset\rangle, \quad \langle n| = \langle\emptyset|\psi_0\psi_1\cdots\psi_{n-1}. \tag{4.3}$$

The current operators,

$$J_n = \sum_{s=0}^{\infty} \psi_{n+s}^+\psi_s, \qquad \bar{J}_n = \sum_{s=0}^{\infty} \psi_s^+\psi_{n+s}, \tag{4.4}$$

will be taken as Hamiltonians as one does for the KP hierarchy. The rôle of these fermions may be understood as follows. Take the case where z is a complex variable. Then the embedding functions f^A are analytic, and each of them is entirely determined by its Taylor expansion around a single point of its analyticity domain. Its behaviour at any other point of its Riemann surface is fixed by analytic continuation. The following free-fermion formalism realizes this continuation automatically. Consider the Taylor expansions at the point z:

$$f^A(z+x) = \sum_{s=0}^{\infty} f^{(s)A}(z)\frac{x^s}{s!}, \quad \bar{f}^{\bar{A}}(\bar{z}+\bar{x}) = \sum_{s=0}^{\infty} \bar{f}^{(s)A}(\bar{z})\frac{\bar{x}^s}{s!}. \tag{4.5}$$

To these developements, we associate the free-fermion operators,

$$\psi_{f^A(z)}=\sum_{s=0}^{\infty}f^{(s)A}(z)\,\psi_s,\quad \psi^+_{\bar f^{\bar A}(\bar z)}=\sum_{s=0}^{\infty}\bar f^{(s)\bar A}(\bar z)\,\psi^+_s. \tag{4.6}$$

The basic property of these operators are

Proposition 1 ***Fermionic representation of chiral functions.***
1) Any change of the Taylor-expansion point z, $\bar z$ can be absorbed by the action of the Hamiltonians J_1, and $\bar J_1$. In particular, one has

$$\psi_{f^A(z)}=e^{-J_1z}\,\psi_{f^A(0)}\,e^{J_1z}\quad \psi^+_{\bar f^{\bar A}(\bar z)}=e^{\bar J_1\bar z}\,\psi^+_{\bar f^{\bar A}(0)}\,e^{-\bar J_1\bar z}. \tag{4.7}$$

2) The embedding functions are represented by the fermion expectation-values

$$f^A(z)=\langle\emptyset|\psi_{f^A(z_0)}e^{J_1(z-z_0)}|1\rangle,\qquad \bar f^{\bar A}(\bar z)=\langle 1|e^{\bar J_1(\bar z-\bar z_0)}\psi^+_{\bar f^{\bar A}(\bar z_0)}|\emptyset\rangle. \tag{4.8}$$

Definition 3 ***KP-parametrization of CP^n.***

Given a chiral surface embedded into CP^n, the associated KP-parameters of the target space are $n+1$ variables $z^{(0)}$, $z^{(1)}=z$, $z^{(2)}$, $\cdots$, $z^{(n)}$, noted $[z]$, and $n+1$ variables $\bar z^{(0)}$, $\bar z^{(1)}=\bar z$, $\bar z^{(2)}$, $\cdots$, $\bar z^{(n)}$, noted $[\bar z]$. The change of coordinates from X^A, $\bar X^{\bar A}$ to $[z]$, $[\bar z]$ is defined by

$$X^A=f^A([z]),\qquad \bar X^{\bar A}=\bar f^{\bar A}([\bar z]) \tag{4.9}$$

where $f^A([z])$, and $\bar f^{\bar A}([\bar z])$, are the solutions of the equations

$$\frac{\partial}{\partial z^{(\ell)}}f^A([z])=\frac{\partial^\ell}{\partial z^\ell}f^A([z]),\quad \frac{\bar\partial}{\partial\bar z^{(\ell)}}\bar f^{\bar A}([\bar z])=\frac{\bar\partial^\ell}{\partial\bar z^\ell}\bar f^{\bar A}([\bar z]) \tag{4.10}$$

with the initial conditions $f^A([z])=f^A(z)$ for $z^{(0)}$, $z^{(2)},\cdots,z^{(n)}=0$, and $\bar f^{\bar A}([\bar z])=\bar f^{\bar A}(\bar z)$ for $\bar z^{(0)}$, $\bar z^{(2)},\cdots,\bar z^{(n)}=0$.

These coordinates coincide with the higher variables of the KP hierarchy. Indeed, their definition is most natural in the free-fermion language, where it is easy to see that

$$f^A([z])=\langle\emptyset|\psi_{f^A}e^{\sum_0^n J_s z^{(s)}}|1\rangle,\quad \bar f^{\bar A}([\bar z])=\langle 1|e^{\sum_0^n \bar J_s\bar z^{(s)}}\psi^+_{\bar f^{\bar A}}|\emptyset\rangle. \tag{4.11}$$

The dependence in $[z]$ and $[\bar z]$ is dictated by the action of the higher currents J, $\bar J$, defined by Eq.4.4,that is, $J_1z\to\sum_{i=0}^n J_iz^{(i)}$, $\bar J_1\bar z\to\sum_{i=0}^n \bar J_i\bar z^{(i)}$ in Eq.4.8. The basic virtue of the KP coordinates is that they allow us to extend the previous discussion away from the W surface (they parametrize at least a neighborhood of it) in such a way that is becomes covariant. In particular, the metric g has the following extension

$$g_{i\bar j}([z],[\bar z])=\sum_A\partial_i f^A([z])\,\bar\partial_{\bar j}\bar f^A([z]),\quad \partial_i\equiv\frac{\partial}{\partial z^{(i)}},\quad \bar\partial_i\equiv\frac{\bar\partial}{\bar\partial\bar z^{(i)}}. \tag{4.12}$$

Now, only first-order derivatives appear. This expression coincides with the *true* Riemannian metric with respect to the KP coordinates.

4.1 W transformations

A general infinitesimal W-transformation is a change of embedding functions which takes the form

$$\delta_W f^A(z) = \sum_{j=0}^{n} w^j(z)\partial^{(j)} f^A(z), \quad \delta_W \bar{f}^{\bar{A}}(\bar{z}) = \sum_{j=0}^{n} \bar{w}^j(\bar{z})\partial^{(j)} \bar{f}^{\bar{A}}(\bar{z}), \tag{4.13}$$

It is not difficult to show there exists a unique extension such that the differential equation 4.10 is left invariant. It is of the form

$$\begin{aligned} \delta_W f^A([z]) &= \sum_r W^r([z])\partial_r f^A([z]), \\ \delta_W \bar{f}^{\bar{A}}([\bar{z}]) &= \sum_r \bar{W}^r([\bar{z}])\partial_r \bar{f}^{\bar{A}}([\bar{z}]) \end{aligned} \tag{4.14}$$

where W^r and $\bar{W}^r$ are functionals of w^j and $\bar{w}^j$ respectively. Only first order derivative appear. Thus the W transformation become extended as reparametrizations

$$\delta_W z^{(r)} = W^r([z]), \quad \delta_W \bar{z}^{(r)} = \bar{W}^r([\bar{z}]). \tag{4.15}$$

They become particular diffeomorphisms of CP^n. Thus they are extended as linear transformations.

4.2 Dynamical equations

The followings topics are discussed in ref.[2]

The above functions τ_a when extended become tau-functions in the sense of Miwa-Jimbo-Sato[7].

The KP coordinates are related with a generalized moving frame, whose integrability condition is equivalent to the the well-known Zakharov-Shabat of the A_n Toda hierarchy.

The extension of the associated WZNW model gives solutions of a 2n dimensional generalization of the WZNW equations where the currents are replaced by the Christoffel symbols of the KP coordinates.

5 SINGULAR POINTS, GLOBAL STRUCTURE

At this point, it is useful to change the viewpoint, and make use of Grassmannians. This part is draws a lot of inspiration from ref.[10]

5.1 Associated mappings

Definition 4 ***Associated mappings.***

Consider the family of osculating hyperplanes with contact of order k denoted $\mathcal{O}_k$ ($k = 1, \cdots, n$) to the original W-surface. With CP^n as the target space, this family defines an embedding into the Grassmannian $G_{n+1,k+1}$, which we call the kth associated mappings to the original W-surface.

This formulation looks different, but is equivalent to the construction of the moving frame and only uses the intrinsic geometries of the induced metrics for $k = 1, \cdots, n$. In practice, what this means is that, instead of forming moving-frame vectors e_k out of $f, \cdots, f^{(k)}$ ($k = 1, \cdots, n$), we consider the nested osculating planes $\mathcal{O}_1 \subset \mathcal{O}_2 \subset \cdots \subset \mathcal{O}_n$. It is obvious that those two have the same information. It is well-known that the Grassmannians are Kähler manifolds. The induced metric on the kth associated surface in $G_{n+1,k+1}$ is simply,

$$g^{(k)}_{z\bar{z}} = \partial\bar{\partial} \ln \tau_{k+1}(z, \bar{z}), \quad g^{(k)}_{zz} = g^{(k)}_{\bar{z}\bar{z}} = 0, \tag{5.1}$$

so that the Toda field $\phi_{k+1} \equiv -\ln(\tau_{k+1})$ appears naturally. Thus $-\phi_{k+1}$ is equal to the Kähler potential of the kth associated surface. At this point, it is very clear that by considering the associated surfaces, we can restrict ourselves to intrinsic geometries. In the discussion of section 4, the Toda equation came out from the Gauss-Codazzi equation. Here, it is equivalent to the local Plücker formula

$$R^{(k)}_{z\bar{z}} \sqrt{g^{(k)}_{z\bar{z}}} = -g^{(k+1)}_{z\bar{z}} + 2g^{(k)}_{z\bar{z}} - g^{(k-1)}_{z\bar{z}}. \tag{5.2}$$

where $R^{(k)}_{z\bar{z}}$ is the only non-vanishing component of the intrinsic Riemann tensor on the kth surface.

5.2 The instanton-numbers of a W-surface

A key point in the coming discussion is to use topological quantities that are instanton-numbers. W-surfaces are instantons of the associated non-linear σ-model. The general situation is as follows. W-surfaces are characterized by their chiral parametrizations which thus satisfy the Cauchy-Riemann relations. These are self-duality equations so that the coordinates of a W-surface define fields that are solutions of the associated non-linear σ model, with an action equal to the topological instanton-number. For a general Kähler manifold M with coordinates ξ^μ and $\bar{\xi}^{\bar{\mu}}$, and metric $h_{\mu\bar{\mu}}$, the action associated with 2D manifolds of M with equations $\xi^\mu = \varphi^\mu(z, \bar{z})$, and $\bar{\xi}^{\bar{\mu}} = \bar{\varphi}^{\bar{\mu}}(z, \bar{z})$ is

$$S = \frac{1}{2} \int d^2x \, h_{\mu\bar{\mu}} \, \partial_j \varphi^\mu \partial_j \bar{\varphi}^{\bar{\mu}}. \tag{5.3}$$

In this short digression we let $z = x_1 + ix_2$, and $\partial_j = \partial/\partial x_j$. The instanton-number is defined by

$$Q = \frac{i}{2\pi} \int d^2x \, \epsilon_{jk} \, h_{\mu\bar{\mu}} \, \partial_j \varphi^\mu \partial_k \bar{\varphi}^{\bar{\mu}}. \tag{5.4}$$

For W-surfaces and their associated surfaces, $\bar\partial\varphi^\mu = \partial\bar\varphi^\mu = 0$, and one has $S = \pi Q$. Q is proportional to the integral of the determinant of the induced metric, that is $Q = \frac{i}{2\pi}\int d^2x \partial\bar\partial \ln \tau_1$.

Moreover we may also apply formula Eq.5.4 to the kth associated surface. This gives

Definition 5 ***Higher instanton-numbers of the W-surface.***
The kth instanton number of the W-surface Q_{k+1} is defined by,

$$Q_{k+1} \equiv \frac{i}{2\pi}\int_\Sigma dz d\bar z g^{(k)}_{z\bar z}, k = 1, \cdots, n-1. \tag{5.5}$$

Its topological nature is obvious from Eq.5.1 which shows that the integrand is indeed a total derivative. The collection of the (k-th) instanton-numbers together with the original one $Q \equiv Q_1$ gives a set of topological quantities which characterize the global properties of the original W-surface.

5.3 Global classification of W surfaces

In section 3, we have constructed the moving frames at the point where the tau-functions are regular. When those functions become irregular, we meet an obstruction to derive the moving frames. In the WZNW language, this signals that there appears a global obstruction to the Gauss decomposition. Toda equations should be modified at these points. In the following, we study the structure of such singularities.

5.4 Gauss-Bonnet Theorem for W-surfaces

For isolated singularities[5], the obstruction to constructing the moving frame may be reduced to the vanishing of certain terms in the Taylor expansion. The latter is characterized by the ramification indices β_k which are integers. Apply the Gauss-Bonnet theorem for each of the kth associated surfaces by computing $\int_{\Sigma_\epsilon} R^{(k)}_{z\bar z}\sqrt{g^{(k)}_{z\bar z}}$. The integral is first computed over Σ_ϵ where small neighborhoods of singularities are removed. The ramification indices at singularity was defined so that at a singular point the induced metric of the kth associated surface behaves as

$$g^{(k)}_{z\bar z} \sim (z-z_0)^{\beta_k(z_0)}(\bar z - \bar z_0)^{\bar\beta_k(\bar z_0)}\tilde g^{(k)}_{z\bar z}, \tag{5.6}$$

where $\tilde g^{(k)}_{z\bar z}$ is regular at z_0, $\bar z_0$. Since we do not assume that $\overline{f(z)} = \bar f(\bar z)$, $\beta_k(z_0)$ and $\bar\beta_k(\bar z_0)$ may be different. By letting $\epsilon \to 0$, one sees, that the contribution of the singularities to the Gauss-Bonnet formula is proportional to the k-th ramification index

$$\beta_k \equiv \frac{1}{2}\sum_{(z_0,\bar z_0)\in\Sigma}\left(\beta_k(z_0) + \bar\beta_k(\bar z_0)\right). \tag{5.7}$$

[5]if there is a cut with a finite number of sheets, one takes the appropriate covering

The contribution of the regular part does not depend upon k, since changing k is equivalent to using a different complex structure, while the result is equal to the Euler characteristic that does not depend upon it. The Gauss-Bonnet theorem for the kth associated surface finally gives

$$\frac{i}{2\pi}\int_\Sigma dz d\bar{z} R^{(k)}_{z\bar{z}} \sqrt{g^{(k)}_{z\bar{z}}} = 2 - 2g + \beta_k. \qquad (5.8)$$

Combining these last relations with Eqs.5.2, one arrives at the

Theorem 3 ***Global Plücker formulae*** *The genus g of a W-surface is related to its instanton-numbers and ramification-indices by the relations*

$$2 - 2g + \beta_k = 2Q_k - Q_{k+1} - Q_{k-1}, \quad \left| \begin{array}{ll} k & = 1, \cdots, n \\ Q_{n+1} & \equiv 0,\ Q_0 \equiv 0 \end{array} \right. \qquad (5.9)$$

5.5 Simple example

Consider the case of Liouville theory, for which one has $n = 1$. The simplest chiral surface corresponds to

$$f^0 = 1,\ f^1 = z; \quad \bar{f}^0 = 1,\ \bar{f}^1 = \bar{z}; \quad \tau_1 = 1 + z\bar{z}, \quad \tau_2 = 1.$$

The instanton number is thus $Q = (i/2\pi)\int dz d\bar{z}/(1 + z\bar{z})^2 = 1$. On the other hand, one has

$$R_{z\bar{z}}\sqrt{g_{z\bar{z}}} = -\partial\bar{\partial}\ln(g_{z\bar{z}}) = \frac{2}{(1 + z\bar{z})^2}$$

so that

$$\frac{i}{2\pi}\int dz d\bar{z} R_{z\bar{z}}\sqrt{g_{z\bar{z}}} = 2Q = 2,$$

and one verifies that the above formulae indeed hold with vanishing genus and ramification index. In this example, the Liouville solution coincides with the metric of the Lobatchevki half-plane. Upon quantization, it corresponds to the $SL(2, C)$ invariant vacuum of the Liouville theory.

1. J.-L. Gervais, Y. Matsuo, *Phys. Lett.* **B274** (1991) 309.
2. J.-L. Gervais, Y. Matsuo, "A_n–W-geometry" preprint LPTENS-95/35, hep-th 9201026, Comm. in Math. Phys. to be published.
3. A. Bilal and J.-L. Gervais, *Phys. Lett.* **B206** (1988) 412; *Nucl. Phys.* **B314** (1989) 646; *Nucl. Phys.* **B318** (1989) 579.
4. J.-L. Gervais, A. Neveu, *Nucl. Phys.* **B199** (59) 1982.
5. J.-L. Gervais, A. Neveu, *Nucl. Phys.* **B202** (125) 1982.
6. J.-L. Gervais, A. Neveu *Nucl. Phys.* **B238** (125) 1984; *Nucl. Phys.* **B238** (396) 1984.

7. M. Sato, *RIMS Kokyuroku* **439** (1981) 30; E. Date, M. Jimbo, M.Kashiwara and T.Miwa, "Transformation Groups for Soliton Equations" in *Proc. of RIMS Symposium on Non-linear Integrable Systems- Classical Theory and Quantum Theory* (Kyoto, Japan, May 1981) (World Scientific Publication Co. Singapore, 1983); G.Segal and G.Wilson, *Pub. Math. IHES* **61** (1985) 5.
8. V.G. Drinfeld and V.V. Sokolov, *Journ. Sov. Math.* **30** (1985) 1975.
9. P.Forgács, A.Wipf, J.Balog, L.Fehér, and L.O'Raifeartaigh, *Phys. Lett.* **227B** (1989) 214; J.Balog, L.Fehér, L.O'Raifeartaigh, P.Forgács and A.Wipf, *Ann. Phys.* (N.Y.) **203** (1990) 76; *Phys. Lett.* **244B** (1990) 435; L.O'Raifeartaigh and A.Wipf, *Phys. Lett.* **251B** (1990) 361; L.O'Raifeartaigh, P.Ruelle, I. Tsutsui and A.Wipf, *W-Algebras for Generalized Toda Theories* Dublin preprint DIAS-STP-91-03, *Comm. Math. Phys.* to appear; L.Fehér, L.O'Raifeartaigh, P.Ruelle, I. Tsutsui and A.Wipf, *Generalized Toda theories and W-algebras associated with integer grading*, Dublin preprint DIAS-STP-91-17, *Ann. Phys.* to appear; L.O'Raifeartaigh, P.Ruelle and I. Tsutsui, *Phys. Lett.* **258B** (1990) 359.
10. P. Griffiths, S. Harris, "Principles of algebraic geometry" Wiley-Interscience (1978).
11. O. Babelon, *Phys. Lett.* **B215** (523) 1988.
12. J.-L. Gervais, *Comm. Math. Phys.* **130** (257) 1990.
13. J.-L. Gervais, *Phys. Lett.* **B243** (85) 1990.
14. J.-L. Gervais, *Comm. Math. Phys.* **138** (301) 1991.
15. J.-L. Gervais, "On the algebraic structure of Quantum gravity in two dimensions", Proceedings of the Trieste Conference on *Topological Methods in Quantum Field Theories*, June 1990, W. Nahm, S. Randjbar-Daemi, E. Sezgin, E. Witten editors, World Scientific.
16. E. Cremmer, J.-L. Gervais, *Comm. Math. Phys.* **134** (619) 1990.
17. J.-L. Gervais, "Quantum group derivation of 2D gravity-matter coupling" Invited talk at the Stony Brook meeting *String and Symmetry 1991*, LPTENS preprint 91/22, Nucl. Phys. B to be published.
18. J.-L. Gervais, A. Neveu, *Phys. Lett.* **151B** (271) 1985.
19. G. Moore, N. Seiberg, *Comm. Math. Phys.* **123** (177) 1989.
20. E. Cremmer, J.-L. Gervais, J.F. Roussel, LPTENS preprint to be published.
21. E. Cremmer, J.-L. Gervais, *Comm. Math. Phys.* **144** (279) 1992.
22. A.N.Leznov, M.V.Saveliev: Acta Appl. Math. **16**,1 (1989).
23. A.N.Leznov, M.V.Saveliev: Lett. Math. Phys. **6**,505 (1982); Comm. Math. Phys. **89**,59 (1983).
24. E. Witten: Phys. Rev. **D44** (1991) 314, *On black holes in string theory*; Lecture at Strings '91, Stonybrook, June 1991.

A brief history of hidden quantum symmetries in Conformal Field Theories*

César Gómez and Germán Sierra
Instituto de Matemáticas y Física Fundamental, CSIC
Serrano 123. Madrid, SPAIN

Abstract

We review briefly a stream of ideas concerning the role of quantum groups as hidden symmetries in Conformal Field Theories, paying particular attention to the field theoretical representation of quantum groups based on Coulomb gas methods. An extensive bibliography is also included.

*Lecture presented in the XXI DGMTP-Conference (Tianjin ,1992) and in the NATO-Seminar (Salamanca,1992).

Integrability, Yang-Baxter equation and Quantum Groups

In the last few years it has become clear the close relationship between conformal field theories (CFT's), specially the rational theories , and the theory of quantum groups.From a more general point of view this is just one aspect of the connection between integrable systems and quantum groups. In fact if one recalls the history, quantum groups originated from the study of integrable models and more precisely from the quantum inverse scattering approach to integrability. The basic relation:

$$R(u-v)\; T_1(u)\; T_2(v) = T_2(v)\; T_1(u)\; R(u-v) \tag{1}$$

which is at the core of this approach may also be taken as a starting point in the definition of quantum groups [FRT].

In statistical mechanics the operator $T(u)$ is nothing but the monodromy matrix which depends on the spectral parameter u and whose trace gives the transfer matrix of a vertex model. Meanwhile the matrix $R(u)$ stands for the Boltzmann weights of this vertex model. The RTT equation guarantees the commutativity of a one parameter family of transfer matrices:

$$[\; tr\; T(u),\; tr\; T(v)\;] = 0 \tag{2}$$

This equation is equivalent to the existence of an infinite number of conserved quantities when taking the thermodynamic limit. In these setting the Yang-Baxter equation:

$$R_{12}(u)R_{13}(u+v)R_{23}(v) = R_{23}(v)R_{13}(u+v)R_{12}(u) \tag{3}$$

which underlies the integrability of the model appears as the associativity condition of the "quantum algebra" (1).

RCFT's, exchange algebras and the braid group : the polynomial equations

Let us turn now to conformal field theories. There, it was soon realized that the braiding matrices of the conformal blocks of a RCFT also satisfy a Yang-Baxter like equation. In fact these matrices provide a representation of the braid group in the space of conformal blocks. The braid group and the exchange algebras appeared suddenly to play an important role in CFT's and more generally in 2D quantum field theories [TK,Ko,Fr,FRS,Rh,MS]. However there are some structural differences with regard to the vertex models in statistical mechanics. In CFT the braiding matrices do not depend on any spectral parameter while in statistical mechanics the Boltzmann weights certainly do depend on it and only when taking the limits $u \to \pm\infty$ one gets non trivial solutions of the Yang-Baxter equation without spectral parameter. In addition ,the Boltzmann weights of a vertex model $R_{ij}^{kl}(u)$ depend on four labels while the braiding matrices $B_{pp'}\begin{bmatrix} k & l \\ i & j \end{bmatrix}$ depend on six labels which are the primary fields of a RCFT and consequently are subject to certain constraints given by the fusion rules. These later conditions are equivalent to the restrictions on the heights of a restricted solid on solid model (RSOS) where the Coxeter diagram is playing the role of the fusion rules in RCFT.

Having in mind the RTT equation of integrable models as well as the braiding properties of the conformal blocks it was proposed in reference [AGS1] that quantum groups should be present in RCFT's as "hidden symmetries" underlying and explaining the origin of the polynomial equations of Moore and Seiberg. The motivation of these authors in writing the polynomial equations was to give an axiomatic definition of RCFT's which would, among another things, give a rigorous proof of the Verlinde conjecture concerning the diagonalization of the fusion rules N_{ij}^{k} by means of the modular matrix S_{ij} which acts on the Virasoro characters [V].

Reference [AGS1] was in this way a first attempt to lay down a "quantum group" interpretation of the duality properties of RCFT's, which are encoded in the polynomial equations. Indeed one of these equations is nothing but the Yang-Baxter equation for the braiding matrices of the conformal blocks mentioned above. A second attempt in the same direction was taken by Moore and Reshetikhin in reference [MR].

A first clue of the problem is suggested by the classical limit of a WZW model based on a group G which is obtained when taking the level k going to infinity. This model has an infinite number of primary fields labelled by their spin $j = 0, 1/2, 1, \dots$ whose braiding and fusion properties can be computed entirely from the representation theory of the classical group G [MS]. Thus for example the fusion matrices F of this classical WZW model become the 6-j symbols of the group G , while the fusion rules are nothing but the usual Clebsch-Gordan decomposition of the tensor product of irreps of G. More generally, what one calls the duality data of a RCFT, namely: fusion rules, braiding matrices, fusion matrices, modular matrices, etc. are computable in the classical limit using ordinary group theory. Of course not all the RCFT's known admit such a classical limit where all the conformal weights of the primary fields vanish and consequently the braiding properties become almost trivial. One example of this is provided by the $c < 1$ theories. However what this limit suggested was that some "deformation" of the "classical" group theory could in principle explain the whole duality data of genuine quantum RCFT's.

Quantum group phenomenology

At that time quantum groups have already taken off from the land of integrable systems. The general notion of quantum groups were developed by Drinfeld and Jimbo in references [Dr1,Ji]. The first non classical example of this definition was a q-deformation of the Lie algebra $Sl(2)$ which was found earlier by Kulish and Reshetikhin [KuR]. Drinfeld and Jimbo generalized this quantum deformation to include all the classical Lie groups affine or not. Later on in a series of papers Kirillov and Reshetikhin [KR,R] constructed the representation theory of $U_q(Sl(2))$ and others groups for a generic value of the deformation parameter q.

After these works it was rather clear that quantum groups were the correct "deformation" of classical groups needed to explain the duality properties of RCFT's. In reference [AGS2] it was considered the case of the WZW model based on the Kac-Moody algebra $\widehat{SU(2)}$ at level k, and it was shown that the braiding matrices , computed by Tsuchiya and Kanie [TK] solving the Knizhnik-Zamolodchikov equation for the conformal blocks [KZ], were in fact given by the representation theory of $U_q(Sl(2))$ with $q = e^{\pi i/k+2}$. This work was later generalized to other groups, where the previous relation between the quantum deformation parameter and the data of the WZW model reads:

$$q = e^{i\pi/k+g} \tag{4}$$

with k the level and g the dual Coxeter number of G. It was also possible to understand the modular properties and in particular the Verlinde theorem using exclusively quantum-group tools [AGS3]. In what concerns modular properties, these lead to the concept of ribbon Hopf algebras [RT].

Similar results were obtained for the minimal models of type (p,p'), where the braiding matrices essentially factorize into the product of two quantum 6-j symbols with deformation parameters: $q = e^{i\pi p/p'}$ and $q' = e^{i\pi p'/p}$ [AGS1, FFK]. Also some orbifold models where shown to have a quantum group structure [DPR] given by quasi-Hopf algebras [Dr2].

The previous examples showed that the deformation parameters coming from RCFT's were given by roots of unity. The representation theory of quantum groups for generic values of q is essentially the same as the classical one [Ro].However quantum groups at roots of unity were shown to display unexpected and interesting properties [L,RA]. The trouble with roots of unity is that many of the formulas in the representation theory for q generic break down or become ambiguous. A solution to this problem, which may look a bit ad hoc, is to restrict the representations to the good ones, so that one has a well defined highest weight theory, and throwing away the bad representations [L,PS,AGS2-3]. For example for $U_q(Sl(2))$ the good representations ,also called regular or of type II, have spin j between zero and k/2, and are in one to one correspondence with the integrable irreps of the WZW model $\widehat{SU(2)}_k$. This truncation was also seen to occur in the representation theory of the Hecke algebra $H_N(q)$ which is the centralizer of the tensor product of the spin 1/2 representation of $U_q(Sl(2))$ [Wz,AGS2]. The relation between fusion rules of RCFT's and quantum groups have been further studied in [GP,FGP,FD,T].

These efforts of trying to match bits of data from different pieces of mathematics and physics were in a way "phenomenological", because it was missing a neat understanding of the close relationship between RCFT's and quantum groups. Further research has deepen and broaden the knowledge of the topic, which cannot yet considered as being closed.

One astonishing aspect of the connection between RCFT's and quantum groups is that on one side one is dealing with an infinite dimensional symmetry i.e. the Virasoro algebra , Kac-Moody algebra ,etc. and on the other the symmetries, although a bit peculiar, are finite dimensional. How could it be possible that some finite dimensional structure would know so much about an infinite dimensional one?. Making a very rough analogy we could say that given a function of the form: $f(z) = z^{\alpha} \sum a_n z^n$ the information that quantum groups are telling us is the integer part of α, which is what determines the monodromy of the function ,i.e. $f(e^{2\pi i}z) = e^{2\pi i\alpha}f(z)$. In RCFT's the role of the function $f(z)$ is played by the conformal blocks while the phase factors $e^{2\pi i\alpha}$ or $e^{\pi i\alpha}$ become the monodromy or braiding matrices respectively. To complete this analogy we could say that the chiral algebra (i.e. Virasoro, Kac-Moody ,etc) would imply a differential equation for the function $f(z)$, equation that would contain in a disguised way the monodromy of their solutions.

In this context the connection between RCFT's and quantum groups is the relation between a differential equation and the monodromy or braiding properties of their solutions. The later problem is deeply connected to the classic and well known Riemann-Hilbert problem , which consist in the characterization of a differential equation and the set of the solutions from the knowledge of their monodromy properties. What the polynomial equations establish are the consistency conditions to be satisfied by the braiding and fusion matrices associated to a set of differential equations which are in principle unknown. The "reconstruction program", advocated by Moore and Seiberg , which tries to determine from every solution to the polynomial equations a set of differential equations whose solutions give the conformal blocks of the theory, is nothing else but a Riemann-Hilbert problem. This program has only been sketched but, as one can imagine, the task of finding a solution to the polynomial equations seems already quite formidable without the help of some guiding principle.

An important thing that we must not forget is the meaning of the differential equations satisfied by the conformal blocks. They guarantee the decoupling of null vectors of the degenerate primary fields. The very existence of these fields is what makes the theory rational (finite number of primary fields) and solvable. We may expect that quantum groups would also have a lot to say about the decoupling of null vectors of the corresponding primary fields.

Vertex formulation of CFT's

From the previous discussion it seemed natural that the ordinary formulation of RCFT's as the representation of a chiral algebra (or rather the product of a holomorphic $\mathcal{A}_L$ times an antiholomorphic chiral algebra $\mathcal{A}_R$) should be enlarged in order to accommodate the quantum group symmetry, which otherwise remains hidden or invisible. A signal of this hidden nature was that objects like 3j-symbols or the R-matrix itself did not show up in RCFT's while for example 6j-symbols did. In general the IRF content of quantum groups was clearly present in RCFT's. This suggested that what was needed, to uncover the quantum group structure of RCFT's, was a "vertex formulation" of conformal field theories. In statistical mechanics it is well known that some models admit both kinds of formulations, either as vertex or as IRF models, in which case there exist a vertex-IRF map which establishes the correspondences [Ba].It happens also, when going from the vertex to the IRF formulation, that one misses along the way the degrees of freedom on which the quantum group is acting.These kind of arguments were also used in [Wi] in an attempt to derive a quantum group structure from the 3D Chern-Simon theory.

In group theoretical language the vertex formulation is like working with tensor products of representations of a group, while the IRF formulation is like performing the tensor product of these irreps and keeping only the spaces on which the centralizer of the group is acting:

$$V_{1/2} \otimes \cdots^{N} \cdots \otimes V_{1/2} = \bigoplus_{j} (W_j^N \otimes V_j) \tag{5}$$

In the vertex formulation the group or q-group is acting on each individual vector space, $V_{1/2}$ in the example above, by means of the comultiplication. The vector space W_j^N is formed by the invariant tensors of the q-group and serves as the representation space of the centralizer $C_{1/2}^N$ which is defined as the set of endomorphisms of $V_{1/2}^{\otimes N}$ that commute with the action of the group. The dimension of W_j^N equals the multiplicity of the spin j irrep into the tensor product $V_{1/2}^{\otimes N}$. In RCFT's W_j^N can be identified with the space of conformal blocks of N legs of spin 1/2 and one leg of spin j, and its dimension can be computed using the fusion rules of the theory.From the previous identification we see that the centralizer of the quantum group is precisely realized on the space of conformal blocks , while they remain invariant under the action of the quantum group. This explains in a neat way the meaning of hidden quantum symmetries when talking about the role of quantum groups in RCFT's.

In more physical terms the centralizer has the structure of a braid group when acting on the space of conformal blocks. In the case of a WZW model based on the group $SU(2)$ it was shown by Kanie and Tsuchiya that the braid group representation that one obtains from the conformal blocks with spin 1/2 primary fields at the external legs gives a representation of the Hecke algebra $H_N(q)$ or more precisely of the Temperley-Lieb-Jones algebra. On the other hand Kirillov and Reshetikhin showed that the centralizer of the quantum group $U_q(Sl(2))$ in the tensor product of spin 1/2 representations was also a Hecke algebra $H_N(q)$ [KR]. This connection was again part of the phenomenology that we mention before, however this time one could see that the missing objects to complete the link with quantum groups were the q-group spaces $V_{1/2}$ and V_j of equation(5). These spaces should not be confused with the Verma modules $\mathcal{H}_j$ of the Kac-Moody algebra. If this were the case then the q-group $U_q(Sl(2))$ would already be contained in the Kac-Moody algebra $\widehat{SU}(2)$. This suggest that the full symmetry algebra should be something like a tensor product $\widehat{SU}(2) \otimes U_q(Sl(2))$. For a general conformal field theory the symmetry would be $\mathcal{A}_L \otimes Q_L$ for the holomorphic degrees of freedom and $\mathcal{A}_R \otimes Q_R$ for the antiholomorphic ones. Q_L and Q_R denote the quantum groups associated to the chiral algebras $\mathcal{A}_L$ and $\mathcal{A}_R$ respectively. Later on we shall discuss in more detail the structure of the tensor products $\mathcal{A}_L \otimes Q_L$ which turns out to be a bit more subtle than just a direct tensor product. The key idea in this respect is that the q-group spaces V_j contain fields or states that cannot be obtained from the primary fields by the action of an operator in the chiral algebra but by that of a genuine quantum group operator. We should then distinguish between two kind of descendants, those of the chiral algebra $\mathcal{A}_L$ and those of the quantum group Q_L.

Having settled these conceptual matters the problem was then to give an "explicit" construction of these quantum group fields which could not be conformal (either primary or descendants)in the usual sense. This technical problem was overcome by the use of the Feigin-Fuchs or Coulomb gas formalism which has proved to be extremely important in the study of conformal field theories [FF].

The Coulomb Gas formalism of CFT's

A Coulomb gas is nothing but a free field realization of a conformal field theory. The first known CFT's admiting a Coulomb gas version are the BPZ $c < 1$ theories [BPZ], which were studied by Dotsenko and Fateev using only one free scalar field [DF]. The CFT's based on the W_n algebras [Z] require the introduction of n-1 scalar fields [FZ,FL], while the WZW models need both scalar fields and $\beta - \gamma$ systems [Wa,FFr]. An open problem is to know wether any conformal field theory admits a free field realization. The fact that a highly interacting field theory allows a free field description is on the other hand a mistery with very deep consequences. Some of them are the ability of a systematic computation of the conformal blocks, the characterization of the null vectors, a straightforward derivation of the Kac's formula, the construction of a BRST formulation *à la* Felder [Fe], and , as we shall show next, the construction of quantum groups.

The basic feature of a Coulomb gas is that the chiral algebra generators can be constructed entirely in terms of a set of free fields which have very simple operator product expansions. Moreover the primary fields, which are essentially vertex operators as in string theory, also satisfy simple OPE's. All this reduces enormously the task of computating correlators which are given by integrals of products of monomials. If we are studying the $c < 1$ or a WZW model we know that these integrals come from the solution of hypergeometric type equations. However in the Coulomb gas these integrals come from the screening charge that one has to introduce in order to balance the background charge. A screening charge is an integral $\int dt J_a(t)$ which has the property of commuting with all the operators of the chiral algebra O_n up to a total derivative:

$$[\, O_n \,, \int dt J_a(t)\,] = \int dt \frac{\partial}{\partial t}(X_n) \tag{6}$$

If there are not boundary contributions to (6) then the screening charges fully commute with the chiral algebra.

Another ingredient of this construction is the use of Fock spaces of free fields instead of Verma modules . The later are recovered as cohomological classes associated to a BRST charge constructed using screening operators [Fe]. The philosophy adopted in references [GS1-2] was to use these Fock spaces to enlarge, rather than to restrict, the number of fields in the theory in order to give room to q-groups. We could say in this sense that quantum groups are hidden in a RCFT in much the same manner that the corresponding free field version is hidden, which lead us in the long run to ask why the solution to the null vector decoupling equations of a RCFT admit an integral representation. This is a highly non trivial fact and a mistery which certainly needs more study.

Field theory formulation of quantum groups

The starting point of the work of reference [GS1] is the definition of a special kind of screened vertex operators which form the basis for the representation spaces of quantum groups. They are defined integrating a collection of screening vertex operators J_a around a primary field ϕ_α:

$$e^{\alpha}_{a_1,\cdots,a_n}(z_P, z_\infty) = \int_{C_1} dt_1 \; J_{a_1}(t_1) \cdots \int_{C_n} dt_n \; J_{a_n}(t_n) \;\; \phi_\alpha(z_P) \tag{7}$$

In the choice of the integration contours $C_1, \cdots, C_n$ we have taken into account the fact that there are branch cuts in the integrand coming from the OPE's between the screening operators with the primary field as well as among the screening themselves.They are choosen as a nested set of Hankel's contours, similar to the ones that appear in the definition of the Γ function, i.e. they go from infinity to infinity encircling the point where the primary field is inserted. Strictly speaking the operators $e^{\alpha}_{a_1,\cdots,a_n}$ depend on two points, one is the point z_P where the primary field is located and the other is the point at infinity z_∞, which plays the role of a base point.

In this construction the primary field itself ϕ_α , is identified with the highest weight vector of a q-group space V_α. The remaining vertices obtained integrating screening operators are identified with q-group descendants. This leads naturally to the interpretation of the screening charges as quantum lowering operators : $F_a = \int_C dt J_a(t)$. This interpretation was already anticipated in references [Sa,BMP]. In the $c < 1$ theory there are two screening operators J_+ and J_- hence we have two lowering operators F_+ and F_- [GS2]. In the Wakimoto construction of $\widehat{SU(n)}$ there are n-1 screenings operators which then lead to n-1 lowering operators each one associated to a positive simple root [RRR]. Under this interpretation each screening charge J_a give rise to a quantum group $U_{q_a}(Sl(2))$, where the deformation parameter q_a is given by the braiding factor among two screenings of the same type. If there are various screening operators they will define in general larger q-groups whose defining relations would follow from their braiding properties. For example for a WZW theory the braiding between screening operators: satisfies:

$$J_a(z)\ J_b(w) = q^{C_{ab}}\ J_b(w)\ J_a(z) \tag{8}$$

where C_{ab} is the symmetrized Cartan matrix and q is given by eq.(4).

Equation (8) reflects the non-local nature of the screening operators and in that sense the quantum deformation parameter q acquires a conformal field theory meaning.

From this equation one can also derive the quantum deformation of the Serre relations satisfied by the lowering operators F_a [BMP,RRR].

After applying a sufficient number of screening operators to a primary field one eventually gets zero or a null vector. To understand why this happens we have to recall that the q-vertex operators do depend in general on the point at infinity z_∞ and the point z_P where the primary field is inserted, however under certain circunstances the branch cut effectively disappears and the contour integrals shrink to the neighbour of the point z_P. This property holds for primary fields within the Kac's formula and it is extensively used in the BRST construction. In our q-group interpretation this means that the primary field give rise to a finite dimensional representation. In other words, a null vector in the CFT is at the same time a null vector for the q-group. This is the core of the connection between quantum groups and RCFT's.

It would be of some interest to find a more intrinsic definition of the q-vertex operators (7) , i.e. a definition independent of the use of the Coulomb gas formalism. A possibility is to consider these vertices as the solutions of the null vector decoupling equations.The primary fields themselves would of course be a solution, the one with conformal properties, while the others will span the rest of a representation space. The order of the equation ,which is nm for the primary field ϕ_{nm} for a $c < 1$ theory, would then give the dimension of the associated representation space of the quantum group.

We conclude that screening charges are lowering operators, similarly the Cartan operators are identified with the Coulomb charges. These two sets of operators form the Borel subalgebra which is traditionally associated with the lower triangular matrices. To complete the RCFT picture of q-groups one needs a definition of the raising operators E_a. These cannot be represented as integrals of screening operators with opposite charge because they do not have the correct conformal properties. What one really needs is an operator that would destroy integrals of screening operators and this can be achieved with the action of the chiral algebra. Take for example the Virasoro operator L_{-1} , acting on a q-vertex operator it integrates each of the screening charges leaving a boundary term as can be seen in eq.(6). Explicit computations show that eq.(6) implies the q-group relation:

$$[O_n, \int dt J_a(t)] = \int dt \frac{\partial}{\partial t}(X_n) \Rightarrow [E_a\ , F_b] = \delta_{a,b} \frac{K_a - K_a^{-1}}{q_a - q_a^{-1}} \tag{9}$$

In this sense the q-group generators E_a are contained in the chiral algebra generators O_n, which shows that the relation between $\mathcal{A}_L$ and Q_L is more intricated than expected. From equation (9) we also see that the shrinking condition for the contours which define a q-vertex operator is equivalent to the statement that this q-vertex operator is a highest weight vector of the quantum group:

$$\frac{\partial}{\partial z_\infty} e^{\alpha}_{a_1,\cdots,a_n} = 0 \Rightarrow E_a\ e^{\alpha}_{a_1,\cdots,a_n} = 0\ \forall a \tag{10}$$

It is also curious to observe some parallelism between this construction and the construction of the quantum double by Drinfeld [GS1,BL1].

Having defined the q-group spaces and the q-group generators, it is easy to obtain the comultiplication, the R-matrix, the 3j-symbols, the 6j-symbols , that is all the ingredients of quantum groups in a field theoretical framework. These ideas and techniques, which were first applied to the $c < 1$ theories, have been extended successfully to WZW models [RRR], W_n algebras [C] , N=1 and 2 superconformal field theories [Jz] and a WZW model based on the supergroup $Gl(1,1)$ [RSa] . Moreover the whole construction has a geometrical and topological flavor [GS3]. Everything follows from contour manipulations involving elementary complex analysis and q-combinatorics. Group theory when quantum deformed becomes a topology of contours (for a more elaborated version of this idea see [FeW,SV]).

Towers of algebras

Until now we have proceed to unravel the quantum group structure of RCFT's from the integral representation of the decoupling equations, we can refer to this line of thought as the analytic approach. A complementary point of view partially inspirated in Pasquier's ethiology of IRF-models [P] consist in starting the construction with the fusion rules and from that piece of information to obtain the duality (braiding and fusion) matrices in the very same way one associates in the Jone's fundamental construction a Temperley-Lieb-Jones algebra to a given graph [GHJ]. In this algebraic approach the RCFT defines a model of the Bratelli diagram in terms of the fusion rules[GS4]. Recent progress by Ocneanu [Oc] indicates ,as it should be expected, that associated with the RCFT tower of algebras it is possible to define in a unique way a quantum group structure. For a different version of the algebraic approach based on operator algebras see [FrK]. We believe that it would be fruitful both from a mathematical and a physical point of view to get a deeper understanding of the interplay between these two apparently different mathematical approaches, namely the monodromy of differential equations with their related Riemann-Hilbert problem and the theory of subfactors.

Cannonical quantization approach to CFT's

The discussions above about the role of quantum groups in CFT's have been done in the general framework of the bootstrap program of BPZ, where actions for the local fields are not strictely needed. In doing so we have omitted another important approach to the role of quantum groups in quantum field theories, namely the one based on the use of local actions or hamiltonians which are quantized in a cannonical way. The traditional methods were applied to Liouville theory, which was shown to exhibit quantum group attributes in a time when quantum groups were not known [GN,FT].In fact in reference [FT] first appeared the defining relations of the quantum Lie group SL_q, not to be confused with the quantum Lie algebra sl_q which is the same as the quantized universal enveloping algebra $U_q(Sl(2))$.

After quantum groups came to fashion this approach was renewed again in connection with Liouville theory [Bb,Ge,ST] , Toda field theory [BG,HM] and the Wess-Zumino-Witten model [AS,Fa,Ga,AF,AFS,AFSV] . The use of a different language has made a bit difficult to stablish the connections with the, let's say, bootstrap approach although they certainly do exist. Particularly interesting are the lattice formulations of the Liouville and the WZW theories which may perhaps allow a Bethe ansantz analysis analogous to the one applied in the study of magnetic chains.We refer to the lectures of prof.L.D.Fadeev for a discussion of these matters.

Integrable quantum field theories

So far we have considered applications of the quantum groups to conformal field theories but their range of applicability includes also massive integrable 2D field theories. An archetype of the later is the well known sine-Gordon theory. In references [RS,BL2] it was shown that there exist a hidden quantum symmetry which governs the scattering of the solitons, antisolitons and bound states of the model, which explains in particular the S-matrix obtained in [ZZ]. This symmetry is affine and can also be given a field theory realization in terms of non-local conserved charges [BL3,FL,BFL].

q-RCFT's: New hidden symmetries ?.Elliptic quantum groups

We would like now to discuss briefly some new directions of research. As we have seen, a particular interesting class of RCFT's are the WZW models based on Kac-Moody algebras. A centrally extended affine Lie algebras can also be deformed in a very non trivial way. These q-RCFT's preserve much of the structure of the ordinary RCFT's and can be interpreted as some kind of massive field theories. In reference [FR] Frenkel and Reshetikhin have found a q-deformation of the Kniniznik-Zamolodchikov equation which governs the correlators of these q-WZW model. This is a finite difference equation whose solutions are q-hypergeometric functions. The connection matrices, or in CFT language, the braiding matrices of this "q-conformal blocks" turns out to be given by the elliptic solution to the Yang-Baxter equation of the RSOS model of Andrews, Baxter and Forrester [ABF]. These authors also pose the question of the possible existence of a hidden quantum symmetry which would underlie the elliptic Boltzmann weights, much in the same way that quantum groups underlie the trigonometric solutions. It has also been suggested as a candidate for this "elliptic quantum symmetry" the Sklyanin algebra which is intimately related to the Baxter's eight vertex model [Sk].

The solution to these questions is not known to us. We would like however to propose a way to attack the problem which consist in following the same steps that were pursued in the unravelling of quantum groups in RCFT's, namely : find a free field realization of the q-Kac-Moody algebra, then look for the q-screening operators and finally define in terms of q-integrals the "elliptic version" of quantum groups. The first two steps of this program have already been achieved in references [FJ,M,Sh,KQS,ABG] where various free field versions of the q deformed Kac-Moody algebra have been given. In [Sh] it is also constructed a q-screening operator whose q-integral *à la* Jackson commutes with the generators of q-Kac-Moody. Therefore one has in principle all the ingredients to unravel a new hidden quantum symmetry.

Spin chains at roots of unity

To finish this brief history we would like to make some comments in connection to recent progress by De Concini and Kac [DK] in the study of the representation theory of quantum groups at roots of unity (see also [DKP]. As we have already mentioned the quantum deformation parameter of the quantum group associated to a RCFT is a root of unity , for example $q = e^{\pi i/k+2}$ for a SU(2) WZW model. This implies that strictely speaking the quantum group associated to this RCFT should be $U_q(SU(2))$ moded out by the central Hopf which appears when q is a root of unity [McSo].

Another phenomena that appears for quantum groups at roots of unity is the existence of non restricted representations, i.e. representations which transform non trivially under the central Hopf subalgebra. In a series of papers [GRS,GS5,BGS,CGS] we have started the study of a family of these representations called semicyclic and nilpotents, which have the property that their tensor product is decomposable [GS5,A] so that one can define 3j and 6j symbols. An important question is wether this new class of braiding and fusion matrices may become the duality data of a new class of decoupling equations defining a new hierarchy of conformal field theories.

Acknowledgements

We would like to thanks Prof.M.L.Ge for his kind invitation to participate in the XXI DGMTP Conference at Tianjin.

References

[ABG] A.Abada,A.H.Bougourzi and M.A.El Gradechi: Deformation of the Wakimoto construction. Preprint (1992).

[AF] A.Yu.Alekseev and L.D.Fadeev: $(T^*G)_t$: A toy model for conformal field theory. Commun.Math.Phys. 141 (1991) 413.

[AFSV] A.Yu.Alekseev,L.D.Fadeev,M.Semenov-Tian-Shansky: Hidden quantum groups inside Kac-Moody algebra.Commun.Math.Phys. 149 (1992) 335.

[AFSV] A.Yu.Alekseev,L.D.Fadeev,M.Semenov-Tian-Shansky and A.Volkov: The unravelling of the quantum group structure in the WZW theory. CERN preprint Th 5981/91 (1991).

[AS] A.Alekseev and S.Shatashvili: From geometric quantization to conformal field theory. Commun.Math.Phys. 128 (1990) 197; Quantum groups and WZNW models. Commun.Math.Phys. 133 (1990) 353 .

[AGS1] L.Alvarez-Gaume,C.Gomez and G.Sierra:Hidden quantum symmetries in rational conformal field theories. Nucl.Phys. B319 (1989) 155.

[AGS2] L.Alvarez-Gaume,G.Gomez and G.Sierra:Quantum group interpretation of some conformal field theories. Phys.Lett. 220B (1989) 142.

[AGS3] L.Alvarez-Gaume,G.Gomez and G.Sierra:Duality and quantum groups.Nucl.Phys. B330 (1990) 347.

[ABF] G.Andrews,R.Baxter and P.Forrester: Eight-vertex SOS model and generalized Rogers-Ramanujan-type identities. J.Stat.Phys. 35 (1984) 193.

[A] D.Arnaudon: Fusion rules and R-matrix for the composition of regular spins with semi-periodic representations of $SL(2)_q$. Phys.Lett. 268B (1991) 217.

[Bb] O.Babelon:Extended conformal algebra and the Yang-Baxter equation. .Phys.Lett. 215B (1988) 523.

[Ba] R.Baxter: Exactly solved models in statistical mechanics.Academic Press (1982).

[BPZ] A.A.Belavin,A.N.Polyakov and A.N.Zamolodchikov: Infinite conformal symmetries in two-dimensional quantum field theory.Nucl.Phys. B241 (1984) 333.

[BGS] A.Berkovich,C.Gomez and G.Sierra: On a new class of integrable models. Inter.J.Mod.Phys. B6 (1992) 1939; q-Magnetism at roots of unity. to appear in J. of Phys.A.

[BFL] D.Bernard,G.Felder and A.LeClair: Non-local conserved currents: the lattice approach.

[BL1] D.Bernard and A.LeClair: The quantum double in integrable quantum field theory. Preprint CLNS 92/1147 (1992).

[BL2] D.Bernard and A.LeClair: Residual quantum symmetries of the restricted sine-Gordon theories. Nucl.Phys. B340 (1990) 721.

[BL3] D.Bernard and A.LeClair: The fractional supersymmetric sine-Gordon models.Phys.Lett. 247B (1990) 309; Quantum group symmetries and non-local conserved currents in 2D QFT .Commun.Math.Phys. 142 (1991) 99.

[BG] A.Bilal and J.L.Gervais: Systematic approach to conformal systems with extended Virasoro symmetries.Phys.Lett. 206B (1988) 412.

[BMP] P.Bouwknegt.J.McCarthy and K.Pilch: Free field realization of WZNW-models; BRST complex and its quantum group structure. Phys.Lett. 234B (1990) 297; Quantum group structure in the Fock space resolutions of $\widehat{sl(n)}$ representations. Commun.Math.Phys. 131 (1990) 125.

[C] R.Cuerno: Quantum symmetries in the free field realization of W_n algebras. Phys.Lett. 271B (1991) 314.

[CGS] R.Cuerno,C.Gomez and G.Sierra:On integrable quantum group invariant antiferromagnets.To appear in Journal of Geometry and Physics.

[DK] C.De Concini and V.G.Kac:Representations of quantum groups at roots of 1. Progress in Math. 92 (1990) 471.

[DKP] C.De Concini,V.G.Kac and C.Procesi: Quantum coadjoint action. Pisa preprint (1991).

[DPR] R.Dijkgraaf,V.Pasquier and P.Roche: Quasi-quantum groups related to orbifold models.Tata Inst. preprint 1990.

[DF] V.S.Dotsenko and V.A.Fateev: Conformal algebra and multipoint correlation functions in 2D statistical models. Nucl.Phys. B240[FS12](1984) 312, B251 (1985) 691.

[Dr1] V.G. Drinfeld: Hopf algebras and the quantum Yang-Baxter equation. Soviet.Math.Dokl.32 (1985) 254; A new realization of Yangians and quantized affine algebras.Soviet. Math.Dokl.36 (1988) 212; Quantum groups. Proc. ICM-86 (Berkeley),Vol.1 AMS (1986) 798.

[Dr2] V.G.Drinfeld: Quasi-Hopf algebras and Knizhnik-Zamolodchikov equation.Preprint ITP-89-43E.

[Fa] L.D.Fadeev: On the exchange matrix for WZNW models. Commun.Math.Phys.132 (1990) 131.

[FRT] L.D.Fadeev, N.Yu.Reshetikhin and L.A.Takhtajan. Quantization of Lie Groups and Lie Algebras. Leningrad LOMI preprint E-14-87. Algebra and Analysis 1 (1989) 178 (in russian).

[FT] L.D.Fadeev and L.Takhtajan: Liouville model on the lattice. Lecture Notes in Physics,vol.246,p.166, Springer (1986).

[FL] V.A.Fateev and S.L.Lykyanov: The models of two-dimensional conformal quantum field theory with Z_n symmetry. Inter.J.Mod.Phys. A3 (1988) 507.

[FZ] V.A.Fateev and A.B.Zamolodchikov: Conformal quantum field theory models in two dimensions having Z_3 symmetry. Nucl.Phys B280[FS18] (1987) 644.

[FFr] B.L.Feigin and E.V.Frenkel: Affine Kac-Moody algebras and semi-infinite flag manifolds. Commun.Math.Phys. 128 (1990) 161.

[FF] B.L.Feigin and D.B.Fuchs: Invariant skew-symmetric differential operators on the line and Verma modules over the Virasoro algebra.Func.Anal.Appl. 16 (1982) 114; Verma modules over the Virasoro algebra. Func.Anal.Appl. 17 (1983) 241; Representations of the Virasoro algebra. In: Topology,Proc. Leningrad 1982, L.D. Fadeev ,A.Malcev (eds.).Lecture Notes in Mathematics,vol 1060. Springer 1984.

[Fe] G.Felder: BRST approach to minimal models. Nucl.Phys. B317 (1989) 215.

[FFK] G.Felder, J.Frohlich and G.Keller: Braid matrices and structure constants for minimal conformal models.Commun.Math.Phys. 124 (1989) 647; On the structure of unitary conformal theory,II: Representation theoretic approach. Commun.Math.Phys. 130 (1990) 1.

[FL] G.Felder and A.LeClair: Restricted quantum affine symmetry of perturbed minimal conformal models. Intern.J.Mod.Phys. A7,Suppl.1A (1992) 239.

[FeW] G.Felder and C.Wieczerkowski: Topological representations of the quantum group $U_q(sl_2)$. Commun.Math.Phys.138 (1991) 583.

[FRS] K.Fredenhagen,K.-H.Rerhen and B.Schroer: Superselection sectors with braid group statistics and exchange algebras, I: General theory. Commun.Math.Phys. 125 (1989) 201.

[FJ] I.B.Frenkel and N.Jing: Vertex representations of quantum affine algebras. Proc.Natl.Acad.Sci.USA 85 (1988) 9373.

[FR] I.B.Frenkel and N.Yu.Reshetikhin: Quantum affine algebras and holonomic difference equations. Commun.Math.Phys.146 (1992) 1.

[Fr] J.Frohlich: Statistics of fields, the Yang-Baxter equation and the theory of knots and links. Cargese lectures, Plenum Press (1987).

[FrK] J.Frohlich and T. Kerler: Preprint ETH-92.

[FD] J.Fuchs and P. van Driel: Some symmetries of quantum dimensions.J.Math. Phys. 31 (1990) 1770 ; WZW fusion rules , quantum groups and the modular matrix S. Nucl.Phys. B346 (1990) 632.

[FGP] P.Furlan,A.G.Ganchev and V.B.Petkova: Quantum groups and fusion rule multiplicities. Nucl.Phys. B343 (1990) 205.

[GP] A.C.Ganchev and V.B.Petkova: $U_q(Sl(2))$ invariant operators and minimal theory fusion rules. Phys. Lett 233B (1989) 374.

[Ga] K.Gawedzki: Classical origin of quantum group symmetries in Wess-Zumino-Witten conformal field theory. Commun.Math.Phys. 139 (1991) 201.

[Ge] J.-L.Gervais: The quantum group structure of 2D gravity and minimal models. Commun.Math.Phys.130 (1990) 257.

[GN] J.-L.Gervais and A.Neveu: New quantum treatment of Liouville field theory.Nucl.Phys. B224 (1983) 329 ; Novel triangle relation and absence of tachyons in Liouville string field theory. Nucl.Phys. B238 (1984) 125.

[GRS] C.Gomez,M.Ruiz-Altaba and G.Sierra: New R-matrices associated with finite dimensional representations of $U_q(sl(2))$ at roots of unity. Phys.Lett. 265B (1991) 95.

[GS1] C.Gomez and G.Sierra: Quantum group meaning of the Coulomb gas.Phys.Lett. 240B (1990) 149.

[GS2] C.Gomez and G.Sierra: The quantum symmetry of rational conformal field theories. Nucl.Phys. B352 (1991) 791.

[GS3] C.Gomez and G.Sierra: Contour representation of quantum groups in RCFT's. Proc.Triestre 1990 Summer School. World Scientific (1991).

[GS4] C.Gomez and G.Sierra: Towers of algebras in rational conformal field theories. Inter.J.Mod.Phys. A6 (1991) 2045.

[GS5] C.Gomez and G.Sierra: A new solution to the star-triangle equation based on $U_q(sl(2))$ at roots of unity. Nucl.Phys. B373 (1992) 761; New integrable deformations of higher spin Heisenberg-Ising chains. Phys.Lett. 285B (1992) 126.

[GHJ] F.M.Goodman,P.de la Harpe and V.F.R.Jones: Coxeter-Dynkin diagrams and towers of algebras.MSRI Publications , Springer-Verlag,1989.

[HM] T.J.Hollowood and P.Mansfield: Quantum group structure of quantum Toda conformal field theories (I). Nucl. Phys. B330 (1990) 720.

[Ji] M.Jimbo:Quantum R matrix for generalized Toda system. Commun.Math.Phys. 102 (1986) 537; A q-difference analogue of $U(g)$ and the Yang-Baxter equation. Lett.Math.Phys. 10 (1985) 63; A q-analogue of $U(gl(N+1)$, Hecke algebra and the Yang-Baxter equation.Lett.Math.Phys. 11 (1986) 247.

[Jz] F.Jimenez: Quantum group structure of N=1 superconformal field theories. Phys.Lett. 252B (1990) 577; The quantized symmetry of N=2 superconformal field theories. Phys.Lett. 273B (1991) 399.

[KQS] A.Kato,Y.H.Quano and J.Shiraishi: Free boson representation of q-vertex operators and their correlation functions. Univ. Tokio preprint UT-618 (1992).

[KR] A.N.Kirillov and N.Yu.Reshetikhin: Representations of the algebra $U_q(sl(2))$, q-orthogonal polynomials and invariants of links. In infinite-dimensional Lie algebras and groups.V.G.Kac (ed.) World Scientific (1989).

[KZ] V.G.Knizhnik and A.B.Zamolodchikov: Current algebra and Wess-Zumino model in two dimensions. Nucl.Phys. B247 (1984) 83.

[Ko] T.Khono: Monodromy representations of braid groups and Yang-Baxter equations. Ann.Inst.Fourier 37 (1987) 139.

[KuR] P.P.Kulish and N.Yu.Reshetikhin: Quantum linear problem for the Sine-Gordon equation and higher representations.Zap.Nauch. Semi. LOMI 101 (1981); J.Sov.Math. 23 (1983) 2435.

[L] G.Lusztig: Modular representations and quantum groups. Contemp.Math. 82 (1989) 58 ;Quantum groups at roots of unity.Geom.Dedicata(1990).

[McSo] G.Mac and V.Schomerus: Quasi-Hopf quantum symmetry in quantum theory. Nucl.Phys. B370 (1992) 185; Quasi-quantum group symmetry and local braid relations in the conformal Ising model.Phys.Lett. 267B (1991) 207.

[M] A.Matsuo: Free field representation of quantum affine algebra $U_q(\widehat{sl(2)})$. Univ.Nagoya preprint (1992).

[MR] G.Moore and N.Yu.Reshetikhin: A comment on quantum group symmetry in conformal field theory. Nucl.Phys. B328 (1989) 557.

[MS] G.Moore and N.Seiberg: Polynomial equations for rational conformal field theories. Phys.Lett. 212B (1988) 451; Naturality in conformal field theory. Nucl.Phys. B313 (1989) 16; Classical and quantum conformal field theory. Commun.Math.Phys. 123 (1989) 177,

[Oc] A. Ocneanu. Lectures at College de France 1992.

[P] V.Pasquier: Etiology of IRF models. Commun.Math.Phys. 118 (1988) 335.

[PS] V.Pasquier and H.Saleur: Common structures between finite systems and conformal field theories though quantum groups. Nucl.Phys. B330 (1990) 523.

[RRR] C.Ramirez,H.Ruegg and M.Ruiz-Altaba:Explicit quantum symmetries of WZNW theories. Phys.Lett. 247B (1990) 499; The contour picture of quantum groups: conformal field theories. Nucl.Phys. B364 (1991) 195.

[Rh] K.-H.Rehren: Locality of conformal fields in two dimensions:Exchange algebra on the light-cone.Commun.Math. Phys.116 (1988) 675.

[R] N.Yu.Reshetikhin: Quantized universal enveloping algebras,the Yang-Baxter equation and invariants of links,I. LOMI preprint E-4-87, 1988; II. LOMI preprint E-17-87,1988.

[RS] N.Yu.Reshetikhin and F.Smirnov: Hidden quantum group symmetry and integrable perturbations of conformal field theories. Commun.Math.Phys. 131 (1990) 157.

[RT] N.Yu.Reshetikhin and V.G.Turaev: Ribbon graphs and their invariants derived from quantum groups. Commun.Math.Phys. 127 (1990) 1.

[RA] P.Roche and D.Arnaudon:Irreducible representations of the quantum analogue of $SU(2)$. Lett.Math.Phys. 17 (1989) 295.

[Ro] M.Rosso: Finite dimensional representations of the quantum analogue of the enveloping algebra of complex simple Lie algebras. Commun.Math.Phys.117 (1988) 581.

[RSa] L.Rozansky and H.Saleur: Quantum field theory for the multi-variable Alexander-Conway polynomial. NuclPhys. B376 (1992) 461.

[Sa] H.Saleur: Lattice models and conformal field theories. Phys.Rep. 184 (1989) 177.

[SV] V.V.Schechtman and A.N.Varchenko: Arrangement of hyperplanes and Lie algebra homology.Invent.Math. 106 (1991) 139; Quantum groups and homology of local systems. IAS preprint (1990).

[Sh] J.Shiraishi: Free boson representation of $U_q(\widehat{sl(2)})$. Univ. Tokio preprint UT-617 (1992).

[ST] F.Smirnov and L.Takhtajan: Towards a quantum Liouville theory with $c < 1$. Colorado preprint 1990.

[Sk] E.K.Sklyanin: Some algebraic structures connected with the Yang-Baxter equation. Func.Anal.Appl. 16 (1982) 27; Some algebraic structures connected with the Yang-Baxter equation. Representations of quantum algebras. Func.Anal.Appl. 17 (1983) 273.

[T] I.T.Todorov: Quantum groups as symmetries of chiral conformal algebras.Lecture Notes in Physics.Vol. 370 p.231,Springer-Verlag 1990.

[TK] A.Tsuchiya and Y.Kanie: Vertex operators in conformal field theory on P^1 and monodromy representations of braid group. Adv.Stud.Pure Math. 16 (1988) 297, Lett.Math.Phys. 13 (1987) 303.

[V] E.Verlinde: Fusion rules and modular transformations in 2D conformal field theories.Nucl.Phys.B300 (1988) 360.

[Wa] M.Wakimoto: Fock representations of the affine Lie algebra $A_1^{(1)}$. Commun.Math.Phys. 104 (1986) 605.

[Wz] H.Wenzl: Hecke algebras of type A_n and subfactors. Invent.Math. 92 (1988) 349.

[Wi] E.Witten: Gauge theories,vertex models and quantum groups. Nucl.Phys. B330 (1990) 285.

[Z] A.B.Zamolodchikov: Infinite additional symmetries in 2-dimensional conformal quantum field theory. Theor.Math.Phys.65 (1986) 1205.

[ZZ] A.B.Zamolodchikov and Al.Zamolodchikov: Factorized S-matrices in two dimensions and the exact solutions to certain relativistic quanttum field theory models.Ann.Phys. 120 (1979) 253.

Z_n Elliptic Sklyanin Algebra and $U_q(\widetilde{sl}(n))$

Bo-yu Hou Kang-jie Shi

Institute of Modern Physics, Northwest University,
Xian 710069, P.R.China

Abstract

We study the limit of the algebra associated with the solution of $Z_n$elliptic YBE. And we give its specific relation with the quantum algebra $U_q(sl(n))$.

(1) Operator YBE and its limit

The R matrix for the elliptical Z_n symmetric Yang-Baxter equation[1-9]is:

$$R_{ij}(u) = \sum_{\alpha \in Z_n^2} W_\alpha(u) I_\alpha^{(i)} I^{-1}{}_\alpha^{(j)} \tag{1}$$

With the n × n matrix $I_\alpha$defined as:

$$I_\alpha = I_{(\alpha_1,\alpha_2)} = h^{\alpha_1} g^{\alpha_2}$$

where $h_{ij} = \delta_{i,j+1}$, $g_{ij} = \omega^i \delta_{ij}$, $\omega = e^{2\pi i/n}$

and $I_\alpha^{(i)} = I \otimes \cdot\cdot\cdot \otimes I_\alpha \otimes I \otimes \cdot\cdot\cdot$

$$W_\alpha(u) = \frac{\sigma_\alpha(u+\eta)}{\sigma_\alpha(\eta)} = \frac{\theta\begin{bmatrix} \frac{\alpha_1}{n}+\frac{1}{2} \\ \frac{\alpha_2}{n}+\frac{1}{2} \end{bmatrix}(u+\eta,\tau)}{\theta\begin{bmatrix} \frac{\alpha_1}{n}+\frac{1}{2} \\ \frac{\alpha_2}{n}+\frac{1}{2} \end{bmatrix}(\eta,\tau)} \tag{2}$$

They satisfy the YBE

$$R_{12}(u-v)R_{13}(u)R_{23}(v) = R_{23}(v)R_{13}(u)R_{12}(u-v) \tag{3}$$

The operator representation for the YBE is

$$L_j(u) = \sum_{\alpha \in Z_n^2} W_\alpha(u) I_\alpha^{(j)} S_\alpha \tag{4}$$

It satisfies $R_{12}(u-v)L_1(u)L_2(v) = L_2(v)L_1(u)R_{12}(u-v)$ (5)
Which is equivalent to

$$\sum_{\gamma \in Z_n^2} \omega^{(\beta_1-\gamma_1)(\gamma_2-\alpha_2)} W_{\alpha\beta\gamma}(u,v) S_{\alpha+\beta-\gamma} S_\gamma = 0 \tag{6}$$

where

$$W_{\alpha\beta\gamma}(u,v) = W_{\gamma-\alpha}(u-v) W_{\alpha+\beta-\gamma}(u) W_\gamma(v) - W_{\beta-\gamma}(u-v) W_\gamma(u) W_{\alpha+\beta-\gamma}(v) \tag{7}$$

It can be shown that[10]

$$W_{\alpha\beta\gamma}(u,v) = C_{\alpha\beta\gamma}(\eta,\tau) \frac{\sigma_0(u-v)\sigma_\beta(u+2\eta)\sigma_\alpha(v)}{\sigma_\beta(2\eta)} = C_{\alpha\beta\gamma}(\eta,\tau) G_{\alpha\beta}(u,v) \tag{8}$$

Consequently we have

$$\sum_{\gamma \in Z_n^2} \omega^{(\beta_1-\gamma_1)(\gamma_2-\alpha_2)} C_{\alpha\beta\gamma}(\eta,\tau) S_{\alpha+\beta-\gamma} S_\gamma = 0 \tag{9}$$

This is a u,v independent equation of n^2 operators S_α. The elliptic quantum algebra thus obtained reduces to Sklyanin algebra[7] when n = 2.

We want to study the limit of this elliptic quantum algebra as $\tau \to i\infty$, We write the matrix element of L(z) as

$$L(z)_{ij} = \sum_{\alpha \in Z_n^2} W_\alpha(z) S_\alpha (I_\alpha)_{ij}, \quad i,j = 1,\ldots,n.$$

$$\text{As } (I_\alpha)_{ij} = \delta_{i-j,\alpha_1} \omega^{j\alpha_2}$$

thus

$$L(z)_{ij} = \sum_{\alpha_2 \in Z_n} W_{(i-j,\alpha_2)}(z) \omega^{j\alpha_2} S_{(i-j,\alpha_2)} \tag{10}$$

As $\tau \to i\infty$, $L(z)$ has the following asymptotic behavior:

$$L(z)_{ij} \to \sum_{\alpha_2 \in Z_n} exp[2i\pi(\frac{i-j}{n} - \frac{1}{2})z] \omega^{j\alpha_2} S_{(i-j,\alpha_2)}, \quad i > j$$

$$\to \sum_{\alpha_2 \in Z_n} exp[2i\pi(\frac{i-j}{n} + \frac{1}{2})z] \omega^{j\alpha_2} S_{(i-j+n,\alpha_2)}, \quad i < j$$

$$\to \sum_{\alpha_2 \in Z_n} \frac{\sin[\pi(z+\frac{w}{n}+\frac{\alpha_2}{n})]}{\sin[\pi(\frac{w}{n}+\frac{\alpha_2}{n})]} \omega^{j\alpha_2} S_{(0,\alpha_2)}, \quad i=j \tag{11}$$

According to Richey and Tracy[12], R_{ik}^{jl} can be written as

$$R_{ij}^{kl} = R_{i+pj+p}^{k+pl+p} = \delta_{i+l,k+j} S^{k-i,l-i} \tag{12}$$

where $\quad ijklp \in Z_n$,

$$S^{ab} = f_1(z) \frac{\sigma_{(-a+b,0)}(z+w,n\tau)}{\sigma_{(-a,0)}(w,n\tau)\sigma_{(b,0)}(z,n\tau)} \tag{13}$$

when $\quad \tau \to i\infty$,

$$S^{00}(zw\tau) \to f_2(z)(2i)\sin\pi(z+w)$$

$$S^{a0}(zw\tau) \to f_2(z)(2i)\sin\pi w \, exp[2i\pi(\frac{1}{2}-\frac{a}{n})z], \quad n>a>0$$

$$S^{0b}(zw\tau) \to f_2(z)(2i)\sin\pi w \, exp[2i\pi(\frac{b}{n}-\frac{1}{2})z], \quad n>b>0$$

$$S^{ab}(zw\tau) \to 0, a \neq 0, b \neq 0 \tag{14}$$

As $\quad f_1(z), f_2(z)$ are independent of a,b, we may ignore them and obtain

$$R_{ii}^{ii} = e^{i\pi z} e^{i\pi w} - e^{-i\pi z} e^{-i\pi w}$$

$$R_{ij}^{ji} = \begin{cases} (e^{i\pi w} - e^{-i\pi w}) e^{2i\pi(\frac{1}{2}+\frac{i-j}{n})z}, & j>i, \; a=j-i. \\ (e^{i\pi w} - e^{-i\pi w}) e^{2i\pi(\frac{-1}{2}+\frac{i-j}{n})z}, & j<i, \; a=j-i+n. \end{cases}$$

$$R_{ij}^{ij} = \begin{cases} e^{2i\pi(\frac{-1}{2}+\frac{j-i}{n})w}(e^{i\pi z} - e^{-i\pi z}), & j>i, \; b=j-i. \\ e^{2i\pi(\frac{1}{2}+\frac{j-i}{n})w}(e^{i\pi z} - e^{-i\pi z}), & j<i, \; b=j-i+n. \end{cases} \tag{15}$$

All other R′s vanish

(2) Operator algebra after similarity transformations

We now take a similar transformation
Let

$$C(z) = \begin{bmatrix} e^{2i\pi\frac{z}{n}} & & & \\ & e^{2i\pi\frac{2z}{n}} & & \\ & & & \\ & & & \\ & & & \cdot \end{bmatrix} \tag{16}$$

$$C^{-1}LC = \overline{L}$$

$$C^{-1}(z_1)\otimes C^{-1}(z_2)R(z_1 - z_2)C(z_1)\otimes C(z_2) = \overline{R}$$

Consequently,As $\tau \to i\infty$,

$$\begin{aligned} \overline{L}(z)_{ij} &\to \sum_{\alpha_2 \in Z_n} exp[-i\pi z]\omega^{j\alpha_2} S_{(i-j,\alpha_2)}, \quad i > j. \\ &\to \sum_{\alpha_2 \in Z_n} exp[i\pi z]\omega^{j\alpha_2} S_{(i-j+n,\alpha_2)}, \quad i < j. \\ &\to \sum_{\alpha_2 \in Z_n} \frac{exp(i\pi z)exp[i\pi(\frac{w}{n} + \frac{\alpha_2}{n})] - exp(-i\pi z)exp[-i\pi(\frac{w}{n} + \frac{\alpha_2}{n})]}{(2i)sin\pi(\frac{w}{n} + \frac{\alpha_2}{n})} \\ &\qquad \times \omega^{j\alpha_2} S_{(0,\alpha_2)}, \quad i = j. \end{aligned} \tag{17}$$

$$\overline{R}^{ii}_{ii}(z) = e^{i\pi z}e^{i\pi w} - e^{-i\pi z}e^{-i\pi w} \tag{18a}$$

$$\overline{R}^{ij}_{ij}(z) = \begin{cases} (e^{i\pi w} - e^{-i\pi w})e^{i\pi z}, & j > i. \\ (e^{i\pi w} - e^{-i\pi w})e^{-i\pi z}, & i > j \end{cases} \tag{18b}$$

$$\overline{R}^{ij}_{ij}(z) = \begin{cases} (e^{i\pi z} - e^{-i\pi z})e^{2i\pi(\frac{-1}{2} + \frac{j-i}{n})w}, & i < j \\ (e^{i\pi z} - e^{-i\pi z})e^{2i\pi(\frac{1}{2} + \frac{j-i}{n})w}, & i > j \end{cases} \tag{18c}$$

All other $\overline{R}$'s vanish.

Rewrite

$$\overline{L}(z) = e^{\pi iz}L_+ + e^{-\pi iz}L_- \tag{19a}$$

$$\overline{R}(z) = e^{\pi iz}R_+ + e^{-\pi iz}R_- \tag{19b}$$

We then have $(q = e^{\pi iw}, \gamma = q^{\frac{2}{n}})$

$$
\begin{aligned}
(L_+)_{ij} &= \sum_{\alpha_2 \in Z_n} \omega^{j\alpha_2} S_{(i-j+n,\alpha_2)}, \quad i<j \\
&= \sum_{\alpha_2 \in Z_n} \frac{e^{\pi i(\frac{w}{n})} e^{\pi i(\frac{(2j+1)}{n})\alpha_2}}{(2i)\sin\pi(\frac{w}{n}+\frac{\alpha_2}{n})} S_{(0,\alpha_2)}, \quad i=j \\
&= 0, \quad i>j
\end{aligned} \tag{20a}
$$

$$
\begin{aligned}
(L_-)_{ij} &= \sum_{\alpha_2 \in Z_n} \omega^{j\alpha_2} S_{(i-j,\alpha_2)}, \quad i>j \\
&= \sum_{\alpha_2 \in Z_n} \frac{e^{\pi i(\frac{-w}{n})} e^{\pi i(\frac{(2j-1)}{n})\alpha_2}}{(2i)\sin\pi(\frac{w}{n}+\frac{\alpha_2}{n})} S_{(0,\alpha_2)}, \quad i=j \\
&= 0, \quad i<j
\end{aligned} \tag{20b}
$$

$$
e^{\pi i(\frac{w}{n})}(L_-)_{jj} = e^{\pi i(\frac{-w}{n})}(L_+)_{j-1j-1} \tag{21}
$$

$$
(L_-)_{11} = \frac{1}{\gamma}(L_+)_{nn}
$$

As $(e_{ij})_{kl} = \delta_{ik}\delta_{jl}$, we have

$$
\begin{aligned}
R_+ &= \sum_i q e_{ii} \otimes e_{ii} + \sum_{i<j} q^{-1}\gamma^{|j-i|} e_{ii} \otimes e_{jj} \\
&+ \sum_{i>j} q\gamma^{-|j-i|} e_{ii} \otimes e_{jj} + \sum_{i<j}(q-q^{-1}) e_{ij} \otimes e_{ji}
\end{aligned} \tag{22a}
$$

$$
\begin{aligned}
R_- &= -\sum_i q^{-1} e_{ii} \otimes e_{ii} - \sum_{i<j} q_{-1}\gamma^{|j-i|} e_{ii} \otimes e_{jj} \\
&- \sum_{i>j} q\gamma^{-|j-i|} e_{ii} \otimes e_{jj} + \sum_{i>j}(q-q^{-1}) e_{ij} \otimes e_{ji}
\end{aligned} \tag{22b}
$$

where $R_{+\,ij}^{\;ij} = -R_{-\,ij}^{\;ij}, \quad i \neq j$

It is easy to show the following:

$$
\hat{R}_+ \hat{R}_- = kI \otimes I, \qquad \hat{R}_+ + \hat{R}_- = k' I \otimes I \tag{23}
$$

where $\hat{R}_i = PR_i$, P is the permutation matrix which satisfies

$$
P(e_i' \otimes e_j') = e_j' \otimes e_i'
$$

Substitute R_+, R_-, L_+, L_- into YBE Eq.(5), remember z_1 and z_2 are independent variables, we obtain

$$
R_+^{(12)} L_+^{(1)} L_+^{(2)} = L_+^{(2)} L_+^{(1)} R_+^{(12)},
$$

$$
R_+^{(12)} L_+^{(1)} L_-^{(2)} = L_-^{(2)} L_+^{(1)} R_+^{(12)},
$$

$$R_{-}^{(12)}L_{-}^{(1)}L_{+}^{(2)}=L_{+}^{(2)}L_{-}^{(1)}R_{-}^{(12)},$$
$$R_{-}^{(12)}L_{-}^{(1)}L_{-}^{(2)}=L_{-}^{(2)}L_{-}^{(1)}R_{-}^{(12)},$$
$$ie.\quad R_{i}^{(12)}L_{i}^{(1)}L_{j}^{(2)}=L_{j}^{(2)}L_{i}^{(1)}R_{i}^{(12)}, \tag{24}$$
$$\text{with}\quad i,j=\pm$$

In fact, only three of the four equations given above are independent.

(3)The limit of elliptic quantum algebra and $U_q(sl(n))$quantum algebra

Let $R_{ik}^{jl}=R(ij,kl), with R=\sum_{ijkl}R(ij,kl)e_{ij}\otimes e_{kl}$,we then have

①$a\neq b,\quad c\neq d$ (24a)

$$R_{(i)}(aabb)L_{(i)ad}L_{(j)bc}-R_{(i)}(ddcc)L_{(j)bc}L_{(i)ad}+R_{(i)}(abba)L_{(i)bd}L_{(j)ac}$$
$$-R_{(i)}(cddc)L_{(j)bd}L_{(i)ac}=0,$$

②$a=b,\quad c\neq d,$ (24b)

$$R_{(i)}(aaaa)L_{(i)ad}L_{(j)ac}-R_{(i)}(ddcc)L_{(j)ac}L_{(i)ad}-R_{(i)}(cddc)L_{(j)ad}L_{(i)ac}=0,$$

③$a\neq b,\quad c=d,$ (24c)

$$R_{(i)}(aabb)L_{(i)ac}L_{(j)bc}-R_{(i)}(cccc)L_{(j)bc}L_{(i)ac}+R_{(i)}(abba)L_{(i)bc}L_{(j)ac}=0,$$

④$a=b,\quad c=d,$ (24d)

$$R_{(i)}(aaaa)L_{(i)ac}L_{(j)ac}-R_{(i)}(aaaa)L_{(j)ac}L_{(i)ac}=0,$$

Remember:

$(L_{+})_{ij}=0,\quad i>j$

$(L_{-})_{ij}=0,\quad i<j$

$(L_{-})_{ii}=\frac{1}{\gamma}(L_{+})_{i-1,i-1},\quad i=j$

Redefine: $e_i=(L_{+})_{ii+1},\ f_i=(L_{-})_{i+1i},\ K_i=(L_{-})_{i+1i+1},$ (25)

We then have

$$q^2e_i^2e_{i+1}-(q^2+1)e_ie_{i+1}e_i+e_{i+1}e_i^{\,2}=0$$
$$e_i^2e_{i-1}-(q^2+1)e_ie_{i-1}e_i+q^2e_{i-1}e_i^{\,2}=0$$
$$q^2f_i^2f_{i+1}-(q^2+1)f_if_{i+1}f_i+f_{i+1}f_i^{\,2}=0$$
$$f_i^2f_{i-1}-(q^2+1)f_if_{i-1}f_i+q^2f_{i-1}f_i^{\,2}=0$$
$$q^2f_iK_i-\gamma K_if_i=0,$$
$$f_iK_j-\gamma K_jf_i=0,\quad i\neq j$$
$$q^2K_ie_i-\gamma e_iK_i=0,$$

$$K_j e_i - \gamma e_i K_j = 0, \quad i \neq j$$
$$[e_i, e_j] = 0, \quad for |i-j| \geqq 2$$
$$[f_i, f_j] = 0, \quad for |i-j| \geqq 2$$
$$[K_i, K_j] = 0,$$
$$\gamma^2 e_i f_{i-1} - f_{i-1} e_i = 0,$$
$$q^{-2}\gamma^2 e_i f_{i+1} - f_{i+1} e_i = 0,$$
$$\gamma^2 e_i f_{i \pm l} - f_{i \pm l} e_i = 0, \ l \geqq 2,$$
$$\gamma^2 e_i f_i - q^2 f_i e_i + (q^2 - 1)(K_{i+1} K_{i-1} - K_i^2) = 0,$$
$$\gamma(1-q^2) K_p (L_+)_{ij} = \gamma^{p-i} (L_+)_{ip} (L_+)_{pj} - \gamma^{j-p} (L_+)_{pj} (L_+)_{ip}$$
$$for \, i < p < j.$$
$$(1-q^{-2}) K_{p-1} (L_-)_{ij} = \gamma^{p-i} (L_-)_{ip} (L_-)_{pj} - \gamma^{j-p} (L_-)_{pj} (L_-)_{ip}$$
$$for \, j < p < i.$$

$$K_i (L_+)_{lm} = \begin{cases} q^{-2}\gamma^{m-l} (L_+)_{lm} K_i, & l \leqq i < m, \\ \gamma^{m-l} (L_+)_{lm} K_i, & all \ others \end{cases}$$
$$K_{i-1} (L_-)_{lm} = \begin{cases} q^2 \gamma^{m-l} (L_-)_{lm} K_{i-1}, & m < i \leqq l, \\ \gamma^{m-l} (L_-)_{lm} K_{i-1}, & all \ others \end{cases} \tag{26}$$

From the above relations, we take the following transformations

Let $E_{ij} = \dfrac{1}{\gamma(1-q^2)} \dfrac{1}{\sqrt[2]{K_i K_j}} (L_+)_{ij}, \quad i < j$

$$E_{ij} = \frac{1}{1-q^{-2}} \frac{1}{\sqrt[2]{K_{i-1} K_{j-1}}} (L_-)_{ij}, \quad i > j$$
$$\hat{e}_i = E_{i,i+1}, \quad \hat{f}_i = E_{i+1,i}, \quad t_i = \sqrt[4]{\frac{K_{i-1} K_{i+1}}{K_i^2}} \tag{27}$$

Substitute Eq.(27) into Eq.(26), we finally have

$$E_{ip} E_{pj} - q E_{pj} E_{ip} = E_{ij}, \quad i < p < j,$$
$$E_{ip} E_{pj} - q^{-1} E_{pj} E_{ip} = E_{ij}, \quad j < p < i,$$
$$[\hat{e}_i, \hat{f}_i] = \frac{1}{q - q^{-1}} (t_i^2 - t_i^{-2}),$$
$$t_i \hat{e}_i t_i^{-1} = q \hat{e}_i$$

$$t_i \hat{f}_i t_i^{-1} = q^{-1} \hat{f}_i,$$

$$t_i \hat{e}_{i\pm 1} t_i^{-1} = q^{\pm(-\frac{1}{2})} \hat{e}_{i\pm 1},$$

$$t_i \hat{f}_{i\pm 1} t_i^{-1} = q^{\pm\frac{1}{2}} \hat{f}_{i\pm 1},$$

$$\left.\begin{array}{l} t_i \hat{e}_j t_i^{-1} = \hat{e}_j \\ t_i \hat{f}_j t_i^{-1} = \hat{f}_j \end{array}\right\} |i-j| \geqq 2,$$

$$\hat{e}_i^2 \hat{e}_{i\pm 1} - (q+q^{-1}) \hat{e}_i \hat{e}_{i\pm 1} \hat{e}_i + \hat{e}_{i\pm 1} \hat{e}_i^2 = 0,$$

$$\hat{f}_i^2 \hat{f}_{i\pm 1} - (q+q^{-1}) \hat{f}_i \hat{f}_{i\pm 1} \hat{f}_i + \hat{f}_{i\pm 1} \hat{f}_i^2 = 0,$$

This is identical with the $U_q(sl(n))$algebra

We still need an extra generator, say $\prod_{i=1}^{n} K_i$, which is a central element, to recover $L_\pm$ from the $n^2 - 1$generators of quantum algebra $U_q(sl(n))$, and Hopf algebra stucture follows subsequently.

(4)Automorphism of Z_n Sklyanin algebra

Because of the $Z_n \otimes Z_n$ symmetry of R matrix, we have

$$(I_\alpha \otimes I_\alpha) R = R(I_\alpha \otimes I_\alpha)$$

Consequently we have

$$(I_\alpha \otimes I_\alpha) RLL = (I_\alpha \otimes I_\alpha) LLR$$

$$R\tilde{L}\tilde{L} = \tilde{L}\tilde{L}R$$

$$R\tilde{L}\tilde{L}(I_{\alpha'} \otimes I_{\alpha'}) = \tilde{L}\tilde{L}R(I_{\alpha'} \otimes I_{\alpha'})$$

$$RL'L' = L'L'R$$

where

$$\tilde{L} = I_\alpha L$$

$$L' = \tilde{L} I_{\alpha'} = I_\alpha L I_{\alpha'}$$

This constitutes the automorphism of the Sklyanin aegebra,we may write L explicitly in terms of generator S

ie.

$$S'_{\beta+\alpha} = (scalar) S_{\beta} e^{-2\pi i \frac{\alpha_1 \beta_2}{n^2}} \omega^{m\beta_1 + n\beta_2} \frac{\theta\begin{bmatrix} \frac{\alpha_1+\beta_1}{n} + \frac{1}{2} \\ \frac{\alpha_2+\beta_2}{n} + \frac{1}{2} \end{bmatrix}(\eta,\tau)}{\theta\begin{bmatrix} \frac{\beta_1}{n} + \frac{1}{2} \\ \frac{\beta_2}{n} + \frac{1}{2} \end{bmatrix}(\eta,\tau)}$$

We have also discussed the fusion of representations[11].

Acknowledgements

The authors are grateful to the Conference organizers at Nankai University for their hospitality . B.H would also like to thank Prof. C.N.Yang for his kind encouragement.This work is partly supported by Natural Science Fund of China.

References

[1] V.G.Drinfel'd, Sov.Math.Dokl.32(1985)254.
M.Jimbo, Lett.Math.Phys.10(1985)63
[2] E.H.Lieb, Phys.Rev 162(1987)162
[3] J.H.H.Perk and C.Schultz, Phys.Lett.84A.(1981)407
[4] R.J.Baxter, Ann.Phys.70(1972)193
[5] G.E.Andrews,R.J.Baxter and P.J.Forrester, J.Stat.Phys. 35(1984)193
M.Jimbo,T.Miwa and M.Okado, Lett.Math.Phys.14(1987)123
[6] B.Y.Hou,D.P.Li and R.H.Yue,Phys.Lett.B229(1989)45
G.Felder,J Frohlich and G.Keller, Commun.Math.Phys.124(1989)717
B.Y.Hou,K.J.Shi,P.Wang and R.H.Yue,Nucl.Phys.B345(1990)659
[7] E.K.Sklyanin,Funct.Anal.Appl.16 (1982)263; 17(1983)273
[8] A.A.Belavin,Nucl.Phys.B180(1980)189.
[9] I.V.Cherednik,Funct.Anal.Appl.19(1983)77.
[10] B.Y.Hou and H.Wei,J.Math.Phys.30(1989)2750
[11] B.Y.Hou and Y.K.Zhou,J.Phys.A.Math.Gen. 23(1990)1147
[12] M.P.Richey, and C.A.Tracy,J.Stat.Phys.42.(1986)311

Quantum Groups and Invariants of 3-Manifolds

by Louis H. Kauffman

Department of Mathematics, Statistics and
Computer Science (M/C 249)
University of Illinois at Chicago, Box 4348
Chicago, Illinos 60680

The purpose of this paper is to indicate a method of defining invariants of 3-manifolds intrinsically in terms of the structure of certain Hopf algebras that are possessed of right integrals. We call such an invariant a *Hennings invariant* [2], as Hennings was the first person to point out that invariants could be defined in this way.

Part of the motivation of this work (joint work with David Radford [5, 5.1]) is to understand the relationship of this type of invariant with the invariants of Reshetikhin-Turaev [11] defined in terms of surgery on framed links and the representation theory of the Hopf algebras. In a sense, the Hennings invariants are closer in spirit to the original Witten definition [13] in terms of functional integration, since that approach gave a direct definition of the invariant in terms of geometrical data (albeit in the context of quantum field theory and at the physical level of rigor). Here we consider invariants defined directly in terms of the geometry of the diagrams and the global structure of the algebra.

I. Algebra

Recall that a *Hopf algebra* A [12] is a bialgebra over a commutative ring k that is associative, coassociative and equipped with a counit, a unit and an antipode. (k is usually taken to be a field, but we may want to tensor the algebra with a ring on which the field acts. Hence it is useful to designate such a ring from the beginning.)

In order to be an algebra, A needs a multiplication $m:A\otimes A \dashrightarrow A$.
It is assumed that M satisfies the associative law:
$m(m\otimes 1) = m(1\otimes m)$.

In order to be a bialgebra, an algebra needs a coproduct $\Delta: A \text{ -----> } A\otimes A$. The coproduct is a map of algebras, and is regarded as the dual of a multiplicative structure. Coassociativity means that $(\Delta\otimes 1)\Delta = (1\otimes\Delta)\Delta$ holds as compositions of maps.

The unit is a mapping from k to A taking 1 in k to 1 in A, and thereby defining an action of k on A. It will be convenient to just identify the units in k and in A, and to ignore the name of the map that gives the unit.

The counit is an algebra mapping from A to k denoted by $E: A \text{ ----->} k$. The following formulas for the counit dualize the structure inherent in the unit: $(E\otimes 1)\Delta = 1 = (1\otimes E)\Delta$. Here 1 denotes the identity mapping of A to itself.

It is convenient to write formally $\Delta(x) = \Sigma\, x_{(1)}\otimes x_{(2)}$ to indicate the decomposition of the coproduct of x into a sum of first and second factors in the two-fold tensor product of A with itself. We shall adopt the summation convention that $\Sigma\, x_{(1)}\otimes x_{(2)}$ can be abbreviated to just $x_{(1)}\otimes x_{(2)}$. Thus we shall write $\Delta(x) = x_{(1)}\otimes x_{(2)}$.

The antipode is a mapping $s: A \text{ -----> } A$ satisfying the equations $m(1\otimes s)\Delta(x) = E(x)1$, and $m(s\otimes 1)\Delta(x) = E(x)1$ where 1 denotes the unit of k as identified with the unit of A in these equations. It is a consequence of this definition that $s(xy) = s(y)s(x)$ for all x and y in A.

A *quasitriangular Hopf algebra A* [1] is a Hopf algebra with an element ρ in $A\otimes A$ satisfying the following equations:

1) $\rho\Delta = \Delta'\rho$ where Δ' is the composition of Δ with the map on $A\otimes A$ that switches the two factors.

2) $\rho_{12}\rho_{13} = (1\otimes\Delta)\rho$, $\rho_{13}\rho_{23} = (\Delta\otimes 1)\rho$.

These conditions imply that $\rho^{-1} = (1\otimes s^{-1})\rho = (s\otimes 1)\rho$.

It follows easily from the axioms of the quasitriangular Hopf algebra that ρ satisfies the Yang-Baxter Equation

$\rho_{12}\rho_{13}\rho_{23} = \rho_{23}\rho_{13}\rho_{12}$.

A less obvious fact about quasitriangular Hopf algebras is that there exists an element u such that u is invertible and $s^2(x) = uxu^{-1}$ for all x in A. In fact, we may take $u = \Sigma\ s(e')e$ where $\rho = \Sigma\ e\otimes e'$.

An element G in a Hopf algebra is said to be *grouplike* if $\Delta(G) = G\otimes G$ and $E(G)=1$ (from which it follows that G is invertible and $s(G) = G^{-1}$.) A quasitriangular Hopf algebra is said to be a *ribbon Hopf algebra* (See [10],[5].) if there exists a special grouplike element G such that (with u as in the previous paragraph) $v = G^{-1}u$ is in the center of A and $s(u) = G^{-1}uG^{-1}$.

Note that the condition of being ribbon implies that $s^2(x) = GxG^{-1}$ for all x in A (since vx=xv implies that $uxu^{-1} = GxG^{-1}$ for all x.). Note also that $s(v) = v$ (Proof: $s(v) = s(G^{-1}u) = s(u)S(G^{-1}) = G^{-1}uG^{-1}G = G^{-1}u = v$.)

II. Diagrammatic Geometry

A function from the Hopf algebra to the base ring k, tr:A -----> k, is said to be a *trace* if tr(xy) = tr(yx) and tr(s(x))= tr(x) for all x and y in the Hopf algebra A. A trace function on a ribbon Hopf algebra yields an invariant of regular isotopy, TR(K), of knots and links in the following way (See [3],[4].): The link diagram is arranged with respect to a vertical direction so that the crossings form the two types indicated below, and so that other than the crossings the only critical points of the height function are maxima and minima. Each crossing is decorated with elements of the Hopf algebra as shown below. (Here $\rho = \Sigma e\otimes e'$ is the Yang-Baxter element in A, and s denotes the antipode.)

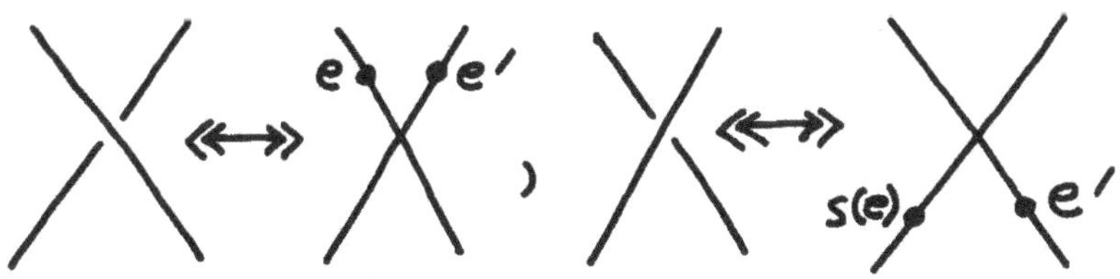

It is implicit in this formalism that there is a summation over all the pairs e,e' for each Yang-Baxter element.

Hopf algebra elements may be moved across maxima or minima at the expense of composition with the antipode:

The link diagram is subject to deformations that generate regular isotopy [3]. We omit these transformations here. An invariant must remain unchanged by these moves. The simplest move is the cancellation of a pair consisting of a maximum and a minimum.

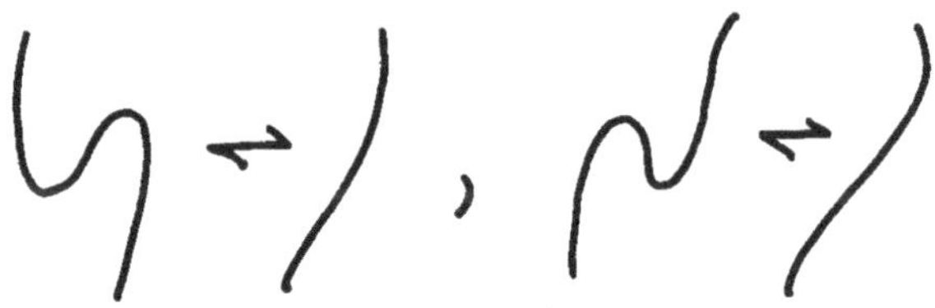

This pair cancellation gives a reformulation of the slide rule for the antipode: The antipode is accomplished by "composition with a maximum and a minimum" and the square of the antipode is diagrammed as "composition with two curls":

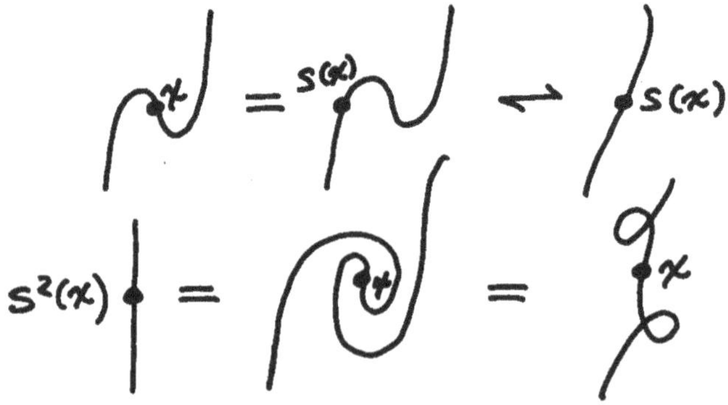

These curls are identified as the special grouplike elements G and G^{-1}.

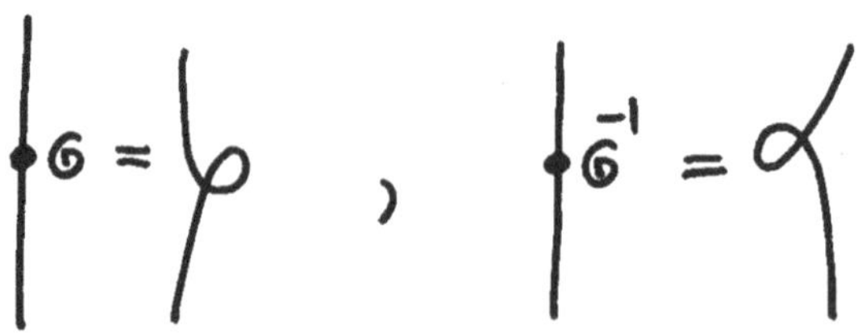

Note that this identification is compatible with the rule $s(G) = G^{-1}$.

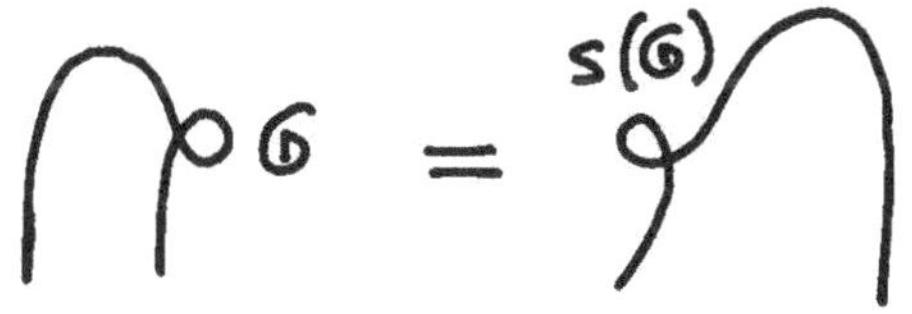

Along a vertical line, algebra elements combine by multiplication.

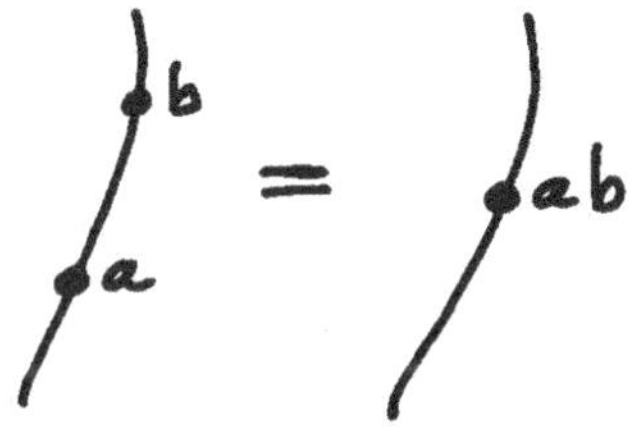

In this way knots on a line can be resolved into algebra elements. For example the twist shown below is equivalent to the ribbon element v. Note how the factorization of v into a product of G^{-1} and $u = \Sigma s(e')e$ is related to the slide convention for the antipode (In the diagrammatic calculation shown below we use the fact that $(s\otimes s)\rho = \rho$.)

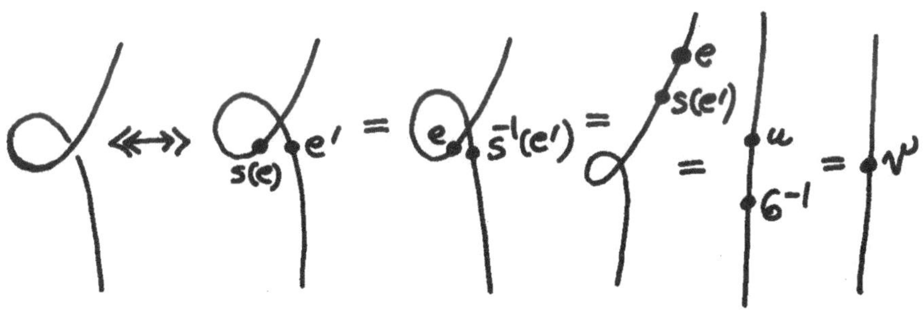

Note also that diagrammatically $s(v) = v$ corresponds to the identification of twists shown below.

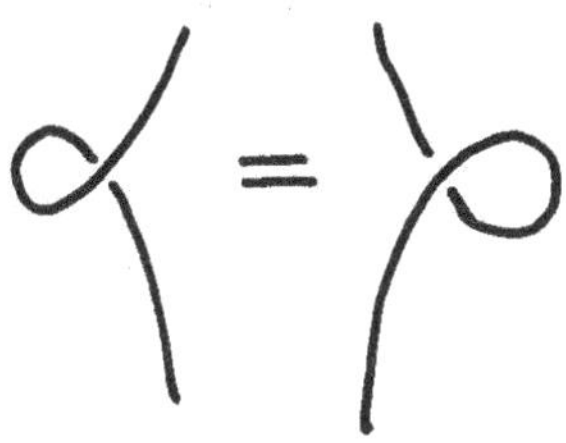

When this identification is added to regular isotopy, the twists catalog only the framing, and the equivalence relation on the link diagrams is equivalent to ambient isotopy of framed links.

Computing TR(K)

To compute the trace TR(K), slide all of the algebra into one vertical portion of the diagram. Amalgamate this algebraic expression according to the rule for multiplying algebra elements on the diagram, as we have done above. Call this localized algebra element w. It is a sum of products, and can be abbreviated as a product where it is understood that there is a sum over all pairs of the type e,e'.

Let d be the Whitney degree of the flat diagram for K that is obtained by traversing K upward from the vertical portion where the algebra has been concentrated. The Whitney degree is the total turn of the tangent vector to the curve as one traverses it in the given direction. For example:

Define TR(K) by the formula $\mathbf{TR(K) = \Sigma tr(wG^{d})}$.
The summation refers to the sum over all the pairs x,x' corresponding to Yang-Baxter elements in w. This defines an invariant of unoriented knots.

In order to define an invariant of unoriented links, concentrate the algebra for each component of the link, and define

$\mathbf{TR(K) = \Sigma tr(w_1 G^{d_1})tr(w_2 G^{d_2})tr(w_3 G^{d_3}) \ldots tr(w_n G^{d_n})}$ where the labels 1,2,...,n refer to the components of the link, and the summation refers to the sum over all the pairs x,x' in these words. The elements $w_1, \ldots, w_n$ are the algebra concentrations for each link component, and the degrees $d_1, \ldots, d_n$ are the Whitney degrees of the components of the link.

Example 1:

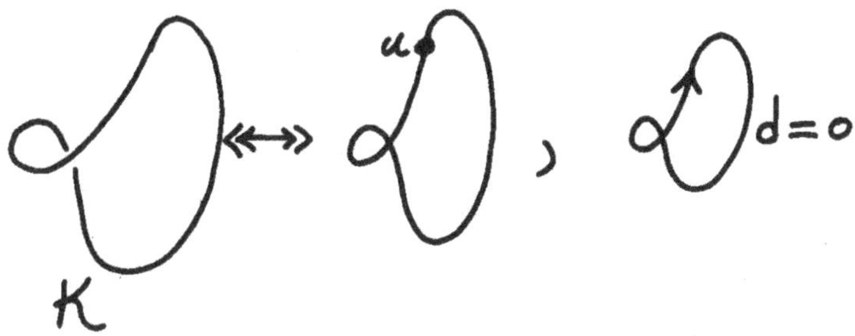

Here $u = \Sigma s(e')e$ and G is the special grouplike element. Thus $TR(K) = tr(uG^0) = tr(u)$.

Example 2. This example points out how the TR(K) is invariant under algebra slides:

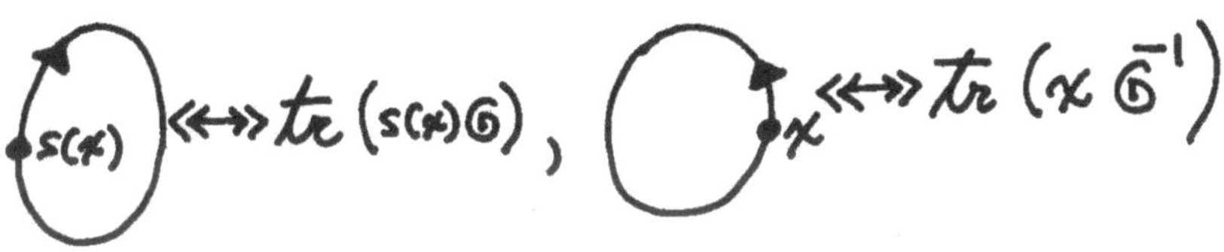

$$tr(s(x)G) = tr(s(s(x)G)) = tr(G^{-1}s^2(x)) = tr(G^{-1}GxG^{-1}) = tr(xG^{-1}).$$

Example 3. Here is the form of calculation for a link.

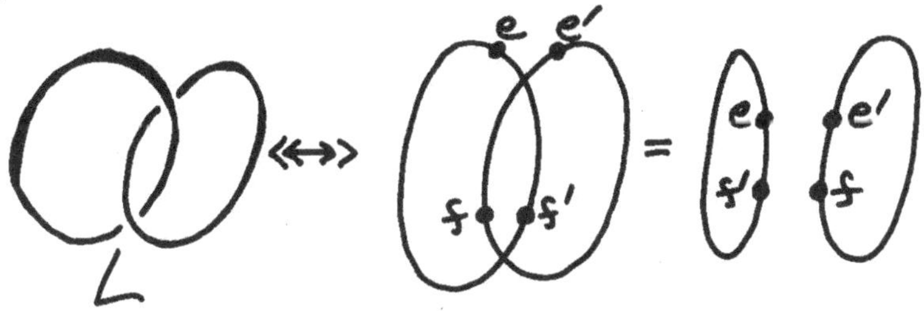

$TR(L) = \Sigma\ tr(f'eG^{-1})tr(fe'G)$.

If $\rho = \sum_{i=1}^{n} x_i \otimes y_i$, then $TR(L) = \sum_{i=1}^{n}\sum_{j=1}^{n} tr(y_j x_i G^{-1}) tr(x_j y_i G)$.

This is how the regular isotopy invariant of the link would look as a specific sum of traces of algebra elements.

III. Categories

Since the description of invariants given in section II may seem a bit ad hoc, the purpose of this section is to give a quick description of our methods in categorical terms. In these terms, everything appears quite functorial, and we see that we have defined functors between categories of diagrams and categories naturally associated with Hopf algebras.

A link diagram, with or without free ends, that is arranged with respect to a vertical direction can be regarded as a morphism in a tensor category with two generating objects V and k. We think of V as the analog of a vector space or module over k. Thus $k \otimes k = k$, $k \otimes V = V \otimes k = V$, and the products $V, V \otimes V, V \otimes V \otimes V, \ldots$ are all distinct. Maxima are morphisms from $V \otimes V$ to k, and denoted $\cap : V \otimes V \text{-----}> k$. Minima are morphisms from k to $V \otimes V$, and are denoted $\cup : k \text{-----}> V \otimes V$. The two types of crossing yield morphisms from $V \otimes V$ to itself, and are denoted by R and $\hat{R}$, respectively: $R, \hat{R} : V \otimes V \text{-----}> V \otimes V$. See the diagrams below.

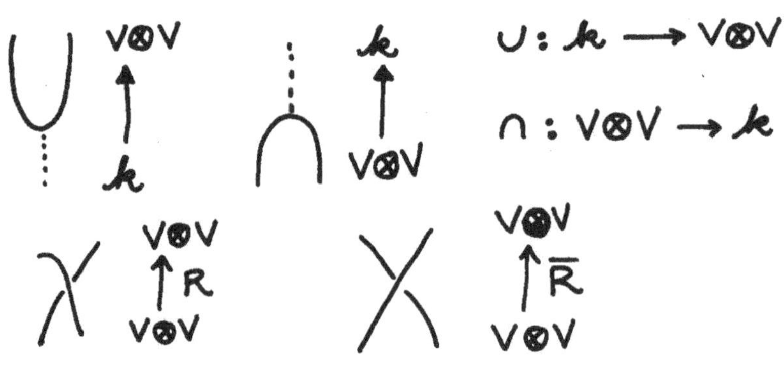

Composition of morphisms corresponds to tying input and output lines together in these diagrams. In this way , each diagram with out any tangle lines becomes a morphism from k to k. Since we call such a diagram a *closed diagram* , we shall call the corresponding morphism from k to k a *closed morphism.* The trace function described in the last section is designed to assign a specific element in a commutative ring (e.g. the complex numbers) to such k to k morphisms. But this puts us ahead of our story.

The morphisms in this regular isotopy category of link diagrams (denoted REG) are assumed to satisfy those relations that correspond to regular isotopy of links with respect to a vertical direction. These identities are indicated below.

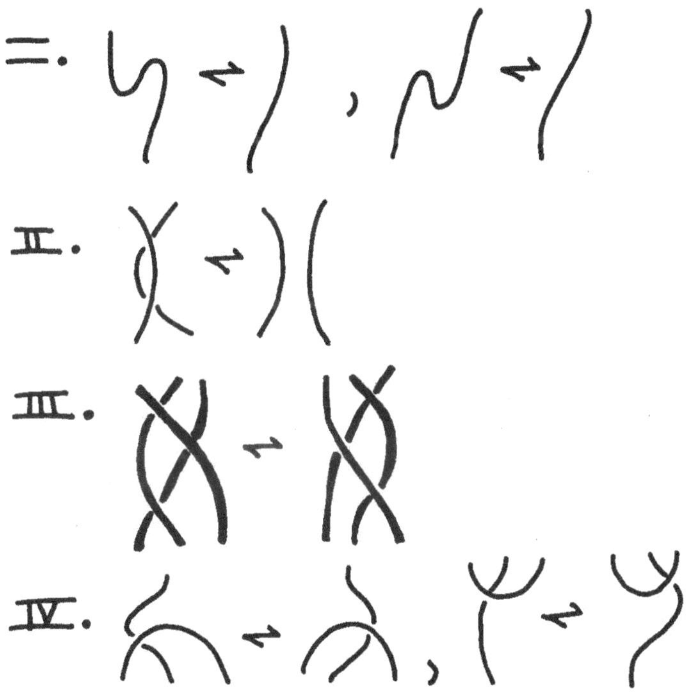

We also use the category of FLAT diagrams. In these the crossing has no distinction between over and under, but there is still a morphism P: V⊗V -----> V⊗V corresponding to a crossing. The rules for the category of FLAT diagrams are exactly the same in form as the rules for the category of REG diagrams, except that there are no restrictions about the types of crossing since there is only one type of crossing.

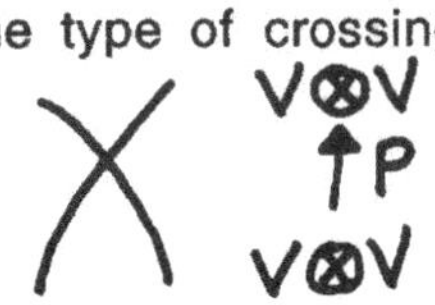

We know from the Whitney-Graustein theorem that FLAT has a particularly simple stucture of closed morphisms. In terms of diagrams any such morphism has the form shown below.

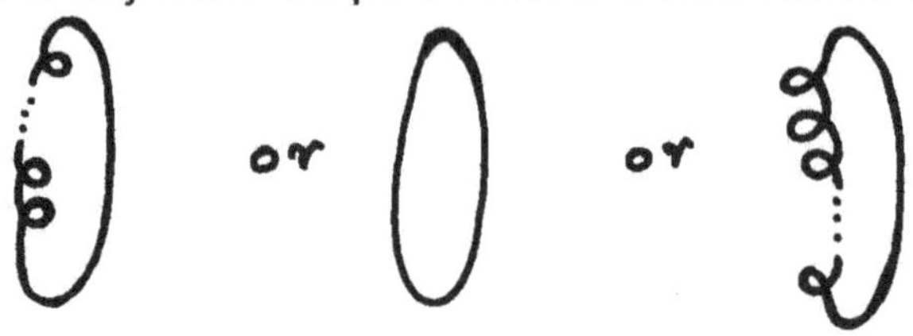

It is for this reason that we have singled out the morphisms G and G^{-1} in the previous section. Here $G = (1\otimes\cap)P(1\otimes\cup)$ as a morphism from V to V. (Note that $V=V\otimes k$.) As we have already remarked, G corresponds to a flat curl, and in the context of labelling the diagram with elements of a quantum group, G corresponds to a grouplike element.

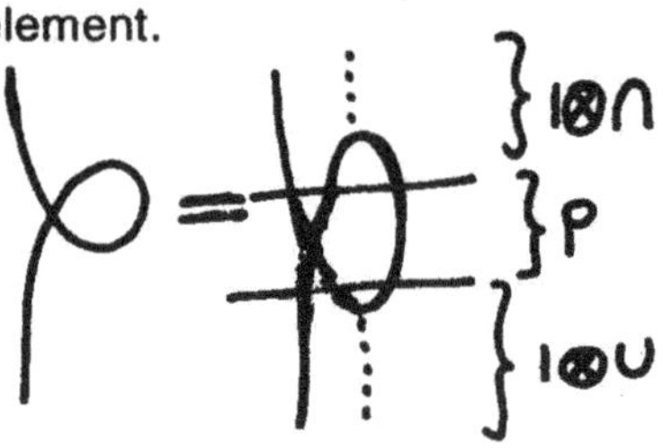

For both categories REG and FLAT we have functors
Δ: REG$\otimes$REG ----->REG and Δ: FLAT$\otimes$FLAT ----->FLAT. On objects W, $\Delta(W) = W\otimes W$. On morphisms, $\Delta(T)$ is obtained by taking the 2-fold parallel cable of each component arc in T.

In order to describe labelled diagrams in categorical language, we begin with a Hopf algebra A, and associate to A a category C(A). The category C(A) has two generating objects V and k with the same properties as the V and k for REG and FLAT. However, k is taken to be identical with the ground field for the Hopf algebra. (This can be generalised if we so desire.) *The morphisms of C(A) are the elements of A.* Each morphism takes V to V and composition of elements corresponds to multiplication in A. The labelled diagrams of simplest form

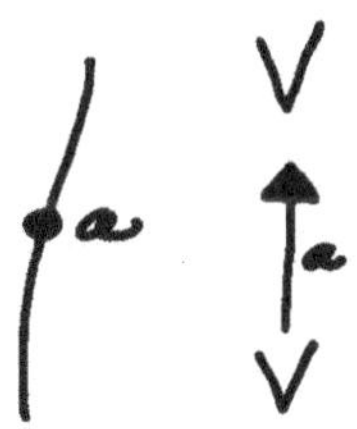

exhibit elements of A in the role of morphisms in C(A).

Omitting technical details, we now define categories C(A)*REG and C(A)*FLAT, by addding the morphisms $\cup$,$\cap$, R, $\hat{R}$, or P to C(A) in the form of diagrams as we have already described them in the previous section. Note how $\cup$ and $\cap$ are defined to carry the structure of the antipode of A.

One then sees that we have, in the previous section, defined a functor from REG to C(A)*FLAT, and then by "moving elements around the diagram, shown how to reduce closed morphisms in C(A)*FLAT to a sum of formal traces of products of elements in A. This functor F: REG -----> C(A)*FLAT is the source of the invariants.

IV. Invariants of 3-manifolds

The structure we have built so far can be used to construct invariants of 3-manifolds presented in terms of surgery on framed links. We sketch here our technique that simplifies an approach to 3-manifold invariants of Mark Hennings [2].

Recall that an element λ of the dual algebra A^* is said to be a *right integral* if $\lambda(x)1 = m(\lambda \otimes 1)(\Delta(x))$ for all x in A. It turns out that for a unimodular (see [7],[8],[12].) finite dimensional ribbon Hopf algebra A (see example at the end of the section) there is a (unique up to scalar multiplication when k is a field) right integral λ satisfying the following properties for all x and y in A:

1) $\lambda(xy) = \lambda(s^2(y)x)$

2) $\lambda(gx) = \lambda(s(x))$ where $g=G^2$, G the special grouplike element for the ribbon element $v=G^{-1}u$.

Given the existence of this λ, define a functional tr:A -----> k by the formula **tr(x) = λ(Gx).**

Theorem. With tr defined as above, then

1) $tr(xy) = tr(yx)$ for all x,y in A.
2) $tr(s(x)) = tr(x)$ for all x in A.
3) $[m(tr\otimes 1)(\Delta(u^{-1}))]u = \lambda(v^{-1})v$ where $v=G^{-1}u$ is the ribbon element.

Proof. The proof is a direct consequence of the properties 1) and 2) of λ. Thus $tr(xy) = \lambda(Gxy) = \lambda(s^2(y)Gx) = \lambda(GyG^{-1}Gx) = \lambda(Gyx) = tr(yx)$, and $tr(s(x)) = \lambda(Gs(x)) = \lambda(gG^{-1}s(x)) = \lambda(s(G^{-1}s(x))) = \lambda(s^2(x)s(G^{-1}))$ $= \lambda(s^2(x)G) = \lambda(GxG^{-1}G) = \lambda(Gx) = tr(x)$. Finally, $[m(tr\otimes 1)(\Delta(u^{-1}))]u = G^{-1}[m(\lambda.G\otimes G)(\Delta(u^{-1}))]u = [m(\lambda\otimes 1)(\Delta(Gu^{-1}))]G^{-1}u$ $= \lambda(Gu^{-1})G^{-1}u = \lambda(v^{-1})v$. This completes the proof. //

The upshot of this Theorem is that for a unimodular finite dimensional Hopf algebra there is a natural trace defined via the existent right integral. Remarkably, this trace is just designed by property 3) of the Theorem to behave well with respect to the Kirby move. (The Kirby move is the basic transformation on framed links that leaves the corresponding 3-manifold obtained by framed surgery unchanged,. See [6], [11].) This means that a suitably normalized version of this trace on framed links gives an invariant of 3-manifolds. To see how this works, here is a sample Kirby move

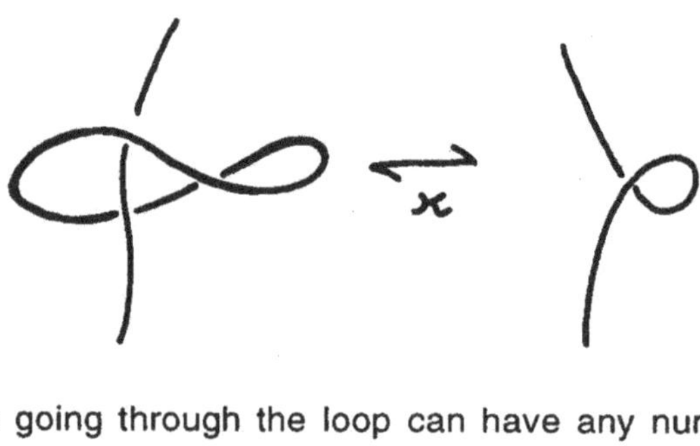

The cable going through the loop can have any number of strands. The loop has one strand and the framing as indicated. The replacement on the right hand side puts a 360 degree twist in the cable with blackboard framing as shown above. Here we calculate the case of a single strand cable:

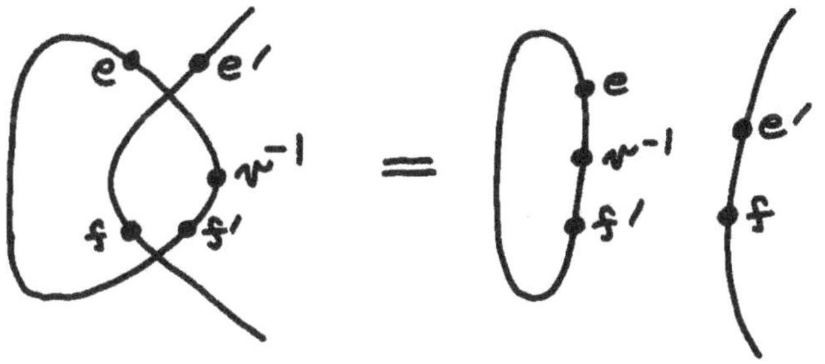

The diagram shows that the trace contribution is (with implicit summation on the repeated primed and unprimed pairs of Yang-Baxter elements)

$\mathrm{tr}(f'v^{-1}eG^{-1})fe' = \mathrm{tr}(f'ev^{-1}G^{-1})fe' = \mathrm{tr}(f'eu^{-1})fe'$

$= [m(\mathrm{tr}\otimes 1)(f'eu^{-1}\otimes fe'u^{-1})]u = [m(\mathrm{tr}\otimes 1)(\rho_{21}\rho_{12}(u^{-1}\otimes u^{-1}))]u$

$=[m(\mathrm{tr}\otimes 1)(\Delta(u^{-1}))]u = \lambda(v^{-1})v.$ $\quad (\Delta(u^{-1}) = \rho_{21}\rho_{12}(u^{-1}\otimes u^{-1}))$

It follows from this calculation that the evaluation of the lefthand picture in the Kirby move is $\lambda(v^{-1})$ times the evaluation of the right hand picture. For an n-strand cable we get the same result with $\lambda(v^{-1})$ raised to the nth power. Thus a proper normalization of TR(K) gives an invariant of the 3-manifold obtained by framed surgery on K. More precisely, (assuming that $\lambda(v)$ and $\lambda(v^{-1})$ are non-zero) let

$$\mathrm{INV}(K) = \{[\lambda(v)\lambda(v^{-1})]^{-c(K)/2}[\lambda(v)/\lambda(v^{-1})]^{-\sigma(K)/2}\}\mathrm{TR}(K)$$

where $c(K)$ denotes the number of components of K, and $\sigma(K)$ denotes the signature of the matrix of linking numbers of the components of K (with framing numbers on the diagonal), then INV(K) is an invariant of the 3-manifold obtained by doing framed surgery on K in the blackboard framing. This is our reconstruction of the form of Hennings invariant [2].

$U_q(sl_2)'$

We realize the invariant INV(K) in terms of a specific right integral for $U_q(sl_2)'$ when q is a root of unity.

Here is the definition of $A = U_q(sl_2)'$.

Let t denote a primitive n-th root of unity over a field k. Let $q=t^2$

and $m = \text{order}(t^4) = \text{order}(q^2)$. Thus we assume that $t^{4m} = 1$.

The algebra A is generated by elements a, e, f with the relations

$$a^n = 1, \quad e^m = 0 = f^m$$

$$ae = qea$$

$$af = q^{-1}fa$$

$$[e,f] = ef - fe = (a^2 - a^{-2})/(q-q^{-1})$$

The formulas for the antipode are
$s(e) = -q^{-1}e$, $s(f) = -qf$, $s(a) = a^{-1}$.

The formulas for the counit are
$E(e) = E(f) = 0$, $E(a) = 1$.

The coproduct is given by
$\Delta(a) = a \otimes a$
$\Delta(x) = x \otimes a^{-1} + a \otimes x, \quad x = e,f.$

The special grouplike element is $G = a^{-2}$.

Proposition. The right integral for $A = U_q(sl_2)'$ is λ **= CHAR{** $a^{2(m-1)}e^{(m-1)}f^{(m-1)}$ **}**, where CHAR{a} is the characteristic function of the algebra element a. Thus λ picks out the coefficient of $a^{2(m-1)}e^{(m-1)}f^{(m-1)}$ in the algebra element to which it is applied.

Proof. This can be verified by a direct calculation.//

The Yang-Baxter element in A (See [6] and [9].) has the formula

$$\rho = \sum_{i,j=0}^{n-1} \sum_{k=0}^{m-1} \{ (t^{-jk+i(j-k)-k} (q-q^{-1})^k)/n[k]_q! \} f^k a^i \otimes e^k a^{-j} .$$

A has linear basis $\{ a^i e^j f^k \mid 0 \le i < n, \ 0 \le j, k < m \}$.

The algebra $A = U_q(sl(2))'$ is a finite dimensional ribbon Hopf algebra. It is our primary object of study for elucidating the invariant INV(K).

References

1. Drinfel'd, V.G. Quantum groups. *Proceedings of the International Congress of Mathematicians*, Berkeley, California, USA (1987), 798-820.

2. Hennings, M.A. Invariants of links and 3-manifolds obtained from Hopf algebras. *preprint*

3. Kauffman, L.H. From knots to quantum groups and back. In **Quantum Groups** (edited by Fairlie and Zachos, World Sci. 1991.), 1-33.

4. Kauffman, L.H. Gauss Codes, Abstract Tensors and Quantum Groups. *to appear*

5. Kauffman, L.H. and Radford, D.E. A necessary and sufficient condition for a finite-dimensional Drinfel'd double to be a ribbon Hopf algebra. *to appear in J. of Algebra*

5.1. Kauffman L.H. and Radford, D.E. Hopf Algebras and Invariants of 3-Manifolds. (Announcement 1992, and paper to appear).

6. Kirby R. and Melvin P. The 3-manifold invariants of Witten and Reshetikhin-Turaev for sl(2,C). *Invent. Math. 105, 473-545 (1991).*

7. Larson, R.G. and Sweedler, M.E. A associative orthogonal bilinear form for Hopf algebras. *Amer. J. Math.* **91** (1969), 75-94.

8. Radford, D.E. The trace function and Hopf algebras. *preprint*

9. Radford, D.E. Generalized double crossproducts associated with quantized enveloping algebras. *preprint*

10. Reshetikhin, N.Yu. and Turaev, V.G. Ribbon graphs and their

invarinats derived from quantum groups. *Commun. Math. Phys.* **127**, 1-26 (1990).

11. Reshetikhin, N.Yu. and Turaev, V.G. Invariants of 3-manifolds via link polynomials and quantum groups. *Invent. Math.* **103**, 547-597 (1991).

12. Sweedler, M.E. **Hopf Algebras.** Mathematics Lecture Notes Series, Benjamin, New York, 1969.

13. Witten,E. Quantum field theory and the Jones polynomial. *Commun. Math. Phys.* 121 (1989), 351-399.

Deformed oscillators, non-commutative geometry, Yang-Baxter and reflection equations

P. P. Kulish

Yukawa Institute for Theoretical Physics, Kyoto, Japan
and
Steklov Mathematical Institute, St Petersburg, Russia

1 Introduction

The Yang-Baxter equation is nowdays a cornerstone of integrability of quantum systems in one dimentional space and two dimentional lattice statistical modeles (cf., e.g.[36,18]). It was used as one of impotent ingredient of the quantum inverse scattering method and it finds applications in many fields of mathematics and theoretical physics: factorizable scattering on a line, classical and quantum soliton equations, quantum Lie groups and Lie algebras, non-commutative geometry, braid groups, link and knot invariants... (cf.[11,36,35,34]). The development of the Yang-Baxter equation (YBE) gives rise to great interest in q-deformed objects, which have many properties and features of the undeformed objects and when a deformation parameter q goes to 1 their limits reproduce the latter ones.
One of the simplest q-deformed object, which appeared many years ago in papers discussing quite different subjects (cf. a review [9]), is the q-oscillator. We can define it as an associative algebra $\mathcal{A}_q$ generated by three elements a, a, N, which satisfy the relations

$$aa^+ - qa^+a = q^{-N}, \quad [N,a] = -a, \quad [N,a^+] = a^+. \tag{1}$$

Treating these three generators as independent ones (it was not the case of usual oscillator, where $N = b^+b$) we found many self-adjoint irreducible representations [16], one of which has vacuum vector and in this manner similar to the Fock representation. Generalization of the q-oscillator to multi-mode case can be done by different approaches. They can be just independent, mutually commuting modes or non-commuting q-oscillators. In the latter case one of the possible criteria to define their commutation relations is the requirement of the quantum group covariance property. It can be written in very compact way using the R-matix formalism, which was apply so fruitfully to the theory of quantum groups [11]. It so happens that in another field of reseach, also related to the quantum groups, non-commutative geoometry (at least for quantum linear spaces), many formulae, being written in the R-matrix form, coinside with the q-oscillator relations. One of the reasons for this fact is the relation and realization of (q-) oscilators by multiplications and differentiation operators (the Bargman-Fock representations). The progress in understanding the YBE has resulted in introducing several new algebraic objects, such as quantum groups and algebras [10,14,11], exchange or Zamolodchikov-Faddeev algebras [5,8,16], Yang-Baxter or quadratic algebras and so on. It seems that the list should be enriched with another item: the reflection equation (RE), which has arisen recently in several independent contexts (cf. [28,1,2,7,12,29,30,17]). This equation is an extension of the YBE and related algebraic structures are discussed at the end of the report.

2 Deformed oscillators

Different deformations of the Heisenberg commutation relations have been discussed in scientific papers many years ago (cf. refs in [9]). In particular, in the end of 60-th the deformed or q-oscillator has been proposed in connection with a generalization of the Veneziano model, which was obtained by substitution in the Veneziano amplitude the gamma-function by q-gamma function. The corresponding q-oscillators were rejuvinated in connection with the quantum group theory recently in few papers and they are used intencively in modern literature.
The deformed oscillators were found also in integrable chain models, in

particular, in descret nonlinear Schrodinger equation [15]. The equations of motion are

$$i(\psi_n)_t = -(\psi_{n+1} + \psi_{n-1})(1 - \psi_n^+ \psi_n) + 2\psi_n, \tag{2}$$

where Bose-like lattice field ψ_n have to satisfy non-canonical Poisson brackets or commutation relations, for Hamiltonian interpretation of this equation:

$$\{\psi_n, \psi_m^+\} = i\delta_{n,m}(1 - \psi_n^+ \psi_n). \tag{3}$$

As result, the corresponding L-operator of the classical and quantum inverse prolem method is very simple one. It is just 2x2 matrix with diagonal (z, z^{-1}) and fields ψ_n, ψ_n^+ off diagonal. Their tensor products intertwine by the XXZ-model R-matrix with broken symmetry. In the multi-mode case the set of n q-oscillators a_i, a_j^+ $; i, j = 1, 2, ..., n$ related to the quantum group $F_q(SU(n))$, which is generated by n^2 elements T_{ij} [11]

$$R_{12}T_1T_2 = T_2T_1R_{12} \ ,$$

is defined by relations $(0 < q < 1)$

$$R_{12}A_1A_2 = qA_2A_1 \ , A_1^+A_2^+R_{12} = qA_2^+A_1^+ \ ,$$

$$A_1 \otimes A_1^+ = qA_2(R_{12}\mathcal{P})A_2^+ + I_1 \ .$$

In these formulas R_{12} is the standard $n \times n$ triangular R-matrix [14,11,33], I_1 is the unit matrix in the first space V_1 and A_k, A_ℓ^+ $; k, \ell = 1, 2$ are columns and rows with entries a_i, a_j^+ in the spaces $V_{1,2}$. It is easy to see that this system of relations is invariant with respect to the transformation (coaction of the quantum group F_q):

$$\psi(A) = \tilde{A} = TA, \quad \psi(A^+) = \tilde{A}^+ = A^+T^+.$$

The case of the deformation parameter q being root of unity is very important for the conformal field theory [3,13] and interesting in frame of representations and realizations of quantum algebras (cf. Ge's report).

3 Non-commutative geometry

In this part of the report we present main formulae of the non-commutative differential geometry (or, better to say, the examples of NCDG, developed in [35,38,24] and refs. therein) of quantum linear spaces in the R-matrix form. The latter ones are defined as associative algebras M_q generated by elements $x_1, x_2, ..., x_n$, which satisfy quadratic relations written in the R-matrix form

$$R_{12}X_1X_2 = qX_2X_1$$

where capital X means a column, entries of which are generators x_j and X_1X_2 means their tensor product $X \otimes X$.

Now one has to define an associative algebra W_q of derivatives (or vector fields) of the algebra M_q generated by elements $\partial_1,\ \partial_2, ..., \partial_n$ which satisfy the relations

$$\bar{R}_{12}D_1D_2 = q^{-1}D_2D_1\ .$$

In this formula D is a column with components ∂_j. The relations among x_j and ∂_k have also the matrix form

$$D_1 \otimes X_1^t = q^{-1}X_2^t\mathcal{P}\bar{R}_{12}D_2 + I_1\ ,$$

where I_1 is the unit matrix in the first space, $\mathcal{P}$ is the permutation operator of two spaces $V_1 \otimes V_2$ and

$$\bar{R}_{12} = \mathcal{P}(R_{12}^{-1})\mathcal{P}.$$

In the same way an algebra of quantum differential forms Λ is generated by q-analogues of 1-forms $w_1, ..., w_n$ which by additional requirement of the existing a nilpotent external differential $d: M_q \rightarrow \Lambda^1$ are proportional to dx_j [35,38]. Three additional set of relations for the generators of q-coordinates M_q, q-derivatives W_q and q-forms Λ can be written consistently in the R-matrix form

$$R_{12}\Omega_1X_2 = q^{-1}X_2\Omega_1, R_{12}\Omega_1\Omega_2 = -q^{-1}\Omega_2\Omega_1, D_1\otimes\Omega_1^t = q^{-1}\Omega_2^t\mathcal{P}\bar{R}_{12}D_2, \quad (4)$$

One can express the external differential d just as in the commuting case in terms of q-derivatives and q-1-forms using a q-trace:

$$d = \Omega^tD = \sum_k \omega_k\partial_k = tr_qD\otimes\Omega^t = trQD\otimes\Omega^t,$$

where the matrix Q is defined by the R-matrix from the requirement of invariance:

$$tr_q(T^t)^{-1}BT^t = tr_qB,$$

for any matrix B and the quantum group matrix T supposing $[B_{ij}, T_{kl}] = 0$. Different kind of q-traces are defined depending on the form of the transformation e.g. $B \to TBT^{-1}$ or $B \to TBT^t$ (cf. [11,38,29,21]).
The similarity of these relations with formulas of previous Sec. on deformed oscillators is obvious (we omit the discussion of real forms or *-structures). In particular, the relations of this Sec. are also covariant w.r.t. the quantum group F_q coaction ϕ

$$\phi(X) = TX,\ \phi(\Omega) = T\Omega,\ \phi(D) = (T^t)^{-1}D,\ \phi(d) = d.$$

The situation with $NCDG$ of the quantum group itself is more cumbersome (cf. [32] and Faddeev's report on this conference).

4 Reflection equations

As was mentioned in the Introduction, that probably in every field, where the Yang-Baxter equation is useful, one can find the application of the corresponding reflection equation. Among these fields let us enumerate few of them: 1. the factorizable scattering on the line (now on a half line) [6], 2. integrable chain models [30,25,26], 3. quantum Lie groups and quantum Lie algebras [1,2], 4. braid groups and link and knot invariants (now in solid handlebody) [19,20,21], 5. quantum homogeneous spaces [27] etc. .

Here we briefly summarize the basic properties of the Reflection Equations (RE) and the Reflection Equation Algebras (REA) in connection with YBE and YBA.

The most natural stage to extend the YBE to the RE is the factorizable scattering with reflection at the end of a line, namely the factorizable scattering on a half line. Then the factorizability conditions consist of two parts, the first part is the YBE itself and the second part contains an additional element, called K-matrix, describing the reflection at the endpoint of the half line.

$$R_{12}^{(1)}K_1R_{12}^{(2)}K_2 = K_2R_{12}^{(3)}K_1R_{12}^{(4)}.$$

Here K is an $N \times N$ matrix and the various (rapidity dependent) R-matrices $R^{(1)}, \ldots, R^{(4)}$ are related to the R-matrix in the YBE and will be specified later. For notational simplicity the rapidity dependence is suppressed. (The suffices of the R-matrices like $R = R_{12}$ indicating the base space $V_1 \otimes V_2$ will be suppressed in most cases.) As in the YBE case we mainly discuss the RE and REA in the rapidity independent form.

The above equation can be considered as a quadratic equation for N^2 c-number entries of K for a given solution R of YBE. On the other hand the reflection equation can be considered as the defining relations of a quadratic algebra (Reflection Equation Algebra) generated by K_{ij} just like the YBA. Then the c-number solutions can be considered as the one-dimensional representations of the algebra.

We require the REA to be closely related with the YBA, namely K to be a "quantum group comodule". In other words the REA should be invariant under either of the following transformations[28]:

$$K \to \delta(K) = \begin{cases} TKT^t \\ TKT^{-1} \end{cases} \qquad [K_{ij}, T_{kl}] = 0.$$

This requirement leads to two groups of RE, called RE1 and RE2 , for which we write the standard forms. For RE1, namely for $\delta(\bar{K}) = T\bar{K}T^t$:

$$R\bar{K}_1 R^{t_1} \bar{K}_2 = \bar{K}_2 \widetilde{R^{t_1}} \bar{K}_1 \widetilde{R^{t_1 \circ t_2}},$$

and for RE2, namely for $\delta(K) = TKT^{-1}$:

$$RK_1 \tilde{R} K_2 = K_2 R K_1 \tilde{R}.$$

In order to distinguish these two groups of RE's we put a bar on the K-matrix appearing in the RE1. Here t_1 means transposition in the first space and $\tilde{A} = \mathcal{P}A\mathcal{P}$ for any matrix on $V_1 \otimes V_2$. The other members of the group are obtained by simply replacing any R by $\tilde{R}^{-1}$ [29]. But we mainly discuss these forms of RE's.

Let us give the explicit forms of the REA's for the $GL_q(2)$ R-matrix given above. For the RE1 :

$$\bar{K} = \begin{pmatrix} \alpha & \beta \\ \gamma & \delta \end{pmatrix}$$

which satisfy the relations [29,17]

$$\begin{array}{lll} [\alpha,\beta] = \omega\alpha\gamma, & [\alpha,\delta] = \omega(q\beta\gamma + \gamma^2), & [\beta,\delta] = \omega\gamma\delta, \\ \alpha\gamma = q^2\gamma\alpha, & [\beta,\gamma] = 0, & \gamma\delta = q^2\delta\gamma. \end{array}$$

This algebra has two central elements, linear and quadratic in generators $\bar{K}_{ij}$,

$$c_1 = \beta - q\gamma, \qquad c_2 = \alpha\delta - q^2\beta\gamma.$$

The c-number solutions are

$$\bar{K} = \epsilon_q = \begin{pmatrix} & 1 \\ -q & \end{pmatrix}, \quad \bar{K} = \begin{pmatrix} \alpha & \beta \\ & \delta \end{pmatrix}.$$

It is well known these days that YBE is related to the braid group B_n, if the R-matrix $\mathcal{P}R$ is identified with a generator σ of B_n. In the same way the RE is related to the braid group in the space with one hole [19].

The relationship between the REA and the braid group for solid handlebody ($g = 1$) can be readily generalized to a braid group $B_n^{(g)}$ for a manifold of an arbitrary genus g. The corresponding REA, to be called "extended REA" consists of g-copies of REA: $K^{(1)}, \ldots, K^{(g)}$,

$$RK_1^{(j)}\tilde{R}K_2^{(j)} = K_2^{(j)}RK_1^{(j)}\tilde{R}, \qquad j = 1, \ldots, g.$$

The mutual relationship among various copies should also be covariant under the "quantum group coaction" $K \rightarrow \delta(K) = TKT^{-1}$. This induces non-trivial relations among them, since the trivial ones (i.e. simply commuting) are not preserved by the transformation

$$[\delta(K_{st}^{(i)}), \delta(K_{uv}^{(j)})] \neq 0, \quad \text{even if} \quad [K_{st}^{(i)}, K_{uv}^{(j)}] = 0.$$

We are led to the following relations between $K^{(i)}$ and $K^{(j)}$ [21,20], which are "linearly ordered" (non-symmetric in i and j),

$$K^{(i)} \succ K^{(j)}, \quad i > j, \qquad \Leftrightarrow \qquad R^{-1}K_2^{(j)}R\,K_1^{(i)} = K_1^{(i)}R^{-1}K_2^{(j)}R.$$

By keeping the "order" in the matrix multiplication we can "compose" $K^{(i,j)} \equiv K^{(i)}K^{(j)}$, $i > j$ which also satisfies the REA,

$$RK_1^{(i,j)}\tilde{R}K_2^{(i,j)} = K_2^{(i,j)}RK_1^{(i,j)}\tilde{R}.$$

It should be remarked that the unit element $K = 1$ always satisfies both equations. The above "composition rule" and the "linear order" and most of the composition rules below are easily understood pictorially by the following identification $K^{(j)} \Leftrightarrow \tau_j$ of the extended REA with $B_n^{(g)}$ $(g \geq 1)$. The above relation can be rewritten in an equivalent form

$$RK_1^{(i)}R^{-1}K_2^{(j)} = K_2^{(j)}R\,K_1^{(i)}R^{-1}.$$

For theses extended reflection equation algebras the central elements are constructed by invariant q-traces [21]. Their physical interpretation connected with a transition from one particle to multiparticle discription [20].

References

[1] Alekseev A. and Faddeev L., *Commun. Math. Phys.* **141** (1991) 413.

[2] Alekseev A., Faddeev L. and Semenov-Tian-Shansky M., Preprint LOMI, E-5-91 (1991); (in [18], 148.)

[3] Alvarez-Gaumé L., Gomez C. and Sierra G., in "Physics and Mathematics of Strings." Brink L., Friedan D. and Polyakov A. (Eds.), World Scientific, Singapore, (1990).

[4] Baxter R. J., *Exactly Solved Models in Statistical Mechanics* Academic Press, (1982).

[5] Babelon O., *Commun. Math. Phys.* **139** (1991) 619 and in [18], 159.

[6] Cherednik I.V., *Teor. Matem. Fiz.* **61** (1984) 55.

[7] Cherednik I.V., *Commun. Math. Phys.* **150** (1992) 109.

[8] Cremmer E. and Gervais J.-L., *Commun. Math. Phys.* **144** (1992) 279; (in [18], 259.)

[9] Damaskinsk E. and Kulish P., *Zap. Nauch. Sem. LOMI* **189** (1991) 41.

[10] Drinfeld V.G., "Quantum groups." In: Proc. ICM-86, **1**, Berkeley, AMS (1987) 798.

[11] Faddeev L., Reshetikhin N. and Takhtajan L., *Alg. Anal.* **1** (1989) 178 (in Russian, English translation: Leningrad Math. J. **1** (1990) 193).

[12] Freidel L. and Maillet J.M., *Phys. Lett.* **B262** (1991) 278; **B263** (1991) 403.

[13] Gomez C. and Sierra G., "contribution to this Proc. of the XXI DGMTP-Conference, Tianjin, 1992, WS (1993).

[14] Jimbo M., *Lett. Math. Phys.* **10** (1985) 63, **11** (1986) 247; *Commun. Math. Phys.* **102** (1986) 537.

[15] Kulish P.P., *Lett. Math. Phys.* **5** (1981) 191.

[16] Kulish P.P., *Teor. Mat. Fiz.* **86** (1991) 157; Phys. Lett. **A161** (1991) 50.

[17] Kulish P., Sasaki R. and Schwiebert C., *J. Math. Phys.* in press.

[18] Kulish P.P. (Ed.), Proc. "Euler Inter. Math. Inst. on Quantum Groups." *Lect. Notes Math.* **1510** Springer, Berlin (1992).

[19] Kulish P.P., "Quantum groups and quantum algebras as symmetries of dynamical systems", Kyoto preprint YITP/K-959 (1991).

[20] Kulish P.P., "Quantum groups, q-oscillators and covariant algebras. Kyoto preprint YITP/K-984, (1992).

[21] Kulish P.P. and Sasaki R., "Covariance properties of reflection equation algebras, Kyoto preprint YITP/U-92-32 (1992).

[22] Kuznetsov V. B., *J. Math. Phys.* **31** (1990) 1167.

[23] Majid S., *Int. J. Mod. Phys.* **A5** (1990) 1; *J. Math. Phys.* **32** (1991) 3426; in [18], 79; preprint DAMTP/92-12.

[24] Manin Yu.I., "Topics in noncommutative geometry." Princeton Univ. Press, (1991); preprint MPI/91-60 (1991).

[25] Mezincescu L. and Nepomechie R., *J. Phys.* **A25** (1992) 2533;

[26] Curtright T., Fairlie D. and Zachos C. (Eds.), "Quantum Groups, World Scientific (1991).

[27] Noumi M. and Mimachi K., *Commun. Math. Phys.* **128** (1990) 521; Noumi M., preprint UT-Komaba (1991).

[28] Sklyanin E.K., *J. Phys. A. : Math. Gen.* **A21** (1988) 2375.

[29] Kulish P.P. and Sklyanin E.K., *J. Phys.* **A25** (1992) 5963.

[30] Kulish P.P. and Sklyanin E.K., J. Phys. **A24** (1991) L 435.

[31] Schirrmacher A., Wess J. and Zumino B., *Z. Phys. C* **49** (1991) 317.

[32] Schupp P., Watts P. and Zumino B., preprint LBL-32314, UCB-PTH-92/13, (1992).

[33] Takhtajan L.A., "Quantum Groups, Lect. Notes. Phys., **370** (1990) 3.

[34] Wadati M., Deguchi T. and Akutsu Y., *Phys. Reports* **C180** (1989) 427.

[35] Wess J. and Zumino B., *Nucl. Phys.B* (Proc. Suppl.) **18B** (1990) 302.

[36] Yang C.N. and Ge M.-L. (Eds.), "Braid group, knot theory and statistical mechanics" World Scientific, Singapore (1989).

[37] Yang C. N. and Yang C. P., *Phys. Rev.* **150** (1966) 321, 327, **151** (1966) 258.

[38] Zumino B., "Introduction to differential geometry of quantum groups, Preprint UCB-PTH-62/91 (1991).

Solving Solvable Lattice Models Using Quantum Affine Symmetry.

Tetsuji Miwa
Research Institute for Mathematical Sciences,
Kyoto University, Kyoto 606, Japan

1. **Introduction** In this talk I will present a new scheme for "solving solvable lattice models". This title might sound funny. I have Baxter's book in mind, the book whose title is "Exactly Solved Models in Statistical Mechanics". I wish our scheme solving unsolved problems in solvable lattice models.

The main part of my talk will be based on our recent works [1–3] with Brian Davies, Omar Foda, Michio Jimbo, Kei Miki and Atsushi Nakayashiki.

2. **CFT and QISM** Conformal field theory (CFT) presents the most successful example of using the infinite-dimensional symmetries. Let us recall the main points of CFT in the context of the chiral Wess-Zumino model [4].

(A1) The symmetry algebra is an affine Lie algebra $\hat{\mathfrak{g}}$ where $\mathfrak{g}$ is a finite dimensional semi-simple Lie algebra.

(A2) The space of states is the direct sum of the irreducible highest weight modules of a given integer level k,

$$\mathcal{F} = \bigoplus_{\langle c,\lambda\rangle = k} V(\lambda).$$

(A3) The local fields are the vertex operators,

$$\varphi(z) : \mathcal{F} \to \mathcal{F} \otimes V_z,$$

where V is a finite dimensional irreducible $\mathfrak{g}$-module and V_z the affinization with the spectral parameter z,

$$V_z = V \otimes F[z, z^{-1}] \quad (F : \text{coefficient field}),$$
$$\mathrm{wt}(v \otimes z^n) = \mathrm{wt}(v) + n\delta \quad (\delta : \text{the imaginary root}).$$

More precisely, $\varphi(z)$ is a formal intertwiner between the two $\hat{\mathfrak{g}}$-modules $\mathcal{F}$ and $\mathcal{F} \otimes V_z$.

(A4) The correlation functions of the local fields solve the Knizhnik-Zamolodchikov equation.

The quantum inverse scattering method is another successful approach to integrable models of infinite degrees of freedom. The Yang-Baxter equation is the basic equation in this approach, which assures the integrability. I would like to give several remarks.

(B1) The applicability of CFT is restricted to the massless continuum theories. This is not the case in QISM.

(B2) The quantum groups underlie the Yang-Baxter equations in many cases.

(B3) The first consequence of the Yang-Baxter equation is the existence of infinite-dimensional abelian symmetries: the commuting transfer matrices or the infinite conservation laws.

(B4) The standard machinery of QISM is the algebraic Bethe Ansatz. It gives the eigenvectors of the transfer matrix on a finite lattice in terms of solutions to the Bethe Ansatz equation.

3. **From Massless to Massive Theories** Following the victory of CFT in massless continuum theory, we wish to make comparable progress in massive theories. If we approach to this problem from the point of view of solvable lattice models, the following are the goals:

(C1) Diagonalization of the transfer matrix.

(C2) Correlation functions of local operators.

(C3) Continuum massive theory in the scaling limit.

These goals were achieved completely for the 2-dimensional Ising model and its relatives even before QISM and CFT. But for other models, progress was made only in (C1) and in a very restricted case of (C2), i.e., the 1 point functions. The catalog of solvable lattice models has been increased by the powerful machinery of quantum groups. But, solving the models was not considered in an explicit connection with the representation theory of quantum groups until recently.

In the last few years there have been remarkable advances in this respect:

(D1) Several auhtors [5–7] realized the quantum affine symmetry in massive continuum integrable models, e.g., the massive sine-Gordon model.

(D2) Theory of "crystals" was invented in [8]. This is the representation theory of $U_q(\hat{\mathfrak{g}})$ at $q = 0$. It was applied [9] to the computation of the 1 point functions on the basis of Baxter's corner transfer matrix method.

(D3) In [10] was established the quantum version of the vertex operators and the KZ equation.

Based on these developments, we have recently proposed a new scheme for solving lattice models [2]. I will proceed to the explanation of the scheme.

Consider the XXZ model on 1-dimensional lattice given by the Hamiltonian

$$H_{\mathrm{XXZ}} = -\frac{1}{2}\sum_{k=-\infty}^{\infty} \sigma_{k+1}^x \sigma_k^x + \sigma_{k+1}^y \sigma_k^y + \Delta \sigma_{k+1}^z \sigma_k^z. \tag{1}$$

We are interested in the massive regime, i.e., $\Delta < -1$. Let V be the 2-dimensional space on which the Pauli matrices act. The Hamiltonian formally acts on the infinite tensor product

$$W_{\mathrm{formal}} = \cdots \otimes V \otimes V \otimes V \otimes V \otimes \cdots. \tag{2}$$

In (A2) the space of states is given a clear mathematical interpretation as a direct sum of irreducible highest weight modules. The infinite tensor product is too wild for an adequate mathematical treatment. In the usual Bethe Ansatz approach, one starts with a finite lattice and, after computation, takes the thermodynamic limit. In our approach, we insist on considering the infinite lattice from the very beginning. The reason is explained below.

4. Quantum Affine Symmetry As noted in [11], the Hamiltonian on a finite lattice, with a suitable boundary condition, acquires the global $U_q(\mathfrak{sl}(2))$ symmetry under the identification $\Delta = (q+q^{-1})/2$.

$$[H^{\text{finite}}_{\text{XXZ}}, U_q(\mathfrak{sl}(2))] = 0.$$

This can be explained as follows: If we consider the space V as the 2-dimensional $U_q(\mathfrak{sl}(2))$ module, then the local interaction term in (1) is nothing but the corresponding R- matrix. Since the R-matrix is an intertwiner, it commutes with the action of $U_q(\mathfrak{sl}(2))$. This symmetry is too small to help us with the diagonalization of the transfer matrix. On the other hand, if we consider the infinite lattice, then the infinite symmetry of the affine quantum algebra $U_q(\widehat{\mathfrak{sl}}(2))$ arises:

$$[H_{\text{XXZ}}, U'_q(\widehat{\mathfrak{sl}}(2))] = 0.$$

Here the prime in $U'_q(\widehat{\mathfrak{sl}}(2))$ means not to include the scaling operator q^d. We will discuss the role of d as the boost operator later. This difference is a reflection of the following fact observed earlier [12].

The degeneracy of the spectrum of the Hamiltonian (1) *increases in the thermodynamic limit.*

The larger symmetry, i.e., the quantum affine Lie algebra, arises in the thermodynamic limit corresponding to this larger degeneracy. It gives us a hint in finding a mathematical formulation of the space of states of our massive (off-critical) model, just as in the Wess-Zumino model. We are led to the following natural question:

What is the $U_q(\widehat{\mathfrak{sl}}(2))$-module relevant to the infinite tensor product?

The answer is:

$$W_{\text{math}} = \bigoplus_{\lambda,\mu} \text{Hom}_{\mathbb{C}}(V(\mu), V(\lambda)) = \bigoplus_{\lambda,\mu} V(\lambda)\hat{\otimes}V(\mu)^*. \tag{3}$$

Here λ, μ are level 1 dominant integral weights. We equip the space (3) with a $U_q(\widehat{\mathfrak{sl}}(2))$-action by using the antipode.

Let us explain why this is the right picture, in connection with the recent developments (D1–3).

(E1) One can expect (3) on the basis of Smirnov's picture for the $\mathfrak{sl}(2)$-invariant Thirring model, which is the continuum limit of the XXZ model. He argues that the Hilbert space of the chiral conformal field theory is downsized from the bigger Hilbert space of the massive theory, and proposes (3) in the Yangian version as the latter space.

(E2) In [1], it is proposed, and checked in a small q expansion, that the eigenstates for the boost operator

$$B = const \sum_{k=1}^{\infty} k(\sigma^x_{k+1}\sigma^x_k + \sigma^y_{k+1}\sigma^y_k + \Delta\sigma^z_{k+1}\sigma^z_k), \tag{4}$$

in the semi-infinite tensor product

$$W_{\text{left}} = \cdots \otimes V \otimes V,$$

form the level 1 highest weight modules. In other words, there is an embedding

$$V(\lambda) \to W_{\mathrm{left}} \tag{5}$$

for each level 1 highest weight module $V(\lambda)$. The operator (4) acts as the derivation of $U_q(\widehat{\mathfrak{sl}}(2))$.

$$[B, e_i] = e_i, [B, f_i] = -f_i.$$

The picture (3) immediately follows from (5), by considering the space of states on the full line to be a tensor product of the spaces of the states on the two half-lines.

(E3) In [2] a conjecture was given that the embedding (5) is given in an appropriate limit of the following composition of the vertex operators:

$$V(\lambda) \to V(\mu) \otimes V \to V(\lambda) \otimes V \otimes V \to \cdots \to W_{\mathrm{left}}. \tag{6}$$

Though the infinite iteration of the vertex operators (6) is not converging in general. we conjecture that by a suitable normalization of the vertex operators it is actually convergent. We have not determined this normalization constant.

5. **Quantum Vertex Operators** To understand (6) more precisely, let us explain some basic facts about the vertex operators and related matters in a general framework for $U_q(\hat{\mathfrak{g}})$, given in [9,10,13]

A vertex operator

$$\tilde{\Phi}_\lambda^{\mu V}(z) = \sum_{k=-\infty}^{\infty} \tilde{\Phi}_{\lambda n}^{\mu V} z^n : V(\lambda) \to V(\mu) \otimes V_z, \tag{7}$$

is a formal generating function of linear operators $\tilde{\Phi}_{\lambda n}^{\mu V}$ from $V(\lambda)$ to $V(\mu)$ such that

$$\tilde{\Phi}_\lambda^{\mu V}(z)(xv) = \Delta(x)\tilde{\Phi}_\lambda^{\mu V}(z)(v)$$

holds where Δ is the coproduct of $U_q(\hat{\mathfrak{g}})$ and the actions on V_z of the Chevalley generators are given by

$$e_i(v \otimes z^n) = e_i v \otimes z^{n+\delta_{i0}}, \quad f_i(v \otimes z^n) = f_i v \otimes z^{n-\delta_{i0}}.$$

The vertex operator is uniquely given once we fix the normalization

$$\tilde{\Phi}_\lambda^{\mu V}(z)(u_\lambda) = u_\mu \otimes v + \cdots, \tag{8}$$

by choosing $v \in V$. Here u_λ denote the highest weight vector in $V(\lambda)$. Obviously, we have $\mathrm{wt}(v) = \lambda - \mu$. This is however not a sufficient condition for the existence of the vertex operator. A necessary and sufficient condition [13] is

$$e_i^{\langle h_i, \mu\rangle + 1} v = 0.$$

In particular, if the finite dimensional module V is perfect of level k as defined in [9], then for each λ there is a unique choice of $\mu = \sigma\lambda$ such that there exists an intertwiner.

Moreover, up to constant multiple there is one and only one intertwiner. If V is perfect and λ is an level k dominant integral weight, we can say heuristically that

$$V(\lambda) = V(\sigma\lambda) \otimes V_z. \tag{9}$$

This suggests the following matching condition for general vertex models of six-vertex type [14].

Let V be perfect of level k and H be the 1-dimensional quantum Hamiltonian obtained from the R-matrix corresponding to V. Then, (3) holds by replacing the level of the highest weight modules to k.

A long list of perfect modules are given in [9]. For $\mathfrak{g} = sl(n)$, the affinization of the irreducible $sl(n)$-module with highest weight $k\Lambda_j$ $(1 \leq j \leq n-1)$ is perfect of level k.

If the level l of a perfect module V in (7), differs from the level of the highest weights λ, μ, then (9) is no more valid. If $l > k$, there is no vertex operator of the form (7). If $l < k$, the uniqueness of the vertex operator for a given λ may not be true. For example, if $\mathfrak{g} = sl(2)$ and $l = 1$ and $k = 2$, then (9) reads as

$$\begin{aligned} V(\Lambda_0 \oplus \Lambda_1) &= V(2\Lambda_0) \otimes V_z = V(2\Lambda_1) \otimes V_z, \\ V(2\Lambda_0) \oplus V(2\Lambda_1) &= V(\Lambda_0 \oplus \Lambda_1) \otimes V_z. \end{aligned}$$

This is a special case of the "fusion rule" as discussed in [13].

Another important subject related to the vertex operators is the quantum KZ equation formulated in [10]. Our scheme uses this as the tool for computation of the two point correlation functions of the vertex operators. Let

$$\Phi_\lambda^{\mu V}(z) = z^{\Delta_\mu - \Delta_\lambda} \tilde{\Phi}_\lambda^{\mu V}(z) : V(\lambda) \to V(\mu) \otimes V_z,$$

where $\Delta_\lambda = (\lambda, \lambda + 2\rho)/2(k + h^\vee)$. Then the correlation function of the vertex operator

$$\Psi(z_1, \ldots, z_n) = \langle u_{\lambda_0} | \Phi_{\lambda_1}^{\lambda_0 V_1}(z_1) \cdots \Phi_{\lambda_n}^{\lambda_{n-1} V_n}(z_n) | u_{\lambda_n} \rangle \in V_1 \otimes \cdots \otimes V_n$$

satifies the following q-difference equation.

$$\begin{aligned} \Psi(z_1, \ldots, pz_k, \ldots, z_n) = & R_{k-1\,k}^{-1}(z_{k-1}/pz_k)^{-1} \cdots R_{1k}^{-1}(z_1/pz_k)^{-1} \times \\ & \times p^{-\phi_k} R_{kn}(z_k/z_n) \cdots R_{k\,k+1}(z_k/z_{k+1}) \Psi(z_1, \cdots, z_n). \end{aligned}$$

Here R_{jk} is a suitable realization of the universal R-matrix on $V_j \otimes V_k$, and ϕ_k is a diagonal matrix acting on the k-th component of $V_1 \otimes \cdots \otimes V_n$.

If V is perfect, the QKZ equation for the n point functions reduces to a single q-difference equation having an explicit solution in an infinite product form. If V is "small" (e.g., $\mathfrak{g} = sl(2)$ and $\dim V = 2$), the QKZ equation for the two-point functions reduces to that of the q-hypergeometric functions [10].

6. Dictionary Once we establish the interpretation (3), the next task is to make a dictionary between the languages of W_{formal} and W_{math}. We have the following:

(F1) The shift operator T, that shifts the k-th component of W_{formal} to the $(k+1)$-th component, is given by the composition of vertex operators.

$$T : V(\lambda) \otimes V(\mu)^* \to V(\sigma(\lambda)) \otimes V \otimes V(\mu)^* \to V(\sigma\lambda) \otimes V(\sigma\mu)^*.$$

(F2) The Hamiltonian H is given as the difference of the boost operator and its shift.

$$H = const\,(T^N B T^{-N} - B).$$

Here, N is the period of σ. Since the boost operator is a derivation, its action on W_{math} is already given.

(F3) A local operator L acting on a local part of W_{formal} is translated to the operator on W_{math} being sandwiched by "open-door" and "close-door" operations.

$$\begin{gathered}\mathcal{L}^{\lambda,\mu}_{z_n,\dots,z_1} : V(\lambda)\otimes V(\mu)^* \to V(\sigma\lambda)\otimes V_{z_n}\otimes V(\mu)^* \to \cdots \\ \to V(\sigma^{n-1}\lambda)\otimes V_{z_n}\otimes\cdots\otimes V_{z_1}\otimes V(\mu)^* \stackrel{\mathrm{id}\otimes L\otimes\mathrm{id}}{\to} V(\sigma^{n-1}\lambda)\otimes V_{z_n}\otimes\cdots\otimes V_{z_1}\otimes V(\mu)^* \to \\ \cdots \to V(\sigma\lambda)\otimes V_{z_n}\otimes V(\mu)^* \to V(\lambda)\otimes V(\mu)^*\end{gathered}$$

Here we fix the constant multiple undetermined in the definition of the vertex operators in such a way that $\mathcal{L}^{\lambda,\mu}_{z_n,\dots,z_1} = \mathrm{id}_{V(\lambda)\otimes V(\mu)^*}$ if $L = \mathrm{id}$.

(F4) The dual coupling between $x \in \mathrm{Hom}_{\mathbf{C}}(V(\lambda), V(\mu))$ and $y \in \mathrm{Hom}_{\mathbf{C}}(V(\mu), V(\lambda))$ that are invariant with respect to the action of $U_q(\hat{\mathfrak{g}})$, is given by

$$\langle y|x\rangle = \frac{\mathrm{tr}_{V(\lambda)} q^{-2\rho} yx}{\mathrm{tr}_{V}(\lambda) q^{-2\rho}}. \tag{10}$$

From this the formula for the 1 point functions in the sense of [15] follows tautologically.

(F5) The vacuum vector $|\mathrm{vac}\rangle$, i.e., the lowest energy state, is given by the identity operator in $\mathrm{Hom}_{\mathbf{C}}(V(\lambda), V(\lambda))$, or equivalently, by the canonical element in $V(\lambda)\otimes V(\lambda)^*$. This is invariant by the shift T^N, and a zero-eigenvector of the boost operator, and hence of the Hamiltonian.

7. **Diagonalization** Let us diagonalize the Hamiltonian on W_{math}. Since the Hamiltonian commutes with the action of $U'_q(\hat{\mathfrak{g}})$, the task is to decompose W_{math} into irreducible pieces with respect to $U'_q(\hat{\mathfrak{g}})$. We say the following.

Finite dimensional irreducible $U'_q(\hat{\mathfrak{g}})$-modules embedded in W_{math} are called "particles".

Consider a finite dimensional irreducible module $V^{(1)}_{z_1}\otimes\cdots\otimes V^{(n)}_{z_n}$ with the spectral parameters $z_1,\dots,z_n$. Having a $U'_q(\hat{\mathfrak{g}})$-linear map

$$V^{(1)}_{z_1}\otimes\cdots\otimes V^{(n)}_{z_n} \to V(\lambda)\otimes V(\mu)^* \tag{11}$$

ia equivalent to having a vertex operator

$$V^{(1)}_{z_1}\otimes\cdots\otimes V^{(n)}_{z_n}\otimes V(\mu) \to V(\lambda). \tag{12}$$

If $\mathfrak{g} = sl(2)$ and $k = 1$, as we have discussed above, such vertex operators exist only if $\dim V^{(j)} = 2$ for all j. Namely, the model has only spin 1/2 particles. This was already observed by [12,16,17]. In general, not all the vertex operators of the form (12) give the particles in the theory [14]. We require the following.

We consider (11) as particles if and only if it preserves the crystal lattice structures.

I have no time to explain in more details on this statement. Intuitively, it means that we consider only those embedding (11) such that "the $q = 0$ limit" is regular. In fact, for $\mathfrak{g} = sl(2)$ and k arbitrary, this restricts the spin of the particles to be not arbitrary but 1/2. In other words, the higher spin XXZ models, say of $k/2$, also have only spin 1/2 particles. This fact was already observed, e.g., see [18]. (Note however that V in the infinite tensor product (2), or in the "open-door" and "close-door" operations, is $(k+1)$-dimensional in that case.)

Summarizing, the creation operators $\varphi^{*\lambda'}_{\lambda v}(z)$ $(v \in V_z)$ are realized by the vertex operator

$$\begin{aligned}&\varphi^{*\lambda'}_{\lambda v}(z) : \mathrm{Hom}_{\mathbf{C}}(V(\mu), V(\lambda)) \ni x = [\cdot \mapsto x(\cdot)] \\ &\mapsto \varphi^{*\lambda'}_{\lambda v}(z)(x) = [\cdot \mapsto \Phi^{\lambda'}_{V\lambda}(v \otimes x(\cdot))] \in \mathrm{Hom}_{\mathbf{C}}(V(\mu), V(\lambda')). \end{aligned} \tag{13}$$

The creation operators create the physical Hilbert space of the eigenstates of the Hamiltonian upon the vacuum discussed in (F5). We denote this space by W_{phys}. The annihilation operator is defined to be the dual operator to the creation operator.

Once the creation and the annihilation operators are understood in this way, the following (G1-3) are the immediate consequences on working with the QKZ equation. Below, we restrict our discussions to the spin 1/2 XXZ model, and denote the creation and the annihilation operators by $\varphi^*_\pm(z)$ and $\varphi_\pm(z)$. Here the sign $\pm$ represents the choice of $v = v_\pm$ in (13).

(G1) Energy and momentum of particles:

$$T\varphi^*_\pm(z)T^{-1} = \tau(z)^{-1}\varphi^*_\pm(z), \quad [H, \varphi^*_\pm(z)] = \epsilon(z)\varphi^*_\pm(z), \tag{14}$$

where

$$\begin{aligned}&\tau(z) = z^{-1/2}\frac{\Theta(qz)}{\Theta(qz^{-1})}, \quad \epsilon(z) = -(q - q^{-1})z\frac{d}{dz}\log\,\tau(z), \\ &\Theta(z) = (z)_\infty(q^4z^{-1})_\infty(q^4)_\infty, \quad (z)_\infty = (z; q^4)_\infty, \quad (z;p)_\infty = \prod_{j=0}^{\infty}(1 - zp^j).\end{aligned}$$

(G2) Zamolodchikov algebra:

$$\begin{aligned}\varphi_{\varepsilon_1}(z_1)\varphi_{\varepsilon_2}(z_2) &= \sum \varphi_{\varepsilon'_2}(z_2)\varphi_{\varepsilon'_1}(z_1)\big(R_{VV}(z_1/z_2)\big)^{\varepsilon_1\varepsilon_2}_{\varepsilon'_1\varepsilon'_2}, \\ \varphi^*_{\varepsilon_1}(z_1)\varphi^*_{\varepsilon_2}(z_2) &= \sum \varphi^*_{\varepsilon'_2}(z_2)\varphi^*_{\varepsilon'_1}(z_1)\big(R_{V^*V^*}(z_1/z_2)\big)^{\varepsilon_1\varepsilon_2}_{\varepsilon'_1\varepsilon'_2}, \\ \varphi_{\varepsilon_1}(z_1)\varphi^*_{\varepsilon_2}(z_2) &= \sum \varphi^*_{\varepsilon'_2}(z_2)\varphi_{\varepsilon'_1}(z_1)\big(R_{VV^*}(z_1/z_2)\big)^{\varepsilon_1\varepsilon_2}_{\varepsilon'_1\varepsilon'_2} + g_\lambda\delta_{\varepsilon_1\varepsilon_2}\delta(z_1/z_2),\end{aligned}$$

where

$$g_\lambda = \mp q^{-1/2}\frac{(q^2)_\infty}{(q^4)_\infty} \quad \text{for } \lambda = \begin{cases}\Lambda_0 \\ \Lambda_1\end{cases}.$$

(G3) The physical picture of the space of the eigenstates of the Hamitonian:

$$W_{\mathrm{phys}} = \left[\sum_{n=0}^{\infty}\int\cdots\int\frac{dz_1}{2\pi i z_1}\cdots\frac{dz_n}{2\pi i z_n}V_{z_1}\otimes\cdots\otimes V_{z_n}\right]_{\mathrm{symm}},$$

where "symm" means the symmetrization with respect to the commutation relations of the Zamolodchikov algebra.

8. **Correlation Functions** As an application of the formula (10), we can obtain an integral formula for arbitrary correlation functions of the spin 1/2 XXZ model [3]. In this case, we are dealing with level 1 highest weight modules and the vertex operators acting on them. Frenkel and Jing [19] constructed a boson realization of the actions of the q-deformed $sl(2)$ currents. We extend their results to the vertex operators, and thereby compute the correlations.

(H1) Bosons:

$$[a_k, a_l] = \delta_{k+l,0}\frac{[2k][k]}{k}.$$

Note that these are not the so-called q-bosons that obey the q-commutation relations.

(H2) Fock space:

$$V(\Lambda_0) \oplus V(\Lambda_1) = F[a_{-1}, a_{-2}, \cdots] \otimes F[P].$$

Here $P = \mathbf{Z}\frac{\alpha}{2}$ (α : the simple root of $sl(2)$) is the weight lattice of $sl(2)$, and $F[P]$ is the group algebra on P. The bosons a_n act on $F[a_{-1}, a_{-2}, \cdots]$ as usual, and e^α and ∂_α act on $F[P]$ by

$$e^\alpha e^\beta = e^{\alpha+\beta}, \quad \partial_\alpha e^\beta = \langle\alpha, \beta\rangle e^\beta.$$

(H3) Currents:

$$X^\pm(z) = \sum_{n\in\mathbf{Z}} x_n^\pm z^{-n-1},$$

where the identification with the Chevalley generators reads as

$$e_1 = x_0^+,\ f_1 = x_0^-,\ e_0t_0 = x_1^-,\ t_0^{-1}f_0 = x_{-1}^+.$$

(H4) Bosonization of the currents:

$$X^+(z) = \exp\Big(\sum_{n=1}^\infty \frac{a_{-n}}{[n]}q^{-n/2}z^n\Big)\exp\Big(-\sum_{n=1}^\infty \frac{a_n}{[n]}q^{-n/2}z^{-n}\Big)e^\alpha z^{\partial_\alpha},$$
$$X^-(z) = \exp\Big(-\sum_{n=1}^\infty \frac{a_{-n}}{[n]}q^{n/2}z^n\Big)\exp\Big(\sum_{n=1}^\infty \frac{a_n}{[n]}q^{n/2}z^{-n}\Big)e^{-\alpha} z^{-\partial_\alpha}.$$

(H5) Vertex operators:

$$\tilde\Phi_{\Lambda_i}^{\Lambda_{1-i}V}(z)(v) = \phi_+^{(i)}(z)(v)\otimes v_+ + \phi_-^{(i)}(z)(v)\otimes v_-,$$
$$\phi_+^{(i)}(z) = \oint \frac{d\xi}{2\pi i\xi}[\phi_-^{(i)}(z), X^-(\xi)].$$

(H6) Bosonization of the vertex operators:

$$\phi_-^{(i)}(z) = \exp\left(\sum_{n=1}^{\infty} \frac{a_{-n}}{[2n]} q^{7n/2} z^n\right) \exp\left(-\sum_{n=1}^{\infty} \frac{a_n}{[2n]} q^{-5n/2} z^{-n}\right) e^{\alpha/2} (-q^3 z)^{(\partial_\alpha + i)/2}.$$

(H7) Correlation functions: Choose signatures $\varepsilon_1, \ldots, \varepsilon_n, \varepsilon'_1, \ldots, \varepsilon'_n$ such that $\varepsilon_1 + \cdots + \varepsilon_n = \varepsilon'_1 + \cdots + \varepsilon'_n$. Let $|\mathrm{vac}\rangle$ be the normalized vacuum in $V(\Lambda_i) \otimes V(\Lambda_i)$. We wish to compute $P^{\varepsilon_n, \ldots, \varepsilon_1}_{\varepsilon'_1, \ldots, \varepsilon'_n}(z_n, \ldots, z_n|i) = \langle \mathrm{vac}|\mathcal{L}^{\Lambda_i, \Lambda_i}_{z_n, \ldots, z_1}|\mathrm{vac}\rangle$ for $L = E_{\varepsilon'_n \varepsilon_n} \otimes \cdots \otimes E_{\varepsilon'_1 \varepsilon_1}$ where E_{ij} is a matrix unit. Using (H1–6) we have the following result: Introduce the following notations

$$\begin{aligned}
&A = \{a_1, \cdots, a_s\} = \{j | \varepsilon'_j = -1\}, \quad B = \{b_1, \cdots, b_t\} = \{j | \varepsilon_j = +1\}, \\
&(s + t = n, \quad a_i < a_j, \quad b_i < b_j \quad \text{for } i < j), \\
&h(z) = (q^2 z; q^4)_\infty (q^4 z; q^4)_\infty (q^2 z^{-1}; q^4)_\infty (q^4 z^{-1}; q^4)_\infty .
\end{aligned}$$

We prepare the integration variables ξ_a $(a \in A)$, ζ_b $(b \in B)$ and set $\eta_j = \xi_{a_j}$ $(1 \le j \le s)$, $= \zeta_{b_{n+1-j}}$ $(s < j \le n)$, $\bar{\eta} = \prod_j \eta_j$ and $\bar{z} = \prod_j z_j$. Then we have

$$\begin{aligned}
&P^{\varepsilon_n, \cdots, \varepsilon_1}_{\varepsilon'_n, \cdots, \varepsilon'_1}(z_n, \cdots, z_1|i) \\
&= (-1)^t q^{\sum_{a \in A} a + \sum_{b \in B} b - n(n+1)/2} \prod_{a \in A} \oint_{C_a} \frac{d\xi_a}{2\pi i(\xi_a - z_a)} \prod_{b \in B} \oint_{C_b} \frac{d\zeta_b}{2\pi i(\zeta_b - z_b)} \\
&\times \prod_{a \in A} \prod_{a < j \le n} \frac{z_j - q^2 \xi_a}{z_j - \xi_a} \prod_{b \in B} \prod_{b < j \le n} \frac{\zeta_b - q^2 z_j}{\zeta_b - z_j} \prod_{j < k} \frac{\eta_k - \eta_j}{\eta_k - q^2 \eta_j} \\
&\times \frac{h(1)^n \prod_{j<k} h(z_j/z_k) h(\eta_j/\eta_k)}{\prod_{j,k} h(\eta_j/z_k)} \frac{\sum_{m \in \mathbb{Z} + i/2} (\bar{z}/\bar{\eta})^{2m} q^{4m^2 - 2m - i}}{(q^4; q^4)_\infty \mathrm{tr}_{V(\mathrm{L}_i)}(q^{-4d - \alpha})}.
\end{aligned} \tag{15}$$

9. **RSOS model** Another class of solvable lattice models to which our diagonalization scheme is immediately applicable is the so-called RSOS models [20,21] in regime III. In [22] it was discussed how the critical RSOS Boltzmann weights arise in the decomposition of the tensor product of $U_q(\mathfrak{g})$-modules. This approach was completed in [10] in off-critical case. They found that the RSOS weights in elliptic parametrization arise as the coefficients in the commutation relations of the $U_q(\hat{\mathfrak{g}})$ vertex operators. To be specific, fix an integer k and a perfect $U'_q(\hat{\mathfrak{g}})$-module of level N, such that $k > N$. Let λ, λ' be level k dominant integral weights. The space of the intertwiners $\mathrm{Hom}_{U'}(V(\lambda), V(\lambda') \otimes V^{\otimes n})$ is spanned by the set of iterated sequences of vertex operators

$$V(\lambda) \to V(\lambda^{(1)}) \otimes V \to V(\lambda^{(2)}) \otimes V \otimes V \to \cdots \to V(\lambda^{(n)}) \otimes V \otimes \cdots \otimes V,$$

such that $\lambda' = \lambda^{(n)}$. The latter is called the set of the restricted paths from λ to λ'. The Boltzmann weights of the RSOS models arise in the commutation relations of the vertex operators and satisfy the Yang-Baxter equation in SOS version.

The 1-point functions of the RSOS models in the so-called regime III, have been studied extensively in connection with the affine Lie algebra characters [21,23]. In [13] was made a systematic study on the basis of the vertex operators and the crystal bases. The general feature of the 1 point functions in the so called regime III is as follows: In this regime, there exist several different sectors corresponding to different boundary conditions. They are labeled by a pair of dominant integral weight (ξ, η) of level $k-N$ and N, respectively. The boundary condition specified by this pair is such that $\lambda^{(n)} = \xi + \sigma^n \eta$ where σ is that of (9). Then, the probability $P(\lambda; \xi\eta)$ of one particular site having the local state λ is given by

$$P(\lambda; \xi\eta) = \frac{\mathrm{ch}\Omega_{\xi\eta\lambda}\mathrm{ch}_\lambda}{\mathrm{ch}_\xi \mathrm{ch}_\eta}.$$

Here ch_λ, etc. are the specialized characters for the irreducible modules $V(\lambda)$, etc., and $\mathrm{ch}\Omega_{\xi\eta\lambda}$ is the character of the space

$$\Omega_{\xi\eta\lambda} = \mathrm{Hom}_{U'}(V(\lambda), V(\xi) \otimes V(\eta))$$

of the highest weight vectors with the highest weight $\lambda \bmod \delta$ in $V(\xi) \otimes V(\eta)$. This result suggests the following mathematical setting of the problem. The semi infinite lattice with the appropriate boundary condition is represented by $V(\xi) \otimes V(\eta)$ modulo the gauge symmetry of $\Delta(U_q(\hat{\mathfrak{g}}))$, or equivalently by the space of the highest weight vectors in $V(\xi) \otimes V(\eta)$. The whole lattice with the appropriate boundary condition is then represented by

$$\begin{aligned}&\mathrm{Hom}_{U'}(V(\xi') \otimes V(\eta'), V(\xi) \otimes V(\eta))\\ &\quad = \oplus_\lambda \mathrm{Hom}_{U'}(V(\lambda), V(\xi) \otimes V(\eta)) \otimes \mathrm{Hom}_{U'}(V(\xi') \otimes V(\eta'), V(\lambda)).\end{aligned}$$

In [23], the details of the working based on this picture is given.

10. **Discussions** I have formulated a new scheme of diagonalizing integrable Hamitonians on one dimensional lattice. The following are interesting questions in this direction.

(i) Generalize the scheme to other cases, e.g., the regime II of the RSOS models, the XYZ model, the chiral Potts model.
(ii) Obtain the correlation functions of the higher spin cases.
(iii) Determine the automorphic property of the correlation functions, if ever exists.
(iv) Study the massive quantum field theory in the scaling limit of the lattice.

I am grateful to C. N. Yang and M. L. Ge, and the Chinese colleagues for the organization of the conference and their warm hospitality.

References

[1] Foda O and Miwa T, *Int. J. Mod. Phys. A*, **7**, Suppl. 1A (1992) 1–53.

[2] Davies B, Foda O, Jimbo M, Miwa T and Nakayashiki A, *RIMS preprint*, **873**, (1992).

[3] Jimbo M, Miki K, Miwa T and Nakayashiki A, *RIMS preprint*, **877**, (1992).

[4] Tsuchiya A and Kanie Y, *Adv. Stud. Pure Math.*, **16**, (1988) 297.

[5] Bernard D, *Commun. Math. Phys.*, **137**, (1991) 191–208.

[6] LeClair A and Smirnov F, *Cornell preprint*, **CLNS 91-1056**, (1991).

[7] Smirnov F, *Int. J. Mod. Phys. A*, **7**, Suppl. 1B (1992) 813–837, 839–858.

[8] Kashiwara M, *Commun. Math. Phys.*, **133**, (1990) 249.

[9] Kang S-J, Kashiwara M, Misra K, Miwa T, Nakashima T and Nakayashiki A, *Int. J. Mod. Phys. A*, **7**, Suppl. 1A (1992) 449–484.

[10] Frenkel I B and Reshetikhin N Yu, *Commun. Math. Phys.*, **146**, (1992) 1–60.

[11] Pasquier V and Saluer H, *Nucl. Phys.*, **B330**, (1990) 523.

[12] Faddeev L and Takhtajan L, *J. Soviet Math.*, **24**, (1984) 241.

[13] Date E, Jimbo M and Okado M, *Osaka Univ. Math. Sci. preprint*, **1**, (1991).

[14] Idzumi M, Iohara K, Jimbo M, Miwa T, Nakashima T and Tokihiro T, *RIMS preprint*, **812**, (1992).

[15] Date, E., Jimbo, M., Kuniba, A., Miwa, T. and Okado, M, *Lett. Math. Phys.*, **17**, (1989) 69–77.

[16] Destri C and Lowenstein J H, *Nucl. Phys.*, **B205**, (1982) 369.

[17] Babelon O, de Vega H J and Viallet C M, *Nucl. Phys.*, **B220**, (1983) 13.

[18] Reshetikhin N Yu, *J. Phys. A: Math. Gen.*, **24**, (1991) 3299-3309.

[19] Freankel I and Jing N-H, *Proc. Natural Acad. Sci. USA*, **85**, (1988) 9373.

[20] Andrews G E, Baxter R J and Forrester P J, *J. Stat. Phys.*, **35**, (1984) 193-266.

[21] Date E, Jimbo M, Kuniba A, Miwa T and Okado M, *Nucl. Phys. B*, **290**, [FS20] (1987) 231–273.

[22] Pasquier V, *Commun. Math. Phys.*, **118**, (1988) 335.

[23] Jimbo M, Miwa T and Ohta Y, *RIMS preprint*, **813**, (1992).

SUPERSYMMETRY AND STABILITY FOR NON–LOCAL INTERACTIONS

KONRAD OSTERWALDER
ETH Zentrum, Department of Mathematics
8092 Zürich, Switzerland

Abstract

A mathematically rigorous construction of supersymmetric quantum field theory models leads to the stydy of hamiltonians with non–local interaction term. Stability problems of such hamiltonians are discussed.

1. Introduction

Supersymmetry is undoubtedly one of the most interesting symmetries to look for in nature. Not only is it the only known tool to bring together in a nontrivial fashion the symmetries of space-time and the internal symmetries of elementary particle physics, but it also leads to interesting new mathematics. As a tool it is helpful in many different contexts and it has led to the discovery of an interesting link between relativistic quantum field theory and non-commutative geometry. See e.g. [1,2,3,4,6] and references in [2].

Supersymmetric quantum field theory models in general exhibit a less singular behaviour than similar models without that extra symmetry. Some (but in general not all) renormalization cancellations turn out to be automatic, i.e. fermionic infinities exactly cancel bosonic infinities. This effect is particularly beautiful in some ($N = 2$) models in two space-time dimensions, where no renormalization is necessary; due to supersymmetry those models become finite, see [3].

A rigorous construction of models usually proceeds in two steps: starting from a formal expression for the hamiltonian or for the Euclidean action of the system a regularization is introduced to make everything well defined and finite. Once the necessary renormalization cancellations have been exhibited explicitly the regularization can be removed in a second step. At that point, of course, it becomes crucial to show that the quantities thus obtained have all the desired properties as required by physics. In particular, the symmetries of the formal hamiltonian now have to be established.

As it often happens, the most straightforward regularization destroys some symmetries and then it becomes very difficult if not impossible to show that in the limit, as the regularization is being removed, the symmetry is restored. Thus one has to look for special regularizations that preserve as much of the symmetries as possible. In the particular case of supersymmetric quantum field theory models, preserving the supersymmetry leads to a non-local interaction term in the hamiltonian. Such hamiltonians can not be dealt with using the standard methods: to show their stability (their boundednes from below) requires some extra work. The purpose of this discussion is to introduce a class of two dimensional supersymmetric models that have been investigated and to show how the problem of non-local interaction arises. Then we explain how to deal with the stability question for a large class of non-local hamiltonians, which of course appear in many other contexts in the physics literature as well. The work reported here has been done jointly with A.Jaffe and A.Lesniewski, [5].

2. Supersymmetric Quantum Field Theory Models in two Dimensions

Two classes of supersymmetric quantum field theory models in 2 space-time dimensions have been studied extensively: (Wess–Zumino) $N=1$ models and $N=2$ holomorphic models, see [3] and references there. For the sake of simplicity we descibe the $N=1$ models only. In that case the supersymmetry algebra is given by

$$\{Q_\alpha, \bar{Q}_\beta\} = 2(\gamma^\mu P_\mu)_{\alpha\beta},$$

where γ^μ are the Dirac matrices, chosen to be

$$\gamma^0 = \begin{pmatrix} 0 & -i \\ i & 0 \end{pmatrix}, \qquad \gamma^1 = \begin{pmatrix} 0 & i \\ i & 0 \end{pmatrix}$$

and Q_α, $\alpha = 1,2$, are the self adjoint supercharges with $\bar{Q}_\alpha = Q_\beta \gamma^0_{\beta\alpha}$.
This implies

$$Q_1^2 = H + P, \qquad Q_2^2 = H - P, \qquad \{Q_1, Q_2\} = 0,$$

or, defining $Q = \frac{1}{2}(Q_1 + Q_2)$ to be the Dirac operator, we find that

$$H = Q^2,$$

for H the hamiltonian of the theory.

The expressions for the Q_α in terms of canonical two component Majorana fields η_α, of a scalar field φ and of its conjugate momentum field π are formally given by

$$\begin{aligned} Q_1 &= \int_{S^1} \eta_2(x)\,(\pi(x) - \partial\varphi(x)) - \eta_1(x) V'(\varphi(x)), \\ Q_2 &= \int_{S^1} \eta_1(x)\,(\pi(x) + \partial\varphi(x)) - \eta_2(x) V'(\varphi(x)). \end{aligned}$$

Here V is any real polynomial, the superpotential of the model, and V' is its derivative. The integration is over the circle S^1, which for simplicity we take to be the unit circle.

A formal computation of the hamiltonian, using the standard (anti)commutation relations of the fields involved, leads to the expression

$$H = \frac{1}{2}\int_{S^1} \bar{\eta}\left(\gamma^1\partial + V''(\varphi)\right)\eta + \pi^2 + (\partial\varphi)^2 + (V'(\varphi))^2 .$$

It is this computation that leads to difficulties, if we do things rigorously, i.e. after the introduction of a regularization.

A regularization consists in replacing the fields η_α and φ by

$$\eta_\alpha^\chi = \chi * \eta_\alpha$$

and

$$\varphi^\chi = \chi * \varphi ,$$

respectively. In the formal calculation of $H = Q^2$ we used the anticommutation relations of the fermi fields to conclude that

$$(\eta_1 - \eta_2)(x)\,(\eta_1 - \eta_2)(y) = \delta(x-y),$$

thus obtaining the local term $\int_{S^1}(V'(\varphi(x)))^2$ in the hamiltonian. Now, instead, we have

$$(\eta_1^\chi - \eta_2^\chi)(x)\,(\eta_1^\chi - \eta_2^\chi)(y) = \chi * \chi(x-y),$$

leading to the non-local term

$$\int_{S^1}\int_{S^1} V'(\varphi^\chi(x))\,\chi * \chi(x-y)\,V'(\varphi^\chi(y)).$$

In the following section we want to neglect the fermionic contributions and to discuss stability problems of purely bosonic hamiltonians with non-local interaction terms of the form obtained above. However, we observe that the infinities introduced by Wick ordering the fermionic part of the full hamiltonian exactly cancel the infinities of the Wick ordering of the bosonic term. We therefore have to take the bosonic interaction in normal ordered form.

3. Non-local Interactions

In this section we look at the general question, whether a hamiltonian of the form

$$\begin{aligned} H^\chi &= H_0 + \int_{S^1}\int_{S^1} : P(\varphi^\chi(x))\,v(x-y)\,P(\varphi^\chi(y)) : \\ &= H_0 + W^\chi \end{aligned}$$

is stable, i.e. bounded from below. Here

$$H_0 = \sum_p \mu(p)\, a^*(p)\, a(p),$$

where $a(p)$ and $a^*(p)$ are the usual annihilation and creation operators,

$$\mu(p) = \sqrt{p^2+1}$$

and the sum runs over $p = 2\pi k, \quad k \in \mathbf{Z}$. Furthermore P is an arbitrary polynomial of order $n \geq 0$ and

$$\varphi^\chi(x) = \frac{1}{\sqrt{2}} \sum_p \mu(p)^{-\frac{1}{2}} \left(a(p)^* + a(-p)\right)\, e^{-ipx}\, \tilde{\chi}(p)$$

is a real, one–component scalar field on the unit circle, regularized with the mollifier function χ.
Two observations are immediate:

- Without the normal ordering $:\,:$ in the interaction term we have

$$W^\chi = \sum_p \left(\widetilde{P(\varphi^\chi)}\right)^2 (p)\, \tilde{v}(p)\ ,$$

  hence $W^\chi \geq 0$ for $\tilde{v}(p) \geq 0$. Wick ordering however makes W^χ unbounded from below (in the limit where χ tends to δ). We conclude that the boundedness of H must depend on a subtle interplay between H_0 and W^χ. This is a classical result of constructive quantum field theory in the local case, $h(x) = \delta(x)$, see [1].

- Some choices of the function v are definitely bad: with $v(x) = 1$ and $P(\xi) = \xi^3$ we obtain $W^\chi =\, :\left(\int \varphi^\chi(x)^3\right)^2:$. As we can easily find field configurations with $\left(\int \varphi^\chi(x)\right)^3$ small but $(\varphi^\chi(x))^2$ large. Using this intuition it can be shown that for some choices of v and of P, H^χ is not bounded from below.

The following assumptions on the cutoff function χ and on the kernel v will allow us to establish stability of H^χ.

$$\begin{array}{llcllcll}
1. & \tilde{\chi}(0) = 1, & 0 & \leq \tilde{\chi}(p) & \leq 1, & \tilde{\chi}(p) & = & \tilde{\chi}(-p) \\
2. & \tilde{v}(0) = 1, & 0 & \leq \tilde{v}(p) & \leq 1, & \tilde{v}(p) & = & \tilde{v}(-p) \\
3. & & (1+\delta|p|)^{-\epsilon} & \leq \tilde{v}(p) & \leq c(1+\gamma|p|)^{-\epsilon} & & &
\end{array}$$

Note that assumption 3 means slow decrease at infinity for $\tilde{v}(p)$, which implies that the delocalization effect of v should not be too strong.

Theorem *Assume the cutoff function χ satisfies assumptions 1 and 2 and there are constants δ, γ, c and ϵ with $0 \leq \epsilon \leq 1/(8n^2)$ such that assumption 3 holds. Then (on a properly chosen domain) the operator*

$$H^\chi = H_0 + W^\chi$$

is self adjoint and bounded from below, uniformly in χ.

Furthermore for $T \geq 0$,

$$Tr(e^{-TH^\chi}) \leq \infty,$$

again with a bound that is uniform in χ.

The basic ingredient for a proof of this theorem is a uniform bound on $Tr(e^{-H^\chi})$, which we prove by using a functional integral representation for this expression.

$$Tr(e^{-H^\chi}) = \int e^{-S}\, d\mu_C(\phi).$$

Here $d\mu_C$ is the Gaussian measure on the space of distributions $\mathcal{D}'(\mathcal{T})$ with mean 0 and covariance $C = (-\Delta + 1)^{-1}$. $\mathcal{T}$ is the two torus $S^1 \times S^1$. The Euclidean action S is given by

$$\begin{aligned} S &= S(\phi^\chi), \\ &= \int_{\substack{x,y \in S^1 \\ t \in S^1}} :P(\phi^\chi(x,t))\, h(x-y)\, P(\phi^\chi(y,t)). \end{aligned}$$

$\phi(x,t)$ is the (Euclidean) field, i.e. an element in path space $\mathcal{D}'(\mathcal{T})$, and

$$\phi^\chi(x,t) = \int_{S^1} \chi(x-y)\phi(y,t)dy\,.$$

To obtain the desired bound we introduce the function

$$f(x) = \mu_C\{S \leq -(1+x)\}.$$

Then we find

$$\int e^{-S} d\mu_C = \int_{-\infty}^{\infty} e^{1+x} f(x) dx\,.$$

Two properties of this function $f(x)$ are obvious:

1. $f(x)$ is monotonically decreasing,
2. $f(-\infty) = \int d\mu_C = Tr(e^{H_0})$,

while the third and most important property requires proof.

Lemma *For $\tilde{v}(p) \geq (1+\delta|p|)^{-\epsilon}, \epsilon$ sufficiently small,*
3. $f(x) \leq e^{-cx^{\alpha}}$, *for some constant c and some* $\alpha \geq 1$.

Properties 1, 2 and 3 immediately lead to the desired bound on $Tr(e^{-H^{\chi}})$.
In order to prove the lemma we introdoce an additional sharp cutoff in the action S. We set

$$\tilde{\phi}_{\kappa}(p,t) = \begin{cases} \tilde{\phi}^{\chi}(p,t), & \text{for } |p| \leq \kappa, \\ 0 \quad, & \text{for } |p| \geq \kappa. \end{cases}$$

and

$$S_{\kappa} = \sum_{p} \int \left| P(\widetilde{\phi_{\kappa}(\cdot,t)}) \right|^2 \tilde{v}(p) dt + \text{Wick ordering terms.}$$

As P is of degree n and in every ϕ_{κ} the momentum is less than κ, the sum in the above formula actually runs over values of p with $|p| \leq n\kappa$ only. Thus we may use the lower bound on $\tilde{v}(p)$ to conclude that

$$S_{\kappa} \geq (1+\delta n\kappa)^{-\epsilon} \sum_{p} \int \left| P(\widetilde{\phi_{\kappa}(\cdot,t)}) \right|^2 dt + \text{Wick ordering terms.}$$

Notice that now the main term on the right hand side is local in x–space, namely, $\int\int (P(\phi_{\kappa}(x,t))^2 dxdt$, and it may therefore be used to dominate the Wick terms in the standard fashion.
This leads to the bound

$$S_{\kappa} \geq \frac{1}{2}(1+\delta n\kappa)^{-\epsilon} \int\int (P(\phi_{\kappa}(x,t))^2 dxdt - M\kappa^{n\epsilon}.$$

While the first term on the right hand side of this inequality can be used to dominate other, lower order, terms in the action, we will use the second one only, to finish the present argument. For the function $f(x)$ we now find

$$\begin{aligned} \mu_C(S \leq -(1+x)) &= \mu_C(S_{\kappa} + (S - S_{\kappa}) \leq -(1+x)) \\ &= \mu_C(S - S_{\kappa} \leq M\kappa^{n\epsilon} - (1+x)). \end{aligned}$$

For the final argument we need an estimate that follows from standard arguments, see [1]. It says that for sufficiently small values of δ and for sufficiently large values of κ,

$$\mu_C(|S - S_{\kappa}| \geq 1) \leq e^{-\kappa^{1/8n}}$$

Choosing κ such that $M\kappa^{n\epsilon} = x$, i.e. $\kappa = \left(\frac{x}{M}\right)^{1/n\epsilon}$ we get

$$\begin{aligned}\mu_C(S \leq -(1+x)) &\leq \mu_C(S - S_\kappa \leq -1)\\ &\leq \mu_C(|S - S_\kappa| \geq 1)\\ &\leq e^{-\kappa^{1/8n}}\\ &\leq e^{-cx^{1/8n^2\epsilon}}.\end{aligned}$$

Choosing $\epsilon \leq 1/8n^2$ as assumed in the theorem we find the desired bound on the function $f(x)$,

$$f(x) = \mu_C(S \leq -(1+x)) \leq e^{-cx^\alpha}$$

for some $\alpha \geq 1$. This proves the lemma.

As we have mentionned above, the bounds are sufficiently strong to show that perturbations of the hamiltonian by polynomials of degree less than $2n$ don't destroy stability.

4. References

1. Glimm, J. and Jaffe, A.: *Quantum Physics*, Second Edition, Springer: New York 1987.

2. Jaffe, A. and Lesniewski, A.: Geometry of Supersymmetry, in *Constructive Quantum Field Theory II*, G.Velo and A.S.Wightman, Editors, Plenum, New York, 1990.

3. Jaffe, A. and Lesniewski, A.: Supersymmetric Field Theory and Infinite Dimensional Analysis, in *Nonperturbative Quantum Field Theory*, G. 't Hooft *et al.*, Editors, Plenum, New York, 1988.

4. Jaffe, A., Lesniewski, A. and Osterwalder, K.: Quantum K–Theory I: The Chern Character, *Commun. Math. Phys.* **178** (1987), 313-329.

5. Jaffe, A., Lesniewski, A. and Osterwalder, K.: Stability for a Class of Bilocal Hamiltonians,*Commun. Math. Phys.*, to appear

6. Witten, E.: Supersymmetry and Morse Theory, *J. Diff. Geom.*17 (1982), 661-692.

QUANTUM MINKOWSKI SPACE AND GENERATORS OF QUANTUM LORENTZ GROUP

XING-CHANG SONG

Department of Physics, Peking University
Beijing 100871, China

ABSTRACT

From the basic 4×4 $\check{R}$ matrix associated to quantum Lorentz group $SL_q(2,C)$ and its various fusion matrices, the covariant differential calculus on the quantum Minkowski space and the $\check{R}$ commutation relation for the covariant generators of quantum Lorentz group are presented.

1. Introduction

Recently much attention has been payed to the investigation on the quantum groups[1,2,3] and the Yang-Baxter equations [4]. In this talk we will report some recent progress in the research of quantum Minkowski space and quantum Lorentz group.

After giving some preliminary results on 4×4 $\check{R}$ matrix associated to $SL_q(2,C)$ and related quantities, we describe the properties of various matrices fusion from $\check{R}$ in the next section. The covariant differential calculus on quantum Minkowski space M_q and the $\check{R}$ commutation relation for the covariant generators of quantum Lorentz group are discussed in Sec. 3 and 4 respectively.

We start from the quantum $\check{R}$ matrix associated with the quantum group $SL_q(2,C)$

$$||\check{R}|| = \left(\check{R}(q)^{\alpha\beta}{}_{\gamma\delta}\right) = q^{-1/2}\begin{pmatrix} q & & & \\ & q-q^{-1} & 1 & \\ & 1 & 0 & \\ & & & q \end{pmatrix} \tag{1.1}$$

which satisfies the Yang-Baxter equation (YBE)

$$\check{R}_{12}\check{R}_{23}\check{R}_{12} = \check{R}_{23}\check{R}_{12}\check{R}_{23} \tag{1.2}$$

and the reduction equation

$$(\check{R}-\mu_1)(\check{R}-\mu_2)=0\ ,\mu_1(q)=q^{1/2},\ \mu_2(q)=-q^{-3/2}. \tag{1.3}$$

The left-acting eigenvectors $t_m(q)^{\alpha\beta}$ and $s(q)^{\alpha\beta}$, as well as the right-acting ones $\bar{t}^m(q)_{\alpha\beta}$ and $\bar{s}(q)_{\alpha\beta}$ are defined by the eigenvalue equations [5]

$$\begin{aligned}&\check{R}(q)^{\alpha\beta}{}_{\gamma\delta}\, t_m(q)^{\gamma\delta}=\mu_2\, t_m(q)^{\alpha\beta}\ ,\quad \check{R}(q)^{\alpha\beta}{}_{\gamma\delta}\, s(q)^{\gamma\delta}=\mu_1\, s(q)^{\alpha\beta}\ ,\\ &\bar{t}^m(q)_{\alpha\beta}\check{R}(q)^{\alpha\beta}{}_{\gamma\delta}=\mu_2\, \bar{t}^m(q)_{\gamma\delta}\ ,\quad \bar{s}(q)_{\alpha\beta}\check{R}(q)^{\alpha\beta}{}_{\gamma\delta}=\mu_1\, \bar{s}(q)_{\gamma\delta}\ .\end{aligned} \tag{1.4}$$

Since $\check{R}$ matrix is symmetric under the transposition, the components of left and right eigenvectors can be chosen as the same

$$\begin{aligned}&t_+(q)^{\alpha\beta}=\begin{pmatrix}1&0\\0&0\end{pmatrix}=\bar{t}^+(q)_{\alpha\beta},\quad t_3(q)^{\alpha\beta}=\begin{pmatrix}0&-q^{1/2}\\-q^{-1/2}&0\end{pmatrix}[2]^{-1/2}=\bar{t}^3(q)_{\alpha\beta}\ ,\\ &t_-(q)^{\alpha\beta}=\begin{pmatrix}0&0\\0&-1\end{pmatrix}=\bar{t}^-(q)_{\alpha\beta},\quad s(q)^{\alpha\beta}=\begin{pmatrix}0&q^{-1/2}\\-q^{1/2}&0\end{pmatrix}[2]^{-1/2}=\bar{s}(q)_{\alpha\beta},\end{aligned} \tag{1.5}$$

In this choice t's are q-symmetric and s q-anti-symmetric, i.e.,

$$t_m(q)^{\beta\alpha}=t_m(q^{-1})^{\alpha\beta},\ s(q)^{\beta\alpha}=-s(q^{-1})^{\alpha\beta}. \tag{1.6}$$

The 2-dim Levi-Civita symbols are related to the singlet eigenvectors

$$\epsilon(q)_{\alpha\beta}=-[2]^{1/2}\bar{s}(q)_{\alpha\beta}\ ,\quad \epsilon(q)^{\alpha\beta}=[2]^{1/2}s(q)^{\alpha\beta} \tag{1.7a}$$

and then

$$\epsilon(q)_{\alpha\beta}\epsilon(q)^{\beta\gamma}=\delta_\alpha{}^\gamma=\epsilon(q)^{\gamma\beta}\epsilon(q)_{\beta\alpha}\ . \tag{1.7b}$$

For latter use we group t's and s into a matrix-valued four vector

$$\begin{aligned}&t_\mu(q)=(t_0(q),\ \ t_m(q))\ ,\quad t_0(q)^{\alpha\beta}=q\ s(q)^{\alpha\beta}\ ;\\ &\bar{t}^\mu(q)=(\bar{t}^0(q),\ \ \bar{t}^m(q))\ ,\quad \bar{t}^0(q)_{\alpha\beta}=q^{-1}\ \bar{s}(q)_{\alpha\beta}\ .\end{aligned} \tag{1.8}$$

They satisfy the orthogonality condition

$$\bar{t}^\nu(q)_{\alpha\beta}\ t_\mu(q)^{\alpha\beta}\ =\delta^\nu{}_\mu\ , \tag{1.9a}$$

and the completeness relation

$$t_\mu(q)^{\alpha\beta}\ \bar{t}^\mu(q)_{\gamma\delta}=\delta^\alpha{}_\gamma\delta^\beta{}_\delta\equiv E^{\alpha\beta}{}_{\gamma\delta}\ , \tag{1.9b}$$

Define the projection operators

$$Q^{(2)}(q)^{\alpha\beta}{}_{\gamma\delta}=t_m(q)^{\alpha\beta}\bar{t}^m(q)_{\gamma\delta}\ ,\quad Q^{(1)}(q)^{\alpha\beta}{}_{\gamma\delta}=s(q)^{\alpha\beta}\bar{s}(q)_{\gamma\delta} \tag{1.10}$$

then

$$Q^{(i)}Q^{(j)} = \delta^{ij}Q^{(j)}\ , \quad Q^{(1)} + Q^{(2)} = E\ . \tag{1.11}$$

Now since $\check{R}$ can be expressed as the linear combination of $Q^{(1)}(q)$ and $Q^{(2)}(q)$

$$\check{R}(q) = \mu_2 Q^{(2)}(q) + \mu_1 Q^{(1)}(q), \tag{1.12}$$

we can obtain

$$Q^{(2)} = \frac{\check{R} - \mu_1 E}{\mu_2 - \mu_1}\ , \quad Q^{(1)} = \frac{\check{R} - \mu_2 E}{\mu_1 - \mu_2}\ , \tag{1.13}$$

from which follow the useful relations

$$Q^{(i)}_{12}\check{R}_{23}\check{R}_{12} = \check{R}_{23}\check{R}_{12}Q^{(i)}_{23}\ , \quad \check{R}_{12}\check{R}_{23}Q^{(i)}_{12} = Q^{(i)}_{23}\check{R}_{12}\check{R}_{23}\ . \tag{1.14}$$

Quantum group $SL(2,C)$ characterized by quantum matrices M and $\bar{M} = (M^{+})^{-1}$. Both of them satisfy the Yang-Baxter relations

$$\check{R}_{12}M_1M_2 = M_1M_2\check{R}_{12},\ \check{R}_{12}\bar{M}_1\bar{M}_2 = \bar{M}_1\bar{M}_2\check{R}_{12} \tag{1.15a}$$

and another mixed one [6]

$$\check{R}_{12}M_1\bar{M}_2 = \bar{M}_1M_2\check{R}_{12}\ . \tag{1.15b}$$

For quantum group $SU_q(2)$, $\bar{M} = M$, then three relations coincide into single one. Eq(1.13) allows one to get a relation similar to (1.14)

$$Q^{(i)}_{12}M_1M_2 = M_1M_2Q^{(i)}_{12},\ Q^{(i)}_{12}\bar{M}_1\bar{M}_2 = \bar{M}_1\bar{M}_2Q^{(i)}_{12}\ . \tag{1.16}$$

All these relations given above can be represented diagrammatically.

2. $\check{R}$ Matrices From Fusion

Various higher dim $\check{R}$ matrices can be obtained from fusion of $\check{R}$ in (1.1). The one associated with the quantum Lorentz group $SO_q(3,1)$ [6,7] is

$$\check{\mathcal{R}}_{(12)(34)} = \check{R}_{23}\check{R}_{12}\check{R}_{34}\check{R}^{-1}_{23},\ \check{\mathcal{R}}^{\alpha\beta,\gamma\delta}{}_{\alpha'\beta',\gamma'\delta'} = \check{R}^{\beta\gamma}{}_{bc}\check{R}^{\alpha b}{}_{\alpha' b'}\check{R}^{c\delta}{}_{c'\delta'}\check{R}^{-1}(q)^{b'c'}{}_{\beta'\gamma'} \tag{2.1}$$

And another one

$$\check{\mathbf{R}}_{(12)(34)} = \check{R}_{23}\check{R}_{12}\check{R}_{34}\check{R}_{23},\ \check{\mathbf{R}}^{\alpha\beta,\gamma\delta}{}_{\alpha'\beta',\gamma'\delta'} = \check{R}^{\beta\gamma}{}_{bc}\check{R}^{\alpha b}{}_{\alpha' b'}\check{R}^{c\delta}{}_{c'\delta'}\check{R}^{b'c'}{}_{\beta'\gamma'} \tag{2.2}$$

represents a reducible $\check{R}$ matrix [5]. The reduction can be shown by multiplying $\bar{t}^{\mu}(q)_{\alpha\beta}$ etc to convert the bispinor indices into four vector ones

$$\begin{aligned} \check{\mathbf{R}}^{\mu\nu}{}_{\kappa\lambda} &\equiv \bar{t}^{\mu}(q)_{\alpha\beta}\bar{t}^{\nu}(q)_{\gamma\delta}\check{\mathbf{R}}^{\alpha\beta,\gamma\delta}{}_{\alpha'\beta',\gamma'\delta'}t_{\kappa}(q)^{\alpha'\beta'}t_{\lambda}(q)^{\gamma'\delta'} \\ &= \check{\mathbf{R}}^{00}{}_{00} \oplus \check{\mathbf{R}}^{m0}{}_{0l} \oplus \check{\mathbf{R}}^{0n}{}_{k0} \oplus \check{\mathbf{R}}^{mn}{}_{kl}\ . \end{aligned} \tag{2.3}$$

Correspondingly the tensor product of two quantum matrices $\Omega_{(12)} = M_1 \otimes M_2$ is converted into

$$\Omega^{\mu}{}_{\nu} = \bar{t}^{\mu}(q)_{\alpha\gamma} M^{\alpha}{}_{\beta} M^{\gamma}{}_{\delta} t_{\nu}(q)^{\beta\delta} = \Omega^{0}{}_{0} \oplus \Omega^{m}{}_{n} \,. \tag{2.4}$$

The reduction comes from (1.14) and (1.16).

It is easy to show that $\check{\mathbf{R}}$ and Ω satisfy their own YBR and YBE

$$\check{\mathbf{R}}_{(12)(34)}\Omega_{(12)}\Omega_{(34)} = \Omega_{(12)}\Omega_{(34)}\check{\mathbf{R}}_{(12)(34)},$$

$$\check{\mathbf{R}}_{(12)(34)}\check{\mathbf{R}}_{(34)(56)}\check{\mathbf{R}}_{(12)(34)} = \check{\mathbf{R}}_{(34)(56)}\check{\mathbf{R}}_{(12)(34)}\check{\mathbf{R}}_{(34)(56)}. \tag{2.5}$$

Or equivalently, their reduction form $(m, n, \cdots = +, 3, -)$

$$\check{\mathbf{R}}^{mn}{}_{kl}\Omega^{k}{}_{i}\Omega^{l}{}_{j} = \Omega^{m}{}_{k}\Omega^{n}{}_{l}\check{\mathbf{R}}^{kl}{}_{ij},$$

$$\check{\mathbf{R}}^{mn}{}_{m'n'}\check{\mathbf{R}}^{n'l}{}_{n''k}\check{\mathbf{R}}^{m'n''}{}_{ij} = \check{\mathbf{R}}^{nl}{}_{n'l'}\check{\mathbf{R}}^{mn'}{}_{in''}\check{\mathbf{R}}^{n''l'}{}_{jk} \,. \tag{2.6}$$

Comparing to the standard form [2], one can see that [5] $\check{\mathbf{R}}^{mn}{}_{kl}$ is nothing but the $\check{R}$ matrix associated with the quantum deformation of $SO(3)$ with parameter q^2. Similar to the eigenequation (1.4), $\check{\mathbf{R}}^{mn}{}_{kl}$ has a singlet eigenvector

$$\mathbf{g}(q^2)^{mn} = (\mathbf{g}(q^2)^{+-}, \mathbf{g}(q^2)^{00}, \mathbf{g}(q^2)^{-+}) = (q^{-1}, 1, q) = \mathbf{g}(q^2)_{kl} \tag{2.7}$$

with the property

$$\mathbf{g}(q^2)_{mn}\Omega^{m}{}_{k}\Omega^{n}{}_{l} = \mathbf{g}(q^2)_{kl}, \;\; \Omega^{m}{}_{k}\Omega^{n}{}_{l}\mathbf{g}(q^2)^{kl} = \mathbf{g}(q^2)^{mn} \,. \tag{2.8}$$

Therefore $\mathbf{g}(q^2)_{mn}$ can be identified as the deformed Euclidean metric in 3-dim quantum orthogonal space.

In a similar manner $\check{\mathcal{R}}_{(12)(34)}$ in (2.1) and the product $\Lambda_{12} = M_1 \otimes \bar{M}_2$ satisfy the YBR and YBE exactly the same as those in (2.5). They can be converted to the vectorial form as in Eqs (2.3) and (2.4). The 16×16 matrix $\check{\mathcal{R}}^{\mu\nu}{}_{\kappa\lambda}$ has three different eigenvalues

$$\lambda_2(q) = q, \;\; \lambda_1(q) = -q^{-1}, \;\; \lambda_0(q) = q^{-3}, \tag{2.9}$$

corresponding to a nonet, a sextet and a singlet respectively. The singlet eigenvector

$$g(q)^{\mu\nu} = (g^{+-}, g^{33}, g^{30}, g^{03}, g^{00}, g^{-+}) = (-q^{-1}, -1, 0, 0, 1, -q) = g(q)_{\mu\nu} \tag{2.10}$$

satisfy [7]

$$g(q)^{\mu\lambda}g(q)_{\lambda\nu} = \delta^{\mu}{}_{\nu} = g(q)_{\nu\lambda}g(q)^{\lambda\mu} \tag{2.11}$$

$$g(q)_{\mu\nu}\Lambda^{\mu}{}_{\kappa}\Lambda^{\nu}{}_{\lambda} = g(q)_{\kappa\lambda}, \;\; \Lambda^{\mu}{}_{\kappa}\Lambda^{\nu}{}_{\lambda}g(q)^{\kappa\lambda} = g(q)^{\mu\nu}. \tag{2.12}$$

This implies that $\Lambda^{\mu}{}_{\nu}$ is the q-deformed Lorentz transformation and $g(q)_{\mu\nu}$ the q-analogue of the Minkowski metric. The sextet eigenvector $u_{ms}(q)^{\mu\nu}$ $(m = +, 3, -, s =$

L, R) and nonet $w_{mn}(q)^{\mu\nu}$ ($m, n = +, 3, -$) can also be constructed from the concrete form $\check{\mathcal{R}}^{\mu\nu}{}_{\kappa\lambda}$. u_{ms} is q-antisymmetric, and the metric q-symmetric

$$u_{ms}(q)^{\nu\mu} = -u_{ms}(q^{-1})^{\mu\nu}, \quad g(q)^{\nu\mu} = g(q^{-1})^{\mu\nu}. \tag{2.13}$$

Now since $\check{\mathcal{R}}$ is not symmetric under the transformation, its right eigenvectors $\bar{u}^{ms}(q)_{\mu\nu}$ and $\bar{w}^{mn}(q)_{\mu\nu}$ have components different to their left counterparts. But still we can choose them all as orthonormal and complete set. Projection operators can be defined [7]

$$\begin{aligned} Q^{(0)}(q)^{\mu\nu}{}_{\kappa\lambda} &= g(q)^{\mu\nu}\bar{g}(q)_{\kappa\lambda}[2]^{-2}, \\ Q^{(1)}(q)^{\mu\nu}{}_{\kappa\lambda} &= u_{ms}(q)^{\mu\nu}\bar{u}^{ms}(q)_{\kappa\lambda}, \\ Q^{(2)}(q)^{\mu\nu}{}_{\kappa\lambda} &= w_{mn}(q)^{\mu\nu}\bar{w}^{mn}(q)_{\kappa\lambda} \end{aligned} \tag{2.14}$$

with

$$Q^{(i)}Q^{(j)} = \delta^{ij}Q^{(j)}, \quad Q^{(0)} + Q^{(1)} + Q^{(2)} = E\,. \tag{2.15}$$

As a matter of fact, we can construct various different "fusion" matrices from quartic product of basic $\check{R}(q)$. Let us define

$$\check{R}_{(12)(34)}(N) = \check{R}(a,b,c,d) = \check{R}^{(a)}_{(23)}\check{R}^{(b)}_{12}\check{R}^{(c)}_{34}\check{R}^{(d)}_{23} \tag{2.16}$$

with $a, b, c, d = \pm$, $\check{R}^{(+)} = \check{R}, \check{R}^{(-)} = \check{R}^{-1}$. There are totally 16 of them, $N = 1$ to 16.

$$\begin{aligned} &\check{R}(1) = \check{R}(++++), && \check{R}(2) = \check{R}(+++-), \\ &\check{R}(3) = \check{R}(++-+), && \check{R}(4) = \check{R}(++--), \\ &\check{R}(5) = \check{R}(+-++), && \check{R}(6) = \check{R}(+-+-), \\ &\check{R}(7) = \check{R}(+--+), && \check{R}(8) = \check{R}(+---), \\ &\check{R}(9) = \check{R}(-+++), && \check{R}(10) = \check{R}(-++-), \\ &\check{R}(11) = \check{R}(-+-+), && \check{R}(12) = \check{R}(-+--), \\ &\check{R}(13) = \check{R}(--++), && \check{R}(14) = \check{R}(--+-), \\ &\check{R}(15) = \check{R}(---+), && \check{R}(16) = \check{R}(----) \end{aligned} \tag{2.17}$$

We see immediately that $\check{\mathbf{R}}_{(12)(34)} = \check{R}(1)$, $\check{\mathcal{R}}_{(13)(24)} = \check{R}(2)$. Pairs of these matrices are inverse of each other

$$\begin{aligned} &\check{R}(1)^{-1} = \check{R}(16), \quad \check{R}(2)^{-1} = \check{R}(8), \quad \check{R}(3)^{-1} = \check{R}(14), \quad \check{R}(4)^{-1} = \check{R}(6), \\ &\check{R}(5)^{-1} = \check{R}(12), \quad \check{R}(7)^{-1} = \check{R}(10), \quad \check{R}(9)^{-1} = \check{R}(15), \quad \check{R}(11)^{-1} = \check{R}(13), \end{aligned} \tag{2.18}$$

and some of them are connected by transposition

$$\check{R}(2)^t = \check{R}(9), \quad \check{R}(4)^t = \check{R}(11), \quad \check{R}(6)^t = \check{R}(13), \quad \check{R}(8)^t = \check{R}(15) \tag{2.19}$$

and others are symmetrical. These 16 matrices fall into six groups, namely

$$(1,16);\ (3,14);\ (5,12);\ (7,10);\ (2,8,9,15);\ (4,6,11,13).$$

A further look shows that some of the matrices have same topology, e.g., $\check{R}(2)$, $\check{R}(14)$, $\check{R}(9)$ and $\check{R}(12)$ have same twine but different direction diagrammatically. They may represent relevant processes, processes with cross symmetry. We have already noticed that $\check{R}(1)$ and $\check{R}(16)$ describe $SO_{q^2}(3)$ quantum group, and $\check{R}(2)$ and $\check{R}(8)$ quantum group $SO_q(3,1)$. One may imagine that the relevant matrices $\{\check{R}(9),\check{R}(15)\}$, $\{\check{R}(3),\check{R}(14)\}$ and $\{\check{R}(5),\check{R}(12)\}$ also have something to do with quantum Lorentz group.

Of all these 16 matrices, $\check{R}(7)$ and $\check{R}(10)$ are special. All others but these two satisfy their own YBE, i.e., for $N \neq 7,\ 10$

$$\check{R}_{(12)(34)}(N)\check{R}_{(34)(56)}(N)\check{R}_{(12)(34)}(N) = \check{R}_{(34)(56)}(N)\check{R}_{(12)(34)}(N)\check{R}_{(34)(56)}(N)\,. \tag{2.20}$$

We have another kind of generalized YBE for $N \neq N'$, i.e.,

$$\check{R}_{(12)(34)}(N)\check{R}_{(34)(56)}(N)\check{R}_{(12)(34)}(N') = \check{R}_{(34)(56)}(N')\check{R}_{(12)(34)}(N)\check{R}_{(34)(56)}(N) \tag{2.21a}$$

$$\check{R}_{(12)(34)}(N)\check{R}_{(34)(56)}(N')\check{R}_{(12)(34)}(N') = \check{R}_{(34)(56)}(N')\check{R}_{(12)(34)}(N')\check{R}_{(34)(56)}(N) \tag{2.21b}$$

Various combinations of N and N' are allowed. And the ones with N, N′ =2,4,6,8 play an important role in discussing the quantum Lorentz group.

It can be easily shown that $\check{R}(4)$, as well as $\check{R}(2) = \check{\mathcal{R}}$, satisfy the same YBR with the quantum Lorentz matrix $\Lambda_{12} = M_1 \otimes \bar{M}_2$

$$\check{R}(N)_{(12)(34)}\Lambda_{(12)}\Lambda_{(34)} = \Lambda_{(12)}\Lambda_{(34)}\check{R}_{(12)(34)}(N), \quad for\ N = 2,4,6,8 \tag{2.22}$$

The 16× 16 matrix $\check{R}(4)^{\mu\nu}{}_{\kappa\lambda}$ has three different eigenvalues, $-q^2$, 1 and $-q^{-2}$. It can be shown that the eigenvectors g, u and w appearing in (2.10) and (2.13) are also the eigenvectors of $\check{R}(4)$

$$\begin{aligned}
\check{R}(4)^{\mu\nu}{}_{\kappa\lambda}g(q)^{\kappa\lambda} &= g(q)^{\mu\nu}\\
\check{R}(4)^{\mu\nu}{}_{\kappa\lambda}u_{mL}(q)^{\kappa\lambda} &= -q^2u_{mL}(q)^{\mu\nu}\\
\check{R}(4)^{\mu\nu}{}_{\kappa\lambda}u_{mR}(q)^{\kappa\lambda} &= -q^{-2}u_{mR}(q)^{\mu\nu}\\
\check{R}(4)^{\mu\nu}{}_{\kappa\lambda}w_{mn}(q)^{\kappa\lambda} &= w_{mn}(q)^{\mu\nu}
\end{aligned} \tag{2.23}$$

So if we decompose the q-antisymmetric projection operator $Q^{(1)}(q)$ into two parts

$$Q^{(1)}(q) = Q^{(1L)}(q) + Q^{(1R)}(q)$$

$$Q^{(1L)}(q)^{\mu\nu}{}_{\kappa\lambda} = u_{mL}(q)^{\mu\nu}\bar{u}^{mL}(q)_{\kappa\lambda}, \quad Q^{(1R)}(q)^{\mu\nu}{}_{\kappa\lambda} = u_{mR}(q)^{\mu\nu}\bar{u}^{mR}(q)_{\kappa\lambda}, \tag{2.24}$$

we have

$$\check{R}(4) = Q^{(0)} - q^2Q^{(1L)} - q^{-2}Q^{(1R)} + Q^{(2)} \tag{2.25a}$$

whereas

$$\check{R}(2) = q^{-3}Q^{(0)} - q^{-1}Q^{(1L)} - q^{-1}Q^{(1R)} + qQ^{(2)} \tag{2.25b}$$

All these eigenvectors can also be constructed by the basic $\check{R}$ matrix (1.1) and basic eigenvectors $t_\mu(q)^{\alpha\beta}$ and $s(q)^{\alpha\beta}$. The results are

$$\begin{aligned} g(q)^{\mu\nu} &= \bar{t}(q)_{\alpha\dot{\beta}}\, \bar{t}(q)_{\gamma\dot{\delta}}\, q^{1/2}\, \check{R}(q)^{\dot{\beta}\gamma}{}_{\beta\dot{\gamma}}\, \epsilon(q)^{\alpha\beta}\epsilon(q)^{\dot{\gamma}\delta}, \\ g(q)_{\mu\nu} &= \epsilon(q)_{\alpha\beta}\, \epsilon(q)_{\dot{\gamma}\dot{\delta}}\, q^{-1/2}\, \check{R}^{-1}(q)^{\beta\dot{\gamma}}{}_{\dot{\beta}\gamma}\, t_\mu(q)^{\alpha\dot{\beta}}\, t_\nu(q)^{\gamma\dot{\delta}}, \\ u_{kL}(q)^{\mu\nu} &= \bar{t}(q)_{\alpha\dot{\beta}}\, \bar{t}(q)_{\gamma\dot{\delta}}\, q^{1/2}\, \check{R}(q)^{\dot{\beta}\gamma}{}_{\beta\dot{\gamma}}\, t_k(q)^{\alpha\beta}\, s(q)^{\dot{\gamma}\dot{\delta}}, \\ u_{kR}(q)^{\mu\nu} &= \bar{t}(q)_{\alpha\dot{\beta}}\, \bar{t}(q)_{\gamma\dot{\delta}}\, q^{1/2}\, \check{R}(q)^{\dot{\beta}\gamma}{}_{\beta\dot{\gamma}}\, s(q)^{\alpha\beta}\, t_k(q)^{\dot{\gamma}\dot{\delta}}, \\ w_{mn}(q)^{\mu\nu} &= \bar{t}(q)_{\alpha\dot{\beta}}\, \bar{t}(q)_{\gamma\dot{\delta}}\, q^{1/2}\, \check{R}(q)^{\dot{\beta}\gamma}{}_{\beta\dot{\gamma}}\, t_m(q)^{\alpha\beta}\, t_n(q)^{\dot{\gamma}\dot{\delta}}. \end{aligned} \tag{2.26}$$

Various properties of g, u_{ks} and w_{mn} can be obtained by these basic relations.

3. Covariant Differential Calculus on M_q

The covariant differential calculus on the quantum plane was first discussed by Wess and Zumino (for the A type case) [8] and then generalized to the cases of quantum orthogonal plane (B,D type) [9,10] or quantum symplectic plane (C type) [10]. For quantum spinor u^α, their components are noncommutative. The same is true for their derivatives $\partial_\alpha = \frac{\partial}{\partial u^\alpha}$

$$u^1\, u^2 = q\, u^2\, u^1, \quad \partial_1\partial_2 = q^{-1}\partial_2\partial_1\ . \tag{3.1}$$

Raising the indices by the Levi-Civita symbol $\partial^\alpha = \epsilon(q)^{\alpha\beta}\partial_\beta$, these two relations can be written as the vanishing of their q-anti-symmetric combination, i.e.,

$$\epsilon(q)_{\alpha\beta}u^\alpha u^\beta = \epsilon(q)_{\alpha\beta}\partial^\alpha\partial^\beta = 0\ . \tag{3.2}$$

The cross commutation relations between the coordinates u^α and derivatives ∂^α determined by the consistency [8] can be put into the form

$$\partial^\alpha u^\beta + \mu_1(q)\check{R}(q)^{\alpha\beta}{}_{\gamma\delta}u^\gamma\partial^\delta = \epsilon(q)^{\alpha\beta} \tag{3.3}$$

or, equivalently

$$\begin{aligned} \bar{t}^m(q)_{\alpha\beta}[\partial^\alpha u^\beta + \mu_1(q)\mu_2(q)u^\alpha\partial^\beta] &= 0, \\ \epsilon(q)_{\alpha\beta}[\partial^\alpha u^\beta + \mu_1(q)u^\alpha\partial^\beta] = \epsilon(q)_{\alpha\beta}\epsilon(q)^{\alpha\beta} &= \mp[2]\ . \end{aligned} \tag{3.4}$$

And from the relation of q-antisymmetric projection operator in (1.13), (3.2) turn out to be

$$[E - \mu_2(q)\check{R}(q)]^{\alpha\beta}{}_{\gamma\delta}\, u^\gamma u^\delta = [E - \mu_2(q)\check{R}(q)]^{\alpha\beta}{}_{\gamma\delta}\, \partial^\gamma\partial^\delta = 0. \tag{3.5}$$

In a similar way for quantum Minkowski coordinates x^μ and their derivatives $\partial_\mu = \frac{\partial}{\partial x^\mu}$, we define $\partial^\mu = g(q)^{\mu\nu}\partial_\nu$. Again their q-antisymmetric combination should vanish.

$$\bar{u}^{ms}(q)_{\mu\nu}x^\mu x^\nu = \bar{u}^{ms}(q)_{\mu\nu}\partial^\mu\partial^\nu = 0\ . \tag{3.6}$$

The consistency condition yields relations between coordinates and derivatives [7]

$$\begin{aligned} &g(q)_{\mu\nu}[\partial^\mu x^\nu + \lambda_1^{-1}(q)\lambda_0^{-1}(q)x^\mu\partial^\nu] = g(q)_{\mu\nu}g(q)^{\mu\nu} = [2]^2 , \\ &\bar{u}^{ms}(q)_{\mu\nu}[\partial^\mu x^\nu + \lambda_1^{-2}(q)x^\mu\partial^\nu] = 0 \\ &\bar{w}^{mn}(q)_{\mu\nu}[\partial^\mu x^\nu + \lambda_1^{-1}(q)\lambda_2^{-1}(q)x^\mu\partial^\nu] = 0 \end{aligned} \tag{3.7}$$

which can be put into the form similar to (3.3)

$$\partial^\mu x^\nu + \lambda_1^{-1}(q)\check{\mathcal{R}}^{-1}(q)^{\mu\nu}{}_{\kappa\lambda}x^\kappa\partial^\lambda = g(q)^{\mu\nu} . \tag{3.8}$$

But now the relations in Eq(3.6) can not be put into the form similar to (3.5) with linear dependence on the matrix $\check{\mathcal{R}} = \check{R}(2)$, since the q-antisymmetric projection operator for Lorentz group is quadratic but not linear in $\check{\mathcal{R}}$. However they can be put into the form

$$[E^{\mu\nu}{}_{\kappa\lambda} - \check{R}(4,q)^{\mu\nu}{}_{\kappa\lambda}]x^\kappa x^\lambda = 0, \quad [E^{\mu\nu}{}_{\kappa\lambda} - \check{R}(4,q)^{\mu\nu}{}_{\kappa\lambda}]\partial^\kappa\partial^\lambda = 0 \tag{3.9}$$

from (2.25a). The consistency among (3.9) and (3.8) can be checked by making use of (2.21) relations for $N = 2, 4, 6, 8$. The details of the differential calculus can be found in our recent paper [7].

4. Covariant Generators of Quantum Lorentz Group

Now similar to the classical case we can define the generators of the $SL_q(2,C)$ in the quantum spinor space as [11]

$$L^{\alpha\beta} = u^\alpha\partial^\beta . \tag{4.1}$$

Using the commutation relations given in (3.4) and (3.5), we can obtain

$$\begin{aligned} &L^{\alpha\beta}L^{\gamma\delta} - q^{-1}\check{\mathcal{R}}^{\alpha\beta,\gamma\delta}{}_{\alpha'\beta',\gamma'\delta'}L^{\alpha'\beta'}L^{\gamma'\delta'} \\ &= q^{1/2}\ \check{R}^{-1}(q)^{\alpha\beta}{}_{\alpha'\beta'}\epsilon(q)^{\beta'\gamma'}\check{R}^{-1}(q)^{\gamma\delta}{}_{\gamma'\delta'}\check{R}^{\alpha'\delta'}{}_{\sigma\tau}L^{\sigma\tau} + \epsilon(q)^{\beta\gamma}L^{\alpha\delta} \end{aligned} \tag{4.2}$$

by a tedious but straightforward derivation. Then by changing the bispinor indices (α,β) into 4-vector index μ and defining

$$L^\mu = \bar{t}^\mu(q)_{\alpha\beta}L^{\alpha\beta} \tag{4.3}$$

we obtain

$$\begin{aligned} &L^\mu L^\nu - q^{-1}\check{\mathcal{R}}(q)^{\mu\nu}{}_{\kappa\lambda}L^\kappa L^\lambda = f^{\mu\nu}_{\rho}L^\rho \\ &= [2]^{1/2}[u_{kL}(q)^{\mu\nu} + u_{kR}(q)^{\mu\nu}]L^k + [2]^{-1/2}(q-q^{-1})g(q)^{\mu\nu}L^0 . \end{aligned} \tag{4.4}$$

In a similar way, the generators in the conjugate spinor space are defined as

$$\bar{L}^{\dot{\alpha}\dot{\beta}} = \bar{u}^{\dot{\alpha}}\bar{\partial}^{\dot{\beta}} \tag{4.5}$$

which satisfy relations similar to those in Eq(4.2)

$$\begin{aligned} &\bar{L}^{\dot{\alpha}\dot{\beta}}\bar{L}^{\dot{\gamma}\dot{\delta}} - q^{-1}\check{\mathcal{R}}^{\dot{\alpha}\dot{\beta},\dot{\gamma}\dot{\delta}}{}_{\dot{\alpha}'\dot{\beta}',\dot{\gamma}'\dot{\delta}'}\bar{L}^{\dot{\alpha}'\dot{\beta}'}\bar{L}^{\dot{\gamma}'\dot{\delta}'} \\ &= q^{1/2}\,\check{R}^{-1}(q)^{\dot{\alpha}\dot{\beta}}{}_{\dot{\alpha}'\dot{\beta}'}\epsilon(q)^{\dot{\beta}'\dot{\gamma}'}\check{R}^{-1}(q)^{\dot{\gamma}\dot{\delta}}{}_{\dot{\gamma}'\dot{\delta}'}\check{R}^{\dot{\alpha}'\dot{\delta}'}{}_{\dot{\sigma}\dot{\tau}}\bar{L}^{\dot{\sigma}\dot{\tau}} + \epsilon(q)^{\dot{\beta}\dot{\gamma}}\bar{L}^{\dot{\alpha}\dot{\delta}}\,. \end{aligned} \tag{4.6}$$

Then if we introduce

$$\bar{L}^{\mu} = \bar{t}^{\mu}(q^{-1})_{\dot{\beta}\dot{\alpha}}\bar{L}^{\dot{\alpha}\dot{\beta}} \tag{4.7}$$

we have

$$\begin{aligned} &\bar{L}^{\mu}\bar{L}^{\nu} - q^{-1}\check{\mathcal{R}}^{\mu\nu}{}_{\kappa\lambda}\bar{L}^{\kappa}\bar{L}^{\lambda} = f^{\mu\nu}{}_{\rho}\bar{L}^{\rho} \\ &= [2]^{1/2}[u_{kL}(q)^{\mu\nu} + u_{kR}(q)^{\mu\nu}]\bar{L}^{k} + [2]^{-1/2}(q-q^{-1})g(q)^{\mu\nu}\bar{L}^{0} \end{aligned} \tag{4.8}$$

The same is true for replacing $q^{-1}\check{\mathcal{R}} = q^{-1}\check{R}(2)$ by $\check{R}(4)$ in Eqs. (4.2), (4.4), (4.6), (4.8). In above definition, the conjugate spinor $\bar{u}_{\dot{\alpha}} \equiv (u^{\alpha})^{*}$ transforms according to the hermitian conjugate (complex transpose) of the quantum matrix M.

$$u^{\alpha} \longrightarrow M(q)^{\alpha}{}_{\beta}u^{\beta}\;, \bar{u}_{\dot{\alpha}} \longrightarrow \bar{u}_{\dot{\beta}}M(q)^{+\dot{\beta}}{}_{\dot{\alpha}}\;. \tag{4.9}$$

The conjugate derivative is $\frac{\partial}{\partial \bar{u}_{\dot{\alpha}}} = \partial^{\dot{\alpha}}$. Raising the dotted index of $\bar{u}_{\dot{\alpha}}$, we see

$$\bar{u}^{\dot{\alpha}} \equiv \bar{u}_{\dot{\beta}}\epsilon(q)^{\dot{\beta}\dot{\alpha}} \longrightarrow \bar{M}^{\dot{\alpha}}{}_{\dot{\gamma}}\,\bar{u}^{\dot{\gamma}}\,. \tag{4.10}$$

The commutation relation among $\bar{u}^{\dot{\alpha}}$ and $\bar{\partial}^{\dot{\alpha}}$ are found to be the same as those for u^{α} and ∂^{α} in (3.3)and (3.5):

$$\begin{aligned} \bar{u}^{\dot{\alpha}}\bar{u}^{\dot{\beta}} &= q^{-1/2}\check{R}^{\dot{\alpha}\dot{\beta}}{}_{\dot{\gamma}\dot{\delta}}\bar{u}^{\dot{\gamma}}\bar{u}^{\dot{\delta}}\;, \\ \bar{\partial}^{\dot{\alpha}}\bar{\partial}^{\dot{\beta}} &= q^{-1/2}\check{R}^{\dot{\delta}\dot{\gamma}}{}_{\dot{\beta}\dot{\alpha}}\bar{\partial}^{\dot{\gamma}}\bar{\partial}^{\dot{\delta}} \\ \bar{\partial}^{\dot{\alpha}}\bar{u}^{\dot{\beta}} &= \epsilon(q)^{\dot{\alpha}\dot{\beta}} + q^{-3/2}\check{R}(q)^{\dot{\alpha}\dot{\beta}}{}_{\dot{\gamma}\dot{\delta}}\bar{u}^{\dot{\gamma}}\bar{\partial}^{\dot{\delta}}\;. \end{aligned} \tag{4.11}$$

And the relations between undotted quantities with dotted ones are

$$\begin{aligned} u^{\alpha}\bar{u}^{\dot{\alpha}} &= q^{1/2}\check{R}^{\alpha\dot{\alpha}}{}_{\dot{\beta}\beta}\bar{u}^{\dot{\beta}}u^{\beta}, \quad & \partial^{\alpha}\bar{\partial}^{\dot{\alpha}} &= q^{1/2}\check{R}^{\alpha\dot{\alpha}}{}_{\dot{\beta}\beta}\bar{\partial}^{\dot{\beta}}\partial^{\beta}, \\ \partial^{\alpha}\bar{u}^{\dot{\alpha}} &= q^{-1/2}\check{R}^{\alpha\dot{\alpha}}{}_{\dot{\beta}\beta}\bar{u}^{\dot{\beta}}\partial^{\beta}, \quad & u^{\alpha}\bar{\partial}^{\dot{\alpha}} &= q^{1/2}\check{R}^{\alpha\dot{\alpha}}{}_{\dot{\beta}\beta}\bar{\partial}^{\dot{\beta}}u^{\beta}, \end{aligned} \tag{4.12}$$

from which we obtain

$$L^{\alpha\beta}\bar{L}^{\dot{\alpha}\dot{\beta}} = \check{\mathbf{R}}^{\alpha\beta,\dot{\alpha}\dot{\beta}}{}_{\dot{\gamma}\dot{\delta},\gamma\delta}\bar{L}^{\dot{\gamma}\dot{\delta}}L^{\gamma\delta} \tag{4.13}$$

Here notice that, the $\check{R}$ matrix is the $\check{\mathbf{R}} = \check{R}(1)$ appearing in (2.4). Transfer to the vector index we see

$$\begin{aligned}\check{\mathbf{R}}(q)^{\mu\nu}{}_{\kappa\lambda} &= \check{\mathbf{R}}^{00}{}_{00} \oplus \check{\mathbf{R}}^{m0}{}_{0l} \oplus \check{\mathbf{R}}^{0n}{}_{k0} \oplus \check{\mathbf{R}}^{mn}{}_{kl} \\ &= 1 \oplus \delta^{m}{}_{l} \oplus \delta_{k}{}^{n} \oplus \check{\mathbf{R}}(q)^{mn}{}_{kl}\end{aligned} \tag{4.14}$$

$\check{\mathbf{R}}^{mn}{}_{kl}$ is the $\check{R}$ matrix for $SO_{q^2}(3)$.

More explicitly, (4.4) can be read as

$$\begin{aligned}&[L^0, L^m] = 0 \ , \\ &q(L^3 - L^0)L^+ - q^{-1}L^+(L^3 - L^0) = q[2]^{1/2}L^+ \ , \\ &qL^-(L^3 - L^0) - q^{-1}(L^3 - L^0)L^- = q[2]^{1/2}L^- \ , \\ &L^+L^- - L^-L^+ + (q - q^{-1})L^3(L^3 - L^0) = q[2]^{1/2}L^3 \ , \\ &-q^{-1}L^+L^- - L^3L^3 + L^0L^0 - qL^-L^+ = q^2[2]^{1/2}L^0 \ .\end{aligned} \tag{4.15}$$

and similar for $\bar{L}^0$ and $\bar{L}^m$. And similarly, (4.13) gives

$$[L^0, \bar{L}^0] = [L^n, \bar{L}^0] = [L^0, \bar{L}^m] = 0 \tag{4.16}$$

$$\begin{aligned}&L^+\bar{L}^+ = q^2\bar{L}^+L^+ \ , \\ &L^+\bar{L}^3 = \bar{L}^3L^+ \ , \\ &L^+\bar{L}^- = q^{-2}\bar{L}^-L^+ \ , \\ &L^3\bar{L}^+ = \bar{L}^+L^3 + (q^2 - q^{-2})\bar{L}^3L^+ \ , \\ &L^3\bar{L}^3 = -q^{-1}(q^2 - q^{-2})\bar{L}^-L^+ + \bar{L}^3L^3 \ , \\ &L^3\bar{L}^- = \bar{L}^-L^3 \ , \\ &L^-\bar{L}^+ = q^{-2}\bar{L}^+L^- - q^{-1}(q^2 - q^{-2})\bar{L}^3L^3 + (1 - q^{-2})(q^2 - q^{-2})\bar{L}^-L^+ \ , \\ &L^-\bar{L}^3 = (q^2 - q^{-2})\bar{L}^-L^3 + \bar{L}^3L^- \ , \\ &L^-\bar{L}^- = q^2\ \bar{L}^-L^- \ .\end{aligned} \tag{4.17}$$

It is easily seen that in the limit $q \to 1$, $\bar{L}^\mu$ and L^ν became two sets of commutative angular momentum with L^0 and $\bar{L}^0$ being their Casimir operators.

All the cross commutation relations given above are explicitly covariant and take the $\check{R}$ commutation form just as those proposed by Woronowicz [3]. All the structure constants are comprised by the relevant $\check{R}$ matrices and their character quantities. The consistency are easily to prove.

Along the same line we can discuss the generators of quantum Poincare group.

References

[1] V.G. Drinfeld, Quantum groups,*Proccedings of ICM*, 1986, Vol.1,798;
M. Jimbo, *Lett. Math. Phys.* **10**(1986)63, **11**(1986)247, *Commun. Math. Phys.* **102**(1986)537;

[2] N.Yu. Reshetikhin, L.A. Takhtajan and L.D. Faddeev, *Leningrad Math. J.* **1**(1990)193.

[3] S.L. Woronowicz, *Publ. RIMS*, Kyoto Univ., **23**(1987)117, *Commun. Math. Phys.* **111**(1987)613; **122**(1989)129.

[4] C.N. Yang, *Phys. Rev. Lett.* **19**(1967)1312; R.J. Baxter, *Exactly Solved Models*, Academic Press, London, 1982.

[5] X.C. Song, *J. Phys.* **A25**(1992)2929.

[6] P. Podles, S.L. Woronowicz, *Commun. Math. Phys.* **130**(1990)381; U. Carow-Watamura, M. Schlieker, M. Scholl and S. Watamura, *Z. Phys.* **C48**(1991)159.

[7] X.C. Song, *Z. Phys.* **C55**(1992)417.

[8] J. Wess and B. Zumino, *Nucl. Phys.* B(Proc. Suppl.) **18B**(1990)302.

[9] U. Carow-Watamura, M. Schlieker and S. Watamura, *Z. Phys.* **C49**(1991)439;

[10] X.C. Song and L. Liao, *J. Phys.* **A25**(1992)623.

[11] X.C. Song, Proc. CCAST Workshop: *Quantum Groups and Low Dimensional Field Theory*, March, 1992, Beijing, p46.

SIMPLE APPROACH TO THERMAL BETHE ANSATZ

H.J. DE VEGA

Laboratoire de Physique Théorique et Hautes Energies * , *Paris* ‡

ABSTRACT

We report on a new approach to the calculation of thermodynamic functions for crossing-invariant models solvable by Bethe Ansatz. In the case of the XXZ Heisemberg chain we derive, for arbitrary values of the anysotropy, a **single** non–linear integral equation from which the free energy can be exactly calculated.These equations are shown to be equivalent to an infinite set of algebraic equations of Bethe type which provide alternatively the thermodinamics. The high–temperature expansion follows in a sistematic and relatively simple way from our non-linear integral equations. For low temperatures we obtain the correct central charge and predict the analytic structure of the full expansion around $T = 0$. Furthermore, we derive a single non-linear integral equation describing the finite–size ground–state energy of the Sine–Gordon quantum field theory.

The computation of thermodynamical functions for integrable models started with the seminal works of C.N. Yang and C.P. Yang [1], of M. Takahashi [2] and M. Gaudin [3]. In Refs. [2,3] the free energy of the Heisenberg spin chain is written in terms of the solution of an infinite set of coupled nonlinear integral equations, derived on the basis of the so–called "string hypothesis". In ref.[4], we proposed a simpler way to solve the thermodynamics by means of a single, rigorously derived, nonlinear integral equation. This method can be applied to a wide class of models solvable by Bethe Ansatz (see ref.[5]).

* *Laboratoire Associé au CNRS UA 280*

‡ *mail address:*
L.P.T.H.E., Tour 16 1er étage, Université Paris VI,
4 Place Jussieu, 75252, Paris cedex 05, France

The XXZ hamiltonian for a periodic chain with $2L$ sites (we assume periodic boundary conditions but generalization to other b.c. is possible) takes the form

$$H_{XXZ} = -J \sum_{n=1}^{2L} \left[\sigma_n^x \sigma_{n+1}^x + \sigma_n^y \sigma_{n+1}^y - \cos\gamma \left(\sigma_n^z \sigma_{n+1}^z + 1 \right) \right] \tag{1}$$

The anisotropy parameter γ is assumed here to lay in the real interval $(0, \pi)$ characteristic of the gapless regime. H_{XXZ} is related to the diagonal–to–diagonal transfer matrix $T_L(\theta)$ of the six–vertex model by

$$T_L(\theta) \overset{\theta\to 0}{=} 1 - \frac{\theta}{2J\sin\gamma} H_{XXZ} + O(\theta^2) \tag{2}$$

where

$$T_L(\theta) = R_{12}R_{34}\dots R_{2L-1\,2L}R_{23}R_{45}\dots R_{2L\,1} \tag{3}$$

and

$$\begin{aligned} R_{nm} &= \frac{a+c}{2} + \frac{a-c}{2}\sigma_n^z\sigma_m^z + \frac{b}{2}(\sigma_n^x\sigma_{n+1}^x + \sigma_n^y\sigma_{n+1}^y) \\ a &= \frac{\sin(\gamma-\theta)}{\sin\gamma}, \quad b = \frac{\sin\theta}{\sin\gamma}, \quad c = 1 \end{aligned} \tag{4}$$

θ is known as spectral parameter while a, b, and c are the conventional six–vertex Boltzmann weights.

The free energy per site is defined as usual

$$f(\beta) = -\frac{1}{\beta} \lim_{L\to\infty} \frac{1}{2L} \log \left[\mathrm{Tr} \left(e^{-\beta H_{XXZ}} \right) \right] \tag{5}$$

and from eq. (2) we read

$$e^{-\beta H_{XXZ}} = \lim_{N\to\infty} \left[T_L(2\tilde{\beta}/N) \right]^N$$

where $\tilde{\beta} = \beta J \sin\gamma$. Hence we can write

$$f(\beta) = -\frac{1}{\beta} \lim_{L\to\infty} \frac{1}{2L} \lim_{N\to\infty} \log Z_{LN}(2\tilde{\beta}/N) \tag{6}$$

where $Z_{LN}(\theta) \equiv \mathrm{Tr}\,[T_L(\theta)]^N$ is the six–vertex partition function on a periodic diagonal lattice with $L \times N$ sites. The two limits in eq. (6) cannot be interchanged since the degeneracy of $T_L(0) = 1$, that is 2^{2L}, is strongly L–dependent. However, the crossing invariance of the six–vertex R–matrix implies that under a rotation by $\pi/2$ of the entire lattice plus the substitution $\theta \to \gamma - \theta$, the numerical value of the partition function does not change (see *e.g.* ref.6). Therefore $Z_{LN}(\theta) = Z_{NL}(\gamma - \theta)$ and

$$f(\beta) = -\frac{1}{\beta} \lim_{L\to\infty} \frac{1}{2L} \lim_{N\to\infty} \log Z_{NL}(\gamma - 2\tilde{\beta}/N) \tag{7}$$

where $Z_{NL}(\theta) \equiv \mathrm{Tr}\,[T_N(\theta)]^L$. Now, as the well-known Bethe Ansatz solution tells us, $T_N(\gamma)$ has a non–degenerate largest eigenvalue for any finite N. Then the two limits in eq. (7) commute and one finds [7]

$$-2\beta f(\beta) = \lim_{N\to\infty} \log \Lambda_N^{max}(\gamma - 2\tilde{\beta}/N) \tag{8}$$

where $\Lambda_N^{max}(\theta)$ denotes the largest eigenvalue of $T_N(\theta)$. Using the results of ref.[8] for the eigenvalues of the diagonal–to– diagonal transfer matrix, the free energy becomes

$$f(\beta) = \frac{1}{2\beta} \lim_{N\to\infty} \left[E_N(\beta) - 2N \log \frac{\sin(2\tilde{\beta}/N)}{\sin\gamma} \right] \tag{9}$$

where

$$\begin{aligned} E_N(\beta) &= i \sum_{j=1}^{N} \epsilon(\lambda_j, \beta) \\ \epsilon(\lambda, \beta) &= \phi(\lambda_j + i\gamma/2, \gamma/2 - \tilde{\beta}/N) - \phi(\lambda_j - i\gamma/2, \gamma/2 - \tilde{\beta}/N) \\ \phi(\lambda, x) &\equiv i \log \frac{\sinh(ix + \lambda)}{\sinh(ix - \lambda)} \qquad (\phi(0,x) = 0) \end{aligned} \tag{10}$$

The real numbers $\lambda_1, \ldots, \lambda_N$ are the roots of the Bethe Ansatz equations

$$N\left[\phi(\lambda_j, \tilde{\beta}/N) + \phi(\lambda_j, \gamma - \tilde{\beta}/N)\right] = \sum_{k=1}^{N} \phi(\lambda_j - \lambda_k, \gamma) + \pi(2j - N - 1) \tag{11}$$

associated to the largest eigenvalue $\Lambda_N^{max}(\gamma - 2\tilde{\beta}/N)$ (we assumed N even). We define as usual the *counting function* [9]

$$z_N(\lambda) = \frac{1}{2\pi} \left[\phi(\lambda, \tilde{\beta}/N) + \phi(\lambda, \gamma - \tilde{\beta}/N) - \frac{1}{N} \sum_{k=1}^{N} \phi(\lambda - \lambda_k, \gamma) \right] \tag{12}$$

which by construction enjoys the properties $z_N(\lambda) = -z_N(-\lambda) = \overline{z_N(\bar{\lambda})}$ and

$$z_N(\lambda_j) = j - (N+1)/2 \tag{13}$$

Our aim is to derive an integral equation for $z_N(\lambda)$. 218z Let us start by assuming that the root distribution becomes the continuus density $\sigma_c(\lambda)$ (normalized to 1) in the limit $N \to \infty$. Then it could be identified with the derivative of $z_N(\lambda)$ and at the same time used to replace the discrete sum in eq. (12) with an integration. For large but finite N we whould therefore obtain the following approximate linear integral equation for $\sigma_c(\lambda)$

$$2\pi\sigma_c(\lambda) = \phi'(\lambda, \tilde{\beta}/N) + \phi'(\lambda, \gamma - \tilde{\beta}/N) - \int_{-\infty}^{+\infty} d\mu\, \phi'(\lambda - \mu, \gamma)\sigma_c(\mu) \tag{14}$$

Fourier transforming leads to the explicit solution

$$\sigma_c(\lambda) = z_c'(\lambda)\,, \qquad z_c(\lambda) = \frac{1}{\pi}\arctan\left[\frac{\sinh(\pi\lambda/\gamma)}{\sin(\pi\tilde{\beta}/\gamma N)}\right] \tag{15}$$

We see that $\sigma_c(\lambda)$ is strongly peaked at $\lambda = 0$ for large N, reflecting the singularity which develops at the origin in the $N \to \infty$ limit of the source term in eqs. (11), (14). The density picture correctly describes the BA roots wherever $\sigma_c(\lambda)$ is of order 1. From eq. (15) we then find as validity interval

$$|\lambda| \leq O(\sqrt{\beta/N}) \tag{16}$$

which shrinks to zero when $N \to \infty$. Roots $|\lambda_j| > O(\sqrt{\beta/N})$ have a spacing of order larger than $O(1/N)$ and cannot be descried by densities. In particular, the roots with largest magnitudes have finite $N \to \infty$ limits spread by $O(1)$ intervals (we checked this fact numerically too). Therefore, contrary to the usual situation [9], we must go beyond the density description to obtain a bulk quantity like the free energy per site. The sum in the last term of eq.(12) can be written as a contour integral

$$\frac{1}{N}\sum_{j=1}^{N}\phi(\lambda - \lambda_j, \gamma) = \int \frac{dz}{2\pi i N}\phi(\lambda - z, \gamma)\frac{d\log[1 + a_N(z)]}{dz} \tag{17}$$

where $a_N = \exp(2\pi i N z_N)$ and the contour in eq.(17) encircles counterclockwise all the real roots $\lambda_k, k = 1, ..., N$. Notice that $a_N(\lambda_k) + 1 = 0, k = 1, ..., N$. Inserting eq.(17) in eq.(12) and integrating by parts yields

$$\log a_N(\lambda)) = iN\left[\phi(\lambda, \tilde{\beta}/N) + \phi(\lambda, \gamma - \tilde{\beta}/N)\right] + i\int_{-\infty}^{+\infty}\frac{dx}{\pi}\phi'(\lambda - x, \gamma)\mathrm{Im}\log[1 + \mathrm{a_N(x + i0)}] \tag{18}$$

Convoluting eq.(18) with the kernel

$$K^{-1} = \delta(\lambda) - p(\lambda)$$

where

$$p(\lambda) = \int_{-\infty}^{+\infty} \frac{dk}{2\pi} \tilde{p}(k) e^{ik\lambda} , \quad \tilde{p}(k) = \frac{\sinh(\pi/2-\gamma)k}{2\sinh(\pi-\gamma)k/2 \cosh(\gamma k/2)} \tag{19}$$

we find

$$\log \frac{a_N(\lambda)}{a_c(\lambda)} = \int_{-\infty}^{+\infty} d\mu\, p(\lambda-\mu) \log \frac{1+a_N(\mu+i0)}{1+\overline{a_N(\mu+i0)}} \tag{20}$$

which is the sought nonlinear integral equation.(Here $a_c = \exp(2\pi i N z_c)$). Notice that $\overline{a_N(\lambda)} = 1/a_N(\bar{\lambda})$. We evaluate the sum in eq. (10) by a similar procedure, with the result

$$L_N(\beta) \equiv E_N(\beta) - E_c(\beta) = \frac{1}{\gamma} \int_{-\infty}^{+\infty} d\lambda \frac{\sinh \pi\lambda/\gamma \cos \pi\tilde{\beta}/\gamma N}{\cosh 2\pi\lambda/\gamma - \cos 2\pi\tilde{\beta}/\gamma N} \log \frac{1+a_N(\lambda+i0)}{1+\overline{a_N(\lambda+i0)}} \tag{21}$$

where $E_c(\beta)$ is that part of the energy that follows from the root density (15), that is

$$E_c(\beta) = \int_{-\infty}^{+\infty} d\lambda\, \epsilon(\lambda,\gamma)\sigma_c(\lambda) = -2N \int_{-\infty}^{+\infty} \frac{dk}{k} \frac{\sinh(\pi-\gamma)k \sinh(\gamma - 2\tilde{\beta}/N)k}{\sinh \pi k \cosh \gamma k} \tag{22}$$

If eqs. (20) and (21) are analytically continued to the axis $\operatorname{Im}\lambda = \pm\gamma/2$ they assume the same structure of those derived in ref. [12] by different methods.

Inserting eqs. (21) and (22) in eq. (9) yields

$$f(\beta) = E_{XXZ}(\gamma) + \beta^{-1} L(\beta) \tag{23}$$

where $E_{XXZ}(\gamma)$ is the ground-state energy of (1), namely

$$E_{XXZ}(\gamma) = 2J \left[\cos\gamma - \sin\gamma \int_{-\infty}^{+\infty} dk \frac{\sinh(\pi-\gamma)k}{\sinh \pi k \cosh \gamma k} \right] \tag{24}$$

while $L(\beta)$ is the $N \to \infty$ limit of (21), that is

$$L(\beta) = \operatorname{Im} \int_{-\infty}^{+\infty} d\lambda \frac{\log[1+a(\lambda+i0)]}{\gamma \sinh \pi(\lambda+i0)/\gamma} \tag{25}$$

Similarly, the function $a(\lambda)$ in this expression is the $N \to \infty$ limit of $a_N(\lambda)$. It is therefore the solution of the $N \to \infty$ limit of eq. (20), that is

$$-i \log a(\lambda) = -\frac{2\pi\tilde{\beta}}{\gamma \sinh \pi\lambda/\gamma} + 2 \int_{-\infty}^{+\infty} d\mu\, p(\lambda - \mu) \,\mathrm{Im}\, \log[1 + a(\mu + i0)] \tag{26}$$

Thus the calculation of the free energy (5) has been reduced to the problem of solving the single nonlinear integral equation (26) and then evaluating the integral (25).

We find from the $N = \infty$ limit of eq.(18) an alternative form for our eq. (26)

$$z(\lambda) = \tilde{\beta} q(\lambda) + \frac{1}{2\pi^2} \mathrm{Im} \int_{-\infty}^{+\infty} d\mu\, \phi'(\lambda - \mu, \gamma) \log \cos \pi z(\mu + i0) \tag{27}$$

where

$$q(\lambda) = \frac{1}{\pi} \left(\frac{\sinh 2\lambda}{\cosh 2\lambda - \cos 2\gamma} - \coth \lambda \right) \tag{28}$$

and $z(\lambda) \equiv \frac{1}{2\pi i} \log a(\lambda)$ is related to the original counting function

$$z(\lambda) = \lim_{N \to \infty} N \left[z_N(\lambda) - \tfrac{1}{2} \mathrm{sign}\, \lambda \right] \quad (\lambda \text{ real}) \tag{29}$$

By the residue theorem we then obtain

$$z(\lambda) = \tilde{\beta} q(\lambda) - \frac{1}{2\pi} \sum_{j \in \mathbb{Z}} [\phi(\lambda - \xi_j, \gamma) - \phi(\lambda, \gamma)] \tag{30}$$

where the real numbers ξ_j, defined by $z(\xi_j) = j - 1/2$, $j \in \mathbb{Z}$, can be identified with the $N \to \infty$ limit of the original BA roots $\lambda_1, \lambda_2, \ldots, \lambda_N$.The new algebraic BA equations $z(\xi_j) = j - 1/2$ embody all the information about the XXZ thermodynamics. Eq. (30) shows that $z(\lambda)$ has periodicity $i\pi$ and has, as unique singularity on the real axis, a simple pole at the origin with residue $-\tilde{\beta}/\pi$. The new algebraic BA equations $z(\xi_j) = j - 1/2, j \in \mathbb{Z}$ embody all the information about the XXZ thermodynamics.They form an infinite set of Bethe Ansatz type algebraic equations equivalent to our non-linear integral equations (26) and (27) . We want to stress that these roots are **discrete** although we have already set $N = \infty$. The root spacing is actually of order T. In the $T \to 0$ limit we recover the continuous distribution of roots associated with the antiferromagnetic ground-state[9]. In terms of the roots ξ_k, the free energy takes the form

$$f(\beta) = J \cos \gamma - \beta^{-1} \ln 2 + \frac{1}{2\beta} \sum_{k \in \mathbb{Z}} \log \left([1 - \cos 2\gamma] / [\cosh 2\xi_j - \cos 2\gamma] \right) \tag{31}$$

The roots ξ_j have an accumulation point at $\xi = 0$. That is, $\xi_j = -\tilde{\beta}/[\pi(j - 1/2]$ for $j \to \infty$. For low temperatures, the largest root, ξ_1, is of order $(\gamma/\pi) \ln \tilde{\beta}$.

Let us now study $f(\beta)$ for high temperatures. When β is small it is convenient to use the alternative form (27) plus the uniform expansion

$$z(\lambda) = \sum_{k=1}^{\infty} \tilde{\beta}^k \, b_k(\lambda) \tag{32}$$

Since the residue of $z(\lambda)$ is linear in $\tilde{\beta}$ and $z(\theta)$ is an odd function, we have

$$b_1(\lambda) \overset{\lambda\to 0}{\simeq} -\frac{1}{\pi\lambda} + O(\lambda)\,, \quad b_k(0) = 0 \quad (k \geq 2) \tag{33}$$

Inserting eq. (32) in eq. (27) we find, up to fourth order in $\tilde{\beta}$

$$\begin{aligned} b_1(\lambda) &= q(\lambda) & b_2(\lambda) &= \frac{1}{4\pi}\phi''(\lambda,\gamma) \\ b_3(\lambda) &= 0 & b_4(\lambda) &= \frac{1}{6\pi}\left(\frac{1}{3} - \frac{1}{\sin^2\gamma}\right)\phi''(\lambda,\gamma) + \frac{1}{144\pi}\phi''''(\lambda,\gamma) \end{aligned} \tag{34}$$

Then, from eqs. (23) and (25),

$$f(\beta) = -\beta^{-1}\ln 2 + J\cos\gamma - \beta J^2\left(1 + \tfrac{1}{2}\cos^2\gamma\right) + \beta^2 J^3\cos\gamma + O(\beta^3) \tag{35}$$

which indeed agrees with the high T expansion [2]. We want to remark that eq. (27) generates the functions $b_k(\lambda)$ *recursively*, with easy integrations which involve only delta functions and derivatives thereof. It is indeed a very efficent way to recover the high–temperature expansion from the Bethe Ansatz solution [5].

We shall consider now the low temperature regime. When $\beta \gg 1$ eq. (26) indicates that $z(\lambda) \sim \beta$, so that $\log[1 + a(\lambda)] \simeq 2\pi i z(\lambda)$, at least as long as $|\lambda| \leq (\gamma/\pi)\log\beta$. At dominant $\beta \gg 1$ order eq. (26) linearizes in the same way as it does for small β. Inserting then $z(\lambda) \simeq \tilde{\beta} q(\lambda)$ in eq. (25), yields $\lim_{\beta\to\infty} L(\beta) = 0$. Therefore eq. (23) tells us that

$$f(\beta) \overset{\beta\to\infty}{=} E_{XXZ}(\gamma) + O(\beta^{-2}) \tag{36}$$

The contributions of order β^{-2} and smaller come from values of λ larger than $(\gamma/\pi)\log\beta$, where the previous assumption $\log a \sim O(\beta)$ does not hold anymore. It is then convenient to introduce the new function

$$A(x) = a(\lambda)\ , \quad x = \lambda - \frac{\gamma}{\pi}\log\frac{4\pi\tilde{\beta}}{\gamma} \tag{37}$$

Then eqs. (25) and (26) reduce, in the $\beta \to \infty$ limit, to

$$L(\beta) = \frac{2}{\pi\tilde{\beta}} \int\limits_{-\infty}^{+\infty} dx\, e^{-\pi x/\gamma} \mathrm{Im}\ \log[1 + A(x + i0)] \tag{38}$$

and

$$-i \log A(x) = -e^{-\pi x/\gamma} + 2 \int\limits_{-\infty}^{+\infty} dy\, p(x-y)\, \mathrm{Im}\ \log[1 + A(y + i0)] \tag{39}$$

Fortunately, it is not necessary to solve this integral equation to calculate the integral (38). In fact, we can use the following lemma.

Lemma. Assume that $F(x)$ satisfies the nonlinear integral equation

$$-i \log F(x) = \varphi(x) + \int\limits_{x_1}^{x_2} dy\, p(x-y)\, \mathrm{Im}\ \log[1 + F(y + i0)] \tag{40}$$

where $\varphi(x)$ is real for real x and x_1, x_2 are real numbers. Then the following relation holds

$$\begin{aligned} \mathrm{Im} \int\limits_{x_1}^{x_2} dx\, \varphi'(x) \log[1 + F(x + i0)] &= \tfrac{1}{2}\mathrm{Im}\ [\varphi(x_2)\log(1 + F_2) - \varphi(x_1)\log(1 + F_1)] \\ &+ \tfrac{1}{2}\mathrm{Re}\ [\ell(F_1) - \ell(F_2)] \end{aligned} \tag{41}$$

where $F_{1,2} = F(x_{1,2})$ and ℓ is a dilogarithm function

$$\ell(t) \equiv \int\limits_0^t du \left[\frac{\log(1+u)}{u} - \frac{\log u}{1+u}\right] \tag{42}$$

To prove the lemma we consider the relation

$$\ell(F_2) - \ell(F_1) = \int\limits_{x_1}^{x_2} dx \left\{\log[1 + F(x)]\frac{d}{dx}\log F(x) - \log F(x)\frac{d}{dx}\log[1 + F(x)]\right\}$$

and then use eq. (40) and its derivative to substitute $\log F(x)$ and $d\log F(x)/dx$. After taking real part and a little algebra this yields eq. (41) (related identities were used in ref. [12]).

The integral in eq. (38) may now be exactly calculated by invoking the lemma with $F(x) = A(x)$, $\varphi = -\exp(-\pi x/\gamma)$ and $x_1 = -\infty$, $x_2 = +\infty$. We have $A(x_1) = 0$, $A(x_2) = 1$ and

$$2\,\mathrm{Im}\int_{-\infty}^{+\infty} dx\, e^{-\pi x/\gamma}\log[1 + A(x + i0)] = -\frac{\gamma}{\pi}\ell(1) = -\frac{\pi\gamma}{6}$$

Then the free energy for low temperature reads

$$f(\beta) = E_{XXZ}(\gamma) - \frac{\gamma}{6J\sin\gamma}\beta^{-2} + o(\beta^{-2}) \tag{43}$$

in perfect agreement with refs. [2,10,11]. As we have just shown, both the high and the low temperature leading behaviors of the free energy can be derived without much effort from our non–linear integral equation (26). The higher order corrections for high T can be obtained in a very sistematic way [5]. The situation for the $o(\beta^{-2})$ terms in the low T expansion (43) is more involved. Qualitatively, however, it is rather easy to establish, from the asymptotic behaviour of the kernel $p(\lambda - \mu)$ in eq. (26) and from the $i\pi$ periodicity implied by eq. (30), that these higher order terms must be integer powers of $T^{\pi\gamma}$ and $T^{\gamma/(\pi-\gamma)}$.

Let us now consider the problem of calculating the (properly subtracted) ground state energy $E(L)$ of 2D integrable massive models of Quantum Field Theory in a finite 1–volume L. This subject recently received much attention in connection with Perturbed Conformal Field Theory. In this contest $E(L)$ is known as ground–state scaling function. Our starting point is the lattice regularization of such models provided by the light–cone approach to integrable vertex models [8]. Specializing to the six–vertex model, the ground state energy, on a ring of length L formed by N sites, takes a form similar to eq. (21):

$$E_N(L) - E_c(L) = \frac{N}{\gamma L}\,\mathrm{Im}\int_{-\infty}^{+\infty} d\lambda\,\left[\mathrm{sech}\,\tfrac{\pi}{\gamma}(\lambda - \Theta) - \mathrm{sech}\,\tfrac{\pi}{\gamma}(\lambda + \Theta)\right]\log[1 + a_N(\lambda + i0)] \tag{44}$$

where

$$E_c(L) = \frac{N^2}{L}\left[-2\pi + \int_{-\infty}^{+\infty} d\lambda\,\frac{\phi(\lambda + 2\Theta, \gamma/2)}{\gamma\cosh\pi\lambda/\gamma}\right] \tag{45}$$

and $a_N(\lambda + i0)$ obeys an equation like (20) with the new source term

$$\log a_c(\lambda) = 2\arctan\frac{\sinh\pi\lambda/\gamma}{\cosh\pi\Theta/\gamma} \tag{46}$$

The L–dependence in these expressions is hidden in the light–cone parameter Θ, which tends to infinity, in the continuum limit $N \to \infty$, as [8]

$$\Theta = \frac{\gamma}{\pi} \log \frac{4N}{mL} \tag{47}$$

with m, the physical mass scale, held fixed. Define now

$$\epsilon(\lambda) = - \lim_{N \to \infty} \log a_N(\gamma\lambda/\pi) \tag{48}$$

then from eqs. (20), (42–45) we find

$$E(L) \equiv \lim_{N \to \infty} [E_N(L) - E_c(L)] = -m \int_{-\infty}^{+\infty} \frac{d\lambda}{\pi} \sinh\lambda \, \mathrm{Im} \log \left[1 + e^{-\epsilon(\lambda + i0)}\right] \tag{49}$$

and

$$\epsilon(\lambda) = mL \sinh\lambda + 2 \int_{-\infty}^{+\infty} d\mu \, G(\lambda - \mu) \, \mathrm{Im} \log \left[1 + e^{-\epsilon(\mu + i0)}\right] \tag{50}$$

where $G(\lambda) = (\gamma/\pi) p(\gamma\lambda/\pi)$. Thanks to the correspondence between the light–cone six–vertex model and the sine–Gordon (or Massive Thirring) model [8], eqs. (49) and (50) give the ground state scaling function $E(L)$ of the sine–Gordon field theory on a ring of length L. This is a rigorous and simpler alternative to the standard Thermodynamic Bethe Ansatz [14]. Notice that all UV divergent terms are contained in $E_c(L)$ (cft. eq. (45)). $E_c(L)$ also contains the finite, scaling part $m^2 L \cot^2 \pi^2/2\gamma$ (also found in ref. [13] by completely different means), which gives the ground–state energy density in the $L \to \infty$ limit.

Let us now study eqs. (49) and (50) for $mL \gg 1$ and $mL \ll 1$. For small values of mL the calculation closely parallels that for low T in the thermodynamics of the XXZ chain. Shifting $\lambda \to \lambda - \log mL$ and applying the lemma leads to

$$E(L) \overset{mL \to 0}{=} -\frac{\pi}{6L} + O(1) \tag{51}$$

showing the expected value $c = 1$ for the UV central charge. The large mL regime easily follows from eq. (50) by iteration. One finds the behavior typical of a massive quantum field theory

$$E(L) \overset{L \to \infty}{\simeq} -\frac{2m}{\pi} K_1(mL) + O(e^{-2mL}) \tag{52}$$

with $K_1(z)$ the modified Bessel function of order 1.

We would like to remark that, contrary to the traditional thermal Bethe Ansatz [2,3], our approach does not relay on the string hypothesis on the structure of the solutions of the BA equations characteristic of the Heisemberg chain. This makes our approach definetly simpler. Most notably, the whole construction of the thermodynamics no longer depends on whether γ/π is a rational or not, unlike Takahashi approach [2]. This applys equally well to the problem of the ground state scaling function of the sine–Gordon field theory, since the standard TBA approach requires the string hypothesis.

Generalizations of the present TBA approach to higher spin chains as well as to magnetic chains and quantum field theories associated to Lie algebras other than A_1 (nested BA solutions) are relatively straightforward provided crossing symmetry holds [5].

REFERENCES

1. C.N. Yang and C.P. Yang, *J. Math. Phys.* **10** (1969 1115.
2. M. Takahashi, *Prog. Theor. Phys.* **46** (1971) 401.
   M. Takahashi and M. Suzuki, *Prog. Theor. Phys.* **48** (1972) 2187.
3. M. Gaudin, *Phys. Rev. Lett.* **26** (1971) 1302.
4. C. Destri and H. J. de Vega,LPTHE Paris preprint, 92-11.
5. C. Destri and H. J. de Vega, in preparation.
6. R.J. Baxter, "*Exactly solved models in Statistical Mechanics*", Academic Press, 1982.
7. J. Suzuki, Y. Akutsu and M. Wadati, *J. Phys. Soc. Jpn.* **59** (1990) 2667.
8. C. Destri and H.J. de Vega, *Nucl. Phys.* **B290** (1987) 363; *J. Phys.* **A22** (1989) 1329.
9. H.J. de Vega, *Int. J. Mod. Phys.* **A** (1989) 2371; *Int. J. Mod. Phys.* **B** (1990) 735.
10. H.J. de Vega and F. Woynarovich, *Nucl. Phys.* **B251** (1985) 439.
11. H.J. de Vega and M. Karowski, *Nucl. Phys.* **B280** (1987) 225.
12. A. Klumper, M.T. Batchelor and P.A. Pearce, *J. Phys.* **A24** (1991) 3111.
13. C. Destri and H.J. de Vega, *Nucl. Phys.* **B258** (1991) 251.
14. Al.B. Zamolodchikov, *Nucl. Phys.* **B342** (1990) 695.

q-Deformed Massive One-Particle States

O. Ogievetsky[1], * M. Pillin[1], † W.B. Schmidke[1],
J. Wess[1,2] and B. Zumino[3] ‡

[1] Max-Planck-Institut für Physik
Werner-Heisenberg-Institut
P.O. Box 40 12 12 , D - 8000 Munich 40, Germany

[2] Sektion Physik, Universität München
Theresienstr. 37, D - 8000 Munich 2, Germany

[3] Department of Physics
University of California, Berkeley
and
Theoretical Physics Group
Physics Division
Lawrence Berkeley Laboratory
1 Cyclotron Road
Berkeley, California 94720

Irreducible representations of the Poincaré algebra are associated with relativistic one-particle states. Quantum groups generalize the concept of symmetry groups. It is in particular the q-deformed Lorentz group ($SL_q(2,\mathrm{C})$) that serves as a standard example for such a quantum group. It is natural to try to generalize this deformation to the inhomogeneous part of the Poincaré group as well and to use this quantum group as the basis for the discussion of q-deformed one particle states.

There is no standard procedure to deform inhomogeneous groups. One possibility is to start from a q-deformed homogeneous group and to apply a group contraction which, in the undeformed case would lead to the inhomogeneous group [1]. We shall follow another procedure [2]. We start from a q-deformed homogeneous group (q-Lorentz group) which we keep as a q-deformed subgroup. For this homogeneous q-group we construct the corresponding quantum spaces that are a q-deformation of the usual

*On leave of absence from P.N.Lebedev Physical Institute, Theoretical Department, 117924 Moscow, Leninsky prospect 53, Russia

†Supported by Studienstiftung des Deutschen Volkes

‡This work was supported in part by the Director, Office of Energy Research, Office of High Energy and Nuclear Physics, Division of High Energy Physics of the U.S. Department of Energy under Contract DE-AC03-76SF00098 and in part by the National Science Foundation under grant PHY-90-21139

representation spaces [3]. Their algebraic structure is to a large extent determined by the R-matrix of the homogeneous q-group. The quantum space that corresponds to the representation by which the inhomogeneous generators transform (q-deformed four vector) is then used as the defining structure for the q-deformed inhomogeneous group. The q-four vectors are now intepreted as q-deformed four-momenta and the momentum length is a Casimir operator. Its eigenvalues in a Hilbert space representation of the q-Poincaré group will characterize the mass of the q-one particle states. An interesting feature of these representations is that for timelike momenta the mass is quantized in the sense that

$$m^2 = d_0^2 q^{2F}. \tag{1}$$

Here q is the real deformation parameter, F is an integer and d_0 is a universal real parameter with dimension of mass.

Instead of the group, we use the Lie algebra because the generators have a direct physical meaning like momenta or angular momenta. The q-deformed Lie algebra of $SU(2)$ is well known [4]:

$$\begin{aligned} q^{-1}T^+T^- - qT^-T^+ &= T^3 \\ q^{\pm 2}T^3T^\pm - q^{\mp 2}T^\pm T^3 &= \pm(q+q^{-1})T^\pm . \end{aligned} \tag{2}$$

For $q=1$ this is the Lie algebra of $SU(2)$. The special feature of the above relations is that they allow a comultiplication that generalizes them to a Hopf algebra. For $q=1$ this comultiplication is

$$\Delta(T^i) = T^i \otimes 1 + 1 \otimes T^i \tag{3}$$

where i stands for $+$, $-$ or 3. It is the basis for the addition of angular momenta. For $q \neq 1$ the rule for comultiplication is

$$\begin{aligned} \Delta(T^\pm) &= T^\pm \otimes 1 + (1-\lambda T^3)^{\frac{1}{2}} \otimes T^\pm \\ \Delta(T^3) &= T^3 \otimes 1 + (1-\lambda T^3) \otimes T^3 \end{aligned} \tag{4}$$

where $\lambda = q - q^{-1}$. It is easy to verify that $\Delta(T^i)$ satisfies the same relations (2) as T^i.

Generators of unitary groups must have well defined properties under hermitian conjugation. A conjugation relation consistent with (2) and (4) is

$$\overline{T^\pm} = q^{\mp 2}T^\mp \qquad \overline{T^3} = T^3. \tag{5}$$

The quantum space that corresponds to the spinor representation of $SU_q(2)$ is known as the Manin plane [3]. Its algebraic properties derived from the R-matrix of $SU_q(2)$ are

$$xy = qyx, \tag{6}$$

the spinor coordinates q-commute. The conjugate spinors satisfy

$$\bar{x}\bar{y} = q^{-1}\bar{y}\bar{x}. \tag{7}$$

We have to specify the relations between spinors and their conjugates. Covariance under the action of $SU_q(2)$ and consistency with (6) and (7) leads to the following relations:

$$\begin{aligned} x\bar{x} &= \bar{x}x - q\lambda\bar{y}y \qquad & y\bar{x} &= q\bar{x}y \\ x\bar{y} &= q\bar{y}x & y\bar{y} &= \bar{y}y. \end{aligned} \tag{8}$$

The generators of $SU_q(2)$ have to act on the spinor space coordinates x, y and $\bar{x}, \bar{y}$ in a way that is consistent with (2), (5) and (6), (7), (8). This consistency condition has the following solution:

$$\begin{aligned} T^+x &= qxT^+ + y & T^+\bar{x} &= q^{-1}\bar{x}T^+ \\ T^+y &= q^{-1}yT^+ & T^+\bar{y} &= q\bar{y}T^+ - q^{-1}\bar{x} \\ \\ T^-x &= qxT^- & T^-\bar{x} &= q^{-1}\bar{x}T^- - q\bar{y} \\ T^-y &= q^{-1}yT^- + x & T^-\bar{y} &= q\bar{y}T^- \\ \\ T^3x &= q^2xT^3 - qx & T^3\bar{x} &= q^{-2}\bar{x}T^3 + q^{-1}\bar{x} \\ T^3y &= q^{-2}yT^3 + q^{-1}y & T^3\bar{y} &= q^2\bar{y}T^3 - q\bar{y}. \end{aligned} \tag{9}$$

From spinors, we construct four vectors:

$$A = \bar{x}y, \qquad B = \bar{y}x, \qquad C = \bar{x}x, \qquad D = \bar{y}y. \tag{10}$$

Their conjugation properties are

$$\overline{A} = B, \qquad \overline{B} = A, \qquad \overline{C} = C, \qquad \overline{D} = D. \tag{11}$$

A real momentum four vector is defined by the linear combinations:

$$\begin{aligned} P_0 &= \frac{q}{q+q^{-1}}(C+D) & P_1 &= \frac{1}{\sqrt{2}}(A+B) \\ P_3 &= \frac{1}{q+q^{-1}}(qD - q^{-1}C) & P_2 &= \frac{i}{\sqrt{2}}(A-B). \end{aligned} \tag{12}$$

From (6), (7) and (8) follow the q-relations for (A, B, C, D):

$$\begin{aligned} AB &= BA - q^{-1}\lambda CD + q\lambda D^2 & BC &= CB - q^{-1}\lambda BD \\ AC &= CA + q\lambda AD & BD &= q^2 DB \\ AD &= q^{-2}DA & CD &= DC. \end{aligned} \tag{13}$$

These relations have the property that they are invertible, an antialphabetic order can be alphabetically rearranged and, most important, the relations are "consistent", they do not create higher order relations. It is this consistency requirement that specifies (13) uniquely.

The action of the $SU_q(2)$ q-Lie algebra (2) on the four momenta can be computed from (9) and (10). It is:

$$\begin{aligned} T^+A &= q^{-2}AT^+ & T^+C &= CT^+ + q^{-1}A \\ T^+B &= q^2BT^+ + qD - q^{-1}C & T^+D &= DT^+ - q^{-1}A \\ \\ T^-A &= q^{-2}AT^- + q^{-1}C - qD & T^-C &= CT^- - qB \\ T^-B &= q^2BT^- & T^-D &= DT^- + qB \\ \\ T^3A &= q^{-4}AT^3 + q^{-1}(q^{-2}+1)A & T^3C &= CT^3 \\ T^3B &= q^4BT^3 - q(q^2+1)B & T^3D &= DT^3. \end{aligned} \tag{14}$$

The relations (13) and (14) define an algebraic system. It is interesting to note that the "energy" P_0 is central in this algebra, it commutes with the other variables. The algebra (2), (13) and (14) can be viewed as a subalgebra of the q-deformed Poincaré algebra, consisting of the q-deformed energy momentum four vector and angular momenta. To arrive at the full q-Poincaré algebra we have to include the boosts of the Lorentz algebra. To this end, we will generalize the ansatz for the generator action on spinors. Consider a set of generators T^i, $i = 1 \ldots n$. The ansatz for the action is:

$$\begin{aligned} T^i x &= a^i{}_j x T^j + b^i{}_j y T^j + \alpha^i x + \beta^i y \\ T^i y &= c^i{}_j x T^j + d^i{}_j y T^j + \gamma^i x + \delta^i y \\ T^i \bar{x} &= \bar{a}^i{}_j \bar{x} T^j + \bar{b}^i{}_j \bar{y} T^j + \bar{\alpha}^i \bar{x} + \bar{\beta}^i \bar{y} \\ T^i \bar{y} &= \bar{c}^i{}_j \bar{x} T^j + \bar{d}^i{}_j \bar{y} T^j + \bar{\gamma}^i \bar{x} + \bar{\delta}^i \bar{y}. \end{aligned} \tag{15}$$

Here $a^i{}_j$, $b^i{}_j$ etc. are $n \times n$ c-number matrices and α^i, β^i etc. are n-dimensional c-number vectors.

Taking into account all the consistency conditions, demanding a closure under q-deformed commutators and choosing a proper normalization we arrive at a minimum set of four new generators. A detailed description of this analysis can be found in ref. [5]. The result is:

$$\begin{aligned} T^2 x &= x T^2 + y \tau^1 & T^2 \bar{x} &= q \bar{x} T^2 \\ T^2 y &= y T^2 & T^2 \bar{y} &= q^{-1} \bar{y} T^2 \\ S^1 x &= x S^1 & S^1 \bar{x} &= q^{-1} \bar{x} S^1 + \bar{y} \sigma^2 \\ S^1 y &= y S^1 & S^1 \bar{y} &= q \bar{y} S^1 \\ \tau^1 x &= q^{-1} x \tau^1 & \tau^1 \bar{x} &= \bar{x} \tau^1 + q \lambda^2 \bar{y} T^2 \\ \tau^1 y &= q y \tau^1 & \tau^1 \bar{y} &= \bar{y} \tau^1 \\ \sigma^2 x &= q x \sigma^2 + q \lambda^2 y S^1 & \sigma^2 \bar{x} &= \bar{x} \sigma^2 \\ \sigma^2 y &= q^{-1} y \sigma^2 & \sigma^2 \bar{y} &= \bar{y} \sigma^2. \end{aligned} \tag{16}$$

For convenience we have used the combinations $\tau^1 = 1 + \lambda T^1$ and $\sigma^2 = 1 + \lambda S^2$. This minimal set of four new generators has been chosen such that the algebra of the generators closes and such that the invariant length of the four vector

$$M^2 = q^{-2} CD - AB \tag{17}$$

is central in the full algebra. The action of the generators on (A, B, C, D) follows from (10) and (16). The existence of such a central element can be deduced from the R-matrix approach to quantum spaces [2]. The decomposition of the R-matrix of the q-Lorentz group into projectors yields a one-dimensional invariant subspace, the invariant length.

The full algebra of the seven q-Lorentz generators is

$$\begin{array}{rcl@{\qquad}rcl}
\tau^1 T^+ &=& T^+\tau^1 + \lambda T^2 & T^+T^2 &=& q^{-2}T^2T^+ \\
\tau^1 T^- &=& q^{-2}T^-\tau^1 - \lambda S^1 & T^-T^2 &=& T^2T^- + \lambda^{-1}(\sigma^2 - \tau^1) \\
\tau^1 T^2 &=& q^2 T^2 \tau^1 & T^+S^1 &=& q^2 S^1 T^+ + \lambda^{-1}(\tau^3\tau^1 - \sigma^2) \\
\tau^1 S^1 &=& S^1 \tau^1 & T^-S^1 &=& S^1 T^- \\
 & & & T^+T^- &=& q^2 T^- T^+ + q\lambda^{-1}(1-\tau^3) \\
\sigma^2 T^+ &=& T^+\sigma^2 - q^2\lambda\tau^3 T^2 & T^2 S^1 &=& S^1 T^2 \\
\sigma^2 T^- &=& q^2 T^-\sigma^2 + q^2\lambda S^1 & & & \\
\sigma^2 T^2 &=& q^{-2}T^2\sigma^2 & \tau^1\sigma^2 &=& \sigma^2\tau^1 + q\lambda^3 T^2 S^1 \\
\sigma^2 S^1 &=& S^1\sigma^2 & \tau^3\tau^1 &=& \tau^1\tau^3 \\
 & & & \tau^3\sigma^2 &=& \sigma^2\tau^3. \\
\tau^3 T^+ &=& q^{-4}T^+\tau^3 & & & \\
\tau^3 T^- &=& q^4 T^-\tau^3 & & & \\
\tau^3 T^2 &=& q^{-4}T^2\tau^3 & & & \\
\tau^3 S^1 &=& q^4 S^1\tau^3 & & &
\end{array} \tag{18}$$

Here we have defined $\tau^3 = 1-\lambda T^3$. For the four new generators a conjugation consistent with (16) and (18) is:

$$\begin{array}{rcl@{\qquad}rcl}
\overline{T^2} &=& -(\tau^3)^{-\frac{1}{2}}S^1 & \overline{\tau^1} &=& (\tau^3)^{-\frac{1}{2}}\sigma^2 \\
\overline{S^1} &=& -q^2(\tau^3)^{\frac{1}{2}}T^2 & \overline{\sigma^2} &=& (\tau^3)^{\frac{1}{2}}\tau^1.
\end{array} \tag{19}$$

The algebra defined by (18) allows a coproduct and a generalization to a Hopf algebra [6]. This algebra appears to have seven generators which in the limit $q \to 1$ turn into the six generators of the Lorentz group. This suggests that the seven generators might not be independent. There is indeed an extra relation in the algebra which allows elimination of one of the seven generators. The quantity

$$Z = \tau^1\sigma^2 - q^2\lambda^2 T^2 S^1 \tag{20}$$

is central. It has eigenvalue one in all representations of q-Lorentz algebra. Therefore with $Z = 1$ one may eliminate τ^1 or σ^2 from the algebra, for example by the substitution $\sigma^2 = (\tau^1)^{-1}(1 + q^2\lambda^2 T^2 S^1)$. The six generator version of the q-Lorentz algebra, however, does not allow q-deformed commutators that are at most quadratic in the generators. It is convenient to keep all seven generators, keeping in mind that they are not independent.

The final step to define the q-Poincaré algebra is to list the algebraic relations between the four new generators and the momenta:

$$\begin{array}{rcl@{\qquad}rcl}
T^2 A &=& qAT^2, & T^2 C &=& qCT^2 + qA\tau^1, \\
T^2 B &=& q^{-1}BT^2 + q^{-1}D\tau^1, & T^2 D &=& q^{-1}DT^2, \\
 & & & & & \\
S^1 A &=& q^{-1}AS^1 + q^{-1}D\sigma^2, & S^1 C &=& (q^{-1}C + q\lambda^2 D)S^1 + qB\sigma^2, \\
S^1 B &=& qBS^1, & S^1 D &=& qDS^1, \\
 & & & & & \\
\tau^1 A &=& qA\tau^1 + q\lambda^2 DT^2, & \tau^1 C &=& (q^{-1}C + q\lambda^2 D)\tau^1 + q\lambda^2 BT^2, \\
\tau^1 B &=& q^{-1}B\tau^1, & \tau^1 D &=& qD\tau^1, \\
 & & & & & \\
\sigma^2 A &=& q^{-1}A\sigma^2, & \sigma^2 C &=& qC\sigma^2 + q\lambda^2 AS^1, \\
\sigma^2 B &=& qB\sigma^2 + q\lambda^2 DS^1, & \sigma^2 D &=& q^{-1}D\sigma^2.
\end{array} \tag{21}$$

The relations (13), (14), (18) and (21) define our q-deformed Poincaré algebra. Whereas the homogeneous part of the algebra allows an extension of the algebra to a Hopf algebra this is not the case for the inhomogeneous part. There a new generator has to be added and the Hopf algebra obtained that way is a deformation of the conformal algebra rather than of the Poincaré algebra [2]. At the moment it is not clear if a q-deformation of the Poincaré algebra in this framework of Hopf algebras exists at all. There is no canonical way to assign a comultiplicaion structure to an algebraic system. This has physical implications. The one-particle states can be defined as Hilbert space representations of the algebra. To define multi-particle states in a Fock space representation a comultiplication has to be known. Thus, at the moment we can only construct the one-particle states of a q-deformed Poincaré algebra and it is only in the conformal case that we can extend this construction to multi-particle states. In the following I shall discuss the q-deformed one-particle states.

A set of commuting observables that we want to diagonalize simultaneously consists of the momentum components P_0 and P_3, the $SU_q(2)$ generator τ^3 and the Casimir operator M^2. It is of advantage to label the states in Hilbert space by the eigenvalues of the operators C, D and τ^3 and to express P_0, P_3 and M^2 in terms of these eigenvalues. We note that A and B shift the eigenvalues of C and D. An n-fold application of A on a state $|c_0, d_0, j_0\rangle$ leads to eigenvalues $c_n = c_0 + d_0 - d_0 q^{2n}$, $d_n = d_0 q^{2n}$ and $j_n = j_0 + n$. Labelling the states with the number n we find that A shifts n to $n+1$ and B shifts n to $n-1$. Normalization and the fact that AB can be expressed in terms of C, D and M^2 gives

$$\begin{aligned} A\,|n,j\rangle &= a_n\,|n+1, j+1\rangle \\ B\,|n,j\rangle &= \overline{a_{n-1}}\,|n-1, j-1\rangle \end{aligned} \tag{22}$$

with $|a_n|^2 = d_n c_{n+1} - m^2$. We now require $q > 1$ and $m^2 > 0$. Then $|a_n|^2$ will become negative for large n. Hence we require a highest state with $n = N$ which is annihilated by A. Analogously, B has to annihilate a state with smallest $n = K$. This can only be achieved if $m^2 = d_0^2 q^{2(N+K)}$. As M^2 is a Casimir operator we find that for an irreducible representation, the quantity $K + N \equiv F$ is fixed. We have used only the relations of (A, B, C, D) to arrive at the following spectrum of P_0 and P_3:

$$\begin{aligned} P_0\,|n,j,N\rangle &= \frac{q}{q+q^{-1}} d_0 \left(q^{2(N+1)} + q^{2K}\right) |n,j,N\rangle \\ P_3\,|n,j,N\rangle &= d_0 \left(q^{2n} - \frac{q^{2(N+1)} + q^{2K}}{q^2+1}\right) |n,j,N\rangle . \end{aligned} \tag{23}$$

The action of the generators of $SU_q(2)$ can be represented in these states. However if we apply τ^1, a non-compact generator of $SL_q(2, \mathrm{C})$, we are moved out of the spectrum as given above. We have to add another representation of (A, B, C, D) whose characteristic length d_0 is $q^{-1} d_0$ of the original one. This is then sufficient to represent all the generators. Summarizing, we have obtained a representation of the q-Poincaré algebra for timelike momenta $m^2 > 0$ and for spin equal zero. The eigenvalues of the energy and the momenta are discrete:

$$\begin{aligned} P_0 &= d_0 \frac{q^{1-r}}{q+q^{-1}} \left(q^{2(N+1)} + q^{2(F-N+r)}\right) \\ P_3 &= d_0 q^{-r} \left(q^{2n} - \frac{q^{2(N+1)} + q^{2(F-N+r)}}{q^2+1}\right) . \end{aligned} \tag{24}$$

Here d_0 is a universal real parameter with dimension of mass. F is an integer, defining the mass of the representation. N and n are integers obeying $F \leq 2N - r$ and $F - N + r \leq n \leq N$ and r takes the value 0 or 1.

The interesting fact is that on these states with a discretized energy-momentum spectrum there is nevertheless a representation of an algebraic system, the q-Lorentz transformations that tend to the usual Lorentz transformations in the limit $q \to 1$. Even on the lattice there is something left that remembers the Lorentz invariance of the continuous theory.

References

[1] J. Lukierski, H. Ruegg, A. Nowicki and V.N. Tolstoy: q-Deformation of Poincaré Algebra. Phys. Lett. B **264**, 331 (1992)

[2] O. Ogievetsky, W.B. Schmidke, J. Wess and B. Zumino: q-Deformed Poincaré Algebra. Commun. Math. Phys. **150**, 495 (1992)

[3] Yu.I. Manin: Quantum Groups and Non-Commutative Geometry. Preprint Montreal University CRM-1561 (1988)

[4] S.L. Woronowicz, "Twisted $SU(2)$ Group. An example of Non-Commutative Differential Calculus", Publ. RIMS-Kyoto **23**, 117 (1987)

[5] W.B. Schmidke, J. Wess and B. Zumino: A q-deformed Lorentz Algebra. Z. Phys. C-Particles and Fields **52**, 471 (1991)

[6] O. Ogievetsky, W.B. Schmidke, J. Wess and B. Zumino: Six Generator q-deformed Lorentz Algebra. Lett. Math. Phys. **23**, 233 (1991)

Mechanical Theorem Proving in Differential Geometries*

WU WEN-TSUN
(Institute of Systems Science, Academia Sinica)

In geometry one usually proves theorems one by one and each individual theorem requires a proof peculiar to that theorem itself. By *mechanical geometry theorem proving* we mean that proving will be carried out by a *universal* method which holds for a whole *class* of theorems instead of single individual ones. It is clear that there is no reason to believe that such a universal method for a given class of theorems necessarily exists. In fact, such non-existence for classes of theorems beyond geometry is abundant as shown by works mainly due to mathematical logicians. On the other hand Tarski in 1950 has proved that for the class of all theorems in elementary euclidean geometry, so far well-defined in some sense not to precise here, such a universal method does exist. A universal method, when it exists for some class of theorems, will be naturally of an *algorithmic* or *mechanical* character and the corresponding theorem asserting such an existence will be called accordingly a *mechanization theorem* for the class of theorems in question or by abuse of language, simply for the domain in question if the class is the class of all theorems in this domain. In such a case one may then turn the algorithm or the mechanical procedure reflecting the universal method into a program and any theorem in that class may then be proved (or disproved) on the computer. The only problem which remains lies in the efficiency or the feasibility of the method. Thus the above mechanization theorem of Tarski, though highly remarkable, suffers from being too inefficient so that no interesting theorem seems to have been ever proved in this way. On the other hand the author has discovered a subclass of the whole class of Tarski and correspondingly a universal or *mechanical* method. The subclass is large enough to contain almost most of interesting theorems known in elementary geometry and the method is efficient enough to be able to prove most of such theorems and even discover a lot of such theorems on the computer in quite simple way. Our task now is to extend the mechanization method of elementary geometries to differential geometries.

To see what *class* of theorems in the case of (local) differential geometries is and what universal mechanical method will permit one to prove any theorem in that *class* let us consider as an example the following Theorema Egregium of Gauss.

For a surface $S : x_i = x_i(u_1, u_2), i = 1, 2, 3$ in the euclidean 3-space let the first and second fundamental forms be

$$I = E * du_1^2 + 2 * F * du_1 * du_2 + G * du_2^2,$$
$$II = L * du_1^2 + 2 * M * du_1 * du_2 + N * du_2^2.$$

*The present paper is partially supported by NSFC Grant JI85312

Set $D = \sqrt{E * G - F^2}$ and let K be the Gaussian curvature of S. Then we have a set of equations $DH_i = 0$ with DH_i given below:

$$DH_1 = E - (\frac{\partial x_1}{\partial u_1})^2 - (\frac{\partial x_2}{\partial u_1})^2 - (\frac{\partial x_3}{\partial u_1})^2,$$

$$DH_2 = F - \frac{\partial x_1}{\partial u_1} * \frac{\partial x_1}{\partial u_2} - \frac{\partial x_2}{\partial u_1} * \frac{\partial x_2}{\partial u_2} - \frac{\partial x_3}{\partial u_1} * \frac{\partial x_3}{\partial u_2},$$

$$DH_3 = G - (\frac{\partial x_1}{\partial u_2})^2 - (\frac{\partial x_2}{\partial u_2})^2 - (\frac{\partial x_3}{\partial u_2})^2,$$

$$DH_4 = D * L - \frac{\partial(x_2, x_3)}{\partial(u_1, u_2)} * \frac{\partial^2 x_1}{\partial u_1^2} - \frac{\partial(x_3, x_1)}{\partial(u_1, u_2)} * \frac{\partial^2 x_2}{\partial u_1^2} - \frac{\partial(x_1, x_2)}{\partial(u_1, u_2)} * \frac{\partial^2 x_3}{\partial u_1^2},$$

$$DH_5 = D * M - \frac{\partial(x_2, x_3)}{\partial(u_1, u_2)} * \frac{\partial^2 x_1}{\partial u_1 \partial u_2} - \frac{\partial(x_3, x_1)}{\partial(u_1, u_2)} * \frac{\partial^2 x_2}{\partial u_1 \partial u_2} - \frac{\partial(x_1, x_2)}{\partial(u_1, u_2)} * \frac{\partial^2 x_3}{\partial u_1 \partial u_2},$$

$$DH_6 = D * N - \frac{\partial(x_2, x_3)}{\partial(u_1, u_2)} * \frac{\partial^2 x_1}{\partial u_2^2} - \frac{\partial(x_3, x_1)}{\partial(u_1, u_2)} * \frac{\partial^2 x_2}{\partial u_2^2} - \frac{\partial(x_1, x_2)}{\partial(u_1, u_2)} * \frac{\partial^2 x_3}{\partial u_2^2},$$

$$DH_7 = D^2 - E * G + F^2,$$

$$DH_8 = D^2 * K - L * N + M^2.$$

The Theorema Egregium of Gauss states that K may be expressed solely in E, F, G as well as their derivatives without intervention of L, M, N. If we like we may write down the conclusion of the theorem in an explicit form of $DC = 0$ where $DC = D^4 * K - DP$, with DP some expression in E, F, G and their derivatives.

We see that these expressions DH_i and DC involved in the hypothesis and conclusion of the theorem are all in the form of ordinary polynomials in some functions $x_1, x_2, x_3, E, L, etc.$ as well as their derivatives w.r.t. some independent variables u_1, u_2. Such polynomials will henceforth be called *differential polynomials* (abbr. *d-pols*). The class of differential geometry theorems considered will be consisting of those for which the hypothesis and conclusion are all expressed in the form of equations in setting some d-pols to 0. Denote this class by **D**. Then we have the following

Mechanization Theorem of Differential Geometries. For the class **D** there is a mechanical procedure following which any theorem in the class **D** may be proved to be *generically true* or not in a finite number of steps.

To see the meaning of *generically true* and to see what is the mechanical procedure let us consider as an example a theorem from elementary geometry,viz.

Feuerbach Theorem (1822). For a triangle the nine point circle is tangent to the inscribed as well as the three escribed circles.

Let us take the coordinates so that vertices of a triangle ABC and the center I, either the in-center or any of the es-center, be given as

$$I = (0, x_1), A = (x_2, 0), B = (x_3, 0), C = (x_4, x_5).$$

Let the center of the nine-point circle be $N = (x_6, x_7)$ and radius be x_8. Denote also the distance between N and I by x_9. Then the geometrical conditions involved will easily be turned into a set of equations $EH_i = 0, i = 1, \cdots, 6$ with *pols* (i.e. polynomials) EH_i given below. Let EPS be the *polset* (i.e. polynomial set) consisting of these pols, then these equations will be abbreviated as $EPS = 0$.

$$\begin{aligned}
EH_1 &= (x_1^2 + x_2 * x_3) * x_4 \; - \; x_1^2 * (x_2 + x_3), \\
EH_2 &= (x_1^2 + x_2 * x_3) * x_5 \; - \; 2 * x_1 * x_2 * x_3, \\
EH_3 &= 4 * x_6 \; - \; (x_2 + x_3 + 2 * x_4), \\
EH_4 &= 4 * x_5 * x_7 \; - \; x_5^2 \; - \; (x_3 - x_4) * (4 * x_6 - 2 * x_2 - x_3 - x_4), \\
EH_5 &= 4 * x_8^2 \; - \; (2 * x_6 - x_2 - x_3)^2 \; - \; 4 * x_7^2, \\
EH_6 &= x_9^2 \; - \; x_6^2 \; - \; (x_7 - x_1)^2.
\end{aligned}$$

The conclusion of the theorem may then be expressed in the form of $EC = 0$ with

$$EC \; = \; (x_9 + x_8 + x_1) * (x_9 + x_8 - x_1) * (x_9 - x_8 + x_1) * (x_9 - x_8 - x_1).$$

Remark that the pols EH_i are *well-arranged* in the sense that they enter the new variables $x_4, \cdots, x_9$ successively one at a time. Such a well-arranged polset EPS will be called an *ascending set* (abbr. *asc-set*) or more exactly an asc-set of *type 0*. The coefficients of the highest degree terms in these new variables, to be called the *initials* will be denoted by $I_1, \cdots, I_6$, e.g. $I_1 = x_1^2 + x_2 * x_3$. That the hypothesis pols may be put in the form of an asc-set is crucial in our method of mechanical proving of theorems in (elementary) geometries. In fact, let us form successive remainders of the conclusion pol EC w.r.t. $EH_6, \cdots, EH_1$ in succession to remove as much as possible the variables $x_9, \cdots, x_4$. The final remainder thus obtained will be denoted by

$$R \; = \; Remdr(EC/EPS),$$

so that we shall have a formula of the form

$$IP * EC \; = \; SUM_i \; EG_i * EH_i \; + \; R,$$

in which IP is some power product of I_i, EG_i are some pols of no interest to us and SUM is over i in the present case from $i = 1$ to $i = 6$. By direct computations it turns out that in the present case we have

$$R \; = \; 0.$$

It follows that $EC = 0$ will follow from $EH_i = 0$ so far $IP \neq 0$ or none of the coefficients I_i is 0. The Feuerbach Theorem is thus proved to be true under the *nondegeneracy conditions*

$$I_i \neq 0, \text{ all } i.$$

We say simply that the theorem is *generically true* under the above *non-degeneracy conditions.*

The above method is of an universal character and any (elementary) geometry theorem will be readily proved to be generically true or not once the hypothesis polset is put into the form of an asc-set (the conclusion being in the form of $EC = 0$ for some pol EC). However, the hypothesis polset given may not be in such a form so that the method is not directly applicable. Thus it arises the problem of how to reduce a polset of pols in confused order to an asc-set with pols in well-arranged order. For this purpose we introduce the following fundamental definition: An asc-set ECS will be called a *characteristic set* (abbr. *char-set)* of a polset EPS if

$$Zero(EPS) \subset Zero(ECS), \text{ and}$$
$$Remdr(EP/ECS) = 0, \text{for all pols } EP \text{ in } EPS.$$

Denote the initials of pols in ECS by I_i and their product by J. Then from the very definition it follows readily

$$Zero(ECS/J) \subset Zero(EPS) \subset Zero(ECS), \text{ and} \tag{E1}$$
$$Zero(EPS) = Zero(ECS/J) + UNION_i\ Zero(EPS + I_i). \tag{E2}$$

Consider now a geometry theorem T with an arbitrary hypothesis polset EHS and a conclusion pol EC. Let ECS be a char-set of the polset $EPS = EHS$. Form the remainder of EC of the theorem T w.r.t. ECS, say

$$R = Remdr(EC/ECS).$$

Suppose that $R = 0$. Then from (E1) and (E2) it follows readily that the theorem T is generically true under the non-degeneracy conditions $J \neq 0$. It may also be proved that the condition $R = 0$ is not only a sufficient condition for a theorem to be generically true under the above non-degeneracy conditions but is also necessary for this to be true in case that the char-set ECS is *irreducible* in some definite sense.

The problem is now reduced to the determination of a char-set of an arbitrary polset in a set of given variables $x_1, \cdots, x_n$. For this purpose let us first introduce a certain partial ordering among all such polsets in variables x_i. The problem is then settled by our main algorithm according to the following scheme (ES):

$$\begin{array}{llllll} EPS & = EPS_0 & EPS_1 & \cdots & EPS_m & \\ & EBS_0 & EBS_1 & \cdots & EBS_m & = ECS \\ & ERS_0 & ERS_1 & \cdots & ERS_m & = empty. \end{array} \tag{ES}$$

In the scheme (ES) each EBS_i is an asc-set of lowest ordering contained in EPS_i, each ERS_i is the set of all non-zero remainders, if there are any, of pols in $EPS_i - EBS_i$ w.r.t. EBS_i, and for each i,

$$EPS_i = EPS_{i-1} + ERS_{i-1}. \tag{E3}$$

It is easily proved that the procedure should end in a finite number of steps with $ERS_m =$ *empty* at a certain stage m. The corresponding asc-set EBS_m is then a char-set ECS of EPS verifying the properties $(E1)$ and $(E2)$ as required.

We remark that the above formulation is the simplest version of the general principles underlying our method of mechanical theorem proving of elementary geometries. In particular, the introduction of the notion of non-degeneracy conditions is not indispensable in our theory. It is introduced only for convenience and for computational sake. Moreover, by deliberately choosing the order of variables involved, the method may be applied to the automatic discovery of new theorems or unknown relations.

How to generalize the above method to the case of differential geometries is now almost self-evident. Instead of variables x_i we have to consider functions x_i and their derivatives w.r.t. to some independent variables u_j. Introduce now an ordering of the independent variables u_j and also an ordering among the functions x_i as well as all their derivatives w.r.t. u_j. All the notions in the ordinary sense will have their corresponding ones in the differential case. These corresponding notions will be called resp. *d-pols, d-polset, d-asc-set, d-remainder, d-char-set, etc.* There are however something new in that for a d-pol DP with leading derivative $DL = \frac{\partial^{j_1+\cdots+j_r} x_k}{\partial u_1^{j_1}\cdots\partial u_r^{j_r}}$, we have to consider besides the *initial* also the *separant* defined as the formal derivative of DP w.r.t. DL. For example, for the d-pol DH_1 given above,the leading derivative is $\frac{\partial x_3}{\partial u_1}$, the *initial* is -1, and the *separant* is $-2 * \frac{\partial x_3}{\partial u_1}$. For most of the concepts we refer to the two classics of Ritt, viz. [Ritt1,2].

For a d-polset DPS and its d-char-set DCS we have now formulas analogous to $(E1)$ and $(E2)$ with slight modifications, viz.

$$d\text{-}Zero(DCS/J) \subset d\text{-}Zero(DPS) \subset d\text{-}Zero(DCS), \text{ and} \tag{D1}$$

$$\begin{aligned} d\text{-}Zero(DPS) &= d\text{-}Zero(DCS/J) + UNION_i\, d\text{-}Zero(DPS + I_i) \\ &\quad + UNION_i\, d\text{-}Zero(DPS + S_i). \end{aligned} \tag{D2}$$

In $(D1)$ and $(D2)$ I_i resp. S_i are now the initials resp. separants of d-pols in the d-char-set DCS, and J is the product of all these initials as well as all these separants. The main algorithm in (ES) is also to be modified to the scheme (DS) below which determines a d-char-set DCS from a given d-polset DPS:

$$\begin{array}{llllll} DPS & = DPS_0 & DPS_1 & \cdots & DPS_m & \\ & DBS_0 & DBS_1 & \cdots & DBS_m & = DCS \\ & DRIS_0 & DRIS_1 & \cdots & DRIS_m & = \textit{empty}. \end{array} \tag{DS}$$

It is to be emphasized that owing to the occurence of integrability conditions when there is more than one independent variables, the relation $(E3)$ has to be replaced by $(D3)$ below:

$$\begin{aligned} DPS_i &= DPS_{i-1} + DRIS_{i-1}, \\ DRIS_i &= DRS_i + DIS_i. \end{aligned} \tag{D3}$$

In $(D3)$ DRS_{i-1} is as in the ordinary case all non-zero d-remainders of d-pols in $DPS_{i-1} - DBS_{i-1}$ w.r.t. DBS_{i-1}, while DIS_{i-1} is consisting of all non-zero d-pols IC_k corresponding to integrability conditions $IC_k = 0$ in DPS_{i-1}. Remark that the latter ones can be determined in a systematic way by some procedure originated from works of Riquier and Janet, cf. [Riq],[J], and [Wu2].

It is easily proved that the mechanical procedure in (DS) should end in a finite number of steps and this proves the Mechanization Theorem of Differential Geometries for the class **D**. For more details cf. [Wu2]. We point out further that, owing to the purely algebraic character of the procedure, we have as a corollary of the Mechanization Theorem the following

Theorem. If a differential geometry theorem in the class **D** is true when all the functions involved are analytic ones, then the theorem will remain to be true so far the functions are differentiable to a sufficient high order.

Consider now the Theorema Egregium of Gauss again. Let us introduce besides D, E, F, G, L, M, N also the Christoffel symbols and arrange them in some definite order. Then the fundamental equations of surface theory will give rise to a d-polset for which a d-char-set may be found by our main algorithm as exhibited in the scheme (DS). The Theorema Egregium is then easily proved by forming the d-remainder of the d-pol $D^2 * K - L * N + M^2$ defining the Gaussian curvature w.r.t. this d-char-set. Cf. the paper [Li].

Our method has been applied mainly to the proving and discovering of theorems in the differential geometry of curves. It has also been applied to the study of some problems in mechanics. Cf. [Wu1] and references there and also [C-G]. On the other hand owing to the high complexity of computations caused by occurence of integrability conditions and the limitation of our equipments, little has yet been done for surfaces or high dimensional cases. Cf. [Wu1] and also [Li].

Bibliography.

[C-G] S.C.Chou & X.S.Gao, Theorems proved automatically using Wu's method — Part on differential geometry and mechanics, MM-Res.Preprints, No.6 (1991) 37-55.

[J] M.Janet, Lecons sur les systemes d'equations aux derivees partielles, Gauthier Villars (1920).

[Li] Z. M. Li, Mechanical theorem-proving of the local theory of surfaces, MM-Res. Preprints, No.6 (1991) 102-120.

[Riq] C.H.Riquier, La methode des fonctions majorantes et les systemes d'equations aux derivees partielles, Gauthier Villars (1928).

[Ritt1] J. F. Ritt, Differential equations from the algebraic standpoint, Amer. Math. Soc., (1932).

[Ritt2] —, Differential algebra, Amer.Math.Soc.,(1950).

[WU1]. Wu Wen-tsun, Mechanical theorem proving of differential geometries and some of its applications in mechanics, MM-Res. Preprints, No.6 (1991) 1-22. Also in J. Automated Reasoning, 7 (1991) 171-191.

[WU2] —, On the foundation of algebraic differential geometry, MM-Res. Preprints, No.3 (1989). Also in Systems Science & Math. Sciences, 2 (1989) 289-312.

ON THE KP HEIRARCHY, $\hat{W}_\infty$ ALGEBRA, AND CONFORMAL $SL(2,R)/U(1)$ MODEL: THE CLASSICAL AND QUANTUM CASES*

Feng Yu† and Yong-Shi Wu‡

†Institute for Theoretical Physics, State University of New York
Stony Brook, New York 11794 USA

‡Center for Theoretical Physics
Laboratory for Nuclear Science and Department of Physics
Massachusetts Institute of Technology
Cambridge, Massachusetts 02139 U.S.A.

and

‡Department of Physics, University of Utah
Salt Lake City, Utah 84112 USA

ABSTRACT

We give a unified description of our recent results on the inter-relationship between the integrable infinite KP hierarchy, nonlinear $\hat{W}_\infty$ current algebra and conformal noncompact $SL(2,R)/U(1)$ coset model both at the classical and quantum levels. In particular, we present the construction of a quantum version of the KP hierarchy by deforming the second KP Hamiltonian structure through quantizing the $SL(2,R)_k/U(1)$ model and constructing an infinite set of commuting quantum $\hat{W}_\infty$ charges (at least at k=1).

INTRODUCTION AND SUMMARY

One of the most exciting recent developments in mathematical physics is the revelation of the close relationship between infinite dimensional (extended) conformal algebras, nonlinear integrable differential systems and 2d exactly solvable field theories, including conformal field models, 2d quantum gravity and string theory. A well-known example is the Virasoro algebra, which is known to be isomorphic to the second Hamiltonian structure of the KdV hierarchy, and appears as the operator product algebra in 2d minimal conformal models, and as constraint algebra in $c < 1$ string theory. Similar relations happen to Zamolochikov's extended W_N algebras

* This work is supported in part by funds provided by the U. S. Department of Energy (D.O.E.) under contract #DE-AC02-76ER03069, and in part by U. S. National Science Foundation grant PHY-9008452.

($N \geq 3$), extension of the Virasoro algebra generated by currents of higher spin ($2 \leq s \leq N$).

In our recent study[1-4], we generalize this profound relationship to the W-infinity algebras, which are generated by currents of all integer spins ($s \geq 1$ or $s \geq 2$) including the energy-momentum tensor and can be viewed as (non-unique) large N limits of the W_N algebras. There are some very deep connection between such algebras and the famous KP hierarchy, which is known to contain all generalized KdV hierarchies, and the conformal $SL(2,R)/U(1)$ coset model that supports the black-hole solution in $D=2$ string theory.

The results reported in our previous papers[1-4] can be summarized as follows: 1) The centerless linear $W_{1+\infty}$ algebra can be identified (also see Ref. [5]) as the first Hamiltonian structure of the KP hierarchy. 2) There exists a unique non-linear deformation[2] of the W_∞ algebra under certain homogeneity assumptions, which we call $\hat{W}_\infty$ and is believed to contain all W_N by reduction. 3) This $\hat{W}_\infty$ algebra is identified as the (modified) second KP Hamiltonian structure (see also Ref. [6]). 4) We have found a two boson realization of the nonlinear $\hat{W}_\infty$ algebra, in closed form for all higher spin currents. 5) Using it, we discovered a field theoretical realization of $\hat{W}_\infty$ as a hidden current algebra in the 2d $SL(2,R)/U(1)$ coset model. 6) The well-known infinite set of KP Hamiltonians, considered as involutive $\hat{W}_\infty$ charges, generate infinitely many KP flows in the coset model, all of which preserve the $\hat{W}_\infty$ current algebra.

The above results 3)-6) hold true not only at the classical but also the *quantum* level: 7) A consistent quantum deformation of the $\hat{W}_\infty$ algebra, together with its two boson realization, is obtained by quantizing the $SL(2,R)_k/U(1)$ coset model. 8) We have been able to construct an infinite set of commuting quantum $\hat{W}_\infty$ charges at level k=1. 9) These quantum charges are used to generate a quantum KP hierarchy, whose Hamiltonian structure is just the quantum $\hat{W}_\infty$.

In this talk we give a unified description of these results. Two long papers containing detailed proofs will be published soon[7,8].

KP BI-HAMILTONIAN STRUCTURE, $\hat{W}_{1+\infty}$ AND $\hat{W}_\infty$

The integrable KP-Sato hierarchy is an infinite set of nonlinear partial-differential equations, which reads in the Lax form

$$\frac{\partial L}{\partial t_m} = [(L^m)_+, L] \tag{1}$$

where L is a first-order pseudo-differential operator (Psd) $L = D + \sum_{r=-1}^{\infty} u_r D^{-r-1}$, with coefficients u_r functions of z and various times $t_m (m = 1,2,3,\ldots)$, and $D \equiv \partial/\partial z$.

The integrability of the KP hierarchy is a direct consequence of its bi-Hamiltonian structure. To put the KP hierarchy into the Hamiltonian form

$$\frac{\partial u_r}{\partial t_m} = \left\{ u_r(z), \oint H_m(z) dz \right\} \quad , \tag{2}$$

one needs to define both the Hamiltonian functions $H_m(z)$ and the Poisson brackets

$$\{u_r(z), u_s(z')\} = k_{rs}(u_p(z))\delta(z-z') \ . \tag{3}$$

To obtain the bi-Hamiltonian structure, we substitute an arbitrary Psd of order N, $P = \sum_{s=-\infty}^{N} D^s p_s$, ($N$ can be arbitrarily large) into

$$\hat{K}(P) \equiv (\hat{L}P)_+\hat{L} - \hat{L}(P\hat{L})_+ = \sum_{r,s=-1}^{\infty} \hat{k}_{rs} p_s D^{-r-1} \ , \tag{4}$$

where $\hat{L} = L + c$ (c is a real parameter).

Proposition 1: For fixed pair (r,s), the operator $\hat{k}_{rs}$ becomes stable (*i.e.* independent of N) when N is sufficiently large. Explicitly, it is given by (for $r, s \geq 0$)

$$\begin{aligned}
\hat{k}_{rs} = & \sum_{\ell=0}^{s+1} \binom{s+1}{\ell} D^\ell \hat{u}_{r+s+1-\ell} - \sum_{\ell=0}^{r+1} \binom{r+1}{\ell} \hat{u}_{r+s+1-\ell}(-D)^\ell \\
& + \sum_{\ell=0}^{\infty} \Bigg[\sum_{t=\ell}^{\ell+s} (-1)^\ell \binom{t-s-1}{\ell} \\
& - \sum_{t=r+1}^{r+s+1} \sum_{k=0}^{t-r-1} (-1)^{\ell+k} \binom{t-k-1}{\ell-k} \binom{s}{k} \Bigg] \hat{u}_{t-\ell-1} D^\ell \hat{u}_{r+s-t} \ , \qquad (5) \\
\hat{k}_{r-1} = & -\sum_{\ell=1}^{r} \binom{r}{\ell} \hat{u}_{r-\ell}(-D)^\ell \ , \\
\hat{k}_{-1s} = & \sum_{\ell=1}^{s} \binom{s}{\ell} D^\ell \hat{u}_{s-\ell} \ , \qquad \hat{k}_{-1-1} = -D \ ;
\end{aligned}$$

with $\hat{u}_{-1} = u_{-1} + c$, $\hat{u}_r = u_r (r \geq 0)$. And the operator matrix k_{rs} defines a one-parameter family of Hamiltonian structures for functions u_r.

Proposition 2: The following limits

$$\{u_r(z), u_s(z')\}_1 = \lim_{c\to\infty} \left(\frac{1}{c} \hat{k}_{rs}(z) \right) \delta(z-z') \qquad r, s \leq 0 \ ; \tag{6}$$

$$\{u_r(z), u_s(z')\}_2 = \lim_{c\to 0} \hat{k}_{rs}(z) \delta(z-z') \qquad r, s \leq -1 \ . \tag{7}$$

coincide, respectively, with the first and second Hamiltonian structures of the KP hierarchy with the Hamiltonian functions

$$H_m^{(1)} = \frac{1}{m+1} Res L^{m+1} \ , \qquad H_m^{(2)} = \frac{1}{m} Res L^m \ . \tag{8}$$

Note the first KP Hamiltonian structure is linear in u_r ($r \geq 0$):

$$\lim_{c\to\infty} \left(\frac{1}{c} \hat{k}_{rs} \right) = \sum_{\ell=0}^{s} \binom{s}{\ell} D^\ell u_{r+s-\ell} - \sum_{\ell=0}^{r} \binom{r}{\ell} u_{r+s-\ell}(-D)^\ell \ . \tag{9}$$

(The u_{-1} current becomes decoupled from the algebra.) Previously we have identified it with the $W_{1+\infty}$ algebra. We call the nonlinear second Hamiltonian structure (7) (with $c=0$) as the $\hat{W}_{1+\infty}$ algebra, in which the current u_r is considered to have spin $r+2$ ($r \geq -1$). It is a universal W-algebra, containing all known W-algebras of the infinite type or of the finite type either by contraction or by reduction. However, u_{-1} does not evolve at all: $\partial u_{-1}/\partial t_m = 0$. One can always impose the constraint $u_{-1}=0$ which is second-class in $\hat{W}_{1+\infty}$.

Proposition 3: With $u_{-1}=0$, the modified second KP Hamiltonian structure is

$$
\begin{aligned}
k_{rs}^{(2D)} = & \sum_{\ell=0}^{s+1} \binom{s+1}{\ell} D^\ell u_{r+s+1-\ell} - \sum_{\ell=0}^{r+1} \binom{r+1}{\ell} u_{r+s+1-\ell}(-D)^\ell \\
& + \sum_{\ell=0}^{\infty} \Bigg[\sum_{t=\ell+1}^{\ell+s} (-1)^\ell \binom{t-s-1}{\ell} \\
& - \sum_{t=r+1}^{r+s} \sum_{k=0}^{t-r-1} (-1)^{\ell+k} \binom{t-k-1}{\ell-k} \binom{s}{k} \Bigg] u_{t-\ell-1} D^\ell u_{r+s-t} \\
& - \sum_{\ell=1}^{r} \sum_{k=1}^{s} (-1)^\ell \binom{r}{\ell} \binom{s}{k} u_{r-\ell} D^{\ell+k-1} u_{s-k} \ .
\end{aligned} \tag{10}
$$

It can be viewed as the unique nonlinear, centerless deformation of W_∞ under certain natural homogeneity requirements. We call it as the $\hat{W}_\infty$ algebra.

A direct consequence is the following interesting relation between $\hat{W}_\infty$ and $W_{1+\infty}$:

$$
\oint \left\{ \hat{W}_{r+1}(z), \hat{W}_{s+1}(z') \right\}_{2D} dz' = \oint \left\{ W_r(z), W_s(z') \right\}_1 dz' \ . \tag{11}
$$

FREE FIELD REALIZATION OF $\hat{W}_\infty$ AND $SL(2,R)/U(1)$

The known realization of $W_{1+\infty}$ with a complex fermion is simply summarized by

$$
L \equiv D + \sum_{r=0}^{\infty} u_r D^{-r-1} = D - \bar{\psi} D^{-1} \psi \ . \tag{12}
$$

Considering only the spin $s \geq 2$ currents, for the W_∞ generators we have

$$
(LD)_- \equiv \sum_{r=0}^{\infty} v_r D^{-r-1} = -(\bar{\psi} D^{-1} \psi D)_- \ . \tag{13}
$$

Meanwhile, W_∞ can be realized in terms of a complex boson

$$
L \equiv D + \sum_{r=0}^{\infty} v_r D^{-r-1} = D + \bar{j} D^{-1} j \ . \tag{14}
$$

where the bosonic currents $\bar{j}(z) = \bar{\phi}'(z)$, $j(z) = \phi'(z)$ satisfy

$$
\{\bar{j}(z), j(z')\} = \partial_z \delta(z-z') \ , \qquad \{\bar{j}(z), \bar{j}(z')\} = \{j(z), j(z')\} = 0 \ . \tag{15}
$$

A nonlinear deformation of this leads to:

Proposition 4: The following gives a realization of all currents u_r of $\hat{W}_\infty$:

$$L \equiv D + \sum_{r=0}^{\infty} u_r D^{-r-1} = D + \bar{j}\frac{1}{D-(\bar{j}+j)}j \\ = D + \bar{j}D^{-1}j + \bar{j}D^{-1}(\bar{j}+j)D^{-1}j + \cdots \quad . \tag{16}$$

This realization of $\hat{W}_\infty$ naturally emerges in the $SL(2,R)/U(1)$ coset model. Recall the bosonized $SL(2,R)_k/U(1)$ parafermion currents: (classically we set $k = 1$)

$$\psi_+ = \bar{j}e^{\bar{\phi}+\phi} \quad , \qquad \psi_- = je^{-\bar{\phi}-\phi} \quad . \tag{17}$$

We propose to expand the bilocal product $\psi_+(z)\psi_-(z')$ in powers of $z - z'$ to all orders.

Proposition 5: The product $\psi_+(z)\psi_-(z')$ generate all two-boson $\hat{W}_\infty$ currents (16):

$$\bar{j}e^{\bar{\phi}+\phi}(z)je^{-\bar{\phi}-\phi}(z') = \sum_{r=0}^{\infty} u_r(z)\frac{(z-z')^r}{r!} \quad . \tag{18}$$

Formally this can also be rewritten as $u_r(z) = \psi_+(z)(-\partial_z)^r\psi_-(z)$, or

$$L = D + \sum_{r=0}^{\infty} u_r D^{-r-1} = D + \psi_+ D^{-1}\psi_- \quad . \tag{19}$$

It follows that the Poisson brackets for $\hat{W}_\infty$ can be generated from

$$\left\{\bar{j}e^{\bar{\phi}+\phi}(z)je^{-\bar{\phi}-\phi}(z-\epsilon), \bar{j}e^{\bar{\phi}+\phi}(w)je^{-\bar{\phi}-\phi}(w-\sigma)\right\} \\ \equiv \{\psi_+(z)\psi_-(z-\epsilon), \psi_+(w)\psi_-(w-\sigma)\} = \sum_{r,s=0}^{\infty} k_{rs}(z)\delta(z-w)\frac{\epsilon^r\sigma^s}{r!s!} \quad . \tag{20}$$

INVOLUTIVE KP CHARGES AND $\hat{W}_\infty$ SYMMETRY IN $SL(2,R)/U(1)$

The KP charges, $Q_m \equiv \oint H_m(z)dz$, are known to be conserved under the KP flows:

$$\frac{\partial}{\partial t_n}\oint H_m(z,t_k)dz = 0 \quad . \tag{21}$$

and they are in involution:

$$\left\{Q_n^{(2)}, Q_m^{(2)}\right\}_{2D} = \left\{Q_{n-1}^{(1)}, Q_m^{(1)}\right\}_1 = 0 \quad . \tag{22}$$

This implies that in the $SL(2,R)/U(1)$ coset model there exist infinitely many commuting $\hat{W}_\infty$ charges, which generate an infinite symmetry

$$\frac{\partial u_r}{\partial t_m} = \{u_r, Q_m\} \quad , \quad \text{or} \quad \delta_m u_r = \epsilon_m\{u_r, Q_m\} \quad . \tag{23}$$

An important property of these symmetry transformations is that the $\hat{W}_\infty$ current algebra for the u_r's is preserved under these flows:

$$\frac{\partial}{\partial t_m}(\{\{u_r(z),u_s(z')\}-k_{rs}(z)\delta(z-z')) = 0 \ . \tag{24}$$

Also these symmetries are abelian to each other, in the sense that the KP flow in one charge direction is invariant under the transformation generated by another $\hat{W}_\infty$ charge:

$$\delta_m\left(\frac{\partial u_r}{\partial t_n} - \{u_r, Q_n\}\right) = \epsilon_m\{u_r,\{Q_n,Q_m\}\} = 0 \ . \tag{25}$$

Note that these results hold for the $W_{1+\infty}$ charges as well.

The free boson realization of $\hat{W}_\infty$ allows us to construct the KP flows of j and $\bar{j}$:

$$\frac{\partial\bar{j}}{\partial t_m} = \{\bar{j}, Q_m\} \ , \qquad \frac{\partial j}{\partial t_m} = \{j, Q_m\} \ . \tag{26}$$

(See also Ref. 9.) Note that the Hamiltonian structure of this hierarchy is simply the fundamental Poisson brackets (15) for $\bar{j}$ and j, which are invariant under the $\bar{j}$-j flows:

$$\frac{\partial\{\bar{j},j\}}{\partial t_m} = \{\{\bar{j},Q_m\},j\} + \{\bar{j},\{j,Q_m\}\} = \{\{\bar{j},j\},Q_m\} = 0 \ . \tag{27}$$

Similarly, the free fermion realization (12) of $W_{1+\infty}$ gives the KP Hamiltonians

$$H_1 = \bar{\psi}\psi \ , \quad H_2 = \frac{1}{2}\left(\bar{\psi}'\psi - \bar{\psi}\psi\right) \ , \quad H_3 = \frac{1}{3}\left(\bar{\psi}''\psi - \bar{\psi}\psi' + \bar{\psi}\psi''\right) \ , \quad etc. \tag{28}$$

Their charges can also be used to generate a compatible and integrable $\bar{\psi}$-ψ hierarchy

$$\frac{\partial\bar{\psi}}{\partial t_m} = \{\bar{\psi}, Q_m\} \ , \qquad \frac{\partial\psi}{\partial t_m} = \{\psi, Q_m\} \ . \tag{29}$$

QUANTUM $\hat{W}_\infty$ AND ITS TWO-BOSON REALIZATION

Because of the connection of the $SL(2,R)/U(1)$ model to the black holes in 2D string theory, it is necessary to generalize what we have obtained above to the quantum theory. We have been succeeded in constructing a quantum version of the nonlinear $\hat{W}_\infty$ algebra, by quantizing the $SL(2,R)_k/U(1)$ coset model, with the bosonized currents

$$\begin{aligned}\psi_+(z;p) &= \frac{1}{2}\left[\left(1+\sqrt{1-2p}\right)\bar{j} + \left(1-\sqrt{1-2p}\right)j\right] e^{\sqrt{p}(\bar{\phi}+\phi)} \ , \\ \psi_-(z;p) &= \frac{1}{2}\left[\left(1-\sqrt{1-2p}\right)\bar{j} + \left(1+\sqrt{1-2p}\right)j\right] e^{-\sqrt{p}(\bar{\phi}+\phi)} \ .\end{aligned} \tag{30}$$

where $\phi(z)$ and $\bar{\phi}$ are a pair of complex bosons, and their currents $\bar{j}(z)$, $j(z)$ satisfy

$$\bar{j}(z)j(z') \sim \frac{1}{(z-z')^2} \ , \qquad \bar{j}(z)\bar{j}(z') \sim j(z)j(z') \sim 0 \ . \tag{31}$$

Note the parameter $p \equiv k^{-1}$ plays the role of the Planck constant as the essential parameter in quantum corrections (or quantum deformation). We have included necessary quantum corrections in the currents. The classical limit is recovered by first rescaling $\phi_i \to \sqrt{k}\phi_i$, $\psi_\pm \to k\psi_\pm$ and then letting $k \to \infty$.

Generalizing the classical equation (18), we use the whole OPE (up to all orders),

$$\psi_+(z;p)\psi_-(z';p) = \epsilon^{-2p}\left\{\epsilon^{-2} + \sum_{r=0}^{\infty} u_r(z;p)\frac{\epsilon^r}{r!}\right\} \tag{32}$$

(with $\epsilon \equiv z - z'$) to generate the quantum $\hat{W}_\infty(p)$ currents $u_r(z;p)$ in the KP basis. (A similar observation was first made in Ref. [10].) A closed expression for all $u_r(z;p)$, in terms of $\bar{j}$ and j only, has been derived in Ref. [8].

The complete structure of the quantum $\hat{W}_\infty(p)$ algebra can be manifested by the OPE's between two currents $u_r(z;p)$ and $u_s(w;p)$, as usual in conformal field theory. In principle, these OPE's can be extracted from the OPE of four $SL(2,R)_k/U(1)$ currents:

$$(\epsilon\sigma)^{2p}(\psi_+(z;p)\psi_-(z-\epsilon;p))(\psi_+(w;p)\psi_-(w-\sigma;p)) \ . \tag{33}$$

The closure of the quantum $\hat{W}_\infty(p)$ algebra is ensured by the closure of the OPE's associated with the enveloping algebra of the $SL(2,R)_k$ currents in the neutral sector in the conformal model. However, once we write the algebra in the form of OPE's among u_r and u_s, the currents u_r may be considered as independent of each other, and the associativity of the OPE's automatically leads to closed Jacobi identities. So our quantum version of the $\hat{W}_\infty$ algebra holds even beyond the context of the coset model.

The $p = 1$ case, corresponding to level $k = 1$, may be interpreted as the "typical" quantum case, For this value of p, a lot of expressions simplify.

Proposition 6: In the case of $p = 1$, we have

$$\begin{aligned} u_0(z)u_s(w) &= \frac{-1}{z-w}u_s'(w) + \text{terms in other powers of } (z-w)^{-1} \ , \\ u_1(z)u_s(w) &= \frac{-2}{z-w}\left[\sum_{l=1}^{s}(-1)^l\binom{s}{l}\left(u_0^{(l)}u_{s-l}\right) + \frac{u_{s+1}'}{(s+1)} + \frac{(-1)^s u_0^{(s+2)}}{(s+1)(s+2)}\right](w) \\ &\quad + \text{terms in other powers of } (z-w)^{-1} \ . \end{aligned} \tag{34}$$

COMMUTING QUANTUM $\hat{W}_\infty$ CHARGES AND QUANTUM KP HIERARCHY

The above OPE's provide sufficient information for constructing an infinite set of commuting quantum $\hat{W}_\infty$ charges, whose densities are quantum deformation of the classical Hamiltonian functions (1.4). For generic p, it is not hard to construct

$$\begin{aligned} H_2(z;p) &= \frac{1}{(1-2p)}u_0(z;p) \ , \qquad H_3(z;p) = u_1(z;p) + \frac{1}{2}u_0'(z;p) \ . \\ H_4(z;p) &= u_2(z;p) + u_1'(z;p) + \frac{(5+4p)}{15}u_0''(z;p) + \frac{p}{(1-2p)}(u_0u_0)(z;p) \ . \end{aligned} \tag{35}$$

In the "typical" quantum case with $p = 1$, we can exploit the two general OPE (2.24) to explicitly construct an infinite number of commuting quantum $\hat{W}_\infty$ charges Q_m.

Let us recall that in conformal field theory the local product $(AB)(z)$ is defined by

$$(AB)(z) = \oint_z \frac{A(w)B(z)}{w-z}dw \tag{36}$$

in which the small contour of integration encircles z. This product is neither commutative or associative. The symmetric local product of N local operators is defined to be the totally symmetrized sum of their multiple local products taken from the left:

$$\langle A_1A_2\cdots A_N\rangle = \frac{1}{N!}\sum_{P\{i\}}(\cdots((A_{i_1}A_{i_2})A_{i_3})\cdots A_{i_N}) \tag{37}$$

where $P\{i\}$ denotes the summation over all possible permutations.

To construct quantum $\hat{W}_\infty$ charges, let us assign a degree 1 to ∂_z and $r+2$ to u_r, and assume that the quantum charge-density $H_m(z)$ is homogeneous and of degree m, and led by the highest-spin current u_{m-2} with unity coefficient. We require their charges commute with each other:

$$\left[\oint_0 H_n(z)dz, \oint_0 H_m(w)dw\right] = 0 \ . \tag{38}$$

Proposition 7: The commutativity with Q_2 is always satisfied:

$$\left[\oint_0 H_2(z)dz, \oint_0 H_m(w)dw\right] = 0 \ . \tag{39}$$

Thus the first set of nontrivial equations in Eq. (38) start with $n = 3$. Here H_3 can only be u_1 plus a derivative of u_0; the latter does not contribute to the charge Q_3. We have shown that charges Q_m are completely determined by the requirement of commutativity with Q_3 alone. Both amusingly and amazingly, the so-determined charges Q_m automatically commute with each other. (A similar situation has happened before for commuting charges for the quantum KdV equation.[11])

Proposition 8: The charges

$$Q_m = \oint_0 \sum_l \sum_{\{i\}} C_{i_1 i_2 \cdots i_l}(m)(u_{i_1} u_{i_2} \cdots u_{i_l})(z)dz \tag{40}$$

with the elegant expression for the coefficients

$$C_{i_1 i_2 \cdots i_\ell}(m) = \frac{(-1)^{\ell-1}(\ell-1)!(m-2)!}{d_1!d_2!\cdots d_k! i_1! i_2! \cdots i_\ell!} \ , \tag{41}$$

provides a unique solution to Eq. (38) for arbitrary m. Here for given number of currents in the product, ℓ, the summation is over all partitions $\{i_k\}$ of $m-2\ell$, satisfying $i_1 + i_2 + \cdots + i_\ell = m - 2\ell$; and here d's denote the degeneracies in the partition: $0 \le i_1 = i_2 = \cdots = i_{d_1} < i_{d_1+1} = \cdots = i_{d_1+d_2} < \cdots = i_{d_1+d_2+\cdots+d_k(=\ell)}$.

We emphasize that in the proof of this Proposition, we have used only the OPE's in Eq. (34); the two boson realization (32) of the $\hat{W}_\infty$ currents u_r is never used.

Now it is straightforward to construct a quantum version of the KP hierarchy in the Hamiltonian form (2). We use the quantum $\hat{W}_\infty$ as the Hamiltonian structure, and the densities of the quantum charges Q_m as corresponding Hamiltonian functions H_m. The latter naturally generate an infinite set of compatible flows in times t_m ($m = 1, 2, \ldots$):

$$\frac{\partial u_r}{\partial t_m} = [u_r, Q_{m+1}] \ . \tag{42}$$

The mutual commutativity of these quantum charges implies that they are conserved charges of the flows (42):

$$\frac{\partial Q_n}{\partial t_m} = [Q_n, Q_{m+1}] = 0 \ . \tag{43}$$

This ensures the complete integrability of the hierarchy (42). We call it the $p=1$ quantum KP hierarchy and view it as a desired quantum version of the classical KP hierarchy. To justify this, we have computed the first three quantum KP charges for generic p and have shown that they coincide with the charges (40) at $p=1$ and reduce to the classical KP charges when $p \to 0$. Also we have calculated the first a few quantum KP flows with generic p:

$$\begin{aligned} &\frac{\partial u_0}{\partial t_1} = (1-2p)u_0' \ , \quad \frac{\partial u_1}{\partial t_1} = (1-2p)u_1' \ , \quad \frac{\partial u_0}{\partial t_2} = (1-2p)(2u_1' + u_0'') \ , \\ &\frac{\partial u_1}{\partial t_2} = (2-3p)u_2' + (1-p)u_1'' + \frac{(1+2p)p}{6}u_0''' + p(u_0u_0)' \ , \\ &\frac{\partial u_0}{\partial t_3} = 3(1-2p)(u_2' + u_1'') + \frac{(3-4p-p^2)}{3}u_0''' + 3p(u_0u_0)' \ . \end{aligned} \tag{44}$$

Manipulating the last three equations, we obtain the first dynamically nontrivial quantum evolution equation for the current $u_0(z;p)$:

$$6(2-3p)\frac{\partial u_0'}{\partial t_3} - 9\frac{\partial^2 u_0}{\partial t_2^2} = (3 - p - 16p^2 + 18p^3)u_0'''' + 18(1-p)p(u_0u_0)'' \ . \tag{45}$$

Rescaling $u_0 \to u_0/p$ and taking $p \to 0$, this reduces to the classical KP equation, as claimed.

It should also be possible to obtain an alternative quantum deformation of the classical KP hierarchy through deforming its much simpler first Hamiltonian structure – the $W_{1+\infty}$. Because of its linearity, quantization of $W_{1+\infty}$ should be straightforward. In the KP basis, the complete structure of the quantum $W_{1+\infty}$ is neatly manifested by the following OPE's:

$$\begin{aligned} u_r(z)u_s(w) = &\sum_{\ell=0}^{r} \frac{r!}{(r-\ell)!} \frac{u_{r+s-\ell}(z)}{(z-w)^{\ell+1}} - \sum_{\ell=0}^{s}(-1)^{\ell}\frac{s!}{(s-\ell)!} \frac{u_{r+s-\ell}(w)}{(z-w)^{\ell+1}} \\ &+ \frac{(-1)^s c r! s!}{(z-w)^{r+s+2}} + O(z-w) \ . \end{aligned} \tag{46}$$

The remaining issue is to construct an complete set of infinitely many commuting quantum charges in accordance to (46), in order for the associated quantum KP hierarchy to be integrable.

To conclude, we remark that it would be interesting to clarify the role of our world-sheet nonlinear $\hat{W}_\infty$ symmetry in physics of 2D black holes, and to invent a formalism for 2D strings in terms of the KP hierarchy or $\hat{W}_\infty$ constraints.

REFERENCES

1. F. Yu and Y.-S. Wu, *Phys. Lett.* **B263** (1991) 220.
2. F. Yu and Y.-S. Wu, *Nucl. Phys.* **B373** (1992) 713.
3. F. Yu and Y.-S. Wu, *Phys. Rev. Lett.* **68** (1992) 2996.
4. F. Yu and Y.-S. Wu, Utah preprint UU-HEP-92/11, May 1992; to be published in *Phys. Lett. B.*
5. K. Yamagishi, *Phys. Lett.* B259 (1991) 436.
6. J. M. Figueroa-O'Farrill, J. Mas and E. Ramos, *Phys. Lett.* **B266** (1991) 298; preprint BONN-HE-92/20, US-FT-92/7 or KUL-TF-92/20.
7. F. Yu and Y.-S. Wu, Utah preprint UU-HEP-92/07, August 1992.
8. F. Yu and Y.-S. Wu, Utah preprint UU-HEP-92/12, September, 1992.
9. D. A. Depireux, Laval preprint LAVAL-PHY-21-92; J. M. Figueroa-O'Farrill, J. Mas and E. Ramos, preprint BONN-HE-92/17, US-FT-92/4 or KUL-TF-92/26.
10. I. Bakas and E. Kiritsis, preprint UCB-PTH-91/44, or LBL-31213 or UMD-PP-91/37.
11. R. Sasaki and I. Yamanaka, it Commun. Math. Phys. 108 (1987) 691; *Adv. Stud. in Pure Math.* **16** (1988) 271.

Parallel Sessions 1

Geometry of Quantum Group and Applications

PURE AFFINE QUANTUM GROUPS

M.CHAICHIAN*
High Energy Physics Laboratory, Department of Physics,
University of Helsinki,
Siltavuorenpenger 20 C, SF-00170 Helsinki, Finland

and

A.P.DEMICHEV
Nuclear Physics Institute, Moscow State University,
119899, Moscow, Russia

Abstract

We show that the appearance of additional scaling transformations in the formerly constructed inhomogeneous q-groups is explained by the properties of the underlying quantum projective space. We construct well defined quantum inhomogeneous groups $IGL(N)_q$ without additional dilatations via the projective approach based on q-spaces with one commuting coordinate.

1.Introduction.

At present there is a well developed theory of quantum deformation of the semisimple matrix Lie groups and Lie algebras [1],[2], [3], [4]. But from the point of view of possible symmetries of the "quantum" spaces, it is interesting to construct and study the inhomogeneous affine-like groups containing the translations. In particular, a quantum analog of Poincaré group has to belong to this class. The first step in such direction was made in [5] and [6] where a q-analog of the group of 2D Euclidean transformations was found. However, this case is rather degenerate because the coaction $\delta(z), \delta(\bar{z})$ on the complex coordinates $z, \bar{z}$ of q-plane has a form

$$\delta(z) = a \otimes z + w \otimes 1,$$

$$\delta(\bar{z}) = \bar{a} \otimes \bar{z} + \bar{w} \otimes 1,$$

where a corresponds to quantum U(1)-rotations and $w, \bar{w}$ are translational parameters. Thus this transformation is factorized and becomes one-dimensional in fact. The problem arises if one try to define inhomogeneous transformations of higher dimensional spaces [7] or of the 2D q-plane with real coordinates. In these cases one is forced to introduce additional scaling transformations [7] which have no analog in the classical limit $q \to 1$. There is the same problem in attempts to construct q-Poincaré algebra [8]. From the other hand there is another approach to the construction of the inhomogeneous universal quantum algebras (without additional dilatations) through contraction of homogeneous algebras, the deformation parameter q being also contracted [9], [10]. This shows that pure affine q-algebras can exist in principle. Unfortunately, the corresponding quantum space transformations and relation with q-derivatives are unknown (one obtains usual commutator for generators of translations in this case). Besides the construction of the universal R-matrix [1] is not consistent with contraction of q [11].

In this letter we give explanation of the appearance of additional dilatations in q-inhomogeneous groups and suggest method for the construction of pure affine q-groups.

*Talk given by M. Chaichian.

2. Affine patch of projective space and dilatations.

Let $\mathbf{CP}^N$ be a projective space with homogeneous coordinates $X = (X^1, ..., X^{N+1}) \in \mathbf{C}^{N+1}$ (see e.g. [12]) and with the equivalence relation $X \sim \lambda X, 0 \neq \lambda \in \mathbf{C}$. The projective space is topologically non-trivial and can be covered by $N+1$ patches $U_1, ..., U_{N+1}$ defined by the conditions

$$X \in U_k \Longleftrightarrow X^k \neq 0.$$

On the patch U_{N+1} one can introduce inhomogeneous coordinates $x^m = X^m/X^{N+1}, m = 1, ..., N$. The action of the group GL(N+1) is deduced from its action in $\mathbf{C}^{N+1}$:

$$X \rightarrow X' = MX; \qquad M \in GL(N+1). \tag{1}$$

On the patch U_{N+1} it is useful to write M in a block-diagonal form,

$$M = \begin{pmatrix} A & \vec{b} \\ \vec{c}^{\,t} & d \end{pmatrix} \tag{2}$$

where $A \in GL(N)$, $\vec{b}$ and $\vec{c}$ are N-dimensional vectors d is a number. The action (1) on the inhomogeneous coordinates $\vec{x} = (x^1, ..., x^m)^t$ takes the form

$$\vec{x} \rightarrow \vec{x}\,' = \frac{A\vec{x} + \vec{b}}{\vec{c} \cdot \vec{x} + d}\,. \tag{3}$$

Let us denote by G the subgroup of GL(N+1)-matrices with $\vec{c} = 0$. The requirement $M \in GL(N+1)$ then gives

$$M = \begin{pmatrix} A & \vec{b} \\ 0 & d \end{pmatrix} \in G \Longleftrightarrow A \in GL(N), d \neq 0. \tag{4}$$

Eq. (3) then reduces to IGL(N) transformations of U_{N+1} given as

$$\vec{x} \rightarrow \vec{x}\,' = d^{-1}(A\vec{x} + \vec{b}) = K\vec{x} + \vec{t}, \tag{5}$$

where $K = d^{-1}A \in GL(N)$ and $\vec{t} = d^{-1}\vec{b}$.

We now identify U_{N+1} with the affine space A^N, and the mapping (5) with the inhomogeneous (affine) transformations. We note that even if we restrict originally the homogeneous part of the affine transformations to a smaller group, e.g. to SL(N), then K is extended by the dilatations. So we are forced to include dilatations into the affine transformations.

Thus the problem of the appearence of additional dilatations in q-inhomogeneous groups can be reduced to a simpler question of the possibility of quantum deformation of the subgroup G of the matrices (4) satisfying the additional condition $d = 1$. Now we shall show that if there is a coordinate among $X^1, ..., X^N$ which does not commutes with X^{N+1} then $d \neq 1$ in the quantum deformation of (4),(5). The coordinates of q-space obey the relation

$$X^i X^j = q_{ij} X^j X^i\,, \qquad i < j$$

and

$$\xi_k \xi_l = q_{kl}^{-1} \xi_l \xi_k\,, \qquad k < l,$$

where ξ_k are the coordinates of the space A^* dual to the space A of formal power series in X^i, X^{N+1} (hereafter i,j,k,l = 1,...,N). The q-matrix transforms these coordinates via coactions

$$\delta : A \rightarrow M \otimes A : \delta(x^i) = \sum_j z^i_j \otimes x^j$$

$$\delta^* : A^* \to M \otimes A^* : \delta^*(\xi_k) = \sum_l \xi_l \otimes z^l{}_k$$

The matrix elements of any 2×2 submatrix of the matrix M

$$\begin{pmatrix} z^m{}_n & z^m{}_q \\ z^p{}_n & z^p{}_q \end{pmatrix} \equiv \begin{pmatrix} a & b \\ c & d \end{pmatrix} \tag{6}$$

have the following commutation relations [3]

$$ab = q_h ba \tag{7}$$
$$cd = q_h dc \tag{8}$$
$$ac = q_v ca \tag{9}$$
$$bd = q_v db \tag{10}$$
$$ad - q_v q_h^{-1} da = q_v cb - q_h^{-1} bc \tag{11}$$
$$ad - q_v^{-1} q_h da = -(q_v^{-1} cb - q_h bc) \tag{12}$$
$$q_v \equiv q_{pm}, \qquad q_h \equiv q_{qn}. \tag{13}$$

Now let $j = N + 1$ and i corresponds to the coordinates X^i such that $q_v = q_{iN+1} \neq 1$. The corresponding submatrix has a form

$$\begin{pmatrix} a & b \\ 0 & d \end{pmatrix} \tag{14}$$

and it is immediately seen from (10) that $z^{N+1}{}_{N+1} = d \neq 1$ in this case. Thus one has to introduce non-trivial dilatations. If $X^{N+1} = e$ is an invertible element of the algebra A, then it is possible to define inhomogeneous coordinates in quantum space also

$$x^i = X^i e^{-1}.$$

In this case the cotransformations

$$\delta(X^i) = X'^{\,i} = \sum_{j=1}^{N} z^i{}_j \otimes X^j + z^i{}_{N+1} \otimes e\ ,$$

$$\delta(e) = e' = d \otimes e\ ,$$

become

$$\delta(x^i) = x'^{\,i} \equiv X'^{\,i}(e')^{-1} = \left(\sum_{j=1}^{N} z^i{}_j \otimes X^j + z^i{}_{N+1} \otimes e\right)\left(d^{-1} \otimes e^{-1}\right) =$$

$$= \left(\sum_{j=1}^{N} z^i{}_j d^{-1} \otimes X^j + z^i{}_{N+1} d^{-1} \otimes 1\right) .$$

This coincides indeed with the transformation which is called q-inhomogeneous one in [7], [8]. But as we have mentioned above they do not reduced to ordinary affine group in the limit $q \to 1$ because of the additional scalings.

3.q-Affine group

From the above discussion it is clear how to avoid the appearance of dilatations in the affine transformations. We have to demand that the last coordinate X^{N+1} of the initial projective space commutes with all the others. As an example, let $q_{ij} = q\ \forall i, j = 1, ..., N;\ i < j;\ q_{iN+1} = 1$. Denote for convenience $z^i{}_{N+1} \equiv t^i$. Then, from (7) -(12), we have for the matrix elements of

$$\begin{pmatrix} Z & \vec{t} \\ 0 & d \end{pmatrix} \tag{15}$$

the following commutation relation

1. For $z^i{}_j(i, j = 1, ..., N)$ we have the usual commutation relations of the homogeneous linear q-transformations [3], [1] (see (7) - (12)).
2. For t^i :

$$z^i{}_j t^i = t^i z^i{}_j,$$
$$t^i t^k = q t^k t^i,$$
$$z^i{}_j t^k - q t^k z^i{}_j = q z^k{}_j t^i - t^i z^k{}_j, \tag{16}$$
$$z^i{}_j t^k - q^{-1} t^k z^i{}_j = -q^{-1} z^k{}_j t^i - t^i z^k{}_j, \tag{17}$$
$$i < k; \qquad j = 1, ..., N .$$

3. For d :

$$z^i{}_j d = d z^i{}_j,$$
$$t^i d = d t^i, \qquad i, j = 1, ..., N .$$

It is easy to verify straightforwardly that the conditions $z^{N+1}{}_i = 0$ do not contradict any of the commutation relation (7) - (12). From the other hand one can not define (at least in this way) the pure translation group as is seen from (16) - (17) (it is impossible to put all $z^i{}_j$ equal to unity and $z^i{}_j = 0$, $i \neq j$). The parameter d and coordinate e commute with all the quantities and can be put equal to unity without any contradiction.

The antipode S and counity ε are defined by one of the well known methods for matrix corepresentation of Hopf algebra. For the homogeneous part its form is well known [1], [4], [3]. For the translations it has a form similar to that of [7] but without the additional scaling parameter:

$$S(t^i) = -S(z^i{}_j) t^j.$$

The form of counity is obvious

$$\varepsilon \begin{pmatrix} Z & \vec{t} \\ 0 & 1 \end{pmatrix} = \begin{pmatrix} 1 & 0 \\ 0 & 1 \end{pmatrix}.$$

4. Conclusion

In conclusion, using one patch of the quantum projective space defined with one commuting coordinate we have managed to define pure affine quantum groups which become the usual affine groups in the limit $q \to 1$. By putting additional conditions on homogeneous matrix elements $z^i{}_j$ [1], one can obtain different inhomogeneous groups with (pseudo)orthogonal, unitary or symplectic homogeneous part. The explicit example of physically interesting 4D Poincaré group and connection with differential calculus on q-spaces [13] will be presented in a later communications.

Acknowledgements

We are grateful to J. de Azcárraga, J.Schulze and V.N.Tolstoy for useful discussions.

References

[1] L.D.Faddeev, N.Yu.Reshetikhin and L.A.Takhtadjan, Algebra i Analis, 1 (1989) 178.

[2] S.L.Woronowicz, Comm.Math.Phys., 111 (1987) 613.

[3] Yu.I.Manin, Comm.Math.Phys., 123 (1989) 163.

[4] T.H.Koornwinder, in Orthogonal Polynomials, ed. P.Nevai, Kluwer Acad.Pub., 1990, p.257.

[5] S.L.Woronowicz, Comm.Math.Phys., 136 (1991) 399.

[6] L.L.Vaksman and L.I.Korogodsky, Dokl.Akad.Nauk SSSR, 304 (1989) 1036 (in Russian).

[7] M.Schlieker, W.Weich and R.Weixler, Inhomogeneous Quantum Groups, pr.LMU-TPW, 1991.

[8] O.Ogievetsky, W.B.Schmidke, J.Wess and B.Zumino, q-Deformed Poincaré Algebra, pr.MPI-Ph/91-98, 1991.

[9] E.Celegnini, R.Giachetti, E.Sorace and M.Tarlini, Jour.Math.Phys., 31 (1990) 2548; *ibid.* 32 (1991) 1155; *ibid.* 32 (1991) 1159.

[10] J.Lukierski, H.Ruegg, A.Nowicki and V.N.Tolstoy, Phys.Lett., 264B (1991) 331.

[11] V.N.Tolstoy, private communication.

[12] W.V.D.Hodge and D.Pedoe, Methods of Algebraic Geometry, v.I, Cambridge, 1947.

[13] J.Wess and B.Zumino, Nucl.Phys.Suppl. 19B (1990) 330.

q - DEFORMED CONFORMAL ALGEBRA AND ITS q - DEFORMED LORENTZ, WEYL AND POINCARÉ SUBALGEBRAS

V.K. Dobrev

International Centre for Theoretical Physics, Trieste, Italy,
on leave from Bulgarian Academy of Sciences, Sofia, Bulgaria

1. $U_q(sl(4,C))$. In this paper we review the example of the q - deformed conformal algebra and its subalgebras given in [1,2] to illustrate a canonical procedure proposed in [1] (cf. also [2]) for the q - deformation of the real forms $\mathcal{G}$ of complex Lie (super-) algebras associated with (generalized) Cartan matrices.

First we need the deformation of the complexification $\mathcal{G}^{\mathbf{C}} = sl(4,\mathbf{C})$ of $\mathcal{G} = su(2,2)$. The q-deformation $\mathcal{U}_q(sl(4,\mathbf{C}))$ is defined in [3,4] as the associative algebra over $\mathbf{C}$ with Chevalley generators $X_j^\pm$, H_j , $j = 1,2,3$ and with relations :

$$[H_j\ ,\ H_k] \ =\ 0,\ \ [H_j\ ,\ X_k^\pm] \ =\ \pm a_{jk} X_k^\pm\ , \tag{1a}$$

$$[X_j^+\ ,\ X_k^-] \ =\ \delta_{jk}\frac{q^{H_j/2} - q^{-H_j/2}}{q^{1/2} - q^{-1/2}} \ =\ \delta_{jk}\,[H_j]_q\ , \tag{1b}$$

$$\begin{gathered}(X_j^\pm)^2 X_k^\pm - [2]_q X_j^\pm X_k^\pm X_j^\pm + X_k^\pm (X_j^\pm)^2 = 0,\ (jk) = (12),(21),(23),(32)\\ [X_1^\pm, X_3^\pm] \ =\ 0\ ,\end{gathered} \tag{1c}$$

where $[x]_q = (q^{x/2} - q^{-x/2})/(q^{1/2} - q^{-1/2})$, $(a_{jk}) = (2(\alpha_j,\alpha_k)/(\alpha_j,\alpha_j))$, $j,k = 1,2,3$, is the Cartan matrix of $sl(4,\mathbf{C})$; $\alpha_1,\alpha_2,\alpha_3$ are the simple roots; the non-zero products between the simple roots are: $(\alpha_j,\alpha_j) = 2$, $j = 1,2,3$, $(\alpha_1,\alpha_2) = (\alpha_2,\alpha_3) = -1$. The non-simple positive roots are : $\alpha_{12} = \alpha_1 + \alpha_2$, $\alpha_{23} = \alpha_2 + \alpha_3$, $\alpha_{13} = \alpha_1 + \alpha_2 + \alpha_3$. The elements H_j span the Cartan subalgebra $\mathcal{H}^{\mathbf{C}}$ of $\mathcal{G}^{\mathbf{C}}$ while the elements $X_j^\pm$ generate the subalgebras $\mathcal{G}^\pm$ of $\mathcal{G}^{\mathbf{C}}$; thus we have the standard triangular decomposition $\mathcal{G}^{\mathbf{C}} = \mathcal{G}^+ \oplus \mathcal{H}^{\mathbf{C}} \oplus \mathcal{G}^-$.

The algebra $U_q(sl(4,\mathbf{C}))$ is a Hopf algebra with co-multiplication δ, co-unit ε and antipode γ defined on the generators as follows [3,4]:

$$\delta(H_j) \ =\ H_j \otimes 1 \ +\ 1 \otimes H_j\ ,\quad \delta(X_j^\pm) \ =\ X_j^\pm \otimes q^{H_j/4} \ +\ q^{-H_j/4} \otimes X_j^\pm\ , \tag{2a}$$

$$\varepsilon(H_j) \ =\ \varepsilon(X_j^\pm) \ =\ 0\ ,\quad \gamma(H_j) \ =\ -H_j\ ,\quad \gamma(X_j^\pm) \ =\ -q^{\pm 1/2}\, X_j^\pm\ . \tag{2b}$$

The Cartan-Weyl basis for the non-simple roots is given by (cf. [4]):

$$X_{jk}^\pm \ =\ \pm q^{\mp 1/4}(q^{1/4} X_j^\pm X_k^\pm - q^{-1/4} X_k^\pm X_j^\pm)\ ,\ \ (jk) \ =\ (12),(23)\ , \tag{3a}$$

$$\begin{aligned}X_{13}^\pm &= \pm q^{\mp 1/4}(q^{1/4} X_1^\pm X_{23}^\pm - q^{-1/4} X_{23}^\pm X_1^\pm) =\\ &= \pm q^{\mp 1/4}(q^{1/4} X_{12}^\pm X_3^\pm - q^{-1/4} X_3^\pm X_{12}^\pm)\ .\end{aligned} \tag{3b}$$

The commutation relations and Hopf algebra relations for the generators in (3) follow from formulae (1),(2).

2. $U_q(su(2,2))$. Let us consider the conformal algebra $su(2,2) \cong so(4,2)$. It has 15 generators $Y_{AB} = -Y_{BA}$, $A,B = 1,2,3,5,6,0$ satisfying

$$[Y_{AB}, Y_{CD}] = \eta_{BC} Y_{AD} - \eta_{AC} Y_{BD} - \eta_{BD} Y_{AC} + \eta_{AD} Y_{BC}\ , \tag{4}$$

where $\eta_{AB} = \text{diag}(- - - - + +)$. It has three non-conjugate classes of Cartan subalgebras represented, say, by $\mathcal{H}^a$, $a = 0, 1, 2$ with a non-compact generators. Thus according to our procedure [1] it has five different deformations - three in the case of $\mathcal{H}^2$ (since there are three nontrivial parabolic subalgebras) and one each for the other two choices of Cartan subalgebras. We shall work with the most noncompact Cartan subalgebra $\mathcal{H} = \mathcal{H}^2$ and with the maximal parabolic subalgebra. Since $su(2,2)$ is the conformal algebra of 4 - dimensional Minkowski space - time we would like to deform it consistently with the subalgebra structure relevant for the physical applications. These subalgebras are : the *Lorentz* subalgebra $\mathcal{M} \cong so(3,1)$ generated by $Y_{\mu\nu}$, $\mu,\nu = 1,2,3,0$, the subalgebra $\tilde{\mathcal{N}}$ of *translations* generated by $P_\mu = Y_{\mu 5} + Y_{\mu 6}$, the subalgebra $\mathcal{N}$ of *special conformal transformations* generated by $K_\mu = Y_{\mu 5} - Y_{\mu 6}$, the *dilatations* subalgebra $\mathcal{A}$ generated by $D = Y_{56}$. The commutation relations besides those for the Lorentz subalgebra are:

$$[D\ ,\ Y_{\mu\nu}] = 0\ ,\quad [D\ ,\ P_\mu] = P_\mu\ ,\quad [D\ ,\ K_\mu] = -K_\mu\ . \tag{5a}$$

$$[Y_{\mu\nu}\ ,\ P_\lambda] = \eta_{\nu\lambda}P_\mu - \eta_{\mu\lambda}P_\nu\ ,\quad [Y_{\mu\nu}\ ,\ K_\lambda] = \eta_{\nu\lambda}K_\mu - \eta_{\mu\lambda}K_\nu\ ,\quad [P_\mu\ ,\ K_\nu] = 2Y_{\mu\nu} + 2\eta_{\mu\nu}D\ . \tag{5b}$$

The algebra $\mathcal{P}_{max} = \mathcal{M} \oplus \mathcal{A} \oplus \mathcal{N}$ (or equivalently $\tilde{\mathcal{P}}_{max} = \mathcal{M} \oplus \mathcal{A} \oplus \tilde{\mathcal{N}}$) is the so called *maximal parabolic subalgebra* of $\mathcal{G}$, where $\tilde{\mathcal{N}}, \mathcal{N}$, resp., are subalgebras forming the positive, negative, resp., root spaces of the root system $(\mathcal{G}, \mathcal{A})$.

For a deformation of $su(2,2)$ one has to express its generators as complex linear combinations of the generators of its complexification $sl(4, \mathbf{C})$ compatible with the reality structure. The deformation introduced in [1,2] is given by the following relations:

$$H = (1/2)(H_1 + H_3)\ ,\ M^\pm = X_1^\pm + X_3^\pm\ ,\ \tilde{D} = (i/2)(H_1 - H_3)\ ,\ N^\pm = i(X_1^\pm - X_3^\pm)\ , \tag{6a}$$

$$D = (1/2)(H_1 + H_3) + H_2\ , \tag{6b}$$

$$P_0 = i(X_{12}^+ + X_{23}^+)\ ,\ P_1 = i(X_{13}^+ + X_2^+)\ ,\ P_2 = X_{13}^+ - X_2^+\ ,\ P_3 = i(X_{23}^+ - X_{12}^+)\ , \tag{6c}$$

$$K_0 = -i(X_{12}^- + X_{23}^-),\ K_1 = i(X_{13}^- + X_2^-),\ K_2 = X_2^- - X_{13}^-,\ K_3 = i(X_{23}^- - X_{12}^-). \tag{6d}$$

The Cartan subalgebra $\mathcal{H}$ of $\mathcal{G}$ is spanned by $H, \tilde{D}, D$ and its complexification is the Cartan subalgebra $\mathcal{H}^{\mathbf{C}}$ of $\mathcal{G}^{\mathbf{C}}$.

Now we can derive the relations in $U_q(su(2,2))$:

1) According to our general scheme the deformed Lorentz subalgebra is a Hopf subalgebra of $U_q(su(2,2))$. Its commutaion relations are given by:

$$[H, M^\pm] = \pm M^\pm\ ,\ [H, N^\pm] = \pm N^\pm\ , [\tilde{D}, M^\pm] = \pm N^\pm\ ,\ [\tilde{D}, N^\pm] = \mp M^\pm\ , \tag{7a}$$

$$[M^+, M^-] = [N^-, N^+] = 2[H]cos(\tilde{D}h/2)\ ,\ [M^\pm, N^\mp] = \pm 2\frac{cosh(Hh/2)sin(\tilde{D}h/2)}{sinh(h/2)} \tag{7b}$$

2) The commutation relations of the generators $H, \tilde{D}, D$ of the Cartan subalgebra $\mathcal{H}$ are not deformed, e.g., (7a) above, and will be omitted further.

3) The deformation of the translations and special conformal transformations subalgebras is given by:

$$P_a(P_0 \pm P_3) = q^{\pm 1/2}(P_0 \pm P_3)P_a\ ,\ a = 1,2;\quad [P_1, P_2] = 0,\quad [P_0 - P_3, P_0 + P_3] = \tilde{\lambda}(P_1^2 + P_2^2) \tag{8}$$

$$K_a(K_0\pm K_3)=q^{\mp 1/2}(K_0\pm K_3)K_a\ ,\ a=1,2;\ [K_1,K_2]=0,\ [K_0-K_3,K_0+K_3]=\tilde\lambda(K_1^2+K_2^2) \tag{9}$$

4) The commutation relations of $M^\pm$ with P_μ are given by:

$$\begin{aligned}
M^+(P_3+P_0)-q^{-1/2}(P_3+P_0)M^+ &= P_1+iP_2\ ,\\
M^+(P_3-P_0)-q^{1/2}(P_3-P_0)M^+ &= q^{1/2}(P_1+iP_2)\ ,\\
M^+(P_1+iP_2)\ -\ \frac{[2]}{2}(P_1+iP_2)M^+ &= \frac{i\tilde\lambda}{2}(P_1+iP_2)N^+\ ,\\
M^+(P_1-iP_2)\ -\ \frac{[2]}{2}(P_1-iP_2)M^+ &= \frac{-i\tilde\lambda}{2}(P_1-iP_2)N^+ +(P_0-P_3)-q^{1/2}(P_3+P_0)
\end{aligned} \tag{10}$$

$$\begin{aligned}
&[M^-,P_3+P_0]=-q^{(i\tilde D+H)/2}(P_1-iP_2),\quad [M^-,P_3-P_0]=-(P_1-iP_2)q^{(i\tilde D-H)/2}\\
&[M^-,P_1+iP_2]=(P_3+P_0)q^{(i\tilde D-H)/2}+q^{(i\tilde D+H)/2}(P_3-P_0),\quad [M^-,P_1-iP_2]=0.
\end{aligned} \tag{11}$$

The commutation relations between $M^\pm$ and K_μ are obtained from the above by the following changes: $M^\pm\mapsto M^\mp$, $N^+\mapsto -N^-$, $H\mapsto -H$, $\tilde D\mapsto \tilde D$, $P_\mu\mapsto(-1)^{\delta_{\mu 1}}K_\mu$, $q^{1/2}\mapsto q^{-1/2}$. These follow from the automorphism of $U_q(\mathcal{G}^{\mathbb{C}})$: $X_1^\pm\longleftrightarrow X_3^\mp$, $H_1\longleftrightarrow -H_3$, $X_2^\pm\mapsto -X_2^\mp$, $H_2\mapsto -H_2$, $q^{1/2}\mapsto q^{-1/2}$, (then $X_{12}^\pm\longleftrightarrow -X_{23}^\mp$, $X_{13}^\pm\mapsto -X_{13}^\mp$). The commutation relations between $N^\pm$ and P_μ, K_μ are obtained from those between $M^\pm$ and P_μ by the changes $M^\pm\longleftrightarrow iN^\pm$, $P_0\longleftrightarrow P_3$, $P_1\longleftrightarrow iP_2$ and from those between $M^\pm$ and K_μ by the changes $M^\pm\longleftrightarrow -iN^\pm$, $K_0\longleftrightarrow K_3$, $K_1\longleftrightarrow -iK_2$.

5) For $[P_\mu,K_\nu]$ we have:

$$\begin{aligned}
[P_3\pm P_0\ ,\ K_3\pm K_0] &= \pm\tilde\lambda\ q^{\pm(H-D)/2}\,(M^+\mp iN^+)(M^-\pm iN^-)\ ,\\
[P_3\pm P_0\ ,\ K_3\mp K_0] &= 4[\mp i\tilde D-D]\ ,\\
[P_1\pm iP_2\ ,\ K_1\mp iK_2] &= 4[\mp H-D]\ ,\\
[P_1\pm iP_2\ ,\ K_1\pm iK_2] &= 0\ ,\\
[P_3+P_0\ ,K_1+iK_2] &= 2\,(M^+-iN^+)\,q^{(H-D)/2}\ ,\\
[P_3+P_0\ ,K_1-iK_2] &= -2\,(M^-+iN^-)\,q^{-(D+i\tilde D)/2}\\
[P_3-P_0\ ,K_1+iK_2] &= 2\,q^{(D-H)/2}\,(M^++iN^+)\ ,\\
[P_3-P_0\ ,K_1-iK_2] &= -2\,q^{(D-i\tilde D)/2}\,(M^--iN^-)
\end{aligned} \tag{12}$$

and four more relations which are obtained from the last four of (12) by the first set of changes described after formulae (11) and by $D\mapsto -D$.

The comultiplication for the Lorentz subalgebra is given by:

$$\begin{aligned}
\delta(M^\pm)=&M^\pm\otimes e^{Hh/4}cos(\tilde Dh/4)-N^\pm\otimes e^{Hh/4}sin(\tilde Dh/4)+\\
&+e^{-Hh/4}cos(\tilde Dh/4)\otimes M^\pm+e^{-Hh/4}sin(\tilde Dh/4)\otimes N^\pm\ ,
\end{aligned} \tag{13a}$$

$$\begin{aligned}
\delta(N^\pm)=&N^\pm\otimes e^{Hh/4}cos(\tilde Dh/4)+M^\pm\otimes e^{Hh/4}sin(\tilde Dh/4)+\\
&+e^{-Hh/4}cos(\tilde Dh/4)\otimes N^\pm-e^{-Hh/4}sin(\tilde Dh/4)\otimes M^\pm\ ;
\end{aligned} \tag{13b}$$

for the Cartan generators it is trivial while for the translations and special conformal transformations we have:

$$\delta(T^\pm) = \begin{cases} T^\pm \otimes q^{(D\pm i\tilde{D})/4} + q^{-(D\pm i\tilde{D})/4} \otimes T^\pm + \delta_1(T^\pm), & T^\pm = P_3 \pm P_0, K_3 \mp K_0 \\ T^\pm \otimes q^{(D\pm H)/4} + q^{-(D\pm H)/4} \otimes T^\pm + \delta_1(T^\pm), & T^\pm = P_1 \pm iP_2, K_1 \mp iK_2 \end{cases} \tag{14a}$$

$$\begin{aligned}
&\delta_1(T^\pm) = \frac{\tilde{\lambda}}{2}(M^\pm \mp iN^\pm)q^{(H-D)/4} \otimes q^{(H\pm i\tilde{D})/4}\tilde{T}^\pm , \\
&\qquad T^+ = P_3 + P_0,\ T^- = K_3 + K_0,\ \tilde{T}^+ = P_1 - iP_2,\ \tilde{T}^- = K_1 + iK_2 , \\
&\delta_1(T^\pm) = \frac{-\tilde{\lambda}}{2}\tilde{T}^\mp q^{(-H\mp i\tilde{D})/4} \otimes q^{(D-H)/4}(M^\mp \pm iN^\mp) , \\
&\qquad T^+ = K_3 - K_0,\ T^- = P_3 - P_0,\ \tilde{T}^+ = P_1 - iP_2,\ \tilde{T}^- = K_1 + iK_2 , \\
&\delta_1(T^\pm) = 0,\ T^+ = P_1 - iP_2, T^- = K_1 + iK_2 , \\
&\delta_1(T^\pm) = \frac{\tilde{\lambda}}{2}(\tilde{T}'^\pm q^{(-H\pm i\tilde{D})/4} \otimes q^{(D\pm i\tilde{D})/4}(M^\pm \pm iN^\pm) - \\
&\qquad - (M^\pm \mp iN^\pm)q^{(-D\pm i\tilde{D})/4} \otimes q^{(H\pm i\tilde{D})/4}\tilde{T}''^\pm) , \\
&\qquad T^+ = P_1 + iP_2,\ T^- = K_1 - iK_2,\ \tilde{T}'^+ = P_3 + P_0 , \\
&\qquad \tilde{T}'^- = K_3 + K_0,\ \tilde{T}''^+ = P_3 - P_0,\ \tilde{T}''^- = K_3 - K_0
\end{aligned} \tag{14b}$$

Consistently with the general scheme [1] formulae (14) tell us that the deformed subalgebras of translations and special conformal transformations are not Hopf subalgebras of $\mathcal{G}$, rather the algebras $U_q(\mathcal{P}_{max})$, $U_q(\tilde{\mathcal{P}}_{max})$ are Hopf subalgebras of $U_q(\mathcal{G})$.

3. q - deformed Poincaré and Weyl algebras. The Poincaré algebra is not a semisimple (or reductive) Lie algebra and our procedure is not directly applicable. One may try to use the fact that it is a subalgebra of the conformal algebra. Indeed, there is a q - deformed Poincaré algebra with generators $M^\pm$, $N^\pm$, H, $\hat{D} = i\tilde{D}$, P_μ , and with commutations relations given by (7), (8), (10), (11) and those obtained from the latter two by the changes $M^\pm \longleftrightarrow iN^\pm$, $P_0 \longleftrightarrow P_3$, $P_1 \longleftrightarrow iP_2$. One can easily see that this q - deformed Poincaré algebra is a commutation subalgebra of $U_q(su(2,2))$, but not a Hopf subalgebra of $U_q(su(2,2))$, cf. formulae (14). Rather the deformation $U_q(\tilde{\mathcal{P}}_{max})$ of the 11 - generator Weyl subalgebra = Poincaré & dilatations = $\tilde{\mathcal{P}}_{max}$ - is a Hopf subalgebra of $U_q(\mathcal{G})$. Another Weyl algebra conjugate to this is $U_q(\mathcal{P}_{max})$ with generators $M^\pm$, $N^\pm$, H, $\hat{D} = i\tilde{D}$, K_μ, D, and with commutations relations given by (7), (9), and those obtained from (10), (11) as explained in the text after formulae (11).

Acknowledgments. The participation of the author in the XXI DGM Conference (Tianjin 1992) was supported by the Organizing Committee of the DGM, by the ICTP, Trieste, Italy, and by the Bulgarian National Science Foundation, Grant $\Phi - 11$.

References

[1] V.K. Dobrev, Göttingen University preprint (July 1991).
[2] V.K. Dobrev, ICTP Trieste preprint IC/92/13 (January 1992).
[3] V.G. Drinfeld, Sov.Math.Dokl. **32** (1985) 254; Proceedings ICM (1987) p. 798.
[4] M. Jimbo, Lett.Math.Phys. **10** (1985) 63; Lett.Math.Phys. **11** (1986) 247.

DEFORMATIONS OF THE CANONICAL COMMUTATION RELATIONS AND THE PROBLEM OF TIME DEPENDENCE

DAVID B. FAIRLIE

Department of Mathematical Sciences
University of Durham, Durham, DH1 3LE, England

ABSTRACT

Deformations of the Canonical Commutation relations which have the effect of altering the spectrum of a standard Hamiltonian, bilinear in creation and annihilation operators are described. The problem of going over from an eigenvalue situation to a theory with time evolution is discussed, and a specific example with deformation parameter an nth root of unity is constructed which possess a consistent time evolution.

This work is an outgrowth of some recent studies of associative deformations of the Heisenberg algebra of several creation and annihilation operators with Jean Nuyts of the University of Mons Hainaut[1,2], and describes some observations of my own concerning the difficulty of implementing time evolution in a quantum group context. It builds on earlier work with Cosmas Zachos (Argonne National Laboratory, U.S.A.)[3,4] which in turn is related to work of Manin[5], and Wess, Zumino and collaborators[6,7,8]. The main idea is that if quantum groups have any role in physics, then they must manifest themselves at the level of deformations of the canonical commutation relations.

Many examples exist in the literature of the use of such deformations in constructing examples of spectra of time independent deformed operators[1], but there is really nothing new in this, as the work of Curtright and Zachos[9] has shown, because for generic q they have demonstrated the existence of invertible transformations which relate transformed co-ordinates to undeformed ones. Thus a simple Hamiltonian, quantised with deformed commutation relations is equivalent to a complicated Hamiltonian quantized canonically. Hence there is no new physical idea in treating merely the spectral problem.

Where something new is possible lies in the realm of time dependent problems. It is my contention that the problem of writing a field theory invariant under a quantum group is a much more difficult problem than has been generally realised, though some authors[10] have also raised this issue. Indeed the problem of simply doing time dependent quantum mechanics, i.e. of solving the usual time dependent Schrödinger equation with deformed quantization rules, is not at all straightforward. When q is a root of unity, the deformation is not invertible and it is possible to construct completely coherent time dependent operators for a certain class of problems which I believe constitute the true generalisation of the idea of fermions to parafermions. The simplest such case will now be illustrated, and this is enough to admit ready generalisation.

1. How to quantize $\frac{d^3}{dt^3}x = x$.

This section addresses the question as to how one might quantize dynamical equations of the form

$$\frac{\partial^n x}{\partial t^n} = (-1)^n x. \tag{1.1}$$

Consider the Hamiltonian:

$$\mathcal{H} = \sum_{i=1}^{n} x_i p_{i+1}, \quad p_0 \equiv p_n.$$

The classical equations of motion for the co-ordinate x_j are

$$\dot{x}_j = \{\mathcal{H}, x_j\} = -x_{j-1}. \tag{1.2}$$

These imply that x_j, $\forall j$ satisfy equation (1). This system may be quantised in the standard fashion, with canonical commutation rules;

$$[x_i, x_j] = 0, \quad [p_i, p_j] = 0, \quad [x_i, p_j] = i\hbar\delta_{ij} \tag{1.3}$$

and with the replacement of (2) by

$$\dot{x}_j = [H, x_j] = -x_{j-1}. \tag{1.4}$$

The solutions are unbounded for $n > 2$.

Now let us consider quantization with quommutators. It is sufficient to consider the case where $n = 3$.

$$\mathcal{H} = x_1 p_2 + x_2 p_3 + x_3 p_1. \tag{1.5}$$

It is easier to start with the representation:

$$x_1 = \begin{pmatrix} \cdot & e^{-\omega t} & \cdot \\ \cdot & \cdot & e^{-t} \\ \cdot & \cdot & \cdot \end{pmatrix}, \; x_2 = \begin{pmatrix} \cdot & e^{-\omega t} & \cdot \\ \cdot & \cdot & \omega e^{-t} \\ \cdot & \cdot & \cdot \end{pmatrix}, \; x_3 = \begin{pmatrix} \cdot & e^{-\omega t} & \cdot \\ \cdot & \cdot & \omega^2 e^{-t} \\ \cdot & \cdot & \cdot \end{pmatrix}. \tag{1.6}$$

$$p_1 = \begin{pmatrix} \cdot & \cdot & \cdot \\ e^{\omega t} & \cdot & \cdot \\ \cdot & -\omega e^{t} & \cdot \end{pmatrix}, \; p_2 = \begin{pmatrix} \cdot & \cdot & \cdot \\ e^{\omega t} & \cdot & \cdot \\ \cdot & -e^{t} & \cdot \end{pmatrix}, \; p_3 = \begin{pmatrix} \cdot & \cdot & \cdot \\ e^{\omega t} & \cdot & \cdot \\ \cdot & -\omega^2 e^{t} & \cdot \end{pmatrix}, \tag{1.7}$$

Here $\omega^3 = 1$ These satisfy the quommutation relations;

$$\begin{aligned} x_1 x_2 &= \omega x_2 x_1, \; x_1 x_3 = \omega^2 x_3 x_1, \; x_2 x_3 = \omega x_3 x_2 \\ p_1 p_2 &= \omega p_2 p_1, \; p_1 p_3 = \omega^2 p_3 p_1, \; p_2 p_3 = \omega p_3 p_2 \end{aligned}$$

$$\begin{aligned} x_1 p_1 - \omega^2 p_1 x_1 &= I \; x_2 p_2 - \omega^2 p_2 x_2 = I, \; x_3 p_3 - \omega^2 p_3 x_3 = I \\ x_1 p_2 - \omega p_2 x_1 &= A, \; x_2 p_3 - \omega p_3 x_2 = A, \; x_3 p_1 - \omega p_1 x_3 = A \\ x_1 p_3 - p_3 x_1 &= B, \; x_2 p_1 - p_1 x_2 = B, \; x_3 p_2 - p_2 x_3 = B \end{aligned}$$

In the above, besides the identity operator, I, two other operators, A, B have occurred. These operators commute, $AB - BA = 0$, since they are realised by diagonal matrices, as is the Hamiltonian;

$$A = \begin{pmatrix} 1 & \cdot & \cdot \\ \cdot & \omega^2 & \cdot \\ \cdot & \cdot & \omega \end{pmatrix}, \quad B = \begin{pmatrix} 1 & \cdot & \cdot \\ \cdot & \omega & \cdot \\ \cdot & \cdot & \omega^2 \end{pmatrix}, \quad \mathcal{H} = \begin{pmatrix} 3 & \cdot & \cdot \\ \cdot & -3 & \cdot \\ \cdot & \cdot & 0 \end{pmatrix}. \tag{1.8}$$

Also

$$x_j I - I x_j = 0, \quad x_j A - \omega A x_j = 0, \quad x_j B - \omega^2 B x_j = 0, \forall j. \tag{1.9}$$

The Poisson Bracket relations are replaced by the quommutators

$$\begin{aligned} \dot{x}_1 &= \frac{1}{3}(\mathcal{H}x_1 - \omega^2 x_1 \mathcal{H}) = -\,3\omega x_3 \\ \dot{x}_2 &= \frac{1}{3}(\mathcal{H}x_2 - \omega^2 x_2 \mathcal{H}) = -\,3\omega x_1 \\ \dot{x}_3 &= \frac{1}{3}(\mathcal{H}x_3 - \omega^2 x_3 \mathcal{H}) = -\,3\omega x_2 \end{aligned} \tag{1.10}$$

The factor 1/3 is introduced as an effective $\hbar$. Now consider, under the assumption that t acts as an ordinary variable,

$$\begin{aligned} \frac{d}{dt}(x_1 x_2 - \omega x_2 x_1) =& \dot{x} y + x \dot{y} - \omega(\dot{y} x + y \dot{x}) \\ =& -\omega(x_3 x_2 + x_1^2 - \omega(x_1^2 + x_2 x_3)) \end{aligned} \tag{1.11}$$

This expression does not vanish for general quommutators (1.6,1.7). However it does vanish in the specific representation given. This is true for all derivatives of these quommutators, that in the given representation they hold consistently.. The representation given may be generalized to all integral n. It cannot be thought of as a faithful representation of an abstract quommutator scheme as clearly not all the operators are linearly independent. There are the relations;

$$\begin{aligned} x_1 + \omega x_2 + \omega^2 x_3 = 0 \\ p_1 + \omega^2 p_2 + \omega p_3 = 0 \end{aligned} \tag{1.12}$$

Nevertheless it provides a consistent scheme for quantization with quommutators which possess a simple time evolution. Note also that all quommutators of any 'position' operator with any 'momentum' operator contain a central term. The time dependence of the solution is consistent with ordinary differentiation with respect to time. In fact, by taking the other linear combinations

$$\begin{aligned} X_1 =& \frac{1}{3}(x_1 + x_2 + x_3) \\ X_2 =& \frac{1}{3}(x_1 + \omega^2 x_2 + \omega x_3) \end{aligned} \tag{1.13}$$

and defining the conjugate momenta P_1, P_2 analogously, a somewhat simpler basis for the problem may be constructed![11] In the case of general n, X_i, P_j, $i, j = 1, \ldots n-1$ are given by monomial matrices

$$\{X_i\}_{kl} = \exp(-w^{-i-1}t)\delta_{i,k}\delta_{i+1,l}, \quad \{P_i\}_{kl} = \exp(w^{-i-1}t)\delta_{i+1,k}\delta_{i,l}. \tag{1.14}$$

These satisfy canonical commutation relationships with each other:

$$P_i X_j - X_j P_i = (\text{diagonal})\ \delta_{i,j}. \tag{1.15}$$

Nevertheless the system is not decoupled, because X_i, X_j fail to commute when $i = j \pm 1$, a situation reminiscent of the braid algebra, where adjacent operators also fail to commute.

2. Conclusions.

This example depends upon a specific finite dimensional representation with a deformation parameter which is a root of unity, so is rather special. When working in the transformed basis of monomial matrices as in the concluding paragraph of the previous section, the commutation rules become almost canonical, but not quite.

The general problem of the introduction of time evolution prompts the question as to what time development is valid anyway. Usually one assumes the time dependent Schrödinger equation and reduces it to an eigenvalue equation by the assumption of an exponential time development $\exp(-\frac{iEt}{\hbar})$. However, we have learned from the Inverse Scattering method for integrable systems that given an eigenvalue equation of Schrödinger type,

$$-\frac{\partial^2}{\partial x^2}y(x,t) + u(x,t)y(x,t) = \lambda y(x,t), \tag{2.1}$$

with eigenvalue λ independent of time, this is compatible with a time evolution of the form

$$\frac{\partial y}{\partial t} = \mathcal{B}y, \tag{2.2}$$

provided the operators $\mathcal{L} = \frac{\partial^2}{\partial x^2} + u(x,t)$, $\mathcal{B}$ satisfy the Lax pair condition:

$$\frac{\partial \mathcal{L}}{\partial t} + [\mathcal{L}, \mathcal{B}] = 0. \tag{2.3}$$

If the potential $u(x,t)$ is constant then the simple exponential time evolution corresponding to constant $\mathcal{B}$ suffices. However, as we know, other possibilities exist e.g.

$$\mathcal{B} = -4\frac{\partial^3}{\partial x^3} + 3u\frac{\partial}{\partial x} + 3\frac{\partial}{\partial x}u. \tag{2.4}$$

This requires that u satisfies the KdV equation;

$$u_t + u_{xxx} - 6uu_x = 0 \tag{2.5}$$

It seems to me that a consistent development of the use of quantum groups in physical applications requires the exploration of systems with more complicated time evolution than the simple exponential. After all it was in the context of quantisation of inverse scattering problems that quantum groups first arose[12]!

References.

1. (with J. Nuyts). Alternative Hamiltonian Quantisation. *J.Phys* **A 24** (1991) L1001-1007.
2. (with J. Nuyts). Neutral and Charged Quommutators with and without Symmetries DTP-92-03 (1992) Submitted to *Zeitschrift für Physik*.

3. (with C.K. Zachos) Multiparameter Associative Generalizations of Canonical Commutation Relations and Quantized Planes *Physics Letters*, **B256** (1991) 43-49.
4. (with C.K. Zachos) Quantized Planes and Multiparameter Deformations of Heisenberg and $GL(N)$ Algebras. Invited talk, to appear in *Proceedings of Nato ARW on Quantum Field Theory*, Coral Gables 1991.
5. Yu.I.Manin, *Quantum Groups and Non-Commutative Geometry*, C.R.M. Montreal Publications (1988).
6. W. Pusz and S.L. Woronowicz, *Rep.Math.Phys.*, **27** (1989) 231.
7. S.P. Vokos, B. Zumino and J. Wess, *Z.Phys. C*, **48** (1990) 317.
8. J.Wess and B. Zumino, Covariant Differential Calculus on the the quantum hyperplane CERN-TH-5697 (1990). To be published in *Nuclear Physics B Supplements; Volume in honour of R. Stora*
9. T. Curtright and C.K. Zachos *Phys.Lett.***B243** (1990) 237.
10. A.Chodos and D.G. Caldi, *J. Phys.***A24** 1991 5505-5515.
11. C.K. Zachos, *Private Communication.*
12. L.D. Faddeev and L.A. Takhtajan, *Russ. Math. Surv.* **34** (1979) 11.

q-Generating Functions and the q-Oscillator

R. J. Finkelstein
Department of Physics, UCLA
Los Angeles, California 90024-1547, USA

Abstract

We construct generating functions which interpolate between the generating functions for Hermitian and Grassmann multinomials. The new functions which arise in this way may be related to state functions of the q-deformed harmonic oscillator.

1. Introduction

We extend the quantum plane relation[1] to two spaces of indeterminates $\{x^\alpha\}$ and $\{t^\alpha\}$:

$$x^\alpha t^\beta = q\, t^\beta x^\alpha \quad \text{all} \quad \alpha,\beta \tag{1.1}$$

In terms of these two sets $\{x^\alpha\}$ and $\{t^\alpha\}$ one may construct a generating function for multinomials in the $\{x^\alpha\}$.

One may also generalize to three spaces. Denoting the α-coordinate of the ith space by $z_i{}^\alpha$ we consider

$$z_i{}^\alpha\, z_j{}^\beta = R^{\alpha\beta}_{ij}\, z_j{}^\beta\, z_i{}^\alpha\,, \qquad j = 1,2,3 \quad \text{all} \quad \alpha,\beta \tag{1.2}$$

where the R_{ii} are restricted by a braid algebra and the R_{ij} $(i\neq j)$ satisfy a Yang-Baxter equation. With these three sets one may again construct new generating functions.

In this framework one obtains new functions which are generalizations of the classical special functions in many variables. We shall here restrict our discussion to (1.1) and the Hermite case. The commuting case $(q=1)$ had been introduced long ago by Hermite[2] while the Grassmannian case $(q=-1)$ naturally turns up in a supersymmetric context.[3,4] Although our work will be based on the Hermitian generating function, the method is general and has been extended to other classical functions.[5]

2. Expansion of Generating Function[6]

Let

$$G = \exp(-A_{\alpha\beta}t^\alpha t^\beta + 2\, B_{\alpha\beta}t^\alpha x^\beta) \tag{2.1}$$

where

$$t^{\alpha}t^{\beta} = q^{tt} t^{\beta} t^{\alpha} \qquad q^{tt} = \pm 1$$
$$x^{\alpha}x^{\beta} = q^{xx} x^{\beta} x^{\alpha} \qquad q^{xx} = \pm 1$$
$$x^{\alpha}t^{\beta} = q^{xt} t^{\beta} x^{\alpha} \qquad q^{xt} = q \tag{2.2}$$

Then

$$G = \sum_{p=0}^{\infty} \frac{(-A+B)^p}{p!} \tag{2.3}$$

By the q-binomial theorem

$$G = \sum_{p=0}^{\infty} \frac{1}{p!} \sum_{s=0}^{p} \left\langle {p \atop s} \right\rangle_{q^2} (-A)^s \, (B)^{p-s} \tag{2.4}$$

where

$$\left\langle {p \atop s} \right\rangle_q = \frac{\langle p \rangle_q !}{\langle s \rangle_q ! \ \langle p-s \rangle_q !} \tag{2.5}$$

and

$$\langle p \rangle_q = \frac{q^p - 1}{q-1} \tag{2.6}$$

We also introduce the corresponding symmetric forms:

$$\begin{bmatrix} p \\ s \end{bmatrix}_q = \frac{[p]_q!}{[s]_q! \ [p-s]_q!} \tag{2.7}$$

where

$$[p]_q = \frac{q^p - q^{-p}}{q-q^{-1}} \tag{2.8}$$

After some rearrangements one finds

$$G = \sum_{n=0}^{\infty} \sum_{\lambda_1 \ldots \lambda_n} t^{\lambda_1} \ldots t^{\lambda_n} H_{\lambda_1 \ldots \lambda_n}(\bar{x}_1 \ldots \bar{x}_n) \tag{2.9a}$$

where

$$H_{\lambda_1 \ldots \lambda_n}(\bar{x}_1 \ldots \bar{x}_n) = \sum_{s=0}^{[n/2]} \frac{(-)^s}{(n-s)!} \begin{bmatrix} n-s \\ s \end{bmatrix}_q \frac{\varepsilon_q(n-s)}{\varepsilon_q(s)}$$
$$\times A_{\lambda_1 \lambda_2} \ldots A_{\lambda_{2s-1} \lambda_{2s}} (2\bar{x})_{\lambda_{2s+1}} \ldots (2\bar{x})_{\lambda_n} \tag{2.9b}$$

and

$$\bar{x}_\gamma = B_{\gamma\delta} x^\delta \tag{2.9c}$$
$$\varepsilon_q(s) = q^{s(s-1)/2}$$

In the derivations of (2.9) no assumption has been made about $q^{(xx)}$. However, (2.9a) implies that the λ-symmetry is determined by the $\{t^\alpha\}$ and consequently the symmetry of the $\{t^\alpha\}$ is passed on to the $\{x^\alpha\}$. Therefore $q^{xx} = q^{tt} = \pm 1$ as anticipated in (2.2).

We are particularly interested in the special case:

$$q^{xx} = q^{tt} = 1 \ , \quad |q^{xt}| \neq 1$$

$$A_{\lambda_1\lambda_2} = B_{\lambda_1\lambda_2} = \delta_{\lambda_1\lambda_2} \quad \text{and} \quad \lambda_1 = \lambda_2 = \ldots \lambda_n \tag{2.10}$$

Then

$$H_n^q(x) = \sum_{s=0}^{[n/2]} \frac{(-)^s}{(n-s)!} \begin{bmatrix} n-s \\ s \end{bmatrix}_q \frac{\varepsilon_q(n-s)}{\varepsilon_q(s)} (2x)^{n-2s} \tag{2.11}$$

If q=1, (2.11) becomes the usual Hermite polynomial. We now turn to the relation of these q-polynomials to the state functions of the q-deformed oscillator.

3. The Quantum Group and the q-Quantum System

We shall here define a q-deformed quantum mechanical system by an unchanged Hamiltonian and the q-modified Heisenberg commutator:

$$Dx - qxD = 1 \,, \quad p = -i\hbar D \tag{3.1}$$

Eq. (3.1) may be rewritten

$$\tilde{z}Qz = 1 \tag{3.2}$$

where

$$Q = \begin{pmatrix} 0 & 1 \\ -q & 0 \end{pmatrix} \qquad z = \begin{pmatrix} D \\ x \end{pmatrix} \tag{3.3}$$

Let $T \in GL_q(2,C)$. Then under $z \to Tz$ (3.2) becomes

$$\tilde{z}Qz = \Delta \tag{3.4a}$$

or

$$Dx - qxD = \Delta \tag{3.4b}$$

where Δ is the quantum determinant.

Since Δ is central, one may put $\Delta = 1$ and if $T \in SL_q(2,C)$, the q-commutators are invariant.

One may satisfy (3.4) by the following choices of D:

$$\text{(a)} \quad q = 1 \quad \Delta = 1 \quad D = \frac{d}{dx} \tag{3.5}$$

$$\text{(b)} \quad q \neq 1 \quad \Delta = 1 \quad D = x^{-1} \langle\theta\rangle_q \tag{3.6}$$

$$\text{(c)} \quad q \neq 1 \quad \Delta \neq 1 \quad D = x^{-1} [\theta]_q \tag{3.7}$$

where

$$\theta = x \frac{d}{dx} \tag{3.8}$$

Here the symbols < > and [ ] are defined in (2.6) and (2.8). In cases (b) and (c) D is a difference operator. In case (c) one finds

$$\Delta = q^{-\theta} \tag{3.9}$$

The wave function of the q-deformed free particle is in case (b)

$$e_q(ipx) = \sum \frac{(ipx)^n}{\langle n\rangle_q!} \tag{3.10}$$

and in case (c) $\langle n\rangle_q \to [n]_q$.

If the q-deformed oscillator is defined by

$$H = \frac{1}{2}(a\bar{a} + \bar{a}a) \tag{3.11a}$$

$$a\bar{a} - q\bar{a}a = q^{-\theta} \tag{3.11b}$$

then in this representation the energy spectrum is

$$E_n = \frac{1}{2}\left([n]_q + [n+1]_q\right) \tag{3.12}$$

while the usual monomial state functions are normalized by $([n]_q!)^{-1/2}$ instead of $(n!)^{-1/2}$.

4. The q-Oscillator in Configuration Space

If the q-oscillator is defined by an unchanged Hamiltonian and a q-commutator, one has

$$H = \frac{1}{2}(p^2+x^2) \tag{4.1}$$

$$Dx - qxD = 1 \tag{4.2}$$

where (4.2) corresponds to (3.1) rather than (3.4). Define

$$D_\pm = D \pm q^{\mp 1/2}\, x \tag{4.3}$$

then

$$H - \frac{1}{2}q^{-1/2} = -\frac{1}{2}D_-D_+ \tag{4.4}$$

If we regard D_+ and D_- as lowering and raising operators, then the ground state may be defined by

$$D_+\psi_o = 0 \tag{4.5}$$

and the lowest energy is then

$$E_o = \frac{1}{2}q^{-1/2} \tag{4.6}$$

Now (4.5) is a functional (difference) equation for the ground state. The exact solution of (4.5), which may be obtained by iteration, is the convergent infinite product:

$$\psi_o(x) = \prod_{s=0}^{\infty}\left[1 + q^{2s}(q^{1/2} - q^{-1/2})\, x^2\right]^{-1}\psi_o(0)\ , \quad |q| < 1 \tag{4.7}$$

and an approximate solution, approximating the usual Gaussian as $q \to 1$, is

$$\psi_o(x) = \exp\left[-f(q)\,\frac{x^2}{2}\right] \tag{4.8}$$

$$f(q) = \left(q^{1/2} - q^{-1/2}\right)(\ell n\ q)^{-1} \tag{4.9}$$

One would like to find a complete set of exact solutions of the deformed oscillator that approach the known eigensolutions of the undeformed system. Part of the difficulty, however, is that there are several reasonable formulations of the deformed problem besides the one defined by Eqs. (4.1) and (4.2). Here are some possibilities for the deformed state functions.

1. The state functions are the Hermitian q-functions (2.11) defined by the generating function (2.1).

2. If one chooses (3.6) then it also is natural to choose $e_q(x)$, given by (3.10), instead of the exponential as the generating function. The new G, replacing (2.1), is then either

$$G = \sum \frac{(-A+B)^n}{\langle n\rangle_q!} \tag{4.10}$$

$$= e_q(-A+B) \tag{4.11}$$

or a similar construction depending on $[n_q]!$ Here q is left free. If, for example, one chooses the new generating function to be $e_{q^2}(-A+B)$ then the new Hermitian q-function $H_n^q(x)'$ is

$$H_n^q(x)' = \sum_{s=0}^{[n/2]} \frac{(-)^s}{\varepsilon_q(s)} \frac{1}{[s]_q!\ [n-2s]_q!} (2x)^{n-2s} \tag{4.12}$$

3. The deformed state functions are solutions of the following integral equation

$$\varphi_n^{(q)}(p) = \lambda_n \int dx\ e_q(ipx)\ \varphi_n^{(q)}(x) \tag{4.13}$$

As $q\to 1$, e_q approaches the usual exponential and $\varphi_n^q(x)$ approaches the familiar Hermite function (which is its own Fourier transform).

All of these q-functions approach the usual Hermite functions in the limit $q\to 1$.

In the q-space of the oscillator there is a complete solution of the dynamical problem not only at q=1 but also at the Grassmann limit, q=-1.[4] In addition, since there is also a complete solution for arbitrary q in the $(a,\bar{a})$ representation, it would be of interest if there were a satisfactory q-transformation theory leading from the $(a,\bar{a})$ to the (x,p) representation There is of course no physical support for a q-quantum mechanics--or any similar modification of quantum theory--at the present time.[7]

5. References

1. Yu. I. Manin, *Ann. Inst. Fourier* **31**, 198 (1987).
2. C. Hermite, *C. R. Acad. Sci. Paris* **58**, 93 (1864).
3. R. Finkelstein and M. Villasante, *J. Math. Phys.* **27**, 6, 1595 (1986).
4. R. Finkelstein and M. Villasante, *Phys. Rev.* **D33**, 6, 1660 (1986).
5. F. L. Chan, R. Finkelstein, and K. M. Lau, UCLA/TEP/14 (1992).
6. R. Finkelstein, UCLA/TEP/18 (1992), J. Math. Phys. (in press).
7. S. Weinberg, Oskar Klein Mem. Lectures 1 (World Scientific, 1991).

QUANTUM GROUPS, SYMPLECTIC GEOMETRY AND GEOMETRIC QUANTIZATION [1]

HAN-YING GUO

Institute of Theoretical Physics, Academia Sinica
P.O.Box 2735, Beijing 100080, China

Abstract. We show that the quantum groups are neither group nor necessarily to be physically quantum. Namely, they can be realized at both classical and quantum levels by means of symplectic geometry and geometric quantization method.

The quantum groups [1-5] are usually thought to appear in certain quantum physical systems and when the quantum systems revert to classical ones the quantum groups reduce to the corresponding Lie groups. We found [6-12], however, this is not the case in principle. Actually, the quantum groups are neither groups (this is well-known) nor necessarilly to be physically quantum. Namely, they can be realized at both classical and quantum levels by means of the symplectic geometry of the phase spaces of certain classical mechanical systems and the geometric quantization theory [13]. In this talk I briefly introduce this approach. We take the case of $SU_q(2)$ as an example, but the approach is available to any quantum groups.

Let us begin with a system of two harmonic oscillators with the Hamiltonian and the symplectic form on the phace space C^2

$$H = \sum_{i=1}^{2} \bar{z}_i z_i, \quad \Omega = -i \sum_{i=1}^{2} d\bar{z}_i \wedge dz_i, \tag{1}$$

where $z_i = \frac{p_i + iq_i}{\sqrt{2}}$, $\bar{z}_i = \frac{p_i - iq_i}{\sqrt{2}}$.

Remarkablely, it can be shown that a set of the q-deformed observables on C^2

$$J'_+ = z_1 \bar{z}'_2, \quad J'_- = z_2 \bar{z}'_1, \quad J'_3 = J_3, \tag{2}$$

[1]Work supported in part by the National Natural Science Foundation of China.

generate the $SU_{q,\hbar\to 0}(2)$ algebra in Poisson brackets

$$[J'_+, J'_-]_{P.B.} = -i\,[2J'_3]_q\,, \quad [J'_3, J'_\pm]_{P.B.} = -i\left(\pm J'_\pm\right), \tag{3}$$

where

$$z'_i = \frac{1}{\sqrt{\gamma \sinh\gamma}} \frac{\sinh(\gamma z_i \bar{z}_i)}{z_i \bar{z}_i} z_i, \quad \bar{z}'_i = \frac{1}{\sqrt{\gamma \sinh\gamma}} \frac{\sinh(\gamma z_i \bar{z}_i)}{z_i \bar{z}_i} \bar{z}_i. \tag{4}$$

When q reaches one, all back to $SU(2)$ via oscillators.

On the $U(1)$ line bundle over the phase space C^2 with curvature Ω and symplectic connection θ, the prequantization operator of the observable f can be given by the map

$$f \longrightarrow \tilde{f} = -i\,(X_f - i\theta(X_f)) + f.$$

Notice that there should be no Planck constant $\hbar$ at this level. From physical point of view, $\hbar$ is of completely quantum characteristic so that it should appear only after suitable polarization is taken.

For z_i and $\bar{z}_i$, their prequantization operators read

$$\tilde{z}_i = -\frac{\partial}{\partial \bar{z}_i} + z_i, \quad \tilde{\bar{z}}_i = \frac{\partial}{\partial z_i},$$

with the commutation relations $[\tilde{z}_i, \tilde{\bar{z}}_j] = -\delta_{ij}$.

To get the prequantization operators of $J'_\pm$ and J'_3, a suitable ordering of $\tilde{z}_i$ and $\tilde{\bar{z}}_i$ is needed. Then we have

$$\begin{aligned} \tilde{J}_+{}' &= \frac{1}{\sqrt{\gamma \sinh\gamma}} \frac{\tilde{z}_1}{\tilde{z}_2} \sinh\left(\gamma \tilde{z}_2 \tilde{\bar{z}}_2\right), \\ \tilde{J}_-{}' &= \frac{1}{\sqrt{\gamma \sinh\gamma}} \frac{\tilde{z}_2}{\tilde{z}_1} \sinh\left(\gamma \tilde{z}_1 \tilde{\bar{z}}_1\right), \\ \tilde{J}'_3 &= \frac{1}{2}\left(\tilde{z}_1 \tilde{\bar{z}}_1 - \tilde{z}_2 \tilde{\bar{z}}_2\right). \end{aligned} \tag{5}$$

with commutation relations of $SU_{q,\hbar\to 0}(2)$ algebra in Lie brackets

$$[\tilde{J}_+{}', \tilde{J}_-{}'] = \frac{1}{\gamma} \sinh\left(2\gamma \tilde{J}'_3\right), \quad [\tilde{J}'_3, \tilde{J}'_\pm] = \pm \tilde{J}'_\pm. \tag{6}$$

With these formulae, we can define the Hopf algebra operations as same as for the usual " quantum algebra" $SU_q(2)$. Then the entire quantum group structure follows as is well known in the quantum group theory[1-5]. But, it has now nothing to do with physical quantization at all.

To quantize a system is to introduce a suitable polarization. Let us take the Kähler polarization on (C^2, Ω) with basis $X = \left\{\frac{\partial}{\partial \bar{z}_1}, \frac{\partial}{\partial \bar{z}_2}\right\}$. Then the quantum operator of a given classical observable f (which is supposed to preserve the polarization) is given by: $f \longrightarrow \hat{f} = -i\hbar\left(X_f - i\theta\left(X_f\right)\right) + f - \frac{i}{2}\hbar a$, where a is determined by the formula $[X_f, X] = aX$. It is easy to get the quantum operators of z_i and $\bar{z}_i$ with commutation relations $[\hat{z}_i, \hat{\bar{z}}_j] = -\hbar\delta_{ij}$.

Then the quantum versions of $J'_\pm$ and J'_3 are given by

$$\begin{aligned} \hat{J}'_+ &= \frac{1}{\sqrt{\gamma \sinh \gamma}} \frac{z_1}{z_2} \sinh\left(\gamma \hbar z_2 \frac{\partial}{\partial z_2}\right) \\ \hat{J}'_- &= \frac{1}{\sqrt{\gamma \sinh \gamma}} \frac{z_2}{z_1} \sinh\left(\gamma \hbar z_1 \frac{\partial}{\partial z_1}\right) \\ \hat{J}'_3 &= -\frac{\hbar}{2}\left(z_2 \frac{\partial}{\partial z_2} - z_1 \frac{\partial}{\partial z_1}\right) \end{aligned} \tag{7}$$

which form the quantum q-deformed algebra $SU_{q,\hbar}(2)$:

$$[\hat{J}'_+, \hat{J}'_-] = \frac{\sinh \gamma\hbar}{\gamma} \frac{\sinh\left(2\gamma \hat{J}'_3\right)}{\sinh \gamma}, \qquad [\hat{J}'_3, \hat{J}'_\pm] = \hbar \hat{J}'_\pm. \tag{8}$$

Obviously, the Hopf algebraic structure of $SU_{q,\hbar}(2)$ is identical to that of $SU_{q,\hbar\to 0}(2)$ and the entire structure of the quantum groups follows as well.

It is worthwhile to point out that the deformations $z_i \to z'_i$ are quasiconformal deformations while q is real. And $(z', \bar{z}')$ describe a q-deformed oscillator which is nonlinear but can be exactly solved at both classical and quantum levels [6-9].

In [10-12], both $SU_{q,\hbar\to 0}(2)$ and $SU_{q,\hbar}(2)$ are realized on a line bundle over a 2-sphere by a set of q-deformed functions satisfying the equation for the q-deformed sphere, which is closely linked to the q-deformed rotor and the q-deformed top, and by corresponding operators on the line bundle over the sphere. The Lie group theory shows that the sphere is the coadjoint orbit of $SU(2)$ so that it is a coadjoint orbit approach to $SU_q(2)$. And this approach can also be used to any "quantum groups".

Our approach also shows that the quantum groups should be the "symmetries" of single deformed dynamical systems at both classical and quantum levels, such as the q-oscillator, the q-rotor as well as the q-top and so on, in addition to that of the solvable models in quantum field theory and statistical mechanics.

I would like to thank Z. Chang, W. Chen, S.M. Fei and H. Yan for creative collaboration on this topic during last two years.

References

[1] V.G.Drinfel'd, Proceedings of the International Congress of Mathematicians, (Berkeley), 798 (1987).

[2] M. Jimbo, Lett. Math. Phys. **10**, 63, (1985); Commun. Math. Phys.**102**, 537 (1986).

[3] S. L. Woronowicz, Commun. Math. Phys.**111**, 613, (1987).

[4] L. D. Faddeev, N. Yu. Reshetikhin and L. A. Takhtajan, preprint of Leningradskogo Otdeleniya Mathenmaticheskogo Insituta, (1987).

[5] Yu. I. Manin, Quantum Groups and Non-commutative Geometry, University of Montreal, 1988.

[6] Z. Chang, W. Chen and H.Y. Guo, J. Phys. **A 23**, 4185 (1990).

[7] Z. Chang, W. Chen, H.Y. Guo and H. Yan, J. Phys. **A 23** 5371 (1990).

[8] Z. Chang, W. Chen, H.Y. Guo and H. Yan, J. Phys. **A24** 1427 (1991).

[9] Z. Chang, S.M. Fei, H.Y. Guo and H. Yan, J. Phys. **A24** 5435 (1991).

[10] S.M. Fei and H.Y. Guo, J. Phys. **A 24** 1 (1991).

[11] S.M. Fei and H.Y. Guo, Comm. Theor. Phys. **16** 79 (1991).

[12] S.M. Fei, J. Phys. **A 24** 5195 (1991)

[13] See, for example, J. Sniatychi, "*Geometric Quantization and Quantum Mechanics*", (Springer-Verlag, New York, 1980), and references therein.

TOWARDS THE CHERN-WEIL HOMOMORPHISM IN NON COMMUTATIVE DIFFERENTIAL GEOMETRY

ANDREAS CAP
PETER W. MICHOR

Institut für Mathematik, Universität Wien,
Strudlhofgasse 4, A-1090 Wien, Austria.

1. Introduction

In this short review article we sketch some developments which should ultimately lead to the analogy of the Chern-Weil homomorphism for principle bundles in the realm of non commutative differential geometry. Principal bundles there should have Hopf algebras as structure 'cogroups'. Since the usual machinery of Lie algebras, connection forms, etc., just is not available in this setting, we base our approach on the Frölicher–Nijenhuis bracket. See [9] for an account of the classical theory using this approach.

In this paper we give an outline of the construction of a non commutative analogy of the Frölicher–Nijenhuis bracket as well as some simple applications. For simplicity we work in a purely algebraic setting but the whole theory can also be developed for topological algebras as well as for the so called convenient algebras (see [5]) which are best suited for differentiation and take care of completed tensor products. For a detailed exposition in the latter setting see [1] and [2].

2. Universal differential forms

2.1. Let A be a unital associative algebra over a commutative field K of characteristic zero. Then the graded algebra $\Omega_*(A)$ of universal differential forms over A is constructed as follows (see [6] and [7]): The tensor product $A\otimes A$ is an A–bimodule and the multiplication map $\mu : A \otimes A \to A$ is a bimodule homomorphism by associativity. Hence the kernel of μ is an A–bimodule which we denote by $\Omega_1(A)$. We define $d : A \to \Omega_1(A)$ by $d(a) = 1 \otimes a - a \otimes 1$.

The map d can be characterized by a universal property as follows: Let M be an A–bimodule. A linear map $D : A \to M$ is called a derivation if and only if for any $a, b \in A$ we have $D(ab) = D(a) \cdot b + a \cdot D(b)$. Obviously d is a derivation and thus for any bimodule homomorphism $\varphi : \Omega_1(A) \to M$ the map $\varphi \circ d : A \to M$ is a derivation. It can be proved that any derivation is of this form:

Supported by Project P 7724 PHY of 'Fonds zur Förderung der wissenschaftlichen Forschung'

2.2. Proposition. *For any A-bimodule M the canonical linear mapping* d^* : $\mathrm{Hom}^A_A(\Omega_1(A), M) \to \mathrm{Der}(A; M)$, *given by* $\varphi \mapsto \varphi \circ d$ *is an isomorphism.*

Clearly the module $\Omega_1(A)$ is determined by this universal property up to canonical isomorphism.

2.3. Now we define the spaces of differential forms of higher degree by $\Omega_k(A) := \Omega_1(A) \otimes_A \cdots \otimes_A \Omega_1(A)$ (k factors). Moreover we put $\Omega_0(A) = A$ and $\Omega(A) = \oplus_{k=0}^{\infty}\Omega_k(A)$. Then $\Omega(A)$ is a graded algebra with the tensor product as multiplication. Next put $\bar{A} := A/K$, where the ground field K is identified with the multiples of the unit of A. Then one proves that the map $a \otimes \bar{b} \mapsto ad(b)$ is an isomorphism between $A \otimes \bar{A}$ and $\Omega_1(A)$. Consequently the map $a_0 \otimes \bar{a}_1 \otimes \cdots \otimes \bar{a}_k \mapsto a_0 d(a_1) \dots d(a_k)$ defines an isomorphism $A \otimes \overbrace{\bar{A} \otimes \cdots \otimes \bar{A}}^{k\text{-times}} \to \Omega_k(A)$, and it turns out that $a_0 d(a_1) \dots d(a_k) \mapsto d(a_0) d(a_1) \dots d(a_k)$ gives a well defined map $d : \Omega_k(A) \to \Omega_{k+1}(A)$ for any k. Then it can be shown that $(\Omega(A), d)$ is a graded differential algebra, i.e. that $d(\omega_p\omega_q) = d\omega_p\omega_q + (-1)^p\omega_p d\omega_q$ for all $\omega_p \in \Omega_p(A)$ and $\omega_q \in \Omega_q(A)$. Again this algebra is characterized by a universal property:

2.4. Proposition. *Let* $(B = \oplus_{k=0}^{\infty} B_k, \delta)$ *be an arbitrary unital graded differential algebra,* $\varphi_0 : A \to B_0$ *a homomorphism of unital algebras. Then there is a unique homomorphism* $\varphi : \Omega(A) \to B$ *of graded differential algebras which restricts to* φ_0 *in degree zero.*

In particular this result shows that the construction of the algebra of universal differential forms is functorial.

3. Construction of the Frölicher–Nijenhuis bracket

The construction is based on the classification of all graded derivations of the graded algebra $\Omega(A)$.

3.1 Definition. The space $\mathrm{Der}_k\,\Omega(A)$ consists of all *(graded) derivations* of degree k, i.e. all bounded linear mappings $D : \Omega(A) \to \Omega(A)$ with $D(\Omega_\ell(A)) \subset \Omega_{k+\ell}(A)$ and $D(\varphi\psi) = D(\varphi)\psi + (-1)^{k\ell}\varphi D(\psi)$ for $\varphi \in \Omega_\ell(A)$.

Lemma. *The space* $\mathrm{Der}\,\Omega(A) = \bigoplus_k \mathrm{Der}_k\,\Omega(A)$ *is a graded Lie algebra with the graded commutator* $[D_1, D_2] := D_1 \circ D_2 - (-1)^{k_1k_2} D_2 \circ D_1$ *as bracket. So the bracket is graded anticommutative,* $[D_1, D_2] = -(-1)^{k_1k_2}[D_2, D_1]$, *and satisfies the graded Jacobi identity* $[D_1, [D_2, D_3]] = [[D_1, D_2], D_3] + (-1)^{k_1k_2}[D_2, [D_1, D_3]]$.

3.2. A derivation $D \in \mathrm{Der}_k\,\Omega(A)$ is called *algebraic* if $D \mid \Omega_0(A) = 0$. Then $D(a\omega) = aD(\omega)$ and $D(\omega a) = D(\omega)a$ for $a \in A$, so D restricts to a bounded bimodule homomorphism, an element of $\mathrm{Hom}^A_A(\Omega_l(A), \Omega_{l+k}(A))$. Since we have $\Omega_l(A) = \Omega_1(A) \otimes_A \cdots \otimes_A \Omega_1(A)$ and since for a product of one forms we have $D(\omega_1 \otimes_A \cdots \otimes_A \omega_l) = \sum_{i=1}^{l} (-1)^{ik} \omega_1 \otimes_A \cdots \otimes_A D(\omega_i) \otimes_A \cdots \otimes_A \omega_l$, the derivation D is uniquely determined by its restriction $K := D|\Omega_1(A) \in \mathrm{Hom}^A_A(\Omega_1(A), \Omega_{k+1}(A))$; we write $D = j(K) = j_K$ to express this dependence. Note the defining equation $j_K(\omega) = K(\omega)$ for $\omega \in \Omega_1(A)$. Since it will be very important in the sequel we will use the notation $\Omega^1_k = \Omega^1_k(A) := \mathrm{Hom}^A_A(\Omega_1(A), \Omega_k(A))$ and $\Omega^1_* = \Omega^1_*(A) = \bigoplus_{k=0}^{\infty} \Omega^1_k(A)$.

It can be shown that for any $K \in \Omega^1_k(A)$ the formula

$$j_K(\omega_0 \otimes_A \cdots \otimes_A \omega_\ell) = \sum_{i=0}^{\ell} (-1)^{ik} \omega_0 \otimes_A \cdots \otimes_A K(\omega_i) \otimes_A \cdots \otimes_A \omega_k$$

for $\omega_i \in \Omega_1(A)$ defines an algebraic graded derivation $j_K \in \mathrm{Der}_k \Omega(A)$ and any algebraic derivation is of this form. Thus $K \mapsto j_K$ is an isomorphism from $\Omega^1_*(A)$ to the space of algebraic graded derivations of $\Omega(A)$ Since the graded commutator of two algebraic derivations is clearly again algebraic we can define a graded Lie bracket $[\ ,\ ]^\Delta$ on the space $\Omega^1_*(A)$ by $j([K,L]^\Delta) := [j_K, j_L]$. This bracket is called the *algebraic bracket*; it is an analogy of the one used in [3].

3.3. The differential d is a graded derivation of $\Omega(A)$ of degree one which is not algebraic. In analogy to the well known formula for the Lie derivative along vector fields we now define the *Lie derivative* along a field valued form $K \in \Omega^1_k(A)$ by $\mathcal{L}_K := [j_K, d] \in \mathrm{Der}_k \Omega(A)$. Then one proves that for any derivation $D \in \mathrm{Der}_k \Omega(A)$ there are unique elements $K \in \Omega^1_k$ and $L \in \Omega^1_{k+1}$ such that $D = \mathcal{L}_K + j_L$. Moreover $L = 0$ if and only if $[D, d] = 0$ and D is algebraic if and only if $K = 0$.

For elements $K \in \Omega^1_k$ and $L \in \Omega^1_\ell$ one immediately verifies that $[[\mathcal{L}_K, \mathcal{L}_L], d] = 0$, so we have $[\mathcal{L}(K), \mathcal{L}(L)] = \mathcal{L}([K, L])$ for a uniquely defined $[K, L] \in \Omega^1_{k+\ell}$. This vector valued form $[K, L]$ is called the *abstract Frölicher-Nijenhuis bracket* of K and L. Clearly this bracket defines a graded Lie algebra structure on the space $\Omega^1_*(A)$.

4. Distributions and Integrability

4.1. Distributions. By a *distribution* in an algebra A we mean a sub-A-bimodule $\mathcal{D}$ of $\Omega_1(A)$.

The distribution $\mathcal{D}$ is called *globally integrable* if there exists a sub algebra B of A such that $\mathcal{D}$ is the subspace generated by $A(d(B))$ and $d(B)A$.

The distribution $\mathcal{D}$ is called *splitting* if it is a direct summand in $\Omega_1(A)$ or equivalently if there is a projection $P \in \Omega^1_1(A) = \mathrm{Hom}^A_A(\Omega_1(A), \Omega_1(A))$ onto $\mathcal{D}$, i.e. $P \circ P = P$ and $\mathcal{D} = P(\Omega_1(A))$. Then there is a complementary sub module $\ker P \subset \Omega_1(A)$.

The distribution $\mathcal{D}$ is called *involutive* if the ideal $(\mathcal{D})_{\Omega_*(A)}$ generated by $\mathcal{D}$ in the graded algebra $\Omega_*(A)$ is stable under d, i.e. if $d(\mathcal{D}) \subset (\mathcal{D})_{\Omega_*(A)}$.

4.2. Comments. One should think of this as follows: In ordinary differential geometry $\mathcal{D}$ should be the A-bimodule of those 1-forms which annihilate the sub bundle of TM. Global integrability then means that it is integrable and that the space of functions which are constant along the leaves of the foliation generates those forms. This is a strong condition: There are foliations where this space of functions consists only of the constants, and this can be embedded into any manifold. So in $C^\infty(M)$ there are always involutive distributions which are not globally integrable. To prove some Frobenius theorem a notion of local integrability would be necessary.

4.3 Curvature and cocurvature. Let $P \in \Omega^1_1(A) = \mathrm{Hom}^A_A(\Omega_1(A), \Omega_1(A))$ be a projection, then the image $P(\Omega_1(A))$ is a splitting distribution, called the *vertical distribution* of P and the complement $\ker P$ is also a splitting distribution, called the *horizontal* one. $\bar{P} := Id_{\Omega_1(A)} - P$ is a projection onto the horizontal distribution.

We consider now the Frölicher-Nijenhuis bracket $[P, P]$ of P and define

$$R = R_P = [P, P] \circ P \qquad \text{the } \textit{curvature}.$$
$$\bar{R} = \bar{R}_P = [P, P] \circ \bar{P} \qquad \text{the } \textit{cocurvature},$$

The curvature and the cocurvature are elements of $\Omega^1_2(A) = \mathrm{Hom}^A_A(\Omega_1(A), \Omega_2(A))$. The curvature kills elements of the horizontal distribution, so it is *vertical*. The cocurvature kills elements of the vertical distribution.

Then one proves:

Proposition. *With P, R and $\bar{R}$ as above we have:*

1. (Bianchi Identity)

$$[P, R + \bar{R}] = 0$$
$$2[R, P] = j_R\bar{R} + j_{\bar{R}}R,$$

where the insertion operators are extended to $\Omega^1_*(A)$ in the obvious way.

2. The curvature R is zero if and only if the horizontal distribution is involutive. The cocurvature $\bar{R}$ is zero if and only if the vertical distribution $P(\Omega_1(A))$ is involutive.

References

1. Cap, Andreas; Kriegl, Andreas; Michor, Peter W.; Vanžura, Jiři, *The Frölicher-Nijenhuis bracket in non commutative differential geometry*, Preprint 1991.
2. Cap, A.; Karoubi, M.; Michor, P. W., *The Chern-Weil homomorphism in non commutative differential geometry*, in preparation.
3. De Wilde, M.; Lecomte, P. B. A., *Formal deformations of the Poisson Lie algebra of a symplectic manifold and star-products. Existence, equivalence, derivations*, Deformation theory of algebras and structures and applications, M. Hazewinkel, M. Gerstenhaber, Eds, Kluwer Acad. Publ., Dordrecht, 1988, pp. 897–960.
4. Frölicher, Alfred; Kriegl, Andreas, *Linear spaces and differentiation theory*, Pure and Applied Mathematics, J. Wiley, Chichester, 1988.
5. Frölicher, A.; Nijenhuis, A., *Theory of vector valued differential forms. Part I*, Indagationes Math **18** (1956), 338–359.
6. Karoubi, Max, *Connexions, courbures et classes caractéristiques en K-theorie algébriques*, Canadian Math. Soc. Conference Proc. Vol 2, 1982, pp. 19–27.
7. Karoubi, Max, *Homologie cyclique des groupes et algébres*, C. R. Acad. Sci. Paris **297** (1983), 381–384.
8. Michor, Peter W., *Remarks on the Frölicher-Nijenhuis bracket*, Proceedings of the Conference on Differential Geometry and its Applications, Brno 1986, D. Reidel, 1987, pp. 197–220.
9. Michor, Peter W., *Gauge theory for fiber bundles*, Monographs and Textbooks in Physical Sciences, Lecture Notes 19, Bibliopolis, Napoli, 1991.

INSTITUT FÜR MATHEMATIK, UNIVERSITÄT WIEN, STRUDLHOFGASSE 4, A-1090 WIEN, AUSTRIA.

E-mail address: cap@awirap.bitnet, cap@pap.univie.ac.at, michor@awirap.bitnet, michor@pap.univie.ac.at

Noncommutative Symplectic Geometry and Quantum Mechanics

A. Dimakis and **F. Müller-Hoissen**

Institut für Theoretische Physik
Bunsenstr. 9, D-3400 Göttingen, Germany

ABSTRACT

We consider differential calculus on the Heisenberg algebra of quantum mechanics. Notions of symplectic geometry are introduced on this noncommutative algebra. It is discussed in which sense quantum mechanics can be formulated as noncommutative symplectic geometry.

Classical mechanics deals with the commutative algebra of functions on a phase space. The structure of classical mechanics is beautifully described in terms of symplectic geometry. Quantum mechanics works with a noncommutative (Heisenberg) algebra. Is it possible to formulate quantum mechanics as noncommutative symplectic geometry using the tools of noncommutative geometry ? This problem has already been addressed by Dubois-Violette, Kerner and Madore.[1,2] Here we approach it from a different point of view, but also point out a relation with the latter work. The most basic structure of noncommutative geometry is the notion of differential calculus on an algebra. Let us first define what we mean by a *differential calculus* on a $*$-algebra $\mathcal{A}$ with unit element $\mathbb{I}$.[3] This is a $\mathbb{Z}$-graded algebra

$$\bigwedge(\mathcal{A}) = \bigoplus_{r\in\mathbb{Z}} \overset{r}{\bigwedge}(\mathcal{A}) \quad , \quad \overset{0}{\bigwedge}(\mathcal{A}) = \mathcal{A} \quad , \quad \overset{r}{\bigwedge}(\mathcal{A}) = \{0\} \quad \forall r < 0 \, .$$

The elements of $\bigwedge^r(\mathcal{A})$ are called r-forms. There is a $\mathbb{C}$-linear "exterior derivative" operator $d : \bigwedge^r(\mathcal{A}) \to \bigwedge^{r+1}(\mathcal{A})$ which satisfies

$$\begin{array}{ll} d^2 = 0 & d(\theta \wedge \theta') = (d\theta) \wedge \theta' + (-1)^r \, \theta \wedge d\theta' \\ (df)^* = d(f^*) & (\theta \wedge \theta')^* = (-1)^{rr'} \, \theta'^* \wedge \theta^* \end{array}$$

where θ and θ' are r- and r'-forms, respectively, and $f \in \mathcal{A}$.

Here we choose for $\mathcal{A}$ the Heisenberg algebra generated by $\mathbb{I}$, q, p satisfying $q^* = q$, $p^* = p$ and the (canonical) commutation relation

$$[\, q \, , p \,] = i \hbar \, \mathbb{I} \, . \tag{CR0}$$

We assume that the differentials dq and dp generate the space of 1-forms. To specify a differential calculus on $\mathcal{A}$, we have to specify commutation relations between algebra elements f and differentials. The general structure is

$$\begin{array}{rcl} f\, dq &=& dq \, \Theta(f)^1_1 + dp \, \Theta(f)^1_2 \\ f\, dp &=& dq \, \Theta(f)^2_1 + dp \, \Theta(f)^2_2 \end{array} \qquad \forall f \in \mathcal{A} \, . \tag{CR1}$$

These relations have to be consistent with the algebraic structure of $\mathcal{A}$ and the rules of differential calculus. More precisely, the operator Θ has to satisfy the *consistency conditions* which arise from the following procedures:

- Apply d to (CR0) and use (CR1).
- Commute dq and dp through (CR0) using (CR1).
- Apply * to (CR1) and use (CR1) again.
- Acting with d on (CR1) leads to commutation relations (CR2) for the differentials. Commute q and p through (CR2).

The problem is to find all consistent differential calculi on $\mathcal{A}$. We were only able to solve it under the rather restrictive assumption that $\Theta(q)$ and $\Theta(p)$ are linear in q and p. The commutation relations between p, q and their differentials are then given modulo $SL(2, \mathbb{R})$ transformations by[4]

$$q\,dq = dq\,q \quad , \quad p\,dq = dq\,p \quad , \quad q\,dp = dp\,q \quad , \quad p\,dp = dp\,p + i\,c\,dq\,q$$

which involves a deformation parameter $c \in \mathbb{R}$. The $\wedge$ product is the classical one:

$$dq \wedge dq = 0 \quad , \quad dq \wedge dp = -dp \wedge dq \quad , \quad dp \wedge dp = 0\ .$$

The parameter c can be transformed away by the nonlinear transformation

$$\begin{pmatrix} q \\ p \end{pmatrix} \mapsto \begin{pmatrix} q \\ p + \frac{c}{6\hbar} q^3 \end{pmatrix} = \begin{pmatrix} q' \\ p' \end{pmatrix}$$

which leaves the canonical commutation relation (CR0) invariant and preserves hermiticity. We are left with the following

Problem: Is it possible to transform *every* consistent differential calculus on $\mathcal{A}$ to the standard one ?

Via

$$df = dp\,\partial_p f + dq\,\partial_q f \qquad (f \in \mathcal{A})$$

we introduce vector fields ∂_p and ∂_q. More generally a *vector field* X is a linear combination of these special vector fields with coefficients in $\mathcal{A}$.

Remark: Dubois-Violette has formulated an approach to noncommutative differential calculus starting from *derivations* (see Refs. 1, 2 and the references cited there). A modification of his formalism has been suggested by Madore[5] and we shall discuss the relation with ours in the following. The first step is to choose as many linearly independent derivations of the algebra $\mathcal{A}$ as there are generators (with the exception of the unit element $\mathbb{1}$). For the Heisenberg algebra this means that we have to choose *two* derivations out of the infinite-dimensional space $\mathrm{Der}(\mathcal{A})$ of derivations (which are all *inner* derivations). Let us consider the following choice:

$$\hat{X}_1 := \frac{i}{\hbar}\,\mathrm{ad}(p + \frac{c}{6\hbar} q^3) \quad , \quad \hat{X}_2 := \frac{i}{\hbar}\,\mathrm{ad}(q)\ .$$

The next step is to introduce a set of 1-forms θ^i dual to $\hat{X}_i$, i.e.

$$\theta^i(\hat{X}_j) = \delta^i_j \, \mathbb{1} \, .$$

Using $df(X) = Xf$ one finds

$$\theta^1 = dq \quad , \quad \theta^2 = -\frac{c}{2\hbar} q^2 \, dq - dp \, .$$

A further rule of the calculus is

$$\theta^i f = f \theta^i \qquad (\forall f \in \mathcal{A})$$

from which we recover our commutation relations between dp, dq and p, q. The differential calculus has a two-dimensional $\mathbb{C}$-vector space of vector fields which are derivations.[4] This space coincides with the space generated by the two derivations $\hat{X}_1, \hat{X}_2$. □

Let us now confine our considerations to the standard differential calculus (i.e. $c = 0$) and formulate symplectic geometry on $\mathcal{A}$. There is a natural candidate for a *symplectic 2-form*:

$$\omega = dp \wedge dq \, .$$

We call a mapping $\phi : \mathcal{A} \to \mathcal{A}$ *canonical* if it leaves ω invariant, i.e.

$$d\phi(p) \wedge d\phi(q) = dp \wedge dq \, .$$

Next we introduce the notion of Hamiltonian vector fields. In order to do this we have to define an inner product of vector fields and differential forms. This is a source of additional freedom[4] but here we will simply choose the analogue of the classical rule (see also Ref. 1):

$$\begin{array}{ll} \partial_q \lrcorner dq = 1 - dq \, \partial_q \lrcorner & \qquad \partial_q \lrcorner dp = -dp \, \partial_q \lrcorner \\ \partial_p \lrcorner dp = 1 - dp \, \partial_p \lrcorner & \qquad \partial_p \lrcorner dq = -dq \, \partial_p \lrcorner \end{array}$$

(note that this is insertion from the left, whereas in the remark above we had insertion from the right). For $H \in \mathcal{A}$ we define the *Hamiltonian vector field* X_H by

$$X_H \lrcorner \omega = -dH$$

in accordance with the classical definition. If $H = (1/2m)p^2 + V(q)$, then

$$dH = V'(q) \, dq + \frac{1}{m} p \, dp \quad , \quad X_H = \frac{1}{m} p \, \partial_q - V'(q) \, \partial_p \, .$$

The vector field X_H generates a flow on the algebra $\mathcal{A}$. Could this flow describe time evolution of quantum mechanics ? The answer is *no* because the latter is an automorphism of $\mathcal{A}$ and the generator of an automorphism is a derivation. But the only vector fields on $\mathcal{A}$ (in the sense defined above) which are derivations are linear combinations of ∂_p and ∂_q with complex coefficients (for the undeformed differential calculus).

There is, however, a remedy. We may associate with a vector field X a *derivation* $\hat{X}$ via

$$\hat{X}q = Xq \quad , \quad \hat{X}p = Xp \, .$$

Since elements of $\mathcal{A}$ are functions of p and q, the action of $\hat{X}$ on any $f \in \mathcal{A}$ is then determined by the derivation property of $\hat{X}$. Via $\hat{X}$, the vector field X generates a family $\hat{\phi}_t$ of endomorphisms of $\mathcal{A}$, the flow of $\hat{X}$. If we want $\hat{\phi}_t$ to describe quantum mechanical time evolution, it has to preserve the canonical commutation relation and the hermiticity of p and q. For X these conditions mean

$$[Xq, p] = [Xp, q] \quad , \quad (Xq)^* = Xq \quad , \quad (Xp)^* = Xp$$

in which case we call X an *evolution vector field.* In particular, the flow of $\hat{X}$ for an evolution vector field X is always canonical.

Our X_H given above is indeed an evolution vector field and from the flow of $\hat{X}_H$ we recover the *Heisenberg equations* of motion. The necessity of going beyond the geometrical framework shows, however, that the correspondence between noncommutative symplectic geometry and quantum mechanics is not perfect.

We have started with an analysis of the possible consistent differential calculi on $\mathcal{A}$. At least for the class of calculi which we found it turned out that deformations relative to the standard differential calculus can be transformed away by canonical transformations. However, such transformations change the Hamiltonian H. If we choose the standard differential calculus on $\mathcal{A}$ and a Hamiltonian H, we may apply a canonical transformation to map H into the free Hamiltonian $H' = (1/2m)p'^2$. The *interaction* (the potential term in the original Hamiltonian) then manifests itself *as a deformation of the differential calculus* :

standard diff. calculus	$\longrightarrow$	deformed diff. calculus
$\vert$	canonical transformation	$\vert$
H	$\longrightarrow$	H' free Hamiltonian

This opens the possibility of a kind of geometrization of interaction and suggests a new way to formulate physical theories.

References

1. M. Dubois-Violette, R. Kerner and J. Madore, "Non-commutative differential geometry of matrix algebras" *J. Math. Phys.* **31** (1990) 316.
2. M. Dubois-Violette, "Noncommutative differential geometry, quantum mechanics and gauge theory", in Differential Geometric Methods in Theoretical Physics, eds. C. Bartocci *et al.* (Springer, 1991).
3. S. L. Woronowicz, "Differential calculus on compact matrix pseudogroups (quantum groups)", *Commun. Math. Phys.* **122** (1989) 125.
4. A. Dimakis and F. Müller-Hoissen, "Quantum mechanics as noncommutative symplectic geometry", to appear in *J. Phys. A.*
5. J. Madore, "The commutative limit of matrix geometry" *J. Math. Phys.* **32** (1991) 332

A realization of Macdonald's symmetric polynomials on quantum homogeneous spaces

MASATOSHI NOUMI

Department of Mathematical Sciences
University of Tokyo

This is a report on our recent result concerning the realization of Macdonald's symmetric polynomials by quantum homogeneous spaces. We introduce quantum analogues of the homogeneous spaces $GL(n)/SO(n)$ and $GL(2n)/Sp(2n)$ in the framework of quantum general linear groups, and investigate their zonal spherical functions associated with finite dimensional representations. As a result, we see that our zonal spherical functions are expressed in terms of Macdonald's symmetric polynomials $P_\lambda(x_1, \cdots, x_n; q, t)$ (of type A_{n-1}) with $t = q^{\frac{1}{2}}$ or q^2 ([M]). Details of the arguments below will be given elsewhere.

This result can be regarded as a generalization of Koornwinder's realization of continuous q-Legendre polynomials on the quantum group $SU_q(2)$ ([K1]). For the quantum analogue of $SL(3)/SO(3)$, it has been known by Ueno-Takebayashi [UT] that Macdonald's symmetric polynomials arise as zonal spherical functions.

Throughout this note, we denote by G the general linear group $GL(N)$ and by $\mathfrak{g}$ its Lie algebra $\mathfrak{gl}(N)$.

1. Recalls on the quantum general linear group.

Taking the field $\mathbb{K} = \mathbb{Q}(q^{\frac{1}{2}})$ of rational functions in $q^{\frac{1}{2}}$ as the ground field, we denote by $U_q(\mathfrak{g}) = U_q(\mathfrak{gl}(N))$ the quantized universal enveloping algebra of Drinfeld and Jimbo, associated to the Lie algebra $\mathfrak{g} = \mathfrak{gl}(N)$. In this note, we will make a systematic use of L-operators, rather than the Chevelley generators for $U_q(\mathfrak{g})$, following the framework of Reshetikhin-Takhtajan-Faddeev [RTF].

Let V the vector representation of $U_q(\mathfrak{g})$ and take the following R-matrices $R^\pm$ acting on the tensor product $V \otimes_{\mathbb{K}} V$:

$$R^\pm = \sum_{i,j} e_{ii} \otimes e_{jj} q^{\pm\delta_{ij}} \pm (q - q^{-1}) \sum_{\pm(j-i)>0} e_{ij} \otimes e_{ji}, \tag{1.1}$$

where the e_{ij}'s are the matrix units with respect to the canonical basis of V. These matrices satisfy the Yang-Baxter equation $R_{12}^\epsilon R_{13}^\epsilon R_{23}^\epsilon = R_{23}^\epsilon R_{13}^\epsilon R_{12}^\epsilon$ $(\epsilon = \pm)$. The corrsponding L-operators will be denoted by $L^\pm = (L_{ij}^\pm)_{1\le i,j\le N}$. The commutation relations among $L_{ij}^\pm \in U_q(\mathfrak{g})$ are described by $R^+ L_1^\epsilon L_2^\epsilon = L_2^\epsilon L_1^\epsilon R^+$ $(\epsilon = \pm)$, $R^+ L_1^+ L_2^- = L_2^- L_1^+ R^+$, and they satisfy $\Delta(L^\pm) = L^\pm \dot{\otimes} L^\pm$, $\varepsilon(L^\pm) = \mathrm{id}_V$. Note that the R-matrices and the L-operators are related by the formulas $R^{\pm} = \sum_{i,j} e_{ij} \otimes \rho_V(L^{\pm}_{ij})$, where $\rho_V : U_q(\mathfrak{g}) \to \mathrm{End}_{\mathbb{K}}(V)$ is the structure homomorphism of the left $U_q(\mathfrak{g})$-module V.

The quantized coordinate ring $A_q(G) = A_q(GL(N))$ is a Hopf algebra generated by the "canonical coordinates" $T = (t_{ij})_{1\le i,j\le N}$ and the inverse of the quantum determinant $\det_q(T)$. The fundamental relations among t_{ij} are given by $R^+T_2T_1 = T_1T_2R^+$ and the Hopf algebra structure of $A_q(GL(N))$ is determined by $\Delta(T) = T\dot{\otimes}T$, $\varepsilon(T) = \mathrm{id}_V$. Note that the quantum determinant $\det_q(T)$ is a central group-like element of $A_q(G)$.

The algebra $A_q(G) = A_q(GL(N))$ has the natural structure of a bimodule over $U_q(\mathfrak{g}) = U_q(\mathfrak{gl}(N))$, representing the regular representation of $GL_q(N)$. It is a $\mathbb{K}$-*algebra with* $U_q(\mathfrak{g})$-*symmetry* in the sense that the multiplication $A_q(G)\otimes_{\mathbb{K}} A_q(G) \to A_q(G)$ and the structure homomorphism $\mathbb{K} \to A_q(G)$ are both homomorphisms of $U_q(\mathfrak{g})$-bimodules. In terms of the L-operators, the $U_q(\mathfrak{g})$-bimodule structure of $A_q(G)$ is described as $L^{\epsilon}_1.T_2 = T_2R^{\epsilon}_{12}$, $T_2.L^{\epsilon}_1 = R^{\epsilon}_{12}T_2$ $(\epsilon = \pm)$.

We denote by $P = \bigoplus_{k=1}^{N} \mathbb{Z}\epsilon_k$ the weight lattice for the Lie algebra $\mathfrak{gl}(N)$ and by P^+ the set of all dominant integral weights, namely, of all $\lambda = \sum_{k=1}^{N} \lambda_k\epsilon_k \in P$ such that $\lambda_1 \ge \lambda_2 \ge \cdots \ge \lambda_N$. For each dominant integral weight $\lambda \in P^+$, we denote by $V(\lambda)$ the irreducible left $U_q(\mathfrak{g})$-module with highest weight λ. Actually each left $U_q(\mathfrak{g})$-module $V(\lambda)$ is obtained as the "differential representation" from the corresponding right $A_q(G)$-comodule structure on $V(\lambda)$. We denote by $W(\lambda)$ the $\mathbb{K}$-vector subspace of $A_q(G)$ spanned by the matrix elements of the right $A_q(G)$-comodule $V(\lambda)$. Then the regular representation $A_q(G)$ has the irreducible decomposition

$$A_q(G) = \bigoplus\nolimits_{\lambda\in P^+} W(\lambda), \quad W(\lambda) \xleftarrow{\sim} \mathrm{Hom}_{\mathbb{K}}(V(\lambda), \mathbb{K}) \otimes_{\mathbb{K}} V(\lambda), \tag{1.2}$$

as a $U_q(\mathfrak{g})$-bimodule, which corresponds to the Peter-Weyl Theorem for the quantum unitary group $U_q(N)$ (see [NYM], for example). We also remark that that the $\mathbb{K}$-subspace $W(\lambda)$ of $A_q(G)$ is characterized as the simultaneous eigenspace of the center of $U_q(\mathfrak{g})$:

$$W(\lambda) = \{\varphi \in A_q(G);\ C.\varphi = \chi_\lambda(C)\varphi \text{ for any central element } C \in U_q(\mathfrak{g})\}, \tag{1.3}$$

where $\chi_\lambda(C)$ stands for the eigenvalue of the central element C acting on the irreducible representation $V(\lambda)$.

2. Quantum analogue of Lie subalgebras of $\mathfrak{gl}(N)$.

In this section, we introduce quantum analogues of some Lie *subalgebras* of the Lie algebra $\mathfrak{g} = \mathfrak{gl}(N)$. The Lie subalgebras $\mathfrak{k}$ to be considered for quantization are:

$$\begin{aligned} &\text{Case (SO):} \quad \mathfrak{k} = \mathfrak{so}(n) = \{X \in \mathfrak{gl}(n);\ {}^tX + X = 0\} \quad (N = n),\\ &\text{Case (Sp):} \quad \mathfrak{k} = \mathfrak{sp}(2n) = \{X \in \mathfrak{gl}(2n);\ {}^tX\,J_n + J_n X = 0\} \quad (N = 2n), \end{aligned} \tag{2.1}$$

where $J_n = \sum_{k=1}^{n} e_{2k-1,2k} - e_{2k,2k-1}$. For each case, we introduce a *coideal* $\mathfrak{k}_q$ of $U_q(\mathfrak{g})$ which "tends" to $\mathfrak{k}$ when $q \to 1$.

We first define the matrix $J \in \mathrm{End}_{\mathbb{K}}(V)$ as follows:

$$\text{(2.2)}\qquad \begin{aligned} &\text{Case (SO):}\ \ J := \sum\nolimits_{k=1}^{n} e_{kk} q^{n-k}, \\ &\text{Case (Sp):}\ \ J := \sum\nolimits_{k=1}^{n} (e_{2k-1,2k} - q e_{2k,2k-1}) q^{2(n-k)}. \end{aligned}$$

By straightforward verification, one can check that the matrix J above satisfies the *reflection equation*

$$\text{(2.3)}\qquad R_{12}^{+} J_2 R_{12}^{+\,t_2} J_1 = J_1 R_{12}^{+\,t_2} J_2 R_{12}^{+}.$$

For the reflection equation and related topics, see [Ku]. We define an $N \times N$ matrix $M = (M_{ij})_{1 \le i,j \le N}$ with entries in $U_q(\mathfrak{g})$ by setting $M = L^+ - J\, S(L^-)^t J^{-1}$, and denote by $\mathfrak{k}_q$ the $\mathbb{K}$-vector subspace of $U_q(\mathfrak{g})$ spanned by the matrix elements M_{ij}:

$$\text{(2.4)}\qquad \mathfrak{k}_q := \sum\nolimits_{1 \le i,j \le N} \mathbb{K} M_{ij} \subset U_q(\mathfrak{g}), \quad M := L^+ - J\, S(L^-)^t J^{-1}.$$

One can easily show that $\mathfrak{k}_q$ becomes a coideal of $U_q(\mathfrak{g})$, namely, $\Delta(\mathfrak{k}_q) \subset \mathfrak{k}_q \otimes U_q(\mathfrak{g}) + U_q(\mathfrak{g}) \otimes \mathfrak{k}_q$ and $\varepsilon(\mathfrak{k}_q) = 0$. By an explicit calculation, one can also check that the elements M_{ij}'s recover the root vectors of $\mathfrak{so}(n)$ or $\mathfrak{sp}(2n)$ when q tends to 1. We remark that, in Case (SO), our coideal $\mathfrak{k}_q$ is closely related to the q-deformed algebra $U_q(\mathfrak{so}(n))$ of Gavrilik-Klimyk [GK].

Regarding this coideal $\mathfrak{k}_q \subset U_q(\mathfrak{g})$ as a quantum analogue of the Lie subalgebra $\mathfrak{k} \subset \mathfrak{g}$, we investigate the $\mathfrak{k}_q$-fixed vectors in $U_q(\mathfrak{g})$-modules. For each dominant integral weight $\lambda \in P^+$, we denote by $V(\lambda)_{\mathfrak{k}_q}$ the $\mathbb{K}$-vector subspace of $V(\lambda)$ consisting of all $\mathfrak{k}_q$-fixed vectors, namely, all elements v with $\mathfrak{k}_q.v = 0$.

Theorem 1. (1) *For each $\lambda \in P^+$, the $\mathbb{K}$-vector subspace $V(\lambda)_{\mathfrak{k}_q}$ of $\mathfrak{k}_q$-fixed vectors is at most one dimensional.*
(2) *The left $U_q(\mathfrak{g})$-module $V(\lambda)$ has a nonzero $\mathfrak{k}_q$-fixed vector if and only if the highest weight $\lambda = \sum_{k=1}^{N} \lambda_k \epsilon_k$ satisfies the following condition:*
Case (SO): $\lambda_k - \lambda_{k+1} \in 2\mathbb{Z}$ $(1 \le k \le n-1)$,
Case (Sp): $\lambda_{2k-1} = \lambda_{2k}$ $(1 \le k \le n)$.

Theorem 1 for Case (SO) is already mentioned in [UT]. Hereafter, we denote by $P_{\mathfrak{k}}^+$ the subset of dominant integral weights λ satisfying the condition above.

3. Quantum homogeneous spaces and zonal spherical functions.

In terms of the coideal $\mathfrak{k}_q$, we can study the quantum analogue of the homogeneous space $K\backslash G$ for the following closed subgroup K of G:

$$\text{(3.1)}\qquad \begin{aligned} &\text{Case (SO):}\ \ K = SO(n) = \{g \in GL(n);\ g^t g = \mathrm{id}_V,\ \det(g) = 1\}, \\ &\text{Case (Sp):}\ \ K = Sp(2n) = \{g \in GL(2n);\ g^t J_n g = J_n,\ \det(g) = 1\}. \end{aligned}$$

For that purpose, we consider the $\mathbb{K}$-vector subspace of $\mathfrak{k}_q$-invariants in $A_q(G)$:

$$A_q(K\backslash G) := \{\varphi \in A_q(G);\ \varphi.\mathfrak{k}_q = 0\}. \tag{3.2}$$

Thanks to the fact that $\mathfrak{k}_q$ is a coideal, this subspace of $\mathfrak{k}_q$-invariants is actually a $\mathbb{K}$-subalgebra of $A_q(G)$. Note also that $A_q(K\backslash G)$ has the structure of a left $U_q(\mathfrak{g})$-module induced from the right regular representation $A_q(G)$. From the irreducible decomposition (1.2) of $A_q(G)$, combined with Theorem 1, we obtain the multiplicity-free irreducible decomposition of $A_q(K\backslash G)$: $A_q(K\backslash G) \cong \bigoplus_{\lambda \in P_\mathfrak{k}^+} V(\lambda)$. As to the structure of this invariant ring, we have

Theorem 2. *Define a matrix $X = (x_{ij})_{1\le i,j\le N}$ with enries in $A_q(G)$ by $X := T^t J^{-1} T$. Then the $\mathbb{K}$-algebra $A_q(K\backslash G)$ is generated by the quadratic elements x_{ij} $(1 \le i, j \le N)$ together with $\det_q(T)^{\pm 1}$.*

Proposition 3. *The elements x_{ij} $(1 \le i, j \le N)$ defined as above satisfy*
Case (SO): $x_{ij} = q x_{ji}$ $(1 \le i < j \le n)$,
Case (Sp): $x_{ii} = 0$ $(1 \le i \le 2n)$, $q x_{ij} + x_{ji} = 0$ $(1 \le i < j \le 2n)$.
They have the commutation relations $R_{12}^+ X_2 R_{12}^{+\,t_2} X_1 = X_1 R_{12}^{+\,t_2} X_2 R_{12}^+$, in each case.

This description of $A_q(K\backslash G)$ corresponds to the realization of the homogeneous space $SL(n)/SO(n)$ and $SL(2n)/Sp(2n)$ as an orbit of symmetric and anti-symmetric matrices, respectively. (Compare the commutation relations above with (2.3).)

The next step is to consider the quantum analogue of the double coset space $K\backslash G/K$. We take the following $\mathbb{K}$-subalgebra of $\mathfrak{k}_q$-biinvariants in $A_q(G)$:

$$\mathcal{H} = A_q(K\backslash G/K) := \{\varphi \in A_q(G);\ \mathfrak{k}_q.\varphi = \varphi.\mathfrak{k}_q = 0\}. \tag{3.3}$$

From the irreducible decomposition of $A_q(G)$ again, we have the direct decomposition of the $\mathbb{K}$-subalgebra $\mathcal{H}$ of $\mathfrak{k}_q$-biinvariants $\mathcal{H} = \bigoplus_{\lambda \in P_\mathfrak{k}^+} \mathcal{H}(\lambda)$, with $\mathcal{H}(\lambda) = \mathcal{H} \cap W(\lambda)$. Note that $\mathcal{H}(\lambda)$ is one-dimensioanl for each $\lambda \in P_\mathfrak{k}^+$. We say that a nonzero element φ of $\mathcal{H}(\lambda)$ $(\lambda \in P_\mathfrak{k}^+)$ is a *zonal spherical function* associated with the representation $V(\lambda)$. This condition is equivalent to saying that

(1) $\mathfrak{k}_q.\varphi = \varphi.\mathfrak{k}_q = 0$,
(2) $C.\varphi = \chi_\lambda(C)\varphi$ for any central element $C \in U_q(\mathfrak{g})$.

For the description of the zonal spherical functions φ, we consider their "restriction" $\varphi|_\mathbb{T}$ to the diagonal subgroup $\mathbb{T} = (\mathbb{K}^*)^N$ of the quantum group $GL_q(N)$. We denote by $z = (z_1, \cdots, z_N)$ the canonical coordinates of the diagonal subgroup. The coordinate ring $A(\mathbb{T})$ is the $\mathbb{K}$-algebra of Laurent polynomials $\mathbb{K}[{z_1}^{\pm 1}, \cdots, {z_N}^{\pm 1}]$, and there exists a Hopf algebra homomorphism $\varphi \mapsto \varphi|_\mathbb{T} : A_q(G) \to A(\mathbb{T})$ such that $t_{ij}|_\mathbb{T} = \delta_{ij} z_j$. We look at the composition $\mathcal{H} \hookrightarrow A_q(G) \to A(\mathbb{T})$.

Theorem 4. *The restriction mapping $\mathcal{H} \to A(\mathbb{T})$ is an injective $\mathbb{K}$-algebra homomorphism; hence, $\mathcal{H}$ is a commutative $\mathbb{K}$-subalgebra of $A_q(G)$. Furthermore the image $\mathcal{H}|_\mathbb{T}$ of $\mathcal{H}$ is given by*

$$\textit{Case (SO)}:\ \mathcal{H}|_\mathbb{T} = \mathbb{K}[{z_1}^2, \cdots, {z_n}^2]^{\mathfrak{S}_n}[(z_1 \cdots z_n)^{-1}],$$

$$\textit{Case (Sp)}:\ \mathcal{H}|_\mathbb{T} = \mathbb{K}[z_1 z_2, \cdots, z_{2n-1} z_{2n}]^{\mathfrak{S}_n}[(z_1 z_2 \cdots z_{2n})^{-1}].$$

Theorem 4 for Case (SO) is also stated in [UT].

4. Macdonald's symmetric polynomials as zonal spherical functions.

For each $\lambda \in P_{\mathfrak{k}}^{+}$, one can find a unique zonal spherical function $\psi_\lambda \in \mathcal{H}(\lambda)$, such that $\mathcal{H}(\lambda) = \mathbb{K}\psi_\lambda$ and $\psi_\lambda(q^\rho) = 1$, where q^ρ is the point $(q^{\frac{N-1}{2}}, q^{\frac{N-3}{2}}, \cdots, q^{\frac{-N+1}{2}})$ of the diagonal subgroup $\mathbb{T} = (\mathbb{K}^*)^N$. The restriction of our zonal spherical functions ψ_λ to $\mathbb{T}$ are expressed explicitly in terms of Macdonald's symmetric polynomials.

Recall that, for each partition $\mu = \sum_{k=1}^{n} \mu_k \epsilon_k$ with $\mu_1 \geq \cdots \geq \mu_n \geq 0$, the ***Macdonald symmetric polynomial*** $P_\mu(x; q, t)$ in n variables $x = (x_1, \cdots, x_n)$ is characterized as the unique symmetric polynomial in $\mathbb{Q}(q,t)[x]$, homogeneous of degree $\sum_{k=1}^{n} \mu_k$, satisfying the following conditions: i) The polynomial $P_\mu(x; q, t)$ has an expression

$$P_\mu(x; q, t) = m_\mu(x) + \sum\nolimits_{\nu < \mu} c_{\mu\nu} m_\nu(x), \tag{4.1}$$

where $m_\mu(x)$ stands for the monomial symmetric polynomial of monomial type μ and the symbol $<$ denotes the dominance order of partitions; ii) $P_\mu(x; q, t)$ satisfies the q-difference equation

$$\sum\nolimits_{k=1}^{n} \frac{\Delta(x_1, \cdots, t x_k, \cdots, x_n)}{\Delta(x_1, \cdots, x_n)} T_{q, x_k} P_\mu(x; q, t) = \left(\sum\nolimits_{k=1}^{n} t^{n-k} q^{\mu_k}\right) P_\mu(x; q, t), \tag{4.2}$$

where $\Delta(x_1, \cdots, x_n)$ stands for the difference product $\prod_{1 \leq i < j \leq n}(x_j - x_i)$ and T_{q,x_k} for the q-shift operator defined by $(T_{q,x_k} f)(x_1, \cdots, x_n) = f(x_1, \cdots, q x_k, \cdots, x_n)$.

Theorem 5. *For each $\lambda \in P_{\mathfrak{k}}^{+}$, the restriction $\psi_\lambda|_{\mathbb{T}}$ of the zonal spherical function $\psi_\lambda \in \mathcal{H}(\lambda)$ is expressed by Macdonald's symmetric polynomials as follows:*

$$\text{Case (SO):} \quad \psi_\lambda|_{\mathbb{T}} = c_\lambda^{-1} P_\mu({z_1}^2, \cdots, {z_n}^2; q^4, q^2)(z_1 \cdots z_n)^\ell,$$

if $\lambda = 2\sum_{k=1}^{n} \mu_k \epsilon_k + \ell \sum_{k=1}^{n} \epsilon_k$ with $\mu_k \geq 0 \, (1 \leq k \leq n)$ and $\ell \in \mathbb{Z}$;

$$\text{Case (Sp):} \quad \psi_\lambda|_{\mathbb{T}} = c_\lambda^{-1} P_\mu(z_1 z_2, \cdots, z_{2n-1} z_{2n}; q^2, q^4)(z_1 z_2 \cdots z_{2n})^\ell,$$

if $\lambda = \sum_{k=1}^{n} \mu_k(\epsilon_{2k-1} + \epsilon_{2k}) + \ell \sum_{k=1}^{2n} \epsilon_k$. with $\mu_k \geq 0 \, (1 \leq k \leq n)$ and $\ell \in \mathbb{Z}$.

The proof of Theorem 5 is based on the calculation of the *radial component* of the central element

$$C_1 = \sum\nolimits_{1 \leq i,j \leq N} q^{N-2i+1} L_{ij}^{+} S(L_{ji}^{-}) \tag{4.3}$$

of $U_q(\mathfrak{g})$, proposed in [RTF]. By analyzing the commutation relations of L-operators described in terms of R-matrices, one can find a q-difference operator $D_1 : A(\mathbb{T}) \to A(\mathbb{T})$, in the variables $z = (z_1, \cdots, z_N)$, such that $(C_1.\varphi)|_{\mathbb{T}} = D_1(\varphi|_{\mathbb{T}})$ for all $\varphi \in \mathcal{H}$. Such a q-difference operator acting on the image $\mathcal{H}|_{\mathbb{T}}$ will be called the radial component of C_1.

Theorem 6. *The radial component of the central element $C_1 \in U_q(\mathfrak{g})$ above is given by the following q-difference operator:*

$$\text{Case (SO):} \quad D_1 = q^{-(n-1)}\sum_{k=1}^{n} \frac{\Delta(x_1,\cdots,q^2x_k,\cdots,x_n)}{\Delta(x_1,\cdots,x_n)} T_{q^2,x_k},$$

where $x_k = z_k{}^2$ for $1 \le k \le n$;

$$\text{Case (Sp):} \quad D_1 = q^{-2(n-1)}(q+q^{-1})\sum_{k=1}^{n} \frac{\Delta(x_1,\cdots,q^4x_k,\cdots,x_n)}{\Delta(x_1,\cdots,x_n)} T_{q^2,x_k},$$

where $x_k = z_{2k-1}z_{2k}$ for $1 \le k \le n$.

Since the q-difference equation $(D_1 - \chi_\lambda(C_1))\psi_\lambda|_{\mathbb{T}} = 0$ determines the restriction $\psi_\lambda|_{\mathbb{T}}$, up to a scalar multiple, we obtain the expression of ψ_λ as in Theorem 5.

Finally we remark that our formulation of qunatum analogues of the homogeneous spaces $GL(n)/SO(n)$ and $GL(2n)/Sp(2n)$ is compatible with the *-structure of $A_q(G)$, conrresponding to the quantum unitary group $U_q(N)$ when q is real. By using the invariant functional on this Hopf *-algebra, we can give an interpretation for the orthogonality relations of Macdonald's symmetric polynomials $P_\lambda(x;q,t)$ for $t = q^{\frac{1}{2}}$ and q^2, under the scalar product $< \, , \, >'_{q,t}$([M]). In our framework, we can also evaluate the scalar products $< P_\lambda, P_\lambda >'_{q,t} \,/\, < 1,1 >'_{q,t}$ for $t = q^{\frac{1}{2}}$ and q^2, to recover Macdonald's result stated in [M] from the Schur orthogonality of our zonal spherical functions.

References

[GK] A.M. Gavrilik and A.U. Klimyk, *q-Deformed orthogonal and pseudo-orthogonal algebras and their representations*, Lett. Math. Phys. **21** (1991), 215–220.

[J] M. Jimbo, *A q-analogue of $U_q(\mathfrak{gl}(N+1))$, Hecke algebra, and the Yang-Baxter equation*, Lett. Math. Phys. **11** (1986), 247–252.

[K1] T.H. Koornwinder, *Continuous q-Legendre polynomials as spherical matrix elements of irreducible representations of the quantum $SU(2)$ group*, CWI Quaterly **2** (1989), 171-173.

[K2] ———, *Orthogonal polynomials in connection with quantum groups*, in "Orthogonal Polynomials: Theory and Practice" (P. Nevai, ed.), NATO ASI Series, Kluwer Academic Publishers, 1990, pp. 257–292.

[Ku] P.P. Kulish, *Quantum groups and quantum algebras as symmetries of dynamical systems*, preprint YITP/K-959 (1991).

[M] I.G. Macdonald, *A new class of symmetric functions*, Actes 20ᵉ Séminaire Lotharingien, Publ. I. R. M. A. Strasbourg, 1988, pp. 131–171.

[N] M. Noumi, *Quantum groups and q-orthogonal polynomials — Towards a realization of Askey-Wilson polynomials on $SU_q(2)$*, in "Special Functions" (M. Kashiwara, T. Miwa, eds.), ICM-90 Satellite Conference Proceedings, Spinger-Verlag, 1991, pp. 260–288.

[NUW] M. Noumi, T. Umeda and M. Wakayama, *A quantum analogue of the Capelli identity and an elementary differential calculus on $GL_q(n)$*, preprint J-Tokyo-Math 91-16 (1991).

[NYM] M. Noumi, H. Yamada and K. Mimachi, *Finite dimensional representations of the quantum group $GL_q(n;\mathbb{C})$ and the zonal spherical functions on $U_q(n-1)\backslash U_q(n)$*, Japanese J. Math. (to appear).

[RTF] N.Yu. Reshetikhin, L.A.Takhtajan and L.D. Faddeev, *Quantization of Lie groups and Lie algebras*, Algebra and Analysis **1** (1989), 178–206; English transl. in Leningrad Math. J. **1** (1990), 193–225.

[UT] K. Ueno and T. Takebayashi, *Zonal spherical functions on quantum symmetric spaces and Macdonald's symmetric polynomials*, in "Quantum Groups" (P.P. Kulish, ed.), Proceedings of Workshops held in the Euler International Mathematical Institute, Lenengrad, Fall 1990, Lecture Notes in Math., vol. 1510, Springer-Verlag, 1992, pp. 142–147.

3-8-1 Komaba, Meguro-Ku, Tokyo 153, Japan

Reflection Equations, c-Number Solutions, Quadratic Algebras and Quantum Groups

R. SASAKI AND C. SCHWIEBERT*

Uji Research Center
Yukawa Institute for Theoretical Physics
Kyoto University, Uji 611, Japan

Basic properties of reflection equations (quantum group comodule), c-number solutions and quadratic algebras defined by them are briefly reviewed. Reflection equations appear under different guises in various places of "integrable physics".

This contribution is based on our work on reflection equations without the spectral parameter [1]. The Reflection Equation (RE) was first discussed in [2] as a condition for factorisable scattering on a half line which extends the Yang-Baxter equation (YBE) (factorisable scattering on a line). Schematically it reads

$$R(\lambda-\nu)K_1(\lambda)R'(\lambda+\nu)K_2(\nu)=K_2(\nu)R''(\lambda+\nu)K_1(\lambda)R'''(\lambda-\nu), \tag{1}$$

where the $n\times n$ K-matrix is a new object describing the reflection at the endpoint of the half line. The parameters λ and ν are the rapidities of the incoming particles or the spectral parameters.

Similar and related equations appeared in different contexts recently. To mention some: Lattice regularized WZWN and integrable models with non-local commutation relations [3] [4 – 7], quantum Liouville theory [8], quantum current algebras and conformal field theory [9], modified Knizhnik-Zamolodchikov equations [10], integrable spin chains with non-periodic boundary conditions [11 – 14], noncommutative differential geometry on quantum groups [15, 16], "braided matrix groups" [17], twisted Yangians [18], algebra of q-Minkowski space coordinates and derivatives [19], etc.

The success of the spectral parameter independent YBE in the description of quantum groups, braid groups etc., motivates the study of spectral parameter independent reflection equations. They have the forms (in the notations of [1])

$$\text{RE1}: \qquad RK_1R^{t_1}K_2=K_2R^{t_1}K_1R, \tag{2}$$

$$\text{RE2}: \qquad RK_1\widetilde{R}K_2=K_2RK_1\widetilde{R}, \tag{3}$$

where t_1 means transposition in the first space, $\widetilde{R}=PRP$ with P the permutation operator. RE2 and YBE, for example, describe the braid group relations of higher genus manifolds [3, 20].

---

* Feodor Lynen and JSPS Fellow. Supported in part by Grant in aid from Ministry of Education, Science and Culture, No. 91081.

The key property of these equations is the covariance under the quantum group, in other words they are comodules w.r.t. the quantum group defined by the corresponding R-matrix [11, 21]. The comodule property means that for a given solution K of the RE

$$K_T = TKT^t \text{ for RE1}, \qquad K_T = TKT^{-1} \text{ for RE2}, \tag{4}$$

is again a solution of the corresponding RE. For this to hold it is necessary that the matrix elements of K and the quantum group matrix T commute.

The comodule property can further be generalised. The following RE

$$R^{(1)}K_1R^{(2)}K_2 = K_2R^{(3)}K_1R^{(4)}, \tag{5}$$

is invariant w.r.t the comodule transformation

$$K \to K_{T,S} = TKS, \tag{6}$$

provided that T and S satisfy the following set of equations

$$\begin{aligned} R^{(1)}T_1T_2 &= T_2T_1R^{(1)}, \qquad & R^{(4)}S_1S_2 &= S_2S_1R^{(4)}, \\ S_1R^{(2)}T_2 &= T_2R^{(2)}S_1, \qquad & T_1R^{(3)}S_2 &= S_2R^{(3)}T_1. \end{aligned} \tag{7}$$

A particular example of this is well known by now, in (7) one may choose the $L^{\pm}$-matrices [21] encoding the commutation relations of the quantum Lie algebra. Also we can replace any R-matrix of the RE by $R \to \widetilde{R}^{-1}$ preserving the comodule property [22].

The RE can be considered to define an associative quadratic algebra generated by the elements of K. Using the standard $SL_q(2)$ R-matrix

$$R = \begin{pmatrix} q & & & \\ & 1 & & \\ & \omega & 1 & \\ & & & q \end{pmatrix}, \quad R^{t_1} = \begin{pmatrix} q & & & \omega \\ & 1 & & \\ & & 1 & \\ & & & q \end{pmatrix}, \quad \omega = q - q^{-1} \tag{8}$$

and writing the K-matrix as $K = \begin{pmatrix} \alpha & \beta \\ \gamma & \delta \end{pmatrix}$, then RE1 gives the commutation relations among them

$$\begin{aligned} [\alpha, \beta] &= \omega\alpha\gamma, \qquad & [\alpha, \delta] &= \omega(q\beta\gamma + \gamma^2), \qquad & [\beta, \delta] &= \omega\gamma\delta, \\ \alpha\gamma &= q^2\gamma\alpha, \qquad & [\beta, \gamma] &= 0, \qquad & \gamma\delta &= q^2\delta\gamma. \end{aligned} \tag{9}$$

This algebra has two central elements [1, 20, 22]

$$c_1 = \mathrm{Tr}(\epsilon_q^t K) = \beta - q\gamma, \qquad c_2 = \mathrm{Tr}(\epsilon_q^t K \epsilon_q K), \tag{10}$$

in which a constant matrix ϵ_q is the quantum invariant metric of $SL_q(2)$ defined below. The RE2 algebra for $SL_q(2)$ is isomorphic to (9) so we don't discuss it further. Truncations of these algebras obtained by putting central elements to a constant leads to quantum homogeneous spaces.

We can easily find the c-number representations of (9), there are two of them

$$K^{(1)} \sim \varepsilon_q = \begin{pmatrix} & 1 \\ -q & \end{pmatrix}, \qquad K^{(2)} = \begin{pmatrix} \alpha & \beta \\ & \delta \end{pmatrix}. \tag{11}$$

Using the universal R-matrix expressed in terms of the generators $J, X_\pm$ of the quantum algebra $sl_q(2)$ (cf.[23])

$$R_U = q^{2J\otimes J} \sum_{n=0}^{\infty} \frac{(1-q^{-2})^n}{[n;q^{-2}]!} q^J X_+ \otimes q^{-J} X_-, \qquad [n;q] = (1-q^n)/(1-q) \tag{12}$$

one can show that both solutions can be extended to universal solutions of the RE. The first one is proportional to the element w of the quantum Weyl group

$$w \simeq q^{-J}\varepsilon_j, \qquad (\varepsilon_j)_{m,m'} = (-1)^{j-m}\delta_{m,-m'}, \quad -j \le m, m' \le j \tag{13}$$

defined in any finite dimensional representation $V_j(\mathrm{dim}V_j = 2j+1)$ [23]. The second one is given by [8]

$$K_U = q^{J^2} \sum_{n=0}^{\infty} \frac{f_n}{[n]!} q^{-nJ}(X_+)^n, \tag{14}$$

where the f_n are related to the q-Hermite polynomials and only f_1 is a free parameter. RE2 has equivalent constant solutions.

To finish we just show the constant solutions of RE1 for $SL_q(3)$ in the defining representation. There are two with non-vanishing determinants

$$K^{(1)} = \begin{pmatrix} a_{11} & & \\ & a_{22} & \\ & & a_{33} \end{pmatrix}, \quad K^{(2)} = \begin{pmatrix} a_{11} & a_{12} & a_{13} \\ & f_{22} & f_{23} \\ & & f_{33} \end{pmatrix}, \tag{15}$$

where the functions f_{ij} depend on the parameters in the first row $f_{ii} = a_{1i}^2/a_{11}X^2$, $f_{ij} = a_{1i}a_{1j}/a_{11}X$ and $X = 1+q^{-1}$. And there are six others with vanishing determinants corresponding to embeddings of the $SL_q(2)$ solutions (11). In contrast to (11) all free parameters in (15) can be transformed away. This is a general feature of $SL_q(N)$ for RE1. For RE2 the solutions are not equivalent to those of RE1. We have analysed much more cases and refer readers to [1] for further details.

REFERENCES

1. Kulish P. P., Sasaki R. and Schwiebert C., "Constant Solutions of Reflection Equations and Quantum Groups." Preprint YITP/U-92-07 (1992) 31p., submitted to J. Math. Phys.
2. Cherednik I.V., *Teor. Mat. Fiz.* **61** (1984) 55.
3. Kulish P.P. (Ed) , Proc. "Euler Inter. Math. Inst. on Quantum Groups." *Lect. Notes Math.* **1510** Springer, Berlin 1992 398pp.
4. Alekseev A., Faddeev L. and Semenov-Tian-Shansky M., in [3] 142.
5. Alekseev A., Faddeev L., Semenov-Tian-Shansky M. and Volkov A., Preprint CERN-TH-5981/91 (1991) 15p.
6. Babelon O., *Commun. Math. Phys.* **139** (1991) 619 and in [3], 121.
7. Freidel L. and Maillet J.M., *Phys. Lett.* **B262** (1991) 278; **B263** (1991) 403.
8. Cremmer E. and Gervais J.-L., in [3] , 181; *Commun. Math. Phys.* **134** (1990) 619.
9. Moore G. and Reshetikhin N. Yu., *Nucl. Phys.* **B328** (1989) 557.
10. Cherednik I.V., Preprint RIMS-862 (1992) 31p.
11. Sklyanin E.K., *J. Phys. A. : Math. Gen.* **21** (1988) 2375.
12. Mezincescu L., Nepomechie R. and Rittenberg V., *Phys. Lett.* **A147** (1990) 70.
13. Mezincescu L. and Nepomechie R., Preprint CERN-TH. 6152/91 14p.
14. Kulish P. and Sklyanin E., *J. Phys.A : Math. Gen.* **24** (1991) L435.
15. Zumino B., Preprint UCB-PTH-62/91 (1991) 17p.
16. Alekseev A. and Faddeev L., *Commun. Math. Phys.* **141** (1991) 413.
17. Majid S., Preprint DAMTP/92-12 (1992) 36p.
18. Olshanskii G., In [3], 104.
19. Carow-Watamura U., Schlieker M., Scholl M. and Watamura S., *Z. Phys. C.* **48** (1990) 159.
    Schmidke W. B., Wess J. and Zumino B., *Z. Phys. C* **52** (1991) 471.
20. Kulish P.P., Preprint YITP/K-959 (1991) 24p.
    Kulish P.P. and R. Sasaki, preprint YITP/U in preparation.
21. Faddeev L., Reshetikhin N. and Takhtajan L., *Alg. Anal.* **1** (1989) 178 (in Russian, English translation: Leningrad Math. J. **1** (1990) 193).
22. Kulish P.P. and Sklyanin E.K., Preprint YITP/K-980 (1992) 18p.
23. Kirillov A.N. and Reshetikhin N.Yu., LOMI preprint, E-9-88, 1988; In: New developments in the theory of knots, ed. T.Kohno, World Scientific, Singapore 1989.

Quantum Differential Calculus and Lie Algebras

A. Sudbery
Department of Mathematics, University of York, Heslington, York, England YO1 5DD

QUANTISED ENVELOPING ALGEBRAS: THE ALTERNATIVES

There are three quite different forms in the literature for the generators and relations of a quantum enveloping algebra: the Jimbo/Drinfel'd form, with relations like

$$[J_+, J_-] = \frac{q^{2J_0} - q^{-2J_0}}{q - q^{-1}}; \tag{1}$$

the Faddeev-Reshetikhin-Takhtajan form, with relations like

$$L^+_{11} L^+_{12} = q L^+_{12} L^+_{11}; \tag{2}$$

and the Woronowicz form, with relations like

$$q^{\mp 2} X_0 X_\pm - q^{\pm 2} X_\pm X_0 = X_\pm. \tag{3}$$

Of these three, the Woronowicz form seems to be closest to the classical concept of a universal enveloping algebra; the fact that its left-hand sides are quadratic in the generators while the right-hand sides are linear means that it is truly an enveloping associative algebra of a finite-dimensional non-associative algebra like a Lie algebra. Woronowicz [15] has given axioms for this generalisation of Lie algebra, but they have been little studied to date. It appears to be difficult to construct the quantum analogues of the classical simple Lie algebras (but see [1]).

In this talk I want to describe a framework in which Woronowicz's finite dimension-preserving quantum deformations of Lie algebras are expected to occur naturally. It is, of course, a development of Woronowicz's version of non-commutative geometry [15], with some essential input from Manin's theory of quantum spaces [4] and Wess and Zumino's quantum differential calculus [14].

QUANTUM GROUPS AS INVARIANCE GROUPS

In Manin's theory [4, 11] quantum groups are defined by copying the definition of classical groups as sets of transformations of a vector space which preserve some structure or structures on the space. The structures in question are algebras satisfied by the coordinates on the space; for example, the classical general linear group GL(n) preserves the commutative algebra of polynomials in n coordinates and the Grassmann algebra with n anti-commuting generators. In the quantum version these two algebras are replaced by the algebras defined by an R-matrix R^{kl}_{ij} with two eigenvalues q and $-q^{-1}$: the commutative law $x_j x_i = x_i x_j$ is replaced by $R^{kl}_{ij} x_k x_l = q x_i x_j$ (which we write as $\mathbf{R}_{12}\mathbf{x}_1\mathbf{x}_2 = q\mathbf{x}_1\mathbf{x}_2$) and the anticommutative law $\xi_j \xi_i = -\xi_i \xi_j$ by $\mathbf{R}_{12}\xi_1\xi_2 = -q^{-1}\xi_1\xi_2$. The quantum group GL$_q(n)$ is determined by these as follows: its algebra of functions is generated by matrix elements a_i^j such that $a_i^j x_j$ and $a_i^j \xi_j$ satisfy the same relations as x_i and ξ_i, respectively, if a_i^j commutes with both x_i and ξ_i. These conditions force the matrix $\mathbf{A} = (a_i^j)$ to satisfy the Faddeev-Reshetikhin-Takhtajan relations $\mathbf{R}_{12}\mathbf{A}_1\mathbf{A}_2 = \mathbf{A}_1\mathbf{A}_2\mathbf{R}_{12}$. The braid relation $\mathbf{R}_{12}\mathbf{R}_{23}\mathbf{R}_{12} = \mathbf{R}_{23}\mathbf{R}_{12}\mathbf{R}_{23}$ guarantees that the resulting algebra has the same number of independent polynomials of any given degree as the commutative algebra if the algebras of the x_i and the ξ_i have similar properties [11]. (This is not, as it is often thought to be, an obvious consequence of the braid relation alone.)

DIFFERENTIAL CALCULUS ON QUANTUM SPACES AND GROUPS

If we are modelling our theory on the classical theory of a group acting on a space, it is not surprising to find the algebra of the coordinates x_i playing a central role. But what is the significance of the other algebra generated by the ξ_i? The answer suggested by the work of Wess and Zumino is that they are differential forms on our space: $\xi_i = dx_i$. If so, they are not a completely distinct algebra but can be multiplied by functions of the coordinates, and we need cross-commutation relations between coordinates and momenta. There are two possibilities for these, namely $d\mathbf{x}_1\mathbf{x}_2 = q^{-1}\mathbf{R}_{12}\mathbf{x}_1 d\mathbf{x}_2$ and $d\mathbf{x}_1\mathbf{x}_2 = q\mathbf{R}_{12}^{-1}\mathbf{x}_1 d\mathbf{x}_2$, both of which define a consistent algebra (i.e. one with the classical numbers of independent polynomials of a given degree in x_i and dx_i) with an exterior derivative d satisfying $d^2 = 0$ and the usual Leibniz rule [12, 13].

What about differentials of the matrix elements? If the algebra of the matrix elements is determined by the requirement that they should give transformations of both coordinates and differentials which preserve their algebras, can the algebra of the da_i^j be determined by a similar requirement? The answer is again yes [3, 5, 12, 13] but it needs a little more work. The requirement is that the transformation $\mathbf{x} \to \mathbf{A}\mathbf{x}$ and its differential $d\mathbf{x} \to (d\mathbf{A})\mathbf{x} + \mathbf{A}d\mathbf{x}$ should preserve the algebras of the x_i and the dx_i in the sense explained above. Now, however, the algebra of the a_i^j and the da_i^j is not completely determined by this requirement; there is a one-parameter family of possible cross-commutation relations between them. For $GL_{p,q}(2)$ these were described by Müller-Hoissen [6] at this conference; they can be written using the R-matrix as

$$d\mathbf{A}_1\mathbf{A}_2 = \mathbf{R}_{12}\mathbf{A}_1 d\mathbf{A}_2\mathbf{R}_{12} + t(\mathbf{R}_{12} + q^{-1})\mathbf{A}_1 d\mathbf{A}_2(\mathbf{R}_{12} + q^{-1}) \tag{4}$$

These possible cross-commutation relations can be narrowed down to just one, namely the one with $t = 0$, by the further very natural requirement that they should yield a consistent algebra, in the sense explained above. It is easy to see that the relations (4) have this property if $t = 0$, for then $d\mathbf{A}_1\mathbf{A}_2\mathbf{A}_3 = \mathbf{R}_{12}\mathbf{R}_{23}\mathbf{A}_1\mathbf{A}_2 d\mathbf{A}_3\mathbf{R}_{23}\mathbf{R}_{12}$ and the requirement of consistency, which is that $\mathbf{R}_{23}$ should still commute with $d\mathbf{A}_1\mathbf{A}_2\mathbf{A}_3$ after one makes the above substitution, is automatically satisfied because $\mathbf{R}$ satisfies the braid relation. It is not so obvious that $t = 0$ is the *only* choice that has this property. This was shown by Maltsiniotis [3] and Manin [5] by heroic calculation; it can also be proved by using the representation theory of the braid group, as in [12]. What is actually proved in [12] is that if $t \neq 0$ then there are some relations that are satisfied by both matrix elements and their differentials; these common relations are extracted from the cubic relations imposed by the above consistency condition. But the relations between differentials, which are obtained by differentiating (4), have zero intersection with the relations $\mathbf{R}_{12}\mathbf{A}_1\mathbf{A}_2 = \mathbf{A}_1\mathbf{A}_2\mathbf{R}_{12}$ between matrix elements. (To see this, note that the relations between matrix elements are equivalent to $\mathbf{P}^{\pm}_{12}\mathbf{A}_1\mathbf{A}_2\mathbf{P}^{\mp}_{12} = 0$, where $P^{\pm} = R \mp q^{\pm 1}$ are the projectors onto eigenspaces of R, while those between differentials are equivalent to $\mathbf{P}^{\pm}_{12}\mathbf{A}_1\mathbf{A}_2\mathbf{P}^{\pm}_{12} = 0$. This holds for all values of t except $t = -q^2(1+q^2)^{-1}$, when the relation with $\mathbf{P}^-$ is absent.) Thus unless $t = 0$ the consistency condition is not satisfied.

QUANTUM LIE ALGEBRAS

Once an algebra of differentials has been defined on a quantum space, it becomes possible to define vector fields on the space. The idea is to copy the classical definition of a vector field as a first-order differential operator, i.e. a map from functions to functions with a derivation property on products, but the meaning of "derivation" has to be modified in a quantum space. The new meaning is provided by the cross-commutation rules between coordinates and differentials. Thus an operator D on functions of the non-commuting variables x_i is a vector field if it is of the form $Df = \hat{D}(df)$ where $\hat{D}$ is a map from differentials to functions with the property $\hat{D}(f\omega) = f\hat{D}(\omega)$. If the coordinates and their differentials satisfy $d\mathbf{x}_1\mathbf{x}_2 = q^{-1}\mathbf{R}_{12}\mathbf{x}_1 d\mathbf{x}_2$ as above, then this definition implies that D is a vector field if it satisfies the derivation rule $D(x_i x_j) = x_i Dx_j + q^{-1}R^{kl}_{ij}x_k Dx_l$. For vector fields describing the infinitesimal action of a quantum group, it follows from Woronowicz's theory [15] that one can define a Lie bracket between them, giving a quantum analogue of the result that the commutator of two derivations is also a derivation. For the case of GL(n), with generators X_i^j defined by $X_i^j(x_k) = \delta^j_k x_i$, this Lie bracket is given by

$$[X_i^j, X_k^l] = X_i^j X_k^l - \Lambda^{jl,ac}_{ik,bd} X^b_a X^d_c \tag{5}$$

where $\Lambda^{jl,ac}_{ik,bd} = R^{jf}_{ke} R^{ga}_{fi} (R^{=1})^{ch}_{bg} (R^{-1})^{el}_{hd}$ and $R^{\sim 1}$ is a second inverse of R, satisfying $R^{kn}_{im}(R^{\sim 1})^{ml}_{nj} = \delta^l_i \delta^k_j$. Using the operator products $X_i^j X_k^l = \delta^j_k X_i^l$, this becomes

$$[X_i^j, X_k^l] = \delta^j_k X_i^l - t^h_g R^{jf}_{ke} R^{ga}_{fi} (R^{-1})^{el}_{hd} \tag{6}$$

where $t_i^j = (R^{\sim 1})^{aj}_{ai}$ is the quantum trace form [2, 8]. From this the quantum structure constants of GL(n) can be obtained.

COPRODUCTS, BRAIDING AND QUANTUM INTERACTION

A disadvantage of the Woronowicz form (3) of the quantised enveloping algebra is that the usual coproducts cannot be expressed easily in terms of the generators $X_{0,\pm}$. This becomes apparent upon using the expressions for the Woronowicz generators in terms of the Jimbo/Drinfel'd generators [10]. However, the Woronowicz relations and their generalisation (5) admit another coproduct of a different type from the usual one. This is a coproduct in a *braided bialgebra* [2]. The usual coproduct assigns to each element X of the enveloping algebra U an element $\Delta(X)$ of $U \otimes X$ so that the coproducts $\Delta(X)$ satisfy the same relations as the elements X, products in $U \otimes U$ being calculated on the assumption that $1 \otimes X$ commutes with $X \otimes 1$. In a braided bialgebra this commutativity is replaced by a braiding, which in our case takes the form

$$(1 \otimes X_i^j)(X_k^l \otimes 1) = \hat{\Lambda}_{ik,bd}^{jl,ac}(X_a^b \otimes 1)(1 \otimes X_c^d) \tag{7}$$

where $\hat{\Lambda}$ differs from Λ in that the final R^{-1} is replaced by R. With this algebra for tensor products, the coproduct can be taken to be the same as in the enveloping algebra of a classical Lie algebra, namely $\Delta(X) = X \otimes 1 + 1 \otimes X$.

This may be unfamiliar in the context of quantum groups, but it makes very good sense in the context of quantum mechanics. If we think of Hopf algebra elements as dynamical variables of a single particle, then the coproduct tells us how to construct the corresponding dynamical variables for a system of two particles. The usual prescription is to form these from one-particle variables, which commute with each other but (if $q \neq 1$) do not simply add to give the two-particle variables, just as the total energy of two interacting particles is not the sum of two one-particle energies. This recalls a slogan of Podles [7]. The braided structure can be interpreted in terms of an old suggestion [9] for a reformulation of quantum mechanics in Newtonian form, using only *kinetic* energy and momentum. These quantities are simply additive; the interaction between two particles is expressed by the fact that their energies and momenta do not commute with each other. It is intriguing to see this idea of Newtonian quantum mechanics emerging in the theory of quantum groups.

References

[1] U.Carow-Watamura, M.Schlieker, S.Watamura and W.Weich, *Commun. Math. Phys.* **142** (1991), 605.

[2] S.Majid, Quantum and braided linear algebra. Preprint (Cambridge, 1992).

[3] G.Maltsiniotis, Calcul différentiel sur le groupe linéaire quantique. Preprint (Paris 7, 1990).

[4] Yu.I.Manin, *Quantum Groups and Non-Commutative Geometry* (CRM, Montreal, 1988).

[5] Yu. I. Manin, Notes on quantum groups and quantum de Rham complexes. Preprint (Max-Planck-Institut, Bonn 1991).

[6] F.Müller-Hoissen, *J. Phys. A* **25** (1992), 1703.

[7] P.Podleś, Quantization enforces interaction. Preprint (RIMS, Kyoto 1991).

[8] P.Schupp, P.Watts and B.Zumino, Differential geometry on linear quantum groups. Preprint (Berkeley, 1992).

[9] A.Sudbery, *Nucl. Phys. B* **44** (1972), 520.

[10] A.Sudbery, in *Quantum Groups*, ed. T.Curtright, D.Fairlie and C.Zachos (World Scientific, Singapore 1990)

[11] A.Sudbery, Matrix-element algebras determined by quadratic coordinate algebras. *J. Algebra*, to be published.

[12] A.Sudbery, The algebra of differential forms on a full matric bialgebra. Preprint (York, 1991).

[13] A.Sudbery, *Phys. Lett. B* **284** (1992), 61 and correction (to be published)

[14] J.Wess and B.Zumino, *Nucl. Phys. B (Proc. Suppl.)* **18B** (1990), 302.

[15] S.L.Woronowicz, *Commun. Math. Phys.* **122** (1989), 125.

YANG-BAXTER MATRIX AND $*$-CALCULI ON QUANTUM GROUPS OF A-SERIES

XIAO-DONG SUN SHI-KUN WANG

Institute of Applied Mathematics,
Academia Sinica, Beijing 100080, P.R.China

ABSTRACT

In this short note, we prove the differential calculi on quantum groups of A-series given in this note and previous papers are $*$-calculi.

It is well known that from Yang-Baxter matrix

$$R_q = q^{1/N}\left(\sum_{i,j=1}^{N} q^{\delta_{ij}} e_{ii}\otimes e_{jj} + (q-q^{-1})\sum_{\substack{i,j=1\\ i>j}}^{N} e_{ij}\otimes e_{ji}\right), \quad q\in \mathbf{C}^*, \tag{1}$$

we can construct quantum group of A-series[1], the coordinate ring of $GL_q(N)$ is generated on $\mathbf{C}$ by t_{ij} $(i,j=1,2,\cdots,N)$ which satisfy the relation:

$$R_q T_1 T_2 = T_2 T_1 R_q. \tag{2}$$

If the relation $Det_q T = \mathbf{1}$ is added, we obtain quantum group $SL_q(N)$.

Recently, differential calculi on quantum planes and quantum groups were discussed by a lot of papers, J.Wess and B.Zumino gave the general methods to study differential calculi on quantum planes[2], S.L.Woronowicz provided basic theory of differential calculi on quantum groups[3] and many other people gave methods to construct differential calculi on quantum groups, but the $*$-calculi have not been discussed very much. In this short note, we give the differential calculi on quantum group $SU_q(N)$ as an extension of the results of Ref.4-6 and prove these differential calculi and differential calculus given in Ref.7 are $*$-calculi. We will point out that Yang-Baxter matrix plays a very important role in the $*$-calculi on quantum groups.

1. Yang-Baxter dual and differential calculi on quantum groups

Let Ω^0 be the coordinate ring of A-series quantum group $\mathcal{A}$, the first order differential calculus is denoted by $\{\Omega^1, d\}$, where Ω^1 is a bimodule of Ω^0 and d is a linear operator from Ω^0 to Ω^1 satisfying

(i) Leibnitz rule $d(xy) = (dx)y + xdy, \qquad \forall x,y\in\Omega^0$,

(ii) for arbitrary element ρ in Ω^1, there always exist some elements $x_k, y_k \in \mathcal{A}$ $(k=1,2,\cdots,M)$ such that $\rho = \sum_{k=1}^{M} x_k \delta y_k$.

On quantum group of A-series, we have two sets of Y-B linear functionals $L^\pm = (l^\pm_{ij})_{1\le i,j\le N}$ defined by

$$< L^+, T >= \lambda_+ P R_q P, \quad < L^-, T >= \lambda_-^{-1} R_q^{-1}, \qquad \lambda_\pm \in \mathbf{C}^*, \tag{3}$$

where P is the permutation matrix. If $\mathcal{A}$ denote quantum group $SL_q(N)$, we must require $\lambda_+^N = \lambda_-^N = 1$. Furthermore, if we introduce two sets of functionals on Ω^0 as follows:

$$\nabla_{ij} = \frac{1}{q-q^{-1}}(S(l_{ik}^-)l_{kj}^+ - \delta_{ij}\varepsilon), \tag{4}$$
$$\theta_{ijkl} = S(l_{ki}^-)l_{jl}^+, \tag{5}$$

where $i,j,k,l = 1,2,\cdots,N$, S is the antipode. Let Ω^1 be the left module generated by ω^{ij} $(i,j = 1,2,\cdots,N)$, therefore the first order differential calculus on $\mathcal{A}$ is given by

$$dx = (\nabla_{ij} * x)\omega^{ij}, \tag{6}$$
$$\omega^{ij}\cdot x = (\theta_{ijkl} * x)\omega^{kl}, \quad \forall x \in \Omega^0, \quad i,j,k,l = 1,2,\cdots,N. \tag{7}$$

From the discussion in Ref.6, we know Eq.6 and Eq.7 in fact give the first order bicovariant differential calculus on quantum group of A-series. Based on the property of Y-B matrix R_q of A-series, we have following important proposition:

Proposition Let ∇_{ij} and θ_{ijkl} $(i,j,k,l = 1,2,\cdots,N)$ be defined by Eq.4 and Eq.5, we have

$$\nabla_{ij}(t_{ab}) = \nabla_{ji}(t_{ba}), \quad \theta_{ijkl}(t_{ab}) = \theta_{jilk}(t_{ba}), \quad i,j,k,l,a,b = 1,2,\cdots,N.$$

Remark: For the definition of bicovariant differential calculus on quantum groups, please see Ref.3.

2. $*$-calculi on quantum group $SU_q(N)$

If we introduce an operator $*$: $\Omega^0 \longrightarrow \Omega^0$ to $SL_q(N)$ satisfying

$$T^* = S(T)^t, \tag{8}$$

where S is the antipode, then quantum group $SU_q(N)$ is obtained. And if the $*$ operator can be extended to an operator on quantum de Rham complex $\Omega^\wedge$ satisfying:

$$(\rho_1 \wedge \rho_2)^* = (-1)^{k_1 k_2}\rho_2^* \wedge \rho_1^*, \quad (d\rho)^* = d(\rho^*), \quad \rho,\rho_1,\rho_2 \in \Omega^\wedge, \tag{9}$$

where k_1 and k_2 are the orders of ρ_1 and ρ_2 respectively, then (Ω^1, d) is called a $*$-calculus. According to the basic theory of quantum matrix group of Woronowicz[3], (Ω^1, d) is a $*$-calculus $\Longleftrightarrow$ $S(\mathcal{H})^* \subseteq \mathcal{H}$, where $\mathcal{H} = ker\varepsilon \cap \{\cap_{i,j=1}^N ker\nabla_{ij}\}$, ε is the counit.

By Ref.6, if we denote the set of the generators of the right ideal $\mathcal{H}$ by Λ, then for quantum group $SU_q(N)$, elements of Λ can be written as

$$\xi_{abcd} = t_{ab}t_{cd} - \nabla_{ij}(t_{ab}t_{cd})(M^{-1})_{kl}^{ij}t_{kl} - C_{abcd},$$

where

$$M^{ij}_{kl} = \nabla_{kl}(t_{ij}),$$
$$C_{abcd} = \varepsilon(t_{ab}t_{cd} - \nabla_{ij}(t_{ab}t_{cd})(M^{-1})^{ij}_{kl}t_{kl}).$$

By Eq.8, we know

$$S(t_{ij})^* = t_{ji}.$$

By the Proposition, we have

$$\nabla_{ij}(t_{ab}t_{cd}) = \theta_{ijkl}(t_{ab})\nabla_{kl}(t_{cd}) - \delta_{ab}\nabla_{ij}(t_{cd}) = \nabla_{ji}(t_{ba}t_{dc}),$$
$$(M^{-1})^{ij}_{kl} = (M^{-1})^{ji}_{lk}.$$

Therefore

$$S(\xi_{abcd})^* = \xi_{badc},$$

i.e. $S(\Lambda)^* = \Lambda$, then we have $S(\mathcal{H})^* \subseteq \mathcal{H}$ straightforward. Hence we have proved the differential calculus on $SU_q(N)$ is $*$-calculus. Therefore we have N differential $*$-calculi on $SU_q(N)$ different from each other by the choice of the product $r = \lambda_+\lambda_-$ ($r^N = 1$).

3. $*$-calculus on quantum Lorentz group

The construction of quantum Lorentz group was first given by P.Podlés and S.L.Woronowicz[8], and then discussed by some other papers. In fact quantum Lorentz group can be treated as $SL_q(2,\mathbb{C})$, the complex version of $SL_q(2)$. The coordinate ring of quantum Lorentz group is generated by 8 elements t_{ij} and $t_{\bar{i}\bar{j}}$ $(i,j = 1,2)$. If we arrange them into two 2×2 matrices as

$$\mathcal{T} = \begin{pmatrix} T & \\ & \hat{T} \end{pmatrix}, \quad T = (t_{ij})_{i,j=1,2}, \quad \hat{T} = (t_{\bar{i}\bar{j}})_{i,j=1,2},$$

then the relations satisfied by these elements can be written as

$$t_{\bar{i}\bar{j}} = (S(t_{ji}))^*,$$
$$Det_q T = t_{11}t_{22} - qt_{12}t_{21} = 1,$$
$$Det_q \hat{T} = t_{\bar{1}\bar{1}}t_{\bar{2}\bar{2}} - qt_{\bar{1}\bar{2}}t_{\bar{2}\bar{1}} = 1,$$

and

$$\mathcal{R}\mathcal{T}_1\mathcal{T}_2 = \mathcal{T}_2\mathcal{T}_1\mathcal{R}, \tag{10}$$

where $\mathcal{R} = (\mathcal{R}^{ab}_{cd})_{a,b,c,d=1,2,\bar{1},\bar{2}}$,

$$\mathcal{R}^{ij}_{kl} = R^{ij}_{kl}, \ \mathcal{R}^{i\bar{j}}_{k\bar{l}} = ((R^+)^{-1})^{ij}_{kl}, \ \mathcal{R}^{\bar{i}\bar{j}}_{\bar{k}\bar{l}} = R^{ij}_{kl}, \ \mathcal{R}^{\bar{i}j}_{\bar{k}l} = ((R^+)^{-1})^{ij}_{kl}, \ i,j,k,l = 1,2,$$

and R is 2×2 Y-B matrix of A-series and other elements of $\mathcal{R}$ are zeros. It can be checked that $\mathcal{R}$ also satisfies Yang-Baxter equation.

We can define two sets of linear functionals by

$$< l^+_{ab}, t_{cd} >= P\mathcal{R}P, \quad < l^-_{ab}, t_{cd} >= \mathcal{R}^{-1}, \; a,b,c,d = 1,2,\bar{1},\bar{2}.$$

We can also define ∇_{ab} and θ_{abcd} $(a,b,c,d = 1,2,\bar{1},\bar{2})$ by Eq.4 and Eq.5 as we have done for $SU_q(N)$.

The differential calculus on quantum Lorentz was discussed in Ref.7, and the elements of the generators set Λ corresponding to quantum Lorentz group can be written as:

$$\xi_{abcd} = t_{ab}t_{cd} - \nabla_{ij}(t_{ab}t_{cd})(M^{-1})^{ij}_{kl}t_{kl} - C_{abcd},$$
$$\xi_{\bar{a}\bar{b}\bar{c}\bar{d}} = t_{\bar{a}\bar{b}}t_{\bar{c}\bar{d}} - \nabla_{\bar{i}\bar{j}}(t_{\bar{a}\bar{b}}t_{\bar{c}\bar{d}})(M^{-1})^{\bar{i}\bar{j}}_{\bar{k}\bar{l}}t_{\bar{k}\bar{l}} - C_{\bar{a}\bar{b}\bar{c}\bar{d}},$$

where

$$M^{ij}_{kl} = \nabla_{kl}(t_{ij}), \quad M^{\bar{i}\bar{j}}_{\bar{k}\bar{l}} = \nabla_{\bar{k}\bar{l}}(t_{\bar{i}\bar{j}}),$$
$$C_{abcd} = \varepsilon(t_{ab}t_{cd} - \nabla_{ij}(t_{ab}t_{cd})(M^{-1})^{ij}_{kl}t_{kl}),$$
$$C_{\bar{a}\bar{b}\bar{c}\bar{d}} = \varepsilon(t_{\bar{a}\bar{b}}t_{\bar{c}\bar{d}} - \nabla_{\bar{i}\bar{j}}(t_{\bar{a}\bar{b}}t_{\bar{c}\bar{d}})(M^{-1})^{\bar{i}\bar{j}}_{\bar{k}\bar{l}}t_{\bar{k}\bar{l}}).$$

After some computation we have

$$\nabla_{ij}(t_{ab}t_{cd})(M^{-1})^{ij}_{kl} = \nabla_{\bar{i}\bar{j}}(t_{\bar{a}\bar{b}}t_{\bar{c}\bar{d}})(M^{-1})^{\bar{i}\bar{j}}_{\bar{k}\bar{l}},$$

and

$$S(\xi_{abcd})^* = \xi_{\bar{b}\bar{a}\bar{d}\bar{c}}, \quad S(\xi_{\bar{a}\bar{b}\bar{c}\bar{d}})^* = \xi_{badc}.$$

Therefore, $S(\Lambda)^* = \Lambda$, and $S(\mathcal{H})^* \subseteq \mathcal{H}$. Thus we have proved that the differential calculus on quantum Lorentz group given in Ref.7 is $*$-calculus.

This work is supported by Natural Science Foundation of Academia Sinica.

References

1. N. Yu. Reshetikhin, L. A. Takhtadzhyan & L. D. Faddeev, Quantization of Lie groups and Lie algebras, *Leningrad Math. J.* Vol. 1, 1(1990), 193-224.
2. J. Wess, & B. Zumino, Covariant differential calculus on the hyperplanes, Preprint CERN-5697, April 1990.
3. S. L. Woronowicz, Differential calculus on compact matrix pseudogroups (quantum groups), *Commun. Math. Phys.* **122**(1989) 125-170.
4. K. Wu & R. J. Zhang, preprint ASITP-91-72, CCAST-91-69, to appear in *Commun. Theor. Phys.*
5. X. D. Sun & S. K. Wang, Bicovariant differential calculus on the two-parameter quantum group $GL_{p,q}(2)$, to appear in *Journal of Math. Phys.*.
6. X. D. Sun & S. K. Wang, Preprint CCAST-92-04, ASIAM-92-07, ASITP-92-12.
7. X. D. Sun, S. K. Wang & K. Wu, Preprint CCAST-92-14, ASITP-92-18, ASIAM-92-11.
8. P. Podlés and S.L. Woronowicz, *Commun. Math. Phys.* **130** (1990) 381.

THE QUANTUM ALGEBRA APPROACH TO q-SPECIAL FUNCTIONS

Roberto Floreanini

Istituto Nazionale di Fisica Nucleare, Sezione di Trieste
Dipartimento di Fisica Teorica, Università di Trieste
Strada Costiera 11, 34014 Trieste, Italy

Luc Vinet

Laboratoire de Physique Nucléaire and Centre de Recherches Mathématiques
Université de Montréal, Montréal, Canada H3C 3J7

Abstract

The quantum algebra interpretation of q-special functions is presented. The connection between these functions and representations of q-algebras is explained and it is shown how generating functions and addition formulas are obtained in this approach. The relation between $\mathcal{U}_q(sl(2))$and the ${}_2\phi_1$ series is used for illustration purposes.

Most special functions of mathematical physics have q-analogues, that is generalizations to a base q. The best-known examples are the basic hypergeometric series [1],

$$
\begin{aligned}
&{}_r\phi_s(a_1,a_2,\ldots,a_r;b_1,\ldots,b_s;q;z) \\
&\qquad = \sum_{n=0}^{\infty} \frac{(a_1;q)_n(a_2;q)_n\ldots(a_r;q)_n}{(q;q)_n(b_1;q)_n\ldots(b_s;q)_n} \left[(-1)^n q^{\frac{n(n-1)}{2}}\right]^{1+s-r} z^n \,,
\end{aligned}
\tag{1}
$$

which are the q-analogues of the (generalized) hypergeometric series ${}_rF_s$. The q-shifted factorials $(a;q)_\alpha$ that enter in the definition (1) are defined as follows for α complex:

$$
(a;q)_\alpha = \frac{(a;q)_\infty}{(aq^\alpha;q)_\infty}\,, \qquad (a;q)_\infty = \prod_{k=0}^{\infty}(1-aq^k)\,, \qquad |q|<1\,. \tag{2}
$$

The q-deformed functions, like the ordinary ones, satisfy various properties; they have appeared in the evaluation of physical quantities and are certainly worthy of interest.

The standard special functions are also known to have Lie interpretations. They arise for instance as matrix elements, basis vectors, etc. of representations of Lie groups and algebras. We have provided (see for instance [2,3,4]) similar interpretations of the q-special functions using quantum algebras. We summarize here some of our results taking as example the relation between the quantum algebra $\mathcal{U}_q(sl(2))$ and the basic hypergeometric series ${}_2\phi_1$.

The following two q-analogues of the exponential play a central role in our approach:

$$
e_q(z) = \sum_{n=0}^{\infty} \frac{1}{(q;q)_n}\, z^n, \qquad E_q(z) = \sum_{n=0}^{\infty} \frac{q^{\frac{1}{2}n(n-1)}}{(q;q)_n}\, z^n\,. \tag{3}
$$

These functions satisfy $e_q(z)E_q(-z) = 1$ and

$$D_z^+ e_q(\lambda z) = \lambda e_q(\lambda z), \quad D_z^- E_q(\lambda z) = -q^{-1}\lambda E_q(\lambda z), \tag{4}$$

with the q-derivatives $D_z^\pm$ defined by $D_z^\pm \varphi(z) = [\varphi(z) - \varphi(q^{\pm 1}z)]/z$.

The basic idea in establishing the connection between quantum algebras and q-special functions is to replace the exponential map from Lie algebras into Lie groups by q-exponentials of quantum algebra generators. To see how this works, let us consider $\mathcal{U}_q(sl(2))$. This quantum algebra is generated by k, k^{-1}, e and f subjected to

$$kek^{-1} = q^{1/2}e, \quad kfk^{-1} = q^{-1/2}f, \quad kk^{-1} = k^{-1}k = e, \quad [e,f] = \frac{k^2 - k^{-2}}{q^{1/2} - q^{-1/2}}. \tag{5}$$

It can be equipped with a coproduct, counit and antipode to define a Hopf algebra. It also admits a module $V^{(\lambda,m_0)}$ with $\lambda, m_0 \in \mathbf{C}$, characterized by the following action of the generators on the basis vectors ξ_j, $j = m_0 + n$, $n \in \mathbf{Z}$:

$$k\xi_j = q^{-j/2}\xi_j, \quad e\xi_j = q^{(1-2\lambda)/4}\frac{1-q^{\lambda+j}}{1-q}\,\xi_{j-1}, \quad f\xi_j = q^{(1-2\lambda)/4}\frac{1-q^{\lambda-j}}{1-q}\,\xi_{j+1}. \tag{6}$$

Given any $a \in \mathcal{U}_q(sl(2))$, its matrix elements $W_{ij}(a)$ in this representation are defined by

$$a\xi_j = \sum_i \xi_j\, W_{ij}(a). \tag{7}$$

In analogy with Lie theory, one considers the following elements in the completion of $\mathcal{U}_q(sl(2))$:

$$U(\alpha,\beta,\gamma) = E_q(\alpha\, e)\, e_q(\beta\, f)\, k^\gamma, \tag{8}$$

with α, β, γ complex parameters. It is then easy to show that

$$\begin{aligned}
W_{ij}\left(U(\alpha,\beta,\gamma)\right) &= q^{-\gamma j/2}\left(\beta\,\frac{q^{(1-2\lambda)/4}}{1-q}\right)^{i-j}\frac{\left(q^{\lambda-i+1};q\right)_{i-j}}{(q;q)_{i-j}} \\
&\quad\times {}_2\phi_1\left(q^{\lambda+i+1}, q^{i-\lambda}; q^{i-j+1}; q, -\alpha\beta\,\frac{q^{(1-2i)/2}}{(1-q)^2}\right), \quad \text{if } i-j \geq 0, \\
W_{ij}\left(U(\alpha,\beta,\gamma)\right) &= q^{-\gamma j/2}q^{(j-i)(j-i-1)/2}\left(\alpha\,\frac{q^{(1-2\lambda)/4}}{1-q}\right)^{j-i}\frac{\left(q^{\lambda+i+1};q\right)_{j-i}}{(q;q)_{j-i}} \\
&\quad\times {}_2\phi_1\left(q^{\lambda+j+1}, q^{j-\lambda}; q^{j-i+1}; q, -\alpha\beta\,\frac{q^{(1-2i)/2}}{(1-q)^2}\right), \quad \text{if } i-j \leq 0.
\end{aligned} \tag{9}$$

This establishes most straightforwardly the connection between $\mathcal{U}_q(sl(2))$ and Heine's ${}_2\phi_1$. We shall now briefly indicate how this connection can be put to use.

It is of course possible to take instead of (8), different combinations of e_q and E_q. In particular for

$$\widetilde{U}(\alpha,\beta,\gamma) = k^\gamma\, E_q(\beta\, f)\, e_q(\alpha\, e), \tag{10}$$

the matrix elements $W_{ij}[\widetilde{U}(\alpha,\beta,\gamma)]$ are again expressed in terms of ${}_2\phi_1$. (See Ref. [4] for the explicit formulas.) Now note that $U(\alpha,\beta,0)\widetilde{U}(-\alpha,-\beta,0) = 1$ or in terms of matrix elements that $\sum_l W_{il}(U(\alpha,\beta,0))\, W_{lj}\left(\widetilde{U}(-\alpha,-\beta,0)\right) = \delta_{i-j,0}$. Substituting in this last relation the expressions for $W(U)$ and $W(\widetilde{U})$one arrives at the following orthogonality relation:

$$\delta_{i-j,0} = \frac{q^{(i^2+j^2)/2}\,(q/z;q)_j}{(q^{\lambda+1};q)_i\,(q^{-\lambda};q)_j} \sum_{l\in \mathbf{Z}+m_0} q^{l(l-1)/2}\left(-zq^{-i-j}\right)^l \frac{(z;q)_l\,(q^{\lambda+1};q)_l\,(q^{-\lambda};q)_l}{(q;q)_{l-j}\,(q;q)_{l-i}} \\ \times {}_2\phi_1\left(q^{\lambda+l+1},q^{l-\lambda};q^{l-i+1};q,zq^{-i}\right)\, {}_2\phi_1\left(q^{\lambda+l+1},q^{l-\lambda};q^{l-j+1};q,zq^{-j}\right). \tag{11}$$

A generating relation is obtained through the construction of a one-variable model for $V^{(\lambda,m_0)}$. Let z be a complex variable and take the basis vectors to be

$$\xi_j = z^n\,, \quad j = m_0 + n \quad n\in \mathbf{Z}\,. \tag{12}$$

Let the generators be the following q-difference operators

$$\begin{aligned} k &= q^{-m_0/2}\,T_z^{-1/2}\,, \\ e &= q^{(1-2\lambda)/4}\left(\frac{1}{1-q}D_z^+ + \frac{1-q^{\lambda+m_0}}{1-q}\,\frac{1}{z}\,T_z\right)\,, \\ f &= q^{(1-2\lambda)/4}\left(\frac{z^2}{1-q}D_z^- + \frac{1-q^{\lambda-m_0}}{1-q}\,z\,T_z^{-1}\right)\,, \end{aligned} \tag{13}$$

where T_z is defined by $T_z\varphi(z) = \varphi(q\,z)$. It is easy to check that the identifications (12) and (13) provide a realization of the representation (λ, m_0). One now acts with $U(\alpha,\beta,0)$ on the basis vectors $\xi_j = z^{j-m_0}$ using for e and f the expressions given above. By construction $U(\alpha,\beta,0)z^{j-m_0} = \sum_i z^{i-m_0} W_{ij}[U(\alpha,\beta,0)]$. On the r.h.s., the expansion coefficients of the series in z are given in terms of ${}_2\phi_1$. The l.h.s. therefore defines a generating function for these ${}_2\phi_1$. This function can be evaluated with the help of the q-binomial theorem $\sum_{n=0}^{\infty}\frac{(a;q)_n}{(q;q)_n}z^n = \frac{(az;q)_\infty}{(z;q)_\infty}$ and after some redefinitions one obtains the following identity between formal power series:

$$(xq^{-2m_0};q)_{2m_0}\,(q/y;q)_{m_0+\lambda}\;{}_2\phi_0\left(q^{m_0+\lambda+1}/y,q^{m_0-\lambda};q,xyq^{-2m_0-1}\right) \\ = \sum_{k\in\mathbf{Z}} \frac{(q^{m_0-\lambda};q)_k}{(q^{2m_0+1}/x;q)_k\,(q;q)_k}\, y^k\;{}_2\phi_1\left(q^{\lambda-m_0+1},q^{-m_0-\lambda};q^{k+1};q,x\right). \tag{14}$$

Addition formulas are arrived at by observing that the matrix elements $W_{ij}[U(\alpha,\beta,0)]$ and $W_{ij}[\widetilde{U}(\alpha,\beta,0)]$ also provide models for the module $V^{(\lambda,m_0)}$. That this is true is shown by presenting operators $\pi^{(i)}(a)$ acting on the variables α and β and such that $\pi^{(i)}(a)W_{ij}[U(\alpha,\beta,0)] = \sum_k W_{ik}[U(\alpha,\beta,0)]W_{kj}(a)$. It suffices of course to give $\pi^i(e)$, $\pi^i(f)$ and $\pi^i(k)$;these are easily obtained by exploiting the properties of the following analogue of the B-C-H formula

$$E_q(\zeta X)\,Y\,e_q(-\zeta X) = \sum_{n=0}^{\infty}\frac{\zeta^n}{(q;q)_n}\,[X,Y]_n\,,, \\ [X,Y]_0 = Y, \qquad [X,Y]_{n+1} = q^n\,X\,[X,Y]_n - [X,Y]_n\,X, \quad \text{for } n = 1,2,\ldots\,. \tag{15}$$

One finds

$$\begin{aligned}\pi^{(i)}(k) &= q^{-i/2}\,T_{\alpha}^{-1/2}\,T_{\beta}^{1/2}\,,\\ \pi^{(i)}(e) &= -qD_{\alpha}^{-} - \frac{q^{1/2}}{(1-q)^2}\,\beta\left(q^{i}T_{\alpha}T_{\beta}^{-1} - q^{-i}T_{\alpha}^{-1}\right)\,,\\ \pi^{(i)}(f) &= D_{\beta}^{+}\,.\end{aligned} \tag{16}$$

This defines a 2-variable model of $V^{(\lambda,m_0)}$ with $\xi_j^{(i)} = W_{ij}\left(U(\alpha,\beta,0)\right)$. From here, one can derive the second- order q-difference equation that the ${}_2\phi_1$ satisfy. The Casimir operator C of $\mathcal{U}_q\left(sl(2)\right)$ is given by $C = \left(q^{1/2} - q^{-1/2}\right)^{-2}\left(q^{1/2}k^2 + q^{-1/2}k^{-2} - 2\right) + fe$ and on $V^{(\lambda,m_0)}$ it takes the value $C^{(\lambda)} = \left(q^{1/2} - q^{-1/2}\right)^{2}\left(q^{\lambda+1/2} + q^{-\lambda-1/2} - 2\right)$. From $\pi^{(i)}(C)\,W_{lj}\left(U(\alpha,\beta,0)\right) = C^{(\lambda)}W_{lj}\left(U(\alpha,\beta,0)\right)$ one sees that ${}_2\phi_1$ obeys:

$$\begin{aligned}\Big\{z(c-abqz)\left(D_z^{+}\right)^2 + \Big[(1-c) + \left((1-a)(1-b) - (1-c)\right)z\Big]D_z^{+}\\ -(1-a)(1-b)\Big\}{}_2\phi_1(a,b;c;q,z) = 0\,.\end{aligned} \tag{17}$$

Now the addition formula. Since the $W_{ij}\left(U(\alpha,\beta,0)\right)$ form models for $V^{(\lambda,m_0)}$ we must have $\pi^{(i)}[\widetilde{U}(\alpha,\beta,0)]W_{ij}\left(U(\alpha,\beta,0)\right) = \sum_k W_{ik}\left(U(\alpha,\beta,0)\right)W_{kj}\left(\widetilde{U}(\alpha,\beta,0)\right)$. Substituting in the r.h.s. the expressions for the matrix elements in terms of ${}_2\phi_1$ and using the specifics of the two-variable model to evaluate directly the l.h.s.,one can derive after some operations (see Ref.[4]) the following relation between formal series

$$\begin{aligned}{}_2\phi_1\left(q^{\lambda+l+1},q^{l-\lambda};0;q,z\right)\,{}_2\phi_0\left(q^{\lambda-j+1},q^{-\lambda-j};q,w/q\right) = \sum_{k\in\mathbf{Z}}\frac{(-1)^k q^{k(k+1)/2}}{(q;q)_{l-j-k}\,(q;q)_k}\\ \times\,{}_2\phi_1\left(q^{\lambda+l+1},q^{l-\lambda};q^{l-j-k+1};q,z\right)\,{}_2\phi_1\left(q^{\lambda-j+1},q^{-j-\lambda};q^{k+1};q,wq^{k}\right)\,.\end{aligned} \tag{18}$$

With this we conclude. We would first like to stress that the examples presented here are far from covering all the applications that the connection between q-special functions and quantum algebras entail. We would also like to point out that our simple quantum algebra approach allows one to recover [3] the results that are obtained in the quantum group picture (see for instance [5]) where one relates the q-special functions to the matrix elements of corepresentations of quantum groups.

REFERENCES

1. Gasper, G. and Rahman, M., *Basic Hypergeometric Series*, (Cambridge University Press, Cambridge, 1990)
2. Floreanini, R. and Vinet, L., Quantum algebras and q-special functions, University of Montreal-preprint, UdeM-LPN-TH54, 1991, to appear in Ann. Phys.
3. Floreanini, R. and Vinet, L., On the quantum group and quantum algebra approach to q-special functions, University of Montreal-preprint, UdeM- LPN-TH86, 1992
4. Floreanini, R. and Vinet, L.,Using quantum algebras in q-special function theory, to appear in Phys.Lett. A
5. Masuda, T., Mimachi, K., Nakagami,Y., Noumi, M., Saburi, Y. and Ueno, K., Unitary representations of the quantum group $SU_q(1,1)$: structure of the dual space of $\mathcal{U}_q(sl(2))$, Lett. Math. Phys. **19**, 187-194 (1990); Unitary representations of the quantum group $SU_q(1,1)$: II-matrix elements of unitary representations and the basic hypergeometric functions, *ibid.* **19**, 195-204 (1990)

Complex Quantum Group, Dual Algebra and Bicovariant Differential Calculus [1]

SATOSHI WATAMURA

Department of Physics, Tohoku University, Sendai 980, Japan

Abstract

Some results about complex quantum group, its dual algebra and bicovariant differential calculus are reported. The motivation of our investigation is to understand deeper the structure of the quantum Lorentz group algebra $\mathcal{U}_q(SL(2,C))$. Infinitesimal q-Lorentz transformations is discussed.

1. Complex Quantum Group

First let me introduce the complex quantum group $Fun_q(SL(N,C))$. Since such a notation is used already, I think that I must explain what is the new point when we say "*complex*". Usually, the quantum group is defined by the algebra generated by the element of $N \times N$ matrix M^i_j. The non-commutativity of the quantum group is given by the $\hat{\mathbf{R}}$ matrix as

$$\hat{\mathbf{R}}^{ij}_{i'j'} M^{i'}_k M^{j'}_l = M^i_{i'} M^j_{j'} \hat{\mathbf{R}}^{i'j'}_{kl} \quad , \tag{1}$$

However, by this definition the matrix elements are the q-analogue of the complex function $Fun(G)$ and we can consider their $*$-conjugation. The $*$-operation defines the conjugate element $\bar{M}^i_j \equiv (M^j_i)^*$ of which commutation relation is defined by the $*$-conjugate of eq.(1):

$$\hat{\mathbf{R}}^{i'j'}_{ij} \bar{M}^l_{j'} \bar{M}^k_{i'} = \bar{M}^{j'}_j \bar{M}^{i'}_i \hat{\mathbf{R}}^{kl}_{i'j'} \quad . \tag{2}$$

Then usually we consider the reality condition which identifies M and $\bar{M}$ such as $\bar{M} = \kappa(M)$ for $Fun_q(SU(N))$ and $M^* = M$ for $Fun_q(SL(N,R))$.

The question is that what happens if we do not impose such conditions. We must have a $*$-Hopf algebra $Fun_q(SL(N,C))$. But we immediately notice that the definition of the algebra is not completed if we do not impose the reality conditions, since we do not know the commutation relation between M^i_j and $\bar{M}^i_j$. Actually this was the problem of the construction of quantum Lorentz group $Fun_q(SL(2,C))$ [PW,CSSW]. And once we know the solution, it is rather straightforward to generalize it to $Fun_q(SL(N,C))$.[Pod1,DSWZ,CW]

Heuristically the commutation relation between M and $\bar{M}$ can be found as follows: Multiplying the antipode on M we can rewrite eq.(1) as $\hat{\mathbf{R}}^{-1}{}^{ij}_{k'l}\kappa(M^s_i)M^{k'}_k = M^j_{j'}\hat{\mathbf{R}}^{-1}{}^{sj'}_{kl}\kappa(M^l_i)$. Recalling that the unitarity condition in the $Fun_q(SU(N))$ is given by $\bar{M} = \kappa(M)$, we replace $\kappa(M)$ by $\bar{M}$. Then, we get

$$\bar{M}^i_{i'} M^{j'}_k \hat{\mathbf{R}}^{-1}{}^{i'j}_{j'l} = \hat{\mathbf{R}}^{-1}{}^{ij'}_{ki'} M^j_{j'} \bar{M}^{i'}_l \quad . \tag{3}$$

We take this relation as the definition of the commutation relation among the M and $\bar{M}$ in the complex quantum group. Note that we can choose $\hat{\mathbf{R}}$ instead of $\hat{\mathbf{R}}^{-1}$. The resulting algebra is isomorphic to the other. By construction it consistently reduces to the $Fun_q(SU(N))$ if we impose the unitarity condition. Note that the condition on the q-deformed determinant $(\det_q M)^* = 1$ must also be imposed.

Now we get the complete definition of the $*$-Hopf algebra $\mathcal{A} = Fun_q(SL(N,C))$: The algebra is defined by $\mathcal{A} \equiv \mathbf{C} < M^i_j, \bar{M}^i_j > /(\text{above relations})$ and by adding for the coalgebra structure the coaction $\Delta(\bar{M}^i_j) = \bar{M}^{i'}_j \otimes \bar{M}^i_{i'}$ and the counit $\epsilon(\bar{M}^i_j) = \delta^i_j$.

For the convenience we introduce the totally q-antisymmetric tensor $\epsilon_{i_1 \dots i_N}$ and the convention raising and lowering suffices with this tensor: $V^{[i]} \equiv V_j \epsilon^{j[i]}$ and $W_{[i]} \equiv \epsilon_{[i]j} W^j$ where $\epsilon^{i[k]}\epsilon_{[k]j} = \delta^i_j$ and $[i] = [i_1 \dots i_{N-1}]$ denotes $N-1$ q-antisymmetrized suffices.

2. Fundamental Bicovariant Bimodule

The bicovariant bimodules over the complex quantum group can be constructed by generalizing the method developed by Woronowicz for the compact quantum groups [Wor2]. The difference is now we also have the conjugate of the functional for the definition.

For the fundamental bimodules $\eta^i_\pm$ and $\bar{\eta}^{[i]}_\pm$, (Here fundamental means it coincides with the fundamental representation in classical limit.) we have four types of functional: $f^i_{\pm j}$ and $\bar{f}^i_{\pm j} : \mathcal{A} \to \mathbf{C}$. These functionals relate the left and right multiplications of bimodules.

$$\eta^i_\pm a = (a * f^i_{\pm j})\eta^j_\pm \quad , \quad \bar{\eta}^{[i]}_\pm a = (a * \bar{f}^{[i]}_{\pm[j]})\bar{\eta}^{[j]}_\pm \tag{4}$$

[1] This talk is based on the collaboration with U. Carow-Watamura.[CW]

Imposing the bicovariance of the above relations as well as the consistency with the quantum group commutation relations, we get the explicit definition of the functionals

Theorem 1 : *The complete definition of the functionals $f^i_{\pm j}$ and $\bar{f}^i_{\pm j}$ are:*

$$f^i_j(\mathbf{I}) = \delta^i_j, \quad \bar{f}^i_j(\mathbf{I}) = \delta^i_j, \tag{5}$$

$$f^i_{\pm j}(M^k_l) = \hat{\mathbf{R}}^{\pm}{}^{ik}_{lj}, \quad f^i_{\pm j}(\bar{M}^{[k]}_{[l]}) = \hat{\mathbf{R}}^{i[k]}_{[l]j}, \quad \bar{f}^{[i]}_{\pm[j]}(M^k_l) = \hat{\mathbf{R}}^{-}{}^{[i]k}_{l[j]}, \quad \bar{f}^{[i]}_{\pm[j]}(\bar{M}^{[k]}_{[l]}) = \hat{\mathbf{R}}^{\mp}{}^{[i][k]}_{[l][j]}, \tag{6}$$

where $\hat{\mathbf{R}}^+ = \hat{\mathbf{R}}$ and $\hat{\mathbf{R}}^- = \hat{\mathbf{R}}^{-1}$ and the $\hat{\mathbf{R}}$ matrix is normalized as $\hat{\mathbf{R}} = q^{-1/N}(q\mathcal{P}_S - q^{-1}\mathcal{P}_A)$ with the projectors $\mathcal{P}_S$ and $\mathcal{P}_A$. Proof : See ref.[CW]

The suffix $\pm$ was introduces to distinguish the two consistent solutions given by $\hat{\mathbf{R}}^{\pm}$ in eq.(6) which define the two different bimodule structures. The new relations comparing to the compact quantum group case are the second and third relations in eq.(6). They define the commutation relation of $\eta(\bar{\eta})$ and $\bar{M}(M$, resp.) and do not depend on the suffix $\pm$.

3. Algebra of Bicovariant Bimodules

We want to construct a bicovariant differential calculus on the complex quantum group which includes the one form of the adjoint representation in classical limit. For this purpose, in analogy with the classical case we consider the following right invariant basis

$$\theta^i_k \equiv \eta^i \eta^{j_1} \ldots \eta^{j_{N-1}} \epsilon_{j_1 \ldots j_{N-1} k} \quad , \tag{7}$$

and its $*$-conjugation $\bar{\theta}^i_j$. Since η^i is the q-analogue of the fundamental representation, the basis θ and $\bar{\theta}$ contain the adjoint representation in classical limit.

Now it is necessary to examine the structure of the algebra of the bicovariant bimodule, since we can take any combination of η_+ and η_- in eq.(7). We did not meet such an ambiguity in the compact quantum group case since the hermiticity condition fixes the choice [CSWW].

First we consider the product of the same basis, the product of η_+ with η_+ and η_- with η_- and their conjugate. By using the projection operators $\mathcal{P}_S$ and $\mathcal{P}_A$, those products can be decomposed into the irreducible representations like in the non-deformed case. For example,

$$\eta^i_+ \eta^j_+ = \mathcal{P}_S{}^{ij}_{kl} \eta^k_+ \eta^l_+ \oplus \mathcal{P}_A{}^{ij}_{kl} \eta^k_+ \eta^l_+ \quad , \tag{8}$$

and the same for the product of η^i_-. The former is the q-symmetric product, whereas the latter one is the q-antisymmetric product. In both cases, the product of the bases becomes again a bimodule.

Note that here we do not impose either the symmetric or the antisymmetric combination to be zero like in the case of the q-plane algebra [Manin]. Such subalgebras contain only a part of the whole representation by definition, and we want to have all kinds of tensor representations in the algebra.

For the different type of the fundamental bimodule, the bicovariance and consistency with the quantum group relations gives the following relations:

Theorem 2 : *Requiring the bicovariance, the commutation relations consistent with the quantum group relations among the right invariant bases which coincide with the (anti-)commuting case in the limit $q \to 1$ are*

$$\eta^i_+ \eta^j_- = \beta \hat{\mathbf{R}}^{ij}_{kl} \eta^k_- \eta^l_+ \quad \text{and} \quad \eta^i \bar{\eta}_j = \gamma \hat{\mathbf{R}}^{-1}{}^{ki}_{lj} \bar{\eta}_k \eta^l \quad , \tag{9}$$

where the second relation holds for any combination of the suffix $\pm$. β and γ are nonzero constants. The other relations among $\bar{\eta}_{\pm i}$ are given by the $*$-conjugate of the first relation. Proof : See ref.[CW].

The point of the theorem is that we can not replace $\hat{\mathbf{R}}$ with $\hat{\mathbf{R}}^{-1}$ as in the quantum plane case (see for example [CSW]), the algebra of comodules. They are fixed as the consequence of the choice of the fundamental commutation relation of M and $\bar{M}$ in eq.(3). The consequence of this theorem is that when we consider the product of the basis $\eta_\pm$ and $\bar{\eta}_\pm$, using the commutation relations in eq.(9) we can reorder the product, since we can not distinguish the corresponding bicovariant bimodules. Therefore, the product depends now only on the number of basis of each type for fixed representation.

4. Dual Algebra and Relation to $Fun_q(SU(N))$

As in the case of the compact quantum groups we can consider the Hopf algebra generated by the functionals $f_\pm$ and $\bar{f}_\pm$, the dual algebra of the $Fun_q(SL(N,C))$ with appropriate commutation relations where the convolution product $f * g = (f \otimes g)\Delta$ is the product of the algebra. The coproduct, counit and antipode of the dual algebra are given as in the case of the compact quantum group with appropriate generalization for the $\bar{f}^i_j$. Therefore, the problem of defining the dual $*$-Hopf algebra $\mathcal{A}'$ is to find the consistent commutation relations among the functionals, i.e. the generators of the dual algebra.

We can show that the following relations among the functionals follow by requiring the consistency with the algebra of the bimodule discussed in the previous section:

$$\widehat{\mathbf{R}}^{ij}_{k'l'} f^{l'}_{\pm l} * f^{k'}_{\pm k} = f^{j}_{\pm j'} * f^{i}_{\pm i'} \widehat{\mathbf{R}}^{i'j'}_{kl} \quad , \widehat{\mathbf{R}}^{ij}_{k'l'} f^{l'}_{+l} * f^{k'}_{-k} = f^{j}_{-j'} * f^{i}_{+i'} \widehat{\mathbf{R}}^{i'j'}_{kl} \quad , \tag{10}$$

which are the same form as in the case of the compact quantum groups [CSWW]. New relations are given by

Proposition 3 : *The commutation relations of the dual algebra are given by (10) and its $*$-conjugation and*

$$\widehat{\mathbf{R}}^{-1}{}^{ki'}_{lj'} \bar{f}^{j'}_{j} * f^{i}_{i'} = f^{l'}_{l} * \bar{f}^{k}_{k'} \widehat{\mathbf{R}}^{-1}{}^{k'i}_{l'j} \quad , \tag{11}$$

where the eq.(11) holds for all combinations of the suffix $\pm$. Proof : See ref.[CW].

In the case of the $Fun_q(SU(N))$ due to the unitarity condition we had a further identity between $f_\pm$ and $\bar{f}_\pm$ (Proposition 3 in ref.[CSWW].) In the complex quantum group case we cannot impose the same relations since there is no unitarity condition. However still we can find similar relations among the functionals. One is due to the fact that the right invariant elements $\eta^{l_1}_{-} ... \eta^{l_{N-1}}_{-} \epsilon_{l_1 ... l_{N-1} j}$ and $\bar{\eta}_{-j}$ can not be distinguished by the bimodule structure. (They transform differently under the coaction.) Therefore we can impose consistently with the bicovariant bimodule algebra a condition

$$\bar{f}^{i}_{-j} = \kappa^{-1}(f^{i}_{-j}) \quad . \tag{12}$$

This is necessary from the point of view of the counting of the generators. We also impose the unitarity condition

$$\det{}_q f_\pm = 1 \tag{13}$$

and its $*$-conjugation. For the explicit definition of the q-determinant see ref.[CW]. These are the possible non-quadratic relations which can be immediately found from the structure of the algebra of the bicovariant bimodules.

We define the commutation relation among the generators of the dual algebra by eqs.(10)-(13) (and their $*$-conjugations). However, from this approach we cannot exclude the possibility that there may still be some additional relations among the functionals.

Nevertheless, we conclude here that the Hopf algebra generated by the functionals f and $\bar{f}$ with the above relations are the dual algebra $\mathcal{A}'$ of the complex quantum group $\mathcal{A}$ defined in section 1. The reason is that the relations listed above are the ones which define the dual quantum group $Fun'_q(SL(N,C)$ as the twisted product of the compact quantum group $Fun_q(SU(N))$ and its dual $\mathcal{U}_q(SU(N))$ and thus this algebra has twice of the independent generators. Let me explain this situation in detail:

Like in the case of $Fun_q(SU(N))$, comparing the Hopf algebra structure with the results obtained in [RTF] the algebra of f_+ and $\bar{f}_+$ can be identified with the dual algebra $\mathcal{U}_q(SU(N))$ by identifying the functionals with L_+ and $L^{*j}_{+i} = \kappa^{-1}(L^{i}_{-j})$ in ref.[RTF], since the values of the functionals f_+ and $\bar{f}_+$ on the generator $\bar{M}^i_j$ given in eq.(6) are consistent with the unitarity condition of $Fun_q(SU(N))$ and thus we can not distinguish the algebra generated by f_+ and $\bar{f}_+$ from the dual algebra of $Fun_q(SU(N))$.(see also [DSWZ]). On the other hand the defining equations of the functionals f_- and $\bar{f}_-$ in eq.(6) are not compatible with the unitarity condition of $Fun_q(SU(N))$. However, the algebra of the functionals f_- and $\bar{f}_-$ can be identified with the $Fun_q(SU(N))$ itself. To see this we identify f^{*i}_{-j} with the quantum matrix M^i_j of $Fun_q(SU(N))$. Then for example eq.(12) is equivalent to the unitarity condition of the quantum matrix.

Therefore the structure of the dual algebra $\mathcal{U}_q(SL(N,C))$ is found to be the twisted product

$$\mathcal{U}_q(SL(N,C)) \sim Fun_q(SU(N)) \tilde{\otimes} \mathcal{U}_q(SU(N)) \quad . \tag{14}$$

It means that $\mathcal{U}q(SL(N,C))$ is generated by the generators of the two subalgebras:

$$\mathcal{U}_q(SL(N,C)) \supset Fun_q(SU(N)) \quad \text{and} \quad \mathcal{U}_q(SL(N,C)) \supset \mathcal{U}_q(SU(N)) \quad . \tag{15}$$

On the other hand, the algebra is not the direct product of the two since the two subalgebras do not commute.

5. Bicovariant Differential Calculus

Using these results we construct now the differential calculus. As a space of one-form Γ^1, we take the bimodule with the right invariant basis $\theta^{i[k]} \equiv \eta_+^i \eta_-^{j_1} \ldots \eta_-^{j_{N-1}} \epsilon_{j_1 \ldots j_{N-1} l} \epsilon^{l[k]}$ and its conjugate $\bar{\theta}^{i[j]} \equiv \epsilon^{i[l]}(\theta^{k[l]})^* \epsilon^{k[j]}$. The bimodule structure is defined by functionals $\mathbf{L}^{i[j]}_{k[l]} \equiv (\kappa(f^i_{+k}) * f^m_{-n})\epsilon^{n[j]}\epsilon_{[l]m}$ and $\widetilde{\mathbf{L}}^{i[j]}_{k[l]} \equiv (\bar{f}^i_{-k} * \kappa(\bar{f}^m_{+n}))\epsilon^{n[j]}\epsilon_{[l]m}$ as: $\forall a \in \mathcal{A}$

$$a\theta^{i[j]} = \theta^{k[l]}(a * \mathbf{L}^{i[j]}_{k[l]}) \quad \text{and} \quad a\bar{\theta}^{i[j]} = \bar{\theta}^{k[l]}(a * \widetilde{\mathbf{L}}^{i[j]}_{k[l]}) \tag{16}$$

There are two left-right invariant one-forms in Γ^1: $\mathbf{X}_{(1)} = \epsilon_{i[j]}\theta^{i[j]}$ and $\mathbf{X}^*_{(1)}$. Using them the exterior derivative can be defined as

Definition 4 : *The exterior derivative* $\mathbf{d}$ *is the map* $\mathbf{d}\ : \mathcal{A} \to \Gamma^1$ *and* $\forall a \in \mathcal{A}$ *it is defined by*

$$\mathbf{d}a \equiv Da + \bar{D}a \quad , \tag{17}$$

where

$$Da \equiv \frac{1}{\mathcal{N}}[a, \mathbf{X}_{(1)}] \quad \text{and} \quad \bar{D}a \equiv (Da^*)^* = \frac{-1}{\mathcal{N}}[a, \mathbf{X}^*_{(1)}] \quad , \tag{18}$$

The normalization constant $\mathcal{N}$ *is real and proportional to* $q - q^{-1}$ *in the limit* $q \to 1$.

Since the exterior derivative is defined as a commutator, it satisfies automatically the Leibniz rule $\mathbf{d}(ab) = (\mathbf{d}a)b + a(\mathbf{d}b)$ where $a, b \in \mathcal{A}$. Under the $*$-conjugation, it behaves as $(\mathbf{d}a)^* = \mathbf{d}a^*$.

The higher order differential forms can be also defined by using the similar methods developed in ref.[CSWW]. There the condition $d^2 = 0$ gives nontrivial consequences in complex quantum group case, contrary to the case of $Fun_q(SU(N))$ where the nilpotency of the exterior derivative is satisfied automatically in our construction in ref.[CSWW]. See for detail ref.[CW].

6. Infinitesimal q-Lorentz Transformation

In the rest of the talk I want to explain how this differential calculus realized on the quantum space algebra $\mathcal{M}$ as a q-analogue of the infinitesimal transformation. This gives the picture of the infinitesimal q-Lorentz transformation on the quantum Minkowski space [CSSW].

Definition 5 : *We define the action of the derivative operator* $\mathbf{d} = D + \bar{D}$ *on an element* $x \in \mathcal{M}$ *by*

$$Dx \equiv \theta^{i[j]}(x * \chi_{i[j]}) \quad \text{and} \quad \bar{D}x \equiv \bar{\theta}^{i[j]}(x * \widetilde{\chi}_{i[j]}) \tag{19}$$

where the $*$-product is the convolution product $(x * f) \equiv (f \otimes id)\Delta_L(x)$.

We also require that the Leibniz rule of $\mathbf{d}$ acting on $\mathcal{M}$ holds as $\mathbf{d}(xy) = (\mathbf{d}x)y + x(\mathbf{d}y)$. This is satisfied by setting the following commutation relations:

$$x\theta^{i[j]} = \theta^{k[l]}(x * \mathbf{L}^{i[j]}_{k[l]}) \quad \text{and} \quad x\bar{\theta}^{i[j]} = \bar{\theta}^{k[l]}(x * \widetilde{\mathbf{L}}^{i[j]}_{k[l]}) \quad . \tag{20}$$

To make the analogy to the infinitesimal transformation, we give the representation of this operation with a commutator of the comodules. We consider a second rank tensor $\nabla_{i[j]}$ and its conjugate $\widetilde{\nabla}_{i[j]}$ together with coordinate functions. The tensor does not commute with the elements of the quantum space algebra and the commutator gives the generalization of the infinitesimal transformation. This tensor $\nabla_{i[j]}$ can be interpreted as a differential operator on the quantum space.

Definition 6 : *We introduce the tensor* $\nabla_{i[j]}$ *as an operator acting on* $\mathcal{M}$ *with the relation*

$$\nabla_{i[j]}x = (x * \chi_{i[j]}) + (x * \mathbf{L}^{k[l]}_{i[j]})\nabla_{k[l]} \quad \text{and} \quad \widetilde{\nabla}_{i[j]}x \equiv (x * \widetilde{\chi}_{i[j]}) + (x * \widetilde{\mathbf{L}}^{k[l]}_{i[j]})\widetilde{\nabla}_{k[l]} \tag{21}$$

With this definition the exterior derivative can be interpreted as the infinitesimal Lorentz transformation. The parameter of the transformation is the bimodule θ and $\bar{\theta}$. The derivative operator $\nabla_{k[l]}$ and $\widetilde{\nabla}_{k[l]}$ is the q-analogue of the generator of the Lorentz transformation represented by the differential operator.

References: The reference cited here can be found in ref.[CW] (U. Carow-Watamura and S. Watamura, "Complex Quantum Group, Dual Algebra and Bicovariant Differential Calculus," Preprint TU-382(1991), to be published in Commun. Math. Physics.)

DUAL DIFFERENTIALS IN QUANTUM SPACE-TIME [1] [2]

ZHAN XU

Department of Physics and Institute of Mordern Physics, Tsinghua University, Beijing 100084

ABSTRACT

The quantum Minkowski space-time is "real" in the sense of C^*-algebra. This seems to be contradictive to the differential calculus in it. Dual differentiations are introduced to solve this problem.

Non-commutative geometry is a developing topic in modern mathematics and also a new concept for constructing spaces and field theories in modern physics [1–10]. The relativistic quantum field theory constructed in the conventional Minkowski space-time is perhaps not applicable when the scale of physics was down to the sub-microscopic world [11]. Therefore the quantum Minkowski space-time based on the q-deformed Lorentz group is suggested [11–16,21,22], which is essentially non-commutative. The differential calculus in this space is of course important in order to formulate field theories, and many authors have worked it [12,13,21,22]. There is, however, still an ambiguous problem to be solved, *i. e.* the contradiction between the reality of the space and the differentiation in the space.

For the differential calculus in quantum spaces, after J. Wess and B. Zumino [17], firstly introduced is the exterior differentiation d satisfying

$$\begin{cases} \mathrm{d}^2 = 0, & \text{(Cartan rule)} \\ \mathrm{d}(fg) = (\mathrm{d}f)g + (-1)^{P(f)} f(\mathrm{d}g), & \text{(Leibniz rule)} \end{cases} \tag{1}$$

where $P(f)$ is the rank of element f in the algebra. And on the other hand, the $*$-conjugation in a C^*-algebra requires

$$\begin{cases} (f^*)^* = f, & \text{(idempotence)} \\ (fg)^* = g^* f^*, & \text{(algebraic antihomomorphism)} \end{cases} \tag{2}$$

for any element f and g in it. If the operations d and $*$ are commutative, *i.e.*

$$\mathrm{d}(f^*) = (\mathrm{d}f)^* \equiv \mathrm{d}f^*,$$

by differentiating the second one in Eq.(2) and taking $*$-conjugation of the second one in Eq.(1) we find that

$$(\mathrm{d}g^*)f^* + (-1)^{P(g)} g^*(\mathrm{d}f^*) = (-1)^{P(f)}(\mathrm{d}g^*)f^* + g^*(\mathrm{d}f^*),$$

and this apparently does not hold for arbitary elements f and g, unless they obey Cartan's commutation relation,

$$fg = (-1)^{P(f)P(g)} gf.$$

[1] Presented at 21th International Conference on Differential Geometry Methods in Theoretical Physics (Parallel Session 1), June 5—9, 1992, Tianjin, China.

[2] Work supported by CCAST (World Lab), Beijing and Institute of Theoretical Physics, Academia Sinica, Beijing and National Natural Science Foundation of China.

Because in quantum space we have different commutations from Cartan's, the two operations become uncompatible with their commutivity. Especially, the quantum Minkowski space-time has real structure, which means that there exists a relation between coordinates $x^\mu(\mu = 0, +, 3, -)$ of the space and their $*$-conjugation,

$$(x^\mu)^* = C^\mu{}_\nu x^\nu, \tag{3}$$

where $C = (C^\mu{}_\nu)$ is a real matrix. If d is commutative with $*$, the coordinates' differentials $\xi^\mu \equiv \mathrm{d}x^\mu$ will have similar conjugation relation,

$$(\xi^\mu)^* = C^\mu{}_\nu \xi^\nu, \tag{4}$$

and this adds unplausible constraints to the differential calculus in the space.

The commutation relations in the quantum Minkowski space, which is regarded a vector representation of the quantum group $SL_q(2, C)$, in Wess-Zumino's form are

$$\begin{cases} x_1 x_2 = \hat{S}_{12} x_1 x_2, \\ \xi_1 \xi_2 = -q\hat{\mathcal{R}}_{12}\xi_1\xi_2, \\ x_1\xi_2 = q\hat{\mathcal{R}}_{12}\xi_1 x_2, \\ \partial_1 x_2 = (G^{-1})_{12} + q(\hat{\mathcal{R}}^{-1})_{12} x_1 \partial_2, \\ \partial_1 \xi_2 = q^{-1}\hat{\mathcal{R}}_{12}\xi_1\partial_2, \\ \partial_1\partial_2 = \hat{S}_{12}\partial_1\partial_2, \end{cases} \tag{5}$$

(also see Ref.18–20). Here we introduced two matrices $\hat{S}$ and $\hat{\mathcal{R}}$ defined by

$$\begin{aligned} (\hat{S}/\hat{\mathcal{R}})^{\mu\nu}{}_{\kappa\lambda} &= K^\mu{}_{a\dot{a}} K^\nu{}_{b\dot{b}} (\hat{\mathbf{S}}/\hat{\mathbf{R}})^{a\dot{a}b\dot{b}}{}_{c\dot{c}d\dot{d}} (K^{-1})^{c\dot{c}}{}_\kappa (K^{-1})^{d\dot{d}}{}_\lambda, \\ &(\mu, \nu, \kappa, \lambda = 0, +, 3, -; \quad a, b, c, d, \dot{a}, \dot{b}, \dot{c}, \dot{d} = 1, 2) \end{aligned} \tag{6}$$

in which K is the transformation matrix from bi-spinor representation to vector representation of $SL_q(2, C)$, $\hat{\mathbf{S}}$ and $\hat{\mathbf{R}}$ are fusions of the $\hat{R}$ matrix of $SL_q(2, C)$ defined by

$$\begin{cases} \hat{\mathbf{S}}_{(12)(34)} = \hat{R}_{23}\hat{R}_{12}\hat{R}_{34}^{-1}\hat{R}_{23}^{-1}, \\ \hat{\mathbf{R}}_{(12)(34)} = q^{-1}\hat{R}_{23}\hat{R}_{12}\hat{R}_{34}\hat{R}_{23}^{-1}, \end{cases} \tag{7}$$

and they satisfy Yang-Baxter equation and the following relation,

$$(E - \hat{S})(E + q\hat{\mathcal{R}}) = 0. \qquad (E : 16 \times 16 \text{ unit matrix}) \tag{8}$$

$G^{-1} = (g^{\mu\nu})$ is the inverse matrix of the metric $g_{\mu\nu}$ of the space. In Eq.(5) the partial differentiation ∂ means ∂^μ which is related to ∂_μ by

$$\partial^\mu = g^{\mu\nu}\partial_\nu, \tag{9}$$

and ∂_μ in its turn to the exterior differentiation d by

$$\mathrm{d} = \xi^\mu \partial_\mu. \tag{10}$$

Commutations Eq.(5) are covariant under quantum Lorentz transformation.

Now we have the following identities for matrices $\hat{S}, \hat{\mathcal{R}}$ and C,

$$C^{\mu}{}_{\tau}C^{\nu}{}_{\rho}(\hat{S}/\hat{\mathcal{R}})^{\rho\tau}{}_{\kappa\lambda} = (\hat{S}/\hat{\mathcal{R}})^{\mu\nu}{}_{\rho\tau}C^{\tau}{}_{\kappa}C^{\rho}{}_{\lambda}, \tag{11}$$

therefore if Eqs.(3) and (4) hold, the first and the second equations in Eq.(5) are invariant under $*$-conjugation, but the third equation changes to

$$\xi_1 x_2 = q\hat{\mathcal{R}}_{12}x_1\xi_2,$$

which gives more relations in the algebra and they are questionable. This forces us to abandon the commutivity of d and $*$, *i.e.* $(\mathrm{d}f)^* \neq \mathrm{d}(f^*)$.

The following solution to this problem is proposed. Redenote the above differential as left one, *i.e.*

$$\mathrm{d}f \equiv \overrightarrow{\mathrm{d}}f, \tag{12}$$

and introduce its dual (or, in some sense, its conjugation) as right differential $\overleftarrow{\mathrm{d}}$ which obeys

$$\begin{cases} \overleftarrow{\mathrm{d}}^2 = 0, \\ (fg)\overleftarrow{\mathrm{d}} = f(g\overleftarrow{\mathrm{d}}) + (-1)^{P(g)}(f\overleftarrow{\mathrm{d}})g, \end{cases} \tag{13}$$

and simultaneously assume they cooperate with $*$-conjugation to the rule

$$\begin{cases} (\overrightarrow{\mathrm{d}}f)^* = (f^*)\overleftarrow{\mathrm{d}}, \\ (f\overleftarrow{\mathrm{d}})^* = \overrightarrow{\mathrm{d}}(f^*). \end{cases} \tag{14}$$

It is easy to prove that the conjectures Eqs.(1), (13), (2) and (14) are consistent. The differential calculus for right action is quite similar to the left one (Eq.(5)), *i.e.*

$$\begin{cases} \eta_1\eta_2 = -q\hat{\mathcal{R}}_{12}\eta_1\eta_2, \\ \eta_1 x_2 = q\hat{\mathcal{R}}_{12}x_1\eta_2, \\ x_1\overleftarrow{\partial}_2 = (G^{-1})_{12} + q(\hat{\mathcal{R}}^{-1})_{12}\overleftarrow{\partial}_1 x_2, \\ \eta_1\overleftarrow{\partial}_2 = q^{-1}\hat{\mathcal{R}}_{12}\overleftarrow{\partial}_1\eta_2, \\ \overleftarrow{\partial}_1\overleftarrow{\partial}_2 = \hat{S}_{12}\overleftarrow{\partial}_1\overleftarrow{\partial}_2, \end{cases} \tag{15}$$

in which $\eta = (\eta^{\mu}) \equiv (x^{\mu}\overleftarrow{\mathrm{d}})$, $\overleftarrow{\partial} = (\overleftarrow{\partial}^{\mu})$,

$$\overleftarrow{\partial}^{\mu} = \overleftarrow{\partial}_{\nu}\, g^{\nu\mu}, \tag{16}$$

and

$$\overleftarrow{\mathrm{d}} = \overleftarrow{\partial}_{\mu}\,\eta^{\mu}. \tag{17}$$

As to the mixed commutation relations we have

$$\begin{cases} \xi_1\eta_2 = \eta_1\xi_2, \\ \xi_1\eta_2 = -q\hat{\mathcal{R}}_{12}\eta_1\xi_2, \\ \overrightarrow{\partial}_1\overleftarrow{\partial}_2 = q^{-1}(\hat{\mathcal{R}}^{-1})_{12}\overleftarrow{\partial}_1\overrightarrow{\partial}_2, \\ \overrightarrow{\partial}_1\eta_2 = q(\hat{\mathcal{R}}^{-1})_{12}\eta_1\overrightarrow{\partial}_2, \\ \xi_1\overleftarrow{\partial}_2 = q(\hat{\mathcal{R}}^{-1})_{12}\overleftarrow{\partial}_1\xi_2, \end{cases} \tag{18}$$

by requirement of consistency. The two sets of differentials ξ^μ and η^μ are dual to one another, *i.e.*

$$\begin{cases} (\xi^\mu)^* = C^\mu{}_\nu \eta^\nu, \\ (\eta^\mu)^* = C^\mu{}_\nu \xi^\nu, \end{cases} \tag{19}$$

and similar are $\overleftarrow{\partial}^\mu$ and $\overrightarrow{\partial}^\mu$,

$$\begin{cases} (\overrightarrow{\partial}^\mu)^* = C^\mu{}_\nu \overleftarrow{\partial}^\nu, \\ (\overleftarrow{\partial}^\mu)^* = C^\mu{}_\nu \overrightarrow{\partial}^\nu. \end{cases} \tag{20}$$

Eqs.(5), (15) and (18) form a complete set of commutation relations for the calculus in quantum Minkowski space. By Eqs.(3), (19), (20), (11) and

$$C^\mu{}_\kappa C^\nu{}_\lambda g^{\lambda\kappa} = g^{\mu\nu}, \tag{21}$$

we can prove that they are self-consistent under $*$-operation.

In summary, the differential calculus defined in quantum Minkowski space should be covariant under quantum Lorentz transformation, compatible with reality of the space and self-consistent, and in order to meet these requirements the introducing of dual differentials is necessary.

References

1. A. Connes and J. Lott, *Nucl. Phys.* (Proc. Suppl.)**18B** (1990)29.
2. M. Dubios-Violette *et.al.*, *J. Math. Phys.* **31**(2)(1990)323.
3. D. Kastler, CPT-90/P2456 (France, Nov. 1990).
4. B. S. Balakrishna *et.al.*, *Phys. Lett.* **B254**(1991)430.
5. V. G. Drinfel'd, in *Proc. ICM*, (Berkeley, 1986), pp.798–820.
6. L. D. Faddeev *et.al.*, in *Algebraic Analysis*, (Academic Press, 1988), pp.129–140.
7. M. Jimbo, *Lett. Math. Phys.* **10**(1985)63; *Comm. Math. Phys.* **102**(1986)537.
8. S. L. Woronowicz, Publ. RIMS, Kyoto Univ. **23**(1987)117; *Comm. Math. Phys.* **111**(1987)613; **122**(1989)125.
9. Yu. I. Manin, "Quantum Group and Non-commutative Geometry", Université de Montrél preprint, CRM-1561(1988); *Ann. Inst. Fourier*, **37**(1987)191; *Comm. Math. Phys.* **123**(1989)163.
10. S. L. Woronowicz and P. Podleś, Warsaw Univ. preprint(1989); Mittag-Leffler Ins. report 20(1988/89).
11. P. Podleś and S. L. Woronowicz, *Comm. Math. Phys.* **130**(1990)381.
12. U. Carow-Watamura *et.al.*, *Z. Phys.* **C48**(1990)159.
13. U. Carow-Watamura *et.al.*, *Int. J. Mod. Phys.* **A6**(1991)3081.
14. A. Schirrmacher, MPI-Ph/91-77(Oct. 1991).
15. W. B. Schmidke *et.al.*, MPI-Ph/91-15(March 1991).
16. O. Ogievetsky *et.al.*, MPI-Ph/91-51(July 1991).
17. J. Wess and B. Zumino, *Nucl. Phys.* (Proc. Suppl.)**18B**(1990)302.
18. U. Carow-Watamura *et.al.*, *Z. Phys.* **C49**(1991)439.
19. J. Birman and H. Wenzl, *Braids, Link Polynomials and a New Algebra*, (Columbia Univ. Press, New York, 1989).
20. J. Murakami, *Osaka J. Math.* **24**(1987)745.
21. X. C. Song, CCAST 91-63 (Beijing, Nov. 1991), to appear in *Z. Phys.* C.
22. Z. Xu, IC/91/366 (ICTP, Miramare-Trieste, Nov. 1991); CCAST 92-16 (Beijing, April 1992).

CONSERVATION LAWS AND QUANTUM GROUP SYMMETRY IN N=1 PERTURBED SUPERCONFORMAL SOLITON THEORY*

Yu Yangzheng
Department of Physics, Xiamen University
Xiamen, Fujian, China
Institute of Theoretical Physics, Academia Sinica
Beijing, China

Tsuneo Uematsu
Department of Physics, College of Liberal Arts and Sciences
Kyoto University, Kyoto 606, Japan

Ken-ichiro Kobayashi
Department of Physics, Purdue University
West Lafayette, IN 47907, USA

1. INTRODUCTION

Recently a great deal of attention has been paid to the study of perturbed conformal field theories (perturbed CFT), i. e. deformation of CFT. It is well known that the two-dimensional CFT describe the critical points of statistical systems that correspond to the renormalization group fixed points in two-dimentional field theories. Now when we apply some perturbation on CFT, the conformal invariance will be lost and we are now in the off-critical region of the original theory.

Zamolodchikov has shown that each model with a relevant perturbation is still solvable as a massive model[1,2]. Eguchi and Yang investigated the models perturbed by $\Phi_{(1,3)}$ operator. It is shown that the sine-Gordon theory can be treated as perturbed CFT[3].

In this talk we would like to discuss the extension of the perturbed conformal models with integrable soliton systems to the N=1 supersymmetric version of sine-Gordon theory.

2. N=1 SUPER SINE-GORDON EQUATION AND CONSERVED QUANTITIES——CLASSICAL LEVEL

For the N=1 supersymmetric extension of sine-Gordon theory, the equation of motion reads

$$\bar{D}D\phi = g\sin\phi \tag{1}$$

Where ϕ is an N=1 superfield which depends on the N=1 supercoordinates $Z=(z,\theta)$ and $\bar{Z}=(\bar{z},\bar{\theta})$, and the corresponding supercovariant derivatives are given by

$$D = \frac{\partial}{\partial\theta} + \theta\partial_z, \qquad \bar{D} = \frac{\partial}{\partial\bar{\theta}} + \bar{\theta}\partial_{\bar{z}} \tag{2}$$

* Talk presented by Y. Z. Yu at XXI DGM, June 5-9, 1992, Tianjin, China.

In the classical level, we have a criterion for A to be a conserved current. It is found to be

Theorem If A satisfies

$$\bar{D}A = DX \tag{3}$$

then A is a conserved current, and the conserved charge is

$$Q = \oint dZ A(Z) \tag{4}$$

Some conserved charges at the classical level are constructed as follows

$$Q_1 = \oint dZ\, T, \qquad Q_3 = \oint dZ (TDT) \tag{5}$$

$$Q_5 = \oint dZ (2T(DT)^2 + \partial T(D^3T))$$

where the super energy-momentum tensor with the Feigin-Fuchs linear term is given as

$$T = \partial\phi D\phi + i\partial D\phi \tag{6}$$

3. N = 1 PERTURBED SUPERCONFORMAL THEORY AND CONSERVATION LAW —— QUANTUM LEVEL

In this section we discuss the quantum case, where the quantum effects are taken into account as a perturbation from the free theory. We add to the N = 1 free action S_{free} a relevant perturbation term as

$$S_{free} \longrightarrow S = S_{free} - \lambda \int d^2 Z \Phi(Z, \bar{Z}) \tag{7}$$

In conformal limit, we assume the factorization

$$\Phi(Z, \bar{Z}) = \Phi(Z)\bar{\Phi}(\bar{Z}) \tag{8}$$

holds and denote the operator product expansion of $A(Z)$ and $\Phi(Z)$ as

$$A(Z_1)\Phi(Z_2) = \sum_k Z_{12}^k \{(A\Phi)_k^{(0)} + \theta_{12}(A\Phi)_k^{(1)}\} \tag{9}$$

where $Z_{12} = Z_1 - Z_2 = z_1 - z_2 - \theta_1\theta_2$ and $\theta_{12} = \theta_1 - \theta_2$ are invariant distances in the superspace. Now we have the conservation condition for a operator A(Z) as follows.

Theorem If a operator A(Z) satisfies

$$(A\Phi)_{-1}^{(1)} = DX \tag{10}$$

then

$$Q = \oint dZ A(Z) \tag{11}$$

is a conserved charge to the lowest order.

We consider a perturbation of the form

$$\Phi = e^{i\beta\phi} + e^{-i\beta\phi} \tag{12}$$

the first two lower dimensional conserved charges are given as

$$Q_1 = \oint dZ T(Z), \qquad Q_3 = \oint dZ (TDT)_0^{(0)}(Z) \tag{13}$$

Here we have written the N = 1 super energy-momentum tensor T in the Feigin-Fuchs-Miura form

$$T(Z) = -\frac{1}{2}\partial\phi D\phi - i\alpha\partial D\phi \tag{14}$$

The central charge is $c = \frac{3}{2}(1 - 8\alpha^2)$ and the conformal weight for the operator $e^{i\beta\phi}$ as a superfield is h

$=\frac{1}{2}\beta^2+\alpha\beta$. We choose the α such that $e^{i\beta\phi}$ be an N=1 screening operator, i.e. h=1/2.

It is worth pointing out that there exist also extra conserved charges of the vertex operator type. It is shown that

$$J^{\pm}(Z)=e^{\pm\frac{i}{\beta}\phi(Z)} \tag{15}$$

is a conserved current. The conservation of this current J follows from

$$(J^{\pm}\Phi)^{(1)}_{-1}=\frac{1}{1-\beta^2}D(e^{\pm i(\frac{1}{\beta}-\beta)\phi}) \tag{16}$$

4. EXTRA CONSERVED CHARGES AND QUANTUM GROUP SYMMETRY

We now turn to discuss the physical implication of the existence of the extra quantum conserved charges. It is shown that the quantum conserved charges of the vertex operator type play an important role in quantum group symmetry in $N=1$ supersymmetric sine-Gordon theories.

As mentioned above, there exist the following extra conserved charges

$$Q^{\pm}=\oint dZe^{\pm\frac{i}{\beta}\phi(Z)}=\oint dzJ^{\pm}(z) \tag{17}$$

It is easy to know

$$J^{\pm}(z)=\pm\frac{i}{\beta}\psi(z)e^{\pm\frac{i}{\beta}\varphi(z)} \tag{18}$$

and the similar expression for the anti-holomorphic part

$$\bar{J}^{\pm}(\bar{z})=\mp\frac{i}{\beta}\bar{\psi}(\bar{z})e^{\mp\frac{i}{\beta}\bar{\varphi}(\bar{z})} \tag{19}$$

here we have written the chiral (anti-chiral) superfield in a power series in θ ($\bar{\theta}$)

$$\phi(Z)=\varphi(z)+\theta\psi(z),\qquad \bar{\phi}(\bar{Z})=\bar{\varphi}(\bar{z})+\bar{\theta}\bar{\psi}(\bar{z}) \tag{20}$$

If we define

$$\Phi^{(\pm)}(w)=\pm i\beta\psi(w)e^{\pm i\beta\varphi(w)},\quad \bar{\Phi}^{(\pm)}(\bar{w})=\pm i\beta\bar{\psi}(\bar{w})e^{\pm i\beta\bar{\varphi}(\bar{w})} \tag{21}$$

then the perturbation term can be rewritten as

$$\lambda\int d^2W\Phi(W,\bar{W})=\lambda\int d^2w\Phi(w,\bar{w}) \tag{22}$$

and

$$\Phi(w,\bar{w})=\sum_{\epsilon=+,-}\Phi^{(\epsilon)}(w)\bar{\Phi}^{(\epsilon)}(\bar{w}) \tag{23}$$

By using following operator product expansion

$$J^{\pm}(z)\Phi^{(\epsilon)}(w)\sim\frac{1}{z-w}\partial_w h^{\pm(\epsilon)}(w),\quad \bar{J}^{\pm}(\bar{z})\bar{\Phi}^{(\epsilon)}(w)\sim\frac{1}{\bar{z}-\bar{w}}\partial_{\bar{w}}\bar{h}^{\pm(\epsilon)}(\bar{w}) \tag{24}$$

with

$$h^{\pm(\mp)}(z)\sim\frac{\pm i\beta}{1-\beta^2}e^{\pm i(\frac{1}{\beta}-\beta)\varphi(z)},\qquad \bar{h}^{\pm(\pm)}(z)\sim\frac{\pm i\beta}{1-\beta^2}e^{\mp i(\frac{1}{\beta}-\beta)\bar{\varphi}(\bar{z})} \tag{25}$$

to the lowest order of perturbation theory we get

$$\partial_{\bar{z}}J^{\pm}=\partial_z H^{\pm},\quad \partial_z\bar{J}^{\pm}=\partial_{\bar{z}}\bar{H}^{\pm} \tag{26}$$

where

$$H^{\pm}=\lambda\sum_{\epsilon}h^{\pm\epsilon}(z)\bar{\Phi}^{(\epsilon)}(\bar{z}),\quad \bar{H}^{\pm}=\lambda\sum_{\epsilon}\Phi^{(\epsilon)}(z)\bar{h}^{\pm(\epsilon)}(\bar{z}) \tag{27}$$

Now we redefine the conserved charges as follows

$$Q_{\pm}=\int dzJ^{\pm}+\int d\bar{z}H^{\pm},\quad \bar{Q}_{\pm}=\int d\bar{z}\bar{J}^{\pm}+\int dz\bar{H}^{\pm} \tag{28}$$

Now it turns out that these quantum conserved charges generate the commutation relations which lead to the q-deformation of the sl(2) affine Kac-Moody algebra denoted as $sl_q(2)$ just like the N=0 case[4,5].

To start with, we derive following quantum group commutation relation where the deformation parameter is

$$q=-e^{-i\pi/\gamma},\quad 1/\gamma=1/2\beta^2-1/2 \tag{29}$$

As in N=0 sine-Gordon theory, there exists a topological current in our N=1 supersymmetric sine-Gordon theory;

$$\mathscr{T}^{\mu}(x,t)=\frac{\beta}{\pi}\varepsilon^{\mu\nu}\partial_{\nu}\phi(x,t) \tag{30}$$

where $\varepsilon^{\mu\nu}$ is the Leve-Civita tensor in two-dimensions. Hence we get the topological charge as follows

$$\mathscr{T}=\frac{\beta}{\pi}[\phi(x=\infty)-\phi(x=-\infty)] \tag{31}$$

Now we can take a soliton configuration in which $\phi(x=\infty)=0$.

With the use of the relations (51) and (53), this leads to the commutation relation

$$Q_{\pm}\bar{Q}_{\mp}-q^{-2}\bar{Q}_{\mp}Q_{\pm}=a(1-q^{\pm 2\mathscr{T}}) \tag{32}$$

where $a=\lambda\beta^2/(1-\beta^2)^2$.

Similarly, we have

$$Q_{\pm}\bar{Q}_{\pm}-q^2\bar{Q}_{\pm}Q_{\pm}=0 \tag{33}$$

$$[\mathscr{T},Q_{\pm}]=\pm 2Q_{\pm},\quad [\mathscr{T},\bar{Q}_{\pm}]=\pm 2\bar{Q}_{\pm}$$

From this commutation relation we can easily see that the extra charges generate the quantum affine algebra $sl_q(2)$.

We have also investigated the N=2 supersymmetric sine-Gordon theory, it is more interesting in physics (see the details in ref. 6).

References

1. A. B. Zamolodchikov, *Sov. J. Nucl. Phys.* 44(1986)529.
2. A. B. Zamolodchikov, *Reviews in Mathematical Physics* 1 (1990)197.
3. T. Eguchi and S. K. Yang, *Phys. Lett.* 224B(1989)373; *Phys. Lett.* 235B(1990)282.
4. D. Bernard and A. LeClair, *Nucl. Phys.* B340(1990)721; *Phys. Lett.* 247B(1990)309.
5. D. Bernard and A. LeClair, *"Quantum Group Symmetries and Non-Local Currents in 2D QFT"*, CLNS—90—1027(1990); SPhT—90/144(1990).
6. K. Kobayashi, T. Uematsu and Y. Z. Yu, KUCP—29/91; PURD—TH—91—03.

Tensor Method In Quantum Superalgebras

Zurong Yu

CCAST (World Laboratory, Beijing)

Department of Physics, Tongji University, Shanghai 200092, China

1. Irreducible Representations of $Su_q(2/1)$

Recently we have developed a technique named tensor method to determine the irreducible representations (irreps) of quantum algebra $su_q(3)$ [1,2] and to calculate the reduced coefficient of $su_q(4) \supset su_q(2) + su_q(2)$ [3]. Here we shall applly the similar technique to the quantum superalgebra $su_q(2/1)$.

We make use the following generators of quantum superalgebra $su_q(2/1)$ as reference [4]

$$J_0 \;,\; J_\pm \;;\; Q \;;\; T_{1/2} \text{ and } V_{-1/2} \tag{1}$$

And, Chichian & Klulish [5] made use symbols h_i, X_i, and Y_i, i=1,2. The relations between them are: $J_0=h_1/2$, $J_+=X_1$, $J_-=Y_1$, $Q=h_1+2h_2$, $T_{1/2}=-Y_2$, $V_{-1/2}=X_2$.

We define two auxiliary operators [4]

$$T_{-1/2} \quad , \quad V_{1/2} \tag{2}$$

These operators obey the following commutation relations

$$[\, Q \,,\, J_0 \,] = [\, Q \,,\, J_\pm \,] = 0 \tag{3a}$$

$$[\, J_0 \,,\, J_\pm] = \pm J_\pm \;, \qquad [\, J_+ \,,\, J_- \,] = [2J_0] \tag{3b}$$

$$[\, J_0 \,,\, T_s \,] = s\,T_s \;, \quad [\, Q \,,\, T_s \,] = T_s \;,\; s=\pm 1/2 \tag{3c}$$

$$[\, J_0 \,,\, V_s \,] = sV_s \;, \quad [\, Q \,,\, V_s \,] = -V_s \;,\; s=\pm 1/2 \tag{3d}$$

$$\{\, T_\pm \,,\, V_\mp \,\} = \mp [\, Q/2 \pm J_0] \tag{3e}$$

and Serre's relations

$$J_-^2 T_{1/2} + T_{1/2} J_-^2 = [2]\, J_- T_{1/2} J_- \tag{4a}$$

$$J_+^2 V_{-1/2} + V_{-1/2} J_+^2 = [2]\, J_+ T_{-1/2} J_+ \tag{4b}$$

and Serre-like relations [4]

$$J_+^2 T_{-1/2} + T_{-1/2} J_+^2 = [2]\, J_+ T_{-1/2} J_+ \tag{4c}$$

$$J_-^2 V_{1/2} + V_{1/2} J_-^2 = [2]\, J_- V_{1/2} J_- \tag{4d}$$

In the following discussion, we suppose that q is not a root of unity.

Now let $|\xi jm\rangle$ be basis vector in the irrep subspace L, and satisfy

$$\langle \xi' j' m' |\ Q\ | \xi j m\rangle = \xi\, \delta_{\xi'\xi}\, \delta_{j'j}\, \delta_{m'm} \tag{5a}$$

$$\langle \xi' j' m' |\ J_\pm\ | \xi j m\rangle = \sqrt{[j \mp m][j \pm m + 1]}\ \delta_{\xi'\xi}\, \delta_{j'j}\, \delta_{m',m\pm 1} \tag{5b}$$

$$\langle\ \ ' j' m' |\ J_0\ |\ j m\rangle = m\ \delta_{\xi'\xi}\, \delta_{j'j}\, \delta_{m'm} \tag{5c}$$

From eqs.(4a) and (4c), the matrix element of T_s can written as

$$\langle \xi+1\ j' m' | T_s |\ j m\rangle = \frac{\langle \xi+1\ j' \| T \| \xi j m\rangle}{\sqrt{[2j'+1]}}\, q^{w/2}\, C_q(jm\tfrac{1}{2}s | j'm') \tag{6a}$$

$$w = (-1)^{s+j-j'}\ (j + 1/2) - m', \quad s = \pm 1/2 \tag{6b}$$

Where $C_q(jm\frac{1}{2}s|j'm')$ is the Clebsch-Gordan coefficient (CGC) of $su_q(2)$)[6]. A similar expression can also obtain for V_s. The expression (5) is a q-analogy of the usual Winger- Eckart Theorem. Then the T_s and V_s may be considered as tensor-like operators of $su_q(2)$.

The elements $\langle \xi' j' \| T \| \xi j\rangle$ and $\langle \xi' j' \| V \| \xi\ j\rangle$ can be calculated by a set of recurrent formula. Thus the irreducible representations of $su_q(2/1)$ may be completely determined[4].

3. Clebsch-Gordan Coefficient of $su_q(2/1)$

CGC appear in a consideration of the tensor product $L \otimes L$. The coproduct of $T_{1/2}$ and $V_{-1/2}$ acting on $L \otimes L$ are

$$\Delta(T_{1/2}) = T_{1/2} \otimes q^{-\frac{1}{2}(\frac{Q}{2}-J_0)} + q^{\frac{1}{2}(\frac{Q}{2}-J_0)} \otimes T_{1/2} \tag{7a}$$

$$\Delta(V_{-1/2}) = V_{-1/2} \otimes q^{-\frac{1}{2}(\frac{Q}{2}-J_0)} + q^{\frac{1}{2}(\frac{Q}{2}-J_0)} \otimes V_{-1/2} \tag{7b}$$

Let the orthonormal basis $|\alpha \varepsilon jm\rangle$ of $L \otimes L$ be

$$|\alpha \varepsilon jm\rangle = \sum_{\varepsilon_a j_a \varepsilon_b j_b} \langle \varepsilon_a j_a , \varepsilon_b j_b \| \alpha \varepsilon j \rangle \, | \varepsilon_a j_a , \varepsilon_b j_b ; jm\rangle \tag{8}$$

where α labels multiplity of the reducation $(\varepsilon_a j_a) \otimes (\varepsilon_b j_b) \rightarrow (\varepsilon j)$. $\langle \varepsilon_a j_a , \varepsilon_b j_b \| \alpha \varepsilon j \rangle$ is called as scalar factor (SF) of the CGC of $su_q(2/1)$, which is invariant for the $su_q(2)$ algebra. And $| \varepsilon_a j_a , \varepsilon_b j_b ; jm\rangle$ is defined as

$$| \varepsilon_a j_a , \varepsilon_b j_b ; jm\rangle = \sum_{m_a m_b} C_q(j_a m_a j_b m_b | jm) \, | \varepsilon_a j_a m_a\rangle \, | \varepsilon_b j_b m_b\rangle \tag{9}$$

Combaining eq. (8) and (9), we can write the CGC of $su_q(2/1)$ as

$$C_q(\varepsilon_a j_a m_a , \varepsilon_b j_b m_b | \varepsilon jm) = \langle \varepsilon_a j_a , \varepsilon_b j_b \| \alpha \varepsilon j \rangle \, C_q(j_a m_a j_b m_b | jm) \tag{10}$$

Due to

$$T_s \, |\alpha \varepsilon_{max} j \, m\rangle = 0 \tag{11}$$

we obtain

$$\sum_{\varepsilon_a j_a \varepsilon_b j_b} \langle \varepsilon_a' j_a' , \varepsilon_b' j_b' ; j' \| T \| \varepsilon_a j_a , \varepsilon_b j_b ; j \rangle \langle \varepsilon_a j_a , \varepsilon_b j_b \| \alpha \varepsilon_{max} j_0 \rangle = 0 \tag{12}$$

Here $|\alpha \varepsilon_{max} j_0 m\rangle$ is a highest weight state in L. Using eq. (7) and the symmetrical properties of the CGC of $su_q(2)$ we have

$$\begin{aligned} &\langle \varepsilon_a' j_a' , \varepsilon_b' j_b' ; j' \| T \| \varepsilon_a j_a , \varepsilon_b j_b ; j \rangle \\ &= (-1)^{\frac{1}{2}+j+j_a'+j_b} \sqrt{[2j+1][2j'+1]} \begin{Bmatrix} 1/2 & j_a & j_a' \\ j_b & j' & j \end{Bmatrix}_q \langle \varepsilon_a' j_a' \| T^{(a)} \| \varepsilon_a j_a \rangle q^{A/2} \\ &+ (-1)^{\frac{1}{2}+j'+j_a+j_b} \sqrt{[2j+1][2j'+1]} \begin{Bmatrix} 1/2 & j_b & j_b' \\ j_a & j' & j \end{Bmatrix}_q \langle \varepsilon_b' j_b' \| T^{(b)} \| \varepsilon_b j_b \rangle q^{B/2} \end{aligned} \tag{13}$$

where

$$A = (-1)^{\frac{1}{2}+j-j'} (j+1/2) - (-1)^{\frac{1}{2}+j_a-j_a'} (j_a+1/2) - \varepsilon_b/2$$

$$B = (-1)^{\frac{1}{2}+j-j'} (j+1/2) - (-1)^{\frac{1}{2}+j_b-j_b'} (j_b+1/2) - \varepsilon_a/2$$

$\begin{Bmatrix} j_1 & j_2 & j_3 \\ l_1 & l_2 & l_3 \end{Bmatrix}_q$ is Racach coefficients of $su_q(2)$ [7]. And, the factor $\langle \varepsilon_i' j_i' \| T^{(i)} | \varepsilon_i j_i \rangle$ (i=a,b) has been known and the expression of $\langle \varepsilon_a' j_a', \varepsilon_b' j_b' ; j' \| V \| \varepsilon_a j_a, \varepsilon_b j_b ; j \rangle$ is similar to eq.(13).

Owing to

$$|\alpha \varepsilon-1\ j'm'\rangle = N(\varepsilon j'j) \sum_{ms} C_q(jm\tfrac{1}{2}s|j'm')\ V_s\ |\alpha jm\rangle \qquad (14a)$$

and

$$\{N(\varepsilon j'j)\} = \frac{\langle \varepsilon-1\ j' \| V \| \varepsilon\ j\rangle}{\sqrt{[2j'+1]}} \sum_{ms} \{C_q(jm\tfrac{1}{2}s|j'm')\}^2 q^{w/2} \qquad (14b)$$

With the help of eqs.(14) and (8), we can obtain finally

$$\begin{aligned} &\langle \varepsilon_a' j_a', \varepsilon_b' j_b' \| \alpha \varepsilon-1, j' \rangle \\ &= \{\langle \varepsilon-1, j' \| V \| \varepsilon., j \rangle\}^{-1} \\ &\times \langle \varepsilon_a j_a, \varepsilon_b j_b \| \alpha \varepsilon j\rangle \langle \varepsilon_a' j_a', \varepsilon_b' j_b' ; j' \| V \| \varepsilon_a j_a, \varepsilon_b j_b ; j \rangle \qquad (15) \end{aligned}$$

Having known $\langle \varepsilon_a j_a, \varepsilon_b j_b \| \alpha\ \varepsilon_{max} j_0 \rangle$, from eq.(15) we can obtain all the SF and CGC of $su_q(2/1)$ in principle.

Note that we need not the coproduct of $T_{-1/2}$ and $V_{1/2}$, so we don't write them here.

Reference

1. Zurong Yu, J. Phys. A: Math. Gen. 24 (1991) L399
2. Zurong Yu, Commun. Theor. Phys. 17 (1992) 369
3. Ya-Ping Yang & Zurong Yu, to be sumitted J. Math. Phys.
4. Zurong Yu,

   High Energy Physics & Nuclear Physics, 16 (1992) 937
5. M. Chaichian & P. Kulish , Phys. Lett. B234 (1990) 72
6. Bo-Yu Hou et al, Commun. Theor. Phys. 13 (1990) 181,341
7. I.I. Kachurik & A.V. Klimyk, J.Phys. A23 (1990) 2717

QUANTUM DEFORMATION OF KDV HIERARCHIES

De-Hai ZHANG

Physics Department, Graduate School, Academia Sinica
P.O.Box 3908, Beijing 100039, P.R.China

Recently, 1+1 dimensional Korteweg-de Vries (KdV) hierarchy[1] arouses the strong interests of the theoretical physicists . A lot of researches demonstrate that the KdV hierarchy is closely related to the matrix models, the non-perturbative 2 dimensional gravity coupled to matter systems and the 2 dimensional topological field theories. In the other hand, the interests of the quantum deformation (so-called q-deformation) of Lie algebra (quantum group) has been growing in the physical and mathematical regions. The q-deformed Virasoro algebra has been studied. Chaichian, Popowicz and Presnajder even researched the q-deformed KdV system[2]. For a long time past it is quite meaningful to discover a new kind of integrable systems. In this paper we shall do a new investigation about the q-deformed KdV hierarchies by defining the suitable q-differentiation. Through building a set of complete operations of the q-deformed pseudo differential operator and using the Lax pair, we present the constructing program of the q-deformed KdV hierarchies and obtain the q-deformed generalization of the ordinary KdV equation. It is believable that the q-deformed KdV hierarchy we gained is a new kind of the integrable system.

At first we define the q-deformed differential operator[3]

$$\tilde{D} = \frac{1}{(1-q^{-2})z}(1-Q^{-2}),$$

where Q is an operator

$$Qf(z) = f(zq) = q^{z\partial} f(z).$$

The q-deformed Leibniz rule of the q-deformed differential operator at high order case is

$$\tilde{D}^n \circ f(z) = \sum_{m=0}^{n} \left[\begin{array}{c} n \\ m \end{array} \right] q^{2m(n-m)} f^{(m,2m-2n)}(z) \tilde{D}^{n-m},$$

where

$$\left[\begin{array}{c} n \\ m \end{array} \right] = \frac{[n]!}{[m]![n-m]!}, \quad [m]! = [m][m-1]\cdots[2][1], \quad [m] = \frac{1-q^{-2m}}{1-q^{-2}},$$

and the symbol $f^{(n,m)}(z) = (\tilde{D}^n Q^m f(z))$ is tntroduced.
One can generalize the above rule to the case of negative n,

$$\tilde{D}^{-n} \circ f(z) = \sum_{m=0}^{\infty} (-1)^m \begin{bmatrix} n+m-1 \\ m \end{bmatrix} q^{-m(m+1)} f^{(m,2m+2n)}(z) \tilde{D}^{-n-m}.$$

All of operations needed for construction of the q-deformed KdV hierarchies is given.

The q-deformed KdV hierarchy is described in terms of the q-deformed pseudo differential operator which is a formal expression

$$K = \sum_{n=-\infty}^{M} k_n \tilde{D}^n,$$

One further introduces the decomposition

$$K_+ = \sum_{n=0}^{M} k_n \tilde{D}^n, \quad K_- = \sum_{n=1}^{\infty} k_{-n} \tilde{D}^{-n}, \quad r\tilde{e}sK = k_{-1},$$

The Nth q-deformed KdV hierarchy consists of an infinite set of commuting q-deformed differential equations for the coefficients $V_n(z, t_p)$ $\quad (n = 0, 1, \cdots, N-1)$ of a q-deformed differential operator L of order N that has been put in the canonical form

$$L = \tilde{D}^N + \sum_{n=0}^{N-1} V_n \tilde{D}^n.$$

In the algebra of q-deformed pseudo differential operators L has an unique N^{th} root $L^{1/N}$, and in the Lax representation [19] the pth flow of the Nth q-deformed KdV hierarchy (called the (p, N) system) is given by[4]

$$\frac{\partial}{\partial t_p} L = [L_+^{p/N}, L],$$

where t_p are called time parameters. Since L commutes with $L^{p/N}$, one has

$$[L_+^{p/N}, L] = [L, L_-^{p/N}].$$

When expanded in powers of $\tilde{D}$ this operator equation gives rise to a single q-deformed differential equation for each of the coefficients V_n.

The simplest system of the (p, N) q-deformed KdV hierarchies must be the (3,2) system which is called q-deformed KdV equations. Let us give this system in order to illustrate the above procedure. This model is obtained by taking L to be the two order q-deformed differential operator

$$K = \tilde{D}^2 + V_1(z,t)\tilde{D} + V_0(z,t).$$

The formal expansion of $L^{1/2}$ in powers of D is given by

$$K^{1/2} = \tilde{D} + \sum_{n=0}^{\infty} W_{-n}\tilde{D}^{-n}.$$

Since one needs only the first five coefficients of W_{-n} in the later q-deformed KdV equations, one gives them in terms of V_1 and V_0 order by order,

$$W_0 = (1+Q^{-2})^{-1}V_1 = \sum_{n=0}^{\infty}(-1)^n V_1^{(0,-2n)},$$

$$W_{-1} = -(1+Q^{-2})^{-1}(-V_0 + W_0^{(1,0)} + W_0^2),$$

$$W_{-2} = -(1+Q^{-2})^{-1}(W_{-1}W_0^{(0,2)} + W_{-1}^{(1,0)} + W_0W_{-1}),$$

$$W_{-3} = -(1+Q^{-2})^{-1}(-q^{-2}W_{-1}W_0^{(1,4)} + W_{-2}W_0^{(0,4)} + W_{-1}W_{-1}^{(0,2)} \\ +W_{-2}^{(1,0)} + W_0W_{-2}),$$

$$W_{-4} = -(1+Q^{-2})^{-1}(q^{-6}W_{-1}W_0^{(2,6)} - q^{-2}[2]W_{-2}W_0^{(1,6)} + W_{-3}W_0^{(0,6)} \\ -q^{-2}W_{-1}W_{-1}^{(1,4)} + W_{-2}W_{-1}^{(0,4)} + W_{-1}W_{-2}^{(0,2)} + W_{-3}^{(1,0)} + W_0W_{-3}).$$

One needs only the first two terms of the q-deformed pseudo differential operator $L_-^{3/2}$,

$$K_-^{3/2} = U_{-1}\tilde{D}^{-1} + U_{-2}\tilde{D}^{-2} + \cdots,$$

where the coefficients U_{-1} and U_{-2} are given by following expressions,

$$U_{-1} = W_{-3} - q^{-2}W_{-1}V_1^{(1,4)} + W_{-2}V_1^{(0,4)} + W_{-1}V_0^{(0,2)},$$

$$U_{-2} = W_{-4} + q^{-6}W_{-1}V_1^{(2,6)} - q^{-2}[2]W_{-2}V_1^{(1,6)} + W_{-3}V_1^{(0,6)} \\ -q^{-2}W_{-1}V_0^{(1,4)} + W_{-2}V_0^{(0,4)}.$$

Now we can obtain the q-deformed KdV equations

$$\frac{\partial V_1}{\partial t} = U_{-1}^{(0,-4)} - U_{-1},$$

$$\frac{\partial V_0}{\partial t} = (U_{-2}^{(0,-4)} - U_{-2}) + V_1(U_{-1}^{(0,-2)} - U_{-1}) \\ -U_{-1}(V_1^{(0,-2)} - V_1) + q^2[2]U_{-1}^{(1,-2)}.$$

When q tends to 1, the above equations return to the ordinary KdV equation.

If $dL/dt = [A, L]$, where L is a q-deformed differential operator, then one has

$$\frac{d}{dt}L^{r/k} = [A, L^{r/k}].$$

Taking the ordinary residue on both sides of the above equation one obtains[5]

$$\frac{d}{dt}resL^{r/k} = res[A, L^{r/k}] = \frac{\partial f}{\partial z}.$$

Integrating it and choosing suitable boundary condition one has

$$\frac{d}{dt}\int dzresL^{r/k} = \int \frac{\partial f}{\partial z}dz = 0.$$

Therefore we see that for any integer r , the ordinary residue

$$H_{r/k} = resL^{r/k}$$

is a density of conservation law for the Lax equation. Of course, nontrivial conservation laws

$$C_{r/k} = \int dzresL^{r/k}$$

correspond only to numbers r not a multiple of k. If one knows the expansion expression of the flows in terms of the q-deformed differential operators

$$L^{r/k} = \sum_{n=-\infty}^{r} a_n \tilde{D}^n,$$

one can obtain the density of the conservation law in terms of the coefficients of their expansion expression

$$H_{r/k} = resL^{r/k} = \sum_{n=1}^{\infty}\left(\sum_{i=0}^{n-1}\frac{(-1)^{n-i-1}q^{-(n-i)(n-i-1)}(1-q^{-2})}{2\ln q\ \ [i]!\ [n-i-1]!}\right)a_{r+n}z^{n-1}.$$

REFERENCES

[1] As a review, see, for example, P.G.Drazin and R.S.Johnson, Soliton: an Introduction, Cambridge University Press 1989.

[2] M.Chaichian, Z.Popowicz and P.Presnajder Phys.Lett.249B(1990)63.

[3] F.H.Jackson, Quart.J.Math.41(1910)193.

[4] P.P.Lax, Comm.Pure Appl.Math.21(1969)467.

[5] V.G.Drinfeld and V.V.Sokolov,Journ.Sov.Math.30(1985)1975.

Parallel Sessions 2

Quantum Group and Knot Theory

INVARIANTS OF COLORED LINKS AND A PROPERTY OF THE CLEBSCH-GORDAN COEFFICIENTS OF $U_q(g)$

TETSUO DEGUCHI and TOMOTADA OHTSUKI †

Department of Physics, Faculty of Science, University of Tokyo, Hongo, Bunkyo-ku, Tokyo 113, Japan

† *Department of Mathematics, Faculty of Science, University of Tokyo, Hongo, Bunkyo-ku, Tokyo 113, Japan*

Abstract.

We show that multivariable colored link invariants are derived from the roots of unity representations of $U_q(g)$. We propose a property of the Clebsch-Gordan coefficients of $U_q(g)$, which is important for defining the invariants of colored links. For $U_q(sl_2)$ we explicitly prove the property, and then construct invariants of colored links and colored ribbon graphs, which generalize the multivariable Alexander polynomial.

1 Introduction

Recently a new family of invariants of colored oriented links and colored oriented ribbon graphs is introduced, which gives generalizations of the multivariable Alexander polynomial.[1,2,6] The new invariants are related to the roots of unity representations of $U_q(sl_2)$ where $(X^{\pm})^N = 0$ and $(K)^N = q^{2Np}$ ($p \in \mathbf{C}$) and $q = \epsilon$.[5,6,7] Here $\epsilon = \exp(\pi i s/N)$, and the integers N and s are coprime. We call the representations nilpotent representations.

The invariants of colored links have a property that they vanish for disconnected links.[2,6] Due to this property a proper regularization method is necessary for definition of the invariants.

In this paper we show an important property of the Clebsch-Gordan coefficients (CGC) of the nilpotent representations of $U_q(sl_2)$, which leads to the definition of the colored link invariants. We consider the nilpotent reps of $U_q(g)$ such that $(X_i^{\pm})^N = 0$, $(K_i)^N = q^{2Np_i}$ ($p_i \in \mathbf{C}$) and $q = \epsilon$. We give a conjecture that the property of CGC also holds for the nilpotent reps of $U_q(g)$ and we can define invariants of colored links for $U_q(g)$ in the same way as $U_q(sl_2)$.

2 CGC of the nilpotent representations

We introduce some symbols for a positive integer n and a complex parameter p.

$$[n]_q = \frac{q^n - q^{-n}}{q - q^{-1}}, \quad [n]_q! = \prod_{k=1}^{n} [k]_q, \quad [p; n]_q! = \prod_{k=0}^{n-1} [p - k]_q. \tag{1}$$

For $n = 0$ we assume $[0]_q! = [p;0]_q! = 1$.

We introduce the nilpotent rep. $V^{(p)}$ [5]: $\pi(X^+)^a_b = \sqrt{[2p-a]_\epsilon[a+1]_\epsilon}\ \delta_{a+1,b}$, $\pi(X^-)^a_b = \sqrt{[2p-a+1]_\epsilon[a]_\epsilon}\ \delta_{a-1,b}$, $\pi(K)^a_b = \epsilon^{(2p-2a)}\delta_{a,b}$. Here $a, b = 0, 1, \cdots N-1$. We note that N is related to the dimensions of the representation. When N is even, ϵ is $2N$-th primitive root of unity, while when N is odd, ϵ may be $2N$-th or N-th primitive root of unity.

In Ref. 1 the explicit matrix representations of the colored braid group were introduced. It was shown that the representations are equivalent to the R matrices of the nilpotent reps, which are derived from the universal R matrix $\mathcal{R}$: $R^{a_1,a_2}_{b_1,b_2} = \pi^{p1} \otimes \pi^{p_2}(\mathcal{R})^{a_1a_2}_{b_1b_2}$. [5]

We can show the fusion rule for the tensor product: $V^{(p_1)} \otimes V^{(p_2)} = \sum_{p_3} N^{p_3}_{p_1p_2} V^{(p_3)}$, where $N^{p_3}_{p_1,p_2} = 1$ for $p_3 = p_1 + p_2 - n$ ($0 \le n \le N-1$ and $n \in \mathbf{Z}$); $N^{p_3}_{p_1,p_2} = 0$, otherwise. The CGC for $V^{(p_i)}$ ($i = 1, 2, 3$) are given by [6]

$$\begin{aligned} & C(p_1, p_2, p_3; z_1, z_2, z_3) = \delta(z_3, z_1 + z_2 - n) \\ & \times \sqrt{[2p_1 + 2p_2 - 2n + 1]_\epsilon} \sqrt{[n]_\epsilon![z_1]_\epsilon![z_2]_\epsilon![z_3]_\epsilon!} \\ & \times \sum_\nu \frac{(-1)^\nu \epsilon^{-\nu(p_1+p_2+p_3+1)} \times \epsilon^{(n-n^2)/2+(n-z_2)p_1+(n+z_1)p_2}}{[\nu]_\epsilon![n-\nu]_\epsilon![z_1-\nu]_\epsilon![z_2-n+\nu]_\epsilon!} \\ & \times \sqrt{\frac{[2p_1-n;z_1-\nu]_\epsilon![2p_1-z_1;n-\nu]_\epsilon![2p_2-n;z_2+\nu-n]_\epsilon![2p_2-z_2;\nu]_\epsilon!}{[2p_1+2p_2-n+1;z_1+z_2+1]_\epsilon!}}. \end{aligned} \quad (2)$$

Here $0 \le z_i \le N-1$, for i=1,2,3, and the sum over the integer ν in (2) is taken under the condition: max $\{0, n - z_2\} \le \nu \le$ min $\{n, z_1\}$. The expression of the CGC was proved through the infinite dimensional representations. [6]

3 A Property of CGC

We consider the following two sums;

$$\begin{aligned} A^n_{p_1p_2} &= \sum_{z_1} [C(p_1, p_2, p_3; z_1, z_2, z_1 + z_2 - n)]^2 q^{2\rho(z_1)}, \\ B^n_{p_1p_2} &= \sum_{z_2} [C(p_1, p_2, p_3; z_1, z_2, z_1 + z_2 - n)]^2 q^{-2\rho(z_2)}. \end{aligned} \quad (3)$$

For the nilpotent reps of $U_q(sl_2)$ we assume $\rho(z) = p - z + Np/2$ and $q = \epsilon$. Due to the irreducibility of the representations the sum $A^n_{p_1p_2}$ ($B^n_{p_1p_2}$) does not depend on z_2 (z_1). We now introduce an important property of CGC.

Proposition 3.1 *The CGC of the nilpotent representations satisfy*

$$A^n_{p_1p_2}/B^n_{p_1p_2} = g(p_1, p_2) = f(p_1)/f(p_2), \text{ independent of } n \quad (p_3 = p_1 + p_2 - n). \quad (4)$$

(For the nilpotent reps we have $g(p_1,p_2) = [2p_2; N-1]_\epsilon!/[2p_1; N-1]_\epsilon!)$.

Proof

We set $z_2 = 0$ for $A^n_{p_1p_2}$ and $z_1 = 0$ for $B^n_{p_1p_2}$, and we define $L^n_{p_1p_2}$ and $R^n_{p_1p_2}$ by $A^n_{p_1p_2} = L^n_{p_1p_2}[2p_3+1]/[2p_3+1+n;N]!$ and $B^n_{p_1p_2} = R^n_{p_1p_2}[2p_3+1]/[2p_3+1+n;N]!$, respectively. Then we can show

$$L^n_{p_1p_2} = (-1)^{nN}[2p_2; N-1]_\epsilon!, \quad R^n_{p_1p_2} = (-1)^{nN}[2p_1; N-1]_\epsilon!. \tag{5}$$

The first eq. in (5) can be shown by using the following recurrence relation on n.

$$L^n_{p_1p_2} = \frac{[2p_1+2p_2-n-N+2]}{[2p_2+1]}\epsilon^{2p_1-n}L^{n+1}_{p_1+1/2,p_2+1/2} - \frac{[2p_1-n]}{[2p_2+1]}L^{n+1}_{p_1p_2+1/2}. \tag{6}$$

The second eq. in (5) (for $R^n_{p_1p_2}$) is derived from that of $L^n_{p_1p_2}$ by exchanging p_1 with p_2, and by setting $\epsilon \to \epsilon^{-1}$.

It is easy to see that the property (4) of CGC also holds for the finite dimensional (spin) representations of $U_q(sl_2)$ with q generic, where we have $g(j_1,j_2) = [2j_1]/[2j_2]$.

We can define invariants of colored links using the proposition 3.1. Let T be a $(1,1)$- tangle. We denote by $\hat{T}$ the link obtained by closing the open strings of T. We note the following proposition. [2]

Proposition 3.2 *Let T_1 and T_2 be two $(1,1)$-tangles. If $\hat{T}_1$ is isotopic to $\hat{T}_2$ as a link in S^3 by an isotopy which carries the closing component of $\hat{T}_1$ to that of $\hat{T}_2$. Then T_1 is isotopic to T_2 as a $(1,1)$-tangle.*

Let us introduce the functor $\phi(\cdot)$ for the tangle diagrams. [2] We denote by $\phi(T,\alpha)^a_b$ the value ϕ for the tangle with variables a and b on the closing component (or edge). It is easy to show that $\phi(T,\alpha)^a_b = \lambda\delta_{ab}$. [2]

We put $L = \hat{T}$ and s is the color of the closing component (or edge) of $\hat{T}$. For a colored link (L,α) and a color s of closing component (or edge), we define Φ by $\Phi(L,s,\alpha) = \lambda$ where L,T,s are as above and $\phi(T,\alpha)^a_b = \lambda\delta_{ab}$. We can show that Φ is well-defined, i.e. $\Phi(L,s,\alpha)$ does not depend on a choice of T. [2] From the proposition 3.1 we have the following.

Proposition 3.3 [2] *For a link L and its color $\alpha = (p_1,\cdots,p_n)$, we have*

$$\Phi(L,s,\alpha)([p_s;N-1]_\epsilon!)^{-1} = \Phi(L,s',\alpha)([p_{s'};N-1]_\epsilon!)^{-1}. \tag{7}$$

Definition 3.4 [2] *For a colored oriented link (L,α), we define an isotopy invariant $\hat{\Phi}$ of (F,α) by*

$$\hat{\Phi}(L,\alpha) = \Phi(L,s,\alpha)([p_s;N-1]_\epsilon!)^{-1}. \tag{8}$$

Thus we have constructed the multivariable invariants from the property (4) of CGC. In the same way we can define the multivariable invariants of colored ribbon graphs. [6]

We now consider CGC of $U_q(g)$, where g is a simple Lie algebra.

$$\begin{aligned} A^{\Lambda_3}_{\Lambda_1\Lambda_2} &= \sum_{\vec{z}_1} \left[C(\Lambda_1,\Lambda_2,\Lambda_n;\vec{z}_1,\vec{z}_2,\vec{z}_1+\vec{z}_2-\vec{n})\right]^2 q^{2\rho(\vec{z}_1)}, \\ B^{\Lambda_3}_{\Lambda_1\Lambda_2} &= \sum_{\vec{z}_2} \left[C(\Lambda_1,\Lambda_2,\Lambda_3;\vec{z}_1,\vec{z}_2,\vec{z}_1+\vec{z}_2-\vec{n})\right]^2 q^{-2\rho(\vec{z}_2)}. \end{aligned} \quad (9)$$

We assume that ρ is given by "half the sum of positive roots". Finally, we propose the following conjecture.

Conjecture 3.5 *(1) The CGC of the nilpotent representations Λ_i of $U_q(g)$ satisfy*

$$A^{\Lambda_3}_{\Lambda_1\Lambda_2}/B^{\Lambda_3}_{\Lambda_1\Lambda_2} = g(\Lambda_1,\Lambda_2), \text{ independent of } \Lambda_3. \quad (10)$$

(2) With a proper normalization of the quantum trace ($\rho(\vec{z}) \to \rho(\vec{z})$ + constant), we can set $g(\Lambda_1,\Lambda_2) = f(\Lambda_1)/f(\Lambda_2)$.
(3) $f(\Lambda)$ is equivalent to the tangle invariant for the Hopf link.

If the conjecture is true, we can construct multivariable invariants of colored links and colored ribbon graphs from $U_q(g)$ in the same way as $U_q(sl_2)$.

It is easy to show that the property (10) holds for CGC of finite dimensional representations of $U_q(g)$ with q generic.

References

[1] Y. Akutsu and T. Deguchi, A New Hierarchy of Colored Braid Group Representations, Phys. Rev. Lett. **67** (1991) 777.

[2] Y. Akutsu, T. Deguchi and T. Ohtsuki, Invariants of Colored Links, Journal of Knot Theory and Its Ramifications **1** (1992) pp. 161-184.

[3] T. Deguchi and Y. Akutsu, A General Formula for Colored Z_N Graded Braid Matrices and Fusion Braid Matrices, J. Phys. Soc. Jpn. **60** (1991) 2559.

[4] T. Deguchi and Y. Akutsu, A New Hierarchy of Colored Vertex Models, J. Phys. Soc. Jpn. **60** (1991) 4053.

[5] T. Deguchi and Y. Akutsu, Colored braid matrices from infinite dimensional representations of $U_q(g)$, Mod. Phys. Lett. **A7** (1992) 767.

[6] T. Deguchi and Y. Akutsu, Colored Vertex Models, Colored IRF Models, and Invariants of Colored Framed Graphs, preprint January 1992.

[7] T. Ohtsuki, Colored ribbon Hopf algebras and universal invariants of framed links, preprint June 1992.

ALGEBRAIC MAPS AND DEFORMATION SYMMETRY [1]

SHAO-MING FEI

Institute of Theoretical Physics, Academia Sinica
P. O. Box 2735, Beijing 100080, China

ABSTRACT

The general algebraic maps among algebras with three generators are systematically investigated in terms of symplectic geometry and geometric quantization on 2-D manifolds. From which the Hamiltonian of Heisenberg spin chain model with $SU_q(2)$ symmetry and arbitrary spin values is given.

Quantum groups[1-4] are usually considered the "quantization" of Lie groups. Nevertheless they have nothing to do with the physical quantization and can be realized both at classical and quantum levels[5,6]. In this talk we briefly introduce the symplectic geometry and geometric quantization approach to the deformation symmetries described by the quantum groups in view of algebraic maps. We will concern on algebras with three generators and take $SU_q(2)$ as an example. The approach may be generalized to general Lie algebras by means of the symplectic geometry on the coadjoint orbits of Lie groups.

From the symplectic geometry we find that there is an one to one correspondence between smooth 2-D manifolds and algebras with three generators up to algebraic isomorphisms[6]. For a general smooth manifold M defined by

$$f(S_1, S_2, S_3) = 0 , \tag{1}$$

where S_i, i=1,2,3, are variables, the corresponding Poisson algebraic relations of S_i's are uniquely given by

$$[S_i, S_j]_{P.B.} = \alpha \sum_k^3 \epsilon_{ijk} \frac{\partial f}{\partial S_k} , \tag{2}$$

where α is a constant.

It is easy to find that with respect to Poisson algebra $SU(2)$,

$$[S_i, S_j]_{P.B.} = \sum_{k=1}^3 \epsilon_{ijk} S_k , \qquad i,j = 1,2,3, \tag{3}$$

the manifold is a 2-D sphere S^2 of radius S_0,

$$S_1^2 + S_2^2 + S_3^2 = S_0^2 \tag{4}$$

[1]Work supported in part by the National Natural Science Foundation of China and LWTZ-1298 of Chinese Academy of Sciences.

with symplectic form

$$\omega = \frac{-1}{2S_0^2}\sum_{ijk}^{3}\epsilon_{ijk}S_i dS_j \wedge dS_k \ . \tag{5}$$

While for the algebra $SU_q(2)$,

$$[S'_+, S'_-]_{P.B.} = [2S'_3]_q \ , \quad [S'_3, S'_\pm]_{P.B.} = \pm S'_\pm \ , \tag{6}$$

there is a q-deformed sphere S_q^2,

$$S_1'^2 + S_2'^2 + \frac{(\sinh\gamma S_3')^2}{\gamma\sinh\gamma} = S_0'^2 \tag{7}$$

with symplectic form

$$\omega' = -\frac{1}{S_0'^2}(S_1'dS_2' \wedge dS_3' + S_2'dS_3' \wedge dS_1' + \frac{\tanh\gamma S_3'}{\gamma}dS_1' \wedge dS_2') \ , \tag{8}$$

where $S'_\pm = S'_1 \pm iS'_2$ and S'_0 is a constant.

Remarkably, the manifolds (4) and (7) are just the guiding equations for the algebraic maps between algebras $SU(2)$ and $SU_q(2)$. One algebra may be realized from another one and the only requirement is that both equations be satisfied. Let $S'_i = S'_i(S_1, S_2, S_3)$, i=1,2,3, be functions of S^2. Then we have conclusion: **the new " observables " $S'_i = S'_i(S_1, S_2, S_3)$ constitute the $SU_q(2)$ algebra in Poisson brackets if and only if S'_i satisfy the eq. (7) of S_q^2.** The proof is straightforward by using eq. (4) and symplectic from (5) on S^2. Here is one kind of algebraic maps between $SU(2)$ and $SU_q(2)$,

$$S'_\pm = \frac{1}{\sqrt{\gamma\sinh\gamma}}S_\pm\frac{\sinh\gamma(S_0 \mp S_3)}{S_0 \mp S_3} \ , \quad S'_3 = S_3 \ , \tag{9}$$

where $S_\pm = S_1 \pm iS_2$.

The Poisson algebra can be realized in Lie commutators from prequantization and geometric quantization method[7]. From which we get the operator equations with respect to eqs. (4) and (7),

$$\hat{S}_+\hat{S}_- + \hat{S}_3^2 = j(j+1), \tag{10}$$

$$\hat{S}'_+\hat{S}'_- + [\hat{S}'_3]_q[\hat{S}'_3 - 1]_q = [j]_q[j+1]_q \ , \tag{11}$$

where j takes integer and half integer values. $S_{3,\pm}$ and $S'_{3,\pm}$ give rise to $SU(2)$ and $SU_q(2)$ algebras in Lie commutators respectively. The non-trivial Hopf algebraic structures for $SU_q(2)$ can also be defined both at prequantization (classical) and quantum levels such as

$$\Delta(\hat{S}'_3) = \hat{S}'_3 \otimes 1 + 1 \otimes \hat{S}'_3 \ , \quad \Delta(\hat{S}'_\pm) = q^{\hat{S}'_3} \otimes \hat{S}'_\pm + \hat{S}'_\pm \otimes q^{-\hat{S}'_3} \ . \tag{12}$$

The eqs. (10) and (11) are again the guiding equations for the operator maps between $SU(2)$ and $SU_q(2)$. For example, an operator map related to (9) is

$$\hat{S}'_+ = \hat{S}_+ \frac{[j-\hat{S}_3]_q}{j-\hat{S}_3} , \quad \hat{S}'_- = \hat{S}_- \frac{[j+\hat{S}_3]_q}{j+\hat{S}_3} , \quad \hat{S}'_3 = \hat{S}_3 , \tag{13}$$

which is the same as the one presented in [8].

Therefore the q-deformation of $SU(2)$ to $SU_q(2)$ is related to the deformations from manifolds (4) to (7) or eq. (10) to eq. (11) which give rise to deformation maps such as (9) and (13). With these formulas one may investigate the deformations of the symmetries described by the deformed algebra. Here as an example we give the construction of Heisenberg chain Hamiltonian of $SU_q(2)$ symmetry and arbitrary spin values.

We consider a spin chain with $N+1$ sites. The total operators of $SU_q(2)$ for this chain is then $\hat{S}'^{tot}_{3,\pm} = \Delta^{(N)}\left(\hat{S}'_{3,\pm}\right)$, where $\Delta^{(N)} = (\Delta \otimes id)\Delta^{(N-1)}$. Let $\mathbf{I}_q = \hat{S}'_+\hat{S}'_- + [\hat{S}'_3]_q[\hat{S}'_3 - 1]_q$. From eq. (11) it is clear that I_q is an invariant of $SU_q(2)$. Hence the Hamiltonian defined by following is of $SU_q(2)$ symmetry,

$$H = \sum_{i=1}^{N} \mathbf{1}^{(1)} \otimes \cdots \otimes \mathbf{1}^{(i-1)} \otimes \Delta\left(C - \mathbf{I}_q\right) \otimes \mathbf{1}^{(i+2)} \otimes \cdots \otimes \mathbf{1}^{(N+1)} , \tag{14}$$

where C is an arbitrary constant. It is straightforward to prove that H commutes with $\hat{S}'^{tot}_{3,\pm}$. However H is not yet a Heisenberg spin chain Hamiltonian as $\hat{S}'_{3,\pm}$ do not constitute the $SU(2)$ algebra. But with maps (13) it really becomes a Heisenberg Hamiltonian of $SU_q(2)$ symmetry for general spin values.

For the case of spin-$\frac{1}{2}$ one gets the usual Hamiltonian[9] from formulas (12), (13) and (14). However for the case of spin-1, we have

$$\begin{aligned}
\Delta(C - \mathbf{I}_q) = \frac{1}{4q^3} \Big[& (q^6 - q^4 - q^2 + 1)\hat{S}_3^2 \otimes \hat{S}_3^2 - (q^6 + 3q^4 + 3q^2 + 1)\hat{S}_3 \otimes \hat{S}_3 \\
& + (-q^6 - q^4 + q^2 + 1)\hat{S}_3^2 \otimes \hat{S}_3 + (q^6 + q^4 - q^2 - 1)\hat{S}_3 \otimes \hat{S}_3^2 \\
& + 2q(q^4 - 2q^3 + 2q^2 - 2q + 1)(\hat{S}_+\hat{S}_3 \otimes \hat{S}_-\hat{S}_3 + \hat{S}_-\hat{S}_3 \otimes \hat{S}_+\hat{S}_3) \\
& + 2q(-q^3 + q^2 - q + 1)\hat{S}_-\hat{S}_3 \otimes \hat{S}_+ + 2q^2(q^3 - q^2 + q - 1)\hat{S}_+ \otimes \hat{S}_-\hat{S}_3 \\
& + 2q(q^3 - q^2 + q - 1)\hat{S}_- \otimes \hat{S}_+\hat{S}_3 + 2q^2(-q^3 + q^2 - q + 1)\hat{S}_+\hat{S}_3 \otimes \hat{S}_- \\
& - 2q(q^2 + 1)\hat{S}_- \otimes \hat{S}_+ - 2q^3(q^2 + 1)\hat{S}_+ \otimes \hat{S}_- \\
& + 2(-q^6 + q^4 + q^2 - 1)(\hat{S}_3^2 \otimes \mathbf{1} + \mathbf{1} \otimes \hat{S}_3^2) \\
& + 2(-q^6 - q^4 + q^2 + 1)\hat{S}_3 \otimes \mathbf{1} + 2(q^6 + q^4 - q^2 - 1)\mathbf{1} \otimes \hat{S}_3 \\
& + 4q^2(-2q^2 + qC - 2))\mathbf{1} \otimes \mathbf{1}\Big] ,
\end{aligned} \tag{15}$$

which gives a Hamiltonian H that is different to the one presented in [9].

The symplectic geomety and geometric quantization approach discussed here is valid for all algebras with three generators such as $SU(1,1)$, $\mathcal{H}(4)$ and their q-deformed counterparts. The algebraic maps among these algebras can be obtained from their related manifolds or operator equations upon prequantization and geometric quantization.

Acknowledgements I would like to thank Professor H.Y. GUO for helpful instructions and discussions.

References

[1] V.G. Drinfeld, *Sov. Math. Dokl.* **32**(1985)254; *Proc. I.C.M. Berkeley* (1986)798-820.

[2] M. Jimbo, *Lett. Math. Phys.* **10**(1985)63; **11**(1986)247; *Comm. Math. Phys.* **102**(1986) 537.

[3] L.D. Faddeev, N. Reshetikhin and L. Takhtajan, LOMI preprint E-14-87, 1987, Leningrad; L.D. Faddeev, in Les Houches Lectures, 1982.

[4] S. Woronowicz, *Comm. Math. Phys.* **111**(1987) 613.

[5] Z. Chang, W. Chen and H.Y Guo, *J. Phys.* **A 23**(1990)4185.
Z. Chang, W. Chen, H.Y Guo and H. Yan, *J. Phys.* **A 23**(1990)5371.
Z. Chang, W. Chen, H.Y Guo and H. Yan, *J. Phys.* **A 24**(1991)1427.
Z. Chang, S.M. Fei, H.Y. Guo and H. Yan, *J. Phys.* **A 24**(1991)5435.

[6] S.M. Fei and H.Y. Guo, *J. Phys.* **A24**(1991)1-10.
S.M. Fei and H.Y. Guo, *Commun. Theor. Phys.* **16**(1991)79-88.
S.M. Fei, *J. Phys.* **A 24**(1991)5195-5214.

[7] J.Sniatychi, *Geometric Quantization and Quantum Mechanics*, Springer-Verlag, 1980.

[8] T.L. Curtright, G.I. Ghandour and C.K. Zachos, *J. Math. Phys.* **32**(1991)3; T.L. Curtright and C.K. Zachos, *Phys. Lett.* **B 243**(1990)237.

[9] F. V. Pasquier and H. Saleur, *Nucl. Phys.* **B 330** (1990), 523.
A.B. Zamolodchikov and V.A. fateev, Sov. *J. Nucl. Phys.* **32**(1980)2.
P.P. Kulish and E.K. Sklyanin, *J . Phys.* **A 24**(1991)L435.
M.T. Batchelor, L. Mezincescu, R.I. Nepomechie and V. Rittenberg, *J. Phys.* **A 23**(1990)L141.

Quantum Group Symmetric Bargmann Fock Construction

Achim Kempf*
Sektion Physik der Ludwig-Maximilians-Universität
Theresienstr.37, 8000 München 2 / FRG
E-mail: kempf@ls-wess.physik.uni-muenchen.dbp.de

Abstract

Usually in quantum mechanics the Heisenberg algebra is generated by operators of position and momentum. The algebra is then represented on an Hilbert space of square integrable functions. Alternatively one generates the Heisenberg algebra by the raising and lowering operators. It is then natural to represent it on the Bargmann Fock space of holomorphic functions. In the following I show that the Bargmann Fock construction can also be done in the quantum group symmetric case. This leads to a 'q- deformed quantum mechanics' in which the basic concepts, Hilbert space of states and unitarity of time evolution, are preserved.

1 Introduction

There are already several approaches to q- deformed algebras of raising and lowering operators in the literature [see e.g. 6,7,8,9,11]. Let us define a 'q- deformed' Heisenberg algebra generated by the operators ${a^\dagger}_i$ and a^i by imposing the following commutation relations which were shown to be preserved under the action of the quantum group $SU_q(n)$ [4,6]:

$$a^r a^s + cR^{sr}_{ji} a^j a^i = 0 \quad , \quad a^r {a^\dagger}_s + 1/cR^{rj}_{is} {a^\dagger}_j a^i = \delta^r_s \quad , \quad {a^\dagger}_r {a^\dagger}_s + cR^{ji}_{sr} {a^\dagger}_j {a^\dagger}_i = 0 \tag{1}$$

The R- matrix reads [for the foundations see e.g. 1,2]:

$$R = q \sum_i e^i_i \otimes e^i_i + \sum_{i \neq j} e^i_i \otimes e^j_j + (q - 1/q) \sum_{i>j} e^i_j \otimes e^j_i \tag{2}$$

*supported by German National Scholarship Foundation

with $e^i_j \in M_n(C)$ matrix units. In the bosonic case we have $c = -1/q$ and in the fermionic case $c = q$, with q a real number. Other commutation relations are possible but the above relations are natural here in the sense that the resulting algebra deviates the least possible from the undeformed case [see 4]. Using the generalized twisting method introduced in [10] it is easy to see, that no further continous parameters can be introduced.

Since the $a^\dagger{}_i$ and a^i will be represented as adjoint operators we can recover hermitean position and momentum operators:

$$x^i := (2m\omega)^{-1/2}(a^\dagger_i + a^i) \quad , \quad p^i := i(2m\omega)^{1/2}(a^\dagger_i - a^i) \tag{3}$$

In the fermionic case one may consider a q- deformed system of n spin 1/2's in a constant magnetic field in z- direction. With similar linear combinations one then gets the x- and y- components of the i'th spin.

We define the ground state as usual $< 0|0 >= 1$ and $a^i|0> = 0$ for $i = 1, ..., n$. The scalar product $<,>$ can then be shown to be still positiv definit.

For the simplest Hamiltonian $H := \omega N$ with $\omega \in \mathbb{R}^+$, which is obviously scalar transforming and hermitean, the energy spectrum is:

$$E_p = \omega(1 + c^{-2} + c^{-4} + ... + c^{-2(p-1)}) = \omega\frac{c^{-2p} - 1}{c^{-2} - 1} \quad \text{with} \quad p = 0, 1, 2, ... \tag{4}$$

2 Bargmann Fock representation

The operators a^i and $a^\dagger_i$ can be represented as differentiation and multiplication operators on a space of (deformed) holomorphic functions [4,5]:

$$\rho : a^i \rightarrow \partial_{\bar{\eta}}{}^i \quad , \quad \rho : a^\dagger{}_i \rightarrow \bar{\eta}_i$$

Wave functions are the polynomials exclusively in $\bar{\eta}$'s , which shall also be denoted holomorphic functions. The bar operation is an antialgebra mapping i.e. we have for example $\overline{\partial_{\bar{\eta}}{}^i\bar{\eta}_j} = \eta^j\partial_{\eta_i}$. The commutation relations among the $\eta^i, \bar{\eta}_i, \partial_{\eta_i}$ and $\partial_{\bar{\eta}}{}^i$ are not unique. I use a choice [4] that is natural in the above mentioned sense, and which is similar but not identical to the differential calculus of Wess and Zumino [3].

The *evaluation* of differentiation is defined as follows: All $\partial_{\bar{\eta}}$'s are to be commuted to the right, the ∂_η's to the left. When they arrive, the corresponding terms are to be set equal zero. What remains is the value of the differentiation.

Let us define the 'c- deformed' exponential function

$$e_c{}^{(\partial_{\eta_i}\partial_{\bar{\eta}}{}^i)} := \sum_r \frac{(\partial_{\eta_i}\partial_{\bar{\eta}}{}^i)^r}{[r]_c!} \quad \text{with} \quad [r]_c := \frac{c^{2r} - 1}{c^2 - 1}. \tag{5}$$

The scalar product $(,)$ of the wave functions ϕ and ψ (which are polynomials in $\bar{\eta}$) is then given [4] by the following "integral":

$$(\phi, \psi) := (\bar{\phi} e_c{}^{(\partial_{\eta_i}\partial_{\bar{\eta}}{}^i)}\psi)_{\text{evaluated at } \eta=0=\bar{\eta}} \tag{6}$$

The evaluation procedure is as follows: At first the differentiations are to be evaluated. Then all terms containing η's and $\bar{\eta}$'s are to be set equal zero. Thus the result is a number. The new integral, which can of course also be used in the undeformed theory, has the same evaluation procedure for the bosonic as for the fermionic case.

It can be shown, that the operators a^k and $a^\dagger_k$ are adjoint in respect to $(,)$ i.e. $(\rho(a^\dagger{}_k)\phi, \psi) = (\phi, \rho(a^k)\psi)$ and that the wave function of the ground state, 1, is normalized. Thus the scalar product $(,)$ coincides with the bracket $<,>$ of the Fock space. The Hilbert space of all wave functions is defined to be the set of power series in $\bar{\eta}$'s that are square integrable in respect to $(,)$. Let us introduce a more familiar notation for the scalar product:

$$\int d\bar{\eta} d\eta \bar{\phi} e_c^{(\partial_{\eta^i} \partial^i_{\bar{\eta}})} \psi := (\bar{\phi} e_c{}^{(\partial_{\eta_i} \partial_{\eta}{}^i)} \psi)_{\text{evaluated at } \eta=0=\bar{\eta}} \tag{7}$$

Like in the undeformed case it is now possible to represent every operator P on the Fock space also as an integral kernel $G_P(\bar{\eta}', \eta)$:

$$\int d\bar{\eta} d\eta G_P(\bar{\eta}', \eta) e_c^{(\partial_{\eta^i} \partial^i_{\bar{\eta}})} \psi(\bar{\eta}) = P\psi(\bar{\eta}') \tag{8}$$

With natural commutation relations between different copies of the function space (e.g. primed and unprimed) the general rule for getting the integral kernel of an arbitrary normal ordered operator $: P(a^\dagger, a) :$ turns out to be as follows: Starting with the integral kernel of the identity operator $e_c^{(\bar{\eta}'_i \eta^i)}$ one writes for each $a^\dagger_i$ a $\bar{\eta}'_i$ to the left of $e_c^{(\bar{\eta}'_i \eta^i)}$ and for all a^i one writes η^i to the right of the $e_c^{(\bar{\eta}'_i \eta^i)}$.

The green function i.e. the integral kernel of the time evolution operator $U = e^{-i(t_f - t_i)H}$ for the simple Hamiltonian of section 1 is found to be

$$G_U = \sum_{r=0}^{\infty} \frac{(\bar{\eta}'_i \eta^i)^r}{[r]_c!} e^{-i\omega(t_f - t_i)[r]_{1/c}} \tag{9}$$

3 Introduction of driving forces

Let us introduce a 'classical' but q-deformed driving force $f(t)g^i$ where $f(t)$ denotes a complex- valued function describing the time dependence of the driving force and g shall be a constant unit vector: $\bar{g}_i g^i = 1$. The Hamiltonian now reads:

$$H = \omega \bar{\eta}_i \partial^i_{\bar{\eta}} - \bar{f}(t) \bar{g}_i \partial^i_{\bar{\eta}} - f(t) \bar{\eta}_i g^i \tag{10}$$

Considering quantum mechanics as quantum field theory with zero space dimensions, our driving forces are q-deformed Schwinger sources. The algebra of the g's and $\bar{g}$'s is now noncommutative, even in the bosonic case [5]. Nevertheless, results that do only depend on the length of the force vector are still ordinary complex numbers for that the usual probability interpretation applies. E.g. the vacuum-vacuum transition amplitude for the switching on of a constant driving force can be calculated to be

$$\begin{aligned} < 0|0(t_f,t_i) > &= 1+\frac{f\bar{f}}{\omega^2}\sum_{z=2}^{\infty}\frac{(-i\omega)^z}{z!}(t_f-t_i)^z \\ &+ \frac{(f\bar{f})^2}{\omega^4}\sum_{z=4}^{\infty}\left\{z-3+\sum_{r=0}^{z-4}(z-3-r)([2]_{1/c})^{r+1}\right\}\frac{(-i\omega)^z}{z!}(t_f-t_i)^z \\ &+ \ldots \end{aligned} \qquad (11)$$

One recognizes that deviations from the undeformed case do not occur before the second order in $(f\bar{f})$. This is plausible because the energy level of the first excitation is not deformed, while the levels (see Eq.4) deviate the more, the higher the excitation is. If it is possible to extend this formalism to a quantum field theory it can be expexted that the ultraviolet behaviour is strongly influenced by the deformation parameter.

Acknowledgement:

I would like to thank very much Prof. Jian-min Shen and Prof. Mo-Lin Ge for the kind invitation to the conference and the generous hospitality.

4 References

1. L.D.Faddeev, N.Yu.Reshetikhin, L.A.Takhtajan, Alg.Anal.1,1,(1989)178-206
2. S.Majid, Int.J.Mod.Phys. A. Vol. 5, No 1 (1990) 1-91
3. J.Wess, B. Zumino, Nucl. Phys. Proc. Suppl. 18B (1991) 302
4. A.Kempf, Preprint LMU-TPW 92-4, submitted to Lett. Math. Phys.
5. A.Kempf, Preprint LMU-TPW 92-7, submitted to J. Math. Phys.
6. W.Pusz, S.Woronowicz, Rep. Math. Phys. 27 (1989) 231.
7. L.Biedenharn, J. Phys. A 22 (1989) L 873
8. A.Macfarlane, J. Phys. A 22 (1989) 4581
9. D.B.Fairlie, C. Zachos, preprint ANL-HEP-CP-91-28 (1991)
10. A.Kempf, in Proc. XX DGM- Conference, eds. S. Catto, A. Rocha (World Scientific, Singapore, 1991) p.546
11. M.Chaichian, P. Kulish, J. Lukierski, Phys. Lett. B 262 (1991) 43

COMMUTANTS OF QUANTUM ALGEBRAS AND THEIR APPLICATION IN PHYSICS

H.C. LEE

Theoretical Physics Branch, Chalk River Laboratories, AECL Research
Chalk River, Ontario, Canada K0J 1J0
and Department of Applied Mathematics, University of Western Ontario
London, Ontario, Canada N6A 5B9

ABSTRACT

We summarize recent results[1] on a functor that maps tangles to commutants of a quasitriangular Hopf algebra. In certain cases, images of *(1,1)*-tangles are new Casimir operators of the algebra whose eigenvalues are link invariants. Some applications to physics are briefly discussed.

1. A Functor Mapping Tangles to a Quantum Algebra

1.1. Tangles. An *oriented* (n, n)*-tangle* T is the disjoint union of n open oriented strands and an arbitrary number of oriented closed strands embedded in a cylinder in a 3-manifold, with all the n tails of the open strands held fixed on the ceiling of the cylinder and all the n tips held fixed on the floor. A *link* is a $(0,0)$-tangle. A *tangle diagram* is a regular two-dimensional projection of T. It is composed of *positive* () and *negative* () *crossings* and *edges* ($\downarrow$). An *ambient (regular, resp.) isotopy class* of tangle diagrams is defined by the equivalence relations generated by the three (two, resp.) Reidemeister moves[2], I, II and III (II and III, resp.) on tangle diagrams, generalized for oriented tangle diagrams. Henceforth the adjectives "oriented" and "diagrams" will normally be suppressed.

1.2. Quasitriangular Hopf algebra. A quasitriangular Hopf algebra[3–5] is a noncommutative, noncocommutative Hopf algebra $\{\mathcal{A},\ S,\ m,\ \triangle,\ \epsilon\}$ (resp. algebra over $\mathbf{C}[\![\eta]\!]$, antipode, multiplication, comultiplication, counit) equipped with a universal invertible R-matrix $\mathcal{R} \in \mathcal{A} \otimes \mathcal{A}$, $\mathcal{R}^{-1} = (S \otimes id)\mathcal{R} = (id \otimes S^{-1})\mathcal{R}$, satisfying $\forall \alpha \in \mathcal{A}$: (i) $m(id \otimes S)\triangle(\alpha) = m(S \otimes id)\triangle(\alpha) = \epsilon(\alpha)\mathbf{1}$; (ii) $(\mathcal{T} \circ \triangle(\alpha))\mathcal{R} = \mathcal{R}\triangle(\alpha)$; (iii) $(\triangle \otimes id)\mathcal{R} = \mathcal{R}_{13}\mathcal{R}_{23}$, $(id \otimes \triangle)\mathcal{R} = \mathcal{R}_{13}\mathcal{R}_{12}$ (iv) $\mathcal{R}_{12}\mathcal{R}_{13}\mathcal{R}_{23} = \mathcal{R}_{23}\mathcal{R}_{13}\mathcal{R}_{12}$. We call the algebra $\mathcal{U} \supset \{\mathcal{A},\ \mathcal{A} \otimes \mathcal{A},\ \mathcal{A} \otimes \mathcal{A} \otimes \mathcal{A},\ \cdots\}$ a quantum algebra, and define $\mathcal{U}_\triangle \supset \{\mathcal{A},\ \triangle\mathcal{A},\ (id \otimes \triangle)\triangle\mathcal{A},\ \cdots\}$. Let $\mathcal{I}$ be the linear subspace spanned by $ab - ba$, $\forall a, b \in \mathcal{A}$. Let $\mathcal{A}_0$ be the quotient space $\mathcal{A}/\mathcal{I}$, then $Sp: \mathcal{A} \to \mathcal{A}_0$ is a linear map such that $Sp(ab) = Sp(ba)$, $\forall a, b \in \mathcal{A}$. We consider Sp as part of the structure of $\mathcal{U}$. For generic η, $\mathcal{R}$ is sometimes an infinite formal sum in $\mathcal{A} \otimes \mathcal{A}$, which is well defined provided $\mathbf{C}[\![\eta]\!]$ is given an η-adic topology[6]. In $\mathcal{U}$, there are two invertible elements $v \equiv m((id \otimes S)\mathcal{R})$ and $v' \equiv S(v) = m((S \otimes id)\mathcal{T} \circ \mathcal{R})$ that satisfy $\forall a \in \mathcal{A}$: $vav^{-1} = S^{-2}(a)$; $v'av'^{-1} = S^{2}(a)$; $[vv' = v'v \equiv \lambda, a] = 0$.

1.3. The functor $\mathcal{V}$. A ***functor $\mathcal{V}$ for tangles*** is a mapping of ambient $[T]$ or regular $[T]_r$ classes of tangles to an algebraic structure. It turns out that, in order to construct $\mathcal{V}$ it is also necessary to identify the *wrong-way* edges in a tangle. For a generic tangle, one way to do this is to examine the Seifert circles in a tangle[1]; on each such circle there is one and only wrong-way edge, denoted by $\downarrow^{\triangleright}$ ($\downarrow^{\triangleleft}$, resp.) if it is on a clockwise (counterclockwise, resp.) Seifert circle. It can be shown[1] that every tangle is regularly isotopic to a partially closed braid, called a ***braid-tangle***. The wrong-way edges on a braid-tangle are just the edges that (partially) close the braid. Henceforth we denote T as a tangle with its wrong-way edges identified and marked as $\downarrow^{\triangleright}$ or $\downarrow^{\triangleleft}$. We treat the edges and crossings as single or crossed ***arrows***. Let α, β, ... be arrows. The tensor product $\alpha \otimes \beta$ is the disjoint union of the two arrows. The multiplication p adjoins two arrows: $p(\alpha \otimes \beta) = \alpha\beta$. Successive multiplications yield an open *string* of arrows: $p_{12}(p_{13}(\alpha \otimes \beta \otimes \gamma)) = p_{12}(\alpha\gamma \otimes \beta) = \alpha\gamma\beta$. Given an open string $\tau = \alpha_1\alpha_2\cdots\alpha_k$, denote a closed string $\hat{\tau}_{\{\alpha\}}$ to be the class $\{\alpha_{s_1}\alpha_{s_2}\cdots\alpha_{s_k} | (s_1, s_2, \ldots, s_k) \in \{\textit{all cyclic perm. of } (1, 2, \ldots, k)\}\}$. We define a functor $\mathcal{V}$ whose actions on the arrows in T are: $\mathcal{V}(\downarrow) = 1$; $\mathcal{V}(\downarrow^{\triangleleft}) = v$; $\mathcal{V}(\downarrow^{\triangleright}) = v'$; $\mathcal{V}(\times) = R$; $\mathcal{V}(\times') = R'$; $\mathcal{V}(\alpha \otimes \beta) = \mathcal{V}(\alpha) \otimes \mathcal{V}(\beta)$; $\mathcal{V}(\alpha\beta) = \mathcal{V}(\alpha)\mathcal{V}(\beta) = m(\mathcal{V}(\alpha) \otimes \mathcal{V}(\beta))$; $\mathcal{V}(\hat{\tau}) = Sp(\mathcal{V}(\tau_*)) \in \mathcal{A}_0$, $\tau_* \in \hat{\tau}$. Let $\{T\}$ be the set of all tangles and $[T_l^{(n)}]_r$ be a regular isotopy of (n, n)-tangles with l closed strands. Then $\mathcal{V}$ is a functor for regular isotopic tangles:

$$\mathcal{V}: \{T\} \to \mathcal{U}, \mapsto \mathcal{V}[T_l^{(n)}]_r = \mathcal{V}(T) \in \mathcal{A}^{\otimes n} \otimes \mathcal{A}_0^{\otimes l}, \forall T \in [T_l^{(n)}]_r. \tag{1}$$

1.4. Representation of $\mathcal{V}$. Let V be the vector space supporting $\mathcal{A}$ and let $\pi \in End(V)$ be an irreducible representation of $\mathcal{A}$. Then $\pi(\lambda) = \lambda_\pi \mathbf{1}_\pi$. Suppose $\lambda_\pi \neq 0$. Define $h_\pi \equiv {\lambda_\pi}^{-1/2}\pi(v)$, $h_\pi' \equiv {\lambda_\pi}^{-1/2}\pi(v)'$, $\mathcal{R}_\pi \equiv (\pi \otimes \pi)\mathcal{R}$, *etc.* Let Tr_π be the matrix trace in π. Then $\mathcal{V}_\pi'$ defined as $\mathcal{V}$ with the set $\{\mathcal{R}, \mathcal{R}', v, v', Sp\}$ substituted by $\{\mathcal{R}_\pi, \mathcal{R}_\pi', h_\pi, h_\pi', Tr_\pi\}$ is obviously a functor that sends $[T_l^{(n)}]_r$ to $(\pi(\mathcal{A}))^{\otimes n}$. Note that $Sp: \mathcal{A} \to \mathcal{A}_0$ but $Tr_\pi: \mathcal{A} \to \mathbf{C}$. Furthermore, if $w(T)$ is the writhe number of a tangle T, then $\mathcal{V}_\pi[T] \equiv {\lambda_\pi}^{w(T)/2}\mathcal{V}_\pi'(T)$ is an invariant for ambient isotopic tangles.

2. Commutants and Tangle Casimir Operators of Quantum Algebras

2.1. The quantum algebra $\tilde{\mathcal{U}}$. Consider a more restrictive type of $\mathcal{U}$ denoted by $\tilde{\mathcal{U}}$, which is a quantum algebra with properties: (*i*) Generated in the Chevalley basis by $\{H_i, X_i^+, X_i^-;\ i = 1, \ldots\}$, with the H_i's generating the Cartan subalgebra;

(ii) $\Delta H_i = H_i \otimes 1 + 1 \otimes H_i$, $\Delta X_i^{\pm} = X_i^{\pm} \otimes q^{H_i/2} + q^{-H_i/2} \otimes X_i^{\pm}$; (iii) $S(H_i) = -H_i$, $S(X_i^{\pm}) = -q^{H_i/2} X_i^{\pm} q^{-H_i/2}$. Examples of $\tilde{\mathcal{U}}$ include $\mathcal{U}_q(g)$ over $\mathbf{C}[\![\eta]\!]$, $q \equiv e^{\eta}$, where g is a simple complex Lie algebra, or is the graded $sl(M|N)$.

2.2. Every $\mathcal{V}[T]_r$ almost commutes with $\tilde{\mathcal{U}}_{\Delta}$. Let $\sigma: T_l^{(n)} \to S_n$ be the natural mapping of tangle diagrams to the symmetry groups. The map does not depend on the number l of the closed strands in $T_l^{(n)}$. Let $\Delta^{\{1\}} = \Delta$ and define, for $n \geq 2$, $\Delta^{\{n\}} = (id \otimes id \cdots \otimes \Delta)\Delta^{\{n-1\}} = (id^{\otimes(n-1)} \otimes \Delta)\Delta^{\{n-1\}}$. By definition, $\mathcal{A}^{\{n\}} \equiv \Delta^{\{n-1\}}\mathcal{A} \in \tilde{\mathcal{U}}_{\Delta} \subset \tilde{\mathcal{U}}$. Then, for any $T_l^{(n)}$,

$$\left(\Delta'^{\{n-1\}}(\alpha) \otimes 1^{\otimes l}\right) \mathcal{V}[T]_r = \mathcal{V}[T]_r \left(\sigma(T) \circ \Delta'^{\{n-1\}}(\alpha)\right) \otimes 1^{\otimes l}, \quad \forall \alpha \in \mathcal{A}. \tag{2}$$

This generalizes the coboundary condition $(\mathcal{T} \circ \Delta(\alpha))\mathcal{R} = \mathcal{R}\Delta(\alpha)$.

2.3. Commutants of $\tilde{\mathcal{U}}_{\Delta}$. Let $\mathcal{P}: [T_l^{(n)}] \to S_n$ with $\mathcal{P}[T_l^{(n)}]$ acting on $V^{\otimes n}$. Then

$$[(\mathcal{T}_n \circ \mathcal{V}[T_l^{(n)}]_r)\mathcal{P}^{-1}(T_l^{(n)}), \Delta^{\{n-1\}}(\alpha) \otimes 1^{\otimes l}] = 0, \qquad \forall \alpha \in \mathcal{A}. \tag{3}$$

2.4. Tangle Casimir operators of $\tilde{\mathcal{U}}_{\Delta}$. For T a *(1,1)*-tangle with l closed strands, we call $\mathcal{V}[T]_r$ a *tangle Casimir operator* of $\tilde{\mathcal{U}}_{\Delta}$; it commutes with $\mathcal{A} \otimes 1^{\otimes l}$. As an example, for $T_{trefoil}$ the tangle that closes to a trefoil, $\mathcal{V}[T_{trefoil}]_r = a_s b_t a_r v b_s a_t b_r$ commutes with $\mathcal{A}$. Generally, in an irreducible representation π of $\tilde{\mathcal{U}}$, for T a *(1,1)*-tangle with any number of closed strands,

$$\mathcal{V}_{\pi}[T] = Q_{\pi}[T]1_{\pi}, \tag{4}$$

where $Q_{\pi}[T]$ is a $\mathbf{C}$-valued eigenvalue.

2.5. Link invariants are eigenvalues of tangle Casimir operators. Let T be a *(1,1)*-tangle and $\hat{T}$ its closure, which is a link. Then $Q_{\pi}[T]$ is a link invariant for $[\hat{T}]$ if (a) $Tr_{\pi}(h_{\pi}) \neq 0$, or (b) the multiplicities in the decomposition of $\pi \otimes \pi$ are not greater than one and $\mathcal{V}_{\pi}$ is sufficiently nontrivial. Condition (a) is met by all representations of $\mathcal{U}_q(g)$, g a simple complex Lie algebra as well as $g = sl(n|m)$, $n \neq m$, for generic q. It is not met by $\mathcal{U}_q(sl(m|m))$. Condition (b) is met by all the fundamental representations of the quantum algebras mentioned above.

3. Some Applications in Physics

3.1. Conserved topological charges in 2d QFT. Recently it has been shown[7,8] that some integrable quantum field theories in two dimensions have explicit quantum algebra symmetries. Such systems should have an infinite number of conserved quantities, called topological charges, one for each distinct tangle Casimir operator. In representations of the quantum algebra that satisfy the conditions in §*2.5*, these conserved topological charges should be link invariants.

3.2. The Alexander-Conway polynomial. For the fundamental 2×2 representation of $\mathcal{U}_q(sl(1|1))$, $Tr_\pi(h_\pi) = 0$ so $\mathcal{V}_\pi[L] = 0$ for all links. But Q_π satisfies[9,10] the skein relation, $Q_\pi[\times] + (q - q^{-1})Q_\pi[)(] - Q_\pi[\times] = 0$, and therefore gives the Alexander-Conway polynomial[11].

3.3. Quantum holonomy and Chern-Simons theory in 3d. Recently[9] it was shown that in a quantum gauge theory, the group-valued quantum holonomy, $\Phi(C_x) = \langle P \exp(i \oint_{C_x} A dx)\rangle$, where C_x is a closed contour with initial point $x \in C$, is gauge invariant and x-independent. This implies that for the Chern-Simons theory in three dimensions, $\Phi(L_x)$ is a link invariant. Earlier Witten[12] has already shown that the the Wilson loop, or character-valued $W_\pi = Tr_\pi(\Phi[L]))$ is a link invariant. Thinking of L_x as a *(1,1)*-tangle, the pair $\{\Phi[L_x], W_\pi[L]\}$ to 3d CST is what $\{\mathcal{V}[L_x], \mathcal{V}_\pi[L]\}$ of §*2.3*, *2.4* is to a quantum algebra. If in an irreducible π,

$\pi(\Phi[L_x]) = P_\pi[L]\mathbf{1}_\pi$, then $W_\pi[L] = P_\pi(L)Tr_\pi(\mathbf{1}_\pi)$. An application of this result is that for the gauge group $SL(M|M)$, $W_\pi[L] \equiv 0$ while $P_\pi[L] \neq 0$ is the Alexander-Conway polynomial[9].

3.4. Quantum spin chains. For a *(2,2)*-tangle $T^{(2)}$, $\check{\mathcal{H}} \equiv \mathcal{P}(T^{(2)})\mathcal{V}_\pi(T^{(2)})$ commutes with $\pi \otimes \pi(\triangle\mathcal{A})$. Consider a tensor space of length N and let $\check{\mathcal{H}}_{i,i+1}$ act on the ith and $i+1^{st}$ spaces. Then the Hamiltonian for an open quantum spin chain with nearest neighbor interaction, $H = \sum_{i=1}^{N-1} \check{\mathcal{H}}_{i,i+1}$, is $\tilde{\mathcal{U}}_\triangle$ invariant and satisfies the first requirement for H to be exactly soluble[13]. (Exact solubility[14,15] also requires that $\check{\mathcal{H}}$ satisfies the Yang-Baxter equation[16,17].) More generally, for any (n,n)-tangle T, define $\check{\mathcal{H}}^{(n)} \equiv (T_n \circ \mathcal{V}_\pi[T])\mathcal{P}^{-1}(T)$, then the Hamiltonian for an open quantum spin chain with $(n-1)^{th}$ nearest neighbor interactions, $H^{(n)} = \sum_{i=1}^{N-n+1} \check{\mathcal{H}}^{(n)}_{i,i+1,\cdots,i+n-1}$, is $\tilde{\mathcal{U}}_\triangle$ invariant. As well, being itself $\tilde{\mathcal{U}}_\triangle$ invariant, $\check{\mathcal{H}}^{(n)}$ also satisfies one of the basic requirements for being a transfer matrix.

For π being the fundamental 2×2 representation of $\mathcal{U}_q(sl(2))$, $\pi \otimes \pi = \underline{3} \oplus \underline{1}$, and $\check{\mathcal{H}}$ has two eigenvalues $\epsilon_{\underline{3}}$ and $\epsilon_{\underline{1}}$. Expressed in terms of Pauli matrices,

$$H_{sl(2)} = \frac{1}{q^{-1}+q}\left((\epsilon_{\underline{3}} - \epsilon_{\underline{1}})H_{PS} + \frac{q^{-1}+q}{4}(3\epsilon_{\underline{3}} + \epsilon_{\underline{1}})\right), \tag{5}$$

where $H_{PS} = \sum_{i=1}^{N-1}(\sigma_i^+\sigma_{i+1}^- + \sigma_i^-\sigma_{i+1}^+ + \frac{q^{-1}+q}{4}\sigma_i^3\sigma_{i+1}^3 + \frac{q^{-1}-q}{4}(\sigma_i^3 - \sigma_{i+1}^3))$ is the soluble Hamiltonian of Pasquier and Saleur[13] for an open spin chain. The infinite classes of *(2,2)*-tangles have, in this case, projected onto two degrees of freedom that affect $H_{sl(2)}$ trivially: (i) an global scale factor; and (ii) a constant term. In particular, H_{PS} has no dependence on the eigenvalues. Hence the Hamiltonian derived from any *(2,2)*-tangle (whose image in $\mathcal{U}_q(sl(2))$ has two nondegenerate eigenvalues) is soluble; changing the Hamiltonian associated with one tangle to a Hamiltonian associated with another tangle amounts to renormalizing the Hamiltonian in such a way that the structure of the energy spectrum is unchanged.

The situation for $\mathcal{U}_q(sl(1|1))$ is similar: $\pi \otimes \pi = \underline{2} \oplus \underline{2}'$ and,

$$H_{sl(1|1)} = \frac{1}{q^{-1}+q}\left((\epsilon_{\underline{2}} - \epsilon_{\underline{2}}')H_{FW} + \frac{q^{-1}+q}{2}(\epsilon_{\underline{2}} + \epsilon_{\underline{2}}')\right), \tag{6}$$

where $H_{FW} = \sum_{i=1}^{N-1}(\sigma_i^+\sigma_{i+1}^- + \sigma_i^-\sigma_{i+1}^+ + \frac{1}{2}(q^{-1}\sigma_i^3 + q\sigma_{i+1}^3))$ is the Hamiltonian for free-fermions[18]. We see that the $\mathcal{U}_q(sl(2))$ and $\mathcal{U}_q(sl(1|1))$ symmetries for open quantum spin chains are extremely robust. This probably will not be the case for higher representations of these quantum algebras and for larger quantum algebras.

This work is supported in part by a grant from the Natural Sciences and Engineering Research Council of Canada. The author thanks the hospitality of the Institute of Theoretical Physics, *CAS*, Beijing, where part of this work was carried out.

Bibliography

1. H.C. Lee, Int. J. Mod. Phys., **A7, Supp. 1B** (1992) 581-610; *Commutants and new Casimir operators of quasitriangular Hopf algebras*, (*to be published*).
2. K. Reidemeister, *Knotentheorie*, (Chelsea, 1948).
3. V.G. Drinfel'd, Proc. Int. Cong. Math., Berkeley, **1** (Academic Press, 1986) 798-820.
4. M. Jimbo, Lett. Math. Phys., **10** (1985) 63-69.
5. L.D. Faddeev, N.Yu. Reshetikhin and L.A. Takhtajan, Algebraic Analysis, **1** (1988) 129-139.
6. V.G. Drinfel'd, Soviet Math. Doklady, **32** (1985) 254-258.
7. C. Gomez and G. Sierra, Nucl. Phys., **B352** (1991) 791-828.
8. B. Bernard and A. LeClair, Comm. Math. Phys., **142** (1991) 99-128.
9. H.C. Lee and Z.Y. Zhu, Phys. Rev., **D44** (1991) 942-945.
10. L. Kauffman and H. Saleur, Comm. Math. Phys, **141** (1991) 293-328.
11. J.W. Alexander, Proc. Nat'l Acad. Sci. (USA), **9** (1923) 93-95; J.H. Conway, *Computational problems in abstract algebra*, (Pergamon, New York, 1970) 329.
12. E. Witten, Comm. Math. Phys., **121** (1989) 351-399.
13. V. Pasquier and H. Saleur, Nucl. Phys., **B330** (1990) 523-556.
14. P.P. Kulish and E.K. Sklyanin, J. Phys., **A24** (1991) L435-L439.
15. L. Mezincescu and R.I. Nepomechie, Mod. Phys. Lett., **A6** (1991) 2497-2508.
16. C.N. Yang, Phys. Rev. Lett., **19** (1967) 1312-1314.
17. R.J. Baxter, *Exactly solved models in statistical mechanics*, (Academic Press, 1982).
18. C. Fan and F.Y. Wu, Phys. Rev., **B2** (1970) 723-733.

SELF YANG-BAXTERIZATION OF BRAID GROUP REPRESENTATION

YOU-QUAN LI
Zhejiang Institute of Modern Physics
Zhejiang University, Hangzhou 310027, China

ABSTRACT

On the basis of Yang-Baxter equation and the initial condition of $\check{R}$-matrix, Self Yang-Baxterization of representations of braid group are discussed systematically. It is shown that there exist many Yang-Baxterization formulae beyond the knowns. New developments of Yang-Baxterization approach are presented. As examples some new solutions of Yang-Baxter equation are given.

1. Introduction

Since Yang-Baxter equation plays a crucial role in the study of solvable models in two-dimensional statistical mechanics, quantum integrable systems and low dimensional quantum field theory etc.[1-4], it is undoubtedly significant to study solutions of Yang-Baxter equation. In principle, it is quite difficult to solve Yang-Baxter equation directly, because they are functional equations.

On the basis of quantum group[5], Jimbo[6] constructed a family of trigonometric solutions of quantum Yang-Baxter equation from the classical r-matrix of Belavin and Drinfeld[7] for the case of non-exceptional affine Lie algebra. In order to understand the connection between knot theory and solvable models, Jones[8] proposed a procedure, he called 'Baxterization', to produce solutions of Yang-Baxter equation from representations of braid group. He also gave explicit formulae[8] to Baxterize to trigonometric solutions for the Hacke case and Birman-Murakami-Wenzl case. Later, [9] is succeeded in constructing Jimbo-type trigonometric $\check{R}(x)$ matrix via a developed approach, called Yang-Baxterization. By employing a tensor product graph and its maximal trees, an explicit formula for Jimbo-type trigonometric solutions of Yang-Baxter equation is derived in [10].

In this paper, further developments of Yang-Baxterization approach are presented. In next section, we give some general considerations on Yang-Baxter equation. It is simple but very helpful for simplifying the discussion of consistency condition. In section 3, we systematically study self Yang-Baxterization of braid group representations. We find that, beyond the formulae in [9,10] there may exist many other formulae of Yang-Baxterization. As an example we derive a formula beyond the simplest case (the knowns). In section 4, we derive the consistency condition for Yang-Baxterization mainly for trigonometric case. Section 5 is the application of the above mentioned formulae, where some solutions of Yang-Baxter equation are given. Finally, in section 6 we give some discussions about the results of the paper.

2. Simple Properties

Conveniently quantum Yang-Baxter equation is written (in this paper we briefly say Yang-Baxter equation) as the following matrix form

$$\check{R}_1(u)\check{R}_2(u+v)\check{R}_1(v) = \check{R}_2(v)\check{R}_1(v+u)\check{R}_2(u) \tag{2.1}$$

where $u,\ v\ \in C$ are additive spectral parameters, $\check{R}_1 := \check{R}\otimes I$, $\check{R}_2 := I\otimes\check{R}$, $\check{R}\in End(C^N\otimes C^N)$ and I stands for an $N\times N$ unit matrix. Owing to the physical significance, whether for the cases of solvable models in two-dimensional statistical mechanics and chain models in one-dimensional quantum mechanics, or in the case of two-dimensional factorized scattering theory, $\check{R}(u)$ is subjected to satisfy the called initial condition.

$$\check{R}(0)\ \propto\ I \tag{2.2}$$

Let us observe (2.1) now. If a $\check{R}$ matrix under consideration satisfies condition (2.2), one can obtain from (2.1) by taking $v=-u$ that

$$\check{R}_1(u)\check{R}_1(-u) = \check{R}_2(-u)\check{R}_2(u) \tag{2.3}$$

This leads to

$$\check{R}(u)\check{R}(-u) = \check{R}(-u)\check{R}(u) = \rho(u)I \tag{2.4}$$

where $\rho(u)$ must be an even scalar function. Furthermore one may notice that (2.1) is an identity either $u=0$ or $v=0$ as long as initial condition (2.2) holds. These characteristics will be useful for the discussion of consistency of Yang-Baxterization in section 4.

It is known that S-matrix of a representation of braid group is a solution of Yang-Baxter relation

$$S_1S_2S_1\ =\ S_2S_1S_2 \tag{2.5}$$

which is the spectral parameter independent case of Yang-Baxter equation.

If there exists two scalar functions $\xi(u)$ and $\zeta(u)$ ($\zeta(u)$ is often a constant) such that

$$\lim_{u\to u_0}\rho(u)\xi(u)\zeta(u)\ =\ \rho,\ \text{ and }\ \lim_{u\to u_0}\check{R}(u)\zeta(u)\ =\ S$$

then

$$\lim_{u\to -u_0}\xi(-u)\check{R}(u)\ =\ \rho S^{-1}.$$

Obviously, when $u_0 = 0$ or ∞ , or $\check{R}(u)$ is parameterized by periodic (or double periodic) functions, it is possible to take (2.5) as a limit case of (2.1).

Our task in the following is to explore some route or procedure (easier than to solve (2.1) directly) to obtain solutions $\check{R}(u)$ of Yang-Baxter equation. Since Yang-Baxter relations (2.5) are algebraic equations and is certainly easier to solve than (2.1), it is significant to determine when it is possible and how to introduce a spectral parameter into a representation of braid group.

3. Yang-Baxterization Formulae

Yang-Baxterization is a procedure originated by Jones[8] to produce solutions of Yang-Baxter equation from a known representation of braid group. The aim of the procedure is to insert a spectral parameter into a given representation of braid group so that Yang-Baxter equation is satisfied and so that the original representation of braid group is the limit in the spectral parameter of Baxterized version.

In this section, we will derive a sequent of formulae starting from the initial condition i.e. (2.2) and (2.4) due to the latter is an immediate consequence of the former.

Given that the S-matrix of a representation of braid group has n distinct eigenvalues λ_i $(i = 1, 2, \cdots, n)$ and obeys a nth order reduction relation

$$\prod_{i=1}^{n}(S - \lambda_i) = 0 \tag{3.1}$$

then we have

$$S = \sum_{i=1}^{n} \lambda_i P_i \tag{3.2a}$$

where P_i is the projector onto subspace of λ_i and is solved uniquely to be

$$P_i = \prod_{j \neq i} \frac{(S - \lambda_j)}{(\lambda_i - \lambda_j)} \tag{3.2b}$$

We restrict ourselves to consider the case that $\check{R}(u)$'s commute among themselves for different values of u, i.e. $[\check{R}(u), \check{R}(u')] = 0$. This means that $\check{R}(u)$'s can be diagonalized (for real symmetric or complex without eigenvector of null norm) 'simultaneously'. This reminds us to consider

$$\check{R}(u) = \sum_{i=1}^{n} \Lambda_i(u) P_i \tag{3.3}$$

where $\Lambda_i(u)$ are scalar functions to be determined so that

$$\begin{aligned} \Lambda_i(u_0) &= \lambda_i \,, \qquad \Lambda_i(0) = \Lambda_j(0) \,, \\ \Lambda_i(u)\Lambda_i(-u) &= \Lambda_j(u)\Lambda_j(-u) \quad \text{for } i \neq j \end{aligned} \tag{3.4}$$

These conditions come from the requirements of $\check{R}(u_0) = S$, (2.2) and (2.4) respectively.

It is found without much difficult that

$$\Lambda_i(u) = \prod_{j=1}^{i-1}(\frac{\lambda_j}{\lambda_{j+1}} f_j(u) + 1) \prod_{k=i}^{n-1}(f_k(u) + \frac{\lambda_k}{\lambda_{k+1}}) \cdot \lambda_n \tag{3.5}$$

obeys (3.4), where $f_l(u)$ $(l = 1,\ 2,\ \cdots,\ n-1)$ are any such functions that

$$f_l(u_0) \ = \ 0, \quad f_l(0) \ = \ 1,$$
$$f_l(-u) = f_l^{-1}(u). \tag{3.6}$$

(i) If consider $f_l(u) = y(u)$ for all l, (3.5) becomes

$$\Lambda_i(u) = \prod_{j=1}^{i-1}(\frac{\lambda_j}{\lambda_{j+1}}y(u)+1)\prod_{k=i}^{n-1}(y(u)+\frac{\lambda_k}{\lambda_{k+1}})\cdot\lambda_n \tag{3.7}$$

This is the well known Jimbo-type case[9]. It is worthwhile mentioning that the permutation of n known concrete eigenvalues of a given S matrix can be arbitrary, i.e. it can be any permutation from a given order via actions of permutation group S_n. But one may notice that (3.7) is invariant under

$$\pi' : \ (\lambda_1,\ \lambda_2,\ \cdots,\lambda_{n-1},\ \lambda_n) \ \mapsto \ (\lambda_n,\ \lambda_{n-1},\ \cdots,\lambda_2,\ \lambda_1) \tag{3.8}$$

So we need not to consider situations related to all elements of S_n, but only need to consider those related to the coset representatives with respect to subgroup $\{e,\ \pi'\}$. From (3.7), (3.3) and (3.2b), one can write out the following formulae

a).The case with two eigenvalues

$$\check{R}(u) \ = \ S + \lambda_1\lambda_2 y(u)S^{-1} \tag{3.9}$$

An explicit formula for this case was early carried out in [8].

b). The case with three eigenvalues

$$\check{R}(u) = L(u)S + M(u)I + N(u)S^{-1} \tag{3.10a}$$

where

$$L(u) = 1 - y(u)$$
$$M(u) = (\lambda_1+\lambda_2)(\lambda_2+\lambda_3)\lambda_2^{-1}y(u)$$
$$N(u) = \lambda_1\lambda_3 y(u)[y(u)-1] \tag{3.10b}$$

c). The case with four eigenvalues

$$\check{R}(u) = K(u)S^2 + L(u)S + M(u)I + N(u)S^{-1} \tag{3.11a}$$

where

$$K(u) = y(u)[\ y(u)-1\ ]\alpha, \quad L(u) = [\ 1-y(u)\ ][\ 1+(\alpha\sigma_1+\beta)y(u)\ ],$$
$$M(u) = y(u)[\ (\alpha\sigma_2+\lambda_1\lambda_4\gamma)y(u)+\beta(\lambda_2+\lambda_3)(y(u)+1)+\lambda_2\lambda_3\gamma\ ],$$
$$N(u) = y(u)[\ y(u)-1\ ][\ \lambda_1\lambda_4 y(u)+\lambda_2\lambda_3\beta\ ],$$

$$\alpha = \frac{\lambda_1\lambda_3 - \lambda_2\lambda_4}{\lambda_2\lambda_3(\lambda_1 - \lambda_4)}, \quad \beta = \frac{(\lambda_2 - \lambda_3)\lambda_1\lambda_4}{\lambda_2\lambda_3(\lambda_1 - \lambda_4)},$$

$$\gamma = \frac{\lambda_1\lambda_2 - \lambda_3\lambda_4}{\lambda_2\lambda_3(\lambda_1 - \lambda_4)}, \quad \sigma_1 = \sum_{i=1}^{4} \lambda_i, \quad \sigma_2 = \sum_{i<j} \lambda_i\lambda_j. \tag{3.11b}$$

Further formulae for more than four eiugenvalues can be easily derived by the following recursive formula

$$\check{R}^{(m)}(x) = (1 + x\frac{\lambda_m}{\lambda_{m-1}})\overline{\check{R}^{(m-1)}(x)} + \Lambda_m P_m \tag{3.12}$$

where $\overline{\check{R}^{(m-1)}(x)}$ stands for what we obtained by changing S^l to be $S^l - \lambda_m^l P_m$ $(l \in Z)$ in the original formula $\check{R}^{(m-1)}(x)$ and

$$\Lambda_m = \prod_{i=1}^{m-1}(1 + x\frac{\lambda_i}{\lambda_{i+1}}) \cdot \lambda_m, \quad P_m = \lambda_m S^{-1}\prod_{j=1}^{m-1}\frac{(S - \lambda_j)}{(\lambda_m - \lambda_j)}.$$

Thus if we have known the formula for $m-1$ eigenvalues $\check{R}^{(m-1)}(x)$, we can obtain the formula for m eigenvalues easily. Acturally starting from the trivial solution of Yang-Baxter equation $\lambda_1 I$ one may use the recursive formula to obtain (3.9), (3.10), (3.11) etc. in turn. When these formulae give rise to solutions of Yang-Baxter equation will be discussed in next section.

(ii) As an example beyond the above simplest case, we consider $f_1(u) = y(u)$, $f_2(u) = y^2(u)$ or $f_1(u) = y^2(u)$, $f_2(u) = y(u)$ for the case with three eigenvalues. For the former case

$$\Lambda_1 = (y(u) + \frac{\lambda_1}{\lambda_2})(y^2(u) + \frac{\lambda_2}{\lambda_3}) \cdot \lambda_3$$

$$\Lambda_2 = (\frac{\lambda_1}{\lambda_2}y(u) + 1)(y^2(u) + \frac{\lambda_2}{\lambda_3}) \cdot \lambda_3$$

$$\Lambda_3 = (\frac{\lambda_1}{\lambda_2}y(u) + 1)(\frac{\lambda_2}{\lambda_3}y^2(u) + 1) \cdot \lambda_3 \tag{3.13}$$

Here (3.13) does not have the symmetry properties like (3.7), so we should consider all the six permutations of eigenvalues related to S_3. Using (3.13), (3.3) and (3.2b), we obtain that

$$\check{R}(u) = L(u)S + M(u)I + N(u)S^{-1} \tag{3.14a}$$

where

$$L(u) = [\ y(u) - 1\ ][\ by(u) - a\ ]a^{-1},$$

$$M(u) = (\lambda_1 + \lambda_2)(\lambda_2 + \lambda_3)y(u)[\ \lambda_1 - \lambda_3 y(u)\ ]a^{-1},$$

$$N(u) = y(u)[\ y(u) - 1\ ][\ a\lambda_3 y(u) + b\lambda_2\ ]\lambda_1 a^{-1},$$

$$a = (\lambda_1 - \lambda_3)\lambda_2, \quad b = (\lambda_1 + \lambda_2)\lambda_3. \tag{3.14b}$$

For the latter case, the result is the same as that of (3.14) by turning $\lambda_1, \lambda_2, \lambda_3$ respectively to be $\lambda_3, \lambda_2, \lambda_1$, which is apparently (because we will consider all six permutations) included in formula (3.14). When (3.14) can give rise to solution of Yang-Baxter equation will be discussed in next section.

4. Consistency Conditions

Now we discuss the consistency condition for the Yang-Baxterization formulae obtained in last section to satisfy Yang-Baxter equation. Let us first consider(3.9). Substituting it into (2.1) and using (2.5), one reach the following equation

$$y(u+v)(S_1S_2^{-1}S_1 - S_2S_1^{-1}S_2) + y(u)y(v)\lambda_1\lambda_2(S_1^{-1}S_2S_1^{-1} - S_2^{-1}S_1S_2^{-1}) = 0 \quad (4.1)$$

For the S matrix with quadratic reduction relation, it can be shown to satisfying the following identity

$$(S_1S_2^{-1}S_1 - S_2S_1^{-1}S_2) + \lambda_1\lambda_2(S_1^{-1}S_2S_1^{-1} - S_2^{-1}S_1S_2^{-1}) = 0 \quad (4.2)$$

Then (4.1) and (4.2) lead to

$$y(u+v) = y(u)y(v) \quad (4.3)$$

which tells that $y(u) = e^{-u}$, i.e. trigonometric parameterization. So (3.9) satisfies Yang-Baxter equation (2.1) identically only for trigonometric case.

In order to discuss consistency condition of Yang-Baxterization formulae for the case of three eigenvalues, we substitute either (3.10a) or (3.14a) into (2.1). After using Yang-Baxter relation (2.5) and cubic reduction relation, one obtain the following equation

$$\begin{aligned}
&\theta_3^+(S_1S_2^{-1}S_1 - S_2S_1^{-1}S_2) + \theta_3^-(S_1^{-1}S_2S_1^{-1} - S_2^{-1}S_1S_2^{-1}) \\
&\quad +\theta_2^+(S_2^{-1}S_1 - S_2S_1^{-1}) + \theta_2^-(S_1^{-1}S_2 - S_1S_2^{-1}) \\
&\qquad +\theta_1^+(S_1 - S_2) + \theta_1^-(S_1^{-1} - S_2^{-1}) = 0
\end{aligned} \quad (4.4a)$$

where

$$\begin{aligned}
&\theta_3^+ = L(u)N(u+v)L(v), \quad \theta_3^- = N(u)L(u+v)N(v), \\
&\theta_2^+ = M(u)[\ N(u+v)L(v) - L(u+v)N(v)\ ], \\
&\theta_2^- = [\ N(u)L(u+v) - L(u)N(u+v)\ ]M(v), \\
&\theta_1^+ = [\ L(u)M(v) + M(u)L(v) + \sigma_1 L(u)L(v) \\
&\quad +\sigma_3^{-1}N(u)N(v)\ ]M(u+v) - M(u)L(u+v)M(v), \\
&\theta_1^- = [\ N(u)M(v) + M(u)N(v) + \sigma_3 L(u)L(v) \\
&\quad +\sigma_2\sigma_3^{-1}N(u)N(v)\ ]M(u+v) - M(u)N(u+v)M(v), \\
&\sigma_1 = \sum_i^3 \lambda_i, \qquad \sigma_2 = \sum_{i<j}\lambda_i\lambda_j, \qquad \sigma_3 = \lambda_1\lambda_2\lambda_3.
\end{aligned} \quad (4.4b)$$

By the way, we talk a few words about (3.6). It is easy to find that exponential function (e.g. $y = e^{-u}$) obeys (3.6) with u_0 at infinite, which gives a trigonometric parameterization of $\check{R}(u)$ matrix. Besides, (3.6) also permits pole type parameterization, for example $y = [x(u_0) - x(u)]/[x(u_0) + x(u)]$, where $x(u)$ is an odd function. General analysis of the latter is worth to study further, here we only discuss the trigonometric case.

If we adopt notations

$$y := e^{-u}, \qquad x := e^{-v} \tag{4.5}$$

then we know (from the discussion in section 2) that Yang-Baxter equation (2.1) holds for the Yang-Baxterization formulae under consideration at least when $x = 0$, $y = 0$, $x = 1$, $y = 1$, and $x = y^{-1}$. This means that the θ's in (4.4b) must have $xy(x-1)(y-1)(xy-1)$ as a common factor.

Substituting (3.10b) into (4.4b), factorizing them and dropping out that common factor, we get one consistency condition

$$\Phi_3 - \lambda_1\lambda_3\Phi_3^{-1} + (\lambda_1+\lambda_2)(\lambda_2+\lambda_3)\lambda_2^{-1}[\Phi_2^{-1} - \Phi_2 + \lambda_2^{-1}\Phi_1 - \lambda_2\Phi_1^{-1}] = 0 \tag{4.6a}$$

where

$$\begin{aligned} \Phi_3 &:= S_1 S_2^{-1} S_1 - S_2 S_1^{-1} S_2 \\ \Phi_2 &:= S_2^{-1} S_1 - S_2 S_1^{-1} \\ \Phi_1 &:= S_1 - S_2 \end{aligned} \tag{4.6b}$$

Therefore if (4.6) holds for some permutations of eigenvalues of S matrix, then relating to the permutation, the $\check{R}$ matrix obtained from (3.10) is a solution of Yang-Baxter equation (2.1).

Similarly, using (3.14b) we obtain seven consistency conditions, which take the following forms

$$\varphi_3^{(i)}\Phi_3 + \tilde{\varphi}_3^{(i)}\Phi_3^{-1} + (\lambda_1+\lambda_2)(\lambda_2+\lambda_3)[\varphi_2^{(i)}\Phi_2 + \tilde{\varphi}_2^{(i)}\Phi_2^{-1} + \varphi_1^{(i)}\Phi_1 + \tilde{\varphi}_2^{(i)}\Phi_1^{-1}] = 0$$

$$i = 1, 2, \cdots 7. \tag{4.7}$$

where Φ's are defined by (4.6b) and the seven sets of coefficients are

(i)

$$\begin{aligned} &\varphi_3^{(1)} = b^2, && \varphi_2^{(1)} = -\lambda_3 b, && \varphi_1^{(1)} = -a\lambda_3/\lambda_2, \\ &\tilde{\varphi}_3^{(1)} = \lambda_1\lambda_3 ab, && \tilde{\varphi}_2^{(1)} = \lambda_3 b, && \tilde{\varphi}_1^{(1)} = -\lambda_3(a\sigma_2/\lambda_2 + (\lambda_2+\lambda_3)b); \end{aligned}$$

(ii)

$$\begin{aligned} &\varphi_3^{(2)} = -ab, && \varphi_2^{(2)} = \lambda_3 a, && \varphi_1^{(2)} = -b, \\ &\tilde{\varphi}_3^{(2)} = \lambda_1\lambda_2 b^2, && \tilde{\varphi}_2^{(2)} = -\lambda_1 b, && \tilde{\varphi}_1^{(2)} = \lambda_2\lambda_3 b; \end{aligned}$$

(iii)

$$\begin{aligned} &\varphi_3^{(3)} = -ab, && \varphi_2^{(3)} = \lambda_1 b, && \varphi_1^{(3)} = -b, \\ &\tilde{\varphi}_3^{(3)} = \lambda_1\lambda_2 b^2, && \tilde{\varphi}_2^{(3)} = -\lambda_3 a, && \tilde{\varphi}_1^{(3)} = \lambda_2\lambda_3 b; \end{aligned}$$

(iv)

$$\varphi_3^{(4)} = \lambda_3 a^3 + \lambda_2 b^3, \quad \tilde{\varphi}_3^{(4)} = \lambda_1\lambda_2^2 b^3 - \lambda_1\lambda_3^2 a^3,$$
$$\varphi_2^{(4)} = \lambda_3(\lambda_3 - \lambda_1)a^2, \quad \tilde{\varphi}_2^{(4)} = \lambda_3(\lambda_3 - \lambda_1)a^2,$$
$$\varphi_1^{(4)} = -(a^2(\lambda_1 - \lambda_3)\lambda_3 + \lambda_2(\lambda_2 + \lambda_3)b^2)/\lambda_2,$$
$$\tilde{\varphi}_1^{(4)} = -(a^3\lambda_3^2 + \lambda_1(\lambda_2 b - \lambda_3 a)b^2)/\lambda_3;$$

(v)

$$\varphi_3^{(5)} = -\lambda_2 b^2, \quad \varphi_2^{(5)} = \lambda_2\lambda_3 b, \quad \varphi_1^{(5)} = -\lambda_1 b,$$
$$\tilde{\varphi}_3^{(5)} = -\sigma_3 ab, \quad \tilde{\varphi}_2^{(5)} = \lambda_1\lambda_3 a, \quad \tilde{\varphi}_1^{(5)} = -\sigma_3 b;$$

(vi)

$$\varphi_3^{(6)} = -\lambda_2 b^2, \quad \varphi_2^{(6)} = -\lambda_1\lambda_3 a, \quad \varphi_1^{(6)} = -\lambda_1 b,$$
$$\tilde{\varphi}_3^{(6)} = -\sigma_3 ab, \quad \tilde{\varphi}_2^{(6)} = -\lambda_2\lambda_3 b, \quad \tilde{\varphi}_1^{(6)} = -\sigma_3 b;$$

(vii)

$$\varphi_3^{(7)} = \lambda_2 ab, \quad \varphi_2^{(7)} = -\lambda_1\lambda_2 b, \quad \varphi_1^{(7)} = \lambda_3(\lambda_1(\lambda_1 + \lambda_2) + \lambda_2(\lambda_2 + \lambda_3)),$$
$$\tilde{\varphi}_3^{(7)} = -\lambda_1\lambda_2^2 b^2, \quad \tilde{\varphi}_2^{(7)} = \lambda_1\lambda_2 b, \quad \tilde{\varphi}_1^{(7)} = -\sigma_3 a.$$

When $\lambda_1 + \lambda_2 = 0$, they reduce to be one simple condition

$$(S_1 S_2^{-1} S_1 - S_2 S_1^{-1} S_2) - \lambda_1\lambda_3(S_1^{-1} S_2 S_1^{-1} - S_2^{-1} S_1 S_2^{-1}) = 0 \tag{4.8}$$

Especially if $(S_1 S_2^{-1} S_1 - S_2 S_1^{-1} S_2) = 0$ meanwhile $\lambda_1 + \lambda_2 = 0$ or $\lambda_2 + \lambda_3 = 0$, the consistency condition (4.7) holds identically.

For the case of four eigenvalues, substituting (3.11) into (2.1), we find that the consistency condition are seven complicated equations. It takes a long tedious careful calculation to obtain that results. For brevity, we omit them here. The detailed discussion as well as concrete examples was given in [12]

5. Examples

In this section, we will apply the previous formulations to give several concrete examples. First we consider the following braid group representation[11]

$$S = q\sum_a E_{aa} \otimes E_{aa} + \sum_{a<b}(tE_{ab} \otimes E_{ba} + t^{-1}E_{ba} \otimes E_{ab}) \tag{5.1}$$

of which the eigenvalues are q, 1 and -1. It is easy to find that (5.1) satisfies

$$S_1 S_2^{-1} S_1 - S_2 S_1^{-1} S_2 = 0$$

Then (4.7) hold for the S matrix given by (5.1) when either $\lambda_1 = \pm 1$, $\lambda_2 = \mp 1$, $\lambda_3 = q$ or $\lambda_1 = q$, $\lambda_2 = \pm 1$, $\lambda_3 = \mp 1$. For the former case, we calculate from (3.14) a trigonometric $\check{R}(u)$ matrix which are new solutions of Yang-Baxter equation

$$\check{R}(x) = (q \mp x^2)\sum_a E_{aa} \otimes E_{aa} + (1 \mp qx^2)\sum_{a<b}(tE_{ab} \otimes E_{ba} + t^{-1}E_{ba} \otimes E_{ab}) \tag{5.2}$$

where $x := e^{-u}$ and the overall scalar factor $1 - x$ has been omited. For the latter case one can get the corresponding $\check{R}(x)$ matrix similarly, which is just equivalent to (5.2).

Now we baxterize the following braid group representation
e.g.

$$S = \begin{pmatrix} 1 & 0 & 0 & q^2/p \\ 0 & q & 1 & 0 \\ 0 & 1 & q & 0 \\ p & 0 & 0 & 1 \end{pmatrix} \tag{5.3}$$

It has three distinct eigenvalues $q+1$, $q-1$ and $1-q$. If we denote them as λ_1, λ_2 and λ_3 respectively, We can find that, for permutations (i) $(\lambda_1, \lambda_2, \lambda_3)$, (ii) $(\lambda_1, \lambda_3, \lambda_2)$, (iii) $(\lambda_3, \lambda_1, \lambda_2)$, (iv) $(\lambda_3, \lambda_2, \lambda_1)$, the consistency conditions are satisfied. Then for case (i) we obtain a new 4×4 solutions of Yang-Baxter equation.

$$\check{R}(x) = \begin{pmatrix} x-1 & 0 & 0 & -(x+1)q^2/p \\ 0 & -(x+1)q & x-1 & 0 \\ 0 & x-1 & -(x+1)q & 0 \\ -(x+1)p & 0 & 0 & x-1 \end{pmatrix} \tag{5.4}$$

here and the following we omitted the overall factors in $\check{R}(x)$ matrix obtained from the Yang-Baxterization formula. For the other cases (ii), (iii) or (iv) the results are nothing new but what one obtained from (5.4) after replacing x by $-x$, $-x^2$, or x^2, respectively. If one introduces a new parameter u such that $x := \exp 2u$, then (5.4) becomes

$$\check{R}(u) = \begin{pmatrix} \mathrm{sh}u & 0 & 0 & -q^2\mathrm{ch}u/p \\ 0 & -q\mathrm{ch}u & \mathrm{sh}u & 0 \\ 0 & \mathrm{sh}u & -q\mathrm{ch}u & 0 \\ -p\mathrm{ch}u & 0 & 0 & \mathrm{sh}u \end{pmatrix} \tag{5.5}$$

This may provide an eight vertex model with trigonometric parameterization.

6. Discussions

In above we studied Yang-Baxterization of representations of braid group. When Baxterizing representations of braid group, the $\check{R}$ matrix to be solved is desired to satisfy (i) initial condition (2.2) and its immediate consequence (2.4); (ii) $[\check{R}(u), \check{R}(v)] = 0$ for $u \neq v$, as well as Yang-Baxter equation. Although the formulation of Yang-Baxterization formulae is based on that the S matrix has n distinct eigenvalues, the results with a slight change are also available to the case beyond that. This is because solutions of Yang-Baxter equation permit an arbitrary overall scalar factor. The formulae (3.9) and (3.10) are valid without any change. One may notice that the denominators of (3.11) and (3.14) would be zero if it were the case $\lambda_1 = \lambda_4$ and $\lambda_1 = \lambda_3$ respectively. However, we can drop out that (i.e. multiply (3.11) by $(\lambda_1 - \lambda_4)$, (3.14) by $(\lambda_1 - \lambda_3)$) as long as the result is consistency with Yang-Baxter equation. At the time, the S matrix is no longer a limit of $\check{R}(u)$.

When the $\check{R}(u)$ matrices calculated from Yang-Baxterization formulae are solutions of Yang-Baxter equation will be answered after checking the consistency condition of Yang-Baxter equation. It has been shown that the formula (3.9) for the case that S obeys a quadratic reduction relation only presents trigonometric solutions and follows Yang-Baxter equation identically. For S obeying a cubic reduction relation, we only discussed the trigonometric case, where the Jimbo's type formula provides solutions of Yang-Baxter equation as long as one consistency condition holds; another type formula provides solutions of Yang-Baxter equation when seven consistency conditions hold.

Acknowledgements

The author would like to thank Prof H Y Guo, Drs. L Y Wang and J Zhang for interesting discussions.

References

[1] R.J. Baxter, *Exactly Solved Models in Statistical Mechanics*, (Academic Press, 1982).
C.N. Yang, *Phys. Rev. Lett.* **19**(1967)1312; *Phys. Rev.* **168**(1968)1920.

[2] L.D. Faddeev, N.Yu. Reshetikhin and L.A. Takhtajan, 1987 preprint LOMI E-14-87.
N.Yu. Reshetikhin, 1987 preprints LOMI E-4-87; E-17-87.
H.J. deVega, *Int. J. Mod. Phys.* **4**(1989)2371.

[3] A.B. Zamolodchikov, Al.B. Zamolodchikov, *Nucl. Phys.* **B133**(1978) 525; *Ann. Phys.* **120**(1979) 253.

[4] L. Alvarez-Gaume, C. Gomez and G. Sierra, *Phys. Lett.* **B220**(1989)142.
G. Moore and N. Seiberg, *Phys. Lett.* **B212**(1988)451.

[5] V.G. Drinfeld, 1986 *Proc. ICM Berkeley* 798; *Zap. Nauch. Sem. LOMI* **155** 18.

[6] M. Jimbo, *Lett. Math. Phys.* **10**(1985)63; *Lett. Math. Phys.* **11**(1986)247; *Comm. Math. Phys.* **102**(1986)537.

[7] B.A. Belavin and V.G. Drinfeld, *Funct. Anal. Appl.* **16**(1982)159; *Soviet. Sci. Rev. Sec.C* **4**(1984)91.

[8] V.F.G. Jones, *Inter. J. Mod. Phys.* **B4**(1990)701.

[9] M.L. Ge, Y.S. Wu and K. Xue, *Inter. J. Mod. Phys.***6**(1991)3735.
L.Y. Wang, *Ph.D thesis of Lanzhou University* 1990.

[10] R.B. Zhang, M.D. Gould and A J Bracken, *Nucl. Phys.* **B354**(1991)625.

[11] Y.Q. Li, preprint ZIMP-91-17.
L.Y. Wang, J.H. Dai and J. Zhang, ICTP internal report IC/91/4.

[12] Y.Q. Li, L.Y. Wang and J. Zhang, preprint ASITP-91-54

QUANTUM CLEBSCH-GORDAN COEFFICIENTS FOR NON-GENERIC q VALUES

BO-YU HOU[1)], BO-YUAN HOU[2)] and ZHONG-QI MA[3)4)]

1) *Institute of Modern Physics, Northwest University, Xian 710069, People's Republic of China*
2) *The Graduate School, Chinese Academy of Sciences, P.O.Box 3908, Beijing 100039, People's Republic of China*
3) *CCAST(World Laboratory), P.O.Box 8730, Beijing 100080, People's Republic of China*
4) *Institute of High Energy Physics, P.O.Box 918(4), Beijing 100039, People's Republic of China*

ABSTRACT

In the decomposition of a coproduct in the direct product of two irreducible representation spaces of the quantum enveloping algebra $U_q sl(2)$, the mixed representations, reducible but indecomposable, appear when q is a root of unity. In this paper, the necessary and sufficient condition for mixture of two irreducible representations is given, the quantum Clebsch- Gordan coefficients which are neither all vanishing nor divergent for the non-generic q values are defined, and the method for computing the new states is presented.

The quantum enveloping algebras $U_q\mathcal{G}$ were firstly presented [1] as a tool for solving the Yang-Baxter equation [2] which plays a crucial role in some completely integrable statistical systems, and now they draw the increasing interests of theoretical physicists and mathematicians.

Recently, in the study of an XXZ spin chain model for the non- generic q values where q is a root of unity [3,4], the structures of the type I representations which are reducible but indecomposable were studied. The necessary condition for appearance of the type I representations was given, but there are several problems that should be studied further. Among them, the sufficient condition, the new states, and the quantum Clebsch-Gordan (q-CG) coefficients for the non-generic q values are the most urgent ones. A state of one IR in the decomposition of a coproduct in the direct product of two irreducible representation (IR) spaces for $U_q sl(2)$ may coincide with a state of another IR when q goes to a non- generic value. When it occurs, we call that two states are degenerate and two IR's are mixed. In this paper we will present the necessary and sufficient condition for mixture of two IR's in the decomposition of a coproduct.

For a given integer p, q_0, called a non-generic value, is defined as

$$q_0^p = \lambda = \begin{cases} 1 \text{ or } -1 & \text{when } p \text{ is odd} \\ -1 & \text{when } p \text{ is even} \end{cases} \tag{1}$$

Let

$$[m] = \frac{q^m - q^{-m}}{q - q^{-1}} \tag{2}$$

In this paper, if without a special notification, a small latin letter, for example n or p etc., denotes a non-negative integer, and a small greek letter, for example α, denotes a non-negative integer less than p: $0 \le \alpha < p$.

For the generic q values, the coproduct $\Delta_q^{j_1 j_2}$ in the direct product of two IR spaces of $U_q s\ell(2)$ is a reducible representation and can be reduced by the quantum Clebsch-Gordan matrix which was computed explicitly in Ref. 5,6.

$$\sum_{m'} (\Delta_q^{j_1 j_2})_{m(M-m)m'(M'-m')} (C_q^{j_1 j_2})_{m'(M'-m')JM'} = (C_q^{j_1 j_2})_{m(M-m)JM} (D_q^J)_{MM'} \tag{3}$$

where $J = j_1 + j_2, j_1 + j_2 - 1, \cdots, | j_1 - j_2 |$. The states in the IR spaces can be combined by q-CG coefficients

$$|J\ M\rangle = \sum_m |j_1\ m\rangle |j_2\ (M-m)\rangle (C_q^{j_1 j_2})_{m(M-m)JM} \tag{4}$$

In the Lusztig's representation [7] the q-CG coefficients are as follows [5,6]

$$\begin{aligned}(C_q^{j_1 j_2})_{m(M-m)JM} &= A_J^{-1} q^{-(j_1+j_2-J)(j_1+j_2+J+1)/2 + m j_2 - (M-m) j_1} \Delta(j_1\ j_2\ J) \left\{ \frac{[2J+1]!}{[2j_1]![2j_2]!} \right\}^{\frac{1}{2}} \\ &\times [j_1+m]![j_1-m]![j_2+M-m]![j_2-M+m]! \sum_n (-1)^n q^{n(J+j_1+j_2+1)} \{[n]![j_1-m-n]! \\ &\times [j_2+M-m-n]![j_1+j_2-J-n]![J-j_1-M+m+n]![J-j_2+m+n]!\}^{-1}\end{aligned}$$

$$\Delta(j_1\ j_2\ J) = \left\{ \frac{[j_1+j_2-J]![j_1-j_2+J]![-j_1+j_2+J]!}{[j_1+j_2+J+1]!} \right\}^{\frac{1}{2}} \tag{5}$$

where A_J is introduced to avoid divergence or all vanish of q-CG coefficients such that the state $|J\ M\rangle$ exists. The explicit form of A_J will be given later.

Without loss of generality, we assume that $j_1 \ge j_2$ and $J' > J$. The mixed condition of two IRs $D_q^{J'}$ and D_q^J is a condition under which the state $|J\ J\rangle$ coincides with the state $|J'\ J\rangle$ up to a constant for a non-generic q value, *i.e.*, when q goes to q_0 the following ratio B_m should be independent of m:

$$B_m \equiv \frac{(C_q^{j_1 j_2})_{m(J-m)J'J}}{(C_q^{j_1 j_2})_{m(J-m)JJ}} \tag{6}$$

The coincidence of the rest states can be proved by the lowering operator f.

Since $e\,|J\ J\rangle = 0$ and $\langle J'\ J|J\ J\rangle = 0$ for the generic q values, we obtain from $|J'\ J\rangle \propto |J\ J\rangle$ when q goes to q_0,

$$\lim_{q\to q_0}\langle J'\ J|J'\ J\rangle = \lim_{q\to q_0} A_{J'}^{-2}\begin{bmatrix} 2J' \\ J'+J \end{bmatrix} = 0$$

$$\lim_{q\to q_0} e\,|J'\ J\rangle = \lim_{q\to q_0}[J'+J+1]\,|J'\ (J+1)\rangle = 0$$

that just is the necessary condition for mixture obtained in the XXZ spin chain model [3,4]:

$$J'+J+1 = 0 \quad mod\ p \tag{7}$$

Introduce the following notations:

$$0 \le j_1 - j_2 = sp+\gamma, \qquad 2j_2 = up+\omega \tag{8}$$

$$J' = tp+\eta, \quad J = (\ell - t)p - \eta - 1 < J' \tag{9a}$$

where both γ and η are integers or half of odd integers spontaneously because j_1 and j_2 may be an integer or half of an odd integer, respectively. Now, the mixed condition is shown in the theorem.

Theorem: In the decomposition of the coproduct $\Delta_q^{j_1 j_2}$, the necessary and sufficient condition for mixture of two IRs $D_q^{J'}$ and D_q^{J} is as follows:

$$\ell = \begin{cases} 2t & \text{if } \eta<\gamma<p-\eta \\ 2t+1 & \text{if } \frac{p-1}{2}<\eta<\gamma \text{ or } \gamma<\frac{p-1}{2}<\eta<p-\gamma \end{cases} \tag{9b}$$

The proof of this theorem was given in our previous paper [8]. Condition (9) guarantees that there is a common r in the following three formulas

$$\begin{gathered} j_2+J'-j_1 = rp+\beta_1, \qquad j_2+J-j_1 = rp+\beta_0 \\ j_2+J-m-n = rp+\beta \end{gathered} \tag{10a}$$

$$0 \le \beta_0 \le \beta \le \beta_1 < p, \qquad r = \begin{cases} t-s & \text{if } \eta>\gamma \\ t-s-1 & \text{if } \eta<\gamma \end{cases} \tag{10b}$$

It is the key point for proving the theorem that there exists a common r which is independent of m and n.

Through tedious calculation, we proved [8] that when q goes to q_0 the ratio B_m tends to a limit independent of m:

1) for the case $\omega \ge \beta_1$ or $\omega < \beta_0$

$$\lim_{q\to q_0} B_m = B, \qquad B = (-1)^{\alpha}\lambda^{\alpha(\ell+r)}q_0^{j_1\alpha}\begin{bmatrix}\beta_1\\ \beta_0\end{bmatrix}_{q=q_0} \ne 0 \tag{11}$$

2) For the case $\beta_0 \le \omega < \beta_1$, we separate the values of m, $J - j_2 \le m \le j_1$, into two groups m_1 and m_2:

$$\begin{array}{ll} m_1 = J - j_2 + v_1 p + \mu_1, & \omega < \mu_1 \le \beta_1 \\ m_2 = J - j_2 + v_2 p + \mu_2, & \mu_2 \le \omega \text{ or } \mu_2 > \beta_1 \end{array} \tag{12}$$

and obtain

$$\begin{aligned} \lim_{q \to q_0} B_{m_1} &= \tilde{B}, \qquad \tilde{B} = \left(1 + \frac{\ell}{u - r}\right) B \\ \lim_{q \to q_0} (C_q^{j_1 j_2})_{m_2 (J - m_2) JJ} &= \lim_{q \to q_0} (C_q^{j_1 j_2})_{m_2 (J - m_2) J' J} = 0 \end{aligned} \tag{13}$$

Owing to this coincidence of the states $|J\ J\rangle$ and $|J'\ J\rangle$ up to a constant, some new states must exist in the linear space spanned by $|j_1 m_1\rangle |j_2 m_2\rangle$. The new state with the highest weight can be computed by a limit process [8].

$$\begin{aligned} |J\ J\rangle_0 &= \sum_m |j_1\ m\rangle |j_2\ (J - m)\rangle (\bar{C}_{q_0}^{j_1 j_2})_{m (J-m) JJ} \\ N(\bar{C}_{q_0}^{j_1 j_2})_{m (J-m) JJ} &= \lim_{q \to q_0} \frac{(C_q^{j_1 j_2})_{m (J-m) J' J} - B (C_q^{j_1 j_2})_{m (J-m) JJ}}{q - q_0} \end{aligned} \tag{14}$$

where N is a normalization factor, and B should be replaced by $\tilde{B}$ for the case $\beta_0 \le \omega < \beta_1$.

This theorem pairs J and J' as partners in the decomposition of the coproduct. When $\eta = (p-1)/2$ or $p - 1/2$, we assume $J' = J$. It was proved that the state $|J\ M\rangle$ exists when $q = q_0$ if we introduce the following A_J where J' is the partner of J:

$$\begin{aligned} A_J &= q^{(J - j_1 - j_2)(J - j_1 + j_2 + 1)/2} \Delta(j_1\ j_2\ J) \\ &\times \left\{ \frac{[2J+1]!}{[2j_1]! [2j_2]!} \right\}^{\frac{1}{2}} \frac{[J' + j_2 - j_1]! [j_1 + j_2 - J']!}{[J + j_2 - j_1]!} \end{aligned} \tag{15}$$

Acknowledgments This work was supported by the National Natural Science Foundation of China through the Nankai Institute of Mathematics.

References

1. V. G. Drinfeld, *Proc. Int. Congr. Math.*, Berkeley, 1986, 798. M. Jimbo, *Lett. Math. Phys.* **10**(1985)63; *Commun. Math. Phys.* **102**(1986)537.
2. C. N. Yang, *Phys. Rev. Lett.* **19**(1967)1312. R. J. Baxter, *Ann. Phys.* **70**(1972)193.
3. V. Pasquier and H. Saleur, *Nucl. Phys.* **B330**(1990)523. D. Levy, *Phys. Rev. Lett.* **64**(1990)499.
4. Bo-Yu Hou, Bo-Yuan Hou and Zhong-Qi Ma, *J. Phys.* **A24**(1991)2847.
5. A. M. Kirillov and N. Yu. Reshetikhin, LOMI preprint E-9-88, 1988.
6. Bo-Yu Hou, Bo-Yuan Hou and Zhong-Qi Ma, *Commun. Theor. Phys.* **13**(1990)181 and 341.
7. G. Lusztig, *Contemp. Math.* 82(1989)59.
8. Bo-Yu Hou, Bo-Yuan Hou and Zhong-Qi Ma, *J. Phys.* **A25**(1992)1211.

INFINITE BRAIDED TENSOR PRODUCTS AND 2D QUANTUM GRAVITY

S. MAJID[1]

Department of Applied Mathematics & Theoretical Physics
University of Cambridge, Cambridge CB3 9EW, U.K.

ABSTRACT

Braided tensor products have been introduced by the author as a systematic way of making two quantum-group-covariant systems interact in a covariant way, and used in the theory of braided groups. Here we study infinite braided tensor products of the quantum plane (or other constant Zamolodchikov algebra). It turns out that such a structure precisely describes the exchange algebra in 2D quantum gravity in the approach of Gervais. We also consider infinite braided tensor products of quantum groups and braided groups.

1. Of central importance in the theory of braided groups initiated in [6][7][8][9][10] is the notion of covariant algebras and their braided tensor products. Here we give a new application of this notion.

Let us recall that a covariant algebra is simply an algebra B on which a quantum group H acts (or a dual quantum group A of function algebra type, coacts) in such a way that the product and unit of B are covariant, i.e. the maps $\cdot : B\otimes B \to B$ and $\eta : \mathbb{C} \to B$ respectively are interwiners for the quantum-group (co)action. This is obviously an important notion if we want to work with physical systems (described by algebras) with quantum-group symmetry. A fancy way to think of covariance is that the algebra B *lives in* the category of (co)-modules of the quantum group[7, Sec. 6]: 'Working in the category' just says that we keep everything manifestly covariant. Thinking about things this way allows us to treat covariance in the same way as we treat super-symmetry (we make everything $\mathbb{Z}_2$-graded). This unification between usual notions of group covariance and super-symmetry is one of the remarkable unifications made possible by quantum groups[7, Sec. 6].

In particular, the (co)modules of a true quantum group (with universal R-matrix or its dual concept for a quantum function algebra) have a braiding Ψ[5, Sec. 7]. This is a coherent collection of maps $\Psi_{V,W} : V\otimes W \to W\otimes V$ allowing the 'transposition' of any two objects in the category (any two representations) with properties like the usual transposition or super-transposition $\Psi_{V,W}(v\otimes w) = (-1)^{|v||w|}w\otimes v$, except that typically we no longer have $\Psi_{V,W} = \Psi^{-1}_{W,V}$ (so they are best represented by braids rather than by permutations).

The main lemma which will concern us, which is the *fundamental lemma of the theory of braided groups* is: if B, C are covariant algebras then there is a *braided tensor product* $B\underline{\otimes}C$ which is again a covariant algebra. Thus, we have a way of combining covariant systems in a covariant way. This braided tensor product algebra is $B\otimes C$ as a space, but product

$$(b\otimes c)\cdot(d\otimes e) = b\Psi_{C,B}(c\otimes d)e, \quad b,d\in B,\ c,e\in C \tag{1}$$

where we mean to first apply $\Psi_{C,B}$ to $c\otimes d\in C\otimes B$ and multiply the result from the left in B and from the right in C. That this is associative follows from functoriality and hexagon identities for Ψ. See [8] for a diagrammatic proof. The braided tensor product $B\underline{\otimes}C$ contains B and C as subalgebras, but unlike the usual tensor product, these subalgebras do not commute. Instead, their exchange is given by the braiding Ψ.

This braided tensor product is like the super-tensor product of super-algebras, leading us to interpret (1) as saying that the elements in the algebras B, C have braid statistics

[1]SERC Fellow and Drapers Fellow of Pembroke College, Cambridge

rather than bose-fermi statistics. A braided group then is a generalized Hopf algebra with $\underline{\Delta} : B \to B\underline{\otimes}B$ an intertwiner (covariant) and an algebra homomorphism provided the elements are treated with braid statistics as in (1).

2. Let $B = \mathbb{C}_q$ (the quantum plane) with generators x_i. It is well-known that this is a covariant algebra for $A = SL_q(2)$ with generators $t^i{}_j$. In this context covariance is equivalent to requiring that the coaction $x_i \mapsto x_j t^j{}_i$ (the tensor product on the right is suppressed) is an algebra homomorphism, i.e. that $x'_i = x_j t^j{}_i$ obey the relations of the quantum plane also. The same applies for any R-matrix with the algebra $B = V^*(R)$ (the Zamolodchikov algebra of covector type) given by $\mathbf{x}_1\mathbf{x}_2 = \mathbf{x}_2\mathbf{x}_1\lambda R$ and A a quantum group obtained from the FRT bialgebra $A(R)$. This was proven already in [5, Sec. 6].

We now compute the infinite braided tensor product of such an algebra B with itself. Infact, we compute the N-fold product and leave the reader to take the $N \to \infty$ limit. The generators of $B^{\underline{\otimes}^N}$ are $\{x_i(m)\}_{m=1}^{m=N}$ say, where $x_i(m) = 1\otimes\cdots\otimes x_i\otimes\cdots 1$ (in the m'th position). Since the element 1 is bosonic (has trivial braiding with everything else) we have from Section 1,

$$x_i(m)\cdot x_j(n) = \begin{cases} 1\otimes\cdots\otimes x_i\otimes\cdots\otimes x_j\otimes\cdots\otimes 1 \\ 1\otimes\cdots\otimes x_i x_j\otimes\cdots\otimes 1 \\ 1\otimes\cdots\otimes\Psi(x_i\otimes\cdots\otimes x_j)\otimes\cdots\otimes 1 \end{cases} = \begin{cases} x_i(m)x_j(n) & \text{if } m<n \\ x_i(m)x_j(n) & \text{if } m=n \\ x_l(n)x_k(m)R^k{}_i{}^l{}_j & \text{if } m>n \end{cases} \quad (2)$$

Here the braiding is $\Psi(x_i\otimes x_j) = x_l\otimes x_k\mathcal{R}(t^k{}_i\otimes t^l{}_j) = x_l\otimes x_k R^k{}_i{}^l{}_j$ as explained in detail in [9]. The expression on the right is in terms of the usual (unbraided) tensor product, relative to which the braided one is a kind of 'ordered product' of the x_i. Also, following a trick in [4], let us write

$$R(n-m) = \begin{cases} R & \text{if } m>n \\ \lambda R & \text{if } m=n\,. \\ R_{21}^{-1} & \text{if } m<n \end{cases} \quad (3)$$

It obeys the parametrized Yang-Baxter equations

$$R_{12}(m-n)R_{13}(m)R_{23}(n) = R_{23}(n)R_{13}(m)R_{12}(m-n) \quad (4)$$

with discrete spectral parameter[4]. Then from (2) we can write $B^{\underline{\otimes}^N}$ as generated by the $\{x_i(m)\}$ with relations

$$\mathbf{x}_1(m)\mathbf{x}_2(n) = \mathbf{x}_2(n)\mathbf{x}_1(m)R(n-m) \quad (5)$$

in an obvious compact notation. Because the braided tensor product preserves covariance, we conclude that $B^{\underline{\otimes}^N}$ and its continuous limit are also covariant. The coaction is the pointwise one $x_i(m)\cdots \mapsto x_j(m)t^j{}_i$ on each generator, extended multplicatively.

This is exactly a discretised version of the kind of exchange algebra that arises in 2D quantum gravity[2]. There, one finds fields $\xi_i(\sigma)$ where $\sigma\in[0,\pi]\subset S^1$ is a continuous parameter and $R(\sigma-\sigma') = \begin{cases} R & \text{if } \sigma<\sigma' \\ R_{21}^{-1} & \text{if } \sigma>\sigma' \end{cases}$. Here R is the usual $SL_q(2)$ matrix and the coaction appears equivalently as an action of $U_q(sl_2)$ by dualizing. The same applies for $U_q(g)$. Note that in this setting σ is a spatial world-sheet co-ordinate and our infinite braided tensor product gives a discretised version of this. Our interpretation of the physical fields then is that they are sections of a vector bundle over $[0,\pi]$ except that the copies of the quantum plane over each point are 'braided-independent' in the sense that they are tensor-producted with braid statistics. In this sense they are classical objects with the quantum non-commutativity being interpreted as braid-statistical non-commutativity.

As well as covector type Zamolodchikov algebras we can just as well have quantum vector algebras $V(R)$ as explained in detail in [9]. The generators are $\{v^i\}$ with relations

$\mathbf{v}_1\mathbf{v}_2 = \lambda R_{12}\mathbf{v}_2\mathbf{v}_1$. One may take their infinite braided tensor product just as above and interpret the continuous limit as some other fields transforming as vectors under $SL_q(2)$. In the case of the quantum plane, the vector and covector algebras are $\mathbf{C}_q$ and $\mathbf{C}_{q^{-1}}$ respectively in our approach[9] (for a fixed choice of λ).

3. Another covariant algebra is the braided group $B(R)$ associated to any bi-invertible R-matrix[7]. In compact matrix form the relations and covariance in [7] can be written as

$$R_{21}\mathbf{u}_1R_{12}\mathbf{u}_2 = \mathbf{u}_2R_{21}\mathbf{u}_1R_{12}, \quad \mathbf{u} \mapsto \mathbf{t}^{-1}\mathbf{u}\mathbf{t}. \tag{6}$$

Here $u^i{}_j$ are the generators. Note that the relations here are *not* the reflection equations arising in lattice models with boundary[11] (these have additional transpositions even in the constant case). In fact, they arise in the theory of braided groups by a systematic process of transmutation of the FRT bialgebras $A(R)$ (hence the name). They have also been used previously in a compact description of $U_q(g)$ in FRT form, for reasons explained in [10].

Setting $B = B(R)$ we can compute $B^{\underline{\otimes}^N}$ in a similar way to that above. Thus $B(R)^{\underline{\otimes}^N}$ has generators $u^i{}_j(m)$ with

$$\mathbf{u}_1(m)\cdot\mathbf{u}_2(n) = \begin{cases} \mathbf{u}_1(m)\mathbf{u}_2(n) & \text{if } m \le n \\ \Psi(\mathbf{u}_1(m)\otimes\mathbf{u}_2(n)) & \text{if } m > n \end{cases}. \tag{7}$$

The braiding between $\mathbf{u}$ and itself is computed in [7]. In compact form it can be written as[9] $\Psi(R_{12}^{-1}\mathbf{u}_1 \otimes R_{12}\mathbf{u}_2) = \mathbf{u}_2R_{12}^{-1} \otimes \mathbf{u}_1R_{12}$ so this is relevant when $m > n$. Comparing this with the relations (6) pertaining when $m = n$, we see that $B(R)^{\underline{\otimes}^N}$ is defined by the relations

$$R_{21}(m-n)\mathbf{u}_1(m)R_{12}(n-m)\mathbf{u}_2(n) = \mathbf{u}_2(n)R_{21}(m-n)\mathbf{u}_1(m)R_{12}(n-m) \tag{8}$$

with $R(n-m)$ as in (3) (with $\lambda = 1$). It has a braided-coproduct $\underline{\Delta}\mathbf{u}(m) = \mathbf{u}(m)\underline{\otimes}\mathbf{u}(m)$.

Once again, by construction we will preserve a global symmetry of these equations under the conjugation (quantum adjoint) coaction $\mathbf{u}(m) \mapsto \mathbf{t}^{-1}\mathbf{u}(m)\mathbf{t}$ at each site. When R is the $SL_q(2)$ matrix then $B(R)$ is the degenerate Sklyanin algebra[10] (after setting $BDET(\mathbf{u}) = 1$ one obtains the algebra of $U_q(sl_2)$). Assuming both covector fields $\mathbf{x}(m)$ and vector fields $\mathbf{v}(m)$ are realized (in the continuum limit) in 2D quantum gravity, it is interesting to ask if this field of degenerate Sklyanin algebras is also realised. The reason to believe this is that locally $u^i{}_j = v^i\underline{\otimes}x_j$ in the braided tensor product $V(R)\underline{\otimes}V^*(R)$ is a realization of $B(R)$ [9] (these are the rank-one braided matrices). Since $BDET(\mathbf{u}) = 0$ in this realization[9], this would *not* be a realization among the fields of $U_q(sl_2)$ but only of the degenerate Sklyanin algebra. Nevertheless, this object is a group-like object (it is a bialgebra with braid statistics, with the matrix coproduct). We have argued in [6] that it is just such group-like objects with braid statistics (not necessarily quantum groups) that should be realised as *internal* symmetries, i.e. among the fields of a general low-dimensional quantum theory. Details will be presented elsewhere.

4. Next, let us consider $B = A(R)$ itself as a covariant algebra under the coaction $\mathbf{s} \mapsto \mathbf{s}\mathbf{t}$ where $\mathbf{s}$ denotes the generators of B (as distinguished from the generators $\mathbf{t}$ of the coacting quantum group A obtained from $A(R)$. Clearly, $A(R)\underline{\otimes}A(R)$ consists of two copies of $A(R)$ with cross-relations given by

$$(1\otimes\mathbf{s}_1)\cdot(\mathbf{s}_2\otimes 1) = \Psi(\mathbf{s}_1\otimes\mathbf{s}_2) = \mathbf{s}_2\otimes\mathbf{s}_1\mathcal{R}(\mathbf{t}_1\otimes\mathbf{t}_2) = \mathbf{s}_2\otimes\mathbf{s}_1R_{12} = (\mathbf{s}_2\otimes 1)\cdot(1\otimes\mathbf{s}_1)R_{12}. \tag{9}$$

Similarly in general the algebra $A(R)^{\underline{\otimes}^N}$ has generators $\mathbf{s}(m)$ and relations

$$\mathbf{s}_1(m)\mathbf{s}_2(n) = \begin{cases} \mathbf{s}_2(n)\mathbf{s}_1(m)R_{21}^{-1} & \text{if } m < n \\ R_{12}^{-1}\mathbf{s}_2(n)\mathbf{s}_1(m)R_{12} & \text{if } m = n \\ \mathbf{s}_2(n)\mathbf{s}_1(m)R_{12} & \text{if } m > n \end{cases}. \tag{10}$$

We note here a striking similarity with the exchange algebra among the fields $u_i, u(x)$ in [1] in their approach to the discretisation of Kac-Moody algebras. To coincide precisely we would need $\mathbf{s}_1\mathbf{s}_2 = R_{21}^{-1}\mathbf{s}_2\mathbf{s}_1R_{12}$ at $m = n$, in contrast to (10). In our case, we preserve the quantum-group symmetry in the form $\mathbf{s}(m) \mapsto \mathbf{s}(m)\mathbf{t}$.

5. For completeness, let us mention a previous iterated construction, not a braided tensor product but a *double cross product* of ordinary Hopf algebras $H_1 \bowtie H_2$. This was introduced in [3, Sec. 3.2] and can be thought of as the quantum analogue of a Manin triple (it factorises into H_1, H_2). It was explicitly constructed as a double semidirect product where each factor acts on the other (hence the notation). We showed that Drinfeld's quantum double is an example by the mutual quantum coadjoint actions[3, Sec. 4].

Moreover, as quite a different example from the quantum double, we showed in [4] that $A(R)$ indeed acts on itself by quantum adjoint actions, so we can form $A(R) \bowtie A(R)$. The actions between the two copies of $A(R)$ with generators $\mathbf{s}, \mathbf{t}$ are[4, Thm. 3.2]

$$\mathbf{s}_1 \triangleright \mathbf{t}_2 = R_{12}^{-1}\mathbf{t}_2R_{12}, \qquad \mathbf{s}_1 \triangleleft \mathbf{t}_2 = R_{12}^{-1}\mathbf{s}_1R_{12} \tag{11}$$

and the resulting structure came out as two copies of $A(R)$ with cross relations

$$R_{12}\mathbf{s}_1\mathbf{t}_2 = \mathbf{t}_2\mathbf{s}_1R_{12}. \tag{12}$$

Recently these double cross products of $A(R)$ have attracted attention as some kind of 'complexification' of $A(R)$. We would like to point out that in [4] we went further and iterated this construction. The infinite double cross product $A(R)^{\bowtie^N}$ comes out[4, Prop. 3.4] as generated by $\{\mathbf{t}(m)\}$ with relations

$$R_{12}(m-n)\mathbf{t}_1(m)\mathbf{t}_2(n) = \mathbf{t}_2(n)\mathbf{t}_1(m)R_{12}(m-n). \tag{13}$$

This is a discrete version of the kind of bialgebra that is useful in solving lattice models. In contrast to the above iterated braided tensor products, it does not preserve a quantum action of $A(R)$.

References

[1] A. Alekseev, L. Faddeev, and M. Semenov-Tian-Shanksy. Hidden quantum group inside Kac-Moody algebras. *LOMI preprint*, E-5-91, 1991.

[2] J.-L. Gervais. The quantum group structure of 2D gravity and minimal models I. *C.M.P.* 130:257, 1990.

[3] S. Majid. Physics for algebraists. *J. Algebra*, 130:17–64, 1990.

[4] S. Majid. More examples of bicrossproduct and double cross product Hopf algebras. *Isr. J. Math*, 72:133–148, 1990.

[5] S. Majid. Quasitriangular Hopf algebras and Yang-Baxter equations. *Int. J. Modern Physics A*, 5(1):1–91, 1990.

[6] S. Majid. Braided groups and algebraic quantum field theories. *L.M.P.* 22:167, 1991.

[7] S. Majid. Examples of braided groups and braided matrices. *J.M.P.* 32:3246, 1991.

[8] S. Majid. Braided groups and braid statistics. In *Q. Probability and Related Topics VIII (Proc. Delhi, 1990)*, World Sci.

[9] S. Majid. Quantum and braided linear algebra. *Preprint*, DAMTP/92-12, 1992.

[10] S. Majid. Braided matrix structure of the Sklyanin algebra and of the quantum Lorentz group. *Preprint*, DAMTP/92-10, 1992.

[11] E.K. Sklyanin. Boundary conditions for the integrable quantum systems. *J.Phys. A*, 21:2375–2389, 1988.

Covariant and Quantized Representation Functions of Quantum Group $SU_q(2)$

Masao NOMURA

Institute of Physics, College of Arts and Sciences
University of Tokyo, Komaba, Tokyo, 153, Japan

Abstract

Quantum group algebras are formulated with the postulate that every relation, e.g. a commutation relation (CR) for operators, should transform in a covariant way according to q-linear transformation. Creation and annihilation operators with a pair of indices are settled. Operator $\vec{J}$ in a coupled system is apparently symmetric, while comultiplication is asymmetric. The representation function (RF or q-d-function) in $SU_q(2)$ and its quantized form (QRF) are discussed.

1. Introduction and Summary

The idea that motivates covariant quantum group algebras[1-4] is q-Wigner theorem i.e. existence of unitary symmetry to ensure invariance of transition probability $A \to B$ under coordinate transformation in q-space. We specify covariance, contravariance and invariance to describe all the quantities and the relations in covariant forms. A consequence is q-Wigner-Eckart theorem.[5,6]

We introduce creation operators a^+_{ij}, which obey covariant transformation, and annihilation operators a_{ij} subject to contravariant transformation:

$$a'^+_{ij} = \sum_k a^+_{ik} x_{kj}\,, \qquad a'_{ij} = \sum_k a_{ik} x^*_{kj}\,. \tag{1}$$

The indices i and j of $a^{(+)}_{ij}$ are taken over $1, 2, \ldots, m$ and $1, 2, \ldots, n$ respectively: m is independent of n and could be generalized into a continuous number. The rectangular matrix $X \equiv \{x_{ij}\}$ describes q-spatial transformation. The $*$-conjugate to x_{ij} is denoted as x^*_{ij}. We assume that $(a^+)^* = a$ and that $x^{(*)}_{kj}$ and $a^{(+)}_{i'j'}$ are commutative. The matrices X form a q-group:[7] $x''_{ij} = \Sigma x'_{ik} x_{kj}$. CR's for $x^{(')}_{ij}$'s are preserved with x''_{ij}'s. We take $q\ (= q^*)$ as a real number $(\neq 0, 1)$.

We settle CR's for x_{ij}, $x^*_{i'j'}$, a^+_{ij} and $a_{i'j'}$ under the following condition:[1-3]
(i) CR's for creation and annihilation operators are preserved under (1).
(ii) Invariants (or, generally, definite-rank tensors) generated from $\{a^+_{ij}\}$ and $\{a_{i'j'}\}$ correspond uniquely to the invariants of $U(n)$ (or $U_q(n)$ with $q \to 1$).

We can regard i of a^+_{ij} as the index specifying one-dimensional sites on which tensor operation is defined. The operator $\vec{J}$ in $U_q(2)$ is apparently symmetric in i.

An extension of (1) in $U_q(2)$ is the relation[1-2] between RF, $d^j_{mk}(x_{ij})$, and QRF, $d^j_{mk}(a^+_{ij})$. Another extension of (1) is q-analog BCS-Bogoliubov transformation.[4] A specific feature of $q \to 0$ limit is also demonstrated.

2. q-Commutation Relations in $U_q(n)$

The elements $\{x_{ij}\}$ and the $*$-conjugates in $U_q(2)$ are specified by[7]

$$x_{21}x_{11}=q^{1/2}x_{11}x_{21},\ x_{22}x_{12}=q^{1/2}x_{12}x_{22},\ x_{12}x_{11}=q^{1/2}x_{11}x_{12},\ x_{22}x_{21}=q^{1/2}x_{21}x_{22},$$
$$x_{12}x_{21}=x_{21}x_{12},\ x_{11}x_{22}-q^{-1/2}x_{12}x_{21}=x_{22}x_{11}-q^{1/2}x_{21}x_{12}\equiv I_2,$$
$$x^*_{11}=x_{22},\ x^*_{12}=-q^{-1/2}x_{21},\ x^*_{21}=-q^{1/2}x_{12},\ x^*_{22}=x_{11}. \tag{2}$$

We do not fix I_2, a center, such as $I_2=1$. In the following are the CR's for q-boson operators, satisfying (i) and (ii):

$$a^+_k a^+_j=q^{1/2}a^+_j a^+_k,\ b^+_k b^+_j=q^{1/2}b^+_j b^+_k,\ b^+_k a^+_k=q^{1/2}a^+_k b^+_k,\ a^+_k b^+_j=b^+_j a^+_k,$$
$$b^+_k a^+_j=a^+_j b^+_k+(q^{1/2}-q^{-1/2})a^+_k b^+_j,\ b_k a^+_j=a^+_j b_k,\ a_k b^+_j=b^+_j a_k,\ a_j a^+_k=q^{1/2}a^+_k a_j,$$
$$b_k a^+_k=q^{1/2}a^+_k b_k+(q^{1/2}-q^{-1/2})\sum_{\ell<k}a^+_\ell b_\ell,\ b_k b^+_j=q^{1/2}b^+_j b_k+(q^{1/2}-q^{-1/2})a^+_j a_k,$$
$$a_k a^+_k=I_2+qa^+_k a_k+(q-1)\sum_{\ell<k}a^+_\ell a_\ell,$$
$$b_k b^+_k=I_2+qb^+_k b_k+(q-1)(a^+_k a_k+\sum_{\ell<k}b^+_\ell b_\ell)+(q^{1/2}-q^{-1/2})^2\sum_{\ell<k}a^+_\ell a_\ell:\quad k>j. \tag{3}$$

These are (31), (34) and (35) of ref. 2 with parameters in them being taken as $t'=t''=-1/2$. ($i>k$ of ref. 2 should read $i<k$.) The relations Hermitian conjugate to (3) are understood: $(a^+)^+=a$. We change notation conventions as

$$a^+_i\to a^+_{i1}\quad\text{and}\quad b^+_i\to a^+_{i2};\quad 1\le i\le m, \tag{4}$$

to find correspondence $x_{ij}\leftrightarrow a^+_{ij}$.

For a general $U_q(n)$ we settle the following relations for $\{x_{ij}\}$:[8]

$$x_{kj}x_{i\ell}=x_{ij}x_{k\ell};\ k>i,\ j<\ell,\quad x_{ik}x_{ij}=q^{1/2}x_{ij}x_{ik};\ k>j,$$
$$x_{kj}x_{ij}=q^{1/2}x_{ij}x_{kj};\ k>i,\quad x_{kj}x_{i\ell}=x_{i\ell}x_{kj}+(q^{1/2}-q^{-1/2})x_{k\ell}x_{ij};\ k>i,\ j>\ell. \tag{5}$$

We understand $*$-conjugate to (5). We settle CR's for $\{x_{ij}\}$ and $\{x^*_{i'j'}\}$ as

$$x^*_{ij}x_{k\ell}=q^{-\delta(i,k)/2}\{q^{\delta(j,\ell)}x_{k\ell}x^*_{ij}+(q^{1/2}-q^{-1/2})(\delta(j,\ell)\sum_{m<j}x_{km}x^*_{im}-\delta(i,k)\sum_{m>i}x^*_{mj}x_{m\ell})\}, \tag{6}$$

together with $$\sum_\ell x^*_{\ell j}x_{\ell k}=I_n\,\delta(j,k). \tag{7}$$

I_n is n-dimensional q-determinant of $|x_{ij}|$. We get CR's for q-boson operators:

$$a^+_{kj}a^+_{i\ell}=a^+_{ij}a^+_{k\ell};\ k>i,\ j<\ell,\quad a^+_{ik}a^+_{ij}=q^{1/2}a^+_{ij}a^+_{ik};\ k>j,$$
$$a^+_{kj}a^+_{ij}=q^{1/2}a^+_{ij}a^+_{kj};\ k>i,\quad a^+_{kj}a^+_{i\ell}=a^+_{i\ell}a^+_{kj}+(q^{1/2}-q^{-1/2})a^+_{k\ell}a^+_{ij};\ k>i,\ j>\ell, \tag{8}$$
$$a_{ij}a^+_{k\ell}=q^{(\delta(i,k)+\delta(j,\ell))/2}a^+_{k\ell}a_{ij}+\delta(i,k)q^{\delta(j,\ell)/2}(q^{1/2}-q^{-1/2})\sum_{m<i}a^+_{m\ell}a_{mj}$$
$$+\delta(j,\ell)q^{\delta(i,k)/2}(q^{1/2}-q^{-1/2})\sum_{n<j}a^+_{kn}a_{in}$$
$$+\delta(i,k)\delta(j,\ell)\{I_n+(q^{1/2}-q^{-1/2})^2\sum_{m<i,n<j}a^+_{mn}a_{mn}\}. \tag{9}$$

We can transcribe (5) into (8), replacing x_{ij} in the former by a^+_{ij} in the latter. The CR's for the q-operators with a single suffix are discussed in refs. 9, 10, etc.

3. Algebras of $U_q(2)$

Let us define A^+ by

$$A^+(jm: z_1 z_2) \equiv q^{(j^2-m^2)/4}\, z_1^{j+m}\, z_2^{j-m} \,/\, ([j+m]![j-m]!)^{1/2}\,. \tag{10}$$

The state $|jm> \equiv q^{-j(2j-1)/4} A^+(jm: a_{i1}^+ a_{i2}^+)|0>$ is normalized to unity: We put $I_2 = 1$ for (11)-(13). We use q-boson operators to express q-analog to angular-momentum operator of a coupled system (in $SU(2)$, $J_m = j_{m1} + j_{m2}$), and find

$$J_+ = q^{1/4}(a_1^+ b_1 + a_2^+ b_2)\,, \qquad J_- = q^{-1/4}(b_1^+ a_1 + b_2^+ a_2)\,,$$
$$J_0 = (-q^{-1/2} a_1^+ a_1 + q^{1/2} b_1^+ b_1) + (-q^{-1/2} a_2^+ a_2 + q^{1/2} b_2^+ b_2)\,. \tag{11}$$

These are symmetric in sites 1, 2. Actions of J_+, etc. on the basis state give

$$\begin{aligned} &J_\pm |(j'j")jm>_q \\ &= q^{j'+j"-1/2}([2j][2j+2]/[2])^{1/2}(1, j, \pm 1, m|jm\pm 1)_q |(j'j")jm\pm 1>_q \\ &= q^{j'+j"-1/2+(m\pm 1)/2}([j\mp m][j\pm m+1])^{1/2}|(j'j")jm\pm 1>_q\,, \end{aligned} \tag{12}$$

$$\begin{aligned} &J_0 |(j'j")jm>_q \\ &= -q^{j'+j"-1/2}([2j][2j+2])^{1/2}(1j0m|jm)_q |(j'j")jm>_q \\ &= q^{-(j'+j"+m-1/2)}(J_+ J_- - J_- J_+)|(j'j")jm>_q\,. \end{aligned} \tag{13}$$

We notice that $A^+(j'm'; a_1^+ b_1^+)$, which are coupled with $A^+(j"m"; a_2^+ b_2^+)|0>$ to produce $|(j'j")jm>_q$, does not commute with b_2.

We extend (1) in various ways. One way is given by the following relation between RF and QRF:[1,2]

$$d^j_{mk}(a'^+) = \sum_\ell d^j_{m\ell}(a^+)\, d^j_{\ell k}(x)\,, \tag{14}$$

in which $(x) = (x_{11}, x_{12}, x_{21}, x_{22})$, and (a^+) means x_{ij} of (x) being replaced by a^+_{ij}. It is shown that RF and QRF are describable as a coupled pair of A^+'s:

$$\begin{aligned} &S_2^{j'+j''-j}\, d^j_{mk}(z)(j'j''j'\ -j''\,|\,jk)_q \\ &= \sum_{m'} (j'j''m'm''\,|\,jm)_q\, ([2j']![2j'']!)^{1/2}\, A^+(j'm'; z_{11}z_{21}) A^+(j''m''; z_{12}z_{22})\,, \end{aligned} \tag{15}$$

where $j'' = j' - k$ and $z = x$ for RF and $z = a^+$ for QRF. We have defined $S = I_2$ and $S = a_{11}^+ a_{22}^+ - q^{-1/2} a_{12}^+ a_{21}^+$, respectively, for RF and QRF.

The present author investigated[3] q-symplecton polynomials obeying transformation (1) that induces mixing of particles and holes: Transformation is such that $a'^+_i = a_i^+ x_{11} + a_i x_{21}$ and $a'_i = a_i^+ x_{21} + a_i x_{22}$. In this formalism of $U_q(2)$, $*$-structure is specified such that

$$x_{12}^* = -q^{-1/2} x_{21}, \ x_{21}^* = -q^{1/2} x_{12}, \ (a^+)^* = q^{1/4} a, \ a^* = -q^{-1/4} a^+. \tag{16}$$

We put $J_+ = -a^+a^+/[2]$ and $J_- = aa/[2]$ satisfying $(J_+)^+ = -J_-$. If q-boson operators are used to realize $SU_q(1,1)$,[11] in which $(J_+)^+ = J_-$, we should specify $*$-structure as

$$x_{12}^* = q^{-1/2}x_{21},\ x_{21}^* = q^{1/2}x_{12},\ (a^+)^* = q^{1/4}a,\ a^* = q^{-1/4}a^+ . \tag{17}$$

We point out that in the $q \to 0$ limit[12] the q-symplecton polynomial $P(j,m;a^+a)$ becomes a monomial:

$$T(j,m) \equiv \lim_{q\to 0} q^{(j^2-m^2)/4} \begin{bmatrix} 2j \\ j-m \end{bmatrix}^{1/2} P(j,m) = a^{+(j+m)}a^{j-m} . \tag{18}$$

Using $aa^+ = 1$, which is $q \to 0$ limit of $aa^+ = q^{1/2}a^+a + 1$, we get

$$T(j,m)T(j'm') = \begin{cases} T(j-m', m+m'); & \text{if } j-j' \geq m+m', \\ T(j'+m, m+m'); & \text{if } j-j' < m+m'. \end{cases} \tag{19}$$

A specific feature is realization of Clebsch-Gordan rule such that

$$\{\sum_m T(j,m)\}\{\sum_{m'} T(j',m')\} = \sum_{J=|j-j'|}^{j+j'} \{\sum_M T(J,M)\}; \qquad M = m+m' . \tag{20}$$

References

1. M. Nomura, J. Phys. Soc. Japan 60,789(1991), 60,3260(1991).
2. M. Nomura, in: Proc. 20th Int. Conf. on "Differential Geometric Methods in Theoretical Physics," S.Catto and A. Rocha ed., World Scientific, Singapore (1992), Vol.1, p.574.
3. M. Nomura, J. Phys. Soc. Japan 60,4060(1991).
   Talk presented at 2nd Int. Wigner Symp. (Goslar, 1991).
4. M. Nomura, J. Phys. Soc. Japan 61,1485(1992).
5. M. Nomura, J. Phys. Soc. Japan 59,439(1991).
6. M. Nomura, J. Phys. Soc. Japan 59,2345(1990).
7. S.L. Woronowicz, Publ. RIMS. Kyoto Univ. 23,117(1987).
8. N.Yu. Reshetikhin, L.A. Takhtadzhyan and L.D. Faddeev, Leningrad Math. J. 1,193(1990).
9. H. Yamada in: in: Proc. 20th Int. Conf. on "Differential Geometric Methods in Theoretical Physics," S.Catto and A. Rocha ed., World Scientific, Singapore (1992), Vol.1, p.661.
10. M. Chaichian, P.Kulish and J. Lukierski, Phys. Lett. B262:43(1991).
    D.B. Fairlie and C.K.Zachos, Phys. Lett. B256:43(1991).
11. P.P. Kulish and E.V. Damaskinsky, J. Phys. A: Math. Gen. 23,L415(1990).
    A.I. Solomon and J. katriel, J. Phys. A: Math. Gen. 23,L1209(1990).
12. M. Kashiwara, RIMS,Kyoto Univ., preprint RIMS-676 (1989).

SUBDUCTION AND INDUCTION COEFFICIENTS OF HECKE ALGEBRAS AND CG COEFFICIENTS OF $SU_q(n)$

Feng Pan

Department of Physics, Liaoning Normal University, Dalian 116022, *P. R. China (Mailing address), and Department of Physics, Nanjing University, Nanjing* 210008, *P. R. China*

and

Jin-Quan Chen

Department of Physics, University of Pennsylvania, PA 19104, *U. S. A.*, *and Department of Physics, Nanjing University, Nanjing* 210008, *P. R. China*

ABSTRACT

A method for evaluation of Subduction Coefficients (SDC) and Induction Coefficients (IDC) of Hecke algebras is outlined. A procedure is formulated for evaluating the CG coeffitients of $SU_q(n)$ in the Gel'fand basis from the IDC's of the Hecke algebras $H_f(q)$ by using the Schur-Weyl duality.

Recently, braid groups and their representations draw evermore attentions of physicsts and mathematicians.[1-6] The Hecke algebras $H_f(q)$ are the special braid group realizations.

In the following we assume that q is real. Let $Y_m^{[\lambda]}$ be the standard Young tableau, and $|Y_m^{[\lambda]})q$ be the orthogonal basis vectors. Let $g_i(Y_m^{[\lambda]})$ be the tableau obtained by interchanging the numbers i and i+1 in $Y_m^{[\lambda]}$. It is understood that if the resultant tableau is not a standard one, the corresponding basis vector $|g_i(Y_m^{[\lambda]}))_q$ is equal to 0. The irrep of $H_f(q)$ in the standard basis is given by

$$g_i|(Y_m^{[\lambda]}))_q = \frac{q^d}{[d]}|Y_m^{[\lambda]})_q + \{\frac{[d+1][d-1]}{[d]^2}\}^{\frac{1}{2}}|g_i(Y_m^{[\lambda]}))_q, \tag{1}$$

where for given x

$$[x] = (q^x - q^{-x})/(q - q^{-1}), \tag{2}$$

and d is the axial distance from i to i+1 in $Y_m^{[\lambda]}$.

The orthogonal subduced basis of $H_f(q) \downarrow H_{f_1}(q) \times H_{f_2}(q)$ is

$$|[\lambda]^{\tau[\lambda_1][\lambda_2]}_{m_1\ m_2})_q \tag{3}$$

where $\tau = 1, 2, \ldots, \{\lambda_1\lambda_2\lambda\}$ is the multiplicity label. The basis in (3) is called the non-standard basis of $H_f(q)$. We can expand the non-standard basis in terms of the standard ones

$$|[\lambda]^{\tau[\lambda_1][\lambda_2]}_{m_1\ m_2}\rangle_q = \sum_{m} |^{[\lambda]}_{m}\rangle_q \langle^{[\lambda]}_{m}|^{\tau[\lambda_1][\lambda_2]}_{[\lambda]m_1\ m_2}\rangle_q \tag{4}$$

The expansion coefficient is called the $[\lambda]\downarrow[\lambda_1]\times[\lambda_2]$subduction coefficient. Using the Linear Equation Method (LEM), we have evaluated subduction coefficients of $H_f(q)$ for $n\leqslant 5$.

The induced rep $([\lambda_1]\times[\lambda_2])\downarrow H_f(q)$, which is called the outerproduct of the irreps $[\lambda_1]$ and $[\lambda_2]$, is a reducible rep of $H_f(q)$. The reduction is determined by the well-known Littelwood rule. The basis vectors of the irrep $[\lambda_1]$and $[\lambda_2]$, is a reducible rep of $H_f(q)$. The reduction is determined by the well-known Littelwood rule. [7]The basis vectors of the irrep $[\lambda]$ can be obtained from a linear combination of the normal ordered umcoupled basis vectors

$$|[\lambda]\tau m\rangle_q = \sum_{m_1 m_2 \omega} C^{[\lambda]\tau m}_{\lambda_1 m_1, \lambda_2 m_2, \omega}(q)\,|\lambda_1 m_1, \lambda_2 m_2; (\omega)\rangle_q \tag{5}$$

where τ is the multiplicity label needed in the outer-product, and the expansion coefficient is the $[\lambda_1]\times[\lambda_2]\downarrow[\lambda]$ induction coefficient. Using the LEM, wh have evaluated induction coefficients of $H_f(q)$ for $n\leqslant 4$.

From the Schur-Weyl duality between the Hecke algebra $H_f(q)$ in the standard basis is also a special Gel'fand basis of $U_q(n)$ for the same irrep. In the following, we will outline a procedure to obtain the general Gel'fand basis of $U_q(n)$ from the standard basis of $H_f(q)$.

Suppose that $|Y^{[\lambda]}_{m}\rangle_q$ is a Yamanouchi basis of the Hecke algebra $H_q(q)$ associated with the Young tableau $Y^{[\lambda]}_{m}$. The general Gel'fand basis vectors of $U_q(n)$ with $n\leqslant f\,|[\lambda](w)\rangle_q$, where (w) is the associated Weyl tableau, canbe obtained from the following method. (a)In the Young tableau $Y^{[\lambda]}_{m}$ the indices $1,2,\cdots,f$ are replaced by the double indices $1i_1, 2i_2, \cdots, fi_f$, respectively. Suppose that we have n indices $a_1, a_2, \cdots, a_n$ which are assigned to be $i_1, i_2, \cdots, i_f (n\leqslant f)$ in the following way

$$i_1 = \cdots = i_{n_1} = a_1, i_{n_1+1} = \cdots = i_{n_1+n_2} = a_2,$$

$$i_{f-n_n+1} = \cdots = i_f = a_n,\ f = \sum_{i=1}^{n} n_i. \tag{6}$$

The replacement has one restriction that no two identical indices a_i appear in the same column. However, there is only one set of $H_f(q)$ generators $g_1, g_2, \ldots, g_n$ at present is pseudo. It is not to be confused with the conventional notation for the $SU(mn)\supset SU(m)\times SU(n)$ labeling scheme.

(b) The indices in $1,2,\ldots,f$ should be symmetrized if the corresponding pseudo indices are the same. The symmetrization procedure is the following.

(1) if i and i+1 are in the same row, they have already been symmetrized, and (2) if i

and $i+1$ are not in the same row of the Young tableau $Y_r^{[\lambda]}$, up to a normalizing factor the basis vector which is symmetrized in i and $i+1$ is equal to

$$|_{\dot{i+1}}\,\dot{i}\rangle_q = [d-1]^{\frac{1}{2}}|Y_r^{[\lambda]}\rangle q + [d+1]^{\frac{1}{2}}|Y_s^{[\lambda]}\rangle q, \tag{7}$$

where the two dots above the indices i and $i+1$ indicates a symmetrization in i and $i+1$, while d is the axial distance between i and $i+1$ with $d>0$, $Y_s^{[\lambda]} = g_i Y_r^{[\lambda]}$, and the Yamanouchi number $r>s$.

By repeatedly using (7) for the adjacent indices to be symmetrized the basis $|[\lambda](w)\rangle_q$ can then be expressed in terms of $|Y_m^{[\lambda]}\rangle_q$

$$|[\lambda](w)\rangle_q = \sum_m{}' c_m^{\lambda w}|Y_m^{[\lambda]}\rangle_q, \tag{8}$$

where $c_m^{\lambda w}$ is the normalized symmetrization coefficients, and the prime in the summation symbol indicates that the sum is restricted to those m which give rise to the same Weyl tableau (w).

Finally, it can be proved that the CG coefficient of $SU_q(n)$ can be expressed in terms of the linear combination of induction coefficients of $H_f(q)$

$$C^{[\lambda](w)}_{[\lambda_1](w_1),[\lambda_2](w_2)}(q) = \sum_{m_1 m_2 m\omega}{}' C_{m_1}^{\lambda w_1} C_{m_2}^{\lambda_2 w_2} f_\omega C^{[\lambda]m}_{\lambda_1 m_1,\lambda_2 m_2,(\omega)}(q) \tag{9}$$

where $c_m^{\lambda w}$'s are the normalized symmetrization coefficients given above, and f_ω is the symmetrization coefficient associated with the two uncoupled basis vectors; and the symmetrization is carried out for those adjacent indices which are in two distinct uncoupled basis with the same corresponding pseudo indices. The evaluation of f_ω is quite similar to that of $c_m^{\lambda w}$'s. The $SU_q(n)$ CGC's for the resulting irreps $[n_1, n_2, n_3]$ with $n_1+n_2+n_3 \leqslant 4$ have been evaluated by using this procedure.

References

1. V. F. R. Jones, Ann. Math. 126(1987)335
2. H. Wenzl, Invent. Math. 92(1988)349; Commun. Math. Phys. 133(1990)383
3. M. Jimbo, Lett. Math. Phys. 10(1985)63; 11(1986)247
4. M. Wadati et al, Phys. Rep. 180(1989)247
5. R. Dipper and G. James, Proc. LMS 52(1986)20; 54(1987)57
6. M. L. Ge et al, in Nonlibear Physics, ed. C. H. Gu et al, (Springer-Verleg, Berlin, 1990)
7. Jin-Quan Chen, Group Representation Theory For Physicsts (World Scoentific, Singapore, 1989)

Solving the Yang-Baxter Equation [1]

Shi He (H. Shi)

Institute of Systems Science, Academia Sinica
Beijing 100080, China

Abstract

The Yang-Baxter equation (YBE) [1,9] plays an important role in many fields of mathematics and theoretical physics such as knot theory, quantum groups, statistical mechanics and integrable systems [3,4].

We set the matrix R and the permutation matrix P to be

$$R = \begin{pmatrix} x_{15} & x_3 & x_1 & x_{11} \\ x_5 & x_{14} & x_{10} & x_2 \\ x_7 & x_9 & x_{13} & x_4 \\ x_{12} & x_8 & x_6 & x_{16} \end{pmatrix}, \quad P = \begin{pmatrix} 1 & 0 & 0 & 0 \\ 0 & 0 & 1 & 0 \\ 0 & 1 & 0 & 0 \\ 0 & 0 & 0 & 1 \end{pmatrix},$$

where x_i are complex variables.

Let $\check{R} = RP$. Then the Yang-Baxter equation is

$$M = (m_{ij}) = \check{R}_{12}\check{R}_{23}\check{R}_{12} - \check{R}_{23}\check{R}_{12}\check{R}_{23} = 0, \tag{YBE}$$

where $\check{R}_{12} = \check{R} \otimes I$ and $\check{R}_{23} = I \otimes \check{R}$.

The YBE is a system of algebraic equations consisting of 64 cubic polynomial equations. The Wu Elimination, established by Wu Wen-tsun, provides a perfect method to solve such a system [6,7,8].

Let K be the complex field. We denote the polynomial set $\{p_1, p_2, ...p_n\}$ by PS where p_i are polynomials in the ring $K[x_1, x_2, ..., x_{16}]$. The system of $p_i = 0, i = 1, 2, ..., n$, is denoted by $PS = 0$. The set of all zeros of PS , i.e., the solution set of $PS = 0$, is denoted by $Zero(PS)$. The set of zeros of PS satisfying $I \neq 0$ is denoted by $Zero(PS/I)$.

The characteristic set of a polynomial set plays a profound role in Wu Elimination. Since a characteristic set is in triangular form, its zeros are well-obtained. Furthermore, the zeros of a polynomial set can be determined by some characteristic sets. The conclusions are

Theorem 1.(Wu Wen-tsun) If $CS = \{C_1, C_2, ..., C_r\}$ is the characteristic set of a polynomial set PS , then the zero set of PS has the following decomposition:

$$Zero(PS) = Zero(CS/I) + \sum_i Zero(PS, I_i), \tag{1}$$

[1] Work supported in part by the National Natural Science Foundation of China and CCAST (World Laboratory).

where I_i are the initials of C_i and I is the product of all I_i .

Theorem 2.(Wu Wen-tsun) The zero set of a polynomial set PS has the following decomposition:

$$Zero(PS) = \sum_j Zero(CS_j/I_j), \tag{2}$$

in which CS_j are some characteristic sets.

Theorem 3.(Wu Wen-tsun) By eliminating, the characteristic set of a polynomial set can be obtained in a finite number of steps.

By using Wu Elimination, the Yang-Baxter equation can be solved. Some multiparameter solutions of YBE are presented in [2]. In this paper we study the YBE for one of the most essential cases. We set

$$x_{14} = x_{13},\ x_{10} = x_9,\ x_8 = x_6,\ x_7 = x_5,\ x_4 = x_2,\ x_3 = x_1. \tag{3}$$

In this case, there are 10 unknowns $x_{16}, x_{15}, x_{13}, x_{12}, x_{11}, x_9, x_6, x_5, x_2, x_1$ in YBE. And the elements m_{ij}, $i = 1,3,6,8; j = 1,3,6,8$ of matrix M are zeros automatically. It is easy to check that the following relations between the polynomials m_{ij} are valid:

$$\begin{aligned}
&m_{12} = -m_{15},\ m_{17} = -m_{14},\ m_{21} = -m_{51},\ m_{41} = -m_{71},\ m_{28} = -m_{58},\\
&m_{48} = -m_{78},\ m_{82} = -m_{85},\ m_{84} = -m_{87},\ m_{22} = -m_{55},\ m_{77} = -m_{44},\\
&m_{54} = -m_{27},\ m_{45} = -m_{72},\ m_{26} = -m_{56},\ m_{43} = -m_{73},\ m_{34} = -m_{37},\\
&m_{62} = -m_{65},\ m_{23} = m_{76} = -m_{53} = -m_{46},\ m_{32} = m_{67} = -m_{35} = -m_{64},\\
&m_{25} = m_{74} = -m_{52} = -m_{47},\ m_{23} - m_{32} - m_{25} = 0,\\
&m_{48} + m_{24} + m_{12} = 0,\ m_{84} + m_{42} + m_{21} = 0,\ m_{15} + m_{57} + m_{78} = 0,\\
&m_{51} + m_{25} + m_{87} = 0,\ m_{27} - m_{37} - m_{78} = 0,\ m_{72} - m_{62} - m_{21} = 0.
\end{aligned}$$

Therefore, in order to solve the YBE with the assumption (3), it is sufficient to solve a system of 16 polynomial equations. These polynomials are

$$\begin{aligned}
m_{12} = {} & x_{15}x_{11}x_5 - x_{15}x_9x_1 + x_{13}x_{11}x_5 - x_{13}x_9x_1 + x_{12}x_{11}x_2\\
& + x_{12}x_{11}x_1 - x_{11}x_9x_6 - x_{11}x_9x_5 - x_6x_1^2 + x_5x_2x_1,\\
m_{21} = {} & x_{15}x_{12}x_1 - x_{15}x_9x_5 + x_{13}x_{12}x_1 - x_{13}x_9x_5 + x_{12}x_{11}x_6\\
& + x_{12}x_{11}x_5 - x_{12}x_9x_2 - x_{12}x_9x_1 + x_6x_5x_1 - x_5^2x_2,\\
m_{17} = {} & x_{16}x_{11}x_9 + x_{16}x_1^2 + x_{15}^2x_{11} - x_{15}x_{11}x_9 + 2x_{15}x_2x_1 - x_{15}x_1^2\\
& - x_{13}^2x_{11} - 2x_{13}x_1^2 - x_{11}x_6x_2 - x_{11}x_6x_1 + x_{11}x_5x_2 + x_{11}x_5x_1,\\
m_{71} = {} & x_{16}x_{12}x_9 + x_{16}x_5^2 + x_{15}^2x_{12} - x_{15}x_{12}x_9 + 2x_{15}x_6x_5 - x_{15}x_5^2\\
& - x_{13}^2x_{12} - 2x_{13}x_5^2 - x_{12}x_6x_2 + x_{12}x_6x_1 - x_{12}x_5x_2 + x_{12}x_5x_1,\\
m_{28} = {} & x_{16}^2x_{11} - x_{16}x_{11}x_9 - x_{16}x_2^2 + 2x_{16}x_2x_1 + x_{15}x_{11}x_9 + x_{15}x_2^2\\
& - x_{13}^2x_{11} - 2x_{13}x_2^2 + x_{11}x_6x_2 + x_{11}x_6x_1 - x_{11}x_5x_2 - x_{11}x_5x_1,\\
m_{82} = {} & x_{16}^2x_{12} - x_{16}x_{12}x_9 - x_{16}x_6^2 + 2x_{16}x_6x_5 + x_{15}x_{12}x_9 + x_{15}x_6^2\\
& - x_{13}^2x_{12} - 2x_{13}x_6^2 + x_{12}x_6x_2 - x_{12}x_6x_1 + x_{12}x_5x_2 - x_{12}x_5x_1,\\
m_{78} = {} & x_{16}x_{11}x_6 - x_{16}x_9x_2 + x_{13}x_{11}x_6 - x_{13}x_9x_2 + x_{12}x_{11}x_2 + x_{12}x_{11}x_1\\
& - x_{11}x_9x_6 - x_{11}x_9x_5 + x_6x_2x_1 - x_5x_2^2,\\
m_{87} = {} & x_{16}x_{12}x_2 - x_{16}x_9x_6 + x_{13}x_{12}x_2 - x_{13}x_9x_6 + x_{12}x_{11}x_6 + x_{12}x_{11}x_5\\
& - x_{12}x_9x_2 - x_{12}x_9x_1 - x_6^2x_1 + x_6x_5x_2,
\end{aligned}$$

$$\begin{aligned}
m_{23} = \; & x_{13}x_{12}x_{11} - x_{13}x_9^2 + x_{13}x_6x_1 - x_{13}x_5x_2 + x_{12}x_1^2 + x_{11}x_6^2 \\
& -x_9x_6x_2 - x_9x_5x_1, \\
m_{32} = \; & x_{13}x_{12}x_{11} - x_{13}x_9^2 - x_{13}x_6x_1 + x_{13}x_5x_2 + x_{12}x_2^2 + x_{11}x_5^2 \\
& -x_9x_6x_2 - x_9x_5x_1, \\
m_{27} = \; & x_{16}x_{13}x_1 + x_{16}x_{11}x_6 + x_{15}x_{13}x_2 + x_{15}x_{11}x_5 - x_{13}^2x_2 - x_{13}^2x_1 \\
& -x_{13}x_9x_2 - x_{13}x_9x_1 + x_{12}x_{11}x_2 + x_{12}x_{11}x_1 - x_{11}x_9x_6 - x_{11}x_9x_5 \\
& -x_6x_2^2 + x_6x_2x_1 + x_5x_2x_1 - x_5x_1^2, \\
m_{72} = \; & x_{16}x_{13}x_5 + x_{16}x_{12}x_2 + x_{15}x_{13}x_6 + x_{15}x_{12}x_1 - x_{13}^2x_6 - x_{13}^2x_5 \\
& -x_{13}x_9x_6 - x_{13}x_9x_5 + x_{12}x_{11}x_6 + x_{12}x_{11}x_5 - x_{12}x_9x_2 - x_{12}x_9x_1 \\
& -x_6^2x_2 + x_6x_5x_2 + x_6x_5x_1 - x_5^2x_1, \\
m_{26} = \; & x_{16}x_{13}x_1 + x_{16}x_{11}x_6 + x_{15}x_{13}x_2 + x_{15}x_9x_1 - x_{13}^2x_2 - x_{13}^2x_1 \\
& -x_{13}x_{11}x_5 - x_{13}x_9x_2 - x_6x_2^2 + x_6x_2x_1 + x_6x_1^2 - x_5x_1^2, \\
m_{73} = \; & x_{16}x_{13}x_5 + x_{16}x_9x_6 + x_{15}x_{13}x_6 + x_{15}x_{12}x_1 - x_{13}^2x_6 - x_{13}^2x_5 \\
& -x_{13}x_{12}x_2 - x_{13}x_9x_5 - x_6^2x_2 + x_6^2x_1 + x_6x_5x_1 - x_5^2x_1, \\
m_{22} = \; & x_{16}x_{12}x_{11} + x_{16}x_5x_1 + x_{15}^2x_9 - x_{15}x_9^2 + x_{15}x_6x_1 + x_{15}x_5x_2 \\
& -x_{15}x_5x_1 - x_{13}^2x_9 - 2x_{13}x_5x_1 + x_{12}x_2x_1 + x_{11}x_6x_5 - x_9x_6x_2 \\
& -x_9x_6x_1 - x_9x_5x_2 + x_9x_5x_1, \\
m_{77} = \; & x_{16}^2x_9 - x_{16}x_9^2 - x_{16}x_6x_2 + x_{16}x_6x_1 + x_{16}x_5x_2 + x_{15}x_{12}x_{11} \\
& +x_{15}x_6x_2 - x_{13}^2x_9 - 2x_{13}x_6x_2 + x_{12}x_2x_1 + x_{11}x_6x_5 + x_9x_6x_2 \\
& -x_9x_6x_1 - x_9x_5x_2 - x_9x_5x_1.
\end{aligned}$$

We denote the set of these polynomials by YS .

The calculation is much simpler when some unknowns are zero. But the discussion is lengthier if we try to list it case by case. In this paper only the most important case is considered. We take the following agreement:

Hypothesis: All unknowns are not zeros.

For the polynomial set YS we have

Proposition. The zero set of YS , i.e., the set of solutions of the system $YS = 0$, can be decomposed into two parts: (I) the zeros of $\{YS, p(11)\}$ and (II) the zeros of $\{YS, p(12), p(13), p(21)\}$, in which

$$\begin{aligned}
p(11) = \; & x_6x_1 - x_5x_2, \\
p(12) = \; & x_{15}x_{12}x_{11} - x_{15}x_9^2 + x_{13}x_{12}x_{11} - x_{13}x_9^2 + x_{12}x_2^2 + x_{12}x_2x_1 + x_{11}x_6^2 \\
& +x_{11}x_6x_5 - 2x_9x_6x_2 - x_9x_6x_1 - x_9x_5x_2, \\
p(13) = \; & x_{16}x_{12}x_{11} - x_{16}x_9^2 + x_{13}x_{12}x_{11} - x_{13}x_9^2 + x_{12}x_2x_1 + x_{12}x_1^2 \\
& +x_{11}x_6x_5 + x_{11}x_5^2 - x_9x_6x_1 - x_9x_5x_2 - 2x_9x_5x_1, \\
p(21) = \; & x_{12}x_{11}^2x_6 + x_{12}x_{11}^2x_5 - x_{12}x_2^2x_1 - x_{12}x_2x_1^2 - x_{11}x_9^2x_6 - x_{11}x_9^2x_5 \\
& -x_{11}x_6^2x_1 - x_{11}x_5^2x_2 + 2x_9x_6x_2x_1 + 2x_9x_5x_2x_1.
\end{aligned}$$

The main result of this paper is

Theorem I. Assume the unknowns x_i are not zeros. All solutions of polynomial set $\{YS, x_6x_1 - x_5x_2\}$ are provided by 30 characteristic sets. Thus, a series of solutions of YBE are obtained.

These 30 characteristic sets are listed in [5].

Example. One of the simplest characteristic sets is

$$CS = \{B_3, B_4, B_7, B_1, B_2\},$$

in which $B_3 = x_{16} + x_{13}$, $B_4 = x_{15} + x_{13}$, $B_1 = x_6 + x_5$, $B_2 = x_2 + x_1$, and $B_7 = x_{13}(x_{12}x_{11} - x_9^2) + x_{12}x_1^2 + x_{11}x_5^2 - 2x_9x_5x_1$. From this characteristic set, two solutions of YBE are provided:

$$\begin{pmatrix} -1 & x_1 & x_1 & y_1^2 - x_1^2 \\ x_5 & 1 & -x_5x_1 \pm y_1y_2 & -x_1 \\ x_5 & -x_5x_1 \pm y_1y_2 & 1 & -x_1 \\ y_2^2 - x_5^2 & -x_5 & -x_5 & -1 \end{pmatrix},$$

in which x_1, x_5, y_1, y_2 are four parameters.

The author wishes to express his gratitude to Prof. M.L. Ge, H.Y. Guo, B.L. Hao, Z.M. Qiu and Dr. S.M. Fei for their helpful suggestions. The author is thankful to CCAST (World Laboratory) for its support.

References

1. R.J. Baxter, *Exactly Solved Models in Statistical Physics*, Academia Press, 1982.
2. S.M. Fei, H.Y. Guo and H. Shi, *Multiparameter solutions of the Yang-Baxter equation*, J. Phys. A: Math. Gen. **25** (1992) 2711-2720.
3. M.L.Ge, C.N.Yang, *Bride group, knot theory and statistical mechanics.* World Scientific, 1989.
4. M.Jimbo, *Yang-Baxter equation in integrable systems.* World Scientific.
5. H.Shi, *On solving of the Yang-Baxter equation.* Proceedings of IWMM'92, International Academic Publishers, Beijing (1992),79-88.
6. Wu Wen-tsün, *Basic Principle of Mechanical Theorem Proving in Geometries (Part on Elementary Geometries)*, Science Press, Beijing, 1984 (in Chinese).
7. Wu Wen-tsün, *Basic Principles of Mechanical Theorem Proving in Elementary Geometries*, J. Sys. Sci. and Math. Scis. **4**(1984) 207-235.Republished in J. of Automated Reasoning **2**(1986)221-252.
8. Wu Wen-tsün, *A Zero Structure Theorem for Polynomial Equation-Solving*, MM Research Preprints, No 1, 1987, Inst. of Systems Science, Academia Sinica.
9. C.N. Yang, Phys. Rev. Lett. **19**(1967)1312-1314.

A NEW SOLUTION OF THE YANG-BAXTER EQUATION WITH COMPLETE Z SYMMETRY

YOUICHI SHIBUKAWA

Department of Mathematics, Hokkaido University, Sapporo 060, Japan

and

KIMIO UENO

Department of Mathematics, Waseda University, 3-4-1 Ohkubo Shinjuku-ku, Tokyo 169, Japan

ABSTRACT

An infinite-dimensional R matrix related to the limiting case $n \to \infty$ of the completely $\mathbb{Z}_n$ symmetric R matrix is discovered. This R matrix is expressed as an operator on $C^\infty(S^1 \times S^1)$.

1. Introduction

The quantum group is closely related to Macdonald's symmetric polynomials. Ueno–Takebayashi [13] and Noumi [9] proved that the zonal spherical functions on some quantum homogeneous spaces are nothing but Macdonald's symmetric polynomials. But, in this case, we can not deal with Macdonald's symmetric functions. Roughly speaking, Macdonald's symmetric functions are the polynomials with infinite variables and two parameters. Then we aim at defining the quantum group $U_{q,t}(gl_\infty)$, a deformation of $U(gl_\infty)$ with two parameters q and t for giving the geometric interpretation of Macdonald's symmetric functions.

On the other hand, Freund–Zabrodin [5] and Zabrodin [15] showed that Macdonald's symmetric functions are related to the limiting case $n \to \infty$ of the completely $\mathbb{Z}_n$ symmetric model.

In this paper, we shall introduce a new infinite-dimensional R matrix, which is related to the limiting case $n \to \infty$ of the completely $\mathbb{Z}_n$ symmetric R matrix (cf. [4] [6] [7]). This R matrix satisfies the Yang-Baxter equation

$$R_{12}(\lambda_1)R_{13}(\lambda_1+\lambda_2)R_{23}(\lambda_2) = R_{23}(\lambda_2)R_{13}(\lambda_1+\lambda_2)R_{12}(\lambda_1). \tag{1.1}$$

Further, by the Fourier transformation, we shall give an R operator on $C^\infty(S^1\times S^1)$. This R operator is also a solution of the Yang-Baxter equation (1.1).

2. Completely $\mathbf{Z}$ symmetric R matrix

Let us consider an infinite-dimensional R matrix

$$R(\lambda) = \sum_{i,j,k,\ell \in \mathbf{Z}} R_{i,j}^{k,\ell}(\lambda) E_{ik} \otimes E_{j\ell}. \tag{2.1}$$

The R matrix is called to be completely $\mathbf{Z}$ symmetric if there exists $S^{ab}(\lambda)$ $(a, b \in \mathbf{Z})$ such that

$$R_{i,j}^{k,\ell}(\lambda) = \delta_{i+j,k+\ell} S^{k-i,\ell-i}(\lambda). \tag{2.2}$$

We find an completely $\mathbf{Z}$ symmetric R matrix, which is related to the limiting case $n \to \infty$ of the completely $\mathbf{Z}_n$ symmetric R matrix ([1] [2] [3] [10]).

Theorem 2.1. *We assume $|q| < 1$, and then the completely $\mathbf{Z}$ symmetric R matrix with*

$$S^{ab}(\lambda) = -\frac{q^a}{e^{2\pi\sqrt{-1}\kappa} - q^a} + \frac{q^b}{e^{2\pi\sqrt{-1}\lambda} - q^b} \tag{2.3}$$

satisfies the Yang-Baxter equation (1.1). *Here κ is an arbitrary constant.*

3. R operator

The Fourier transformation of the weight $R_{ij}^{k\ell}(\lambda)$ is as follows (see [6] [8]).

Proposition 3.1. *For $x, y \in \mathbf{R}$, $|Im\,\lambda| < Im\,\tau$, and $|Im\,\kappa| < Im\,\tau$,*

$$\begin{aligned} &\sum_{i,j \in \mathbf{Z}} R_{i,j}^{k,\ell}(\lambda) e^{2\pi\sqrt{-1}(ix+jy)} \\ &= G(x-y:\lambda) e^{2\pi\sqrt{-1}(\ell x+ky)} - G(x-y:\kappa) e^{2\pi\sqrt{-1}(kx+\ell y)}. \end{aligned} \tag{3.1}$$

Here

$$G(x:\lambda) = \frac{1}{2\pi\sqrt{-1}} \frac{\vartheta_1'(0)\vartheta_1(\lambda+x)}{\vartheta_1(\lambda)\vartheta_1(x)}, \tag{3.2}$$

and $\vartheta_1(z)$ is the elliptic theta function

$$\vartheta_1(z) = 2q^{\frac{1}{8}} \sin \pi z \prod_{m=1}^{\infty} (1-q^m)(1-2q^m \cos 2\pi z + q^{2m}). \tag{3.3}$$

By the above proposition, the R matrix is expressed as an operetor on $\mathcal{V} \otimes \mathcal{V}$, where $\mathcal{V} = C^\infty(S^1)$:

$$\begin{aligned} &(R(\lambda)(\varphi_k \otimes \varphi_\ell))(x,y) \\ &\stackrel{\text{def}}{=} G(x-y:\lambda)(\varphi_k \otimes \varphi_\ell)(y,x) - G(x-y:\kappa)(\varphi_k \otimes \varphi_\ell)(x,y). \end{aligned} \tag{3.4}$$

Here the set $\{\varphi_k(x) = e^{2\pi\sqrt{-1}kx};\, k \in \mathbf{Z}\}$ is a basis of $\mathcal{V}$.

Let $\mathcal{V}^{\hat{\otimes} N} = \overbrace{\mathcal{V}\hat{\otimes}\cdots\hat{\otimes}\mathcal{V}}^{N times} \stackrel{\text{def}}{=} C^\infty(S^1 \times \cdots \times S^1)$ for $N \geq 2$. We define the operator $R_{ij}(\lambda) \in End(\mathcal{V}^{\hat{\otimes} N})$ $(1 \leq i, j \leq N, i \neq j)$.

Definition 3.1. For $\varphi \in \mathcal{V}^{\hat{\otimes} N}$, let

$$(G_{ij}(\lambda)\varphi)(x_1, x_2, \ldots, x_N) = G(x_i - x_j : \lambda)\varphi(x_1, x_2, \ldots, x_N), \tag{3.5}$$

$$(\sigma_{ij}\varphi)(\ldots, x_i, \ldots, x_j, \ldots) = \varphi(\ldots, x_j, \ldots, x_i, \ldots). \tag{3.6}$$

Then we define

$$R_{ij}(\lambda) = G_{ij}(\lambda)\sigma_{ij} - G_{ij}(\kappa). \tag{3.7}$$

When $N = 2$, $i = 1$, and $j = 2$, we denote $R_{12}(\lambda)$ by $R(\lambda)$ and call it the R operator.

Theorem 3.2. *The R operator $R(\lambda)$ satisfies the Yang-Baxter equation* (1.1) *in $End(\mathcal{V}^{\hat{\otimes} 3})$.*

The proof of the theorem above leads us to the more general one.

Theorem 3.3. *If an analytic function $\theta(x)$ satisfies the three term equation*

$$\begin{aligned} &\theta(x+y)\theta(x-y)\theta(z+w)\theta(z-w) \\ &+ \theta(x+z)\theta(x-z)\theta(w+y)\theta(w-y) \\ &+ \theta(x+w)\theta(x-w)\theta(y+z)\theta(y-z) \\ &= 0, \end{aligned} \tag{3.8}$$

then the operator

$$R(\lambda) \stackrel{\text{def}}{=} G(\lambda)\sigma - G(\kappa) \tag{3.9}$$

is a solution of the Yang-Baxter equation (1.1). *Here*

$$(G(\lambda)\varphi)(x, y) = \frac{\theta'(0)\theta(\lambda + x - y)}{\theta(\lambda)\theta(x - y)}\varphi(x, y), \tag{3.10}$$

$$(\sigma\varphi)(x, y) = \varphi(y, x) \tag{3.11}$$

for a function $\varphi(x, y)$.

The analytic solutions of the three term equation above are given by

$$\theta(x) = \vartheta_1(x)\exp(\frac{1}{2}Ax^2 + B), \tag{3.12}$$

$$\theta(x) = \sin(\pi x)\exp(\frac{1}{2}Ax^2 + B), \tag{3.13}$$

$$\theta(x) = x\exp(\frac{1}{2}Ax^2 + B), \tag{3.14}$$

where A and B are arbitrary constants (see [14] p.461).

For this R operator (3.7), one can obtain the first inversion relation and can apply the fusion procedure to this case ([11]).

Let $\mathfrak{F}$ be a finite index set, and let $V^{\mathfrak{F}} = \oplus_{\alpha\in\mathfrak{F}} \mathbf{C} f_\alpha$ be a finite-dimensional subspace of $\mathcal{V}$ with a basis $\{f_\alpha : \alpha \in \mathfrak{F}\}$. If the R operator preserves $V^{\mathfrak{F}} \otimes V^{\mathfrak{F}}$, then we call $V^{\mathfrak{F}} \otimes V^{\mathfrak{F}}$ a finite-dimensional representation of $R(\lambda)$, and define $R^{\mathfrak{F}}(\lambda)$ to be the matrix representation of $R(\lambda)|_{V^{\mathfrak{F}}\otimes V^{\mathfrak{F}}} \in End(V^{\mathfrak{F}} \otimes V^{\mathfrak{F}})$ with respect to the basis $\{f_\alpha \otimes f_\beta\}$. We should remark that $R^{\mathfrak{F}}(\lambda)$ automatically becomes a solution of the Yang-Baxter equation (1.1).

We can construct finite-dimensional representations of the trigonometric case (3.13) with $A = B = 0$.

Proposition 3.4. *For $n \geq 1$, we set $\mathfrak{F} = \{0, 1, \ldots, n-1\}$, $f_\alpha(x) = e^{2\pi\sqrt{-1}\alpha x}$ $(\alpha \in \mathfrak{F})$, and $V^{\mathfrak{F}} = \oplus_{\alpha\in\mathfrak{F}} \mathbf{C} f_\alpha$. Then $V^{\mathfrak{F}} \otimes V^{\mathfrak{F}}$ is a finite-dimensional representation.*

These finite-dimensional R matrices $R^{\mathfrak{F}}(\lambda)$ are trigonometric solutions of the Yang-Baxter equation, but not triangular. We notice that one can get the trigonometric, triangular R matrices from these matrices $R^{\mathfrak{F}}(\lambda)$ (see [11]).

4. Acknowledgements

The authors express their gratitude to Professor M. L. Ge for inviting them to this conference. The second author is partially supported by Grant-in-Aid for Scientific Research on Priority Areas 231 "Infinite Analysis".

References

1. A. A. Belavin, Nucl. Phys. **B180** [FS2] (1981) 189–200.
2. I. V. Cherednik, Soviet Math. Dokl. **33** No.2 (1986).
3. ———, J. Soviet Math. **38** No.3 (1987) 1989–2027.
4. V. A. Fateev, A. B. Zamolodchikov, Phys. Lett. **92A** No.1 35–36.
5. P. G. O. Freund, A. V. Zabrodin, $\mathbb{Z}_n$ Baxter models and quantum symmetric spaces, Chicago preprint EFI 92–11.
6. M. Gaudin, J. Phys. France **49** (1988) 1857–1865.
7. C. Gómez, G. Sierra, A new solution to the star-triangle equation based on $U_q(sl(2))$ at roots of unit, to appear in Nucl. Phys. B.
8. C. Jordan, Fonctions elliptiques, Springer 1981.
9. M. Noumi, Macdonald's symmetric polynomials as zonal spherical functions on some quantum homogeneous spaces, preprint (1992).
10. M. P. Richey, C. A. Tracy, J. Stat. Phys. **42** 3/4 (1986) 311–348.
11. Y. Shibukawa, K. Ueno, to appear in Lett. Math. Phys.
12. E. K. Sklyanin, Funct. Anal. Appl. **16** (1983) 263–270; **17** (1984) 273–284.
13. K. Ueno, T. Takebayashi, Zonal spherical functions on quantum symmetric spaces and Macdonald's symmetric polynomials, Lecture Notes in Mathematics 1510 Springer (1992) 142–147.
14. E. T. Whittaker, G. N. Watson, A course of modern analysis, Cambridge.
15. A. V. Zabrodin, to appear in Mod. Phys. Lett. A.

Quantized enveloping algebras associated to simple Lie superalgebras and their universal R-matrices

Hiroyuki Yamane

Faculty of Science, Osaka Univ.,
Toyonaka, 560, Japan.

Introduction

Let G be a simple Lie superalgebra of type A-G. Let Π and p: $\Pi \to \{0,1\}$ be a set of simple roots and a parity function of G. (We assume that (Π,p) is of the distinguished type if G is of type F_4 or G_3.) We define a (Jimbo type) quantized enveloping superalgebras $U_q(\Pi,p)$ of $U(G)$ by Chevalley generators and defining relations. (Moreover, letting $q\to 1$ in the defining relations, we obtain defining relations of $U(G)$. This result also seems to be new.) A remarkable fact is that the defining relations are not exhausted by binary relations such as the q-Serre relations; we also need trinomial and quadrinomial relations.

Let σ be the involution of $U_q(\Pi,p)$ defined by $\sigma(x) = x$ (resp. $-x$) for an even (resp. odd) element x. We introduce the "Hopf algebraization (Bosonaization)" $U_q(\Pi,p)^\sigma$ of $U_q(\Pi,p)$; as an algebra, $U_q(\Pi,p)^\sigma$ is the twisted product of $U_q(\Pi,p)$ and $\langle\sigma\rangle = \mathbb{Z}/2\mathbb{Z}$. Our main result describes explicitly the universal R-matrices $\widetilde{R}_q$ of $U_q(\Pi,p)^\sigma$ by using certain q-root vectors and the q-exponential. Our construction is due to Drinfeld's quantum double construction. In the process of quantum double construction, we also show a P.B.W. theorem for $U_q(\Pi,p)$.

We also introduce a finite dimensional quantized algebras $u_\zeta(\Pi,p)^\sigma$ at roots ζ of unity and their universal R-matrices $\widetilde{R}_\zeta$.

In [Y1-2], we introduced a (Drinfeld type) quantized enveloping superalgebras $U_h(\Pi,p)$, showed $U_h(\Pi,p)$ is an h-adic topologically free C[[h]]-Hopf superalgebras, and gave an explicit formula of the universal R-matrices $\widetilde{R}_h$ of $U_h(\Pi,p)^\sigma$. The arguements used in this paper are the essentially same arguements in [Y1-2].

1. Root systems of simple Lie superalgebras.

Let G be a simple Lie superalgebra of type A_{N-1}, B_N, C_N, D_N, F_4, G_3. In this paper, we treat $(\Pi = \{\alpha_1,\ldots,\alpha_n\}$, p: $\Pi \to \{0,1\})$ associated to following Dynkin diagrams. The i-th dot stands for ✕ for ○, ⊗ if $(\alpha_i,\alpha_i) \neq 0$ (resp. $= 0$). If i-th dot are ○, ⊗, ●, then we define $p(\alpha_i) = 0, 0, 1$ respectively.

(A_{N-1}) 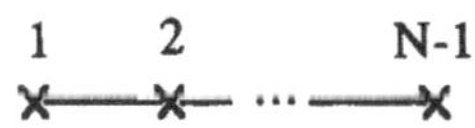 ,

(B_N)  ,  ,

(C_N) 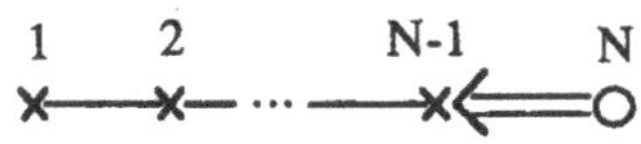 ,

(D_N) 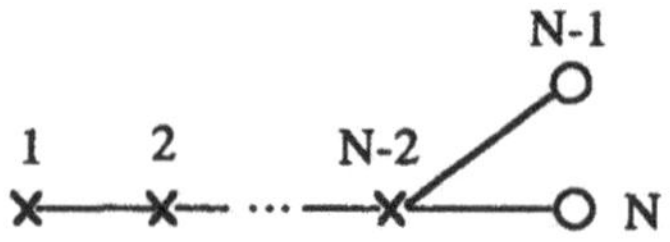 , 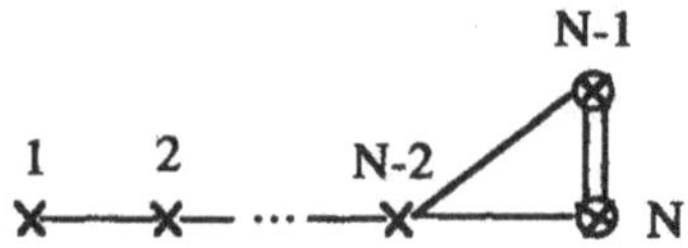 ,

(F_4) 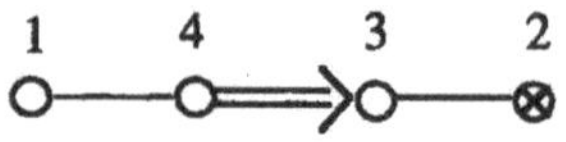 ,

(G_3) 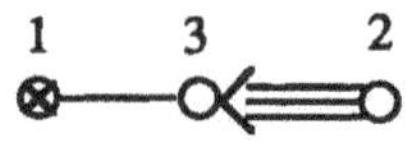 .

2. The defining relations of $U_q(\Pi,p)$ and $U(G)$.
Let E_i, F_i, K_i be Chevally generators of $U_q(\Pi,p)$. Let X_λ be an element satisfying $K_iX_\lambda K_i^{-1}=q^{(\lambda,\alpha_i)}X_\lambda$. Put $[X_\lambda , X_\mu] = X_\lambda X_\mu - (-1)^{p(\lambda)p(\mu)}q^{-(\lambda,\mu)}X_\mu X_\lambda$ and $[X_\lambda , X_\mu] = X_\lambda X_\mu - (-1)^{p(\lambda)p(\mu)}X_\mu X_\lambda$. Let $(\mathrm{ad}_{[\ ]}X_\lambda)(X_\mu) = [X_\lambda , X_\mu]$. Then defining relations of the C-algebra $U_q(\Pi,p)$ are given by :

(2.1) $K_iK_j = K_jK_i$, $K_iK_i^{-1}= K_i^{-1}K_i= 1$,

(2.2) $K_iE_jK_i^{-1}= q^{(\alpha_i,\alpha_j)}E_j$, $K_iF_jK_i^{-1}= q^{-(\alpha_i,\alpha_j)}F_j$,

(2.1) $[E_i , F_j] = \delta_{ij}(K_i-K_i^{-1})/(q^{d_i}- q^{-d_i})$ (Here d_i's are some integers defined in [Y1-2].)

(2.4) (i) $[E_i , E_j] = 0$ if $(\alpha_i,\alpha_j) = 0$,

(ii) $(\mathrm{ad}_{[\ ]}E_i)^{m_{ij}}(E_j) = 0$ if $(\alpha_i,\alpha_j) \neq 0$ and $m_{ij} = 2(\alpha_i,\alpha_j)/(\alpha_i,\alpha_i)$,

(iii) $[[[E_i , E_j], E_k], E_j] = 0$

if 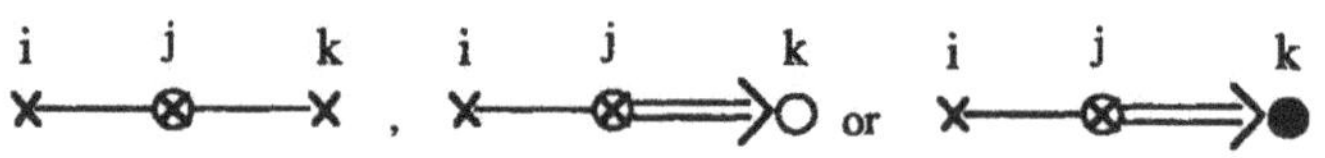 ,

(iv) $[[E_{N-2} , E_{N-1}], E_N] = [[E_{N-2} , E_N], E_{N-1}]$ if 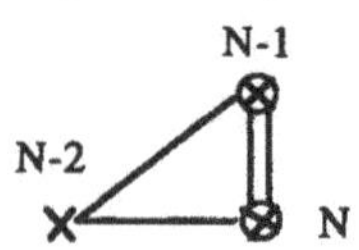 ,

(v) $[[E_{N-2} , E_{N-1}], [[E_{N-2} , E_{N-1}], E_N]] , E_{N-1}] = 0$ if 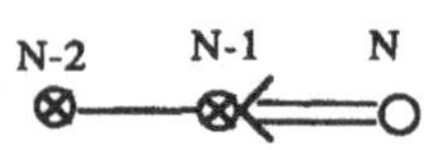 ,

(vi) $[[[[[[E_{N-3} , E_{N-2}], E_{N-1}], E_N], E_{N-1}], E_{N-2}], E_{N-1}] = 0$ if 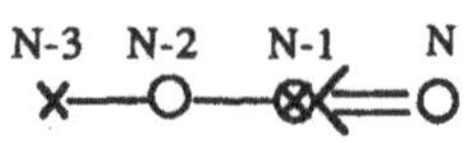 .

Theorem (Serre's theorem of $U(G)$). Letting $q \to 1$ in the defining relations, we obtain defining relations of $U(G)$.

This result also seems to be new.

$U_q(\Pi,p)$ is a Hopf superalgebra with a coproduct Δ defined by $\Delta(K_i) = K_i \otimes K_i$, $\Delta(E_i) = E_i \otimes 1 + K_i \otimes E_i$, $\Delta(F_i) = F_i \otimes K_i^{-1} + 1 \otimes F_i$.

3. Main results

Let Δ_+ be the positive roots with respect to (Π,p) in 2 and $\Delta_+^{red} = \{\beta \in \Delta_+ | \beta/2 \notin \Delta_+\}$. For $\beta = c_1\alpha_1 + \cdots + c_N\alpha_N$, put $ht(\beta) = c_1 + \cdots + c_N$, $g(\beta) = \min\{i \mid i \neq 0\}$, $c_\beta = c_{g(\beta)}$ and $\underline{ht}(\beta) = ht(\beta)/c_\beta$. For $\alpha, \beta \in \Delta_+^{red}$, we say that $\alpha < \beta$ if they satisfy one of the following: (1) $g(\alpha) < g(\beta)$, (2) $g(\alpha) = g(\beta)$ and $\underline{ht}(\alpha) < \underline{ht}(\beta)$, (3) if (Π,p) is of type D_N and $p(\alpha_i + \cdots + \alpha_{N-1}) = 1$, we put

$$\alpha_i + \cdots + \alpha_{N-1} < 2\alpha_i + \cdots + 2\alpha_{N-2} + \alpha_{N-1} + \alpha_N < \alpha_i + \cdots + \alpha_{N-2} + \alpha_N .$$

We define q-root vectors E_β and $F_\beta \in U_q(\Pi,p)$ $(\beta \in \Delta_+^{red})$ as follows: (1) Put $E_{\alpha_i} = E_i$ $(\alpha_i \in \Pi)$. (2) For α, $\alpha + \alpha_i \in \Delta_+^{red}$, $\alpha_i \in \Pi$ such that $g(\alpha) < i$, we put $E_{\alpha+\alpha_i} = [E_\alpha, E_i]$. (3) For α, $\alpha+\beta$, $\beta \in \Delta_+^{red}$ such that $g(\alpha) = g(\beta)$, $\alpha < \beta$, $\underline{ht}(\beta) - \underline{ht}(\alpha) \leq 1$, we put $E_{\alpha+\beta} = [E_\alpha, E_\beta]$. (4) F_β is wrriten similar to (1),(2),(3) by replacing generators E_i's by F_i's. In the process of quantum double construction (and a similer argument to [Y0]), we have:

Theorem (P.B.W Theorem of $U_q(\Pi,p)$). If $q^m \neq 1$ for $m >> 0$, then the elements

$$\Big(\prod^{<}_{\alpha \in \Delta_+^{red}} E_\alpha^{m_\alpha}\Big)\Big(\prod_{\alpha_i \in \Pi} K_i^{l_i}\Big)\Big(\prod^{<}_{\beta \in \Delta_+^{red}} F_\beta^{n_\beta}\Big)$$

(m_α, $n_\beta \in \mathbb{Z}_+$, $l_i \in \mathbb{Z}$; $\prod\limits^{<}_{\alpha \in \Delta_+^{red}}$ denotes the product taken with a total order on Δ_+^{red} compatible with the partial order $<$.) form a C-basis of $U_q(\Pi,p)$.

Let $t_0 = \Sigma_i \varepsilon_i \otimes \varepsilon_i \in \boldsymbol{H} \otimes \boldsymbol{H}$ $((\varepsilon_i, \varepsilon_j) = \delta_{ij})$ be the canonical element of the Cartan part $\boldsymbol{H}$ of $\boldsymbol{G}$. Let $U_q(\Pi,p)^\sigma$ be a twisted product of $U_q(\Pi,p)$ and $<\sigma>$ such that $\sigma^2 = 1$ and $X = (-1)^{p(X)}X$ $(X \in U_q(\Pi,p))$. $U_q(\Pi,p)^\sigma$ is a Hopf algebra with a coproduct $\Delta^{(\sigma)}$ such that $\Delta^{(\sigma)}(X) = X^{(1)}_i \sigma^{p(X^{(2)}_i)} \otimes X^{(2)}_i$ $(X \in U_q(\Pi,p);\ \Delta(X) = X^{(1)}_i \otimes X^{(2)}_i)$, $\Delta^{(\sigma)}(\sigma) = \sigma \otimes \sigma$. By using Drinfeld's quantum double construction[D] (and Tanisaki's argument[T]), we have:

Theorem(Universal R-matrix of $U_q(\Pi,p)^\sigma$). Assume that q is not a root of unity. Then the universal R-matrix of $U_q(\Pi,p)^\sigma$ is given by

$$\widetilde{R}_q = \Big(\prod^{<}_{\alpha \in \Delta_+^{red}} e((q^{d_\alpha} - q^{-d_\alpha})V(\alpha)E_\alpha \otimes F_\alpha ; (-1)^{p(\alpha)}q^{(\alpha,\alpha)})\Big) \cdot q^{t_0} \cdot \Big(2^{-1} \cdot \prod_{0 \leq c_1, c_2 \leq 1} (-1)^{c_1 c_2} \sigma^{c_1} \otimes \sigma^{c_2}\Big)$$

where $e(X;T) = \sum\limits_{n=0}^{\infty} \Phi_n(T)^{-1} X^n$ and $\Phi_n(T) = \Big(\prod\limits_{p=1}^{n} (T^p - 1)/(T - 1)\Big)$.

Here d_α are some integers and $V(\alpha)^{-1}$ are some Lourent polynomials in q (see [Y1-2]).

4. Finite dimensional quantized superalgebra at a root of unity.
Let $\{ \bar{\varepsilon}_i \in H \}$ be an orthogonal basis defined in [Y1-2]. Put $\bar{d}_i = (\bar{\varepsilon}_i, \bar{\varepsilon}_i)$. By using the quantum double construction, we have:

Theorem. Let ζ be a primitive root of unity for large m .

Put $g_\alpha = \min\{ g > 0 \mid \Phi_n((-1)^{p(\alpha)}\zeta^{(\alpha,\alpha)}) = 0\}$ and $g_i = \min\{ g > 0 \mid \Phi_n(\zeta^{\bar{d}_i})) = 0\}$.

(i) There is a finite dimensiona Hopf superalgebra $u_\zeta(\Pi,p)$ satisfying:

(4,1) There exists a short exact seaquience:

$$(0) \to (E_\alpha^{g_\alpha} = F_\alpha^{g_\alpha} = 0, K_{\bar{\varepsilon}_i}^{g_i} = 1) \to U_\zeta(\Pi,p) \to u_\zeta(\Pi,p) \to (0),$$

(4,2) $$\Big(\prod^{<}_{\alpha \in \Delta_+^{red}} E_\alpha^{m_\alpha}\Big)\Big(\prod_i K_{\bar{\varepsilon}_i}^{l_i}\Big)\Big(\prod^{<}_{\beta \in \Delta_+^{red}} F_\beta^{n_\beta}\Big)$$

$(0 \le m_\alpha, n_\beta < g_\alpha, 0 \le l_i < g_i)$ form a C-basis of $u_\zeta(\Pi,p)$.

(ii) The universal R-matrix of $u_\zeta(\Pi,p)^\sigma$ is given by

$$\widetilde{R}_\zeta = \Big(\prod^{<}_{\alpha \in \Delta_+^{red}} \sum_{n=0}^{g_\alpha - 1} \Phi_n((-1)^{p(\alpha)}q^{(\alpha,\alpha)}))^{-1}((\zeta^{d_\alpha} - \zeta^{-d_\alpha})^n V(\alpha)^n E_\alpha^n \otimes F_\alpha^n \Big)$$

$$\cdot \Big(\prod_i g_i^{-1} \cdot \prod_{0 \le p_1, p_2 \le g_i} \zeta^{\bar{d}_i p_1 p_2} K_{\bar{\varepsilon}_i}^{p_1} \otimes K_{\bar{\varepsilon}_i}^{p_2}\Big)$$

$$\cdot \Big(2^{-1} \cdot \prod_{0 \le c_1, c_2 \le 1} (-1)^{c_1 c_2} \sigma^{c_1} \otimes \sigma^{c_2}\Big).$$

References

[D] D.G. Drinfeld: Quantum Groups, Proc. of ICM, Berkeley (1986).

[T] T. Tanisaki: Killing form, Harish-Chandra isomorphisms, and the universal R-matrices for quantum algebras, Proc. of Project 91, RIMS (1992).

[Y0] H. Yamane: A Poincaré - Birkhoff - Witt Theorem for quantized universal enveloping algebras of type A_N, Publ. RIMS Kyoto Univ., 25 (1989), 503-520.

[Y1] H. Yamane: Universal R-matrices for quantum groups associated to simple Lie superalgebras, Proc. Japan Acad., 67, Ser. A (1991), 108-112.

[Y2] H. Yamane: Quantized enveloping algebras associated to simple Liesuperalgebras and their universal R-matrices, submitting.

ALTERING THE SYMMETRY OF WAVEFUNCTIONS IN QUANTUM ALGEBRAS AND SUPERSYMMETRY

C.K. ZACHOS

High Energy Physics Division
Argonne National Laboratory, Argonne, IL 60439-4815, USA
zachos@anlhep

Abstract

The statistics-altering operators $\mathfrak{Y}$ present in the limit $q = -1$ of multiparticle $SU_q(2)$-invariant subspaces parallel the action of such operators which naturally occur in supersymmetric theories. In a consideration of ref.1, I illustrate this heuristically by comparison to a toy $N = 2$ superymmetry algebra, and ask whether there is a supersymmetry structure underlying $SU_q(2)$ in that limit. I remark on the relevance of such alternating-symmetry multiplets to the construction of invariant hamiltonians.

Even though the suitability of quantum algebras[2] to particle physics has not yet been fully understood, there has been a spate of speculation in non-co-commutative contexts in solid state, molecular, and nuclear physics. The new generalized permutation symmetries involved in quantum algebras[3] have led to several connections[4,5,6] to anyons and parastatistics. In previous work[7] on deformed multioscillator algebras, we have stressed the difference between mere anticommutativity of some wavefunctions (easy to arrange through deforming maps of boson operators) versus nilpotence (the Pauli exclusion principle), harder to arrange, which remains the genuine hallmark of fermionic behavior. In fact, it might appear that q-deformations provide a smooth interpolation between bosons and fermions[4,5]: For the deformed algebra $bb^\dagger - qb^\dagger b = 1$, deformation to conventional bosonic oscillators $b^\dagger = \sqrt{(q^N-1)/((q-1)N)}\, A^\dagger$ suggests a zero in the state $b^\dagger b^\dagger|0\rangle$ for $q = -1$, which amounts to the exclusion principle. At the algebraic level, the relation following from the above algebra: $bb^\dagger b^\dagger = q^2 b^\dagger b^\dagger b + (q+1)b^\dagger$, reduces to $bb^\dagger b^\dagger = b^\dagger b^\dagger b$ for $q = -1$; so, acting on the vacuum state with $bbb^\dagger b^\dagger$ likewise yields $b^\dagger b^\dagger|0\rangle = 0$. However, this interpolation has not been achieved in general, and, as a result, q-deformations do not automatically blend graded and conventional bosonic algebras into the same one-parameter structure, and normally each has to be deformed separately—the noncommutative plane and quantum superplane are distinguishable. (Contrast to H.C. Lee's *two*-parameter unification, Chalk River preprint CRL-TP-91-0228).

Nevertheless, investigating the $\mathcal{R}$-matrix which codifies the symmetries of composite states, Saleur[8] has noted intriguing connections of conventional $SU_q(2)$-invariant models to supersymmetry, for generic parameter q. Specifically, he observed a "hidden supersymmetry" $Osp_q(1|2)$ in, e.g., vertex models, leading to a quasi-graded classification of their states: $SU_q(2)$-symmetric vertex models map to $Osp_{-q^2}(1|2)$ graded models for representations of the same dimension. Very recently, R.B. Zhang[8] has further connected $Osp_q(1|2n)$ to $SO_{-q}(2n+1)$ for non-spinorial representations.

Here, in a different vein, we focus on statistics-altering operators $\mathfrak{Y}$ in q-algebras in the particular extreme case $q = -1$, and briefly note their formal similarity with operators of this type that arise naturally in extended supersymmetric theories, illustrated with a toy $N = 2$ supersymmetry algebra. Recall Drinfeld–Jimbo's standard $SU_q(2)$ quantum deformation:

$$[J_0, J_+] = J_+ \qquad [J_+, J_-] = \frac{1}{2}\,[2J_0]_q \qquad [J_-, J_0] = J_- \,, \tag{1}$$

where $[x]_q \equiv (q^x - q^{-x})/(q - q^{-1})$, the "$q$-deformation" of x. The Casimir invariant of this algebra is

$$C_q \equiv 2J_+J_- + [J_0]_q[J_0 - 1]_q = 2J_-J_+ + [J_0]_q[J_0 + 1]_q \,. \tag{2}$$

Recall that the classical limit $q \to 1$ yields SU(2). Further note that, for generic q, the irreducible representations of dimension $2j+1$ for j integer or half-integer yield the eigenvalue $[j]_q[j+1]_q$ for this Casimir invariant.

The relevant feature of this deformed algebra is that the standard[2] coproduct algebra homomorphism

$$\Delta(J_0) = J_0 \otimes \mathbb{1} + \mathbb{1} \otimes J_0 \qquad \Delta(J_\pm) = J_\pm \otimes q^{J_0} + q^{-J_0} \otimes J_\pm \,, \tag{3}$$

is non-cocommutative, in contrast to that of the classical limit (addition of angular momenta), i.e. it differentiates between leading and trailing representation vector spaces. Since the coproduct does not commute with the permutation group, the representations of the deformed algebra will not be expected to be characterized by conventional permutation symmetries. (cf. ref.3 for q-modifications of permutations).

Apply the above coproduct to combine two spin-1/2 doublets, $\mathbf{2} \otimes \mathbf{2} = \mathbf{3} \oplus \mathbf{1}$: The singlet is $\alpha = |q^{1/2} \uparrow\downarrow - q^{-1/2} \downarrow\uparrow\rangle$, while the triplet is

$$\beta = |\uparrow\uparrow\rangle \qquad \Delta(J_-)\beta = \frac{1}{\sqrt{2}}\,|q^{1/2} \uparrow\downarrow + q^{-1/2} \downarrow\uparrow\rangle \qquad \Delta(J_-)^2\beta = |\downarrow\downarrow\rangle. \tag{4}$$

The normalization of α and $\Delta(J_-)\beta$ is $1/\sqrt{[2]_q}$ for real q^2, but we ignore inessential normalizations for simplicity, on account of the complex cyclotomic limits to follow presently.

Now consider the $q = -1$ limit. Polychronakos[4] has noted that this case, among p-th roots of unity, provides a singular exception to the dimensionality restriction of representations[9], essentially because $[p/2]_q$ is nonvanishing for $p = 2$. In fact, the q-algebra itself reduces to classical SU(2), the coproduct remaining non-cocommutative all the same. (The raising/lowering generators are hermitean conjugates, $J_+^\dagger = J_-$, for half-integral spin representations, whence unitarity; but antihermitean conjugates—"negative parity" in the terminology of ref.10—for integral spins, comprising an SU(1,1); renormalizing to hermitean ones in this case reverts to SU(2) unitary representations.) For integral spins, the above Casimir eigenvalue is negative, $-j(j+1)$, and the q-dimension is $2j+1$; but for half-integral spins, C_q diverges to $4/(q-1/q)^2 + 1/2 + j(j+1)$, and the q-dimension is $-2j-1$. The discussion to follow may thus be thought of as the exploration of this unconventional composition law for SU(2)-representations.

The above states become: $\alpha = |i \uparrow\downarrow + i \downarrow\uparrow\rangle$, and

$$\beta = |\uparrow\uparrow\rangle \qquad \Delta(J_-)\beta = \frac{1}{\sqrt{2}}\,|i \uparrow\downarrow - i \downarrow\uparrow\rangle \qquad \Delta(J_-)^2\beta = |\downarrow\downarrow\rangle. \tag{5}$$

By $\Delta(J_+)|\downarrow\downarrow\rangle = -\Delta(J_-)\beta$ and $\Delta(J_+)^2|\downarrow\downarrow\rangle = \beta$, it is evident in this case that $\Delta(J_+)^\dagger = -\Delta(J_-)$, so that renormalizing $\Delta(J_+) \to -\Delta(J_+)$ leads to the conventional spin-1 representation of SU(2). In general, the unitary representation of SU(2) is provided by $(-)^{2j+1}J_+$, J_-, J_0.

Now note that the above singlet wavefunction α is symmetric, and the members of the triplet are symmetric, antisymmetric, symmetric, respectively. The symmetry of the wavefunction alternates *within* a multiplet, but the dimensionality of multiplets has not changed. The raising and lowering operators in the coproduct act as statistics-altering operators $\mathfrak{Y}$. It is crucial for this alternation of symmetry that the spins entering into the coproduct be half-integral.

Have the constituents of the states been converted to fermions? No, as is apparent in the multi-isodoublet wavefunctions below. Nevertheless, this alteration of the symmetry of the wavefunction is reminiscent of supersymmetry. To illustrate this, introduce the elementary supersymmetry algebra consisting of the (graded) direct product of two "supersymmetric quantum mechanics" algebras:

$$ss^\dagger + s^\dagger s = \mathbb{1} \qquad SS^\dagger + S^\dagger S = \mathbb{1} \tag{6}$$

$$S^\dagger S^\dagger = s^\dagger s^\dagger = SS = ss = 0 \tag{7}$$

$$sS + Ss = 0 \qquad s^\dagger S^\dagger + S^\dagger s^\dagger = 0 \tag{8}$$

$$sS^\dagger + S^\dagger s = 0 \qquad s^\dagger S + Ss^\dagger = 0. \tag{9}$$

This supersymmetry algebra may be realized on two boson states $|B\rangle$, $|b\rangle$, and two fermion states $|F\rangle$, $|f\rangle$, so that: $S|B\rangle = |F\rangle$, $s|b\rangle = |f\rangle$, $S^\dagger|F\rangle = |B\rangle$, $s^\dagger|f\rangle = |b\rangle$, the remaining actions being null: $S|F\rangle = S|b\rangle = s|B\rangle = s^\dagger|F\rangle = s^\dagger|b\rangle = S^\dagger|f\rangle = 0$, etc. Then:

$$s|Bb + bB\rangle = |Bf + fB\rangle , \qquad Ss|Bb + bB\rangle = |Ff - fF\rangle. \tag{10}$$

So $\Delta(J_-)$ switches the symmetry of the wavefunction like the even (bosonic) operator $Ss = -sS$, but only the latter and not the former is nilpotent. Morevover, there appear no odd (fermionic) states or operators in the former case ($SU_q(2)$), although a less direct connection cannot be excluded. It may be tempting to ask whether some sort of underlying supersymmetry could become manifest after appropriate fermionization.

Just as in the case of conventional SU(2), combining n doublets for any q yields the Clebsch-Gordan decomposition series (the number preceeding the boldface representation dimensionality label indicates multiplicity of that representation in the reduction):

$$\mathbf{2}^{\otimes n} = \bigoplus_{k=0}^{\lfloor n/2 \rfloor} \left(\frac{n+1-2k}{n+1} \binom{n+1}{k} \right) \ (\mathbf{n+1-2k}) , \tag{11}$$

where $\lfloor n/2 \rfloor$ is the integer floor function. The coproduct used is the iterated composition of Eq.3, unique by co-associativity. For instance, for spins $\mathbf{2} \otimes \mathbf{2} \otimes \mathbf{2} = \mathbf{4} \oplus \mathbf{2} \oplus \mathbf{2}$, we have, unnormalized, a Quartet:

$$|\uparrow\uparrow\uparrow\rangle \tag{12}$$

$$|q \downarrow\uparrow\uparrow + \uparrow\downarrow\uparrow + q^{-1} \uparrow\uparrow\downarrow\rangle \tag{13}$$

$$|q^{-1} \uparrow\downarrow\downarrow + \downarrow\uparrow\downarrow + q \downarrow\downarrow\uparrow\rangle \tag{14}$$

$$|\downarrow\downarrow\downarrow\rangle, \tag{15}$$

and two different doublets. As a result, for $q=-1$, all wavefunctions are of mixed symmetry. This is reminiscent of fermions, but, again, aligned isospins do not annihilate, and hence the representations do not chop up to smaller ones. From the structure of the statistics-altering operator $\mathfrak{Y}$ coproduct it is evident that, in ex-symmetric representations, any aligned isospins are symmetric to each other, while any two anti-aligned ones are antisymmetric with respect to each other. The basic structure and dimensionality of multiplets remains the same as in the classical Lie algebra limit. This contrasts to the general feature of graded algebra representations, whose dimensions differ from those of the corresponding representations of bosonic Lie algebras (the symmetric representations shrink in size, while the antisymmetric ones expand)—but *not*[8] for $\mathrm{Osp}_q(1|2n)$!

The features discussed here are expected to also occur in more general contexts and, by relaxing standard exchange-operator restrictions, may well be useful in constructing new invariant hamiltonians. The effect could also be reversed by *in addition* introducing a supersymmetric structure on top of the $\mathrm{SU}_q(2)$. This may help counter-act the well-known difficulty in constructing $\mathrm{SU}_q(2)$-invariant field theoretic lagrangians for generic q: Bose or Fermi symmetry mix (via symmetrization or antisymmetrization) the singlet α with the triplet, and hence spoil invariance. Nevertheless, for more involved contexts such as spin-chains or anyonic systems, the alternate exchange symmetry structures may allow new constructions, Boltzmann-weight arrangements, and whence partition functions, which merit further attention.

We thankfully acknowledge conversations with P. Freund, H. Yan, A. Polychronakos, J. Uretsky, and S. Vokos. Work supported by the U.S. Department of Energy, Division of High Energy Physics, Contract W-31-109-ENG-38.

References

1. C. Zachos, Mod.Phys.Lett. **A7** (1992) 1595.
2. V. Drinfeld, Sov.Math.Dokl. **32** (1985) 254; *Proc.Int.Cong. Mathematicians*, Berkeley 1986, (1987) 798. M. Jimbo, Lett.Math.Phys. **10** (1985) 63.
3. W. Pusz & S. Woronowicz, Rep.Math.Phys. **27** (1989) 231. T. Curtright & G. Ghandour, unbublished. M. Salam & B. Wybourne, J.Phys. **A24** (1991) L317; **A25** (1992) 2297.
4. A. Polychronakos, Mod.Phys.Lett. **A5** (1990) 2325.
5. Z. Chang, H.-Y. Guo, and H. Yan, Comm.Theor.Phys. **14** (1990) 475.
6. V. Spiridonov, in *Quarks '90*, V. Matveev et al. (eds), 1991, World Scientific, p. 233; M. Chaichian & P. Kulish, Phys.Lett. **234B** (1990) 72; O. Greenberg, Phys.Rev. **D43** (1991) 4111.
7. C. Zachos in *Differential Geometric Methods in Theoretical Physics XX*, S. Catto & A. Rocha (eds.), World Scientific, 1992, p. 426.
8. H. Saleur, Nucl.Phys. **B336** (1990) 363. R. B. Zhang, Canberra preprint, June 1992.
9. P. Roche & D. Arnaudon, Lett.Math.Phys. **17** (1989) 295. For a concise review, see C. Zachos, "Paradigms of Quantum Algebras", in *Symmetries in Science V*, B. Gruber et al. (eds.), Plenum, 1991, p. 593.
10. L. Mezincescu & R. Nepomechie, Phys.Lett. **246B** (1990) 412.

Parallel Sessions 3
Quantum Fields

NON-PERTURBATIVE ANALYSIS, GRIBOV HORIZONS AND THE BOUNDARY OF THE FUNDAMENTAL DOMAIN †

PIERRE VAN BAAL *
Institute for Theoretical Physics, P.O.Box 80.006, NL-3508 TA Utrecht, The Netherlands

and

R.E. CUTKOSKY
Physics Department, Carnegie Mellon University, Pittsburgh, PA 15213, USA

ABSTRACT

In this contribution to the proceedings we will describe some of the details for constructing the Gribov horizon and the boundary of the fundamental modular domain, when restricting to some low energy modes of pure SU(2) gauge theory in a spherical spatial geometry. The fundamental domain is a one-to-one representation of the set of gauge invariant degrees of freedom, in terms of transverse gauge fields. Boundary identifications are the only remnants of the Gribov copies.

1. Introduction

At the conference one of us gave an overview of the finite volume analysis on a torus, relevant for comparison with lattice Monte Carlo results [1]. Recently, for technical reasons, this analysis was extended to a spherical spatial geometry [2], results of which were only briefly touched upon during the talk. The other author has, from a different perspective, been interested in this geometry for quite some years now [3]. Thus we found some common interest, whose hitherto unpublished fruits we will discuss in these proceedings. It involves computing the Gribov horizon and the boundary of the fundamental domain in a spherical spatial geometry. The remaining part of the talk is summarized in the contribution to the Shanxi conference [4]. For the applications and further references see [1-4].

In gauge theories one can fix to the transverse or Coulomb gauge, $\partial_i A_i = 0$, using a functional method, which allows one to pick from the different Gribov copies [5] (transverse gauge fields that are nevertheless gauge equivalent) an (almost) unique representative. The collection of these configurations should thus form a fundamental modular domain, in other words it should form a one-to-one mapping with the Yang-

† Contributed talk presented at 21st Conference on Differential Geometric Methods in Theoretical Physics (XXI DGM), June 5-9, Nankai University, Tianjin, P.R. China.

* KNAW fellow

Mills configuration space. The relevant functional is the L^2-norm of the gauge field: $\| A_i \|^2 = \int_M \mathrm{Tr}(A_i^\dagger A_i)$. For each gauge invariant field configuration this gives a Morse functional on the gauge orbit. One easily verifies that stationary points of this Morse functional satisfy the Coulomb gauge condition and that the Hessian (second order derivative) at the stationary point is precisely given by the Faddeev-Popov operator, whose determinant measures the volume of the gauge orbit. It is natural to choose among the various Gribov copies the one with lowest norm. The collection of transverse fields thus obtained has a boundary. Points on this boundary are usually degenerate in norm and gauge equivalent to at least one other point on the boundary [6]. This gauge equivalence induces boundary identifications that make the set into a fundamental modular domain Λ (no Gribov copies occur at the interior). In general Λ is well contained in the Gribov region Ω, which is by definition the collection of transverse potentials for which the norm functional (when considered as a function of the gauge orbit) has a local minimum, i.e. the Hessian or Faddeev-Popov operator $FP(A) = -\partial_i D_i(A)$ is positive semi-definite ($D_i(A)$ is the usual adjoint covariant derivative). There can [6], however, be points at the boundary of Λ that coincide with the boundary of Ω. Since the latter is the Gribov horizon, which is where the lowest eigenvalue of the Faddeev-Popov operator (and hence the Faddeev-Popov determinant) vanishes, these points will still require some extra care. They have, so far, not been considered in the subsector of the theory we will be studying here.

2. Defining the subspace

The subsector we will study is given by 18 modes that are degenerate in energy, to lowest (quadratic) order. There are only twelve modes with a lower energy, also all degenerate. These 18 modes contain degrees of freedom relevant for the tunnelling from the $A = 0$ vacuum to the two vacua that have Chern-Simons number $Q(A) = \frac{1}{8\pi^2}\int_{S^3} \mathrm{Tr}(A \wedge dA + \frac{2}{3} A \wedge A \wedge A)$ one or minus one. Here $A = A_i dx_i = iA_i^a \tau_a dx_i/2$ is the connection one-form for a SU(2) vector potential A_i^a on the three-sphere. In other words the two vacua with $Q(A) = \pm 1$, are transverse vector potentials, that are pure gauge ($A = gdg^{-1}$) with a gauge function g that has a winding number $n(g) = \pm 1$, where $n(g) = \frac{1}{24\pi^2}\int_{S^3} \mathrm{Tr}((g^{-1}dg)^3)$. The tunnelling between the degenerate vacua is of course described by instantons, easily obtained from the well-known instanton solutions on $\mathbb{R}^4$ by the conformal transformation that relates $S^3 \times \mathbb{R}$ to $\mathbb{R}^4$. To be precise, the 18 modes contain the tunnelling paths that describe the transition over the lowest barrier separating the nearest-neighbour vacua. This particular saddle point of the energy functional is also known as a sphaleron.

One way of describing these modes is by using [2] the 't Hooft η symbols (useful in giving an explicit expression for the instanton vector potentials), to define a framing of the three-sphere. It is easy to specify the η symbol in terms of a basis (and its dual) of the unit quaternions σ_μ ($\bar{\sigma}_\mu$) with $\sigma_4 = \bar{\sigma}_4 = 1$ and $\sigma_a = -\bar{\sigma}_a = i\tau_a$. One has $2i\eta^a_{\mu\nu}\tau_a = \sigma_\mu\bar{\sigma}_\nu - \sigma_\nu\bar{\sigma}_\mu$ and if we parametrize the three-sphere by the unit vectors n_μ in four dimensions, one can define a dreibein by $e^i_\mu = \eta^i_{\mu\nu}n_\nu$. The 18 modes now split in two categories: there are 9 modes described by constant components $A_i^a(n) = c_i^a$ (it is essential that the index i refers to a flat index with respect to the above defined framing), whereas the other 9 modes are constant up to a coordinate-dependent rotation, $V_i^b(n) = \frac{1}{2}\mathrm{Tr}(n^\mu\sigma_\mu\sigma_i n^\nu\bar{\sigma}_\nu\sigma_b)$, of the dreibein, i.e. $A_i^a(n) = -V_i^b(n)d_b^a$. One easily verifies that both vector potentials are transverse. It is furthermore not too difficult to show that (minus) the rotated frame is precisely the one obtained by replacing η by $\bar{\eta}$ (defined by $2i\bar{\eta}^a_{\mu\nu}\tau_a = \bar{\sigma}_\mu\sigma_\nu - \bar{\sigma}_\nu\sigma_\mu$), which is used to express the anti-instantons. The instanton is represented by $c_i^a = -2\delta_i^a/(1+e^{-2t})$ and $A_0 = 0$, with an identical expression for the anti-instanton in terms of d_i^a. In these proceedings we will restrict ourselves to the two-dimensional cross-section of the field space corresponding to the direction of these particular instanton and anti-instanton configurations which, since they describe tunnelling through the sphaleron (the lowest barrier separating two nearest-neighbour vacua), were called sphaleron modes [2]. Thus we will consider $A_i^a(n) = vV_i^a(n) - u\delta_i^a$.

The classical vacua in the (u, v) plane are located at $(0,0)$, $(0,2)$ and $(2,0)$, whereas the sphalerons can be found at $(0,1)$ and $(1,0)$. Furthermore, applying the gauge transformation $g(n) = n^\mu\bar{\sigma}_\mu$ (which has winding number $n(g) = -1$) to the configuration $(u, 0)$ can be shown [2] to yield $(0, 2-u)$. As it should, this maps the vacua $(2,0)$ and $(0,0)$ (which have respectively $Q(A) = 1$ and $Q(A) = 0$) to the vacua $(0,0)$ and $(0,2)$ (where $(0,2)$ has $Q(A) = -1$). Furthermore it maps the sphaleron at $(1,0)$ (with $Q(A) = \frac{1}{2}$) to the sphaleron at $(0,1)$ (with $Q(A) = -\frac{1}{2}$).

3. The Gribov horizon

To diagonalize the Faddeev-Popov operator $FP(A)$ in the subspace (u, v) it is convenient to introduce angular momentum operators [2] $L_1^a = \frac{i}{2}e_\mu^a\partial_\mu = -\frac{i}{2}\eta^a_{\mu\nu}n^\mu\partial_\nu$, $L_2^a = -i\frac{1}{2}\bar{\eta}^a_{\mu\nu}n^\mu\partial_\nu$, $T^a = \frac{1}{2}\mathrm{ad}(\tau_a)$, and $\vec{J} = \vec{L}_1 + \vec{L}_2 + \vec{T}$. It is not too difficult to check that $FP(u, v)$ commutes both with $\vec{L}_1^2 = \vec{L}_2^2$ and with $\vec{J}$. For this it is convenient to write $FP(u,v) = 4\vec{L}_1^2 + 2u\vec{T}\cdot\vec{L}_1 - v\ \mathrm{ad}(\bar{\sigma}_\mu n^\mu\vec{\tau}\sigma_\nu n^\nu)\cdot\vec{L}_1$; only the last term in this expression requires some care in computing the commutators with $\vec{J}$ and $\vec{L}_1^2$. Thus $FP(u, v)$ can be diagonalized in the subspace defined by $\ell \neq 0$, j and j_z. There is an obvious degeneracy in j_z and the eigenvalues will be denoted by $\lambda_{2\ell,j}(u, v)$. Note that the Coulomb gauge does not fix the constant gauge transformations, which means that the constant modes ($\ell = 0$) are eliminated from the spectrum of $FP(A)$. The remaining invariance under constant gauge transformations is easily taken into account without further gauge fixing. As the Gribov horizon is defined as the set of configurations where the lowest eigenvalue of $FP(A)$ vanishes, it suffices to diagonalize $FP(u, v)$ for $\ell = \frac{1}{2}$. There are 12 eigenfunctions in this sector that split in one $j = 0$ singlet ($n_a\tau_a$), two $j = 1$ triplets ($n_0\tau_a$ and $\varepsilon_{abc}n_b\tau_c$) and one $j = 2$ quintet ($n_a\tau_b + n_b\tau_a - \frac{2}{3}n_c\tau_c\delta_{ab}$). For the singlet and quintet the problem of diagonalizing $FP(u, v)$ becomes one-dimensional, whereas for the two triplets one has to diagonalize a 2×2 matrix. One easily checks that $\lambda_{1,0} = 3 - 2s$, $\prod\lambda_{1,1} = (9 - 3s - 2p^2)^3$ and $\lambda_{1,2} = 3 + s$ (for convenience we introduced the scalar and pseudoscalar (even and odd) helicity combinations $s \equiv u + v$ and $p \equiv u - v$). In fig. 1 the full curves give the solutions of $\lambda_{1,j}(u, v) = 0$, whereas the Gribov horizon is indicated by fat sections. The result is in accordance with the convexity of Ω. This convexity is a simple consequence of the linear dependence of $FP(A)$ on the vector potential. Note that the sphaleron configurations are well within the Gribov region, but that the vacua nearest to $A = 0$ are outside (but not on) the Gribov horizon. Nevertheless, it is easily proved [6] that at these vacua the Faddeev-Popov determinant has to vanish. It turns out that the relevant eigenvalues (there are actually 3 of them) that vanish at the vacua nearest to $A = 0$ have $\ell = 1$. So it is instructive to diagonalize $FP(u,v)$ in this sector too. One finds one $j = 0$ singlet, three $j = 1$ triplets, two $j = 2$ quintets and one $j = 3$ septet. We thus have to diagonalize at most a 3×3 matrix to obtain the eigenvalues for $FP(u,v)$, with the following result $\lambda_{2,0} = 8 - 2s$, $\prod \lambda_{2,1} = 512((8-2s)^2 - s^2 + (2s-7)p^2)^3$, $\prod \lambda_{2,2} = (64 - s^2 - 3p^2)^5$ and $\lambda_{2,3} = 8 + 2s$. This allows us to determine the location of the zero's for the Faddeev-Popov determinant in this sector, indicated in fig. 1 by the dashed curves.

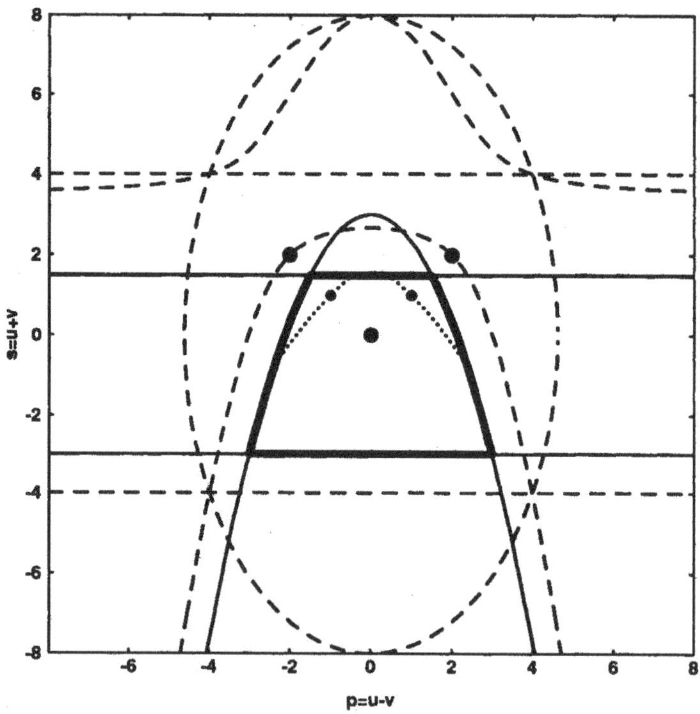

Figure 1: Location of the classical vacua (large dots), sphalerons (smaller dots), zeros of the Faddeev-Popov determinant (full curves at $\ell = \frac{1}{2}$, dashed curves at $\ell = 1$), the Gribov horizon (fat sections) and part of the boundary of the fundamental domain (dotted curves) in the plane specified by $A_i^a = vV_i^a(n) - u\delta_i^a$.

4. The boundary of the fundamental modular domain

To construct the boundary of the fundamental modular domain requires one to find transverse gauge copies that are degenerate in norm. This is unfortunately a rather non-local problem. We can, however, make use of the fact that the two sphalerons are transverse, have the same norm and are gauge copies of each other, as we discussed above. We can therefore start from $(0,1)$ and construct $(u,r(u))$, expanding in powers of u, such that the gauge transformation of $A(u,r(u))$ (coinciding with the sphaleron configuration at $(1,0)$) is again transverse and has the same norm (note that for $u \neq 0$, the gauge transformed potential is *not* part of the (u,v) plane). Obviously, also $(r(v),v)$ is part of the boundary of the fundamental modular domain. These two branches are indicated by the dotted curves in figure 1. To find the desired gauge transformation one writes $g(n)=n^\mu\sigma_\mu\exp(X(n))$ and one minimizes $M(u,v)\equiv\|g(n)\vec{A}(u,v)g(n)^{-1}+g(n)\vec{\partial}g(n)^{-1}\|^2-\|\vec{A}(u,v)\|^2$ with respect to X. It can be verified a posteriori (by checking transversality of the gauge transformed vector potential) that the solution is of the form $X(n)=if(n_0)\vec{n}\cdot\vec{\tau}$. Using this as an ansatz, reduces the problem of finding the stationary solution to solving an ordinary 2nd order differential equation. Its solution can be written as $f(x)=x\sum_{j=1}\sum_{k=0}^{j-1}a_{j,k}(v)u^jx^{2k}$. One finds: $a_{1,0}(v)=2(2+v)^{-1}$, $a_{2,0}=-2(v^2+6v-16)(2+v)^{-3}(10+v)^{-1}$, $a_{2,1}(v)=4(6+v)(v+2)^{-2}(v+10)^{-1}$, etc. By solving $M(u,r(u))=0$ we find

$$r(u)=1-\frac{1}{9}u^2-\frac{2}{81}u^3-\frac{25}{2673}u^4-\frac{1238}{264627}u^5-\frac{172442}{66950631}u^6-\frac{687429956}{457339760361}u^7+\mathcal{O}(u^8).$$

The result is exhibited in figure 1 by the dotted curves. Similar to the Gribov region, the fundamental modular domain is convex and its boundary has "corners", where different branches of copies that are degenerate in norm intersect. Usually this will keep $\partial\Lambda$ from touching the Gribov horizon [6].

Acknowledgements This research has been made possible by a fellowship of the Royal Netherlands Academy of Arts and Sciences and by the U.S. Dept. of Energy under grant No. DE-FG02-91ER40682.

References

1. J. Koller and P. van Baal, *Nucl. Phys.* **B302** (1988) 1; P. van Baal, *Phys. Lett.* **B224** (1989) 397.
2. P. van Baal and N.D. Hari Dass, "The theta dependence beyond steepest descent", to appear in *Nucl. Phys.* **B**, Utrecht preprint THU-92/03, January 1992.
3. R.E. Cutkosky, *J. Math. Phys.* **25** (1984) 939; R.E. Cutkosky and K. Wang, *Phys. Rev.* **D37** (1988) 3024; R.E. Cutkosky, *Czech. J. Phys.* **40** (1990) 252.
4. P. van Baal, "Topology of Yang-Mills configuration space", Utrecht preprint THU-92/15, to appear in the proceedings of ISATQP, June 11-16, Shanxi University, (Science Press of China).
5. V. Gribov, *Nucl. Phys.* **B139** (1978) 1.
6. P. van Baal, *Nucl. Phys.* **B369** (1992) 259.

Dirac operators and exceptional Lie groups

Frank Flaherty
Department of Mathematics
Oregon State University
Corvallis, OR, 97331-4605, USA

Abstract

In dimension eight there are three basic representations for the spin group. These representations lead to a concept of triality and hence lead to the construction of two exceptional commutative algebras: The Chevalley algebra $\mathcal{A}$ of dimension twenty-four and the Albert algebra $\mathcal{J}$ of dimension twenty-seven. All of the exceptional Lie groups can be described using the octonians and these two algebras.

On a complex four-manifold, with triality a parallel field, the Dirac and associated twistor operators can be constructed on either bundle of exceptional algebras. A vanishing theorem will be proved under a postive curvature condition.

In nine and ten dimensions there are notions of triality and multiplication, though not as robust as eight dimensions, which may prove to have application to string theory.

Introduction

Cartan's classification of simple Lie groups yields all Lie groups with some exceptions. In his thesis Cartan described these exceptional groups but, save the one of smallest rank G_2, he was unable to describe the geometry of the groups. In 1950 Chevalley and Schafer successfullly identified F_4 and E_6 as structure groups for the exceptional Jordan algebra and the the Freudenthal multiplication on $\mathcal{J}$, respectively. Later Freudenthal successfully identified the geometry associated with the remaining exceptional groups, E_7 and E_8. All of the exceptional groups have low rank and are, in one way or another, related to the octonians (Cayley numbers), the principle of triality, and the exceptional algebras.

With the advent of superstring theory and the work of Witten and others, it is necessary to understand the exceptional Lie groups and the corresponding differential operators. To this end, the Dirac operator is described explicitly while the twistor operator is merely indicated without detail.

Dirac operators

To define the Dirac operator on a manifold one needs the global notion of a spin manifold. Without going into details this is a (C^∞) orientable Riemannian manifold M for which the second Stiefel-Whitney class vanishes.

The bundle of Clifford Algebras over M is then constructed using $T_p(M)$ and the Riemannian quadratic form $g_p(x,x)$. The resulting vector bundle of Clifford algebras will be denoted by $Cl(M)$. The resulting spinor bundle will be denoted by S, with S_o, S_1 the even and odd spinors (if dim $M = 2m$).

Denoting the Riemannian connection on M by D, it can be shown that on a spin manifold D extends as a derivation to the bundle S, regarded as $Cl(M)$ module. In other words,

$$D(xu) = (Dx)u + xDu$$

$$u \in \Gamma(Cl(M)), \quad x \in \Gamma(S)$$

The notation Γ means take the sections of the bundle that follows.

There is then a first order differential operator $\mathcal{D} : \Gamma(S) \to \Gamma(S)$ which locally is defined by $\mathcal{D}s = \sum e_i D_{e_i} s$ for (e_i) a local orthonormal frame field, $x \in \Gamma(U, S)$. The sections are local sections of S over the chart U. Note that the symbol of $\mathcal{D} = \sum e_i.\zeta_i$ at the cotangent vector, $\zeta = (\zeta_i)$ is $\zeta.$, left Clifford multiplication by ζ, and hence, it follows that

$$symb(\mathcal{D}^2) = -\zeta.\zeta = \| \zeta \|$$

The self-adjoint first order operator $\mathcal{D}$ is called the Dirac operator.

Theorem. *Let $\mathcal{D}$ be the Dirac operator on a compact spin manifold M then*

$$ker\mathcal{D} = ker\mathcal{D}^2$$

and $ker\mathcal{D}$ is finite dimensional.

Proof. Finite dimensionality follows from ellipticity.

$$\mathcal{D}^2 u = 0 \Rightarrow \| \mathcal{D}u \| = 0 \Rightarrow \mathcal{D}u = 0.$$

In the next section, the Dirac operator is extended to the bundle of algebras $\mathcal{J}$.

The exceptional group F_4

The exceptional Lie group F_4 is the group of automorphisms of the exceptional Jordan algebra $\mathcal{J}$. The Chevalley algebra, $\mathcal{A}$ sits as a subalgebra of $\mathcal{J}$. There is an exact sequence displaying ordinary 3-space as a coextension of $\mathcal{J}$ via $\mathcal{A}$. The Lie algebra structure of F_4

can then be displayed as a vector space sum of *spin*(8), and the three basic representations in dimension eight. The bracket is defined in terms of the triality operator on $\mathcal{A}$.

More precisely, the Lie algebra of F_4 can be written in the form:

$$\mathfrak{f}_4 = \mathfrak{g} \oplus V \oplus S_o \oplus S_1$$

in which $\mathfrak{g} = \text{spin}(8)$. Using triality, the bracket operation may be defined as:

$$[V, V] = [S_o, S_o] = [S_1, S_1] = \mathfrak{g}$$

while $[\mathfrak{g}, V] = V$ and cyclically. Finally, $[V, S_o] = S_1$, and cyclically. The Lie algebra of E_8 can be given in a similar manner. I am indebted to Bert Kostant for this description of F_4 and E_8. The group F_4 serves as the group of the bundle of exceptional Albert algebras.

Dirac operators on the exceptional algebras

The Dirac operator can be introduced on the bundle of Albert (Chevalley) algebras on a complex four-fold using the corresponding multiplications in place of the Clifford multiplication. For the Chevalley algebra, this construction is due to Troy Warwick. In addition, the twistor operator can be defined. The question of integrabilty of complex stuctures on twistor space can then phrased in terms of vanishing of curvature. Finally, Bochner and Lichnerowicz type identities are derived, giving rise to a vanishiing theorem for harmonic sections of the bundle of Albert algebras. A sketch of these ideas follows.

Consider a Kaehler manifold M of complex dimension four on which the triality map is a parallel filed. Over M there is the bundle of exceptional algebras $\mathcal{J}$. A section of $\mathcal{J}$ is of the form

$$X = \begin{pmatrix} \xi_1 & x_3 & \overline{x_2} \\ \overline{x_3} & \xi_2 & x_1 \\ x_2 & \overline{x_1} & \xi_3 \end{pmatrix}$$

in which the the ξs are real and the xs are octonians. The basic operation on these matrices is the Jordan multiplication defined by:

$$X \circ Y = \frac{1}{2}(XY + YX)$$

in which the matrix multiplication on the right hand side uses octonian multiplication of the entries. It can be verified that $\mathcal{J}$ is a Jordan algebra, that is, in addition to being commutative, the multiplication rule $\circ$ satisfies the identity

$$X \circ (Y \circ X^2) = X^2 \circ (Y \circ X)$$

for all $X, Y \in \mathcal{J}$. Note that in physicists notation the exceptional Jordan algebra $\mathcal{J}$ is usually denoted by M_3^8 after Jordan, von Neumann, and Wigner.

The Dirac operator can then defined by

$$\mathcal{D} = \sum e_i \circ D_i.$$

The *e*s are embedded in the right corner of the matrix in the off-diagonal position.

Theorem. *If X is a harmonic section of $\mathcal{J}$ for which ξ_1 vanishes and if the Cayley curvature of $\mathcal{J}$ is non-negative then X is parallel. Further, if the Cayley curvature of $\mathcal{J}$ is positive somewhere on M then the any harmonic section of $\mathcal{J}$ in which $\xi = 0$ must vanish.*

Proof. A Bochner-Lichnerowicz formula for $\mathcal{D}$ yields

$$\langle \mathcal{D}^2(X), X\rangle = \text{Laplacian terms in X} \oplus \text{Cayley curvature terms in X}$$

The argument then follows in a trivial manner.

Note that the curvature terms must be made more precise. It is not clear what role any vanishing theorem plays in physics although they contribute to Penrose's twistor theory in a basic way. The notion of a Killing section of $\mathcal{J}$ is obvious and these sections generate the solutions to the twistor equation. Although the twistor operator equation has not been explicitly defined here, it follows from representation theory that there is a projection as in dimension four and the definition is clear.

SOLUTIONS OF THE KNIZHNIK - ZAMOLODCHIKOV EQUATION AND MINIMAL MODEL CORRELATION FUNCTIONS

P. Furlan[a,b], A.Ch. Ganchev[b*], R. Paunov[c*] and V.B. Petkova[d]

[a] Dipartimento di Fisica Teorica, dell'Universitá di Trieste, Italy
[b] Istituto Nazionale di Fisica Nucleare (INFN), Sezione di Trieste, Italy
[c] International School for Advanced Studies (SISSA), 34014 Trieste, Italy
[d] Institute for Nuclear Research and Nuclear Energy, 1784 Sofia, Bulgaria

Abstract

In a $\hat{sl}(2,\mathbb{C})$ WZNW conformal theory a set of chiral n-point functions is found which are solutions of the Knizhnik-Zamolodchikov equations with rational levels and isospins. They have the property of reducing to the corresponding chiral n-point functions of Virasoro minimal models by a kind of quantum Hamiltonian reduction procedure.

The observation of Polyakov [1], that two-dimensional gravity induced by conformal field theory possesses a hidden chiral $\hat{sl}(2,\mathbb{R})$ symmetry, has been the inspiring idea for several people to look for relations between Virasoro algebra and $\hat{sl}(2,\mathbb{R})$ Kac-Moody (KM) algebra and, in general, between W_n-algebras and $\hat{sl}(n,\mathbb{R})$ KM algebras. The authors of ref.[2] follow a classical approach essentially, while those of ref.[3] use a quantum analogue of the classical Hamiltonian reduction of Drinfeld and Sokolov [4]. Despite their success Bershadsky and Ooguri limit their analysis to the Hilbert space. But the description of the full quantum field theory is given by the Hilbert space plus operators or equivalently by all the correlators.

In the work [5] (we want to report on here and to which we refer for notations and conventions) we have implemented the quantum Hamiltonian reduction from $\hat{sl}(2,\mathbb{C})$ [1] to Virasoro algebra directly on the correlation functions of the conformal field theory introduced by Knizhnik and Zamolodchikov (KZ) [6] which has $\hat{sl}(2,\mathbb{C})$ as its infinite-dimensional symmetry.

In the KZ theory the Sugawara representation of the Virasoro generators is used; as a consequence the chiral n-point correlation functions of the $\hat{sl}(2,\mathbb{C})$ primary fields $\Psi^{\Lambda}(z)$ (with $sl(2,\mathbb{C})$ weight Λ) must satisfy the KZ differential equation

$$\left(\frac{\partial}{\partial z_a} - \frac{1}{(k+h)}\sum_{\substack{b=1\\ b\neq a}}^{n}\frac{\Omega_{ab}}{(z_a - z_b)}\right) W^{(n)}(z_1,\Lambda_1;\ldots;z_n,\Lambda_n) = 0\,, \tag{1}$$

with $a = 1,2,\ldots.n$, where k is the level of $\hat{sl}(2,\mathbb{C})$, h is the dual Coxeter number, $\Omega_{ab} = q_{\alpha\beta}S_a^{\alpha}S_b^{\beta}$, with q the inverse of the Killing - Cartan form and S_a^{α}, $\alpha = 0,\pm$ acting on the $a^{\rm th}$ field, are generators of a representation C_{Λ_a} of $sl(2,\mathbb{C})$. Let us put the isospin and coordinate $sl(2,\mathbb{C})$ symmetry on equal footing by introducing a new complex variable x, such that the generators $S^{0,\pm}$ of (isospin) $sl(2,\mathbb{C})$ are realized as differential operators

$$S^0 = 2x\partial_x - 2J, \quad S^- = -\partial_x, \quad S^+ = x^2\partial_x - 2Jx \tag{2}$$

* Permanent address: INRNE, Sofia

[1] We deal here with $\hat{sl}(2,\mathbb{C})$ instead of $\hat{sl}(2,\mathbb{R})$ because our two-dimensional "space-time" is Euclidean instead of Minkowskian like in [3].

and the primary fields acquire an x-dependence

$$\Psi^J(x,z) = \sum_{\tau=0} \binom{2J}{\tau} x^{2J-\tau}\, \Psi^J_{J-\tau}(z) . \tag{3}$$

The two- and three-point chiral correlation functions are given by the factorized expressions

$$W^{(2)}(x_1,z_1,J;x_2,z_2,J) = (x_{12})^{2J}\,(z_{12})^{-2\Delta_J} , \tag{4}$$

$$W^{(3)}(x_1,z_1,J_1;...;x_3,z_3,J_3;) = C_{J_1J_2J_3}\,\frac{x_{12}^{J_1+J_2-J_3}\,x_{23}^{J_2+J_3-J_1}\,x_{31}^{J_1+J_3-J_2}}{z_{12}^{\Delta_{J_1}+\Delta_{J_2}-\Delta_{J_3}}\,z_{23}^{\Delta_{J_2}+\Delta_{J_3}-\Delta_{J_1}}\,z_{31}^{\Delta_{J_1}+\Delta_{J_3}-\Delta_{J_2}}} , \tag{5}$$

where $\Delta_J = \frac{J(J+1)}{k+2}$ is the Sugawara scale dimension. Let us consider rational levels and isospins, $J = j-(k+2)j' \equiv J_{j,j'}$, $2j, 2j' \in \mathbb{Z}_+ , k+2 = p/p' , p, p'$ - coprime (positive) integers, and j, j' - restricted furthermore to $1 \le 2j+1 \le p-1$, $1 \le 2j'+1 \le p'-1$ [7]. If we now let $x_a \to z_a$ in (4),(5), we obtain just the two- and three-point chiral correlation functions $\langle 0|\Psi^{h_J}(z_1)\Psi^{h_J}(z_2)|0\rangle$ and $\langle 0|\Psi^{h_1}(z_1)\Psi^{h_2}(z_2)\Psi^{h_3}(z_3)|0\rangle$ of the Virasoro primary fields Ψ^h with conformal dimension $h(j,j') \equiv h_J = \Delta_J - J$ belonging to the minimal model with central charge $c_k = 13 - 6(k+2+\frac{1}{k+2})$.

This is the manifestation of quantum Hamiltonian reduction at the level of correlators (see also ref.[8] for a connection between isospin and space-time degrees of freedom). The aim of our work is to show that the same procedure is valid also for n-point correlation functions with $n \ge 4$, even if they do not have anymore a factorized form like Eqs. (4) , (5) . Exploiting the free-field (or Coulomb-gas) realization of the minimal models, the correlators of n Virasoro primary fields $\Psi^{h_a}(z_a)$, $h_a = h(j_a, j'_a)$, are represented by a multiple contour integral [9]

$$\langle 0|\Psi^{h_1}(z_1)\Psi^{h_2}(z_2)\ldots\Psi^{h_n}(z_n)|0\rangle_\Gamma = (prefactor) I_\Gamma^{(s,s')}(\underline{z}) , \tag{6}$$

where Γ are the integration contours, $\underline{z}_a$ are the anharmonic ratios and

$$I_\Gamma^{(s,s')}(\underline{z}) = \int_\Gamma du_1 \ldots du_s\, dv_1 \ldots dv_{s'}\; \Phi_J^{(s,s';\alpha_-^2)}(u_i, v_{i'}; \underline{z}_a) \tag{7}$$

where $\alpha_\pm = \pm(k+2)^{\pm 1/2}$ and the multivalued function Φ is given by

$$\Phi_J^{(s,s';\nu)}(u_i,v_{i'};z_a) = \Phi_J^{(s;\nu)}(u_i;z_a)\,\Phi_{-\nu J}^{(s';1/\nu)}(v_{i'};z_a)\prod_{i=1}^{s}\prod_{i'=1}^{s'}\frac{1}{(u_i - v_{i'})^2} , \tag{8}$$

with

$$\Phi_J^{(s;\nu)}(u_i;z_a) = \prod_{i<j}^{s}(u_i-u_j)^{2\nu}\prod_{i=1}^{s}\prod_{a=1}^{n-1}(u_i - z_a)^{-2J_a\nu} . \tag{9}$$

If the free-field bosonization tecnique is used also in building up the correlators of $\hat{sl}(2,\mathbb{C})$ KM algebra, it is easy to see [10] , [5] that it is impossible for rational level and generic (half-) integer values of j and j' to recover the minimal model correlators starting from those ones of $\hat{sl}(2,\mathbb{C})$. Therefore we try to find *directly* a solution of the general KZ-ZF (ZF for Zamolodchikov-Fateev [11]) equation which is obtained by inserting into Eq. (1) the isospin $sl(2,\mathbb{C})$ generators (2). Assuming that the solution is represented as an (infinite in principle) power series in $(x-z)$, we get, for the correlators with the last point sent to infinity

$$U^{(n)}(x,z) \equiv \frac{1}{(2J_n)!}\lim_{\substack{x_n\to\infty\\ z_n\to\infty}} x_n^{-2J_n}\,z_n^{2\Delta_n}\,\langle 0|\Psi^{J_1}(x_1,z_1)\ldots\Psi^{J_n}(x_n,z_n)|0\rangle \tag{10}$$

$$= \prod_{1\le b<a\le n-1} z_{ab}^{2J_aJ_b/(k+2)}\, G_J^{(n)}(x,z) :$$

$$G_J^{(n)}(x,z) = \sum_{t=0} \sum_{\tau:\,|\tau|=t,} \prod_{a=1}^{n-1} (x_a - z_a)^{\tau_a} B_\tau\, I_\tau^{(s,s')}(z)\,, \tag{11}$$

where

$$B_\tau = \frac{(-1)^t \beta_J}{(S-t)!} \frac{(c_t)!}{(c_s)!} \prod_{a=1}^{n-1} \binom{2J_a}{\tau_a}, \quad t = |\tau|\,, \tag{12}$$

and

$$I_{\Gamma,\sigma}^{(s,s')} = \int_\Gamma du_1 \dots du_s\, dv_1 \dots dv_{s'}\, \varphi_\sigma^{(s,s')}\, \Phi_J^{(s,s';\nu)}\,, \tag{13}$$

with $\Phi_J^{(s,s';\nu)}$ given by the minimal model integrand (8) and

$$\varphi_\sigma^{(s,s')}(\{u_i, v_{i'}, z_a\}) = \frac{(s'K)!s!}{S!} \sum_\tau \prod_a \binom{\sigma_a}{\tau_a} \binom{S-|\sigma|}{s-|\tau|} \varphi_{\sigma-\tau}^{[K]}(\{v_{i'}, z_a\}) \varphi_\tau^{[1]}(\{u_i, z_a\})\,, \tag{14}$$

where $K \equiv -k-2$, $S = s + s'K$ and the sum runs over $\{\tau_a = 0,1,...,s, \quad |\tau| \le s\}$, with

$$\varphi_\mu^{[L]}(\{u_i, z_a\}) = \frac{(L!)^s (mL - |\mu|)!}{(mL)!} \sum_{\{N_{ia}\}} \frac{\prod_a \mu_a!}{\prod_i [(L - \sum_a N_{ia})! \prod_a N_{ia}! (u_i - z_a)^{N_{ia}}]}\,, \tag{15}$$

where the sums are finite running over $N_{ia} = 0,1,...,\mu_a$, $i = 1,2,...,m$, $a = 1,2,...,n-1$, subject to the constraints

$$\sum_{i=1}^s N_{ia} = \mu_a\,. \tag{16}$$

Here we want to emphasize the *meromorphic* structure of the expressions (14),(15), in contrast to that one obtained by bosonization in the general case.

The main technical tool and the hardest part to prove is the following

Proposition. *The integrals (13) satisfy the following relations:*

the linear relations

$$\sum_a (2J_a - \sigma_a) I_{\sigma+e_a}^{(s,s')} = 0, \tag{17}$$

the recursion relations

$$\sum_a (2J_a - \sigma_a)\, z_a\, I_{\sigma_a+e_a}^{(s,s')} + c_{|\sigma|}\, I_\sigma^{(s,s')} = 0\,, \tag{18}$$

$$c_{|\sigma|} = \sum_{a=1}^n J_a - |\sigma| - k - 1\,, \tag{19}$$

and the differential equations

$$\begin{aligned}(k+2)\,\partial_a I_\sigma^{(s,s')} \;&+\; \sum_{b(\ne a)} \frac{1}{z_{ab}} [\sigma_a(2J_b - \sigma_b)(I_\sigma^{(s,s')} - I_{\sigma-e_a+e_b}^{(s,s')}) + \sigma_b(2J_a - \sigma_a)(I_\sigma^{(s,s')} - I_{\sigma+e_a-e_b}^{(s,s')})] \\ &=\; (S - |\sigma|)(2J_a - \sigma_a) I_{\sigma+e_a}^{(s,s')}\,. \end{aligned} \tag{20}$$

This statement holds true for arbitrary $k+2 \ne 0$ and any combination of $\{J_a = j_a -$

$j'_a(k+2)\}$ producing nonnegative integers s, s'. Having this proposition in hand, it is almost strightforward to show that the correlators (10) together with Eqs. (11),(12),(13),(14),(15) satisfy the KZ-ZF equation. Furthermore from Eq.(11) we see that in the limit $x_a \to z_a$ only the term with $t = 0$ survives, providing just the correct minimal model correlator. Not only correlators go into correlators in this limit; also differential equations governing them are transformed ones into the others. In fact the KZ-ZF first-order matrix differential equation for the coefficient functions of the $(x-z)$-expansion of the $\hat{sl}(2,\mathbb{C})$ correlators can be reduced to a scalar (BPZ)[12] equation for the correlators of the minimal models – the latter reflecting the decoupling of the Virasoro null-vectors. In the general case the KZ-ZF system of equations is infinite-dimensional (as a result of the infinite dimensionality of the $sl(2,\mathbb{C})$ irreducible representations for non-integer isospins), but there exists an infinite set of algebraic relations, leaving at the end only a *finite* set of independent coefficient functions. These algebraic equations for the $\hat{sl}(2,\mathbb{C})$ primary field correlation functions are provided by the presence of (an infinite number of) singular vectors of the KM Verma modules [13].

References

[1] A.M. Polyakov, *Mod. Phys Lett.* **A 2** (1987) 893,
V.G. Knizhnik, A.M. Polyakov and A.B. Zamolodchikov, *Mod. Phys. Lett.* **A 3** (1988) 819.

[2] J.Balog, L. Fehér, L. O'Raifeartaigh, P. Forgács and A. Wipf, *Phys. Lett.*B**227** (1989) 214, *Ann. Phys.* **203** (1990) 76.

[3] M. Bershadsky and H. Ooguri, *Comm. Math. Phys.* **126** (1989) 49,
B. Feigin and E. Frenkel, *Phys. Lett.* **B 246** (1990) 75.

[4] V. Drinfeld and Sokolov, *J. Sov. Math.* **30** (1984) 1975.

[5] P. Furlan, R. Paunov, A.Ch. Ganchev and V.B. Petkova, *Phys. Lett.* **B 267** (1991) 63; "Solutions of the Knizhnik-Zamolodchikov equation with rational isospins and the reduction to the minimal models", preprint CERN-TH.6289/91, SISSA 120/FM/91.

[6] V.G. Knizhnik and A.B. Zamolodchikov, *Nucl. Phys.* **B 247** (1984) 83.

[7] V.G. Kac and M. Wakimoto, *Proc. Natl. Acad. Sci. USA* **85** (1988) 4956.

[8] A.M. Polyakov, *Int. J. Mod. Phys.* **A 5** (1990) 833.

[9] V.S. Dotsenko and V.A. Fateev, *Nucl. Phys.* **B 240** (1984) 312; **B 251** (1985) 691.

[10] Vl.S. Dotsenko, *Nucl. Phys.* **B358** (1991) 67.

[11] A.B. Zamolodchikov and V.A. Fateev, *Sov. J. Nucl. Phys.* **43** (1986) 657.

[12] A. Belavin, A. Polyakov and A. Zamolodchikov, *Nucl. Phys.* **B 241** (1984) 333.

[13] V. G. Kac and D. A. Kazhdan, *Adv. Math.* **34** (1979) 97.

POISSON LOOP GROUPS, CLASSICAL AND QUANTUM GENERALIZED PERIODIC TODA LATTICES*

GUO MAOZHENG
Department of Mathematics, Peking University
Beijing 100871, P.R. CHINA

and

QIAN MIN
Department of Mathematics, Peking University
Beijing 100871, P.R. CHINA

and

WANG BAILING
Department of Mathematics, Peking University
Beijing 100871, P.R. CHINA

ABSTRACT

The solutions of integrable systems related to loop group are described as geodesics on an infinite dimensional Riemannian manifolds. The generalized periodic Toda lattices are described as finite dimensional reduction of these geodesics.

§1. Introduction

*work supported by NNSFC Grand. to the first two authers

For a given classical r-matrix, the integrable systems can be constructed in principle, and the solutions of Hamiltonian equations can be obtained by Konstant's representation theory. Ten years ago, A.M. Perelomov and M.A. Olshanetsky (cf [3]) posed and systematically studied the question whether the solutions corrrspond to geodsics on some homogeneous spaces for some integrable systems. They illustrated this property in the case of several kinds of finite dimensional integrable systems. They showed that the integrable systems of generalized nonperiodic Toda lattics associated with any Lie group G describe the special geodesic motions on the Riemannian non-compact symmetric space which is the quotient of the normal real form of G by its maximal compact subgroup. L.A. Ferreia and D.I. Olive(cf [1]) illustrated this picture further by adding more geometric ingedient and relating the Lax operator to the r-matrix of the solutions of the Yang-Baxter equation.

In this paper, we consider the integrable systems of periodic Toda lattices associated with any semisimple Lie group G. We show that the solutions of Lax equation related to loop group LG can be described as geodesics on an infinite dimensional Riemannian homogeneous manifold. As an application to generalized periodic Toda lattics, we obtain the reduction to a finite dimensional symplectic manifold. The quantum pictures of the theory are given through geometric quantization procedure of this symplectic manifold.

2. Lax equation on $\mathcal{LG}$

Let G be a complex seisimple Lie group with Lie algebra $\mathcal{G}$. Let K be maximal compact subgroup of G with trivial center and $\mathcal{K}$ be Lie algebra of K. Consider loop group $LG = C^\infty(S^1, G)$, $LK = C^\infty(S^1, K)$, their Lie algebras are $\mathcal{LG}$ and $\mathcal{LK}$ respectively.

Theorem 1. Let $G = KAN$ be the Iwasawa decomposition. Denote S the subgroup of LG such that for $r \in S$, r can be extended to the unit disk holomorphically with $r(0) \in AN$, then

$$LG = LK \cdot S \tag{1}$$

Denote $M = LG/LK$, which is an infinite dimensional Riemannian symmetric space. Let $\Phi : S \to M$ by $\Phi(s) = [s]$, this is one to one and onto mapping.

Consider the root space decomposition of $\mathcal{LG} = \mathcal{G} \otimes C[\lambda, \lambda^{-1}]$

$$\mathcal{LG} = \oplus_{k\in Z}\lambda^k\mathcal{H} \oplus \sum_{k\in Z,\ \alpha\in\Delta} \lambda^k\mathcal{G}_\alpha \tag{2}$$

where $\mathcal{H}$ is the Cartan subalgebra of $\mathcal{G}$, Δ is the root system of $(\mathcal{G}, \mathcal{H})$. We define the generalized Koszual operator on $\mathcal{LG}$, which is linear and $R|_{\mathcal{H}} = o$, $R|_{\mathcal{G}_\alpha} = \text{sign}(\alpha)Id$, for $\alpha \in \Delta$, and $R|_{\lambda^k \mathcal{G}} = \text{sign}(k)Id$, for $k \neq 0$. We know that R is a r-matrix on $\mathcal{LG}$.

Now construct the following Lax equation on $\mathcal{LG}$:

$$\begin{cases} \dfrac{dA}{dt} = [R(A), A] \\ A(0) = A_0 \in \mathcal{LP} \end{cases} \tag{3}$$

where $\mathcal{P}$ is the orthogonal complement of $\mathcal{K}$ with respect to the killing form of $\mathcal{G}$ and $\mathcal{LP} = C^\infty(S', \mathcal{P})$. Then we have

Theorem 2 Adjoint-orbits and Geodesic Picture of the Solutions

(a) The Lax equation (3) has a unique solution, expressed in adjoint-orbit form:

$$A(t) = Ad(s(\exp tA_0))A_0 = Ad(k^{-1}(\exp tA_0))A_0$$

where $s(t) = s(\exp tA_0)$ and $k(t) = k(\exp tA_0)$ are defined by $\exp(tA_0)$ in the decomposition of (1), i.e $\exp tA_0 = k(t)s(t)$

(b) $x(t) = \Phi(s(t))$ is the geodesic flow in M.

As application, we reduce the Lax equation to the appropriate finite dimensional manifold of $\mathcal{LG}$ and show that the solution of the integrable systems of the generalized periodic Toda lattices induce the geodesic flow.

References

1. Ferreire, L.A., and Olive, D.I., Non-compact symmetric spaces and Toda molecule equations, *Math. Phys.* **99**(1986) 365–84.
2. Helgason, S., *Differential Geometry, Lie Groups and Systemmetric Spaces*, Academic Press, New York, 1978.
3. Olshanetsky, M.A., and Perelomov, M.A., Classical integrable finite dimensional systems related to Lie algebra, *Phys. Reports* **71**(1981), No. 5, 313–400.
4. Pressley, A., and Segal, G., *Loop Groups*, Oxford, 1986.

NON-RELATIVISTIC CONFORMAL SYMMETRIES

P. A. HORVÁTHY

Département de Mathématiques, Université,
Parc de Grandmont, F-37200 TOURS (France)

1. Introduction.

Twenty years ago Niederer and Hagen [1] found that the maximal invariance group of the Schrödinger equation of a free, non-relativistic particle was larger then just the Galilei group. It containes two more 'conformal' generators, namely

$$D = tH - \tfrac{1}{4}\{\mathbf{p},\mathbf{r}\} \quad \text{and} \quad K = tH^2 - tD + \tfrac{m}{2}r^2 \tag{1.1}$$

dilatations — expansions

These operators generate, with the Hamiltonian $H = \mathbf{p}^2/2m$, an $o(2,1)$ symmetry algebra. Adding the Galilei group yields the Schrödinger group (2.4).

That the Schrödinger equation has a larger symmetry group then the Galilei group is in contrast with the relativistic case where the free Klein-Gordon has the Poincaré group as symmetry, and the conformal group $o(4,2)$ only occurs for *massless* particles.

The $o(2,1)$ in Eq. (1.1) is also a symmetry for a charged, spin 0 particle in a Dirac monopole field [2] and even for the Pauli Hamiltonian

$$H = \frac{1}{2m}\left[\pi^2 - q\frac{\mathbf{r}.\sigma}{r^3}\right] \qquad \pi = -i\partial - q\mathbf{A}_D \tag{1.2}$$

of a non-relativistic spin $\frac{1}{2}$ particle in the field of a Dirac monopole [3]: Eq. (1.2) has furthermore a superconformal symmetry: the fermionic charges

$$Q = \frac{1}{\sqrt{2m}}\pi.\sigma \quad \text{and} \quad S = \sqrt{\frac{m}{2}}\,\mathbf{r}.\sigma - tQ \tag{1.3}$$

close into the symmetry algebra $osp(1,1)_{superconf}$ (see eqn. (3.6) below).

It was argued [4-5], that the proper arena for non-relativistic physics is extended 'Bargmann' spacetime M, endowed with a metric g with signature $(-,+,+,+,+)$, and with a covariantly-constant null vector ξ. Classical motions are *massless geodesics* in M and hence *conformally invariant.* Our construction differs from Kaluza-Klein theory in that the extra dimension is null, rather then space-like.

2. Spinless particles.

For a spinless, free particle the extended space is $M = \{t,\mathbf{r},s\} = \mathbf{R}\times\mathbf{R}^3\times\mathbf{R}$, endowed with the flat metric $g_{\mu\nu}dx^\mu dx^\nu = d\mathbf{r}^2 + 2dt\,ds$, and with the null Killing vector $\xi = \partial_s$. Viewing the wave function as an equivariant function on extended space, $\partial_s\Phi = im\Phi$, the free Schrödinger equation $-\Delta\varphi = 2mi\partial_t\varphi$ (where Δ is the Laplacian on ordinary 3-space) can be written as

$$\Delta_g\Phi = 0. \tag{2.1}$$

g being the Laplacian on M and $\Phi = e^{ims}\varphi$. Consider now those ξ-preserving conformal isometries C of M,

$$C^*g = \Omega^2 g, \qquad C_*\xi = \xi. \tag{2.2}$$

called ***non-relativistic conformal transformations.*** For a free particle for example, the conformal diffeomorphisms form $O(5,2)$ which is reduced by the ξ-constraint to a 13 dimensional subgroup. Its infinitesimal action on extended spacetime $\{t, \mathbf{r}, s\}$ is

$$\Big(\kappa t^2 + \delta t + \epsilon,\ \omega \times \mathbf{r} + (\frac{\delta}{2} + \kappa t)\mathbf{r} + \beta t + \gamma,\ -(\frac{1}{2}\kappa\, r^2 + \beta.\mathbf{r} + \eta)\Big), \tag{2.3}$$

where $\omega \in so(3)$, $\beta, \gamma \in \mathbf{R}^3$, $\epsilon, \delta, \kappa, \eta \in \mathbf{R}$. This is the ***extended Schrödinger algebra***, with ω representing rotations, β Galilei boosts, γ space-translations, ϵ time-translations, δ dilatations, κ expansions, and η translations in the vertical direction.

The associated conserved quantities $\mathbf{L}$, $\mathbf{b}$, $\mathbf{p}$, H, D, K and m satisfy the commutation relations of the extended Schrödinger algebra,

$$\begin{array}{llll} [\mathbf{L},\mathbf{L}] = i\mathbf{L} & [\mathbf{L},\mathbf{b}] = i\mathbf{b} & [\mathbf{L},\mathbf{p}] = i\mathbf{p} & [\mathbf{L},H] = 0 \\ [\mathbf{b},H] = i\mathbf{p} & [\mathbf{b},\mathbf{b}] = 0 & [\mathbf{b},\mathbf{p}] = im & [\mathbf{p},\mathbf{p}] = [\mathbf{p},H] = 0 \\ [H,D] = iH & [H,K] = 2iD & [D,K] = iK & \\ [\mathbf{p},D] = \frac{i}{2}\mathbf{p} & [\mathbf{b},D] = -\frac{i}{2}\mathbf{b} & [\mathbf{p},K] = -i\mathbf{b} & [\mathbf{b},K] = 0 \\ [\mathbf{L},K] = 0 & [\mathbf{L},D] = 0 & [\mathbf{b},K] = 0 & \end{array} \tag{2.4}$$

and all generators commute with m. The action of conformal transformations on a wavefunction can be deduced from that of $O(2,5)$ on M, yielding a representation of the extended Schrödinger group.

Another application is provided by the harmonic oscillator. The Bargmann space is again $\mathcal{M} = \mathbf{R}\times\mathbf{R}^3\times\mathbf{R}$, endowed with the metric $g = d\mathbf{x}^2 + 2ds\,dt - \omega^2 r^2\,dt^2$. Now the transformation $f : (t, \mathbf{x}, s) \mapsto (\tau, \boldsymbol{\xi}, \sigma)$ where

$$\tau = \frac{1}{\omega}\tan\omega t, \qquad \boldsymbol{\xi} = \frac{\mathbf{x}}{\cos\omega t}, \qquad \sigma = s - \frac{\omega r^2}{2}\tan\omega t \tag{2.5}$$

satisfies $f^*(d\boldsymbol{\xi}^2 + 2d\sigma d\tau) = \frac{1}{\cos^2\omega t}g$, so that (every half period of) the harmonic oscillator is mapped onto a free motion. The two systems have therefore the same symmetries [6].

Electromagnetic fields can be included as external fields. We only consider the theory obtained by minimal coupling, $\partial_t \mapsto \partial_t + ieV$, $\mathbf{p} \mapsto \boldsymbol{\pi} = \mathbf{p} - e\mathbf{A}$. Those conformal space-time symmetries which preserve the electromagnetic fields will still act as symmetries.

For example, the Bargmann space of a Dirac monopole is $\{(t, \mathbf{r}, s)\} = \mathbf{R} \times \mathbf{R}^3 \setminus \{0\} \times \mathbf{R}$, with the flat metric above. Its conformal symmetries form the subgroup of the Schrödinger group which preserves the origin and are readily identified with $SO(3) \times SL(2, \mathbf{R})$, generated by the angular momentum $\mathbf{L} = \mathbf{r} \times \boldsymbol{\pi} - q\hat{r}$, and by $H = \boldsymbol{\pi}^2/2m$, and D and K in (1.1). Since they also preserve the electromagnetic field, they are symmetries.

3. Spinning particles and Supersymmetry.

Chosing Dirac matrices to satisfy $\{\gamma^\mu, \gamma^\nu\} = 2g^{\mu\nu}$ on 5-dimensional extended space, spin $\frac{1}{2}$ particles are described by the massless minimally coupled Dirac equation,

$$
\not{D}\Psi \equiv \gamma^\mu D_\mu \Psi = 0, \tag{3.1}
$$

Using the particular form of the Bargmann metric, one gets the Lévy-Leblond [7] equation

$$
\begin{aligned}
\sigma.\boldsymbol{\pi}\,\varphi \quad &+ \quad 2m\chi = 0 \\
(i\partial_t - eV)\varphi \quad &+ \quad \sigma.\boldsymbol{\pi}\,\chi = 0
\end{aligned} \tag{3.2}
$$

Since

$$
\not{D}^2 = \Big[2m(i\partial_t - eV) - \boldsymbol{\pi}^2 - e\sigma.\mathbf{B}\Big], \tag{3.3}
$$

each component satisfies the *Pauli equation*.

Since we are working with the massless Dirac equations, all ξ-preserving conformal transformations of extended space are symmetries. For a free particle, we get an irreducible representation of the extended Schrödinger group.

For the Dirac monopole, the origin-preserving subgroup $SO(3) \times SL(2, \mathbf{R})$ of the Schrödinger group yields the bosonic symmetry algebra for the Lévy-Leblond equation and thus also for its square, the Pauli equation.

For an arbitrary static magnetic field $\mathbf{A}$, the *helicity operator*

$$
Q = \frac{1}{\sqrt{2m}}\,\sigma.\boldsymbol{\pi} \tag{3.4}
$$

is conserved, $[Q, \not{D}] = 0$. Commuting the free helicity operator $Q = \sigma.\mathbf{p}$ with the generators of the Schrödinger group yields two new supercharges,

$$
\boldsymbol{\Sigma} = i[Q, \mathbf{b}] = \sqrt{\frac{m}{2}}\,\sigma \qquad \text{and} \qquad S = -i[K, Q] = \frac{1}{\sqrt{2m}}\,\sigma.\mathbf{b}. \tag{3.5}
$$

The commutation relations are deduced from those of the extended Schrödinger group. This yields to the 'super-Schrödinger - algebra' [8] with commutation relations (2.4), supplemented with

$$\begin{array}{llll}[Q,D]=\frac{i}{2}Q & [K,Q]=iS & [S,H]=iQ & [S,D]=-\frac{i}{2}S\\ [Q,\mathbf{L}]=0 & [Q,\mathbf{b}]=-i\mathbf{\Sigma} & [Q,\mathbf{p}]=0 & [Q,H]=0\\ [S,\mathbf{L}]=0 & [S,\mathbf{p}]=0 & [S,K]=0 & \\ \{Q,Q\}=2H & \{Q,S\}=-2D & \{S,S\}=2K & \\ \{\mathbf{\Sigma},\mathbf{\Sigma}\}=m, & \{\mathbf{\Sigma},Q\}=\mathbf{p} & \{\mathbf{\Sigma},S\}=\mathbf{b} & \end{array} \tag{3.6}$$

Having Q act on the $o(2,1)$ subalgebra yields an $osp(1,1)$ sub-superalgebra spanned by H, D and K and by the odd charges Q and S. These are the symmetries which remain unbroken when a Dirac monopole is added [3].

I am indebted to C. Duval and G. W. Gibbons in collaboration with whom these results were obtained.

References

[1] U. Niederer, Helv. Phys. Acta **45**, 802 (1972); C. R. Hagen, Phys. Rev. **D5**, 377 (1972)

[2] R. Jackiw, Ann. Phys. (N. Y.) **129**, 183 (1980)

[3] E. D'Hoker and L. Vinet, Phys. Lett. **137B**, 72 (1984)

[4] C. Duval, G. Burdet, H. P. Künzle and M. Perrin, Phys. Rev. D31, 1841 (1985); C. Duval, in *Proc. XIVth Int. Conf. Diff. Geom. Meths. in Math. Phys.*, Salamanca '85, Garcia, Pérez-Rendón (eds). Springer LNM **1251**, p. 205 Berlin (1987); in Proc.'85 Clausthal Conf., Barut and Doebner (eds), Springer LNPhysics **261**, p. 162 (1986); C. Duval, G. Gibbons and P. Horváthy, Phys. Rev. **D43**, 3907 (1991)

[5] W. M. Tulczyjew, J. Geom. Phys. **2**, 93 (1985); M. Omote, S. Kamefuchi, Y. Takahashi, Y. Ohnuki, in *Symmetries in Science III*, Proc '88 Schloss Hofen Meeting, Gruber and Iachello (eds), p. 323 Plenum : N. Y. (1989)

[6] U. Niederer, Helv. Phys. Acta **46**, 192 (1973); G. Burdet, C. Duval and M. Perrin, Lett. Math. Phys. **10**, 255 (1985)

[7] J-M. Lévy-Leblond, Comm. Math. Phys. **6**, 286 (1967)

[8] J. P. Gauntlett, J. Gomis and P. K. Townsend, Phys. Lett. **B248**, 288 (1990)

Electromagnetic Tensor Fields and Semicomplex Structures

by Hou Zixin (Dept. of Math. Nankai Univ. Tianjin, People's Republic of China)
and Zhao Dun (Dept. of Math. Lanzhou Univ. Lanzhou People's Republic of China)

Let V_4 be the space time of general relativity with the gravitational symmetric tensor field g, which is of index of inertia 2, i.e. the signature is $+\ +\ +\ -$ or $-\ -\ -\ +$, rank $(g) = 4$.

Let $\tilde{k}$ be the electromagnetic tensor field of the type (0,2), then $\tilde{k}$ is skew-symmetric. Put

$$g(k(X), Y) = \tilde{k}(X, Y) \quad \forall X, Y \in \mathfrak{s}(V_4) \tag{1}$$

Then k is a tensor field of type (1,1), called the Maxwell fields. The characteristic equation for k is

$$k^4 + 2Kk^2 + |k| I_4 = 0 \tag{2}$$

where $K = \frac{1}{4}$ trace (k^2) and $|k|$ is the determination of k [1]. The eigenvalues for k are

$$\lambda_1 = -\lambda_2 = \sqrt{-1} \cdot \sqrt{\sqrt{D} + K}$$
$$\lambda_3 = -\lambda_4 = \sqrt{\sqrt{D} - K}$$

where $D = K^2 - |k|$.

The electromagnetic tensor field is said to be the first kind if $K|k| \neq 0$. In this case, (3) can be written as

$$(k^2 - p^2)(k^2 + q^2) = 0 \tag{2}$$

put $J = qk$, then J satisfies the follow equation

$$(J^2 - \lambda^2)(J^2 + 1) = 0 \tag{3}$$

and (3) defines three complementary distributions θ_1, θ_1' and θ_2 respectively as follows

$$\theta_1(p)T_p(V_4) = \{X \in T_p(V_4) | J_p X = \lambda X\},$$
$$\theta_1'(p)T_p(V_4) = \{X \in T_p(V_4) | J_p X = -\lambda X\},$$
$$\theta_2(p)T_p(V_4) = \{X \in T_p(V_4) | J_p^2 X = -X\}.$$

When these three distributions are all completely integral, θ_2 determines an almost complex structure on its integral submanifold. Under this assumption, (3) defines a $G-$ structure $GL(r, \mathbb{R}) \times GL(r, \mathbb{R}) \times GL(s, \mathbb{C})$, where $r = dim\theta_1 = dim\theta_1'$ and $2s = dim\theta_2$.

It is obvious that the $G-$structure is independent on the choice of λ, in particular, we can choose the tensor field J which fits for $J^4 = id$ to define the $G-$structure.

Definition 1 Let M be a 2n-dimensional smooth real manifold, a $(1,1)$ - tensor field J on M is called an almost semicomplex structure on M if the followings hold:

(J1) $J^4 = id$;

(J2) For each point $p \in M$, the dimensions of ± 1 - eigenspaces of J_p are the same.

From [3], we can define the integrability of an almost semicomplex structure.

Definition 2 An almost semicomplex structure J is said to be integrable, if the Nijenhuis tensor field N of J vanishes, i.e, for any $X.Y \in \mathfrak{s}(M)$.

$$J^2[X,Y] + [JX,JY] - J[JX,Y] - J[X,JY] = 0$$

holds. An integrable almost semisomplex structure is called a semicomplex structure.

If there exists a (almost) semicomplex structure J on M, then (M,J) is called a (almost) semicomplex manifold. It should be noted that the conception of (almost) semicomplex structure is exact the generalization of the conceptions of (almost) complex structure $(J^2 = id)$ and (almost) paracomplex structure $(J^2 = ik)$[2].

Because of [1], we give

Definition 3 Let (M,J) be a (almost) semicomplex manifold, g be a pseudo - Riemannian metric on M. g is called a (almost) semihermitian metric, if

$$g(JX,Y) + g(X,JY) = 0$$

holds for any $X,Y \in x(M)$, and then (M,J,g) is called a (almost) semihermitian manifold.

Definition 4 Let (M,J,g) be a (almost) semihermitian manifold, for any $X.Y \in x(M)$, put

$$\omega(X,Y) = g(X,JY)$$

if $dw = 0$, then g is called a (almost) semikähler metric and (M,J,g) is called a (almost) semikähler manifold.

If (M,g,J) is semikählerian, it is obvious that (M,Ω) is a symplectic manifold.

Proposition 1 Let (M,I,g) is an almost semihermitian manifold, ∇ is the Levi - Civita connection of g, then (M,I,g) is semikahlerian off $\nabla J = 0$.

Definition 5 Let (M,J) and (M',J') be (almost) semicomplex manifolds, a smooth map

$$f : M \longrightarrow M'$$

is called a (almost) semiholomorphic map, if

$$(df)_{p*} J_p = J'_{f(p)*}(df)_p$$

holds for each $p \in M$.

The group consisting of all semiholomorphic trousformations on M is denoted by Aut (M,J). For a given connected (almost) semihermitian or semikähler manifold (M,J,g), we can define its automorphism group

$$\text{Aut}\,(M,J,g) = \text{Aut}\,(M,J) \cap I(M,g).$$

If the action of Aut (M,J,g) is transitive, then (M,J,g) is called homogeneous.

Definition 6 A connected semihermitian manifold (M,J,g) is said to be a semihermitian symmetric space, if for every $p \in M$, there exists a semiholomorphic isometry $s_p \in$ Aut (M,J,g) such that

1) $s_p^2 = id_M$;
2) p is an isolated fined points of s_p.

A semihermitian symmetric space is of course a pseudo - Riemannian symmetric space. In general, an almost semihermitian metric may not be an semikähler metric, but we have

Proposition 2 A semihermitian symmetric space is a homogeneous semikähler space and therefore a homogeneous symplectic space.

Definition 7 Let G be a connected Lie group, H be a closed subgroup of G. if G/H is a pseudo - Riemannian symmetric coset space with an invariant almost semicomplex struclure J and a G - invariant almost semihermitian metric g, then $(G/H,J,g)$ is called a semihermitian symmetric coset space.

Proposition 3 A semihermitian symmetric space is isometric semiholomorphic to a semihermitian symmetric coset space $(G/H,J,g)$. On the other hand, a semihermitian symmetric coset space $(G/H,J,g)$ is a semihermitian symmetric space.

Examples:

1. Minkowski space-time V^4 has a natural pseudo-Riemannian metric

$$g = dx_1^2 + dx_2^2 + dx_3^2 - dx_4^2$$

we define J on V^4 by

$$I\frac{\partial}{\partial x_1} = \frac{\partial}{\partial x_4} \qquad I\frac{\partial}{\partial x_2} = \frac{\partial}{\partial x_3}$$

$$I\frac{\partial}{\partial x_3} = -\frac{\partial}{\partial x_2} \qquad I\frac{\partial}{\partial x_4} = \frac{\partial}{\partial x_1}$$

then J is a semicomplex structure on V^4 and g is a semihermitian metric about J. We know that V^4 can be represented as symmetric coset space O(1,3)/SO(3), so V^4 is a semihermitian symmetric space with constant curvature 0.

2. Anti-de Sitter space-time [4] is the universal coverd manifold of H, where

$$H = \{x \in \mathbf{R}^5 \mid -x_1^2 - x_2^2 - x_3^2 + x_4^2 + x_5^2 = 1\}.$$

We know that H can be represented as symmetric coset space $\mathrm{SO}^0(3,2)/\mathrm{SO}^0(3,1)$, we shall introduce a pseudo-Riemannian metric on H later. Let $G = \mathrm{SO}^0(3,2)$, we can define an automorphism σ on G by

$$\sigma(A) = E_{4,1} A E_{4,1}$$

then σ is an involutive automorphism. The Lie algebra of G is

$$\begin{aligned} \mathfrak{g} &= \{A \in gl(5,\mathbf{R}) \mid {}^tA \cdot E_{3,2} + E_{3,2} \cdot A = 0\} \\ &= \left\{ A = \begin{pmatrix} B & \alpha \\ -{}^t\alpha E_{3,1} & 0 \end{pmatrix} \middle| {}^tBE_{3,1} + E_{3,1}B = 0 \right\}, B \in gl(4,\mathbf{R}) \end{aligned}$$

we still denote the induced involutive automorphism on $\mathfrak{g}$ by σ, then we can give the canonical decomposition of $\mathfrak{g}$ as follows:

$$\mathfrak{g} = \mathfrak{h} + \mathfrak{m}$$

where

$$\mathfrak{h} = \left\{ X = \begin{pmatrix} B & 0 \\ 0 & 0 \end{pmatrix} \middle| {}^tBE_{3,1} + E_{3,1}B = 0 \right\}$$

$$\mathfrak{m} = \left\{ X = \begin{pmatrix} 0 & \alpha \\ -{}^t\alpha E_{3,1} & 0 \end{pmatrix} \middle| \alpha \in \mathbf{R}^4 \right\}$$

It is easy to see that $\mathfrak{m}$ has a base $\{\varepsilon_1, \varepsilon_2, \varepsilon_3, \varepsilon_4\}$, here

$$\varepsilon_1 = \begin{pmatrix} 0 & e_1 \\ -{}^te_1 & 0 \end{pmatrix} \qquad \varepsilon_2 = \begin{pmatrix} 0 & e_2 \\ -{}^te_2 & 0 \end{pmatrix}$$

$$\varepsilon_3 = \begin{pmatrix} 0 & e_3 \\ -{}^te_3 & 0 \end{pmatrix} \qquad \varepsilon_4 = \begin{pmatrix} 0 & e_4 \\ {}^te_4 & 0 \end{pmatrix}$$

put

$$Z = \begin{pmatrix} 0 & 1 & & & \\ -1 & 0 & & & \\ & & 0 & 1 & \\ & & 1 & 0 & \\ & & & & 0 \end{pmatrix}$$

then $Z \in \mathfrak{h}$. Under the base given above, we get

$$ad_m Z = \begin{pmatrix} 0 & -1 & & \\ 1 & 0 & & \\ & & 0 & 1 \\ & & 1 & 0 \end{pmatrix}$$

Let B be the Killing form of $\mathfrak{g}$, put $J_0 = ad_m Z$ and $Q = B|_m$, then we can prove that J_0 and Q can uniquely define a semicomplex structure J and a pseudo-Riemannian metric g of signature (3,1) on H respectively such that (H, J, g) is a semihermitian symmetric space. Furthermore. we know that (H, J, g) has constant curvature -1.

3. The dual space of H

$$S = \{x \in \mathbf{R}^5 | -x_1^2 + x_2^2 + x_3^2 + x_4^2 + x_5^2 = +1\}$$

is called sitter space-time it can be represented by $SO^0(4,1)/SO^0(3,1)$. In the same way as in example 2, we can prove that S is a semihermitian symmetric space with constant curvature 1.

We can also give a structure theorem on almost all the semisimple homogeneous manifolds with invariant symplectic structure by using semicomplex structure, we shall disscuss the problem in another paper.

References:

[1]. R. S. Miskra, Tensor N. S. Vol.30(1976) 345 - 356.
[2]. S. Kaneyuki & M.Kozai, Tokyo J. Math. 8(1985) 81 - 98.
[3]. E. Kobayashi, Parific J. Math. 12(1962) 963 - 977.
[4]. J. K. Beem & P. E. Ehrtich, ≪Global Lorentzian Geometry≫.

QUANTIZATION ON RIEMANNIAN SPACES FROM GROUPOID C^*-ALGEBRAS

N. P. LANDSMAN[1]
Department of Applied Mathematics and Theoretical Physics, University of Cambridge, Silver Street, Cambridge CB3 9EW, United Kingdom

ABSTRACT

Let (P, Q, H) be a principal H-bundle over a Riemannian space Q. It is shown that the Poisson algebra $C_c^\infty((T^*P)/H)$ is quantized by the C^*-algebra of the gauge groupoïd of P. Mackey's quantization on $Q = G/H$ is a special case, the groupoïd C^*-algebra in that case reducing to the crossed product $C^*(G, G/H)$. For H trivial one obtains a generalization of Weyl quantization to any Riemannian space. Physical applications include the motion of a particle on Q coupling to a Yang-Mills field with gauge group H, and quantization on multiply connected spaces.

1. Deformation quantization and operator algebras

Quantization may be understood as some (functorial) correspondence between Poisson algebras and operator algebras. We here consider Poisson algebras of the type $C_c^\infty(P)$, to be completed into the commutative C^*-algebra $C_0(P)$, which is topologized by the sup-norm. Its 'quantization' $\mathcal{A}$ is taken to be a C^*-algebra, which is non-commutative in general. A further essential ingredient is a family of maps $\{Q_\hbar\}_{\hbar\in I}$ from a dense subalgebra $\mathcal{A}_0$ of $C_0(P)$, on which the Poisson bracket is defined, to $\mathcal{A}$; here I is an interval containing 0. This family has to be real (i.e., $Q_\hbar(\overline{f}) = Q_\hbar(f)^*$), and should satisfy the usual asymptotic correspondence between the algebraic structure of classical and quantum mechanics. That is, $\lim_{\hbar\to 0}(1/2[Q_\hbar(f), Q_\hbar(g)]_+ - Q_\hbar(fg)) = 0$ and $\lim_{\hbar\to 0}(i/\hbar[Q_\hbar(f), Q_\hbar(g)]_- - Q_\hbar(\{f,g\})) = 0$ in norm. Continuity in $\hbar$ may be imposed by demanding that either the maps $\hbar \to Q_\hbar(f)$ from I to $\mathcal{A}$, or else the functions $\hbar \to \| Q_\hbar(f) \|$, are continuous for all $f \in \mathcal{A}_0$, cf. [9, 1].

This notion of quantization fits in well with both the modern geometric theory of classical mechanics [7] and the algebraic theory of superselection sectors in quantum mechanics [3, 5], and, in fact, attempts to connect the two [6]. It has been tested on a system of physical relevance, namely a particle with a nonabelian charge moving on a homogeneous configuration space $Q = G/H$; the charge couples to an external Yang-Mills field with gauge group H. The classical mechanics of this system is described by the Poisson manifold $P = (T^*G)/H$, whose symplectic leaves correspond to co-adjoint orbits in $\mathcal{H}^*$ (the dual of the Lie algebra of H), which may be identified with classical charges the particle may have (see [7], and

[1]Supported by an S.E.R.C. Advanced Research Fellowship

refs. therein to the original work of Sternberg, Weinstein, and others). Its quantization is the crossed product $\mathcal{A} = C^*(G,Q)$, whose irreducible representations correspond to irreducible unitary representations of H [5], which are the quantum charges the particle may have. The deformation of $C^*(G,Q)$ into $C_0((T^*G)/H)$ (that is, the maps $Q_\hbar$) was constructed in [6].

That $C^*(G,Q)$ is indeed the correct quantization of $C_0((T^*G)/H)$ could only be ascertained with hindsight; a similar comment applies to the special case $H = G$, previously considered by Rieffel [9]; here the Poisson algebra $C_c^\infty(\mathcal{G}^*)$ (with the Lie-Kirillov Poisson bracket) is quantized by the group algebra $C^*(G)$. This case is of special interest in connection with quantum groups, which may be arrived at by deforming the Poisson structure on $\mathcal{G}^*$, and subsequently quantizing in the way sketched above.

The purpose of the present note is to generalize these results to arbitrary Riemannian configuration spaces Q (and principal H-bundles over Q, where H is taken to be an amenable Lie group, for simplicity), and this is achieved by reformulating them in terms of the so-called gauge groupoïd of the bundle (P, Q, H).

2. From groupoïds to quantization

A groupoïd Γ [8, 2] is a set op points (the base space or unit space B of Γ), plus a collection of arrows between some of these points. These arrows are invertible, and may be composed if their endpoints match. Γ is fibered over B, the fiber F_b at a point $b \in B$ being the set of arrows in Γ that end at b. An element $x \in \Gamma$ is both an arrow from $s(x)$ to $t(x)$, and an isomorphism I_x between $F_{s(x)}$ and $F_{t(x)}$. We assume that Γ is a manifold. Given a measure on each fiber, such that the family of measures thus obtained is invariant under all isomorphisms I_x, one can construct a C^*-algebra $C^*(\Gamma)$, whose representations correspond to the representations of Γ itself [8]. $C^*(\Gamma)$ is the norm-closure of the function space $C_c^\infty(\Gamma)$, equipped with a generalized convolution product, and a natural norm and involution.

Now regard $\Gamma_0 = (T^*P)/H$ as a trivial groupoïd (that is, $B = \Gamma_0$ and every point is only connected to itself by a single arrow), with each point having measure 1. Then $C^*(\Gamma_0) = C_0(\Gamma_0)$, the commutative algebra of classical observables of a charged particle moving on Q. This system is quantized by replacing Γ_c by the groupoïd $\Gamma_q = P \times_H P \equiv \Gamma_H(P)$, the gauge groupoïd of the principal bundle (P, Q, H). This has base space Q, and its elements are equivalence classes $[p_1, p_2] \equiv [p_1 h, p_2 h]$, interpreted as classes of H-equivariant paths in P [2]. If $\pi : P \to Q$ is the bundle projection on P, the groupoïd projection $\pi_g : \Gamma_H(P) \to Q$ is given by $\pi_g([p_1, p_2]) = \pi(p_1)$. The arrow $x = [p_1, p_2]$ starts at $s(x) = \pi(p_2)$ and ends at $t(x) = \pi(p_1)$, from which the composition rule follows.

The reason that the passage $\Gamma_c \to \Gamma_q$ quantizes the system is that $\mathcal{A} = C^*(\Gamma_q)$ turns out to be exactly the right quantum algebra of observables of the given physical system. It may be shown that $\mathcal{A} \simeq \mathcal{K}(L^2(P))^H \simeq \mathcal{K}(L^2(Q)) \otimes C^*(H)$, where $\mathcal{K}(\mathcal{H})$ is the C^*-algebra of compact operators on a Hilbert space $\mathcal{H}$, $L^2(M)$ is the Hilbert space of square-integrable half-densities on a manifold M, and $\mathcal{K}(L^2(P))^H$ are the compact operators which are invariant under the natural unitary representation of H on $L^2(P)$. Hence $\mathcal{A}$ is Morita-equivalent to $C^*(H)$, and its irreducible representations yield the correct superselection sectors of the system, cf. [5] for the case $P = G$. In that case, $C^*(\Gamma) = C^*(G, Q)$, which has been exhaustively analyzed as a quantum algebra of observables of a particle moving on Q [5, 6].

With hindsight, the passage from Γ_c to Γ_q has hereby been justified, but it may be motivated in itself in various ways (none of which is compelling). Firstly, there is a canonical construction due to Connes (unpublished, but reported in [4]), which, given an injection $i : \Gamma_1 \hookrightarrow \Gamma_2$ of groupoïds (which are manifolds) canonically produces a continuous field of C^*-algebras over the interval [0, 1], that is, a deformation quantization (the C^*-algebra defined by this field is the C^*-algebra of the so-called normal groupoïd corresponding to the injection i). If we apply this to the injection $i : Q \hookrightarrow \Gamma_H(P)$ (with Q regarded as a trivial groupoïd), defined by $i(q) = [s(q), s(q)]$ (with $s : Q \to P$ an arbitrary section), we obtain the C^*-algebras Γ_c and Γ_q above, as well as a full family of cross-sections interpolating between them. Hence apart from the construction of the quantization maps $Q_\hbar$, which is a separate problem discussed below, deformation quantization on Riemannian manifolds is equivalent to Connes' construction of a certain normal groupoïd. *(I am very grateful to R. Nest and G. Skandalis, who pointed this out in the special case $P = G$ in connection with the deformation given in [6]).*

Alternatively, the fact that $\Gamma_c = (T^*P)/H$ is 'quantized' by $\Gamma_q = P \times_H P$ appears natural if one actually constructs the quantization maps $Q_\hbar$. To make this more transparent, we put $H = \{e\}$ for the moment, and give the argument for $\Gamma_c = T^*P$ and $\Gamma_q = P \times P$; the result will quantize H-invariant functions on T^*P by H-invariant compact operators on $L^2(P)$, so that one can simply quotient by H to cover the general case.

Hence consider the diagonal embedding $i : P \to P \times P$. The corresponding normal bundle is the tangent bundle TP, and by the tubular neighbourhood theorem there exists a diffeomorphism φ from a neighbourhood $U \subset TP$ containing P (identified with the zero section) to a neighbourhood $V \subset P \times P$ containing P (identified with $i(P)$). Using the Riemannian metric on P, we define φ by $\varphi^{-1}(X) = (\exp(X/2), \exp(-X/2)$. Now choose a Lebesgue measure on each of the fibers of TP, so that the resulting collection is invariant under parallel transport. This defines a bijection between $C^*(T^*P) = C_0(T^*P)$ and $C^*(TP)$ (the C^*-algebra of the groupoïd TP with base space P, and all arrows closed, composition being addition in a given fiber) by a fiberwise Fourier transform $\mathcal{F}$. For $\hbar > 0$ fixed, one thus obtains a diffeomorphism $\varphi_\hbar$ between $U/\hbar$ and V given by $\varphi_\hbar^{-1}(X) = \varphi(\hbar X)$. As will become clear shortly, this diffeomorphism is essentially quantization, which is thus seen to consist of 'stretching out' the tangent bundle TP into the path space $P \times P$, the stretching factor being $\hbar$. Other diffeomorphisms would correspond to different operator ordering prescriptions.

The dense subalgebra $\mathcal{A}_0 \subset C_0(T^*P)$ of quantizable functions consists of those functions f whose 'Fourier transform' $\mathcal{F}f$ is in $C_c^\infty(TP)$. Such f has a quantization $Q_\hbar(f)$ for those $\hbar$ for which $\hbar(\text{supp}f) \subset U$. It is given by $Q_\hbar(f) = \Delta\,\varphi_\hbar^* f$, where Δ is an f-independent function on $P \times P$ which depends on the precise choice of Lebesgue measures mentioned above (for example, if $P = G$ and these measures are related by the action of the group on itself then Δ is constant). $Q_\hbar(f)$ is in $C_c(P \times P)$, and should be identified with the Hilbert-Schmidt operator defined by this kernel. An easy argument using the majorization of the uniform operator norm by the Hilbert-Schmidt norm shows that this quantization satisfies the conditions stated in sect. 1. Application of this scheme to $P = \mathbf{R}^n$ reproduces the Weyl quantization of functions on $T^*\mathbf{R}^n$, and the multiplication of Hilbert-Schmidt kernels is equivalent to the Weyl calculus of pseudo-differential operators, restricted to the compacts.

The connection between e.g. traditional geometric quantization and our method is not straightforward, as we do not quantize symplectic spaces as such, but entire Poisson manifolds. The algebraic correspondence between $C_0((T^*P)/H)$ and its quantization $C^*(\Gamma_H(P))$ is clear (and given by deformation), but their respective representation theories (on symplectic leaves and Hilbert spaces, respectively [6]) are not so clearly related. Recent techniques to quantize Poisson algebras using symplectic groupoïds [2] take a quite different route, too (for example, the groupoïd structure on T^*P is not taken to be the trivial one, as in our method, but the one we use on TP).

References

[1] A. Connes and N. Higson, *C. R.* **311** (1990) 101-106.

[2] A. Coste, P. Dazord, and A. Weinstein, *Publ. Dépt. Math. Univ. C. Bernard-Lyon I* **2A** (1987) 1-62.

[3] R. Haag, *Local Quantum Physics*, (Springer, Berlin, 1992).

[4] M. Hilsum and G. Skandalis, *Ann. scient. Éc. Norm. Sup. 4e série* **20** (1987) 325-390.

[5] N.P. Landsman, *Rev. Math. Phys.* **2** (1990) 45-72, 73-104.

[6] N.P. Landsman, preprints DAMTP 92-14, 92-44.

[7] J.E. Marsden, *Lectures on Mechanics*, (Cambridge Univ. Press, Cambridge, 1992).

[8] J. Renault, *A groupoïd approach to C^*-algebras*, SLNM **793**.

[9] M.A. Rieffel, *Am. J. Math.* **112** (1990) 657-686.

An Exact Classical Solution of the Dirac Field Equation

Qi-keng Lu

Institute of Mathematics,
Academia Sinica, Beijing 100080, P.R.China

ABSTRACT

The Dirac equation is written into a 2×2 matrix form and a classical exact solution is constructed.

As what we have done for Maxwell equation [1], we use spinor analysis to write the Dirac field equation firstly into spinor form and then into matrix form in order that some special exact solutions would be more easily constructed than by usual method.

Let

$$\epsilon = (\epsilon_{AB}) = (\epsilon^{AB}) = \begin{pmatrix} 0 & 1 \\ -1 & 0 \end{pmatrix}.$$

For a 2×2 matrix H, its spinor conjugate matrix H^* is defened by

$$H^* = \epsilon \overline{H} \epsilon^T$$

where $\bar{H}$ is the complex conjugate matrix of H and ϵ^T is the transposed matrix of ϵ.

The 2×2 Pauli matrices are denote by

$$\sigma^j = \left(\sigma^j_{AB}\right)_{1\le A,B\le 2}, \; j = 0,1,2,3.$$

The great Latin indices $A, B, C, \cdots$ run from 1 to 2 and the small Latin indices $j, h, l, \cdots$ run from 0 to 3. Moreover the summation convention is used.

Let the 4×4 Dirac matrices be

$$\beta^0 = \begin{pmatrix} I & 0 \\ 0 & -I \end{pmatrix}, \quad \beta^\lambda = \begin{pmatrix} 0 & \sigma^\lambda \\ -\sigma^\lambda & 0 \end{pmatrix}, \lambda = 1,2,3$$

The usual Dirac equation is

$$\beta^j \frac{\partial}{\partial x^j} \Psi = \mathrm{im} \Psi, \quad \Psi = \begin{pmatrix} \Psi_0 \\ \Psi_1 \\ \Psi_2 \\ \Psi_3 \end{pmatrix}. \tag{1}$$

Let

$$\phi = (\phi_A) = \begin{pmatrix}\phi_1\\ \phi_2\end{pmatrix}, \psi = (\psi_A) = \begin{pmatrix}\psi_1\\ \psi_2\end{pmatrix}$$

be a pair of spinors and

$$U_0 = \frac{1}{\sqrt{2}}\begin{pmatrix} I & I \\ I & -I \end{pmatrix}, I = \begin{pmatrix} 1 & 0 \\ 0 & 1 \end{pmatrix}.$$

Setting

$$\Psi = U_0\begin{pmatrix}\phi^*\\ \psi\end{pmatrix}, \gamma^j = U_0\beta^j U_0 = \begin{pmatrix} 0 & (\sigma^j)^* \\ \sigma^j & 0 \end{pmatrix},$$

one can see that equation (1) can be written into the form

$$\gamma^j \frac{\partial}{\partial x^j}\begin{pmatrix}\phi^*\\ \psi\end{pmatrix} = \text{im}\begin{pmatrix}\phi^*\\ \psi\end{pmatrix} \tag{2}$$

or

$$\sigma^j \frac{\partial}{\partial x^j}\phi^* = \text{im}\psi, \; (\sigma)^* \frac{\partial}{\partial x^j}\psi = \text{im}\phi^*. \tag{3}$$

Denote the 2×2 matrices

$$D = \sigma^j \frac{\partial}{\partial x^j} = 2\begin{pmatrix} \frac{\partial}{\partial u} & \frac{\partial}{\partial \bar{z}} \\ \frac{\partial}{\partial z} & \frac{\partial}{\partial v} \end{pmatrix}$$

where

$$u = x^0 + x^3, \; v = x^0 - x^3, z = x^1 + ix^2, \bar{z} = x^1 - ix^2,$$

and

$$\Phi = (\phi, \psi).$$

Then the Dirac equation (3) in spinor form can be writtten into matrix form

$$D^*\Phi = \text{im}\Phi^*. \tag{4}$$

Put

$$\Phi = f(\sigma)I + g(\sigma)Z \tag{5}$$

where f and g are differentiable functions,

$$\sigma = uv - |z|^2 \quad \text{and} \quad Z = \begin{pmatrix} u & \bar{z} \\ z & v \end{pmatrix}.$$

Equation (4) reduces to the ordinary differential equations

$$\begin{aligned} 2\sigma g' + 4g &= \text{im}\bar{f} \\ 2f' &= \text{im}\bar{g} \end{aligned} \tag{6}$$

where $f' = \frac{df}{d\sigma}$ and $g' = \frac{dg}{d\sigma}$.

It is proved that the general solution of equation (6) is

$$\begin{cases} g(\sigma) = \frac{1}{\sigma}\left[c_1 J_2(\mathrm{im}\sqrt{\sigma}) + c_2 Y_2(im\sqrt{\sigma})\right] \\ f(\sigma) = \frac{\mathrm{im}}{2}\int_{\sigma_0}^{\sigma}\frac{1}{\tau}\left[\bar{c}_1\overline{J}_1(\mathrm{im}\sqrt{\tau}) + \bar{c}_2\overline{Y}_2(\mathrm{im}\sqrt{\tau})\right]d\tau + c_3 \end{cases} \quad (7)$$

where c_1, c_2, c_3 are arbitrary constants and

$$J_2(\tau) = \sum_{k=0}^{\infty}\frac{(-1)^k(\frac{\tau}{2})^{2k+2}}{k!\Gamma(2+k+1)}$$

and

$$\begin{aligned} Y_2(\tau) &= \tfrac{2}{\pi}J_2(\tau)log\tfrac{\tau}{2} - \tfrac{1}{\pi}(\tfrac{2}{\tau})^2 - \tfrac{1}{\pi} - \tfrac{2}{\pi} \\ &-\tfrac{1}{\pi}\sum_{k=0}^{\infty}\tfrac{(-1)^k(\frac{\tau}{2})^{2k+2}}{k!(2+k)!}\left[\tfrac{\Gamma'(2+k+1)}{(2+k)!} + \tfrac{\Gamma'(k+1)}{(k)!}\right] \end{aligned}$$

are the Bessel functions of first and second kind respectively.

References

1. Qi-keng Lu, A study of Maxwell equation by spinor analysis. Dedicated to Prof. Pei-Yuan Chou for his 90th birthday (preprint).

QUANTUM GROUP σ MODELS - $q^N = 1$ CASE

YITZHAK FRISHMAN

Department of Physics, Weizmann Institute
Rehovot, Israel

and

JERZY LUKIERSKI*

Institute of Theoretical Physics, University or Wrocław,
pl. Maxa Borna 9, 50-204 Wrocław, Poland

and

WOJCIECH J. ZAKRZEWSKI

Department of Mathematical Sciences
University of Durham, Durham DH1 3LE, England

1. Introduction. In this talk we will consider fields taking values in quantum groups and we will discuss the corresponding σ models. Let us recall that the usual σ-field $\phi(x)$ describes the mappings from the coordinate manifold S into the target space M (see *e.g.* [1]). In principle we can "q-deform" the target space M ($M \to M_q$) as well as the coordinate manifold S ($S \to S_q$), *i.e.* we can introduce three kinds of σ models:

a) with quantum deformation of the target manifold

$$\phi_a^q(x): \qquad x \in S \;\to\; \phi_a^q \in M_q, \tag{1}$$

where the index a ennumerates the local coordinates on M_q.

b) with quantum deformation of the base manifold

$$\phi_a(x_q): \qquad x_q \in S_q \;\to\; \phi_a \in M, \tag{2}$$

c) with both manifolds deformed

$$\phi_a^q(x_q): \qquad x_q \in S_q \;\to\; \phi_a^q \in M_q. \tag{3}$$

We believe that in the near future, the quantum group σ models of all three types will be studied. In the present paper we will discuss mainly the case (1).

⋆ The talk presented by J. Lukierski

The description of quantum group σ models can be presented in two different ways:
i) We can consider quantum group σ fields as fields satisfying at a point $x \in S$ the quantum algebra, and so we can study the general properties which follow from the basic formulae of noncommutative differential geometry on quantum groups. Such an approach, which we shall call algebraic, was recently used by Arefeva and Volovich[2] [3] . In the algebraic formulation of a quantum group σ model one can repeat the major part of the geometric formulation of the standard approach to the σ models (Cartan forms, Cartan structure relations, algebra of the covariant derivatives *etc.*), provided that the exchange relations between quantum group valued σ fields and their derivatives are properly introduced. It appears that the effective use of the algebraic formulation of quantum group σ-models requires new developments.
ii) We can assume that the quantum group G_q σ fields $\phi^q_A(x)$ are products of "ordinary" functions $f_A(x) \in H$ and of the x independent algebra A related to the quantum group algebra. This approach gives us expressions which belong to the tensor product $H_G \otimes T_A$, where for an $n \times n$ matrix quantum group σ model the first part H_G is parametrised by an $n \times n$ matrix of "classical" fields (suitably constrained standard $GL(n)$ σ fields), and T_A carries the realisation of the algebra $A = f(G_q)$, solutions satisfying the separability condition described above. The realisation T_A can be described by finite matrices in one case: when q is complex and $q^N = 1$ ($q = e^{i\frac{2\pi}{N}}$). It is worth mentioning that some realisations of the quantum groups for q being the N-th root of unity have recently been found to have physically relevant applications (see *e.g.* ref [4,5,6]). It appears, however, that the quantum deformations $O_q(n)$ and $SU_q(n)$ of the semisimple groups $O(n)$ and $SU(n)$ which are the natural candidates for quantum group σ fields, do not permit complex q (see *e.g.* [7]).
In this talk we shall consider the simplest example of a compact quantum group with $q^N = 1$, the quantum group $U_q(2)$. We will show firstly that the $U_q(2)$ for complex values of q can be obtained as a special case of a two-parameter deformation $GL_{q,p}(2, C)$ of the 2×2 general linear group. Then, using the results on the matrix realisations of $GL_q(n)$ for $q^N = 1$[8] [9] [10] we will embed the $U_q(2)$ σ model with $q^n = 1$ ("anyonic σ model") into the conventional $U(2n)$ σ model. It appears that this embedding is described by conventional $U(2)$ σ-fields.

2. Quantum Group $U_q(2)$. We shall define the quantum group $U_q(2)$, for complex q, as a real form of a two parameter deformation of $GL(2, C)$, denoted by $GL_{p,q}(2)$[10] We have

$$\begin{aligned} ab &= pba \quad ac = qca \quad cd = pdc \\ pbc &= qbc \qquad bd = qdb \\ ad - da &= (p - q^{-1})bc \end{aligned} \tag{4}$$

with the coproduct $\Delta(U) = U \otimes U$, where U is 2×2 matrix describing $U_q(2)$. If we now introduce the determinant

$$\begin{aligned} D &= ad - pbc = ad - qcb \\ &= da - p^{-1}cb = da - q^{-1}bc \end{aligned} \tag{5}$$

then one can check from (4) that the following relations hold:

$$\begin{aligned} [D, a] = 0, &\qquad [D, d] = 0 \\ qDb = pbD, &\qquad pDc = qcD. \end{aligned} \tag{6}$$

We see that only if $q = p$ we can put $D = 1$, *i.e.* we have the quantum group $SL_q(2)$.
The quantum group $GL_{p,q}(2)$ is a genuine Hopf algebra for any complex q and p. In particular, the antipode of U is given by the formulae

$$S\begin{pmatrix} a & b \\ c & d \end{pmatrix} = D^{-1}\begin{pmatrix} d & -q^{-1}b \\ -qc & a \end{pmatrix} = \begin{pmatrix} d & -p^{-1}b \\ -pc & a \end{pmatrix} D^{-1}, \tag{7}$$

where we have used $DD^{-1} = D^{-1}D = 1$. If we now impose the unitarity condition $U^\dagger = S(U)$, *i.e.*

$$\begin{aligned} a^\star &= D^{-1}d = dD^{-1} \\ b^\star &= -qD^{-1}c = -pcD^{-1} \\ c^\star &= -q^{-1}D^{-1}b = -p^{-1}bD^{-1} \\ d^\star &= D^{-1}a = aD^{-1}, \end{aligned} \tag{8}$$

we find as the consistency conditions, that

$$D^\star = D^{-1}, \qquad p = q^\star \tag{9}$$

and so obtain the following $U_q(2)$ algebra

$$\begin{aligned} ac = qca, &\qquad ac^\star = q^\star c^\star a, \\ &c^\star c = cc^\star. \end{aligned} \tag{10}$$

Moreover, from (5) and (8) we find that

$$a^\star a + c^\star c = 1, \qquad aa^\star + |q|^2 cc^\star = 1. \tag{11}$$

The quantum matrix $U_q(2)$ given by

$$U = \begin{pmatrix} a & -q^\star c^\star D \\ c & a^\star D \end{pmatrix}. \tag{12}$$

It describes the generators of the Hopf algebra with standard comultiplication rule and the antipode condition (using $qD^{-1}c = q^\star cD^{-1}$) $U^+ = S(U)$. It should be stressed that even for $|q| = 1$ but with $q \neq 1$, we cannot put $D = 1$ as in this case (11) would give

$$a^\star a + c^\star c = 1 \qquad a^\star a = a a^\star. \tag{13}$$

3. 2N × 2N matrix realization of $\mathbf{U_q(2)}$ for $\mathbf{q^N = 1}$. If $q^N = 1$ the elements a, c, $a^\star$ and $c^\star$ can be represented by $N \times N$ matrices. Following ref. [8,9] we introduce the following matrices

$$Q = \begin{pmatrix} q & .. & 0 \\ 0 & \ddots & 0 \\ 0 & .. & q^N \end{pmatrix} \qquad P = \begin{pmatrix} 0 & 1 & .. & 0 \\ 0 & 0 & \ddots & 0 \\ 0 & 0 & .. & 1 \\ 1 & 0 & .. & 0 \end{pmatrix}. \tag{14}$$

These matrices satisfy $PQ = qQP$, and with $q = exp(\frac{2\pi i}{N})$ they generate the algebra of all $N \times N$ matrices. Moreover, as $Q^N = P^N = 1$ we find that $Q^\dagger = Q^{N-1}$ and $P^\dagger = P^{N-1}$ and, as is easy to see,

$$q^{ij} Q^i P^j = P^j Q^i. \tag{15}$$

We choose the following realization of the algebra (10) - (11)

$$\begin{aligned} a &= sin\alpha e^{i\psi} P, \quad a^\star = sin\alpha e^{-i\psi} P^{N-1} \\ c &= cos\alpha e^{i\phi} Q, \quad c^\star = cos\alpha e^{-i\phi} Q^{N-1}, \end{aligned} \tag{16}$$

where α and ψ are real numbers. Then it is easy to check that the conditions (10) and (11) (with $|q| = 1$) are satisfied if we choose

$$D = e^{i\xi} P^2. \tag{17}$$

Hence a $2N \times 2N$ matrix (12) with its entries given by (16) and (17) is a representation of $U_q(2)$ for $q = exp(\frac{2\pi i}{N})$.

In the following we shall express the matrix (12) as a product of two matrices $U = T\tilde{D}$, $U^\dagger = \tilde{D}^\dagger T^\dagger$, where

$$T = \begin{pmatrix} sin\alpha e^{i\psi} P & -q^\star cos\alpha e^{-i\phi} Q^\dagger \\ cos\alpha e^{i\phi} Q & sin\alpha e^{-i\psi} P^\dagger \end{pmatrix} \qquad \tilde{D} = \begin{pmatrix} 1 & 0 \\ 0 & e^{i\xi} P^2 \end{pmatrix}. \tag{18}$$

The basic Lagrangian then becomes

$$L = -\frac{1}{2} \mathrm{Tr} \, (U^\dagger \partial_\mu U)(U^\dagger \partial^\mu U) \tag{19}$$

with the trace taken with respect to the $U_q(2)$ matrix indices as well as the ones describing the realizations (18). One can write

$$\Omega^\mu = U^\dagger \partial^\mu U = \begin{pmatrix} V^\mu & Z^\mu D \\ D^{-1}K^\mu & -\tilde{V}^\mu \end{pmatrix} \tag{20}$$

where ($\psi^\mu, \equiv \frac{\partial\psi}{\partial x^\mu}$, $\alpha^\mu, \equiv \frac{\partial\alpha}{\partial x^\mu}$ etc.)

$$\begin{aligned} V^\mu &= i(\psi,^\mu sin^2\alpha + \phi,^\mu cos^2\alpha) \\ Z^\mu &= Q^\dagger P^\dagger y^{\dagger\mu} = Q^\dagger P^\dagger e^{-i\psi - i\phi}[\alpha,^\mu - i(sin\alpha cos\alpha)(\psi,^\mu - \phi,^\mu)] \\ K^\mu &= -PQy^\mu = -PQe^{i\psi+i\phi}[\alpha,^\mu + i(sin\alpha cos\alpha)(\psi,^\mu - \phi,^\mu)] = -Z^{\mu\dagger} \end{aligned} \tag{21}$$

$$\tilde{V}^\mu = V^\mu - D^{-1}\partial^\mu D = V^\mu - i\partial^\mu\xi \tag{22}$$

The Lagrangian (19) is given by the formula

$$L = -\frac{1}{2}V^\mu V_\mu + y^\mu y^\dagger_\mu + \frac{1}{2}\xi_{,\mu}\xi,^\mu - iV^\mu\xi_{,\mu} \tag{23}$$

where $\xi(x)$ describes the $U(1)$ field extending $SU(2)$ to $U(2)$. Indeed, one can show that if we put $\xi = 0$ we recover the classical action for the $SU(2)$ σ model. On the other hand, we can generalise the Lagrangian (19) to

$$L = -\frac{1}{2}\,\mathrm{Tr}\,(U^\dagger \nabla_\mu U)(U^\dagger \nabla^\mu U) = -\,\mathrm{Tr}\,\frac{1}{2}\tilde{\Omega}_\mu\tilde{\Omega}^\mu, \tag{24}$$

where we have introduced the $U(1)$ covariant derivative

$$\nabla_\mu = \begin{pmatrix} \partial_\mu \cdot 1_N & 0 \\ 0 & (\partial_\mu - A_\mu)\cdot 1_N \end{pmatrix}, \tag{25}$$

and so

$$\tilde{\Omega}^\mu = \begin{pmatrix} V^\mu & Z^\mu D \\ D^{-1}K^\mu & -(\tilde{V}^\mu + A^\mu) \end{pmatrix}. \tag{26}$$

Then, if in particular, we choose the pure gauge mode for the A_μ field

$$A_\mu = D^{-1}\partial_\mu D = i\xi_{,\mu}\cdot 1_N, \tag{27}$$

we find that the action (24) reduces to the one for the conventional $SU(2)$ σ model. We see that the $U(1)$ gauge field A_μ leads to the appearance of the gauge invariance which allows us to set $D = 1$ in (22).

4. Final Remark Our paper can be treated as a preliminary study of some aspects of sigma models taking values in a quantum group. We consider here the case $q^N = 1$. Additional aspect related with the algebraic formulation of quantum group σ-model for generic q were discussed in our recent paper [11].

5. References

1. W.J. Zakrzewski - *Low dimensional σ models* - Adam Hilger (1989)
2. I.Y. Arefeva and I.V. Volovich - *Mod. Phys. Lett.* **6A**, 893 (1991)
3. I.Y. Arefeva and I.V. Volovich - *Phys. Lett.* **264B**, 62 (1991)
4. C. de Concini and V.G. Kac - "Representations of quantum groups at root of 1", *Progress in Math.* Vol **92**, 471 (1990); ed. Birkhauser (Basel)
5. E. Date, M. Jimbo, K. Miki and T. Miwa - *Com. Math. Phys.* **137**, 133 (1991)
6. D. Arnaudon and A. Chakrabarti - *Com. Math. Phys.* **139**, 605 (1991)
7. L.D. Faddeev, N.Yu. Reshetikin and L.A. Takhtajan - *Algebra and Analysis* **1**, 178 (1989)
8. E.G. Floratos - *Phys. Lett.* **B233**, 395 (1989)
9. J. Weyers - *Phys. Lett.* **B240**, 396(1990)
10. A. Schirrmacher, J. Wess and B. Zumino - *Zeit. f. Phys.* **C49**, 317 (1991)
11. Y. Frishman, J. Lukierski and J.W. Zakrzewski - *Journ. Phys. A*, in press

On the bootstrap structure of Yangian-invariant factorized S-matrices

Niall J. MacKay*

Department of Mathematical Sciences, Durham, DH1 3LE, England

The message of this talk is that there is some very rich, largely unremarked structure in the representation theory of the Yangian (a quasitriangular Hopf algebra discovered by Drinfeld[1] in 1985).

We begin by taking the point of view[2,3] of a physicist investigating the charge algebra of integrable field theories in 1+1 dimensions which have a Lie algebra valued, curvature-free, conserved current (of which notable examples are the principal chiral model and the Gross-Neveu model and its generalizations). Such theories thus contain a charge Q_0^a, valued in a Lie algebra $\mathcal{A}$, which acts additively on tensor products of asymptotically independent particle states: defining its action on a two-particle state to be $\Delta(Q_0^a)$, we have

$$\left[Q_0^a, Q_0^b\right] = i\hbar f^{abc} Q_0^c \quad \text{and} \quad \Delta(Q_0^a) = Q_0^a \otimes 1 + 1 \otimes Q_0^a \ . \tag{1}$$

Such theories also have higher *non-local* conserved charges. The first of these, Q_1^a, satisfies

$$\left[Q_0^a, Q_1^b\right] = i\hbar f^{abc} Q_1^c \quad \text{and} \quad \Delta(Q_1^a) = Q_1^a \otimes 1 + 1 \otimes Q_1^a + \frac{1}{2} f^{abc} Q_0^c \otimes Q_0^b \ . \tag{2}$$

The missing commutator of two Q_1's is fixed by requiring that asymptotic states carry a representation of the charge algebra (*i.e.* that Δ be a homomorphism), and is ($\mathcal{A} \neq sl(2)$)

$$f^{d[ab}[Q_1^{c]}, Q_1^d] = \frac{i\hbar}{12} f^{api} f^{bqj} f^{crk} f^{ijk} Q_0^{(p} Q_0^q Q_0^{r)} \ . \tag{3}$$

This complicated relation may safely be ignored for the moment, although we shall return to it later. The Lorentz boost does not act trivially on these charges. If we boost the rapidity (defined by $\mathbf{p} = (m\cosh\theta, m\sinh\theta)$) of a particle by θ, it is found that

$$L_\theta: \ Q^{(0)} \mapsto Q^{(0)} \ , \quad Q^{(1)} \mapsto Q^{(1)} - \frac{\hbar C_{Adj}}{4\pi} \theta\, Q^{(0)} \qquad (\text{where } C_{Adj}\delta_{ad} = f^{abc} f^{cbd} \ ). \tag{4}$$

This has important implications for the S-matrix, which is heavily constrained by conservation of Q_0 and Q_1. The S-matrix for the interaction of two particles of rapidities θ_1 and θ_2 must satisfy

$$[\, S(\theta_1 - \theta_2)\, , Q_0^a \otimes 1 + 1 \otimes Q_0^a\,] = 0$$

$$S(\theta_1 - \theta_2)\, (L_{\theta_1} \otimes L_{\theta_2} \Delta(Q_1^a)) = (L_{\theta_2} \otimes L_{\theta_1} \Delta(Q_1^a))\, S(\theta_1 - \theta_2) \ . \tag{5}$$

*Supported by a UK SERC studentship.

These equations, together with unitarity and crossing-symmetry, determine the S-matrix up to an overall scalar factor. The classic result implied by these equations is that the S-matrix is *factorized*: that the S-matrix for the interaction of a large number of particles may be written as a product of two-particle S-matrices. That this works consistently follows from the Yang-Baxter equation (YBE),

$$S_{V_2V_3}(\theta')S_{V_1V_3}(\theta+\theta')S_{V_1V_2}(\theta) = S_{V_1V_2}(\theta)S_{V_1V_3}(\theta+\theta')S_{V_2V_3}(\theta') \tag{6}$$

(where θ and θ' denote the appropriate rapidity differences), which follows from (5): the S-matrix does not depend on the order in which the particles interact, and its only action is to exchange quantum numbers within the multiplets V_i of particles of equal mass m_i. Factorization was actually deduced some years ago from other considerations, and such S-matrices have been studied for some time[4,5]. In fact, as Bernard[3] pointed out, the algebra (1,2,3) is precisely Drinfeld's Yangian $Y(\mathcal{A})$, which is a quasitriangular Hopf algebra with an R-matrix satisfying (5,6) (where $S = \mathbf{P}R$ with $\mathbf{P}$ the transposition map), and has a one-parameter automorphism (4). Now suppose that we wish to construct S-matrices without using the underlying Yangian structure. In order to do so we may use the ***bootstrap principle***: that, at appropriate poles, intermediate states of the S-matrix should be regarded as physical particles. For example, where **1** is the particle in the vector representation of $SO(N)$ we have[4]

$$S_{11}(\theta) = s(\theta)\left[(\theta_0-\theta)(i\pi-\theta)P_S + (\theta_0+\theta)(i\pi-\theta)P_A + (\theta_0+\theta)(i\pi-\theta)P_0\right]$$

where P_S, P_A and P_0 project onto the traceless symmetric, antisymmetric and trace components of $\mathbf{1}\otimes\mathbf{1}$, $\theta_0 = \frac{2i\pi}{N-2}$, and $s(\theta)$ is a scalar function. The only value of θ in the physical strip $0 \leq \mathrm{Im}\theta < \pi$ at which any of these components vanish is $\theta = \theta_0$, and we therefore choose the ambiguous factor $s(\theta)$ to have a simple pole at θ_0, and interpret $P_A + P_0$ as a particle state **2**, so that we have a 'fusing' $\mathbf{11} \to \mathbf{2}$. Conservation of momentum tells us the relative masses of these particles, and we can construct S_{12} and S_{22} according to the principle[6]

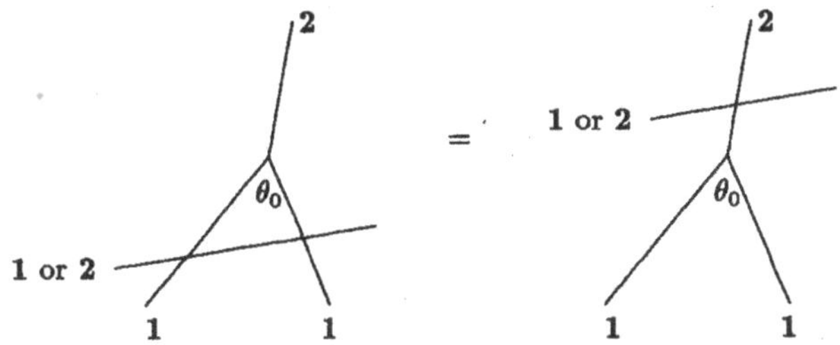

We can now continue the bootstrap, examining S_{12} and S_{22} for poles, interpreting these as third and higher particles, and so on. In principle, we would hope that eventually all appropriate poles would have been interpreted as physical states, so that the bootstrap would have closed on a finite set of particles. The mass ratios of the particles would be known, as would the representation (possibly reducible) of $\mathcal{A}$ carried by the particle, and the set of particle 'fusings' $\mathbf{ab} \to \mathbf{c}$. Unfortunately, calculating the S-matrices involves a great deal of computation, and calculations beyond $S_{\mathbf{22}}$ are unfeasible in practice. However, we can already extract one important point. In all cases hitherto computed, it has been possible to assign a particle $\mathbf{i}$ to a reducible representation of $\mathcal{A}$ containing the ith fundamental representation V_i of $\mathcal{A}$. In some cases (*e.g.* for $\mathcal{A} = a_n$ and for vector and spinor representations of $SO(N)$) there are no other components. A general method exists for calculating the S-matrices in irreducible representations[7], and the bulk of the S-matrices so far constructed have been of this type[5,7], and led to the conjecture that the fusings are in precise agreement with the Clebsch-Gordan (CG) decomposition, *i.e.* that $\mathbf{ab} \to \mathbf{c}$ if and only if $V_a \otimes V_b \supset V_c$. However, when the particle multiplet is reducible, this fails: for example, when we construct $S_{\mathbf{22}}$ for the $SO(N)$ theories[8], we find that $\mathbf{22} \not\to \mathbf{2}$ in general even though $V_2 \otimes V_2 \supset V_2$.

Now let us compare these S-matrices with another type of factorized S-matrix. In *purely elastic scattering theories* or PESTs (of which the prime examples are affine Toda theories[9] and integrable deformations of CFTs[10]), there is no mass-degeneracy: the particles do not form multiplets, and (6) is trivial. The S-matrix is a scalar factor determined by unitarity, analyticity, crossing-symmetry and the bootstrap, which is now much easier to implement. The result is a closed bootstrap on a full spectrum of massive particles and their fusings. For simply-laced $\mathcal{A}$ these match precisely the known results for the YBE-dependent S-matrices (including the CG 'holes')†; moreover, the PEST structure has a beautiful description in terms of root systems of Lie algebras[11]. It therefore seems likely that similar structure exists in the YBE-dependent S-matrices - it is simply that the bootstrap, whose implementation is specific to the $\mathcal{A}$ under consideration, is unable to perceive it.

Let us therefore shift viewpoint and consider the particle multiplets $\mathbf{i}$ as fundamental representations of $Y(\mathcal{A})$. We can use (4) to boost $\mathbf{i} \mapsto \mathbf{i}(\theta)$; the fusion procedure for $S_{\mathbf{ab}}(\theta)$ is then that of decomposing $\mathbf{a}(\theta) \otimes \mathbf{b}$ into its $Y(\mathcal{A})$-irreducible components. An alternative approach to constructing S-matrices is to construct the action of Q_0 and Q_1 explicitly and solve (5). This has been done, as far as I am aware, for two cases only. The first is when $\mathbf{i} = V_i$ is irreducible and $\rho_{\mathbf{i}}(Q_1) = 0$: this only works when the action of the right-hand side of (3) is zero. The second is the extension of the adjoint representation of $\mathcal{A}$ by a scalar representation to the $Y(\mathcal{A})$-irreducible representation $\mathcal{A} \oplus \mathbf{C}$: the action

†For non-simply-laced $\mathcal{A}$ there does seem to be a relationship, but it is much less clear.

of Q_0 is the adjoint action on $\mathcal{A}$ and zero on $\mathbf{C}$, and the action of Q_1 is then constructed to satisfy (2,3). Both of these types of representation were discovered by Drinfeld[1], and the corresponding R-matrices have been calculated[7,12]. Unfortunately this is where the story currently ends. The construction of other representations of $Y(\mathcal{A})$ seems to me to be an interesting problem for mathematicians: for example, can we construct ρ_i by taking Q_0 to have the natural action, and then fixing the action of Q_1 so that (2,3) are satisfied? Beyond this, solving (5) for S is certain to be a tricky process, and may yield little insight into the information obtained; at all levels, it seems, generality is lacking.

In conclusion, there is every prospect of interesting structure in the representation theory of $Y(\mathcal{A})$ alongside a dearth of general techniques for its discovery. For mathematical physicists, I should perhaps point out in addition the wide range of 1+1 dimensional integrable theories to which $Y(\mathcal{A})$ is relevant - apart from those mentioned at the beginning, the close relationship with PESTs has even led Belavin[13] to suggest that the underlying symmetry of affine Toda theories may be $Y(\mathcal{A})/\mathcal{A}$. I can only echo this, and hope to have convinced people that the subject deserves more attention than it has hitherto received.

References

[1] V. G. Drinfeld, Sov.Math.Dokl. 32 (1985) 254.

[2] M. Lüscher, Nucl.Phys. B135 (1978) 1.

[3] D. Bernard, Comm.Math.Phys. 137 (1991) 191.

[4] A. B. Zamolodchikov and A. B. Zamolodchikov, Ann.Phys. 120 (1979) 253.

[5] E. Ogievetsky and P. Wiegmann, Phys.Lett. B168 (1986) 360.

[6] M. Karowski, Nucl.Phys. B153 (1979) 244.

[7] P. P. Kulish, E. K. Sklyanin and N. Y. Reshetikhin, Lett.Math.Phys. 5 (1981) 393.

[8] N. J. MacKay, Nucl.Phys. B356 (1991) 729.

[9] H. W. Braden, E. Corrigan, P. E. Dorey and R. Sasaki, Nucl.Phys. B338 (1990) 689.

[10] A. B. Zamolodchikov, Int.J.Mod.Phys A4 (1989) 4235.

[11] P. E. Dorey, Nucl.Phys. B358 (1991) 654.

[12] V. Chari and A. Pressley, J.für die reine und ang. Math. 417 (1991) 87.

[13] A. A. Belavin, Phys.Lett. 283B (1992) 67.

Q-Trace and the Yang-Mills Theory

Z. Popowicz

Institute of Theoretical Physics
University of Wroclaw
Wrocław, Pl. Maxa Borna 9, 50–205 Wrocław, Poland

In the Lagrangian formulation of the gauge theory four important objects appear:gauge group,covariant derivative,invariants and the space–time. If the attempt of utilization of the quantum group appear reasonable then one can expect that these four objects should be modified compare to the usual case. Here I would like to present the possible way of a deformation of the gauge theory. Let me mention that recently Aref'eva and Volovich [1] has considered the deformation of a chiral field theory and our approach is in this direction. We also mention some open problems This talk presents my results obtained in the collaborations with A.P. Isaev [2].

First we investigate the notion of the quantum trace. We show that the Witten type quantum algebra $U_q(SL(2))$ yields the "adjoint representation" of the $SL_q(2)$ quantum group. Second we define the quantum generalization of the Lagrangian of the Yang-Mills fields as the quantum group invariant. We write down this Lagrangian in the terms of the q-deformed covariant derivatives. The gauge potentials of these q-deformed covariant derivatives are the elements of the quantum universal enveloping algebra related to the quantum gauge group. The deformed Lagrangian takes values in the set of the real numbers only after additional averaging. For the q-deformed $SU_q(2) \otimes U(1)$ gauge theory the "abelian gauge field" and the third component of the "nonabelian field" are mixed with the mixing angle which is related with the deformation parameter q. It is interesting to mention that the mixing of operator-valued gauge fields in the q-deformed case coincides with the mixing of gauge fields in the Weinberg-Salam model if we choose the parameter of deformation q as

$$q^2 = (1 - tan(\theta))/(1 + tan(\theta)), \tag{1}$$

where θ is the Weinberg angle.

A well known example of a quantum group is $GL_q(2)$ which is described by 2×2 matrix with the noncommutative matrix elements a, b, c and d. The quantum group $U_q(2)$ can be obtained from $GL_q(2)$ by assuming the "unitarity" condition which can be reduced to

$$d = a^{*}det_q(T), \qquad b = -qc^{*}det_q(T) \tag{2}$$

where det $q(t) := ad - qbc$. In this case the parameter q must be real number. Now let us introduce the quantum 2×2 matrix E with noncommutative elements E such that

$$[E_{ij}, T_{kl}] = 0 \tag{3}$$

In order to find the proper analogy of the concept of the trace (or character) let us consider the following $GL_q(2)$ transformation of the matrix E:

$$E_{ij} = T_{ik}E_{kl}T^{-1}_{lj} \tag{4}$$

Now we should look for the combinations of the matrix elements that are invariant under (4). It can be shown straightforwardly that

$$Tr_qE = q^{-1}E_{11} + qE_{22} = Tr_q\left(TET^{-1}\right) \tag{5}$$

In the next we call $Tr_q(E)$ as the quantum (or q-deformed) trace. Let us now rewrite the matrix E in the form

$$E = \frac{1}{q+q^{-1}}\left[\begin{pmatrix} Tr_qE & 0 \\ 0 & Tr_qE \end{pmatrix} + \begin{pmatrix} qE_0 & (q+q^{-1})E_{12} \\ (q+q^{-1})E_{21} & -q^{-1}E_0 \end{pmatrix}\right] \tag{6}$$

where $E_0 = E_{11} - E_{12}$,.

We see that (4) (similarly as in the nondeformed case) defines the three-dimensional "representation" of the quantum group $GL_q(2)$. Indeed the following formula

$$\begin{pmatrix} E'_0 \\ E'_{12} \\ E'_{21} \end{pmatrix} = \frac{1}{det_qT}\begin{pmatrix} det_qT + (q+q^{-1})bc, & -(a+q^{-1})ac, & (q+g^{-1}db \\ -ba, & a^2, & -q^{-1}b^2 \\ cd, & -qc^2, & d^2 \end{pmatrix}\begin{pmatrix} E_0 \\ E_{12} \\ E_{21} \end{pmatrix} \tag{7}$$

follows directly from (4) and (6) Now we would like to show that this "adjoint representation" (7) for the $SL_q(2)$ group can be related to the quantum algebra $U_q(sl(2))$. Let the coordinates A_i of the quantum plane generate the quantum oscillator

$$A_1A_2 - qA_2A_1 = 1 \tag{8}$$

It can be shown ([3]) that the algebra (8) is invariant under the left action of the quantum group $SL_q(2)$

$$A_i \Longrightarrow A'_i = T_{ij}A_j, \tag{9}$$

On the other hand (8) is invariant under the right transformation of the dual quantum plane also

$$\tilde{A}_i \longrightarrow \tilde{A}'_i = \tilde{A}_jT^{-1}_{ji} \tag{10}$$

$$\tilde{A}_i = (-qA_2, A_1) \tag{11}$$

Now let us consider the matrix

$$E_{ij} = A_i\tilde{A}_j \tag{12}$$

Under the transformation (9) and (10) the matrix E_{ij} transforms as

$$E'_{ij} = T_{ik}E_{kl}T^{-1}_{ij} \tag{12}$$

Now it is easy to prove that:

$$[E_0, E_\pm]_{q^\pm, q^\mp} = q^\pm E_0 E_\pm - q^\mp E_\pm E_0 = \pm (r_+)^{2E_\pm} \tag{14}$$

$$[E_+, E_-] = \frac{r}{q+q^{-1}}\left(r_+ E_0 - r_- E_0^2\right) \tag{15}$$

where

$$r_\pm = q^{1/2} \pm q^{-1/2}, \qquad E_{12} = E_+, \qquad E_{21} = E_1 \tag{16}$$

It is possible to rescale the generators E_0, E_+, E_- in such a way that the algebra (14) and (15) will reproduce the Witten type quantum algebra $U_q(sl(2))$ [4]. Thus we can say that the Witten deformation of $sl(2)$ is in some sense the "adjoint representation" of the quantum group $SL_q(2)$.

Now let us discuss the four major objects which appear in the gauge theory. First we do not change the space - time. The alternative way where one consider the q–deformation of space time has been recently considered in [5]. The remaining objects in the usual case are ;

Covariant Derivative :

$$\nabla_\mu = \partial_u + \mathcal{A}_\mu(x) = \partial_\mu + A^a_\mu\sigma^a + B_\mu\sigma^0; \tag{17}$$

where $x_\mu = (x_0, x_1, \dots x_d)$ are the coordinates of the d–dimensional euclidean (or Minkowski) space-time.

Invariants

$$F_{\mu\nu} = [\nabla_\mu, \nabla_\nu] = \partial_\mu A_\nu - \partial_\nu A_\mu + [A_\mu A_\nu]; \tag{18}$$

$$L = Tr\ F_{\mu\nu}F_{\mu\nu}, \tag{19}$$

From the properties of the usual trace it is clear that the lagrangian (18) is invariant under the gauge transformation

$$\nabla_\mu \to \mathcal{G}\nabla_{mu}\mathcal{G}^{-1} \tag{20}$$

$$F_{\mu\nu} \to \mathcal{G}F_{\mu\nu}\mathcal{G}^{-1} \tag{21}$$

Now we would like to define the q–analog of the Yang-Mills lagrangian in terms of the q-deformed curvature $F_{\mu\nu}$ using the q–trace. Then one have to know what the meanings are of the q–deformed covariant derivative and the q–deformed gauge field. To define these objects let us consider potentials $(I = 0, 1, 2, 3)$ as some operators while the gauge group elements $G(x)$ belong to the quantum group $U_q(2)$. Now we would like to illustrate the situation with q–deformed gauge fields. Let $\{1, \epsilon^i\}$ are the generators of $U_q(u(2))$, then we have the expansion of the gauge field operator $A^I = a^I \cdot 1 + a^I_i \epsilon^i + a^2_{ij}\epsilon^i\epsilon^j + \ldots$ and we can interpret a^I as the usual vector fields and a^I_i as the composite matter fields. Following the ideology of the quantum group approach we consider class of potential for which $[A^I_\mu, G_{ij}] = 0$ bearing in mind that now A^I_μ is an element of the quantum universal enveloping algebra. We assume for the q-deformed covariant derivative the same formula as in the nondeformed case. Thus we found the expression for Yang-Mills curvature

$$F_{\mu\nu} = B_{\mu\nu} \otimes 1 + F^a_{\mu\nu} \otimes \sigma^a \tag{22}$$

To extract the invariant "abelian" component $F^0_{\mu\nu}$ of $F_{\mu\nu}$ we apply the q-trace and we obtain

$$F^0_{\mu\nu} = B_{\mu\nu} \cos\theta + F^3_{\mu\nu} \sin\theta \tag{23}$$

where θ is defined by (1). At this moment we cannot consider L as the Lagrangian for the field theory because $B_{\mu\nu}$ and $F^a_{\mu\nu}$ are operators. Let us assume that $B_{\mu\nu}$ and $F^a_{\mu\nu}$ could be represented as the finite dimensional matrices i9n the fundamental representation and then consider the usual trace. In that sense we can construct the usual lagrangian for the field theory. Finally let me mention that the assumption $[A^I_{\mu\nu}, G_{ij}] = 0$ is very restrictive. Indeed the matrix entries of the transformed $F'_{\mu\nu}$ need not similarly commute with the entries of G_{ij} but it means that the product of two transformation does not belong to the class under consideration Indeed if $M \in GL_q(2)$ then $M \times M \in GL_{q^2}(2)$ [6]. The solution of this problem remains open up to now.

Acknowledgement: I would like to thank to Prof. Ge-Mo-Lin for the kind invitation and hospitality in Tianjin. I am grateful to rof. A. Kulish for the discussion and presentation me the paper of A. Fadeev where the q-trace was discussed. I would like to thank to Foundation of S. Batory for supporting my travel to China.

References

[1] I.Ya. Aref'eva and I.V. Volovich, Phys. Lett. **B 264** (1991) , 62.

[2] A.P. Isaev and Z. Popowicz, Phys. Lett **B 281** (1992), 271.

[3] B. Zumino, Mod. Phys. Lett **A6** (1991), 1225.

[4] E. Witten, Nucl. Phys. **B 330** (1990), 285.

[5] T. Brzezinski, S. Majid, "Quantum Group Gauge Theory on Classical Spaces", preprint DAMPT/92-51, August 1992.

[6] E. Corrigan, D. Fairle, P. Fletcher and R. Sasaki J. Math. Phys. **31** (1990), 776.

Generalized Quantization on Principal Bundles over Homogeneous Spaces and WZW Model (A Summary of a Paper with Guo M.Z. and Qian Z.H.)

Qian Min
Department of Mathematics
Peking university
Beijing 100871,China

In this note, we set up a framework of generalized quantization by starting from a single structure on the principal bundle P associated with the Lie group pair (G, H), which takes advantage of the Ehresmann connection induced from the G-actions on P.

Conditions under which our suggested procedure can really be carried out are given in this paper. This is the main result. Illustrating examples which show the usefulness of our scheme are, among others, the following:

1.a short and transparent treatment of geometric quantization and Borel-Weil theorem for compact groups(other finite dimensional cases could easily be found also satisfying our conditions).

2.Borel-Weil theorem for loop groups $LG(M = LG/T)$ as treated in [6],

3.geometric quantization of Wess-Zumino-Witten(WZW) model whose correct phase space should be LG/G instead of being LG/T as most people seem to take for granted ([3], [6], [9], [10]), and

4.the modle spaces of loop groups and the modle space of WZW modle.

1. Generalized Quantization on Principal Bundles over Homogeneous Spaces.

Let G be a Lie group, and H be a close Lie subgroup of G. Then G can be identified with a principal boundle on the homogeneous space G/H of G (the left coset space of H) taking H as both its fibre and its structure group. The projection mapping $\pi : G \to G/H$ is just the natural quotient mapping induced by H. The left actions of G on G induced the actions of G on G/H, and they commute with π.

The right actions of H on G are just the actions of structure group H on the principal bundle $\pi : G \to G/H$. Thus at each point g of G, we identify the Lie algebra $\mathcal{H}$ of H with the subspace of the tangent space T_gG as follows:

$$\sigma_g : \mathcal{H} \longrightarrow T_gG; \qquad \xi \mapsto \sigma_g(\xi) \tag{1.1}$$

where $\sigma_g(\xi)$ is the tangent vector of T_gG corresponding to the one-parameter current $g \exp -t\xi$ of G for $t \in R^1$.([8])

THEOREM 1.1. **(The connection α and the curvature Ω of α on the principal bundle $\pi : G \to G/H$)** *Let $\mathcal{G}$, $\mathcal{H}$ be Lie algebras of the Lie groups G, H respectively. Suppose that there exist a direct sum decomposition of $\mathcal{G}$ which satisfies that*

$$\mathcal{G} = \mathcal{H} \oplus \mathcal{M}; \qquad [\mathcal{H}, \mathcal{M}] \subset \mathcal{M} \tag{1.2}$$

We define a $\mathcal{H}$-valued 1-form α on G as

$$\alpha(L_\xi)(g) \stackrel{def}{=} -\xi_{\mathcal{H}} \text{ for all } \xi \in \mathcal{G}, \text{ and all } g \in G, \tag{1.3}$$

where L_ξ is the left-invarient vector field on G with $L_\xi(e) = \xi$, and $\xi_{\mathcal{H}}$ is the $\mathcal{H}$-component of ξ under the decomposition of $\mathcal{G}$ in (1.2). Then:

(1).α is an Ehresmann connection in the principal bundle $\pi : G \longrightarrow G/H$([14]), such that

$$T_gG = H_g \oplus V_g, \tag{1.4a}$$

$$\alpha(\sigma(\xi)) = \xi; \qquad \alpha(R_{h*}X)(g) = Ad_{h^{-1}}\alpha(X)(g) \tag{1.4b}$$

for all $\xi \in \mathcal{H}$, all $g \in G$, and all tagent vectors $X \in T_gG$, where the mapping σ is defiend in (1.1), $R_h$denotes the right action of h on G, and R_{h*} is the tangent mapping of R_h. (1.4a) is the horizontal space decomposition of T_gG.

(2).The curvature Ω of the connection α is a $\mathcal{H}$-valued 2-form on G such that

$$\Omega(L_\xi, L_\eta) = [\xi_{\mathcal{M}}, \eta_{\mathcal{M}}]_{\mathcal{H}}; \quad \Omega(R_{h*}X, R_{h*}Y) = Ad_{h^{-1}}\Omega(X,Y)(g) \tag{1.5}$$

for all $\xi, \eta \in \mathcal{G}$, all $h \in H$, all $g \in G$, and all tagent vectors $X, Y \in T_gG$, where $\xi_{\mathcal{M}}$ denotes the $\mathcal{M}$-component of ξ.

Remark Generally speaking, the curvature Ω on the principal bundle $\pi : G \longrightarrow G/H$ is not the pullback of a $\mathcal{H}$-valued 2-form ω on G/H by the mapping π. However, in the following discussed case, the curvature Ω is really the pullback of a $\mathcal{H}$-valued 2-form ω on G/H, which could be regarded as the basic structure of generalized prequantization of the principal bundle $\pi : G \longrightarrow G/H$ (the counterpat of ordinary symplectic structures of classical phase spaces).

By the left actions of G on G, and the induced action of G on G/H, we define the infinitesimal actions of $\mathcal{G}$ on $C^\infty(G)$, and $C^\infty(G/H)$ respectively as

$$\begin{aligned} \xi_G(F)(g) &\stackrel{def}{=} \frac{d}{dt}F(\exp -t\xi g)|_{t=0} \\ \xi_{G/H}(f)(\pi(g)) &\stackrel{def}{=} \frac{d}{dt}f \circ \pi(\exp -t\xi g)|_{t=0} \end{aligned} \tag{1.6}$$

for all $\xi \in \mathcal{G}$, all $F \in C^\infty(G)$, all $f \in C^\infty(G/H)$.

In fact, we can relate $\xi_{G/H}$ to ξ_G with the connection α defined in Theorem 1.1. The details are in the following theorem.

THEOREM 1.2 (GENERALIZED KONSTANT'S MAPPINGS).

Suppose that the Lie algebras $\mathcal{G}, \mathcal{H}$ of the Lie groups $\mathcal{G}, H$ satisfy (1.2), and that

$$[\mathcal{M}, \mathcal{M}] \subseteq \mathcal{M} \oplus Z; \quad \mathcal{H} = Z \oplus \mathcal{Y}; \quad [\mathcal{H}, \mathcal{Y}] \subseteq \mathcal{Y} \tag{1.7}$$

where Z is the center of $\mathcal{H}$.

We define two linear mappings $\delta_G : \mathcal{G} \longrightarrow C^\infty(G, \mathcal{H})$, and $\delta_{G/H} : \mathcal{G} \longrightarrow C^\infty(G/H, Z)$ respectively as

$$\begin{aligned} \delta_G(\xi)(g) &\stackrel{def}{=} -(Adg.^{-1}\xi)_{\mathcal{H}} \\ \delta_{G/H}(\xi)(\pi(g)) &\stackrel{def}{=} -(Adg.^{-1}\xi)_Z \end{aligned} \tag{1.8}$$

for all $\xi \in \mathcal{G}$, and all $g \in G$. Then:

(1). The curvature Ω of the connection α defined in Theopem 1.1 is the pullback of a Z-valued 2-form ω on G/H, which satisfies that

$$\omega(\pi_*(L_\xi(g)), \pi_*(L_\xi(g)(\pi(g)) = [\xi, \eta]_Z \tag{1.9}$$

for all $\xi, \eta \in \mathcal{M}$, and all $g \in G$.

(2). For all $\xi \in \mathcal{G}$, and all $g \in G$,

$$\xi_G(g) = \tilde{\xi_{G/H}}(g) + L_{\delta_G(\xi)}(g) = \tilde{\xi_{G/H}}(g) - \sigma(\delta_G(\xi))(g) \tag{1.10}$$

$$d\delta_{G/H}(\xi) = i(\xi_{G/H})\omega \tag{1.11}$$

where $\tilde{\xi}_{G/H}$ denotes the horizontal lift of the vector field $\xi_{G/H}$ on G/H determined by the horizontal decomposition of T_gG in (1.4a) corresponding to the Ehresmann connection α. The identity (1.11) makes sense to call the vector field $\xi_{G/H}$ on G/H as a generalized Hamiltonian vector field of $\delta_{G/H}$ corresponding to ω.(Cf.[7], [11]). Due to the identities (1.10) and (1.11), we call the mappings δ_G, and $\delta_{G/H}$ as generalized Konstant's mappings of the principal bundle $\pi : G \longrightarrow G/H$.(Cf.[4], [7], [11].)

COROLLARY 1.3. *Suppose that the Lie algebras $\mathcal{G}, \mathcal{H}$ satisfy the conditions in (1.2), and H is abelian. Then the mappings $\delta_G, \delta_{G/H}$ defined in (1.8) satisfy that*

$$\delta_G(\xi) = \delta_{G/H}(\xi) \circ \pi \text{ for all } \xi \in \mathcal{G} \tag{1.12}$$

$$\begin{aligned}\xi_G(g) &= \tilde{\xi_{G/H}}(g) + L_{\delta_{G/H}(\xi)\circ\pi(g)}(g) \\ &= \tilde{\xi_{G/H}}(g) - \sigma(\delta_{G/H}(\xi)(\pi(g))(g).\end{aligned} \tag{1.13}$$

Remark If the Lie algebras $\mathcal{G}, \mathcal{H}$, of the Lie groups G, H satisfy the conditions assumed in Theorem 1.2, then we can associate the left infinitesimal actions of $\mathcal{G}$ on $C^\infty(G)$ with their induced actions of $\mathcal{G}$ on $C^\infty(G/H)$ through the Ehresmann connection α in Theorem 1.2. This procedure is similar to ordinary prequantization of a complex line bundle on a symplectic manifold ([7],[11]). However, since the Z-valued 2-form ω on G/H (corresponding to the curvature Ω of the connection α) in (1.9) may be degenerate, and the dimensions of the fibre H of the principal bundle $\pi : G \longrightarrow G/H$ may be large than 1, we would like to call this procedure(Theorem 1.1 and Theorem 1.2) as generalized prequantization of the principal bundle G on its homogeneous space G/H.

DEFINITION 1.4. **(Poisson struture of a subspace of $C^\infty(G/H, Z)$,and generalized prequantization operators)**

Assume that the Lie algebras $\mathcal{G}, \mathcal{H}$ satisfy the conditions (1.2),(1.6) and (1.7). We define a subspace $C^\infty_{\mathcal{G}}(G/H, Z)$ of $C^\infty(G/H, Z)$ as

$$C^\infty_{\mathcal{G}}(G/H, Z) \stackrel{def}{=} \{\delta_{G/H}(\xi) | \xi \in \mathcal{G}\} \tag{1.14}$$

Then

(1).For all $\xi, \eta \in \mathcal{G}$, we define the Poisson bracket of two Z-valued smooth functions $\delta_{G/H}(\xi)$ and $\delta_{G/H}(\eta)$ as

$$\{\delta_{G/H}(\xi), \delta_{G/H}(\eta)\} \stackrel{def}{=} \omega(\eta_{G/H}, \xi_{G/H}) \tag{1.15}$$

where ω is a Z-valued 2-form on G/H defined in (1.9), then $\xi_{G/H}$ defined in (1.6) is the generalized Hamiltonian vector field of $\delta_{G/H}(\xi)$ which satisfies (1.11).

(2).For all prequantization operator $\hat{\delta}_{G/H}(\xi)$ on $C^{\infty}(G)$, one has

$$\hat{\delta}_{G/H}(\xi) \stackrel{def}{=} \tilde{\xi}_{G/H} + L_{\delta_G(zi)} = \xi_G, \tag{1.16}$$

where ξ_G is defined in (1.6), $\delta_G(\xi)$ is in (1.8), and $\tilde{\xi}_{G/H}$ is the horizontal lift of $\xi_{G/H}$.

THEOREM 1.5. **(Isomorphisms Theorem)**

Suppose that the Lie algebras $\mathcal{G}, \mathcal{H}$ of the Lie groups G, H satisfy the conditions (1.2), (1.6), (1.7).

Then:

(1). For all $\xi, \eta \in \mathcal{G}$,

$$\{\delta_{G/H}(\xi), \delta_{G/H}(\eta)\} = \delta_{G/H}([\xi, \eta]), \tag{1.17}$$

where $\{,\}$ is the Poisson bracket defined in (1.15). Moreover, the Lie algebra

$$(C^{\infty}_{\mathcal{G}}(G/H, Z), \{,\})$$

is Lie-isomorphic to $\mathcal{G}$.

(2). The generalized prequantization mapping $\mathcal{Q} : \delta_{G/H}(\xi) \mapsto \hat{\delta}_{G/H}(\xi)$ is a Lie homomorphism from $(C^{\infty}_{\mathcal{G}}(G/H, Z), \{,\})$ to the space of linear differential operators on $C^{\infty}(G)$.

2. Applications of the Formulatoin.

2.1 Loop Groups and Their Central Extensions.

Suppose that G is a connected, simpply connected and compact semisimple Lie group. Let LG be the loop group of G, and $\tilde{L}G$ be the central extension of LG determined by the basic 2-cocycle ω of LG in [6].

$\tilde{L}G$ can be regarded as different principal bundles on different homogeneous spaces. In this paper, we would like to discuss the following two fundamental homogeneous spaces ([6]):

1. LG/T: the principal bundle $\mathbf{T} \times T \longrightarrow \tilde{L}G \longrightarrow LG/T$, where $\mathbf{T}, T$ denote the center of $\tilde{L}G$, and the maximal torus of G respectively.

2. LG/T:the principal bundle $\mathbf{T} \times G \longrightarrow \tilde{L}G \longrightarrow \Omega G$, where ΩG is the subgroup of LG whose element γ satisfies that $\gamma(1) = e$ where $1 \in S^1$, and e is the unit element of G. In fact, $\Omega G \cong LG/G$.

2.2 The Principal Bundle $\pi : \tilde{L}G \longrightarrow LG/T$.

By the structure of the Lie algebra $\tilde{L}\mathcal{G}$ of $\tilde{L}G$, it follows directly that the principal bundle $\mathbf{T} \times T \longrightarrow \tilde{L}G \longrightarrow LG/T$ satisfies the conditions supposed in Theorem 1.2 and Corollary 1.3. Then this principal bundle $\pi : \tilde{L}G \longrightarrow LG/T$ can be equipped with the structures of the connection α, and generalized Konstant's mapping $\delta_{LG/T}$ through Theorem 1.1, Theorem 1.2 and Corollary 1.3 apply.

If $\vec{\lambda}$ is weight of $\tilde{L}G$ ([6]), then we can induce a complex line bundle $\mathbf{C} \longrightarrow L_{\vec{\lambda}} \longrightarrow LG/T$ on LG/T from this principal bundle, and impose the induced connection and induced Konstant's mapping on the complex line bundle. Thus we regard $L_{\vec{\lambda}}$ as generalized prequantization line bundle on LG/T. Moreever, by the general theorem of Pressley and Segal[10], the fundamental homogeneous space LG/T can identified with a homogeneous complex mamifold $LG_{\mathbf{C}}/B^+G_{\mathbf{C}}$, where $B^+G_{\mathbf{C}}$ consists of the boundary values of holomorphic maps

$$\gamma : \{z : |z| < 1\} \longrightarrow G_{\mathbf{C}}$$

such that $\gamma(0)$ belongs to the Borel subgroup B^+ of $G_{\mathbf{C}}$ corresponding to the positive roots of $\mathcal{G}_{\mathbf{C}}$.

We choose the polarization P to the determined by the Lie subalgebra $N^+\mathcal{G}_{\mathbf{C}}$ of $\tilde{L}\mathcal{G}_{\mathbf{C}}$ which consists of the boundary values of holomorphic maps

$$\xi : \{z : |z| < 1\} \longrightarrow \mathcal{G}_{\mathbf{C}}$$

such that $\xi(0)$ belongs to the nipotent Lie subalgebra $\mathcal{N}^+$ of $\mathcal{G}_{\mathbf{C}}$ corresponding to the positive roots of $\mathcal{G}_{\mathbf{C}}$. By the complex structure of LG/T, it follows that the polarization space

$$\Gamma_0^P(L_{\vec{\lambda}} \overset{def}{=} \{F \in \Gamma(L_{\vec{\lambda}}) | \nabla^{(\vec{\lambda})}_{X(\pi(\gamma))} F(\gamma) = 0 \text{ for all } \gamma \in \tilde{L}G,$$
$$\text{and all tangent vector} X(\pi(\gamma)) \in P(\pi(\gamma))\}$$

is just the holomorphic section space $\Gamma_{\vec{\lambda}}$ of the holomorphic line bundle $L_{\vec{\lambda}}$ on the homogeneous complex manifold $LG/T(\cong LG_{\mathbf{C}}/B^+G_{\mathbf{C}})$. Being similar to compact groups, Borel-Weil theorem for for loop groups [10] comletely describes the space $\Gamma_{\vec{\lambda}}$.

2.3 WZW Model. The Principal Bundle $\pi : \tilde{L}G \longrightarrow LG/G(\cong \Omega G)$.

From the discussion in [6] and [10], we know the classical phase space of WZW model is $\Omega G \times \Omega G$. Since the left copy and the right copy of ΩG in this classical phase space are homeomorphic, it is enough to investigate geometric quantization of ΩG.

With the argument above, it is natural to take the principal bundle $\mathbf{T} \times G \longrightarrow \tilde{L}G \longrightarrow \Omega G$ as the starting point of geometric quantization of ΩG. It follows that this principal bundle satisfies the conditions (1.2), and (1.17) in Theorem 1.2. By Theorem 1.1 and Theorem 1.2, we obtain an Ehresmann condition and generalized Konstant's mapping.

If $\vec{\lambda} = (0, \lambda, h)$ is an antidominant weight of $\tilde{L}G$ ([6], where the level h (a positive integer) is the character of the center $\mathbf{T}$ of $\tilde{L}G$, then λ is an antidomiant weight of G. By Borel-Weil theorem for compact groups, there exists one and only one irreducible unitary representation $(V_\lambda, \rho_\lambda)$ with the lowest weight λ, where the representation space V_λ is the holomorphic section space Γ_λ of the induced holomorphic line bundle L_λ on the homogeneous complex manifold G/T. With $\vec{\lambda}$, and the irreducible representation $(V_\lambda, \rho\lambda)$ of G, the complex vector bundle $V_\lambda \longrightarrow E^{(r)}_{\vec{\lambda}} \longrightarrow \Omega G$ can be induced from the principal bundle $\mathbf{T} \times G \longrightarrow \tilde{L}G \longrightarrow \Omega G$. Then we can also impose the structures of the induced connection and induced generalized Konstant's mappings on this complex vector bundle. Thus we regard the complex vector bundle $E^{(r)}_{\vec{\lambda}}$ equipped with the structures stated above as the prequantization vector bundle of ΩG.

In order to obtain the polarization of the prequantization vector bundle of ΩG, we first briefly investigate the structure of the homogeneous space $\Omega G \cong LG/G$. As we know from [6], ΩG can be identified with the homogeneous complex manifold $LG_{\mathbf{C}}/L^+G_{\mathbf{C}}$, where $L^+G_{\mathbf{C}}$ consists of the boundary values of the holomorphic maps

$$\gamma : \{z : |z| <\} \longrightarrow G_{\mathbf{C}}.$$

Moreever, ΩG is Kähler manifold whose Kähler form is just the basic 2-cocycle ω of LG.([9].)

Now, we choose the polarization P of ΩG as

$$P(\pi(\gamma)) \stackrel{def}{=} \{\pi_*(L_\gamma(\xi))|\xi \in L_0^+\mathcal{G}_{\mathbf{C}}\} \tag{2.1}$$

for all $\gamma \in \tilde{L}G$ where the Lie subalgebra $L_0^+\mathcal{G}_{\mathbf{C}}$ consists of the boundary values of holomorphic maps

$$\xi : \{z : |z| < 1\} \longrightarrow \mathcal{G}_{\mathbf{C}}$$

such that $\xi(0) = 0$. The polarization space is defined as

$$\Gamma_0^P(E_{\tilde{\lambda}}^{(r)}) \stackrel{def}{=} \{F \in \Gamma(E_{\tilde{\lambda}}^{(r)})|\ \nabla_{X(\pi(\gamma))}^{(\tilde{\lambda})} F(\gamma) = 0 \text{ for all } \gamma \in \tilde{L}G,$$

$$\text{and all tangent vector } X(\pi(\gamma)) \in P(\pi(\gamma))\} \tag{2.2}$$

where $\Gamma(E_{\tilde{\lambda}}^{(r)})$ denotes the smooth section space of the vector bundle $E_{\tilde{\lambda}}(r)$, which can be identified with the following subspace of $C^\infty(\tilde{L}G, V_\lambda)$

$$\Gamma(E_{\tilde{\lambda}}^{(r)}) \cong \{F \in C^\infty(\tilde{L}G, V_\lambda)|F(\gamma t) = \tilde{\lambda}(t)^{-1}F(\gamma), F(\gamma g) = \rho_\lambda(g).^{-1}F(\gamma) \text{ for all}$$

$$\gamma \in \tilde{L}G, \quad \text{all } t \in \mathbf{T}, \quad \text{and all } g \in G\}. \tag{2.3}$$

LEMMA 2.2. **(The explicit form of $\Gamma_0^P(E_{\tilde{\lambda}}^{(r)})$)**

Suppose that $\tilde{\lambda}(= (0, \lambda, h))$ is an antidominant weight of $\tilde{L}G$. Let $(V_\lambda, \rho_\lambda)$ be the irreducible unitary representation of $G_{\mathbf{C}}$ induced by the antidomanant weight λ of G. We extent ρ_λ as the homomorphism $\rho_{\tilde{\lambda}}$ from $L^+G_{\mathbf{C}} \times \mathbf{C}^\times$ to $Hom(V_\lambda, V_\lambda)$ which satisfies that

$$\rho_{\tilde{\lambda}}|_{L_1^+G_{\mathbf{C}}} = 1, \rho_{\tilde{\lambda}}|_{G_{\mathbf{C}}} = \rho_\lambda,\ \rho_{\tilde{\lambda}}|_{\mathbf{C}^*} = h \tag{2.4}$$

where $\mathbf{C}^\times$ denotes the complexification of the center $\mathbf{T}$ of $\tilde{L}G$, and the subgroup $L_1^+G_{\mathbf{C}}$ of $LG_{\mathbf{C}}$ consists of the boundary value of the holomorphic maps

$$\gamma : \{z : |z| < 1\} \longrightarrow G_{\mathbf{C}}$$

such that $\gamma(0) = e \in G$.

We define $E_{\tilde{\lambda}}$ as the induced holomorphic vector bundle from the complex principal bundle $L^+G_{\mathbf{C}} \times \mathbf{C}^\times \longrightarrow \tilde{L}G_{\mathbf{C}} \longrightarrow \Omega G(\cong LG_{\mathbf{C}}/L^+G_{\mathbf{C}})$ by the extended homomorphism $\rho_{\tilde{\lambda}}$: $L^+G_{\mathbf{C}} \times \mathbf{C}^\times \longrightarrow Hom(V_\lambda, V_\lambda)$.

Then the polarization space $\Gamma_0^P(E_{\tilde{\lambda}}^{(r)})$ of the prequantization vector bundle $E_{\tilde{\lambda}}^{(r)}$ is just the holomorphic section space $\Gamma_{hol}(E_{\tilde{\lambda}}^{\lambda})$ of the holomorphic vector bundle $V_{\tilde{\lambda}} \longrightarrow E_{\tilde{\lambda}} \longrightarrow \Omega G$ on ΩG.

PROOF:

(1).Since for all $F \in \Gamma_0^P(E_{\tilde{\lambda}}^{(r)})$, all $\xi \in L_0^+\mathcal{G}_{\mathbf{C}}$, and all $\gamma \in \tilde{L}G$, by (2.2),

$0 = \nabla_{\pi_*(L_\xi(\gamma))}^{(\tilde{\lambda})}(F)(\gamma) = L_\xi(F)(\gamma) = \frac{d}{dt}F(\gamma \exp t\xi)|_{t=0}$, it follows that $F(\gamma b) = F(\gamma)$ for all $b \in L_1^+G_{\mathbf{C}}$. This implies that $F \in \Gamma_{hol}(E_{\tilde{\lambda}})$, i.e, $\Gamma_0^P(E_{\tilde{\lambda}}^{(r)}) \subseteq \Gamma_{hol}(E_{\tilde{\lambda}})$.

(2). By the complex structure of ΩG, and the definition of the polarization space $\Gamma_0^P(E_{\bar{\lambda}}^{(r)})$ in (2.2), it follows directly that $\Gamma_{\text{hol}}(E_{\bar{\lambda}}) \subseteq \Gamma_0^P(E_{\bar{\lambda}}^{(r)})$. From (1) and (2) above, the identity that $\Gamma_0^P(E_{\bar{\lambda}}^{(r)}) = \Gamma_{\text{hol}}(E_{\bar{\lambda}})$ holds.
The existance of non-zero section can be discussed as in [6]. Q.E.D.

Supported by NNSFC.

References

[1]. Belavin, A.A, Polyakov, A.M., and Zamolochikov, A.B., *Infinite conformal symmetry in two-dimensional quantum field theory*, Nucl.Phys **B241** (1984), p. 333-80.

[2]. Bott, R., *Homogeneous vector bundles*, Ann.of Math. **66** (1957), p. 203-48.

[3]. Gawędzki, K. Wess-Zumino-Witten, *Conformal field theory*, in Constructive field theory II Erice, Italy, July (1988).

[4]. Kirillov, A.A., *Elements of the theory representations*, Grundlehren der mathematischen Wissenschaften **220**. Springer-Verlag, Berlin-Heidelberg-New, (1976).

[5]. Newlander, A., and Nirenberg, L., *II.Ser.*, Ann.of Math **65** (1957), p. 391-404.

[6]. Pressley, A.N, and Segal, G.B., *Loop groups*, Oxford Mathematical Monographs **12**, Clarendon Press, Oxford (1986).

[7]. Sniatycki, J., *Geometric quantizatation and quantum mechnics*, Applied Mathematical Science **30** Springger-Verlag, New York (1980).

[8]. Spivak, M., *Acomprehensive introduction to differential geometry*, Second edition, Vol. II, Ch.8, Publish or Perish, Inc., Berkeley (1979).

[9]. Tsuchiya, A., and Kanie, K., *Vertex operators in conformal field theory on P^1 and monodromy representation of braid groups*, Lett. Math. Phys. **13** (1987), p. 303.

[10]. Witten, E., *Non-abelian bosonization in two dimensions*, Commu.Math.Phys **92** (1984), p. 455-72.

[11]. Woodhouse, E., *Geometric Quantization*, Oxford Mathematical Monographs, 4, Clarendon Press, Oxford.

A FEW BRST BICOMPLEXES

RUDOLF SCHMID

ABSTRACT. There are many double complexes in the mathematics and physics literature which are related to BRST transformations and anomalies, e.g. Variational-, BRST-, Faddeev-, Koszul-Tate-, Weil-, Gelfand-Fuks-, semi infinite-, Cech-DeRham-, foliation-, Lie group/algebra- bicomplex. The goal is to identify them and to establish relations. We list and identify a few of them.

1. VARIATIONAL BICOMPLEX

Of all the bicomplexes mentioned above, the variational bicomplex is the "smallest" in the following sense. Let $\pi : P \to M^n$ be a fiber bundle and consider the jet bundle $J^\infty(\pi)$. Its exterior forms are bigraded by horizontal (r) and vertical (s) degrees

$$\Omega^p(J^\infty(\pi)) = \bigoplus_{r+s=p} \Omega^{r,s}(J^\infty(\pi)) , \qquad d = d_H + d_V : \Omega^{r,s} \to \Omega^{r+1,s} \oplus \Omega^{r,s+1}.$$

Horizontal and vertical derivatives satisfy Poincare lemmas and $d_H^2 = 0$, $d_V^2 = 0$, $d_H d_V = -d_V d_H$, hence $d^2 = 0$. With $\mathcal{F}^s = \{\omega \in \Omega^{n,s} | I(\omega) = \omega\}$ (functional forms), I the Euler operator we have the augmentad variational bicomplex [1]:

$$\begin{array}{ccccccccccccc}
 & & \uparrow d_V & & & & & & \uparrow d_V & & \uparrow \delta_V & & \\
0 & \longrightarrow & \Omega^{0,3} & & & \cdots & & & \Omega^{n,3} & \xrightarrow{I} & \mathcal{F}^3 & \longrightarrow & 0 \\
 & & \uparrow d_V & & & & & & \uparrow d_V & & \uparrow \delta_V & & \\
0 & \longrightarrow & \Omega^{0,2} & \xrightarrow{d_H} & \Omega^{1,2} & \xrightarrow{d_H} \cdots & \Omega^{n-1,2} & \xrightarrow{d_H} & \Omega^{n,2} & \xrightarrow{I} & \mathcal{F}^2 & \longrightarrow & 0 \\
 & & \uparrow d_V & & \uparrow d_V & & \uparrow d_V & & \uparrow d_V & & \uparrow \delta_V & & \\
0 & \longrightarrow & \Omega^{0,1} & \xrightarrow{d_H} & \Omega^{1,1} & \xrightarrow{d_H} \cdots & \Omega^{n-1,1} & \xrightarrow{d_H} & \Omega^{n,1} & \xrightarrow{I} & \mathcal{F}^1 & \longrightarrow & 0 \\
 & & \uparrow d_V & & \uparrow d_V & & \uparrow d_V & & \uparrow d_V & & & & \\
\mathbf{R} & \longrightarrow & \Omega^{0,0} & \xrightarrow{d_H} & \Omega^{1,0} & \xrightarrow{d_H} \cdots & \Omega^{n-1,0} & \xrightarrow{d_H} & \Omega^{n,0} & & & &
\end{array}$$

The *local* cohomologies $H^p(\pi) = ker\ d\ /im\ d$ can be computed explicitly using spectral sequences [1]. These are *local* cohomologies because for any $\omega \in \Omega^{r,s}(J^\infty(\pi))$

The autor was supported in part by Emory University Research Grant #2-31220.

we have $\omega(j, X_1, \cdots, X_s) \in \Omega^r(M)$, for any jet $j \in J^\infty(\pi)$ and vector fields X_i on $J^\infty(\pi)$.

2. BRST BICOMPLEX

The BRST bicomplex described in [2],[8],[9] is related to the variational bicomplex as follows: Let $\pi : P \to$ M be a principal $G-$ bundle and $\pi^p : \Omega^p(P, LieG) \to M$ Lie algebra - valued equivariant p-forms. Let $\mathcal{G}$ denote the Lie group of gauge transformations and $Lie\mathcal{G}$ its Lie algebra. Set $\mathbf{C}^{q,p}_{loc} = \Lambda^q(Lie\mathcal{G}, \Omega^{p,0}(J^\infty(\pi^p)))$ (local q - cochains) and define $\delta_{loc} : \mathbf{C}^{q,p}_{loc} \to \mathbf{C}^{q+1,p}_{loc}$ to be the Chevalley-Eilenberg coboundary operator with respect to the adjoint representation:

$$\begin{aligned}(\delta_{loc}\phi)(\xi_0, \cdots, \xi_q) &= \sum_{i=0}^{q}(-1)^i \rho'(\xi_i)\phi(\xi_0, \cdots, \hat{\xi}_i, \cdots, \xi_q) \\ &+ \sum_{i<j}(-i)^{i+j}\phi(\rho'(\xi_i)\xi_j, \cdots, \hat{\xi}_i, \cdots, \hat{\xi}_j, \cdots, \xi_q),\end{aligned}$$

where ρ' is the derived representation of $Lie\mathcal{G}$ on $\Omega(P, LieG)$ induced by a representation ρ of G . We have $\delta^2_{loc} = 0$. Then we define the BRST operator $\mathbf{s} : \mathbf{C}^{q,p}_{loc} \to \mathbf{C}^{q+1,p}_{loc}$ as

$$\mathbf{s} \equiv \frac{(-1)^{p+1}}{q+1}\delta_{loc} .$$

It is clear that $\mathbf{s}$ is nilpotent, $\mathbf{s}^2 = 0$. We call the associated *local* cohomology of this bicomplex $\{\mathbf{C}^{q,p}_{loc}, \mathbf{s}\}$ the *BRST cohomology* of the gauge algebra $Lie\mathcal{G}$, denoted by $H^*_{BRST}(Lie\mathcal{G})$. In [8],[9] we derived the classical BRST transformations using the Chevalley-Eilenberg constriction for the *adjoint* representation:

$$\mathbf{s}A = d\eta + [A, \eta] , \qquad \mathbf{s}\eta = -\frac{1}{2}[\eta, \eta] , \qquad \mathbf{s}\bar{\eta} = b , \qquad \mathbf{s}b = 0 ,$$

where the vector potential $A \in \mathbf{C}^{0,1}_{loc}$ and the ghost field $\eta \in \mathbf{C}^{1,0}_{loc}$ is the Maurer Cartan form on $\mathcal{G}$. We derive a homotopy formula on this bicomplex and with the introduction of Chern-Simons type forms $\omega^{i-1}_{2q-i} = a_i p(A, [A, A]^{i-1}, F^{q-1}_A)$ we obtain the associated descent equations $\delta\omega^0_{2q-1} = -d\omega^1_{2q-2}$, $\delta\omega^1_{2q-2} = -d\omega^2_{2q-3}$, $\cdots$, $\delta\omega^{2q-1}_0 = 0$. We identify the non-Abelian anomaly, which automatically satisfies the Wess-Zumino consistency condition, as a cohomology class in $H^1_{loc}(Lie\mathcal{G})$ represented by ω^1_{2q-2} in $n = 2q - 2$ dimensions.

For example, for $q = 2, q = 3$ we get the 2- and 4-dimensional non-Abelian anomaly respectively, represented by $\omega^1_2 = Tr(\eta\delta_{loc}\tilde{A})$, and $\omega^1_4 = Tr(\eta\delta_{loc}(\tilde{A}\delta_{loc}\tilde{A} + \frac{2}{3}\tilde{A}^3))$ resp., where $\tilde{A} = A + \eta$.

3. Faddeev's bicomplex

Let $\pi : (P, G) \to M$ be a principal bundle and consider $G^p = \{f : S^p \to G | \infty \to 1\}$, the space of p-loops. We have the exterior derivative $d : \Omega^q(P \times G^p) \to \Omega^{q+1}(P \times G^p)$ and the simplicial group coboundary operator $\Delta : \Omega^{q-p}(P \times G^p) \to \Omega^{q-p}(P \times G^{p+1})$ induced by $\Delta_i : P \times G^{p+1} \to P \times G^p : (x, g_1, \cdots, g_{p+1}) \mapsto (x, g_1, \cdots, g_i g_{i+1}, \cdots, g_{p+1})$.

For example for S^3 the Chern-Simon form is $\omega_5^0 = Tp(A)$, where $dTp(A) = p(F) = Trace\ F^3$ (p=invar. polynomial, T= transgression). We get the stair case equations [10]:

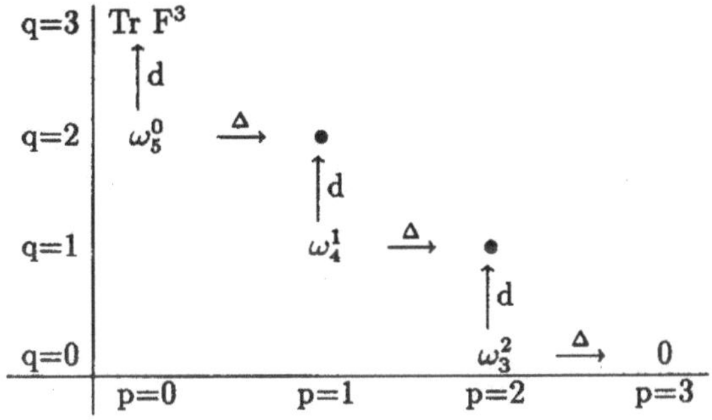

ω_3^2 represents the anomaly.

4. Koszul-Tate complex

Let M be a Poisson manifold with a Hamiltonian G action. Extend the momentum map $J : LieG \to C^\infty(M)$ to a super derivative δ and extend the Lie algebra d, $d : \Lambda LieG \otimes C^\infty(M) \to Lie^*G \otimes (\Lambda LieG \otimes C^\infty(M))$ defined by $dk(\xi) = \xi \cdot k$,($\cdot$ = repres. of $LieG$ on $\Lambda LieG \otimes C^\infty(M)$) to $\tilde{d}$ such that we have

$$\begin{array}{ccc} \Lambda^p Lie^*G \otimes \Lambda^q LieG \otimes C^\infty(M) & \xrightarrow{\delta} & \Lambda^p Lie^*G \otimes \Lambda^{q-1} LieG \otimes C^\infty(M) \\ \tilde{d}\Big\downarrow & & \\ \Lambda^{p+1} Lie^*G \otimes \Lambda^q LieG \otimes C^\infty(M) & & \end{array}$$

δ and $\tilde{d}$ being defined by

$$\delta(\omega \otimes \xi \otimes 1) = \omega \otimes 1 \otimes J(\xi)\ ,\ \delta(\omega \otimes 1 \otimes f) = \omega \otimes 0\ ,\ \tilde{d}(\omega \otimes k) = d\omega \otimes k + (-1)^p \wedge dk.$$

We have $\delta^2 = 0, \tilde{d}^2 = 0$ and $\tilde{d}\delta = \delta\tilde{d}$. The total differential defines the BRST operator $D = \tilde{d} + (-1)^p 2\delta : C^k \to C^{k+1}$, satisfying nilpotency $D^2 = 0$, where $C^k = \sum_{p-q=k} \Lambda^p Lie^*G \otimes \Lambda^q LieG \otimes C^\infty(M)$.

The functions on the reduced phace space are given by the cohomology [7]

$$C^\infty(J^{-1}(0)/G) = H_D(\Lambda Lie^*G \otimes \Lambda LieG \otimes C^\infty(M)) .$$

which equals the space $E_2^{0,0}$ of the associated spectral sequence.

5. Weil complex

Let $\Lambda(Lie^*\mathcal{G})$ be the exterior algebra and $\mathbf{S}(Lie^*\mathcal{G})$ the symmetric algebra of $Lie\mathcal{G}$, the Lie algebra of infinitesimal gauge transformations. The Weil algebra $W(Lie\mathcal{G}) \equiv \Lambda(Lie^*\mathcal{G}) \otimes \mathbf{S}(Lie^*\mathcal{G})$ is a graded differential $\mathcal{G}$ algebra

$$W(Lie\mathcal{G}) = \sum_r W^r \ , \ W^r = \sum_{p+2q=r} \Lambda^p(Lie^*\mathcal{G}) \otimes S^q(Lie^*\mathcal{G}).$$

Let $\{e_\alpha\}$ be a basis of $Lie\mathcal{G}$ and $\{\theta^\alpha\}$ its dual basis of $Lie^*\mathcal{G}$, and let $\{u^\alpha\}$ be a basis of $\mathbf{S}(Lie^*\mathcal{G})$. The antiderivation δ_W of degree 1 on $W(Lie\mathcal{G})$ is given by $\delta_W = \delta_\Lambda + \delta_S + h$, where $\delta_\Lambda : W^p \to W^{p+1}$ is given by: for $\phi \in \Lambda^p(Lie\mathcal{G})$, $x_i \in Lie\mathcal{G}$

$$(\delta_\Lambda \phi)(x_0, \cdots, x_p) = \sum_{\nu<\mu} (-1)^{\nu+\mu} \phi([x_\nu, x_\mu], x_0, \cdots, \hat{x}_\nu, \cdots, \hat{x}_\mu, \cdots, x_p),$$

or $\delta_\Lambda = \frac{1}{2} \sum_\alpha \mu(\theta^\alpha) L_\Lambda(e_\alpha)$, where L_Λ is the Lie derivative and $\mu(a)b \equiv a \wedge b$. We have $\delta_\Lambda^2 = 0$. Moreover $\delta_S = \sum_\alpha \mu(\theta^\alpha) L_S(e_\alpha) : W^p \to W^{p+1}$. Note that $\delta_s^2 \neq 0$ but $(\delta_\Lambda + \delta_S)^2 = 0$, so $\delta_S^2 = -(\delta_\Lambda \delta_S + \delta_S \delta_\Lambda)$. The operator $h : W^p \to W^{p+1}$ is defined by $h = \sum_\alpha \mu_S(\theta^\alpha) \otimes i_A(e_\alpha)$ and is and antiderivative of degree 1 (i_A = interior product). The BRST operator is the total differential $\delta_W = \delta_\Lambda + \delta_S + h$.

References

1. I.M. Anderson, *The Variational Bicomplex*, preprint, 1991.
2. L. Bonora, P. Cotta-Ramusino, *Some Remarks on BRS Transformations, Anomalies and the Cohomology of the Lie Group of Gauge Transformations* , Comm. Math. Phys. **87** (1983), 589-603.
3. J.A. Dixon, *Calculations of BRS Cohomology with Spectral Sequences*, Comm. Math. Phys. **139** (1991),495-526.
4. M. Dubois-Violette, *The Weil-B.R.S. Algebra of a Lie Algebra and the Anomalous Term in Gauge Theories*, J. Geom. Phys. **3** (1986) 525-565.
5. J. Fisch, M. Henneaux, J. Stasheff, C. Teitelboim, *Existence, Uniqueness and Cohomology of the Classical BRST Charge with Ghosts of Ghosts* , Comm. Math. Phys.**120** (1989), 379-.
6. I.B. Frenkel, H. Garland, G.J. Zuckerman, *Semi-Infinite Cohomology and String Theory* , Proc. Natl. Acad. Sci. USA **83** (1987), 8442-8446.
7. B. Kostant, S. Sternberg, *Symplectic Reduction, BRS Cohomology, and Infinite Dimensional Clifford Algebras* , Ann. Phys. **176** (1987), 49-113.
8. R. Schmid, *Local Cohomology in Gauge Theories BRST Transformations and Anomalies*, Differential Geometry and its Applications, (1992) to appear.
9. R. Schmid, *BRST Cohomology and Anomalies,* Geometrical and Algebraic Aspects of Nonlinear Field Theory, 159-172, S. De Filippo et al (eds.), North - Holland, 1989.
10. J. Stasheff, *The De Rham Bar Construction as a Setting for the Zumino, Faddee'v, etc. Descent Equations*, Proc. Symp. anomalies, Geometry, Topology, Argonne Natl. Lab. 1984. W.E. Bardeen, A. R. White (eds.), World Scientific, Singapore (1985), 220-233. [Related papers can be found in the same proceedings].

Department of Mathematics, Emory University, Atlanta, GA 30322
E-mail: rudolf@mathcs.emory.edu

QUANTUM SYMMETRY IN LOCAL QUANTUM FIELD THEORY

GERHARD MACK and VOLKER SCHOMERUS *
II. Institut für Theoretische Physik, Universität Hamburg

ABSTRACT

It is well known that many low dimensional quantum field theories show signs of internal symmetries more general than groups. We give a general definition of a quantum symmetry, such that symmetry operations act on the Hilbert space $\mathcal{H}$ of physical states and notions of unitarity, invariance and covariance are defined. Weak quasi quantum groups are described as a distinguished class of quantum symmetries. Their deep connection with locality is discussed in the context of local braid relations among field operators. Minimal conformal models provide examples of local quantum field theories with a nontrivial weak quasi quantum group symmetry.

1. Quantum symmetry in quantum theory

In the following let $(\mathcal{G}^*, \Delta, \epsilon, *)$ denote a non-co-associative bi-*-algebra. In detail this means that $\mathcal{G}^*$ is a *-algebra with unit e and $\Delta : \mathcal{G}^* \to \mathcal{G}^* \otimes \mathcal{G}^*$ (*coproduct*), $\epsilon : \mathcal{G}^* \to \mathbf{C}$ (*counit*) are *-homomorphisms. For Δ the notion of a *-homomorphism involves a definition of $*$ on $\mathcal{G}^* \otimes \mathcal{G}^*$ [1]. Finally, the coproduct Δ and the counit ϵ satisfy

$$(\epsilon \otimes id)\Delta = id = (id \otimes \epsilon)\Delta \quad . \tag{1.1}$$

The coproduct Δ determines a tensor product $\tau \bigotimes \tau'$ for representations τ, τ' of $\mathcal{G}^*$

$$(\tau \bigotimes \tau')(\xi) = (\tau \otimes \tau')(\Delta(\xi)) \quad \text{for all} \quad \xi \in \mathcal{G}^* \quad . \tag{1.2}$$

With respect to this tensor product of representations, the counit ϵ furnishes a trivial 1-dimensional representation. Triviality refers to the property $\epsilon \bigotimes \tau = \tau = \tau \bigotimes \epsilon$ for all representations τ of $\mathcal{G}^*$, which follows from Eq. (1.1).

Consider some quantum mechanical system $(\mathcal{H}, \{\Psi\}, |0\rangle, H)$ within a second quantized formalism. The Hilbert space $\mathcal{H}$ of physical states should contain a unique ground state $|0\rangle$ with respect to the Hamiltonian H. We assume that $\mathcal{H}$ is generated from $|0\rangle$ by multiplets of field operators $\psi_i^I(\mathbf{r}, t)$ (I labels multiplets,i labels fields in the multiplet I) which create particles or excitations.

Definition 1.1: [1] *A bi-*-algebra* $(\mathcal{G}^*, \Delta, \epsilon, *)$ *is called* quantum symmetry *of the system* $(\mathcal{H}, \{\Psi\}, |0\rangle, H)$, *if there exits a representation* $\mathcal{U} : \mathcal{G}^* \to \mathcal{B}(\mathcal{H})$, *such that*
i) $\mathcal{U}$ *is unitary in the sense that* $\mathcal{U}(\xi^*) = \mathcal{U}(\xi)^*$ *for all* $\xi \in \mathcal{G}^*$.
ii) the Hamiltonian H *and the ground state* $|0\rangle$ *are invariant, i.e.*

$$[H, \mathcal{U}(\xi)] = 0 \quad , \quad \mathcal{U}(\xi)|0\rangle = |0\rangle\epsilon(\xi) \quad \textit{for all} \quad \xi \in \mathcal{G}^* \tag{1.3}$$

*E-mail: I02VOL@DHHDESY3.BITNET

iii) field operators $\Psi_i^I(\mathbf{r},t)$ transform covariantly according to the representation τ^I of $\mathcal{G}^*$,

$$\mathcal{U}(\xi)\Psi_i^I(\mathbf{r},t) = \sum_p \Psi_j^I(\mathbf{r},t)\tau_{ji}^I(\xi_p^1)\mathcal{U}(\xi_p^2)\ , \ \text{if } \Delta(\xi) = \sum_p \xi_p^1 \otimes \xi_p^2\ . \tag{1.4}$$

Note that the quantum symmetries we introduced in this definition do not act on the space–time variable of the fields, i.e. they are internal symmetries.

Special examples of bi-*-algebras $(\mathcal{G}^*.\Delta, *, \epsilon)$ are furnished by compact groups resp. by their group algebras $\mathcal{G}^*$. When Definition 1.1 is applied to group algebras, it reduces to the standard definition of a group symmetry in quantum mechanics.

2. Weak quasi Quantum Groups

Within the large class of quantum symmetries discussed in the preceding section there is a distinguished subclass that we want to describe next. Before we state the assumptions leading to this subclass, let us emphasize that we do not assume $\Delta(e) = e \otimes e$. In general $\Delta(e)$ will be some projector $P \in \mathcal{G}^* \otimes \mathcal{G}^*$ so that we can accommodate for truncated tensor products, i.e. the tensor product of two representations possibly vanishes on a true subspace of the full representation space.

Assumption 1: *Δ is quasi–coassociative in the sense that there is $\varphi \in \mathcal{G}^* \otimes \mathcal{G}^* \otimes \mathcal{G}^*$ and quasi–inverse φ^{-1} [i.e. $\varphi\varphi^{-1} = (id \otimes \Delta)\Delta(e), \varphi^{-1}\varphi = (\Delta \otimes id)\Delta(e)$] such that*

$$\varphi(\Delta \otimes id)\Delta(\xi) = (id \otimes \Delta)\Delta(\xi)\varphi \quad \text{for all } \xi \in \mathcal{G}^*\ . \tag{2.1}$$

This in turn implies that $(\tau^I \otimes \tau^J) \otimes \tau^K$ and $\tau^I \otimes (\tau^J \otimes \tau^K)$ are equivalent representations (but not equal). Following Drinfel'd [2] we postulate

$$(id \otimes id \otimes \Delta)(\varphi)(\Delta \otimes id \otimes id)(\varphi) = (e \otimes \varphi)(id \otimes \Delta \otimes id)(\varphi)(\varphi \otimes e) \tag{2.2}$$

Assumption 2: *The tupel $(\mathcal{G}^*, \Delta, \epsilon, \varphi)$ is quasitriangular in the sense that there is $R \in \mathcal{G}^* \otimes \mathcal{G}^*$ and quasi–inverse R^{-1} [i.e. $RR^{-1} = \Delta'(e), R^{-1}R = \Delta(e)$] such that*

$$R\Delta(\xi) = \Delta'(\xi)R \quad \text{for all } \xi \in \mathcal{G}^*\ . \tag{2.3}$$

If $\Delta(\phi) = \sum_p \phi_p^1 \otimes \phi_p^2$ then $\Delta'(\phi) = \sum_p \phi_p^2 \otimes \phi_p^1$ by definition. Quasitriangularity implies $\tau^I \otimes \tau^J \cong \tau^J \otimes \tau^I$. We postulate the following two relations.

$$(id \otimes \Delta)(R) = \varphi_{231}^{-1} R_{13} \varphi_{213} R_{12} \varphi^{-1}\ , \tag{2.4}$$

$$(\Delta \otimes id)(R) = \varphi_{312} R_{13} \varphi_{132}^{-1} R_{23} \varphi\ . \tag{2.5}$$

We used the standard notation. If $R = \sum r_a^1 \otimes r_a^2$ then $R_{13} = \sum r_a^1 \otimes e \otimes r_a^2$ etc. If s is a permutation of 123 and $\varphi = \sum \varphi_\sigma^1 \otimes \varphi_\sigma^2 \otimes \varphi_\sigma^3$ then $\varphi_{s(1)s(2)s(3)} = \sum \varphi_\sigma^{s^{-1}(1)} \otimes \varphi_\sigma^{s^{-1}(2)} \otimes \varphi_\sigma^{s^{-1}(3)}$. These relations imply validity of quasi Yang Baxter equations,

$$R_{12}\varphi_{312}R_{13}\varphi_{132}^{-1}R_{23}\varphi = \varphi_{321}R_{23}\varphi_{231}^{-1}R_{13}\varphi_{213}R_{12}\ . \tag{2.6}$$

If in addition existence of an antipode S is assumed, we call these structures *weak quasi quantum groups*. To understand this name, one should regard them as generalizations of group algebras. In the group situation we know that $\Delta(e) = e \otimes e =$

$R, \varphi = e \otimes e \otimes e$ leading to a commutative and associative tensor product of representations. Quantum groups are obtained from groups by droping the restriction $R = e \otimes e$. For Drinfel'ds quasi quantum groups [2] we keep only $\Delta(e) = e \otimes e$. In a last step of generalization we give up $\Delta(e) = e \otimes e$ to get weak quasi quantum groups. It can be shown that all algebraic structures that satisfy the above assumptions determine a representation of the braid group [3]. This has been applied to construct knot invariants from quasi quantum groups [4].

The restriction to weak quasi quantum group symmetries should be regarded as a consequence of locality. Some aspects of this topic will be touched in the next section. For a wide class of local quantum field theories - including the conformal quantum field theories - quantum symmetries (in the sense of definition 1.1) can be shown to satisfy assumption 1,2. The proof requires that the observables of the theory generate the set of invariant fields (i.e. the fields commuting with all symmetry operations $\mathcal{U}(\xi)$) [5].

3. Quantum Symmetry, Statistics and Locality

In quantum field theory, permutation group statistics is implemented through quadratic relations among the field operators, namely canonical (anti–)commutation relations for Bosons (Fermions). The spin statistics theorem states that Fermions have spin $s = \frac{1}{2}, \frac{3}{2}, \ldots$, whereas Bosons have integer spin. More general values for the spin (remember that the spin labels representations of the rotation group,e.g. SO(2) in 2 space dimensions) are possible in low dimensional quantum field theory. They are associated with braid group statistics. It has been proposed to implement braid group statistics through local braid relations [6].

$$\Psi_i^I(x,t)\Psi_j^J(y,t) = \Psi_k^J(y,t)\Psi_l^I(x,t)\mathcal{R}_{kl,ij}^{IJ>}\omega^{IJ} \quad x > y \, . \tag{3.1}$$

Here $x > y$ refers to some ordering prescription and ω^{IJ} should be complex numbers generalizing the $\pm$ for Bose/Fermi commutation relations. In contrast with [6] we do not restrict the $\mathcal{R}$–matrix to have C–number entries, but the matrix elements should assume values in $\mathcal{U}(\mathcal{G}^*)$ instead.

In a theory with weak quasi quantum group symmetry, consistency of Eq. (3.1) with the quantum symmetry $(\mathcal{G}^*, \Delta, \epsilon, *)$ and locality may be exploited to obtain constraints on the coefficients $\mathcal{R}_{kl,ij}^{IJ>}$. The analysis suggests

$$\mathcal{R}_{kl,ij}^{IJ>} = (\tau^I \otimes \tau^J \otimes \mathcal{U})_{kl,ij}(\varphi_{213}(R \otimes e)\varphi^{-1}) \quad . \tag{3.2}$$

To gain some insight into the structure of this expression we demonstrate that *local braid relations Eq. (3.1) with $\mathcal{R}$ given by Eq. (3.2) are at least consistent with the transformation law of fields*, i.e. that both sides of the equation transform in the same way. The products of covariant fields which appear in Eq. (3.1) are in general not covariant. However, one may use φ to construct a nonassociative (but still quasi-associative) covariant product $\times$ [1,3].

$$(\Psi^I \times \Psi^J)_{ij} = \sum \Psi_m^I \Psi_n^J \tau_{mi}^I(\varphi_\sigma^1)\tau_{nj}^J(\varphi_\sigma^2)\mathcal{U}(\varphi_\sigma^3) \tag{3.3}$$

The importance of this definition derives from the fact that $\Psi^I \times \Psi^J$ transforms covariantly according to the tensor product representation $\tau^I \otimes \tau^J$. If this covariant product is used to rewrite the local braid relations, Eq. (3.1,3.2), consistency with the transformation law is evident from the intertwining properties of R, Eq. (2.3). There are various other consistency checks which lead to a deeper justification of Eq.(3.2) [1]. Assuming validity of Eq. (3.1), methods from the algebraic theory of superselection sectors can be used to prove Eq. (3.2) [5].

Once Eq. (3.2) is established, it follows that group symmetries (i.e. $R = e \otimes e, \varphi = e \otimes e \otimes e$) are not consistent with nontrivial braid group statistics implemented through Eq. (3.1) (with $\mathcal{R} \neq 1$). Moreover, giving up the requirement $\varphi = e \otimes e \otimes e$ enforces us to encounter non-numerical $\mathcal{R}$-matrices (and vice versa).

It has been argued that the quantum group algebras $U_q(sl_2), q^p = 1$, are the quantum symmetries of the minimal conformal models. The arguments rely on a nonassociative truncated tensor product of the "physical" representations of $U_q(sl_2)$ - i.e. its irreducible representations with nonzero quantum dimension - which is not defined directly in terms of a coproduct. However, it was shown [1] that there exists a truncated quantum group algebra $U_q^T(sl_2)$ associated with $U_q(sl_2)$ which is a nontrivial weak quasi quantum group algebra, i.e. $\varphi, \mathcal{R}, \Delta(e)$ are nontrivial. The indecomposable representations of $U_q^T(sl_2)$ are precisely the physical representations of $U_q(sl_2)$ and the tensor product of representations which is furnished by the coproduct agrees with the truncated tensor product of physical representations of $U_q(sl_2)$. We propose these truncated quantum group algebras to be the true quantum symmetries of the minimal conformal models. The difference between quantum group and quasi quantum group symmetries becomes relevant for possible braid relations of local fields. Investigations of 4-point functions show that local braid relations with a numerical $\mathcal{R}$-matrix cannot hold on all states of the Hilbert space $\mathcal{H}$ [7](this rules out quantum group symmetry). For the conformal Ising model we have constructed a Hilbert space of physical states on which the symmetry $U_q^T(sl_2)$ acts together with covariant fields [8]. These fields obey local braid relations of the type discussed in this section with a nonnumerical $\mathcal{R}$-matrix.

4. References

1. G. Mack, V. Schomerus, Nucl. Phys. B370(1992)185
2. V.G. Drinfel'd, in *Problems of modern quantum field theory*, Proceedings Alushta 1989, Springer Verlag Heidelberg 1989
3. G. Mack, V. Schomerus, DESY 91–144, to be publ. in *New symmetry principles in quantum field theory*, J. Fröhlich et al.(eds.), Lectures given at the 1991 Cargese Summer School
4. D. Altschuler, A. Coste, *Quasi quantum groups, knots, three manifolds and topological field theory*, CERN-TH-6360- 92-REV.
5. G. Mack, V. Schomerus,in preparation
6. J.Fröhlich, in *Nonperturbative quantum field theory*, G.t'Hooft et al.(eds.), Plenum Press 1988
7. D.Buchholz,G.Mack,I.T.Todorov, as reported by I. Todorov in Proceedings of Conf. on Quantum Groups, Clausthal Zellerfeld, July 1989
8. G. Mack, V. Schomerus, Phys.Lett. B267(1991)207

q-DEFORMED CONFORMAL FIELD THEORY

C.H. Oh and K. Singh

Department of Physics
National University of Singapore
Lower Kent Ridge, Singapore 0511
Republic of Singapore

ABSTRACT

We examine some of the standard features of conformal fields in the framework of a q-deformed theory.

Recently, Curtright and Zachos [1] constructed a q-analogue of the centreless Virasoro algebra by generalizing a differential realization of $SU_q(1,1)$ (see also Ref.[2]). The central extension to this algebra has been furnished by Aizawa and Sato [3] and they have also found a q-deformed operator product expansion (OPE) between two energy momentum tensors which realizes this algebra.

In this paper we study the properties of primary fields in the spirit of Ref.[3]. Here we re-examine some of the well known issues pertaining to standard conformal field theory (CFT) [4] in the context of such a q-deformed theory. In particular, we introduce a q-OPE of the energy momentum tensor with a primary field which extends the q-OPE of Ref.[3] to primary fields of arbitrary conformal weights. Using arguments paralleling those used in standard CFT, we obtain the q- analogue of the conformal Ward identity and the projective Ward identities for correlation functions of primary fields. In particular, for the two-point function, it is shown that these Ward identities do not uniquly determine it when $|q| = 1$. Finally, the q-analogues of the descendant fields are furnished.

We begin by defining the q-analoque of a primary field. For this purpose we consider the deformation as defined by Chaichan et. al.[5]:

$$\delta_n^q \Phi(z) = [z\partial + h(n+1) - n] z^n \Phi(z) \tag{1}$$

where

$$[x] = \frac{q^x - q^{-x}}{q - q^{-1}}. \tag{2}$$

To implement this variation as an 'equal-time' commutator, we introduce a q-OPE of $T(z)$

with $\Phi(w)$, which we write as,

$$(T(z)\Phi(w))_q = \frac{[h/2]}{(z-w)}\left\{\frac{\Phi(wq^{-1})}{zq^{h/2}-wq^{-h/2}} + \frac{\Phi(wq)}{zq^{-h/2}-wq^{h/2}}\right\} + \frac{1}{(z-w)}\partial_w^q\Phi(w) + \text{regular terms} \tag{3}$$

where ∂_w^q is the q-analogue of the derivative:

$$\partial_w^q f(w) = \frac{f(wq)-f(wq^{-1})}{w(q-q^{-1})}. \tag{4}$$

Using this definition for the q-derivative, we can also rewrite the q-OPE as

$$(T(z)\Phi(w))_q = \frac{1}{w(q-q^{-1})}\left\{\frac{\Phi(wq)}{z-wq^h} - \frac{\Phi(wq^{-1})}{z-wq^{-h}}\right\} + \text{regular terms} \tag{5}$$

which shows that it is singular at the points $z = wq^{\pm h}$. It is easy to verify that the above OPE leads to the correct variation in Φ, i.e. eqn.(1) by evaluating the integral

$$\oint_C \frac{dz}{2\pi i} z^{n+1}(T(z)\Phi(w))_q, \tag{6}$$

with C taken as a contour encircling the points wq^h and wq^{-h}. Before proceeding further, let us make a few observations:

(1) It is evident from the above expression that there are poles present at two points rather than one. These poles are both of order 1, unlike the undeformed case where the $z = w$ pole is of order 2. It is interesting to note, however, that in the limit $q \to 1$ these poles will coalesce to form a pole of order 2 at $z = w$. In fact, in the limit $q \to 1$, our q-OPE reduces to the standard one.

(2) Recall that the the 'equal-time' commutator in standard CFT [7] is evaluated as the difference of two integrals with contours which are concentric and close to the $|z| = |w|$ contour but one having radius $|z| > |w|$ and the other $|z| < |w|$. They combine into a single contour which is taken to be a small circle centred around the singular point ($z = w$). For this scheme to be applicable here, we must require that the two poles lie on the $|z| = |w|$ contour, since otherwise this poles will not make any contributions to the integral. This means that we must restrict ourselves to the case when $|q| = 1$, i.e. q should be taken as a pure phase ($q = e^{i\alpha\pi}$).

(3) When $h = 2$, our expression is similar to the one given by Aizawa and Sato [3] for the OPE of two energy momentum tensors when the central charge in their expression is taken as zero.

With the q-OPE defined as above, we can write down the q-Ward identities for the correlation functions of primary fields. We begin by considering the action of the generator of infinitisimal conformal transformations on the correlation of n primary fields $\{\Phi_i(w_i)\}$ with correponding conformal weights $h_i (i = 1, 2 \ldots n)$:

$$< \oint_{C_0} \frac{dz}{2\pi i} \epsilon(z) T(z) \Phi_1(w_1) \ldots \Phi_n(w_n) >_q . \tag{7}$$

(The contour C_0 encircles all the points $\{w_i q^{hk} | k = \pm 1\}_{i=1,2,\ldots n}$.) In the above expression, the correlation function $< \ldots >_q$ is taken relative to the 'in' ($|0>_q$) and the 'out' ($_q < 0|$) vacuums which are defined by requiring that

$$L_m |0>_q = 0 \qquad m \geq -1 \qquad _q < 0| L_m = 0 \qquad m \leq 1 \tag{8}$$

where

$$L_m = \oint_{C_0} \frac{dz}{2\pi i} z^{m+1} T(z), \qquad m \in Z, \tag{9}$$

are modes in the expansion,

$$T(z) = \sum_{m \in Z} L_m z^{-m-2}. \tag{10}$$

Note that conditions (8) ensure the regularity of $T(z)|0>_q$ and its adjoint at $z = 0$ and $z = \infty$. By analyticity, the contour C_0 in expression (7) can be deformed to a sum of n contours with each contour C_i surrounding the points, $\{w_i q^h, w_i q^{-h}\}$. Then as a consequence of the q-OPE, and the fact that $\epsilon(z)$ is arbitrary, we have

$$< T(z) \Phi_1(w_1) \ldots \Phi_n(w_n) >_q = \sum_{i=1}^{n} \mathcal{L}^{h_i}_{z;w_i} < \Phi_1(w_1) \ldots \Phi_n(w_n) >_q , \tag{11}$$

where the differential operator $\mathcal{L}^{h_i}_{z;w_i}$ is given by

$$\mathcal{L}^{h_i}_{z;w_i} \equiv \frac{1}{(z - w_i)} \left\{ [h_i/2] \left(\frac{q^{-w_i \partial_{w_i}}}{z q^{h_i/2} - w_i q^{-h_i/2}} + \frac{q^{w_i \partial_{w_i}}}{z q^{-h_i/2} - w_i q^{h_i/2}} \right) + \partial^q_{w_i} \right\}. \tag{12}$$

This is the unintegrated form of the q-Ward identity. It is worth noting that it reduces to the usual one as $q \to 1$.

Next let us consider the q-analogue of the projective Ward Identities. From (7), it is easy to see that the generators $L_{0,\pm 1}$ annhilate both the 'in' and the 'out' vacuums. On multiplying $\epsilon(z) = z^{m+1}$ on both sides of (11) for $m = -1, 0, 1$ and integrating, we have

$$\sum_{i=1}^{n} w_i^m [w_i \partial_{w_i} + h_i(m+1)] < \Phi_1(w_1) \ldots \Phi_n(w_n) >_q = 0 \tag{13}$$

for any n-point function. These are the q-analogues of the projective Ward identities. Now, it is well known that in standard CFT the two-point and three-point functions are severely constrained by the Ward identities. In fact, they are uniquely determined up to a normalization constant. The situation for the q-deformed case is not quite the same. Here when $|q| = 1$ the q- Ward identities do not uniquely specify them as we will illustrate below. For this purpose assume an ansatz for the correlation function of two primary fields $\Phi_1(w_1), \Phi_2(w_2)$ with conformal weight h_1, h_2 respectively to be of the form

$$< \Phi_1(w_1)\Phi_2(w_2) >_q = \frac{1}{(w_1 - w_2)_q^n} \qquad |w_1| > |w_2|, \tag{14}$$

where

$$(w_1 - w_2)_q^n = \prod_{k=1}^{n}(w_1 - w_2 q^{n-2k+1}) = \sum_{k=1}^{n} \frac{[n]!}{[n-k]![k]!} w_1^{n-k}(-w_2)^k \tag{15}$$

is the q-analogue of the distance function $(w_1 - w_2)^n$. On substitution into (13), we obtain the following conditions:

$$[h_1 - n] + [h_2] = 0, \qquad [h_2 - n] + [h_1] = 0, \qquad [2h_2] - [2h_1] = 0$$
$$[2h_1 - n] = 0, \qquad [2h_2 - n] = 0. \tag{16}$$

Apart from the obvious solution

$$h_1 = h_2 = n/2, \tag{17}$$

we also have for $q = e^{i\pi\alpha}$,

$$h_1 = n/2 + k/\alpha, \qquad h_2 = n/2 + l/\alpha \tag{18}$$

where k and l are arbitrary integers which are either both even or both odd. Adding the two we have

$$n = h_1 + h_2 - (k + l)/\alpha \tag{19}$$

and this means that n which characterizes the solution is not unique by virtue of the fact that k and l are arbitrary.

Now, a few words about descendant fields. As in the standard case, we obtain them from the regular part of the OPE of $T(z)$ with a primary field $\Phi(w)$ (in this case the q-OPE) of conformal weight h. Denoting the regular terms of (3) and (5) by $\mathcal{F}_q(z;w)$, we can expand it by using the q-analogue of the Taylor expansion:

$$\mathcal{F}_q(z;w) = \sum_{k=0}^{\infty} \phi_{2k}(wq)(z-w)_q^{2k} + \sum_{k=0}^{\infty} \phi_{2k+1}(w)(z-wq)_q^{2k+1}. \tag{20}$$

The descendant fields $\phi_{2k}(wq)$ and $\phi_{2k+1}(w)$ ($k = 0, 1, 2, ...$) can then be expressed as integrals of $\mathcal{F}_q(z;w)$:

$$\phi_{2k}(wq) = \oint_C \frac{dz}{2\pi i} \frac{\mathcal{F}_q(z;w)}{(z-wq)_q^{2k+1}}, \qquad \phi_{2k+1}(w) = \oint_C \frac{dz}{2\pi i} \frac{\mathcal{F}_q(z;w)}{(z-w)_q^{2k+2}} \tag{21}$$

where the contour C in both cases encloses the $z = |w|$ contour. Note that the the term $\mathcal{F}_q(z;w)$ in the above integrands can be replaced with the q-OPE $(T(z)\Phi(w))_q$ since the singular part of this expansion makes no contribution to the integrals. Hence we have

$$\phi_{2k}(wq) = \oint_C \frac{dz}{2\pi i} \frac{(T(z)\phi(w))_q}{(z-wq)_q^{2k+1}}, \qquad \phi_{2k+1}(w) = \oint_C \frac{dz}{2\pi i} \frac{(T(z)\phi(w))_q}{(z-w)_q^{2k+2}} \tag{22}$$

which are the q-analogues of the descendant fields.

Finally, one can also compute the correlation functions of these descendant fields from the correlation functions of primary fields. As an example, the two-point function between the descendant field $\phi_{2k}(w_1)$ of a primary field with conformal weight h_1 and a primary field $\Phi_2(w_2)$ of conformal weight h_2 is given by,

$$< \phi_{2k}(w_1)\Phi_2(w_2) >_q = \oint_C \frac{dz}{2\pi i} \frac{1}{(z-w_1 q)_q^{2k+1}} < T(z)\Phi_1(w_1)\Phi_2(w_2) >_q$$

$$= \frac{1}{[2k]!}(\partial_z^q)^{2k} \frac{\{[h_2/2](\frac{q^{-w_2\partial_{w_2}-h_2/2}}{(z-w_2q^{-h_2})} + \frac{q^{w_2\partial_{w_2}+h_2/2}}{(z-w_2q^{h_2})}) + \partial_{w_2}^q\} < \Phi_1(w_1)\Phi_2(w_2) >_q}{(z-w_2)}\Bigg|_{z=w_1 q}. \tag{23}$$

REFERENCES

[1.] T. Curtright and C. Zachos, *Phys. Lett.* **B243** (1990) 237.

[2.] M. Chaichan, P. Kulish and J. Lukierski, *Phys. Lett.* **B237** (1990) 401.

[3.] N. Aizawa and H. Sato, *Phys. Lett.* **B256** (1991) 185.

[4.] A.A. Belavin, A.M. Polyakov and A.B. Zamolodchikov, *Nucl. Phys.* **B241** (1984) 333.

[5.] M. Chaichan, A.P. Isaev, J. Lukierski, Z. Popowicz and P. Prešnajder, *Phys. Lett.* **B262** (1991) 32.

[6.] C.H. Oh and K. Singh, *J. Phys.* **A25** (1992)L149.

[7.] D. Lüst and S. Theisen, *Lecture Notes in Physics* Vol 346 (Springer-Verlag, Berlin 1989).

Z-ALGEBRA CONSTRUCTION OF W-ALGEBRAS *

PEI WANG
Institute of Modern Physics,
Northwest University 710069 Xian China

ABSTRACT

We propose a new construction-the Z-algebra construction for W-algebras which include the Virasoro algebra as a special case. We develope a calculating program for Z-algebras. Applying it we deduce the usual conformal properties of parafermions and construct the W-algebras for $SU(2)_k$ and $SU(3)_k$ parafermions.

The talk is a result of cooperation with C.Y.Zhu, K.J.Shi, H.Fan and X.M.Ding.[1,2]

Since Zamolodchikov suggested an extension of the conformal algebra[3] , W-algebra becomes one of central interest in two-dimensional physics.It encompasses many areas of theoretical physics,and is also interested by mathematicians.

Without using the current free field realization we derive W-algebras from the parafermionic model[4,5](In a recent paper Narganes-Quijano derived W-algebras also from parafermions[6], but in a two boson expression which is of course a free field realization too).Definitely we propose a new construction-the Z-algebra construction for the W-algebras which includs the Virasoro algebra as a special case. It is a natural generalization for conventional Sugawara construction of the stress-energy tensor field operator in the WZWN model(cf,ref.[7]).

Because of the mutually semilocal property[4,5] between two parafermions $\psi_\alpha(Z)$, $\psi_\beta(W)$ (α,β are roots of the underlying group) the radial ordering product is a multivalued functions, so we can define

$$R(\psi_\alpha(z)\psi_\beta(w))=\begin{cases}\psi_\alpha(z)\psi_\beta(w),\ \omega^{\alpha\cdot\beta}\psi_\alpha(z)\psi_\beta(w),\cdots,\ \omega^{(k-1)\alpha\cdot\beta}\psi_\alpha(z)\psi_\beta(w), & |z|>|w|,\\ \omega^{\alpha\cdot\beta/2}\psi_\beta(w)\psi_\alpha(z),\ \omega^{3\alpha\cdot\beta/2}\psi_\beta(w)\psi_\alpha(z),\cdots,\ \omega^{(k-1/2)\alpha\cdot\beta}\psi_\beta(w)\psi_\alpha(z), & |z|<|w|\end{cases} \tag{1}$$

where

$$\omega=\exp(2\pi i/k). \tag{2}$$

Especially we have

$$R(\psi_\alpha(z)\psi_\beta(w))(z-w)^{\alpha\cdot\beta/k}=R(\psi_\beta(w)\psi_\alpha(z))(w-z)^{\alpha\cdot\beta/k}. \tag{3}$$

We drop the R symbol hereafter and consider the operator product expansion with higher regular terms for parafermion

* Work supported in part by National Science Fundation of China.

$$\psi_\alpha(z)\,\psi_\beta(w)(z-w)^{\alpha\cdot\beta/k}$$
$$= \delta_{\alpha,-\beta}/(z-w)^2 + [\,\ell_{\alpha\beta}/\sqrt{k}\,(z-w)]\ \psi_{\alpha+\beta}(w) + \sum_{n=0}^{\infty}(z-w)^n(\psi_\alpha\psi_\beta)^n(w) \tag{4}$$

in which k is an integer defining the cyclic symmetry of parafermion, and can also be identified with the level in the current algebra, $\ell_{\alpha\beta}$ is the structure constant of the $[E_\alpha, E_\beta]$ in the Cartan–Weyl basis if $\alpha+\beta$ is a root,otherwise $\ell_{\alpha\beta}$ is zero. Continuing the r. h. s. of eq.(3) to the left we then get the recurrence relations

$$(\psi_\alpha\psi_\beta)^n(w) - (-1)^n(\psi_\beta\psi_\alpha)^n(w)$$
$$= [\,\ell_{\alpha\beta}/\sqrt{k}(n+1)\,!\,]\ \partial_w^{n+1}\psi_{\alpha+\beta}(w) + \sum_{r=0}^{n-1}\ [(-1)^r/(n-r)\,!]\ \partial_w^{n-r}(\psi_\beta\psi_\alpha)^r(w). \tag{5}$$

Let $\varphi_{\lambda\bar\lambda}(z,\bar z)$ be the fields having pair of charges which take values in the weight lattice. The Fourier coefficients of the parafermion are

$$A_m^{\alpha,\lambda}\varphi_{\lambda\bar\lambda}(w,\bar w) = \oint (dz/2\pi i)(z-w)^{m+\alpha\cdot\lambda/k}\,\psi_\alpha(z)\,\varphi_{\lambda\bar\lambda}(w,\bar w), \tag{6}$$

and

$$(\psi_\alpha\psi_\beta)_m^{n,\lambda}\,\varphi_{\lambda\bar\lambda}(w,\bar w)$$
$$= \oint (dz/2\pi i)(z-w)^{m+n+1+(\alpha+\beta)\cdot\lambda/k}\,(\psi_\alpha\psi_\beta)^n(z)\ \varphi_{\lambda\bar\lambda}(w,\bar w). \tag{7}$$

The so-called Z-algebras are defined through the following fundamental relations of these Fourier coefficients (cf. ref. [8]) :

$$A_n^{\alpha,\beta}\psi_\beta(w) = 0\ (n>1)\ ;\quad \delta_{\alpha,-\beta}(n=1);$$
$$(\ell_{\alpha\beta}/\sqrt{k})\ \psi_{\alpha+\beta}(w)\ (n=0)\ ;\quad (\psi_\alpha\psi_\beta)^{-n-1}(w)\ (n<0), \tag{8}$$

and

$$\sum_{l=0}^{\infty} C^{(l)}_{-p-1+\alpha.\beta/k}\ [\,A^{\alpha,\beta+\lambda}_{m-l-p+q}\ A^{\beta,\lambda}_{n+l+p-q} + (-1)^p\,A^{\beta,\alpha+\lambda}_{n-l-q-1}\ A^{\alpha,\lambda}_{m+l+q+1}\,]$$
$$= C^{(p+2)}_{m+q+1+\alpha\cdot\lambda/k}\ \delta_{\alpha,-\beta}\ \delta_{m,-n} + C^{(p+1)}_{m+q+1+\alpha\cdot\lambda/k}(\ell_{\alpha\beta}/\sqrt{k})\ A^{\alpha+\beta,\lambda}_{m+n}$$
$$+ \sum_{r=0}^{p} C^{(p-r)}_{m+q+1+\alpha.\lambda/k}\ (\psi_\alpha\psi_\beta)^{r,\lambda}_{m+n},\quad p=2q\ \text{ or }\ 2q+1\,, \tag{9}$$

where $C^{(l)}_x = (-)^l x(x-1)\ldots(x-l+1)\ /\ l\,!$ is the binomial coefficients. By using these formulae and recurrence relations we establish a calculation program. Applying it we deduce the usual conformal properties of parafermion,and construct the W–algebras.

Define the stress–energy tensor of parafermion

$$T(z) = T^{(2)}(z) = \sum_{\alpha} [k/2(k+g)] \, (\psi_{\alpha} \psi_{-\alpha})^0 \, (z), \quad g\text{=dual Coexter number} \tag{10}$$

Because of the complication of the situation, we consider only $SU(2)_k$ and $SU(3)_k$ parafermions. In these two cases

$$T^{(2)}(z) = [k/2(k+2)] \, [(\psi_1 \psi_{-1})^0(z) + (\psi_{-1} \psi_1)^0(z)], \quad \text{for SU(2)}, \tag{11a}$$

$$T^{(2)}(z) = \sum_{i=1}^{3} T^{(\alpha_i)}(z) = [k/2(k+3)] \sum_{i=1}^{3} [(\psi_{\alpha_i} \psi_{-\alpha_i})^0 (z) + (\psi_{-\alpha_i} \psi_{\alpha_i})^0 (z)] \quad \text{for SU(3)}, \tag{11b}$$

respectively. Where the order of SU(3) roots is indicated in the Figure.

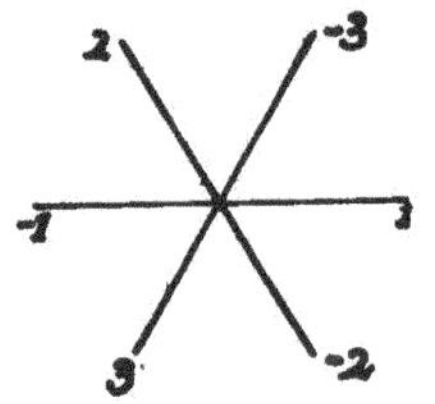

Define a new operator

$$T^{(3)}(z) = B_3 \, [(\psi_1 \psi_{-1})^1 (z) - (\psi_{-1} \psi_1)^1 (z)], \quad (B_3)^2 = [k^3/2(k+2)(3k+4)] \quad \text{for SU(2)}, \tag{12a}$$

and

$$\begin{aligned} T^{(3)}_{(z)} &= B_3 \sum_{i=1}^{3} [(\psi_{\alpha_i} \psi_{-\alpha_i})^1 (z) - (\psi_{-\alpha_i} \psi_{\alpha_i})^1 (z)] \\ &= B_3 (k^{3/2}/(k+2)) \, \epsilon_{ijk} \, [(\psi_{-\alpha_i} (\psi_{-\alpha_j} \psi_{-\alpha_k})^0)^0 + (\psi_{\alpha_i} (\psi_{\alpha_j} \psi_{\alpha_k})^0)^0] \, (z), \\ (B_3)^2 &= k^3/6(k+1)(k+3) \quad \text{for SU(3)}. \end{aligned} \tag{12b}$$

From these operators we can calculate operator product expansions between any two of $T^{(r)}$s (including equal $T^{(r)}$) and $T^{(r)}$ and ψ for SU(2) and SU(3) respectively. Especially we have

$$\begin{aligned} T^{(3)}(z) \, T^{(3)}(w) &= (k-2) \{ c/3(z-w)^6 + 2T(w)/(z-w)^4 \\ &+ \partial_w T(w)/(z-w)^3 + [1/(z-w)]^2 \, [2b^2 \Lambda(w) + (3/10) \partial_w^2 T(w)] \\ &+ (1/(z-w)) \, [b^2 \partial_w \Lambda(w) + (1/15) \partial_w^3 T(w)] \} \\ &+ C_4^{33} [1/(z-w)^2 + 1/2(z-w) \, \partial_w] \, T^{(4)}(w) + \cdots \end{aligned} \tag{13}$$

in which

$$\Lambda(w) = (T \cdot T)^0 (W) - (3/10) \partial_w^2 T(w), \quad b^2 = 16/(22+5c) \, ; \tag{14}$$

$$\begin{aligned} (C_4^{33})^2 &= 128 \, (2k+3)^2 \, (2k-1)/(k+2) \, (3k+4) \, (16k+17) \quad &\text{for SU(2)}, \\ &\ 16(2k+3) \, (4k+3) \, (8k-1) \, / \, 3(k+1) \, (k+3) \, (13k+9) \quad &\text{for SU(3)}, \end{aligned} \tag{15}$$

and

$$T^{(4)}(z)$$

$$= B_4 \{ (\psi_1 \psi_{-1})^2 (z) + (\psi_{-1} \psi_1)^2 (z) - [3(k+2)/10k] \partial_z^2 T(z) - [b^2 (k+2)(6k-5)/8k^2] \Lambda (z) \},$$

$$(B_4)^2 = k^4 (16k+17)/8(k+2)(2k+1)(3k+4), \qquad \text{for } SU(2), \tag{16a}$$

$$T^{(4)}(z) = B_4 \{ \sum_{i=1}^{3} [(\psi_{\alpha_i} \psi_{-\alpha_i})^2 (z) + (\psi_{-\alpha_i} \psi_{\alpha_i})^2 (z) - (3(k+2)^2/k^2 (4k+3)) \Lambda^{(\alpha_i)}(z)]$$

$$- (k-3)(k+3)^2 (12k+11)/k^2(4k+3)(13k+9) \Lambda (z) - [3(k+3)/10k] \partial_z^2 T(z) \},$$

$$\Lambda^{(\alpha_i)}(z) = ((k+3)/(k+2))^2 (T^{(\alpha_i)} \cdot T^{(\alpha_i)})^0 (z) - [3(k+3)/10(k+2)] \partial_z^2 T(z),$$

$$(B_4)^2 = k^4(4k+3)(13k+9)/12(k+3)(2k+3)(k+1)(8k-1) \qquad \text{for } SU(3). \tag{16b}$$

The point is that one can show two point functions

$$< T^{(n)} (z), T^{(n)} (w) > = (z-w)^{-2n}(k-2)(k-3) \cdots (k-n+1)\, c/n. \tag{17}$$

(Although we didn't prove eq. (17) for higher n (>8 for SU(2) , >4 for SU(3)) we conjecture it holds.) It is obvious that the operator series $\{ I, T, T^{(3)}, \ldots \}$ will be trancated at $T^{(n)}(n=k+1)$, if we choose a finite k. We may renomalize $T^{(n)}$ to $W_n(z) = [\prod_{m=2}^{n-1} (k-m)]^{-1/2}\, T^{(n)}(z)$, then the series $\{ I, T, W_3, \ldots, W_k \}$ form a W_k-algebra.

It is easy to find the central charges of parafermions

$$C = (kD/(k+g)) - r, \text{ D–dimension and r–rank of the group.} \tag{18}$$

For $SU(2)_k$ and $SU(3)_k$ they are $2(k-1)/(k+2)$ and $6(k-1)/(k+3)$ respectively. The former agrees with a special case in the Fateev Lykyanov's W_k-algebra series[9]

$$C = (k-1)(1-k(k+1)/p(p+1)) \, |_{p=k+1}$$

but the latter doesn't agree with any known W_k-algebra.

References

1. P.Wang and C.Y.Zhu, IMP–NWU–910525, 1991 submitted to J.Math.Phys.
2. X.M.Ding, H.Fan, K.J.Shi, P.Wang and C.Y.Zhu, IMP–NWU–920601.
3. A.B.Zamolodchikov, TMP 65(1985)1205;
   V.A.Fateev and A.B.Zamolodchikov, Nucl.Phys.B280(1987)644.
4. A.B.Zamolodchikov and V.A.Fateev, JETP 62(1985)215;
   D.Gepner and Z.Qiu, Nucl, Phys.B285(1987)423.
5. D.Gepner, Nucl.Phys.B290(1987)10.
6. F.J.Narganes–Quijano, Int .J.Mod.Phys, A6(1991)2611.
7. F.Bais, P.Bouwknegt, M.Surridge and K.Schoutens, Nucl.Phys.B304(1988)348;371.
8. J.Lepowski and G.Wilson, Ivent, Math.77(1984)199.
9. V.A.Fateev and S.L.Lykyanov, Int.J.Mod.Phys.A3(1988)507.

Parallel Sessions 4

Fractional Statistics and Lattice Models

NEW GENERALIZED LIE STRUCTURES FROM RECENT DEVELOPMENTS IN QUANTUM MECHANICS

Jules BECKERS
Theoretical and Mathematical Physics, Institute of Physics (B.5)
University of Liege
B-4000 LIEGE 1 (Belgium)

and

Nathalie DEBERGH
University of Liege
B-4000 LIEGE 1 (Belgium)

ABSTRACT

Symmetries of fundamental relativistic wave equations such as the Dirac and Kemmer ones generate very well-known Lie algebras. The recent study of their supersymmetries and parasupersymmetries leads to generalized Lie structures.

Supersymmetry has implied at the level of nonrelativistic quantum mechanics a new field [1] called, in particular, N=2-supersymmetric quantum mechanics (SSQM) dealing with usual bosons and fermions and a superpotential. Here the diagonal Hamiltonian is made of two superpartners, it is an even operator while the supercharges are odd ones leading in a natural way to graded structures such as the Lie superalgebras. If paraparticles [2] are considered, N=2-parasupersymmetric quantum mechanics (PSSQM) takes place [3,4,5] and consists in the superposition of bosons and parafermions (of order 2 of paraquantization). Here we are then dealing with two superpotentials and they enter into two SSQM-parts which are not independent [3,4]. The associated Hamiltonian and parasupercharges do also generate closed structures actually called Lie parasuperalgebras, so that interesting inclusions such as are effectively pointed out inside nonrelativistic quantum mechanics through its important (para)(super)symmetries.

The main questions we ask here are : "What about the situation at the level of relativistic quantum mechanics and its implications in quantum field theory ?". In a certain sense, such questions come back on welcome information in connection with particle physics !

A. Spin 1/2 particles and the Dirac equation

A first answer is given by the general connection existing between the Dirac theory and some supersymmetric developments. Beside the ten well-known Poincaré symmetries [6] of the free Dirac equation (they generate the famous Poincaré algebra under the usual Lie bracket understood as a commutator), there are extended Lie symmetries [7]. Altogether they can be collected in covariant notations as follows [8]

$$\{P_\mu, M_{\mu\nu}, I, W_\mu, W_{\mu\nu}, A_\mu, B\}, \quad (\mu, \nu = 0, 1, 2, 3) \tag{1}$$

where we recognize the ten Poincaré generators associated with translations (P_μ) and Lorentz rotations ($M_{\mu\nu}$), the identity and fifteen further ones forming a 4-vector (the Bargmann-Wigner 4-vector W_μ), a 2d order-antisymmetric tensor ($W_{\mu\nu}$), another 4-vector (A_μ) and a scalar B. These twenty-six operators do not close under commutation but some of them combined with the ten second order operators $P_\mu P_\nu$ can be selected in order to get a closed structure. Indeed, the following thirty-one generators

$$\{P_\mu, M_{\mu\nu}, I, P_\mu P_\nu \; ; \; W_\mu^D, W_{\mu\nu}^D\} \tag{2}$$

close but under a Lie superbracket including anticommutators : we thus get a Lie superalgebra associated with the Dirac context (twenty-one operators are even and ten ones are odd). Such a connection has interesting consequences [9,10] and can be generalized if we remember that Dirac particles are nothing else but usual fermions, i.e. parafermions corresponding to the p=1-order of paraquantization [2]. Supersymmetry, superalgebras and superfields have to deal with Dirac characteristics [11].

B. Mesons and the Kemmer equation

Another answer can be obtained by the connection between the Kemmer theory [12] and some parasupersymmetric developments. The (free) Kemmer equation is given by

$$(i\beta^{\mu}\partial_{\mu} - m)\Psi_{\kappa}(\vec{x}, t) = 0 \tag{3}$$

where the β-matrices form a 126-dimensional Kemmer algebra characterized by only three irreducible representations (of dimensions 1, 5 and 10) and are such that

$$\beta^{\mu}\beta^{\nu}\beta^{\lambda} + \beta^{\lambda}\beta^{\nu}\beta^{\mu} = g^{\mu\nu}\beta^{\lambda} + g^{\lambda\nu}\beta^{\mu}. \tag{4}$$

Such a covariant formulation evidently admits the <u>ten</u> Poincaré symmetries (associated with P^{μ} and $M^{\mu\nu}$) but also the <u>sixteen</u> supplementary ones, so that up to some specificities the set (1) is still valid for spin 1-particles. By adding the ten second order operators $P^{\mu}P^{\nu}$, it is now once again possible to determine a closed structure between the thirty-one generators (2) <u>but</u> under the new Lie bracket we refer to by double commutators : we thus get Lie parasuperalgebras in this spin 1-context. This can also be illustrated through pararelativistic oscillators [13] or parasuperfields [14] when Poincaré invariance is required. Hidden parasupersymmetry of order $p = 2$ of paraquantization is also present in the Kemmer theory, etc.

In conclusion, through the above results of the cases A and B where we have considered respectively

$$p = 1 = 2s\big|_{s=\frac{1}{2}}, \qquad p = 2 = 2s\big|_{s=1}, \tag{5}$$

we can extend [14] all the above properties to arbitrary spins s <u>or</u>, equivalently, to arbitrary orders p of paraquantization such that $p = 2s$.

References

1. E. WITTEN, *Nucl.Phys.* **B188** (1981) 513.
2. E.P. WIGNER, *Phys.Rev.* **77** (1950) 711 ;
   H. GREEN, *Phys.Rev.* **90** (1953) 270 ;
   Y. OHNUKI and S. KAMEFUCHI, *Quantum Field Theory and Parastatistics* (University of Tokyo Press, Tokyo, 1982).
3. V.A. RUBAKOV and V.P. SPIRIDONOV, *Mod.Phys.Lett.* **A3** (1988) 1337.
4. J. BECKERS and N. DEBERGH, *Nucl.Phys.* **B340** (1990) 767.
5. S. DURAND and L. VINET, *J.Phys.* **A23** (1990) 3661.
6. E.P. WIGNER, *Ann.Math.* **40** (1939) 149.
7. N. KAMRAN and R.G. McLENAGHAN, *Phys.Rev.* **D30** (1984) 357 ;
   W.I. FUSHCHICH and A.G. NIKITIN, *J.Phys.* **A20** (1987) 537 ;
   J. BECKERS and N. DEBERGH, *J.Phys.* **A24** (1991) 2889.
8. J. BECKERS, N. DEBERGH and A.G. NIKITIN, *Lie Extended Symmetries and Relativistic Particles*, to be published in *J.Phys.* **A** (1992).
9. J. BECKERS and N. DEBERGH, *Phys.Rev.* **D42** (1990) 1255.
10. J. BECKERS, N. DEBERGH and A.G. NIKITIN, *Phys.Lett.* **B279** (1992) 333.
11. E. SOKATCHEV and D.T. STOYANOV, *Mod.Phys.Lett.* **A1** (1986) 577.
12. N. KEMMER, *Proc.Roy.Soc. (London)* **173A** (1939) 91.
13. J. BECKERS, N. DEBERGH and A.G. NIKITIN, *On Pararelativistic Quantum Oscillators*, to be published in *J.Math.Phys.* (1992).
14. J. BECKERS and N. DEBERGH, *Poincaré Invariance and Quantum Parasuperfields*, Liège preprint PTM-92/05 (April 1992).

Exact Solution, Jump Function, Fractal Dimension and Entropy Formation of the p-adic Delay Functional Equation $x(t) = x(t-1) + x(t-\theta)$

CHIN C.H.[1] and JENG H.T.[2]

[1] Institute of Applied Mathematics and Institute of Electrophysics, National Chiao Tung University, Hsin Chu, Taiwan, R.O.C.
[2] Institute of Physics, National Tsing Hwa University, Hsin Chu, Taiwan, R.O.C.

Introduction

Over the past few years, Ghil and Mullhaupt observed log-periodic oscillations in the scaling of jump functions of solutions of Boolean delay equation[1],[2]. The interesting results are encounted such basic concept as generalized Hardy-Littlewood tauberian theorem for polynomial log-periodic growth laws in number theory and functional analysis[3]. Apart from their intrinsic interest, the simplest delay functional equation can also illustrate such ideas as recurrence, solvability, completeness, consistency, provability and axiomatic deduction in logic [4],[5],[6]. The correspondance between physical process and computation idea related closely to undecidability and interactability in theoretical physics is indeed clarified[7],[8]. As an example of semidiscrete dynamical systems, the spatio-temporal chaos and multi-fractal structures can be understood, so that the new measure theory can be constructed. Fractal dimension and entropy creation in the system are thus calculated without difficulty[9].

For this particular functional equation, the more generalized form p-adic case is considered in stead of the dyadic case. The corresponding theorems may be further improved. In this short paper, four mean results are presented therein as theorems. Namely, each of the theorems is concerning the exact solution, jump function, r-level-jump function and fractal entropy formation.

Exact Solution

Let $x : R^+ \to Z_p$, and satisfy the delay funcitonal equation $x(t) = x(t-1) + x(t-\theta)$, where $\theta \in (0,1)$ with initial data belonging to $\{1, 2, ..., p-1\}$. To find the exact solution of $x(t)$, the recursive property comes in the scheme naturally.

Theorem 1. The exact solution is

$$x(t) = \left[\sum_{\substack{\mu>0,\nu>0 \\ t-1<\mu+\nu\theta\leq t}} \binom{\mu+\nu}{\mu}_p \right]_p , \qquad \mu, \nu \in Z^+,$$

where $\binom{m}{n}_p$ denotes the residue of $\binom{m}{n}$ modulo p.

Jump Function and Fractal Dimension

To describe the complexity of $x(t)$, the jump function $J(t)$ is defined to be the total number of jumps of $x(t)$ before t. Under the restriction of p=2,

$$J(t) = \sum_{\substack{\mu>0,\nu>0 \\ \mu+\nu\theta<t}} \binom{\mu+\nu}{\mu}_2 + \eta(t),$$

where $\eta(t) = \frac{1}{2}$, as $t = \mu + \nu\theta$, and $\binom{\mu+\nu}{\mu}_2 = 1$. Therefore, it ensures the condition that

$$J(t) = \frac{1}{2} \left[\lim_{s \nearrow t} J(s) + \lim_{s \searrow t} J(s) \right].$$

The jump function is derived as follows.
Lemma 1. (Mullhaupt) In Z_2

$$J(t) = J(\frac{t}{2}) + J(\frac{t-\theta}{2}) + J(\frac{t-1}{2}).$$

Theorem 2. In Z_p

$$J(t) = \sum_{q=0}^{p-1}\sum_{\nu=0}^{q} J\left[\frac{t-(q-\nu)\theta-\nu}{p}\right].$$

Proof.

$$\begin{aligned} Q(u) &= \left(1+u^\theta+u\right)^{-1} \\ &= \sum_{q=0}^{p-1}\sum_{\mu=0}^{\infty}\left(u^\theta+u\right)^{p\mu+q} \\ &= \sum_{q=0}^{p-1}\left(u^\theta+u\right)^q \sum_{\mu=0}^{\infty}\left[\sum_{\nu=0}^{p}\binom{p}{\nu} u^{(p-\nu)\theta}u^\nu\right]^\mu. \end{aligned}$$

Since

$$\binom{p}{\nu}_p = \begin{cases} 1, & \nu = 0, p \\ 0, & \text{otherwise,} \end{cases}$$

it implies that

$$\begin{aligned} (Q(u))_p &= \sum_{q=0}^{p-1}\left[\left(u^\theta+u\right)^q\right]_p \sum_{\mu=0}^{\infty}\left(u^{p\theta}+u^p\right)^\mu \\ &= \sum_{q=0}^{p-1}\sum_{\nu=0}^{q}\binom{q}{\nu}_p u^{(q-\nu)\theta}u^\nu Q(u^p). \end{aligned}$$

Let $u = e^{-s}$, then

$$\begin{aligned} X(s) &= Q(u) \\ &= \sum_{q=0}^{p-1}\sum_{\nu=0}^{q}\binom{q}{\nu}_p e^{-s\theta(q-\nu)}e^{-s\nu}. \end{aligned}$$

By the Perron summation formula and integrating over the contour which includes all the real singularities, it follows that

$$x(t) = \frac{1}{2\pi i}\int_{c-i\infty}^{c+i\infty} e^{st}\sum_{q=0}^{p-1}\sum_{\nu=0}^{q}\binom{q}{\nu}_p e^{-s\theta(q-\nu)}e^{-s\nu}Q(e^{-ps})\frac{ds}{s}.$$

Let $\tilde{s} = ps$, then

$$\begin{aligned} x(t) &= \sum_{q=0}^{p-1}\sum_{\nu=0}^{q} \binom{q}{\nu}_p \{\frac{1}{2\pi i}\int_{c-i\infty}^{c+i\infty} exp\left[\tilde{s}\left(\frac{t-(q-\nu)\,\theta-\nu}{p}\right)\right] Q\left(e^{-\tilde{s}}\right)\frac{d\tilde{s}}{\tilde{s}}\} \\ &= \sum_{q=0}^{p-1}\sum_{\nu=0}^{q} \binom{q}{\nu}_p x\left[\frac{t-(q-\nu)\,\theta-\nu}{p}\right]. \end{aligned}$$

Since $\nu \leq q < p$ and $\binom{q}{\nu}_p \neq 0$ for all q, ν, there are $\frac{p(p+1)}{2}$ terms being summed up and the theorem is proved.

Definition. The r-level-jump function is defined as

$$J_r(t) =: \sum_{\substack{\mu>0,\nu>0 \\ \mu+\nu\theta<t}} \Xi_p^r(t) \quad + \eta(t),$$

where $\eta(t) = \frac{1}{2}$ as $t = \mu + \nu\theta$ and $\binom{\mu+\nu}{\mu}_p = r$, and

$$\Xi_p^r(t) = \begin{cases} 1, & \mu+\nu\theta<t, \quad \binom{\mu+\nu}{\mu}_p = r \\ 0, & \text{otherwise} \end{cases}.$$

Theorem 3. The r-level-jumps of $x(t)$ is

$$J_r(t) = \sum_{\tau=1}^{p-1}\sum_{q=0}^{p-1}\sum_{\nu=0}^{q} \xi_{\tau r} J_\tau\left[\frac{t-(q-\nu)\,\theta-\nu}{p}\right],$$

where $\xi_{\tau r} = \begin{cases} 1 & , \ \left(\tau\binom{q}{\nu}\right)_p = r \\ 0 & , \ \text{otherwise} \end{cases}$

Proof: In view of the equation

$$(Q(u))_p = \sum_{q=0}^{p-1}\sum_{\nu=0}^{q} \binom{q}{\nu}_p u^{(q-\nu)\theta}u^{\nu}Q\left(u^p\right).$$

and taking the rth power of $(Q(u))_p$,

$$\begin{aligned} (Q(u))_p^r &= \left[\sum_{q=0}^{p-1}\sum_{\nu=0}^{q} \binom{q}{\nu}_p u^{(q-\nu)\theta}u^{\nu}Q\left(u^p\right)\right]_p^r \\ &= \sum_{\tau=1}^{p-1}\sum_{q=0}^{p-1}\sum_{\nu=0}^{q} \binom{q}{\nu}_p^{\lambda} u^{(q-\nu)\theta}u^{\nu}\left(Q\left(u^p\right)\right)_p^{\tau}, \end{aligned}$$

where $(\lambda\tau)_p = r, \quad \tau \in \{1, 2, ..., p-1\}$.
Therefore,

$$\begin{aligned} x_r(t) &= \sum_{\tau=1}^{p-1}\sum_{q=0}^{p-1}\sum_{\nu=0}^{q} \binom{q}{\nu}_p^{\lambda} x_\tau\left[\frac{t-(q-\nu)\,\theta-\nu}{p}\right] \\ &= \sum_{\tau=1}^{p-1}\sum_{q=0}^{p-1}\sum_{\nu=0}^{q} \left(\tau\binom{q}{\nu}\right)_p^{r} x_\tau\left[\frac{t-(q-\nu)\,\theta-\nu}{p}\right], \end{aligned}$$

and the theorem is proved.

Lemma 2. $lim_{t\to\infty} J_r(t) = K(\theta)t^\sigma \phi_r(t)/2$, where $\sigma = \log_3 6$, $r = 1,2$.

Proof: By the identities

$$J(t) = J_1(t) + J_2(t),$$

$$J_r(t) = K_r(\theta)t^\sigma \phi_r(t), \qquad r = 1,2,$$

$$J(t) = K(\theta)t^\sigma[\phi_1(t) + \phi_2(t)],$$

and

$$J(3t) = 3^\sigma(t),$$

it concludes that $K_1(\theta) = K_2(\theta) = K(\theta)/2, \quad \sigma = log_3 6$.

The asymptotic forms of p-1 r-level functions, therefore, are

$$lim_{t\to\infty} J_r(t) = p^\sigma J_r(\frac{t}{p}) = K(\theta)t^\sigma \phi_r(t)/(p-1), \quad r \; = \; 1,2,...,p-1,$$

with $\sum_{r=1}^{p-1} \phi_r(t) = \phi(t) = \phi(pt), \quad \sigma = \log_p \frac{p(p+1)}{2}$

The r-level-jump functions are all log-periodic functions with polynomial growth. The fractal dimension is $\sigma = \log_p \frac{p(p+1)}{2}$.

Entropy Formation

Now it is ready to achieve the value of entropy of the system at time t. Let the trajectory $\vec{x}(t) = [x_1(t), x_2(t), ..., x_d(t)]$ of a dymanical system in d-dimensional phase space be partitioned into boxes of size l^d. The state of the system is measured in time interval τ. Let $P_{i_0...i_n}$ be the joint probability so that $\vec{x}(t+n\tau)$ is in the box i_n.

Shannon entropy is defined to be $FH_n = -\sum_{i_0...i_n} P_{i_0...i_n} \log \, P_{i_0...i_n}$. The fractal entropy FH is defined to be

$FH = lim_{\tau\to 0} lim_{l\to 0} lim_{N\to\infty} \frac{1}{N\tau} \sum_{n=0}^{N-1}(FH_{n+1} - FH_n)$,

which measures the loss of information.

Lemma 3. The probability P_{r_n} for the system of in r-level-jump state in t $\in$ [n, n+1) is

$$P_{r_n} = \theta\sigma K(\theta)n^{\sigma-2}/(p-1), \quad r = 1,2,...,p-1, \; where \; \sigma = \log_p \frac{p(p+1)}{2}$$
$$= 1 - \theta\sigma K(\theta)n^{\sigma-2}, \quad r \; = 0$$

Proof:

The number of total lattice points under the line $t = \mu + \nu\theta$ is $T = \frac{t^2}{2\theta}$. In time interval [t, t+1), the number of lattice points on the line segment $T_d = \frac{t}{\theta}$ is being counted. The jump density function $j_r(t) \sim K(\theta)\sigma t^{\sigma-1}/(p-1)$, $r = 1,2,...,p-1$.

$$P_{r_n} \; = \; \frac{j_r(n)}{T_d} = \theta\sigma K(\theta)n^{\sigma-2}/(p-1),$$

$$P_{o_n} = \frac{j_o(n)}{T_d} = 1 - \theta\sigma K(\theta)n^{\sigma-2}.$$

Theorem 4. $\lim_{t\to\infty} FH \sim t^{\sigma-1}\log t$, $\sigma = \log_p[p(p+1)/2]$.
Proof. Assume we have known $P_{r_0\cdots r_{n-1}}$ at time $t = n-1$, then $P_{r_0\cdots r_n} = P_{r_0\cdots r_{n-1}}P_{r_n}$ at time $t = n$.

$$\begin{aligned} FH_n &= -\sum_{r_0\cdots r_n} P_{r_0\cdots r_n}\log P_{r_0\cdots r_n} \\ &= -\sum_{r_0\cdots r_{n-1}}\sum_{r_n} P_{r_0\cdots r_{n-1}}P_{r_n}\log P_{r_0\cdots r_{n-1}}P_{r_n} \\ &= -\sum_{r_0\cdots r_{n-1}} P_{r_0\cdots r_{n-1}}\left(1-\theta\sigma K(\theta)n^{\sigma-2}\right) \\ &\qquad\cdot\log\left[P_{r_0\cdots r_{n-1}}\left(1-\theta\sigma K(\theta)n^{\sigma-2}\right)\right] \\ &\quad -\sum_{r_0\cdots r_{n-1}}\sum_{r_n\neq 0} P_{r_0\cdots r_{n-1}}\frac{\theta\sigma K(\theta)}{p-1}n^{\sigma-2}\log\left(P_{r_0\cdots r_{n-1}}\frac{\theta\sigma K(\theta)}{p-1}n^{\sigma-2}\right). \end{aligned}$$

Since $\sigma \in [\log_2 3, 2)$, $\lim_{n\to\infty} n^{\sigma-2} = 0$,

$$\begin{aligned} \lim_{n\to\infty}\triangle FH &= \lim_{n\to\infty}(FH_n - FH_{n-1}) \\ &= -\lim_{n\to\infty}\sum_{r_0\cdots r_{n-1}}\sum_{r_n\neq 0}\frac{\theta\sigma K(\theta)}{p-1}n^{\sigma-2}P_{r_0\cdots r_{n-1}} \\ &\qquad\cdot\log\left(P_{r_0\cdots r_{n-1}}\frac{\theta\sigma K(\theta)}{p-1}n^{\sigma-2}\right) \\ &= -(\sigma-2)\frac{\theta\sigma K(\theta)}{p-1}n^{\sigma-2}\log(n)\sum_{r_0\cdots r_{n-1}}\sum_{r_n\neq 0}P_{r_0\cdots r_{n-1}}. \end{aligned}$$

Finally,

$$\begin{aligned} \lim_{t\to\infty} FH &= \lim_{t\to\infty}\int^t d(FH) \\ &\sim t^{\sigma-1}\log t. \end{aligned}$$

In conclusion, this study gives some results to investigate the p-adic functional delay equation. The practical applications may be known to the problems in telecomunication, computer logic design, neural network, chemical reaction, biological immune system, quantum groups as well [10],[11],[12].

References

1. Mullhaupt A.P., "Boolean Delay Equations, Debruijn Oscillations and Intermittency", Phys. Lett. A, Vol.124, No.3, (1987), pp.151-158.

2. Ghil M. and Mullhaupt A.P., "Boolean Delay Equation, Periodic and Aperiodic Solutions", Joul. Stat. Phys., Vol.42, (1985), pp.125-173.

3. Schikhof W.H., "Ultrametric Calculus, An Introduction to p-adic Analysis", Camb. Studies in Adv. Math. 4, Camb. Univ. Press, (1984).

4. Kleene S.C., "Introduction to Mathematics", D. Van Nostrand Comp., INC., New York, (1962).

5. Soare R.I., "Recursively Enumerable Sets and Degrees", Springer-Verlag, (1987).

6. Sacks G.E., " Degrees of Unsolvability ", Ann. of Math. Studies, No.55, Princeton Univ. Press, (1966).

7. Wolfram S., "Undecidability and Interactibility in Theoretical Physics", Phys. Rev. Lett., Vol.54, No.8, (1985), pp.735-738.

8. Urias J., "One Dimensional Cellular Automata as Arithmetic Recursions", Physica D, Vol.36, (1989), pp.109-110.

9. Chin C.H., "Invariant in Chaos and Fractal", Mathmedia, Vol.15, No.3, (1991), pp.16-21.

10. Gutowitz H., " Cellular Automata : Theory and Experiment ", Physica D, Vol.45, No.1-3, (1990).

11. Biedenharn L.C. and Gould M., "The Pattern Calculus for q Tensor Operators in the q-Group U_q (u(n))", Preprint, (1992).

12. Chin C.H., "Order, Hierarchy and Logic, Truth", Mathmedia, (1992).

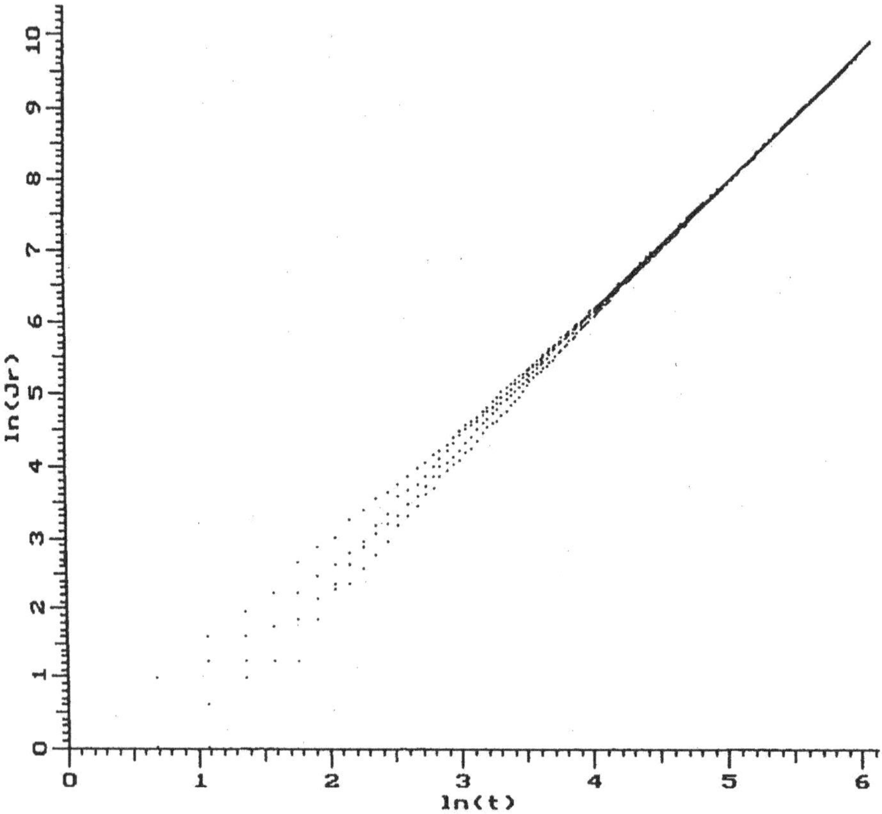

Figure Log scheme of the r-level-jump function in Z_5.

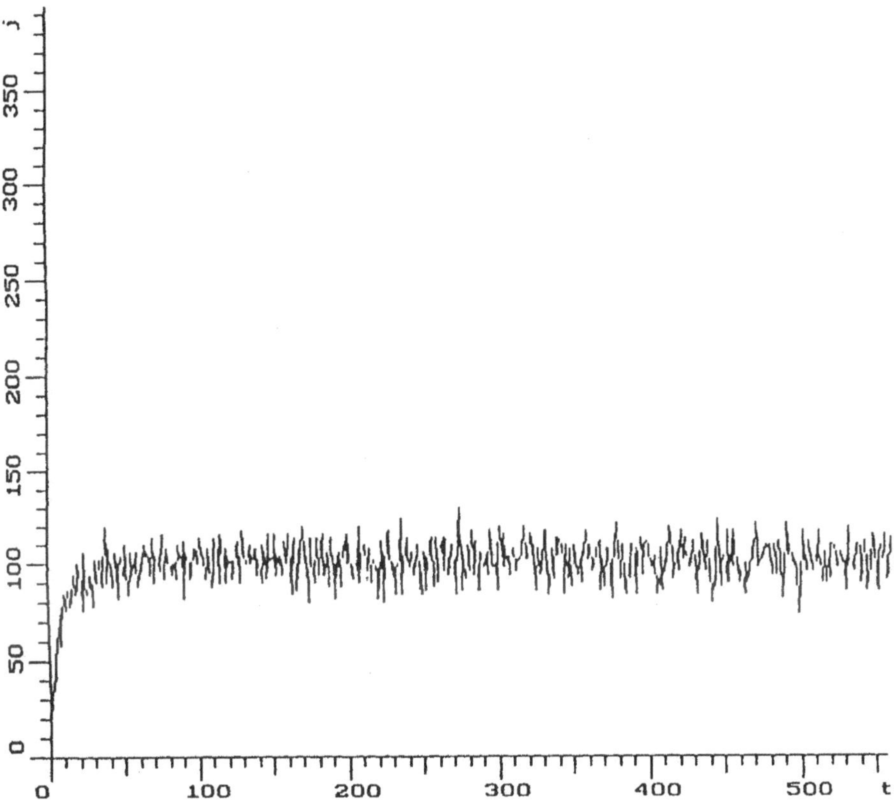

Figure For rational $\theta = 0.11$, the jump density function oscillates but not grows for large t.

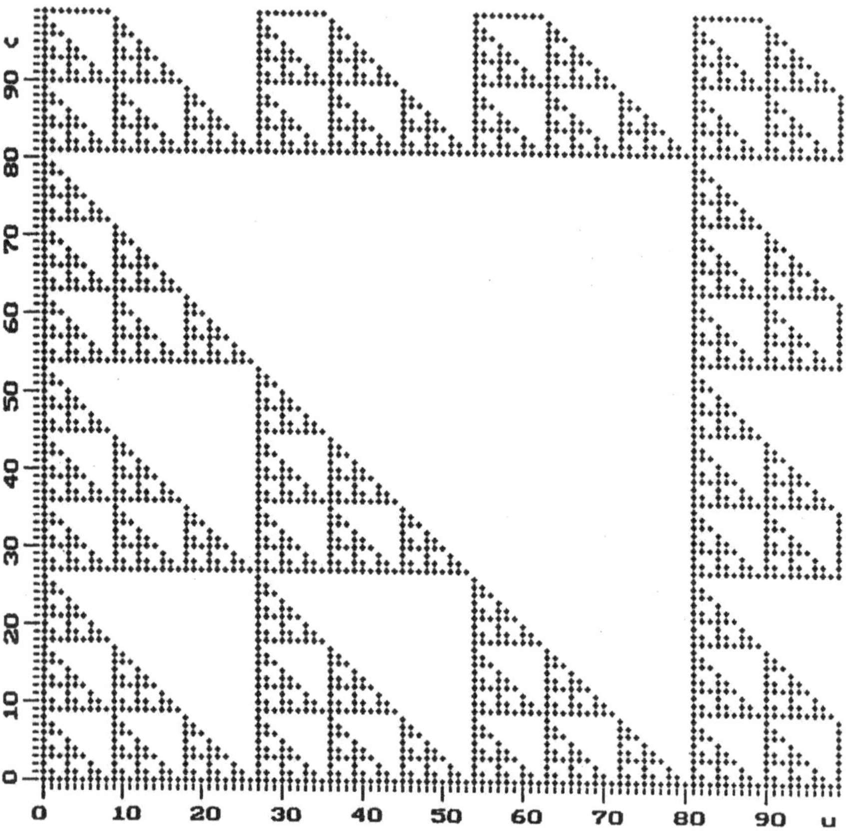

Figure 3. The $\binom{\mu+\nu}{\mu}_3$ matrix.

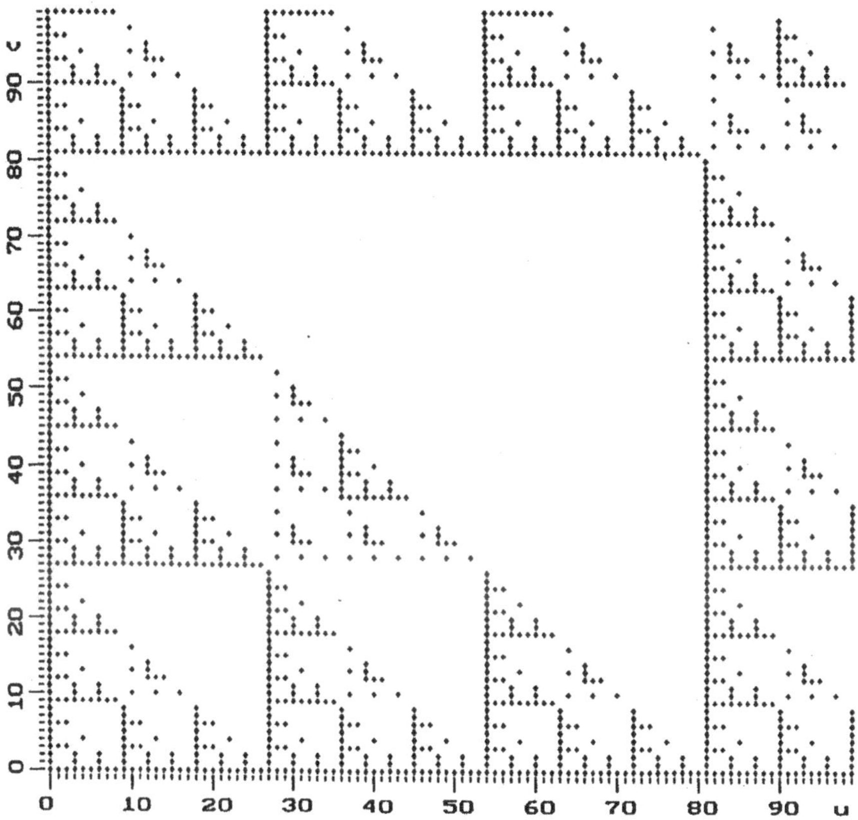

Figure The $\binom{\mu+\nu}{\mu}_3^1$ matrix.

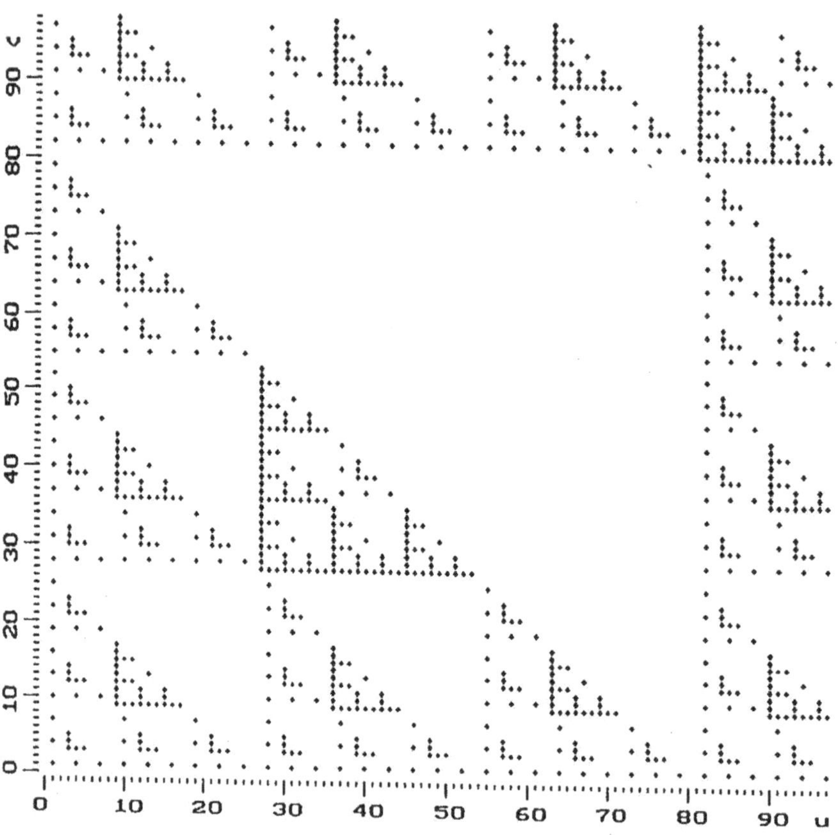

Figure The $\binom{\mu+\nu}{\mu}_3^2$ matrix.

Anyons, Non-Integrable Phases, and Braid Group on a Torus

Choon-Lin Ho
Department of Physics, Tamkang University, Tamsui, Taiwan 25137, R.O.C.
and
Yutaka Hosotani
School of Physics and Astronomy, University of Minnesota, Minneapolis, MN 55455, U.S.A.

Abstract

We show how the braid group structures are realized in a Chern-Simons theory defined on a torus. Emphasis is placed on the important rôle of the dynamics of the non-integrable phases of the two Wilson line integrals.

It is realized recently that the theory of anyons, particles obeying fractional statistics in $1+2$ dimensions, could be of relevance to condensed matter systems, such as the fractional quantum Hall effects and the high temperature superconductivity [1]. One elegant way to implement fractional statistics is by introducing a Chern-Simons interaction among matter field. It has been demonstrated that when the coefficient of the Chen-Simons term is quantized in such a way that the anyon has statistics phase parameter $\theta_s = \pi/N$ for integral N, the system of anyon gas becomes superconducting.

There is no theoretical reason as to why anyon should assume the special quantized values of the Chern-Simons term on a flat two-dimensional space. On the other hand, this arises very naturally when the space is a Riemann surface of genus $g \geq 1$ together with the requirement of manifest invariance under large gauge transformations [2].

It turns out that anyon systems on such Riemann surfaces possess many interesting features not shared by the coresponding planar systems. The analysis [3] based on representation theory of braid group, which is essential in formulating quantum mechanics of anyons, has revealed that the global topology of the Riemann surfaces gives constraints on the possible values of θ_s, the number of anyons, and the number of components of the wavefunctions. For a Riemann surface of $g \geq 1$, it is shown that fractional statistics is consistent only with multi-component wavefunctions.

In this talk we shall show how the general results from braid group representation theory manisfest themselves in a Chern-Simons field theory defined on a torus, based on work presented in [4].

We consider a Chern-Simons theory with non-relativistic electron field given by the Lagrangian

$$\begin{aligned} \mathcal{L} = \frac{N}{4\pi}\, \varepsilon^{\mu\nu\rho} a_\mu \partial_\nu a_\rho + \frac{i}{2}\left(\psi^\dagger D_0 \psi - (D_0\psi)^\dagger \psi\right) - \frac{1}{2m}|D_k\psi|^2 \ , \\ D_0 = \partial_0 + i a_0 \quad , \quad D_k = \partial_k - i a^k \end{aligned} \tag{1}$$

on a torus T^2 with funamental domain defined by $0 \leq x_j \leq L_j, j = 1,2$. Boundary conditions of the fields are given by

$$\begin{aligned} a_\mu[T_j(x)] &= a_\mu[x] + \partial_\mu \beta_j(x) \ , \\ \psi[T_j(x)] &= e^{-i\beta_j(x)}\ \psi(x) \ , \\ T_j : x_j &\rightarrow x_j + L_j \quad , \quad j = 1,2. \end{aligned} \tag{2}$$

That is, the fields return to their original values up to gauge transformations after translations along the non-contractible loops. The two operators T_1 and T_2 need only to commute with each other up to gauge transformation. Furthermore, the requirement of smoothness of the field operator $\psi(x)$ in its covering space, $\psi[T_1 \cdot T_2(x)] = \psi[T_2 \cdot T_1(x)]$, leads to quantisation of the Chern-Simons flux, $\Phi = \int dx\ f_{12} = 2\pi m$ (m : integers), and a constraint on the β_j's:

$$\left\{\beta_1(T_2 x) - \beta_1(x)\right\} - \left\{\beta_2(T_1 x) - \beta_2(x)\right\} = -2\pi m. \tag{3}$$

Typical β_j's which solve (3) are $\beta_j(x) = -\epsilon^{jk}\pi m x_k/L_k$, which will be taken in the rest of the talk. The flux Φ is related , through one of the Chern-Simons equation of motions to the total charge $Q = \int dx\ j^0$ ($j^0 = \psi^\dagger\psi$) by

$$\Phi + \frac{2\pi}{N}\, Q = 2\pi\left(m + \frac{Q}{N}\right) = 0. \tag{4}$$

This becomes a constraint condition on the physical states when the theory is quantized. Hence the eigenvalues of Q are constrained by $q = -Nm$. q is also the number of particles living on the torus. As we shall see below, requirement of large gauge invariance leads to integral values of N.

The Chern-Simons equations of motion can be solved in the $\nabla \cdot \mathbf{a} = 0$ gauge. The solution of the spatial components of a_ν have the form $a^j(x) = \frac{\theta_j(t)}{L_j} + \frac{\pi q}{NL_1L_2}\epsilon^{jk}x_k + \hat{a}^j(x)$, where $\hat{a}^j(x)$ is expressible in terms of the matter field operators only. The zero modes of $a^j(x)$, θ_j, are the non-integrable phases of the Wilson line integrals on the torus. They cannot be eliminated by residual gauge transformations, and play an important rôle in giving rise to the braid group structures of the systems.

In the quantum theory, the physical degrees of freedom are the matter fields ψ and $\psi^\dagger$ (taken to be fermionic), and the non-integrable phases θ_j's. θ_1 and θ_2 are canonical conjugate of each other and satisfy $[\theta_1, \theta_2] = 2\pi i/N$. There are two sets of important operators in the theory, namely, the Wilson line operators $W_j = e^{i\theta_j}$, and the generators of large gauge transformation U_i. In the vacuum sector U_i can be expressed in terms of θ_i: $U_i = e^{i\epsilon^{ik}N\theta_k}$. U_i and W_j satisfy

$$\begin{aligned} W_1W_2 &= e^{-\frac{2\pi i}{N}}W_2W_1 \ , \\ U_1U_2 &= e^{-2\pi i N}U_2U_1 \ , \\ [W_k, U_j] &= 0. \end{aligned} \tag{5}$$

If we demand that the system be invariant under large gauge transformations, then U_1 and U_2 must commute with each other. This implies that N must be an integer. This leads to the special statistics parameter $\theta_s = \pi/N$ mentioned previously. It also follows that the vacua are N-fold degenerate. We shall denote the a-th vacuum by $|0_a\rangle$ $(a = 0, \ldots, N-1)$. Actions of W_j on $|0_a\rangle$ are:

$$\begin{aligned} W_1^{\pm 1}|0_a\rangle &= e^{\pm i(\alpha_1 + 2\pi a)/N}|0_a\rangle \ , \\ W_2^{\pm 1}|0_a\rangle &= e^{\pm i\alpha_2/N}|0_a\rangle \ , \end{aligned} \tag{6}$$

where α_i are the vacuum angles.

We can now define the first quantized many-body Schrödinger wavefunction from the field operators. The wavefunction is required to be invariant under large transformation. Since the vacuum is N-fold degenerate, the wavefunction must have N components. One candidate is

$$\begin{aligned} \phi_a(\mathbf{x}_1, \ldots, \mathbf{x}_q; t) = \frac{1}{\sqrt{q!}}\langle O_a|\Omega\psi(x_1)\ldots\psi(x_q)|\Psi_q\rangle \ , \\ (a = 0, \ldots, N-1) \ , \end{aligned} \tag{7}$$

where $\Omega \equiv exp\left\{-i\sum_{j=1}^{q}\left(\frac{x_1^{(j)}}{L_1}\theta_1+\frac{x_2^{(j)}}{L_2}\theta_2\right)\right\}$ and $|\Psi_q\rangle$ is the q-particle physical state. Gauge invariance requires the presence of θ_i in the definition of the Schrödinger wavefunction. With the aid of eq.(6), it can be shown that ϕ_a transform among themselves when the particles are translated along the two contractible loops of the torus in a way that realizes the braid group representation.

Acknowledgment

This work is supported in part by the R.O.C. Grant NSC 81-0208-M032-502 (C.-L.H.) and by the U.S. Department of Energy under Contract No. DE-AC02-83ER-40105 (Y.H.).

References

1. For a review, see eg. E. Fradkin, *Field Theories of Condensed Matter Systems*, Addison-Wesley, 1991.

2. Y. Hosotani, *Phys. Rev. Lett.* **62** (1989) 2785; *ibid.* **64** (1990) 1691.

3. T. Einarsson, *Phys. Rev. Lett.* **64** (1990) 1995; *Mod. Phys. Lett.* **B5** (1991) 675; T.D. Imbo and J. March-Russell, *Phys. Lett.* **252B** (1990) 84; Y.S. Wu, Y. Hatsugai and M. Kohmoto, *Phys. Rev. Lett.* **66** (1991) 659; Y. Hatsugai, M. Kohmoto and Y.S. Wu, *Phys. Rev.* **B43** (1991) 2661, 10761.

4. C.-L. Ho and Y. Hosotani, "*Anyon Equation on a Torus*", Minnesota preprint UMN-TH-935/91, *Int. J. Mod. Phys.* **A** (in press).

QUANTUM INVERSE SCATTERING METHOD AND SEPARATION OF VARIABLES

Igor V. KOMAROV

St. Petersburg University, Institute of Physics
St. Petersburg, Russia 198904

ABSTRACT

Status of multiplicative variable separation in the QISM is reviewed.

The Quantum Inverse Scattering Method (QISM) gives the most general approach to the theory of integrable systems[1]. In classical mechanics Liouville's definition of integrability claims the number of degrees of freedom N should be equal to number of independent constants of motion I_j commuting with respect to the Poisson brackets. Then the Liouville theorem proves the existence of invariant tori. This also means that at fixed values of integrals of motion there is an additive separation of variables for the action

$$S = \Sigma_{i=1}^{N} S_i(y_i, \{I\}) \tag{1}$$

where each item depends on one variables y_i. For Abelian set $\{I_j\}$ the Liouville theorem gives no regular way to construct $\{y_i\}$. One should study I_j and equations of motion together to get the Eq.(1). In quantum mechanics it is natural to apply Liouville's definition of integrability with minimal changes. Instead of the Poisson brackets one can use commutators and by functional independence of quantum integrals of motion one can mean the independence of their classical limits. As a quantum counterpart of the Eq.(1) the common eigenfunctions of I_j can be represented in a factorized form

$$\Psi = \Pi_{i=1}^{N} \varphi^{(i)}(y_i) \tag{2}$$

In classical limit it should be equivalent to semiclassical quantization of Eq.(1).

There are many examples of constructing the representation (1). The general ideas follow from the Classical Inverse Scattering Method, where both equations and integrals of motion can be find with the help of the Lax matrix $L(u)$ depending on a free complex parameter u. The motion is constrained by the algebraic curve $\Gamma(u,\mu;\{I_j\}) = \{\det(L(u)-\mu I)\}=0\}$. The Baker-Akhiezer function ψ satisfying to $L(u)\psi=\mu\psi$, $\psi_t=A(u)\psi$ is analytic on $\Gamma(u,\mu)$. For hyperelliptic curve ψ can be expressed in terms of the Riemann θ - functions. The separation variables in the case are roots of the d-th component [a]

$$\psi_d(u)=0 \qquad u \Rightarrow \{y_i\}_1^N \tag{4}$$

and this gives rise the representation (1).

Similar analysis can be done in the QISM [2]. In standard notations (d=2) the basic quadratic algebra holds

$$R(u-v)\overset{1}{T}(u)\overset{2}{T}(v) = \overset{2}{T}(v)\overset{1}{T}(u)R(u-v)$$

where R is a solution of the Yang-Baxter equation. For the XXX model $R(u)=u+\eta P$, where η is the Plank constant and P is a permutation in $C^2 \oplus C^2$. The quantum determinant $\Delta(u)$ generates the center of the QISM algebra $[\Delta(u),T(v)]=0$. Commuting integrals of motion are extracted from $t(u)=\mathrm{tr}T(u)$. The common eigenfunction Ψ of integrals of motion

$$t(u)\,\Psi = \tau(u)\,\Psi \tag{5}$$

does not depend on u . The standard way to find Ψ is given by the Algebraic Bethe Ansatz (ABA). There is an independent on u pseudovacuum state $|0\rangle$, that is killed by C(u)

$$C(u)\,|0\rangle = 0$$

Then on $|0\rangle$ T(u) becomes triangular

$$T(u)|0\rangle = \begin{pmatrix} \Delta_-(u) & * \\ 0 & \Delta_+(u) \end{pmatrix} |0\rangle$$

where $\Delta_\pm(u)$ are related to the quantum determinant

$$\Delta_+(u+\eta/2)\,\Delta_-(u-\eta/2) = \Delta(u)$$

For the Mth excited state eigenfunction reads $|M\rangle = B(u_1)B(u_2)\,.....B(u_M)|0\rangle$, if rapidities u_α satisfy to the algebraic Bethe equations

$$\frac{\Delta_-(u_\alpha)}{\Delta_+(u_\alpha)} = \prod_{\beta=1}^{M}\left[\frac{u_\alpha-u_\beta-\eta}{u_\alpha-u_\beta+\eta}\right] \qquad \alpha=1,\,.....M \tag{6}$$

The ABA often provides an effective solution to the spectral problem Eq.(5), but *i*) its connection to classical solution Eq.(3) is not obvious, *ii*)there is only one index M labeling Ψ, *iii*) when M is large we need a lot of computer time to solve Eq.(6) and for other M' to repeat all again from the very beginning, *iiii*) sometimes there is no vacuum state $|0\rangle$ as for the periodic Toda chain and the Goryachev-Chaplygin top. This ambiguities are clarified by the Functional Bethe Ansatz (FBA) that provides the separation of variables in the QISM [2].The basic algebra for $T(u)=\sum_{k=1}^{N} u^{k-1}\begin{pmatrix} A_k B_k \\ C_k D_k \end{pmatrix}$ implies $[C(u),C(v)]=0$. Instead of C_k a new set of operator variables is defined by

$$C(u)=0 \qquad \Rightarrow \qquad u=\{y_i\}_1^N \tag{7}$$

that is a quantum analog of zeros of the Baker-Akhiezer function Eq.(4). Instead of A_k , D_k one introduces

$$Y_i^- = A(u = y_i) = \sum_{k=1}^{N} y_i^{k-1} A_k \ , \qquad Y_i^+ = D(u = y_i) = \sum_{k=1}^{N} y_i^{k-1} D_k$$

where the left ordering of y_i and A_k, D_k is fixed.

Theorem (Sklyanin). The operators y_i and $Y_i^{\pm}$ satisfy the relations

$$[\, y_i \, , \, y_k \,] = [\, Y_i^{\pm} \, , \, Y_k^{\pm} \,] = 0, \quad [\, Y_i^{\pm}, \, y_k \,] = \pm \, \eta \, \delta_{ik} Y^{\pm}{}_k$$

$$Y_i^{\pm} \, Y_i^{\mp} = \Delta \, (y_i \pm \eta/2) \tag{8}$$

Matrix elements of T(u) are determined by the quantum analog of the Lagrange interpolation

$$\begin{aligned} C(u) &= \Pi_{i=1}^{N} (u - y_i) \, C_N \\ A(u) &= C(u) \, A_N/C_N + \Sigma_{i=1}^{N} P_i(u) \, Y_i^- \\ D(u) &= C(u) \, D_N/C_N + \Sigma_{i=1}^{N} P_i(u) \, Y_i^+ \\ B(u) &= C^{-1}(u+\eta) \, (D(u)A(u+\eta) - \Delta(u+\eta/2)) \end{aligned} \tag{9}$$

Here A_N , C_N , D_N are c-numbers and $P_i(u) = \Sigma_{k=1}^{N} (u-y_k)/(y_i-y_k)$ is the Lagrange projector $P_i(y_k) = \delta_{ik}$. Thus we changed variables in T(u) to $\{y_i \ , \ Y_i^{\pm}\}$.

We will treat the spectral problem Eq.(5) in the y_i representation . It is supposed that the spectrum $\mathcal{Y}^{(i)}$ of each y_i is known, thus $\Psi \Rightarrow \Psi(y)$ .Action t(u) on $\Psi(u)$ follows from Eqs.(8),(9). Taking Ψ in multiplicative form Eq.(2) we arrived at a set of onedimensional spectral problems with a common set of eigenvalues $\{I\}$

$$\begin{aligned} &\Delta_+(y_i)\varphi^{(i)}(y_i-\eta) + \Delta_-(y_i)\varphi^{(i)}(y_i+\eta) = \tau(y_i, \{I_j\})\varphi^{(i)}(y_i) \\ &\Delta_+(y_i+\eta/2)\Delta_-(y_i-\eta/2) = \Delta(y_i) \quad ,i=1,\ldots N, \ \ y_i \in \mathcal{Y}^{(i)} \end{aligned} \tag{10}$$

Eqs.(10) at $y_i \Rightarrow u$ coincide to the Baxter equation [1] . The latter is a functional equation with uncertain class of solution and the ABA is equivalent to seeking the solution in the polynomial form $\varphi(u) = \Pi(u-u_\alpha)$, u_α being its zeros. The former is the set of the second order finite difference linear equations on Y_i with the step η, or the system of spectral problems for tridiagonal matrices

$$X^{(i)}(\eta, \{I_j\})\, \Phi^{(i)} = 0 \qquad i=1,\ldots N$$

i)Each function φ has a label α_i, thus Ψ is labeled by the multiindex $(\alpha_1, \alpha_2, \ldots \alpha_N)$. To compare with the ABA the excitation M is decomposed $M=\Sigma\, \alpha_i$. For spin 1 su(2) magnet $M = Nl - J$, J being total angular momentum. *ii*) Spectrum of integrals is a solution of algebraic system

$$\det X^{(i)}(\eta, \{I_j\}) = 0 \qquad i=1,\ldots N$$

This generalizes the Bethe Eqs(6) of the ABA . When y has finite spectrum , X is a finite matrix with the entries that are analytical functions of η . Therefore integrals are multivalued functions of complex η . There are branch points in complex η-plane walking around each we come from one state to another. Thus quantum numbers labeling sheets of the spectral surface have topological meaning. *iii*) Connection to classical mechanics can be readily seen. In separated equation $(Y^+ + Y^- - \tau(y))\varphi = 0$ after substitutions $Y^+ \triangleright w$, $Y^- \triangleright w^{-1}\Delta$, $y \triangleright \eta$ and multiplication by the w from the left we get $(w^2 - w\tau(y) + \Delta(u))\varphi = 0$, where the operator coincides to the classical algebraic curve.

This approach was applied to the Goryachev-Chaplygin and other tops , the periodic Toda chain, The XXX magnet and the generalized polyspheroidal functions, connected to reflection equation [see Ref.in [2]]. Recently Sklyanin[4] extended this approach to the classical sl(3) magnet and Kuznetsov[5] has shown that all results on variables separation for the Laplace operator in Riemannian spaces of constant curvature follow from this scheme.

REFERENCES

1. L.D.Faddeev, in *Nankai Lectures on Mathematical Physics, Integrable Systems*, ed.X.C.Song, (World Scientific, Singapore 1990)pp.23-70.
2. E.K.Sklyanin, *Quantum Inverse Scattering Method. Selected topics*, (Preprint HU-TFT-91-51,Helsinki 1991).
3. S.P.Novikov,S.V.Manakov,L.P.Pitaevski,V.E.Zakharov, *Theory of solitons - Inverse Scattering Method*, (Plenum, New York, 1984).
4. E.K.Sklyanin, *Separation of variables in the classical integrable SL(3) magnetic chain*, (Preprint RIMS-871,Kyoto, 1992).
5. V.B.Kuznetsov, *J.Math.Phys.* 32 (1992) August

Rogers Dilogarithm in Integrable Systems

ATSUO KUNIBA
Department of Mathematics, Kyushu University
Fukuoka 812 JAPAN

and

TOMOKI NAKANISHI
Lyman Laboratory of Physics, Harvard University
Cambridge, MA 02138 USA

ABSTRACT

We discuss some curious aspects of the Rogers dilogarithm appearing in integrable systems in two dimensions.

1. Introduction

This note is a brief exposition of the appearance of the Rogers dilogarithm function in relation to integrable lattice models and conformal field theory (CFT) in two dimensions. The content is mainly the known facts in refs.[1,2] but also includes a few new informations based on a collaboration with Junji Suzuki.

The Rogers dilogarithm is a function of a variable x defined by

$$L(x) = -\frac{1}{2}\int_0^x \left(\frac{log(1-y)}{y} + \frac{log y}{1-y}\right)dy \quad (0 \leq x \leq 1). \tag{1}$$

The following identity is know essentially due to refs.[3]

$$\frac{6}{\pi^2}\sum_{m=1}^{\ell} L\left(\frac{sin^2\frac{\pi}{\ell+2}}{sin^2\frac{\pi(m+1)}{\ell+2}}\right) = \frac{3\ell}{\ell+2}, \quad \text{for } \ell \in \mathbb{Z}_{\geq 1}. \tag{2}$$

In the above, the rhs is the well known value of the central charge for the level ℓ $A_1^{(1)}$ WZW model in conformal field theory. On the other hand, the argument of the dilogarithm has the form $(Q_m)^{-2}$, where Q_m is the $m+1$ dimensional irreducible A_1 character specialized to some "rational point". Thus Eq. 2 is connecting the two important quantities in the A_1-related theory, i.e., the central charge and the specialized character.

In fact, there is a conjectural generalization of the identity Eq. 2 into arbitrary classical simple Lie algebra X_r[4,1].

$$\frac{6}{\pi^2}\sum_{a=1}^{r}\sum_{m=1}^{t_a \ell} L(f_m^{(a)}) = \frac{\ell \dim X_r}{\ell + g}, \qquad \text{for } \ell \in \mathbb{Z}_{\geq 1}. \tag{3}$$

Here, g denotes the dual Coxeter number and t_a is the integer defined as the ratio of the a-th Kac and dual Kac label. The argument $0 \leq f_m^{(a)} \leq 1$ arises through thermodynamic Bethe ansatz (TBA) analysis. Here we shall give its definition only for $X_r = A_r$. (See refs.[1,4,5] for the general case.)

$$f_m^{(a)} = f_m^{(a)}(z=0), \qquad f_m^{(a)}(z) = 1 - \frac{Q_{m+1}^{(a)}(z)Q_{m-1}^{(a)}(z)}{Q_m^{(a)}(z)^2}, \tag{4}$$

where $Q_m^{(a)}(z)$ is the irreducible A_r character with the highest weight $m\Lambda_a$. ($z \in$ dual space of the Cartan subalgebra and Λ_a is the a-th fundamental weight. Nodes on the Dynkin diagram are enumerated according to ref.[1].) In general, the quantity $Q_m^{(a)}(z)$ is a Yangian character and we adopt the z dependence as given in ref.[2]. The conjecture (4) was firstly systematically used in the TBA analysis in refs.[6,1].

2. Scaling Dimensions from Dilogarithm

There is a generalization[2] of Eq. 3 so as to include the parafermion scaling dimensions (modulo integer) in CFT $\Delta_\lambda^\Lambda = \frac{(\Lambda|\Lambda+2\rho)}{2(\ell+g)} - \frac{|\lambda|^2}{2g}$, where Λ is a level ℓ dominant integral weight of $X_r^{(1)}$ and $\lambda \in \Lambda +$ root lattice. This is achieved by considering the specialization $f_m^{(a)}(z=\Lambda)$ instead of the principal one in Eq. 4. Now that the $f_m^{(a)}(\Lambda)$ is a complex number in general, one can consider various analytic continuations $L_{a,m}(x)$ of $L(x)$. Leaving all the technical points[2], we have a conjecture

$$\frac{6}{\pi^2}\sum_{a=1}^{r}\sum_{m=1}^{t_a \ell - 1} L_{a,m}(f_m^{(a)}(\Lambda)) + \text{Logarithmic terms} = \frac{\ell \dim X_r}{\ell+g} - r - 24(\Delta_\lambda^\Lambda + \text{integer}). \tag{5}$$

Here, λ depends on the integration contour along which the $L(x)$ is analytically continued. Its explicit form and the logarithmic terms can be found in ref.[2]. Sec also refs. [7,8,9,10] for some physical aspects. Here we shall only present a conjecture on the value $Q_{t_a \ell}^{(a)}(\Lambda)$, under which the congruence $\lambda \equiv \Lambda$ (modulo root lattice) can be verified directly. Put $k =$ number of the Kac labels a_i $(0 \leq i \leq r)$ equal to 1. Then we conjecture that $Q_{t_a \ell}^{(a)}(\Lambda)$ is a k-th root of unity as

$$\begin{aligned} Q_{t_a \ell}^{(a)}(\Lambda) &= \exp(-2\pi i c(\Lambda)\overline{\gamma}_a/k) \quad \text{for } X_r \neq D_r, \\ &= \exp(2\pi i c_2(\Lambda)\gamma_a^{(2)}/k) \quad \text{for } X_r = D_r,\ r = \text{odd}, \\ &= \exp(2\pi i((c_2(\Lambda) - rc_1(\Lambda))\gamma_a^{(1)} + c_1(\Lambda)\gamma_a^{(2)})/k) \quad \text{for } X_r = D_r,\ r = \text{even}. \end{aligned} \tag{6}$$

Here, for $\Lambda = \sum_{a=1}^{r} \mu_a \Lambda_a$, we have set $c(\Lambda) = \sum_a \gamma_a \mu_a \mod k$ for $X_r \neq D_r$ and $c_i(\Lambda) = \sum_a \gamma_a^{(i)} \mu_a \mod k/(3-i)$ for $X_r = D_r$, $i = 1, 2$ and γ is the rank-dimensional integer vector given by

$$\begin{aligned} &A_r, B_r, C_r, E_6 : \gamma_a = a, \\ &D_r : \gamma^{(1)} = (0, \ldots, 0, 1, 1), \quad \gamma^{(2)} = (2, 4, 6, \ldots, 2(r-2), r-2, r), \\ &E_7 : \gamma = (0, 0, 0, 1, 0, 1, 1), \\ &E_8, F_4, G_2 : \gamma = (0, \ldots, 0). \end{aligned} \tag{7}$$

Finally, $\overline{\gamma} = -\gamma$ if $X_r = A_r$ and $\overline{\gamma} = \gamma$ of the dual algebra of X_r if $X_r \neq A_r, D_r$. Eq. 6 is a special solution of $\prod_b Q_{t_b \ell}^{(b)}(\Lambda)^{C_{ab}} = 1$ where C is the Cartan matrix of X_r. From numerical tests it also seems valid that $Q_m^{(a)}(\Lambda) = Q_{t_a \ell}^{(a)}(\Lambda) Q_{t_a \ell - m}^{(a)}(\Lambda)^*$ for $-1 \leq m \leq t_a \ell + 1$ and $Q_{t_a \ell+1}^{(a)}(\Lambda) = 0$ for any level ℓ dominant integral weight Λ. These are interesting arithmetic properties of the specialized Yangian characters. We note especially that $Q_m^{(a)}(0)$ appears[1] as the high temperature limit of $log(entropy)$ per site in the TBA system connected to X_r. This implies that $Q_m^{(a)}(0)$ yields the largest eigenvalue of the incidence matrix for a fusion $X_r^{(1)}$ RSOS model.

3. Functional Relations

The Yangian character $Q_m^{(a)}(z)$ is known to satisfy interesting recursion relations[4,5]. For example in $X_r = A_r$ case,

$$Q_m^{(a)}(z)^2 = Q_{m+1}^{(a)}(z) Q_{m-1}^{(a)}(z) + Q_m^{(a+1)}(z) Q_m^{(a-1)}(z) \tag{8}$$

and similar relations are known for all the other algebras. Furthermore there is a "spectral parameter dependent version (or Yang-Baxterization)" of these relations. Below we shall describe it briefly for $X_r = A_r$. In ref.[6] Bazhanov and Reshetikhin wrote down a system of functional relations among the row to row transfer matrices for the fusion $A_r^{(1)}$ model[11].

$$T^{\xi,\eta}(u) = \det \left(T^{\xi,(\eta_i - i + j)\Lambda_1}(u + \eta_1 + i - \eta_i - 1) \right)_{1 \leq i,j \leq \eta'_1} \tag{9a}$$

$$= \det \left(T^{\xi,\Lambda_{\eta'_i - i + j}}(u + \eta_1 - j) \right)_{1 \leq i,j \leq \eta_1}, \tag{9b}$$

where ξ (resp. η) is the Young diagram representing the fusion type in the horizontal (resp. vertical) direction and the $T^{\xi,\eta}(u)$ is transferring the states into the vertical direction. $\eta' = [\eta'_1, \ldots, \eta'_{\eta_1}]$ is the transpose of $\eta = [\eta_1, \ldots, \eta_{\eta'_1}]$ and u is the spectral parameter entering the solution of the Yang-Baxter equation that underlies the model[11]. Eq. 9 is a "quantum analogue" of the 2nd Weyl character formula (Jacobi-Trudi's formula)[6]. (In fact Eq. 9 is different from that in ref.[6] and the alternation is based on a private communication with V.V. Bazhanov. Note that $T^{\xi,\eta}(u)$ may be regarded as representing an eigenvalue thanks to the commutatibity.) Now let η be the a by m rectangular Young diagram corresponding to $m\Lambda_a$ and write the $T^{\xi,\eta}(u)$ as $T^{(a)}_m(u)$. Then from Eq. 9 one can prove the functional relation

$$T^{(a)}_m(u)T^{(a)}_m(u+1) = T^{(a)}_{m+1}(u)T^{(a)}_{m-1}(u+1) + T^{(a+1)}_m(u)T^{(a-1)}_m(u+1), \tag{10}$$

which is a "Yang-Baxterization" of Eq. 8. From this we find that the combination

$$y^{(a)}_m(u + \frac{a+m}{2}) = \frac{T^{(a)}_{m+1}(u)T^{(a)}_{m-1}(u+1)}{T^{(a+1)}_m(u)T^{(a-1)}_m(u+1)} \tag{11}$$

solves essentially the following $U_q(A^{(1)}_r)$ functional relation proposed in ref.[2]

$$y^{(a)}_m(u+\frac{1}{2})y^{(a)}_m(u-\frac{1}{2}) = \frac{(1+y^{(a)}_{m+1}(u))(1+y^{(a)}_{m-1}(u))}{(1+y^{(a+1)}_m(u)^{-1})(1+y^{(a-1)}_m(u)^{-1})}. \tag{12}$$

As noted in ref.[2], the above equation is also satisfied by $e^{\epsilon^{(a)}_m(u)}$ ($\epsilon^{(a)}_m(u)$: pseudo energy) in the TBA at ∞ temperature. In this way, one can roughly say that the TBA equation has a solution in terms of "Yang-Baxterized" Yangian characters. It is interesting to note that Eq. 10 can be viewed as a simplest example of the Plücker relation under Eq. 9.

References

1. A.Kuniba, "Thermodynamics of the $U_q(X^{(1)}_r)$ Bethe ansatz system with q a root of unity" , to appear in Nucl.Phys.B.
2. A.Kuniba and T.Nakanishi, "Spectra in conformal field theories from the Rogers dilogarithm", ANU preprint (1992) SMS-042-92.
3. A.N.Kirillov and N.Yu.Reshetikhin, J.Phys.**A20** (1987) 1587.
4. A.N.Kirillov, Zap.Nauch.Semin.LOMI **164** (1987) 121.
5. A.N.Kirillov and N.Yu.Reshetikhin, Zap.Nauch.Semin.LOMI **160** (1987) 211.
6. V.V.Bazhanov and N.Yu.Reshetikhin, J.Phys.**A23** (1990) 1477.
7. M.J.Martins, Phys.Rev.Lett. **67** (1991) 419.
8. T.R.Klassen and E.Melzer, Nucl.Phys.**B370** (1992) 511.
9. P.Fendley, Nucl.Phys. **B374** (1992) 667.
10. A.Klümper and P.A.Pearce, Physica **A183** (1992) 304.
11. M.Jimbo, A.Kuniba, T.Miwa and M.Okado, Commun.Math.Phys. **119** (1988) 543.

Topological Ideas in Simple Quantum Systems

Li Hua Zhong

Physics Department and Advanced Research Centre

Zhongshan University, Guangzhou, China

510275

《Abstract》

Application of topological idea to physics has a long history. However, it is only recent years topological ideas in simple quantum systems attract much attention. What we mean by "Simple" quantum systems are systems of a small number of degrees of freedom, for example, particle motion,nuclear, atomic and molecular system which in contrast to the quantum field system need no description with field theories. The recent interest in applications of topological idea were initiated by the celebrated paper of M. V. Berry, [1], after whom a new concept of Berry's phase anholonomy were widely known. From physical point of view, this concept is important because it is based on experimentally verified facts [2] .From mathematical point of view, this is one example in which fibre bundle theoretic concepts are found in a physical system which can be prepared in ordinary laboratories. The following presentations are the works in which the present author was involved.In the present talk the relationships between geometry and wave functions of quantum systems are illustrated. How and why geometry of quantum state space is related to properties of the wave funcnion are shown by examples. Then by purely geometrical method the Berry's phases are calculated from Riemann connection on certain fibre fundle constructed from some physical models [3] .It turns out that non-abelian gauge potential of magetic monopole type are relevent to the phase anholonomy. The mangnetic monopole type fields do have physical effect to trigger transmutation from isospin degrees of freedom into spin degrees of freedom [4] . As is well known, the nonintegral phase factors can be used to define gauge fields.A non-integral phase factor of Berry's type can be meaningful for any path of evolution of the physical system and it gives measurable physical effects [5] . This concept of topological phase for non-cyclic processe was also verified experimentally quite recently [6] .

An interesting question is that, while topological structure (fibre bundle structure) of a single magnetic monopole potential was well-known, is there any topology associated with the simplest system, a single electron? The question is explored. An idea that the spin of electron would possibly be • the manifestation of the topological structure of certain space associatcd with the point electron is conjecture. This idea stemedfrom the recent contribution of Polyakov [7] ,who showed that when a scalar particle is coupled to Chern-Simns fields the whole system behaves as a spin 1/2 particle. The dynamics in this system would be a scalar particle moving in a magnetic monopole field. This mechanism can be provided by some kind of topological structure descri bed in [3] [4] of an internal space associated with the electron.

[References] :

[1] M. V. Berry proc. Roy. Soc. (London) . A 392, (1984) 45.
B. Simon Phys. Rev. Lett. 51. (1983) 2167.

[2] Sec, for exrmple, in "Geomefrical phases in physics" , ed by A. Shapere and F. Wilczek. 1989. World Scientific Pub.

[3] H. Z. Li Phys Rev. Lett. 58, (1987) 539.

[4] H. Z. Li Phys Rev. D35, (1987) 2615.

[5] Y. S. Wu and H. Z. Li Phys. Rev. B 38, (1988) 11907.

[6] H. Weinfurter and G, Badurek phys. Rev. Lett. 64, (1990) 1318.

[7] A. M. Polyakov. Mod. Phys. Lett. A3, (1988) 325.

PERTURBATIVE GRAND PARTITION FUNCTION FOR BRAID STATISTICS

Stéphane OUVRY [1]
Division de Physique Théorique [2], IPN, Orsay Fr-91406

Abstract

We describe a perturbative computation of the grand partition function of the non relativistic abelian anyon gas. The anyons are confined in a weak harmonic potential in place of the usual finite box or periodic boundary conditions. Close to Bose statistics, standard perturbation theory is ill defined. Removing the diagonal of the configuration space, -which allows for braid group statistics-, yields a perturbative problem equivalent to the original one, and now correctly defined. At first order, only the second virial coefficient gets corrected close to Bose statistics. At second order the complexity of the model appears.

In [1] $2+1$ dimensions intermediate (or fractionnal) statistics which interpolate between Bose and Fermi statistics are available. The quantum wave function of two identical particles take a phase $\exp(i\pi\alpha)$ when their positions are interchanged. This nontrivial phase appears as a unitary one-dimensional representation of the abelian braid group [2], which is the first homotopy group of the configuration space of a system of identical particles (defined as the product of each individual configuration space R^2 quotiented by the permutation group since all particles are identical). A crucial point is the removal of the diagonal of the configuration space in order to accomodate braid group statistics. (The diagonal is defined has any configuration such that two particles are located at the same point.)

The anyon model [1] is a quantum mechanical realization of the above nontrivial braiding phase. One has the physical representation of particles carrying a charge e and a flux tube $\phi = e/\kappa$ where κ is the Chern- Simons coupling constant of the underlying topological dynamics [3], and interacting à la Aharonov- Bohm. The N-anyon Hamiltonian reads

$$H = \frac{1}{2m}\sum_{i=1}^{N}(\vec{p}_i - \alpha\vec{A}_i)^2 \tag{1}$$

where the statistical gauge field $\vec{A}_i = \sum_{j\neq i}\frac{\vec{k}\times\vec{r}_{ij}}{r_{ij}^2}$ ($\vec{k}$ is the unit vector perpendicular to the plane, $\alpha = e\phi/\pi$). In the Aharonov- Bohm effect, the charge and the flux are electromagnetical; in the anyon model they are statistical. The interaction is purely quantal, as it should to describe non trivial statistics.

Up to now, the particles are either bosons or fermions. A singular gauge transformation $\vec{A}_i = \sum_{j\neq i}\frac{\vec{k}\times\vec{r}_{ij}}{r_{ij}^2} = \vec{\partial}_i\sum_{k\leq l}\theta_{kl}$ erases the flux tubes, yielding a free Hamiltonian with a punctured plane, the essence of braid statistics. Indeed, the phase factor $\exp(i\alpha\sum_{k\leq l}\theta_{kl})$ is responsible for the nontrivial braiding phases.

---

[1] *and LPTPE, Tour 16, Université Paris 6 / electronic e-mail: OUVRY@FRCPN11*

[2] *Unité de Recherche des Universités Paris 11 et Paris 6 associée au CNRS*

One is interested by the equation of state of the anyon model. First of all, the system has to be regularized by confining the particles, and only then the thermodynamic limit should be taken. We have proposed [4] to confine the anyons in a harmonic potential, instead of the usual box or periodic boundary conditions. In the 2-anyon case, adding a harmonic regulator to the system, one gets [1] a discrete spectrum $E_{nm} = (2n + |m - \alpha| + 1)\omega$. One notes a linear interpolation between the Bose and Fermi spectra; when the relative orbital angular momentum $m = 0$, the spectrum is nonanalytic in α. Important efforts have been devoted to the study of the N-anyon spectrum [5], but little is known about it. One finds linear states that generalize the 2-anyon states. But contrary to the latter, they are only a part of the spectrum.

Since a complete knowledge of the N-anyon spectrum seems out of reach, we have proposed to perturb the Bose or Fermi systems with a small anyonic interaction. The fact that such a perturbative analysis is at all possible [6] is not obvious. Indeed let us compute at first order the 2-anyon spectrum in a harmonic well. The Hamiltonian has two terms, one linear in α, and one quadratic, α^2/r^2. The linear perturbation yields finite energyshift. One has to check that the quadratic corrections $\alpha^2 < \psi^0_{nm}(r,\theta)|1/r^2|\psi^0_{nm}(r,\theta) >$ are properly defined. The zeroth order wave function $\psi^0_{nm}(r,\theta) \to r^{|m|}$ as $r \to 0$. Thus for $m = 0$ a logarithmic divergence shows up due to the singular nature of the interaction. The $m = 0$ wave function does not vanish when the two particles are at the same point, a characteristic of Bose statistics.

There is a way to circumvent this perturbative obstruction. Let us define $\psi(r,\theta) = r^{|\alpha|}\tilde{\psi}(r,\theta)$. Then $\tilde{\psi}(r,\theta)$ satisfies an eigenvalue equation without the divergent α^2/r^2 singularity. Note that the above redefinition imposes to the anyonic wavefunction to vanish when $r \to 0$, which is nothing but excluding the diagonal of the configuration space. Similarly in the N-anyon case, the singular nature of the α^2/r^2_{ij} interactions requires that the N-anyon wave function $\psi(\vec{r}_1, \cdots, \vec{r}_N)$ be rewritten as

$$\psi(\vec{r}_1, \cdots, \vec{r}_N) = \prod_{i<j} r_{ij}^{|\alpha|}\tilde{\psi}(\vec{r}_1, \cdots, \vec{r}_N) \tag{2}$$

The singular terms disappear in the new Hamiltonian acting on $\tilde{\psi}$

$$\tilde{H} = \sum_{i=1}^{N}\left(\frac{\vec{p}_i^2}{2m} + \frac{i\alpha}{m}\sum_{j\neq i}\frac{\vec{k}\times\vec{r}_{ij}}{r^2_{ij}}\vec{\partial}_i - \frac{|\alpha|}{m}\sum_{j\neq i}\frac{\vec{r}_{ij}}{r^2_{ij}}\vec{\partial}_i\right) \tag{3}$$

In (2) nonperturbative effects in $|\alpha|$ are encoded by hand in the anyonic wave function. As a bonus, the interaction in (3) is a sum of two-body interactions.

One is now in position to address the perturbative expansion [6,7] of the equation of state. We drop the "tilde" symbol to simplify the notations. One wants to compute the thermodynamical potential $\Omega = -\beta PV = -\ln Tr e^{-\beta(\mathcal{H}-\mu\mathcal{N})}$ where $\mathcal{H}$ and $\mathcal{N}$ stand respectively for the second quantized anyon Hamiltonian and number operator of particles, and μ is the chemical potential. More precisely one looks for a perturbative expansion of the cluster coefficients b_l defined by $-\beta P = \sum b_l z^l$, where $z = \exp(\beta\mu)$ is the fugacity. The virial coefficients a_l are directly deduced from the cluster coefficients. In terms of the second quantized fields ψ, $\mathcal{H}$ reads $\mathcal{H} = \mathcal{H}^o + \frac{1}{2}\int d\vec{r}_1 d\vec{r}_2 \psi^+(\vec{r}_1)\psi^+(\vec{r}_2)V(\vec{r}_1 - \vec{r}_2)\psi(\vec{r}_2)\psi(\vec{r}_1)$ where

$$V(\vec{r}_1 - \vec{r}_2) = \frac{(i\alpha\vec{k}\times \quad -|\alpha|)}{m}\frac{\vec{r}_{12}}{r^2_{12}}(\vec{\partial}_1 - \vec{\partial}_2). \tag{4}$$

$\mathcal{H}^o$ stands for the unperturbed Hamiltonian, i.e. independant particles bound to the origin by a harmonic interaction. The thermodynamic limit is understood as $\omega \to 0$, with the correspondance [6] $1/(\beta^2\omega^2) = V/\lambda^2$. Care has to be taken regarding overall normalization. For example letting $\omega \to 0$ in the grand canonical partition function of a system of bosonic or fermionic oscillators does not reproduce the usual Bose or Fermi grand canonical partition function. At a given power s of z, one has to multiply the harmonic result by s in order to get the usual infinite volume limit.

At first order [6], one has two diagrams corresponding to the two possible contractions in the Wick's expansion

$$\begin{aligned}\Omega = \Omega_o + \int_0^\beta d\beta_1 \int d^2\vec{r}_1 d^2\vec{r}_2 \frac{i\alpha\vec{k}\times - |\alpha|}{m}\frac{\vec{r}_{12}}{r_{12}^2}\vec{\partial}_1 \quad & (\{\psi^+(\vec{r}_1,\beta_1)\psi(\vec{r}_1,\beta_1)\}\{\psi^+(\vec{r}_2,\beta_1)\psi(\vec{r}_2,\beta_1)\} \\ & \pm\{\psi^+(\vec{r}_1,\beta_1)\psi(\vec{r}_2,\beta_1)\}\{\psi^+(\vec{r}_2,\beta_1)\psi(\vec{r}_1,\beta_1)\})\end{aligned} \tag{5}$$

In (5), the differential operator $\vec{\partial}_1$ is acting on the field $\psi(\vec{r}_1,\beta_1)$ and not on the field $\psi^+(\vec{r}_1,\beta_1)$. Computing these diagrams, one immediately sees the importance of a correct regularization. Indeed, the 2-anyon spectrum yield the second virial coefficients [4,8] $a_2^b = -\frac{\lambda^2}{4}(1-4|\alpha|+2\alpha^2)$ and $a_2^f = +\frac{\lambda^2}{4}(1-2\alpha^2)$ where λ is the thermal wave length. Perturbation theory at first order should give a $-|\alpha|$ correction close to Bose statistics, and no correction close to Fermi statistics. A direct computation in the thermodynamic limit uncorrectly implies a vanishing of the first diagram of (5), and thus a $\pm$ sign between the Bose and Fermi first order corrections to the second virial coefficient.

To see the effect of the regularization in a transparent way, let us integrate by parts (5) by using the identity $(i\alpha\vec{k}\times - \ |\alpha|)\vec{r}_{12}/r_{12}^2\vec{\partial}_1 = \vec{\partial}_1(i\alpha\vec{k}\times - \ |\alpha|)\vec{r}_{12}/r_{12}^2 + 2|\alpha|\pi\delta(\vec{r}_1,\vec{r}_2)$. Then one subtitutes the right hand side of the last equation in (5), where the differential operator acts both on $\psi^+(\vec{r}_1,\beta_1)$ and $\psi(\vec{r}_1,\beta_1)$. The total divergence leads to a surface term when integrated. This surface term does not vanish if one uses a plane wave basis, but it does vanish if one considers the system in a box or harmonically bound to the origin, since the wavefunctions are either zero on the boundary in the box case, or decreasing exponentially in the harmonic case. Thus, if one correctly takes into account boundary conditions, one gets, at first order in perturbation theory, a $|\alpha|\pi\delta(\vec{r}_1,\vec{r}_2)/m$ repulsive pointlike interaction. Once more, the exclusion of the diagonal of the configuration space appears naturally in the perturbative framework, close to Bose statistics. At first order, the "strength" of the exclusion is proportionnal to $|\alpha|$. Both diagrams in (5) are equal, and thus cancel in the fermionic case due to the relative $-$ sign. In the bosonic case, they add and one obtains [6]

$$\Omega = \Omega_o + 2|\alpha|\frac{\pi\beta}{m}\int d\vec{r}_1\{\psi^+(\vec{r}_1)\psi(\vec{r}_1)\}^2 = \Omega_o + |\alpha|\frac{V}{\lambda^2}\ln^2(1-z) \tag{6}$$

Lastly we consider the second order contribution [7]. Using Wick's theorem one obtains twenty connected diagrams. Each graph is regularized as in the first order case. A tadpole on which a differential operator acts is always replaced by a δ interaction. Again a naive computation with plane wave basis would lead in certain cases to the incorrect conclusion that a diagram vanishes. The harmonic regulator breaks the symmetry responsible for the vanishing of the integral in the plane wave basis, and gives instead a finite result, in the limit $\omega \to 0$. Contrary to the first order case

(6), each cluster coefficient has to computed on its own, since Bose- Einstein or Fermi-Dirac distributions have to be expanded in power series of the fugacity, to make the computations tractable. A simple symmetry argument involving parity shows that contributions of the type $\alpha|\alpha|$ cancel.

The third cluster coefficient reads $b_3^{b,f} = \frac{1}{\lambda^2}(\pm\frac{3}{2} + (1 - \frac{1}{24}))\alpha^2$ implying that the third virial coefficient is [7]

$$a_3^{b,f} = 4(a_2^{b,f})^2 - 2\frac{b_3^{b,f}}{b_1^3} = \frac{\lambda^4}{36}(1 + 3\alpha^2) \tag{7}$$

In the same way, α^2 corrections to all cluster coefficients have been obtained, from which corrections to all virial coefficients can be deduced. The corrections to a_4, a_5, a_6 are given in [7] where more technical details can be found. It seems however impossible at second order to sum the series in powers of z to get a closed expression as in (6).

Finally let us mention a recent numerical quantum analysis [9] where (7) has been reproduced for small α close to Fermi statistics.

Acknowledgments

I would like to thank my collaborators A. Comtet, A. Dasnieres de Veigy and J. McCabe, with whom the material presented here has been elaborated.

References

1. J.M. Leinaas and J.Myrheim, Nuovo Cimento B **37** (19770 1; F. Wilczek, Phys. Rev. Lett. **49** (1982) 957; Y.S.Wu, Proc. 2nd Int. Symp. Foundations of Quantum Mechanics, Tokyo (1986) 171

2. see Y.S. Wu in [2]

3. Y.H. Chen, F. Wilczek, E. Witten, and B. Halperin, Int. Journ. Mod. B3 (1989) 1001

4. A. Comtet, Y. Georgelin and S. Ouvry, J. Phys. A : Math. Gen. **22** (1989) 3917 see also K. Olaussen, "On the Harmonic Oscillatror Regularization of Partition Functions", Trondhiem University Preprint

5. G.V. Dunne, A. Lerda and C.A. Trugenberger, Mod. Phys. Lett. A **6** (1991) 2891; ibid, Int. Jour. Mod. Phys. B **5** (1991) 1675; G.V. Dunne, A. Lerda , S. Sciuto and C.A. Trugenberger, Nucl. Phys. B **370** (1992) 601 A.P. Polychronakos, Phys. Lett. B **264** (1991) 362; C. Chou, Phys. Lett. A **155** (1991) 245; A. Khare, J. McCabe and S. Ouvry, Phys. Rev. D **46** (1992) 2714

6. J. McCabe and S. Ouvry , Phys. Lett. B **260** (1991) 113; A. Comtet, J. McCabe and S. Ouvry , Phys. Lett. B **260** (1991) 372; see also D. Sen, Nucl. Phys. B **360** (1991) 397; G. Amelino-Camelia, Phys. Lett. B **286** (1992) 97

7. A. Dasnières de Veigy and S. Ouvry, "Perturbative Equation of State for a Gas of Anyons : Second Order", IPNO/TH 91-96 Report, to be published in Phys. Lett. B; "Perturbative Anyon Gas", IPNO/TH 92-54 Report, submitted to Nuclear Physics

8. D.P. Arovas, R. Schrieffer, F. Wilczek and A. Zee, Nucl.Phys. B **251** (1985) 117; J.S. Dowker, J. Phys. A **18** (1985) 3521

9. J. Law, A. Suzuki, and R. K. Bhaduri, "Quantum Virial Coefficient of a Fractional Statistics Gas" McMaster Preprint MCM-TP-92-1

THE $N \to \infty$ LIMIT OF THE CHIRAL POTTS MODEL

HELEN AU-YANG and JACQUES H.H. PERK*
Department of Physics, Oklahoma State University
Stillwater, OK 74078-0444, USA†

ABSTRACT

We consider the $N \to \infty$ limits of the N-state chiral Potts model. We find new weights that satisfy the star-triangle relations with spin variables either taking all the integer values or having values from a continous interval. The models provide chiral generalizations of Zamolodchikov's Fishnet Model.

1. Introduction

In the integrable N-state chiral Potts model[1] the Boltzmann weights $W(a-b)$ and $\overline{W}(a-b)$ — corresponding to the horizontal and vertical pair interactions between spins in states a and b — satisfy the "star-triangle" equation[2,3]

$$\sum_{d=1}^{N} \overline{W}_{qr}(b-d)\, W_{pr}(a-d)\, \overline{W}_{pq}(d-c) = R_{pqr}\, W_{pq}(a-b)\, \overline{W}_{pr}(b-c)\, W_{qr}(a-c). \quad (1)$$

The resulting weights have a product form.[2,3] Here we rewrite[4] these results as

$$\frac{W_{pq}(n)}{W_{pq}(0)} = \left(\frac{\sin\theta_p \sin\phi_q}{\sin\theta_q \sin\phi_p}\right)^{n/2N} \prod_{j=1}^{n} \frac{\sin[\pi(j-\frac{1}{2})/N - (\theta_q-\phi_p)/2N]}{\sin[\pi(j-\frac{1}{2})/N + (\phi_q-\theta_p)/2N]}, \quad (2)$$

$$\frac{\overline{W}_{pq}(n)}{\overline{W}_{pq}(0)} = \left(\frac{\sin\theta_p \sin\theta_q}{\sin\phi_p \sin\phi_q}\right)^{n/2N} \prod_{j=1}^{n} \frac{\sin[\pi(j-1)/N + (\phi_q-\phi_p)/2N]}{\sin[\pi j/N - (\theta_q-\theta_p)/2N]}, \quad (3)$$

using the substitutions

$$\frac{b_p}{c_p} = \omega^{\frac{1}{2}} e^{i\theta_p/N}, \quad \frac{a_p}{d_p} = e^{i\phi_p/N}, \quad \omega = e^{2\pi i/N}, \quad \frac{c_p}{d_p} = \left(\frac{e^{i\phi_p}\sin\phi_p}{e^{i\theta_p}\sin\theta_p}\right)^{1/2N}, \quad (4)$$

where θ_p and ϕ_p differ by a factor N from those of Baxter.[4] The last equality in (4) follows from the integrability conditions[2,3] or

$$k = \frac{\sin\frac{1}{2}(\theta_p - \phi_p)}{\sin\frac{1}{2}(\theta_p + \phi_p)} = \frac{\sin\frac{1}{2}(\theta_q - \phi_q)}{\sin\frac{1}{2}(\theta_q + \phi_q)}. \quad (5)$$

*Supported in part by NSF Grants Nos. DMS 91-06521 and INT 91-12563

†Work done at RIMS, Kyoto University, Sakyo-ku, Kyoto 606, Japan

The Fourier transforms, which are the weights after duality transformation,[2,3]

$$W_{pq}^{(f)}(n) = N^{-1}\sum_{j=0}^{N-1}\omega^{-jn}W_{pq}(j), \qquad \overline{W}_{pq}^{(f)}(n)) = N^{-1}\sum_{j=0}^{N-1}\omega^{-jn}\overline{W}_{pq}(j), \tag{6}$$

can be rewritten as

$$\frac{W_{pq}^{(f)}(n)}{W_{pq}^{(f)}(0)} = e^{in(\theta_q-\phi_q+\theta_p-\phi_p)/2N}\prod_{j=1}^{n}\frac{\sin[\pi(j-1)/N+(\bar\theta_q-\bar\theta_p)/2N]}{\sin[\pi j/N-(\bar\phi_q-\bar\phi_p)/2N]}, \tag{7}$$

$$\frac{\overline{W}_{pq}^{(f)}(n)}{\overline{W}_{pq}^{(f)}(0)} = e^{in(\theta_q-\phi_q-\theta_p+\phi_p)/2N}\prod_{j=1}^{n}\frac{\sin[\pi(j-\frac{1}{2})/N-(\bar\theta_q-\bar\phi_p)/2N]}{\sin[\pi(j-\frac{1}{2})/N+(\bar\phi_q-\bar\theta_p)/2N]}, \tag{8}$$

where

$$\bar\phi_p = \tfrac{1}{2}(\theta_p+\phi_p)+\tfrac{1}{2}i\log\frac{\sin\phi_p}{\sin\theta_p}, \quad \bar\theta_p = \tfrac{1}{2}(\theta_p+\phi_p)-\tfrac{1}{2}i\log\frac{\sin\phi_p}{\sin\theta_p}. \tag{9}$$

By direct substitution we can show that if the weights satisfy the star-triangle equation (1) then their Fourier transforms satisfy the star-triangle equation

$$\frac{N}{R_{pqr}}\overline{W}_{qr}^{(f)}(a)\,W_{pr}^{(f)}(b)\,\overline{W}_{pq}^{(f)}(a+b) = \sum_{d=0}^{N-1} W_{pq}^{(f)}(b-d)\,\overline{W}_{pr}^{(f)}(a+b-d)\,W_{qr}^{(f)}(d). \tag{10}$$

This equation has the exact same form as equation (1), as can be seen replacing $a \to a-b$, $b \to b-c$, $a+b \to a-c$, and $c+d \to d$.

For $\theta_p = \phi_p$, $\theta_q = \phi_q$ we recover the self-dual Fateev and Zamolodchikov[5] solution with $W^{(f)}(n)/W^{(f)}(0) = \overline{W}(n)/\overline{W}(0)$, $\overline{W}^{(f)}(n)/\overline{W}^{(f)}(0) = W(n)/W(0)$.

2. The $N \to \infty$ Limit of the Boltzmann Weights

We shall now obtain the $N \to \infty$ limit of the Boltzmann weights (2) and (3) or their dual weights (7) and (8). Note that these all have the product form

$$\frac{W(n)}{W(0)} = A^{n/N}\prod_{j=1}^{n}\frac{\sin\left(\pi(j+\alpha-1)/N\right)}{\sin\left(\pi(j+\beta-1)/N\right)}, \quad A = \frac{\sin(\pi\beta)}{\sin(\pi\alpha)}, \tag{11}$$

with α and β given constants depending on parameters θ_p, θ_q, ϕ_p, and ϕ_q satisfying (5). Also, the condition on A guarantees that $W(n+N) = W(n)$. We can rewrite formula (11) in terms of a convergent series in powers of $1/N$, i.e.

$$\log\frac{W(n)}{W(0)} = \log\left(A^{n/N}\prod_{j=1}^{n}\frac{j+\alpha-1}{j+\beta-1}\right) + \sum_{l=0}^{\infty}\frac{B_{l+1}(\alpha)-B_{l+1}(\beta)}{(l+1)!}\left(\frac{2\pi}{N}\right)^{l}\left(\frac{d}{dz}\right)^{l}\log\left(\frac{\sin z}{z}\right)\Bigg|_{z=\pi n/N}, \tag{12}$$

where $0 < n < N$ and we used the functional relations of Bernoulli polynomials[6] $\sum_{m=0}^{p} \binom{p}{m} B_m(x) y^{p-m} = B_p(x+y)$ and $B_p(x+1) - B_p(x) = px^{p-1}$.

There are three regimes for the limit, the first having $N \to \infty$, while n remains finite. In this case, (12) results in

$$\text{I.} \quad \frac{W(n)}{W(0)} = \prod_{j=1}^{n} \frac{j+\alpha-1}{j+\beta-1} = \prod_{j=1}^{-n} \frac{j-\beta}{j-\alpha} = \frac{\Gamma(n+\alpha)\Gamma(\beta)}{\Gamma(n+\beta)\Gamma(\alpha)}, \quad -\infty < n < \infty. \tag{13}$$

The second regime has N, $n \to \infty$ such that $2\pi n/N \to x$ remains finite. Consequently, the weights $W(n)$ in (11), which originally took N different values and which were periodic modulo N, now depend on the continuous spin values x and they are periodic modulo 2π. Using the asymptotic formula[7] for $\log \Gamma(x)$, and $B_1(x) = x - \frac{1}{2}$ we find that (12) in this limit gives

$$\text{II.} \qquad W(x) = C\, A^{\frac{x}{2\pi} - \left[\frac{x}{2\pi}\right]} \left|\sin \tfrac{1}{2}x\right|^{\alpha-\beta}, \tag{14}$$

where $[x]$ stands for integral part of x and

$$C = W(0)\,(N/\pi)^{\alpha-\beta}\,\Gamma(\beta)\,/\Gamma(\alpha). \tag{15}$$

The results in Regimes I and II are consistent with duality transformation. More precisely, if the limiting weights are in Regime II, their Fourier transforms are in Regime I, and vice versa. Indeed, the infinite sum in the Fourier transform can be summed using the formula[8]

$$\sum_{n=-\infty}^{\infty} e^{inx} \prod_{j=1}^{n} \frac{j+\alpha-1}{j+\beta-1} = \frac{2^{\beta-\alpha-1}\,\Gamma(1-\alpha)\Gamma(\beta)}{\Gamma(\beta-\alpha)}\, e^{\frac{1}{2}i(1-\alpha-\beta)(x-\pi)} \left|\sin \tfrac{1}{2}x\right|^{\beta-\alpha-1}, \tag{16}$$

for $0 < x < 2\pi$ (and periodically extended) and with α and β in (13) calculated from (2) or (3), to give an identical formula for the weights as in (14) for Regime II, with new values of α and β corresponding to (7)-(9). This generalizes the large-N limit of Fateev and Zamolodchikov.[5]

A third intermediate regime can also appear with N, $n \to \infty$ such that $n/f(N) \to x$ for some function $f(N)$ that blows up slower than N. We have

$$\text{III.} \qquad W(x) = D\, A^{-\frac{1}{2}\mathrm{sign}(x)} \, |x|^{\alpha-\beta}, \quad -\infty < x < \infty, \tag{17}$$

which is a chiral generalization of Zamolodchikov's Fishnet Model.[10] In (17), $D = C\,A^{\frac{1}{2}}$, with C given by (16). The sign function in (17) emerges if we rewrite (11) for negative n in the form (12), see also (13), and compare constants in (17) for $x > 0$ and $x < 0$; then we can use $\Gamma(1-\alpha)/\Gamma(1-\beta) = A\,\Gamma(\beta)/\Gamma(\alpha)$, which follows from $\Gamma(x)\Gamma(1-x) = \pi/\sin(\pi x)$, leading to an extra factor A for $x < 0$. The sign function relates directly to the effect of the integer part in (14) near $x = 0$; in fact, by the identical reasoning for Regime II, we need the same extra factor A for $-2\pi < x < 0$ in (14).

We note that in the previously known cases[5,10] $\alpha + \beta = 1$ and $A = 1$. We have only one condition (11) on A allowing the deformations (13), (14), and (17) to provide integrable field theories with chirality.

3. The $N \to \infty$ Limit of the Chiral Potts Model

Having obtained explicit prescriptions on how to take the $N \to \infty$ limit of the Boltzmann weights of the N-state chiral Potts model, we are now in a position to examine the various limits of the star-triangle equation (1). The summation over d must be split in several pieces as we must choose from which regimes to take the three weights in the summand. As $N \to \infty$, we can rigorously show that most pieces can be ignored.

The least complicated (but also a most interesting) case occurs when all the $\beta - \alpha$ in the formulae of section 2 are between 0 and 1, defining a principal domain for the spectral parameters. In this case, the three types of large-N behavior do not mix. More precisely, if we take the three spin states a, b, and c in (1) mutually separated according to one and the same regime in the sense of the previous section — so that we are taking the same type of limit for all three weights in the right-hand side of (1) — the dominant part of the sum over spin state d comes from the piece with all three weights in the left-hand side of (1) from the same regime. This is easily proved as the formulae in section 2 give full control over the leading term and the corrections in the large-N limit.

For example, taking Regime II, the star-triangle equations become

$$\int_0^{2\pi} dw\, \overline{W}_{qr}(y-w)\, W_{pr}(x-w)\, \overline{W}_{pq}(w-z) = R\, W_{pq}(x-y)\, \overline{W}_{pr}(y-z)\, W_{qr}(x-z), \tag{18}$$

after a suitable renormalization of R. This has the solution

$$\begin{aligned} W_{pq}(x) &= C_{pq}\, e^{(\gamma_p-\gamma_q)(\frac{x}{2\pi} - [\frac{x}{2\pi}])} \left|\sin \tfrac{1}{2}x\right|^{\lambda_p-\lambda_q}, \\ \overline{W}_{pq}(x) &= \overline{C}_{pq}\, e^{(\gamma_p+\gamma_q)(\frac{x}{2\pi} - [\frac{x}{2\pi}])} \left|\sin \tfrac{1}{2}x\right|^{\lambda_q-\lambda_p-1}, \end{aligned} \tag{19}$$

with

$$\lambda_p = \frac{1}{\pi} \arctan\left(\frac{\sin\theta_p}{\cos\theta_p + k}\right), \quad \gamma_p = \tfrac{1}{2}\log\left(\frac{1+2k\cos\theta_p + k^2}{1-k^2}\right), \tag{20}$$

and similar formulae for λ_q, γ_q, λ_r, and γ_r, as follows from the prescriptions given in the previous two sections. If $\lambda_p < \lambda_q < \lambda_r < 1 + \lambda_p$ all six Boltzmann weights in (18) are real and positive and the parameters are in the principal domain.

We remark that it is straightforward to show that the weights obtained by just dropping the integral part $[x/2\pi]$ in (19) also satisfy the same star-triangle equation (18). This solution can be viewed as the Fateev-Zamolodchikov solution[5] with a site-dependent gauge transformation. We emphasize that we have both numerically verified and analytically proved that the "chiral" weights (19) also satisfy (18).

Since $W_{pq}(x)$ and $\overline{W}_{pq}(x)$ are now functions of x modulo 2π, their Fourier transforms

$$W_{pq}^{(f)}(j) = \frac{1}{2\pi}\int_0^{2\pi} dx\, e^{-ijx}\, W_{pq}(x), \qquad \overline{W}_{pq}^{(f)}(j) = \frac{1}{2\pi}\int_0^{2\pi} dx\, e^{-ijx}\, \overline{W}_{pq}(x), \tag{21}$$

are over all the integer value j, ranging from $-\infty$ to ∞. Substituting (21) into (18), we find that the Fourier transforms satisfy different star-triangle equations,

$$\frac{2\pi}{R}\,\overline{W}^{(f)}_{qr}(a)\,W^{(f)}_{pr}(b)\,\overline{W}^{(f)}_{pq}(a+b)=\sum_{d=-\infty}^{\infty}W^{(f)}_{pq}(b-d)\,\overline{W}^{(f)}_{pr}(a+b-d)\,W^{(f)}_{qr}(d), \quad (22)$$

in which the sum is over all integer values of d.

Taking Regime I instead of Regime II, leads to the proof of a star-triangle equation of the form (21) instead of (18). This is an interesting identity in the theory of generalized hypergeometric series generalizing the Dougall-Ramanujan identity.[9]

Choosing Regime III, we can prove a star-triangle equation of the form (18), but with integration over $(-\infty,+\infty)$. This generalizes Symanzik's integral,[11] which Zamolodchikov used to prove the star-triangle equation for the Fishnet Model[10] and which further provided the proof for the Fateev-Zamolodchikov model[5] via a conformal transformation, $\xi=\tan\frac{1}{2}w$.

Finally, as in our joint work with Baxter,[2] we can let the R-matrix be the product of four weights of any of the above types I, II, or III, i.e.

$$R(a,b,c,d)=\overline{W}_{p_1q_1}(a-c)\,W_{p_1q_2}(c-b)\,\overline{W}_{p_2q_2}(d-b)\,W_{p_2q_1}(a-d). \quad (23)$$

Then any such infinite-dimensional R-matrix satisfies the Yang-Baxter equation. But our solutions are very different from those of Gaudin,[12] Shibukawa and Ueno.[13]

References

1. H. Au-Yang, B. M. McCoy, J. H. H. Perk, S. Tang, and M.-L. Yan, *Phys. Lett.* **123A** (1987) 219.
2. R. J. Baxter, J. H. H. Perk, and H. Au-Yang, *Phys. Lett.* **128A** (1988) 138.
3. H. Au-Yang and J. H. H. Perk, *Adv. Stud. Pure Math.* **19** (1989) 57.
4. R. J. Baxter, *J. Stat. Phys.* **52** (1988) 639.
5. V. A. Fateev and A. B. Zamolodchikov, *Phys. Lett.* **92A** (1982) 37, see also 35.
6. *Higher Transcendental Functions* Vol. 1, *H. Bateman Manuscript Project*, eds. A. Erdélyi et al., (McGraw-Hill, New York, 1953), eqs. 1.13 (2), (6).
7. Ref. 6, eq. 1.18 (1).
8. Ref. 6, eqs. 2.1.2 (6), 2.9 (1), (13), (22), (27).
9. Ref. 6, eq. 4.5 (7).
10. A. B. Zamolodchikov, *Phys. Lett.* **97B** (1980) 63.
11. K. Symanzik, *Lett. Nuovo Cim.* **3** (1972) 734.
12. M. Gaudin, *J. Phys. France* **49** (1988) 1857.
13. Y. Shibukawa and K. Ueno, *Lett. Math. Phys.* **25** (1992) 239; *Waseda University preprint.*

QUANTUM GROUPS IN QUANTUM OPTICS

Allan I. Solomon *

Faculty of Mathematics, The Open University, Milton Keynes, MK7 6AA, UK

Abstract

We review some of the recent application of quantum groups to quantum optics. We concentrate on models involving exotic *states* rather than the dynamics, and show how these states relate to the conventional descriptions of coherent and squeezed states.

Introduction

Although their use is ubiquitous in physics, group theoretical methods have not been very widely applied in quantum optics. Nevertheless, the recent introduction of quantum groups has given some impetus to the use of theory as a means of producing new exotic states of light. This provides a fertile playground for the theorist; it remains to be seen if the resulting systems are of more than theoretical interest.

Two general areas of application are:

- Exotic States
- Atomic Systems

The latter approach usually involves some variant of the Jaynes-Cummings Model and depends on postulating a hamiltonian for the interaction between the light quanta and the atomic levels. The former does not rely on the introduction of a hamiltonian; in its absence one has to forego a discussion of the evolution of the system. However, as there is ambiguity as to the choice of a deformed hamiltonian, in this brief review we shall restrict our discussion to states.

A general method for the production of coherent states corresponding to any group G - and this also includes the *squeezed states* which are a special case of coherent states - is to write down a realisation of G in terms of boson creation and annihilation operators $a, a^\dagger$ which satisfy the 3-dimensional Heisenberg-Weyl Algebra hw

$$[a, a^\dagger] = 1 \tag{1}$$

Then the action of an element of G on the vacuum state $|0>$, which satisfies $a|0>= 0$, gives a coherent state of G. In the case of the group corresponding to the hw algebra of

*Talk presented at the XXI International Conference on Differential Geometric Methods in Theoretical Physics, Tianjin, China June 5–9,1992

Eq. (1) we obtain conventional *coherent* states. When $G = SU(1,1)$ we obtain *squeezed* states (which the author obtained over twenty years ago [1] without, however, recognizing their squeezing, that is, *noise-reduction*, properties [2]). The physically useful property of these states, which gives rise to the term "squeezed", derives from the fact that in them the quantum dispersion (variance) $(\Delta x)^2$ of $x \equiv (a + a^\dagger)/\sqrt{2}$ (which in optical applications corresponds to a component of a single mode of the quantized electromagnetic field) may attain values below its vacuum (or coherent) state value of 1/2. Such states have been produced experimentally. As another example, we may use the standard two-mode (Schwinger) realisation of $SU(2)$ to produce $SU(2)$ squeezed states. I first review briefly the conventional group theoretical approach to some quantum optics states.

1. States of Conventional Groups

(i) Vacuum state $|0>$. This may be considered as associated in a somewhat trivial sense with the $u(1)$ algebra.
(ii) Conventional coherent states: These states are defined as eigenstates of the operator a, $a|\lambda>= \lambda|\lambda>$. They are obtained by the action of the displacement operator $D(\lambda)$ on the vacuum, $D(\lambda)|0>= |\lambda>$, where D is an element of a representation of the group of the hw algebra Eq. (1). Up to a normalization factor we have $|\lambda>= \exp(\lambda a^\dagger)|0>$.
(iii) $SU(1,1)$ states: These are obtained by the action of $U(\xi)$, an element of the group of the $su(1,1)$ algebra generated by $\{a^2, (a^\dagger)^2, (aa^\dagger + a^\dagger a)\}$.
(iv) Squeezed states: Conventional squeezed states are obtained by the action of the squeezing group $\mathcal{S}$ whose algebra s is a semi-direct sum of the algebras of (ii) and (iii) above. (The solvable algebra hw is an ideal of s; this is a real-life example of the Levi decomposition theorem.) Thus a typical squeezed state may be written

$$|\xi, \lambda>= U(\xi)D(\lambda)|0> . \tag{2}$$

For example, choosing $U(\xi) = \exp(-\frac{1}{2}\xi a^{\dagger 2})$, $D(\lambda) = \exp(\lambda a^\dagger)$ $|\xi| < 1$ it is easy to show that

$$(a + \xi a^\dagger)|\xi, \lambda>= \lambda|\xi, \lambda> . \tag{3}$$

An equivalent definition of squeezed states arises from the observation that the automorphism group of the hw algebra is $SU(1,1)$; thus a unitary transformation $U \in SU(1,1)$ acting on a gives

$$a \mapsto U\ a\ U^\dagger = \mu a + \nu a^\dagger \tag{4}$$

where $|\mu|^2 - |\nu|^2 = 1$. A conventional squeezed state is then defined as an eigenstate of the transformed photon annihilation operator

$$(\mu a + \nu a^\dagger)|\xi>= \xi|\xi> . \tag{5}$$

These states may be generalized, still within the framework of conventional photons, in the following ways:

- Multiphoton states
- Multimode states
- Fractional photon states

- States corresponding to other groups

Multiphoton states [3] make use of the properties of generalized boson operators A_k, first introduced by Brandt and Greenberg [4] which are defined to satisfy an algebra isomorphic to the hw algebra Eq. (1). We may carry over unchanged, except for the substitution capital A for small a, all the algebraic properties of one-photon states. The properties of these generalized multiphoton coherent states, and the analogous $SU(1,1)$ and $S(2)$ states, have been extensively studied [3, 5].

Multimode states may be produced from an $n \times n$ matrix realisation of a given group, analogously to the 2-mode realization (Schwinger realization) of $SU(2)$; this latter has recently been discussed [6, 7] in an experimentally feasible context. Multimode, multi-photon states [8] may be constructed in a way analogous to the method described above for the k-photon operator; namely, by defining multi-mode operators which satisfy Eq. (1).

Fractional photon states arise if we regard the transformations from single photon a to multiphoton operators A_k as forming a semigroup of canonical transformations. These transformations, which are defined for positive integral k only, may be extended formally to a group, isomorphic to the positive rationals under multiplication. The resulting objects are formally identified with fractional photon operators, but may arise quite naturally as states of conventional photons. Fractional photon states were first defined in Reference [9]. The fractional index r may be shown to give a direct measure of the ratio of number-variance to phase-angle-variance in number-phase minimum uncertainty states [10].

States corresponding to *other groups* have been introduced largely through the Holstein-Primakoff realization [11]. By this means the properties of states related to k-photon realisations of $SU(2)$ and $SU(1,1)$ have been analyzed [3], as well as $SU(n)$ [12].

In the next section we show how the introduction of q-bosons modifies the above considerations, and how q-analogues of these quantum optics states may be defined.

2. States of Quantum Groups

A deformation of the standard boson operator a was introduced some years ago by Arik and Coon [13]. Their deformed bosons satisfy

$$a\tilde{a} - q\tilde{a}a = 1. \tag{6}$$

More recently, the Quantum Heisenberg-Weyl Algebra

$$\begin{aligned} aa^\dagger - q'a^\dagger a &= q'^{-N} \\ \left[N, a^\dagger\right] &= a^\dagger \end{aligned} \tag{7}$$

has been introduced [14, 15]. This may be obtained from Eq.(6) by writing $\tilde{a} = a^\dagger q'^N$ and $q = q'^2$. The introduction of this second form enables a Schwinger-type realization of the quantum group $SU_{q'}(2)$ to be given.

Just as in the conventional case we may define *q-Coherent states* as eigenstates of the annihilation operator [13], [15]. The deformed boson operators Eq. (6) satisfy

$$[a, \tilde{a}] = [N+1] - [N] \tag{8}$$

where $[x]$ is the *basic number* defined by $(q^x - 1)/(q-1)$. It is easy to show by induction that

$$[a, \tilde{a}^r] = \tilde{a}^{(r-1)}([N+r] - [N]). \tag{9}$$

In general we have

$$[a, F(\tilde{a})]\,|0> = {}_qD_{\tilde{a}}F(\tilde{a})|0> . \tag{10}$$

where ${}_qD_xF(x) \equiv (F(qx) - F(x))/(x(q-1))$ is the q-derivative of classical q-analysis [16].

Acting on the vacuum with each side of Eq.(9) shows that an eigenstate of a is given by $E_q(\lambda\tilde{a})|0>$ where $E_q(x)$ is the classical q-exponential function satisfying ${}_qD_xE_q(\lambda x) = \lambda E_q(x)$. In the absence of time evolution, the q-coherent states defined above do not give rise to squeezing, just as in the case of the conventional coherent states. In fact, it may be shown [17] that the term which gives rise to squeezing is, in general, $< a_q^2 > - < a_q >^2$which is zero for eigenstates of a_q. However, unlike the conventional coherent states, if we relax our reticence concerning dynamics, it may be shown that there is time-dependent squeezing, by choice of a suitable analogue of the usual Hamiltonian [18, 19]. We may generalize the conventional $SU(1,1)$ squeezed states, defined in (i) of the preceding section, to the quantum group context.by using a q-boson realization of the analogous $su_q(1,1)$ algebra; one may then define the analogous q-squeezed states by the exponential action of the resulting operators on the vacuum (or on the q-coherent states). A realization of $su_q(1,1)$ was given by Kulish and Damaskinsky [20]; it is essentially the Hopf algebra generated by a^2 and its hermitian conjugate, where $a, a^\dagger$ satisfy Eq. (7) above. A suitable candidate $|\xi,\lambda>$ for a q-analogue of the squeezed state of (iv) of the previous section may be obtained as

$$|\xi,\lambda> = E_{q^2}(-\frac{1}{[2]_q}\xi\tilde{a}^2)E_q(\lambda\tilde{a})|0> . \tag{11}$$

Not only does this give a normalizable state for $q < 1$, unlike the corresponding more usual exp_q [14], [15], but use of the q-chain rule [21]

$$ {}_qD_xF(G(x)) =_{\frac{G(qx)}{G(x)}} D_GF(G)\,{}_qD_xG(x) \tag{12}$$

and the Product (Leibniz) Rule

$$ {}_qD_x(F(x)G(x)) = {}_qD_xF(x).G(x) + F(qx).{}_qD_xG(x) \tag{13}$$

shows that $|\xi,\lambda>$ satisfies

$$(a + \xi\tilde{a})|\xi,\lambda> = \lambda|\xi,\lambda>$$

provided $\xi\lambda = q^{-2}\lambda\xi$. This gives a direct q-analogue of Eq.(3), and also shows the necessity for the introduction of quantum plane variables at the algebra level. All the states mentioned in Section 1 may be generalized to the quantum group case; ***multiphoton q-states*** [17], ***multimode q-states*** [23], ***fractional q-photon*** states [24] and realizations of ***other q-groups***, which last can be obtained by use of Holstein-Primakoff transformations [17]. I refer the interested reader to the references given for further details concerning the construction of these states.

References

[1] A.I. Solomon, J.Math.Phys.**12**, 390 (1971)

[2] H.P.Yuen, Physics Letters **56A**,105(1976)

[3] J.Katriel, A.I.Solomon, G.d'Ariano and M.Rasetti, Phys.Rev.D**34**,2332(1986)

[4] R.A.Brandt and O.W.Greenberg, J.Math.Phys.**10**,1168(1969)

[5] J.Katriel, A.I.Solomon, G.d'Ariano and M.Rasetti, J.Opt.Soc.Am.**B** **10**,1728(1987)

[6] V.Buzek and T.Quang, J.Opt.Soc.Am.**B6**, 2448(1989)

[7] W.K.Lai,V.Buzek and P.L.Knight, Phys.Rev.**A43**,6323(1991)

[8] J. Katriel and A.I.Solomon, Quantum Optics,**1**,85 (1989)

[9] J.Katriel, M.Rasetti and A.I.Solomon, Phys.Rev.D**35**,1248(1987)

[10] G.d'Ariano, Phys.Rev.D,**41**,2636- 2644(1990)

[11] T.Holstein and H.Primakoff, Phys.Rev.**58**,1048(1940)

[12] J.Katriel, M.Rasetti and A.I.Solomon, Phys.Rev.D**35**,2601(1987)

[13] M. Arik and D.D. Coon, J.Math.Phys.**17**,524(1976)

[14] A.J.MacFarlane, J.Phys.A:Math.Gen.**22**,4581(1989)

[15] L.C.Biedenharn, J.Phys.A:Math.Gen.**22**,L873(1989)

[16] FH Jackson, Quart.J.Pure Appl.Math., **41**,193 (1910)

[17] J.Katriel and A.I. Solomon, J.Phys.A:Math.Gen.**24**,2093(1991)

[18] V.Buzek, J.Mod.Opt.**38**,801(1991)

[19] E.Celeghini, M.Rasetti and G.Vitiello, Phys.Rev.Lett.**66**,2056–2091(1991)

[20] P.P.Kulish and E.V.Damaskinsky, J.Phys.A:Math.Gen.**23**,L415–419(1990)

[21] A.I.Solomon and J.Katriel, Unpublished.

[22] A.I.Solomon and J.Katriel, J.Phys.A:Math.Gen.**23**,L1209(1990)

[23] A.I.Solomon and J. Katriel, to be published.

[24] A.I.Solomon and J. Katriel, Proceedings of the Second International Wigner Symposium, Goslar,1991. (To be published by Springer.)

GENERALIZED XYZ MODEL ASSOCIATED TO SKLYANIN ALGEBRA

TAKASHI TAKEBE

Department of Mathematical Sciences
the University of Tokyo
Hongo 7-3-1, Bunkyo-ku, Tokyo 113, Japan

ABSTRACT. The free energy of a lattice model, which is a generalization of the Heisenberg XYZ model with the higher spin representation of the Sklyanin algebra, is calculated by the generalized Bethe Ansatz of Takhtajan and Faddeev.

§0. Introduction.

The XYZ model, a quantum chain system with Hamiltonian

$$H = -\frac{1}{2}\sum_{n=1}^{N}(J_x\sigma_n^x\sigma_{n+1}^x + J_y\sigma_n^y\sigma_{n+1}^y + J_z\sigma_n^z\sigma_{n+1}^z),$$

was first solved by Baxter in early 70's by means of the Bethe Ansatz. This work was interpreted from the view point of the quantum spectral transformation method (QSTM) by Takhtajan and Faddeev in 1979 [TF]. Their starting point is the fundamental relation of the L operators

$$R^{12}(\lambda-\mu)L^{13}(\lambda)L^{23}(\mu) = L^{23}(\mu)L^{13}(\lambda)R^{12}(\lambda-\mu), \tag{0.1}$$

where

$$L(\lambda) = \sum_{a=0}^{3} W_a^L(\lambda)\sigma^a \otimes \sigma^a, \tag{0.2}$$

$$W_0^L(\lambda) = \frac{\theta_{11}(\lambda)}{\theta_{11}(\eta)}, W_1^L(\lambda) = \frac{\theta_{10}(\lambda)}{\theta_{10}(\eta)}, W_2^L(\lambda) = \frac{\theta_{00}(\lambda)}{\theta_{00}(\eta)}, W_3^L(\lambda) = \frac{\theta_{01}(\lambda)}{\theta_{01}(\eta)},$$

σ^a's are Pauli matrices, $\theta_{ab}(\lambda) = \theta_{ab}(\lambda;\tau)$ (we follow the notations of [M] for the theta functions), and the R matrix is the Baxter's R matrix defined by

$$R(\lambda) = \sum_{a=0}^{3} W_a^R(\lambda)\sigma^a \otimes \sigma^a, W_a^R(\lambda) := W_a^L(\lambda+\eta).$$

Sklyanin found the algebraic structure underlying the above relation, replacing the L operator by

$$L(\lambda) = \sum_{a=0}^{3} W_a^L(\lambda)\sigma^a \otimes S^a,$$

and writing down the relations among S^a's equivalent to (0.1) [Sk1]. The algebra Q generated by S^a's is called the Sklyanin algebra. He also constructed series of representations in [Sk2], among which we use the spin l representation ρ_l : $Q \to \mathrm{End}(V_l)$, $\dim V_l = 2l+1$. The algebra Q is a deformation of the universal enveloping algebra $U(sl(2,\mathbb{C}))$, and ρ_l is corresponding deformation of the spin l representation of $sl(2,\mathbb{C})$.

As the solvability of the XYZ (or 8 vertex) model comes from the fundamental relation (0.1), it is natural to consider the generalization of this model by using the above representations of the Sklyanin algebra, and calculate its free energy.

§1. Description of the model. Now we construct a generalization of the XYZ model by replacing the L operators of the form (0.2) with

$$L_n(\lambda) = \sum_{a=0}^{3} W_a^L(\lambda)\,\sigma^a \otimes \rho_l^{(n)}(S^a),$$

(1.1)

$$\rho_l^{(n)}(S^a) = 1 \otimes \cdots \otimes 1 \otimes \underbrace{\rho_l(S^a)}_{n\text{-th component}} \otimes 1 \otimes \cdots 1,$$

which act on the space $\mathbb{C}^2 \otimes V_l^{\otimes N}$. The transition matrix of the higher spin generalization of the 8 vertex model is defined by:

$$T(\lambda) = L_N(\lambda)\cdots L_2(\lambda)L_1(\lambda).$$

The transition matrix of the corresponding chain model is

$$t(\lambda) = \mathrm{tr}_{\mathbb{C}^2}\, T(\lambda),$$

which acts on the space $V_l^{\otimes N}$. The free energy of the model under consideration is, by definition, the Perron-Frobenius eigenvalue of the transfer matrix $t(\lambda)$, Λ_N, or its thermodynamic limit,

$$f = \lim_{N\to\infty} \frac{1}{N} \ln \Lambda_N.$$

Similarly to the choice of [TF], we assume that the elliptic modulus τ is of the form $\tau = i/t$, where $t > 0$ and the parameters η, λ belong to the domain $0 < \lambda + \eta < \max(\eta, 2l\eta) < \frac{1}{2}$ (antiferromagnetic region).

In the previous paper [Take], the author showed that the generalized Bethe Ansatz of Takhtajan and Faddeev is applicable to this model. According to this result, if $M = Nl$ parameters, $\lambda_1, \dots \lambda_M$, satisfy the Bethe equation,

(1.2)
$$\left(\frac{\theta_{11}(\lambda_j - 2l\eta)}{\theta_{11}(\lambda_j + 2l\eta)}\right)^N = \prod_{k=1,k\neq j}^{M} \frac{\theta_{11}(\lambda_k - \lambda_j - 2\eta)}{\theta_{11}(\lambda_k - \lambda_j + 2\eta)},$$

for $j = 1, \dots, M$, then the transfer matrix $t(\lambda)$ has an eigenvalue

(1.3)
$$(2\theta_{11}(\lambda - 2l\eta))^N \prod_{k=1}^{M} \frac{\theta_{11}(\lambda - \lambda_k + 2\eta)}{\theta_{11}(\lambda - \lambda_k)} + (2\theta_{11}(\lambda + 2l\eta))^N \prod_{k=1}^{M} \frac{\theta_{11}(\lambda + \lambda_k + 2\eta)}{\theta_{11}(\lambda - \lambda_k)}.$$

§2. Thermodynamic limit. Now we take the thermodynamic limit $N \to \infty$. From the results of the XXX and XXY models [Takh], [B] and [S], we assume that the following *string hypothesis* holds also for this model: assymptotically as $N \to \infty$, the solutions of the Bethe equation (1.2) corresponding to the ground state cluster in complexes, so called $2l$-strings:

$$\{\lambda_k^0 + 2\alpha\eta \,|\, \alpha = -l + \tfrac{1}{2}, \dots, l + \tfrac{1}{2}\}_{k=1,\dots,N/2}$$

(assuming that N is even), and centres of these strings λ_k^0 distribute in a purely imaginary interval $\{ix| -1/2t < x < 1/2t\}$ and have a smooth density $\rho(\lambda')d\lambda'$ as $N \to \infty$ ($\lambda' = \lambda/\tau$).

Under these hypotheses, the Bethe Ansatz equation (1.2) reduces to an integral equation

$$(2.1) \quad \sum_{\alpha=0}^{2l-1} \Psi(\lambda, (2\alpha+1)\eta) =$$

$$= -2\pi\rho(\lambda') + \int_{-1/2}^{1/2} \Big(\sum_{\beta=1}^{2l-1} \Psi(\lambda-\mu, 2\beta\eta) + \sum_{\beta=0}^{2l-1} \Psi(\lambda-\mu, 2(\beta+1)\eta)\Big)\rho(\mu')\,d\mu',$$

where $\lambda' = \lambda/\tau$, $\mu' = \mu/\tau$ and

$$\Psi(\lambda,\eta) = \frac{\tau}{i}\frac{d}{d\lambda}\ln\frac{\theta_{11}((\lambda-\eta)/\tau; -1/\tau)}{\theta_{11}((\lambda+\eta)/\tau; -1/\tau)}.$$

We can solve this equation, expanding $\rho(\lambda')$ into Fourier series, and obtain

$$(2.2) \qquad \rho(\lambda') = \sum_{n\in\mathbf{Z}} \frac{e^{2\pi i n\lambda'}}{2\cos 2\pi n\eta'}.$$

Using this density of the solutions of the Bethe equation, we can take the limit of the eigenvalue (1.3) and obtain the free energy of the model $f = \lim_{N\to\infty} \ln\Lambda_N/N$ as

$$(2.3) \quad e^f \propto \theta_{11}(\lambda + 2l\eta)e^{2\pi t(\lambda-\eta)(2l\eta-\frac{1}{2})}\times$$

$$\times \exp\sum_{n=1}^{\infty} \frac{\sinh 2\pi nt(2l\eta - \frac{1}{2})\,\sinh 2\pi nt(\lambda-\eta)}{n\sinh\pi nt\,\cosh 2\pi nt\eta} =$$

$$= \theta_{11}(\lambda + 2l\eta)\,q^{-(\lambda-\eta)(2l\eta-\frac{1}{2})}\frac{\sigma(\lambda;\eta,l|q)}{\sigma(-\lambda;\eta,l|q)},$$

where $q = \exp(-2\pi t)$ and

$$\sigma(\lambda;\eta,l|q) = \prod_{k=0}^{\infty} \frac{\Gamma_q(\lambda + 4\eta k + 2(l+1)\eta)\,\Gamma_q(\lambda + 4\eta k - 2l\eta + 1)}{\Gamma_q(\lambda + 4\eta k + 2l\eta)\,\Gamma_q(\lambda + 4\eta k - 2(l-1)\eta + 1)},$$

$$\Gamma_q(x) = \frac{(q;q)_\infty}{(q^x;q)_\infty}(1-q)^{1-x}, \qquad (a;q)_\infty = \prod_{n=0}^{\infty}(1 - aq^n).$$

This result is consistent with that of [TF] which corresponds to the case $l = \frac{1}{2}$.

Details and the further results concerning the thermodynamic properties and the excited states of this model will be published in the forthcoming paper.

The author expresses his great thanks to Professors E. K. Sklyanin, H. J. de Vega, M. Wadati, E. Date, Kimio Ueno, M. Noumi and Dr. T. Deguchi for their comments, advices, interests and encouragement.

REFERENCES

[B] Babujian H.M., Phys.lett.A **90** (1982), 479–482.

[M] Mumford, D., *Tata Lectures on Theta I*, Birkhäuser, 1982.

[S] Sogo K., Phys.lett.A **104** (1984), 51–54.

[Sk1] Sklyanin E.K., Func. Anal. Appl. **16-4** (1982), 263-270.

[Sk2] ———, Func. Anal. Appl. **17-4** (1983), 273-284.

[Take] Takebe T., J.Phys.A **25** (1992), 1071–1083.

[Takh] Takhtajan L.A., Phys.lett.A **87** (1982), 479–482.

[TF] Takhtajan L.A., Faddeev L.D., Uspekhi Mat. Nauk **34:5** (1979), 13–63; English transl. in Russ. Math. Surv. **34** (1979), 11–68.

EXISTENCE THEOREM OF THE FOCK-LIKE SPACE FOR QUONS

TING YU* and ZHAOYAN WU
Center for Theoretical Physics, Jilin University, Changchun, Jilin 130023, P. R. China

ABSTRACT

It is pointed out that there were fatal flaws in an earlier attempt to prove the existence theorem of the Fock-like space for quons. The theorem is proven by using construction. It turns out that all the Fock-like spaces for quons with $q\in(-1,1)$ are isomorphic to the one with $q=0$, which allows the Boltzmann statistics.

The intermediate statistics has been the focus of extensive interest recently[1]. One of the several approaches to it uses deformations of the bilinear Bose and Fermi commutation relations. Particles obeying this type of statistics are called quons[2]. The Fock-like space for quons refers to a Hilbert space F_q satisfying the following conditions:

(i) There is a set of linear operators $\{a_1(q), a_2(q), \cdots, a_N(q)\}$ (or $\{a_j(q)\,|\,j\geqslant 1\}$) in F_q such that

$$a_j(q)a_k^+(q) - qa_k^+(q)a_j(q) = \delta_{jk}, \tag{1}$$

where q is a fixed number.

(ii) There is a vector in F_q, say $|\rangle$, with a unit norm, $\langle|\rangle=1$, killed by all $a_j(q)$'s, $a_j(q)|\rangle = 0$, and such that F_q is spanned by $|\rangle$ and all $a_{j_1}^+(q)a_{j_2}^+(q)\cdots a_{j_m}^+(q)|\rangle$'s.

The existence of the Fock-like space is of great importance to the concept of quons. Fivel gave a proof of it[3]. However, there were fatal flaws in his reasoning (see the appendix). We will prove the exitence theorem by using construction.

Theorem The Fock-like space exists if and only if $-1\leqslant q\leqslant 1$. Any two Fock-like spaces with the same $q\in[-1,1]$ and the same dimension of the single particle state subspace are isomorphic to each other, and the correspondence

$$a_{j_1}^+(q)\cdots a_{j_m}^+(q)|\rangle \;\longmapsto\; a'^+_{j_1}(q)\cdots a'^+_{j_m}(q)|\rangle',$$

is an isomorphism preserving the inner product (uniqueness).

To prove the necessity and uniqueness, suppose F_q is a Fock-like space. We can obtain

$$\langle|a_{j_m}\cdots a_{j_2}a_{j_1}a_{k_1}^+a_{k_2}^+\cdots a_{k_n}^+|\rangle = \begin{cases} 0, & \text{if } m\neq n;\\ \sum\limits_{P\in S_m}\Big(\prod\limits_{t=1}^{m}\delta_{j_t k_{P(t)}}\Big)q^{\nu(P)}, & \text{if } m=n, \end{cases} \tag{2}$$

*Permanent address: Department of Applied Mathematics, Jilin University of Technology, Changchun, Jilin 130025, P. R. China.

where $\nu(P)$ is the inversion number of P. We see that the inner product of any two generating vectors is completely determined by the number q, hence the uniqueness follows. Calculating the squared norms of the following vectors

$$|\psi_\pm\rangle = (a_1^+ a_2^+ \pm a_2^+ a_1^+)|\rangle,$$

we find

$$\langle\psi_\pm|\psi_\pm\rangle = 2(1 \pm q).$$

The positive definiteness of the inner product leads to the necessary condition

$$-1 \leqslant q \leqslant 1, \text{ for } N \geqslant 2.$$

From the equation

$$\langle|a_1 a_1 a_1^+ a_1^+|\rangle = 1 + q,$$

we obtain the necessary condition

$$q \geqslant -1, \text{ for } N = 1.$$

To prove the sufficiency we start with noting that the Fock spaces for Bosons and for Fermions are the Fock-like spaces for quons with $q = \pm 1$. We will construct Fock-like spaces for quons with $-1 < q < 1$.

Suppose F is a complex vector space of countably infinite dimensions, and B is a basis of F. We can always label the base vectors in B as $|\rangle$ and $|j_1 j_2 \cdots j_m\rangle (1 \leqslant m, 1 \leqslant j_t)$. Let q be a complex number. Define a sesquilinear form f: $F \times F \to C$, such that

$$f(|j_1\cdots j_m\rangle, |k_1\cdots k_n\rangle) \equiv \langle j_1\cdots j_m|k_1\cdots k_n\rangle = \begin{cases} 0, & \text{if } m \neq n \\ \sum_{P\in S_m}\left(\prod_{t=1}^{m}\delta_{j_t k_{P(t)}}\right)q^{\nu(P)}, & \text{if } m = n. \end{cases} \quad (3)$$

Since $\nu(P) = \nu(P^{-1})$, the Gram matrix is symmetrical. And it is real if and only if q is real. Henceforth we will suppose q is real, thus f is Hermitian. We are going to prove the Hermitian form f is positively definite for $-1 < q < 1$. For $q = 0$, the Gram matrix is the unit matrix. Hence it suffices to show that the Gram determinant does not vanish for $-1 < q < 1$. Define an Equivalence relation in B: $|j_1\cdots j_m\rangle \sim |k_1\cdots k_n\rangle$, iff $m = n$ & $\exists\ P \in S_m$ such that $|j_1\cdots j_m\rangle = |k_{P(1)}\cdots k_{P(m)}\rangle$. Obviously the Gram matrix elecments between nonequivalent base vectors vanish. For a base vector $|j_1\cdots j_m\rangle$, denote by F^J the subspace of F spanned by all its equivalent base vectors, and by H^J the following subgroup of S_m:

$$H^J = \left\{P \in S_m \,\middle|\, |j_1\cdots j_m\rangle = \left|j_{P(1)}\cdots j_{P(m)}\right\rangle\right\}. \quad (4)$$

It is easy to show that $|j_{P^{-1}(1)}\cdots j_{P^{-1}(m)}\rangle = |j_{Q^{-1}(1)}\cdots j_{Q^{-1}(m)}\rangle$, if and only if $PH^J = QH^J$. The dimension of F^j is equal to the index of H^J. The nonvanishing elements of the Gram matrix are

$$\langle j_{R^{-1}(1)}\cdots j_{R^{-1}(m)}|j_{S^{-1}(1)}\cdots j_{S^{-1}(m)}\rangle = \sum_{P\in RH^J S^{-1}} q^{\nu(P)} = \sum_{P\in H^J} q^{\nu(RPS^{-1})}, \quad (5)$$

where RH^JS^{-1} is the two-sided coset of H^J in S_m.

$|j_1\cdots j_m\rangle$ contains at most m different indices. For simplicity we assume that $1 \leqslant j_t \leqslant m$, $\forall\ 1 \leqslant t \leqslant m$. Let R_m be the subspace of F spanned by m^m base vectors with all indices no greater than m. Let us define two representations, D of S_m and T of U_m, in R_m:

$$D(P)|k_1\cdots k_m\rangle = |k_{P^{-1}(1)}\cdots k_{P^{-1}(P)}\rangle, \ \forall \ P \in S_m.$$

$$T(\alpha)|k_1\cdots k_m\rangle = |l_1\cdots k_m\rangle \alpha^{l_1}_{k_1}\cdots \alpha^{l_m}_{k_m}, \ \forall \ \alpha \in U_m.$$

According to the well-established result, there exists a new basis $\{e^{\lambda}_{bg}\}$ of R_m such that

$$D(P)e^{\lambda}_{bg} = \sum_{h=1}^{d\lambda} e^{\lambda}_{bh} D^{\lambda}_{hg}(P), \ \forall \ P \in S_m, \tag{6}$$

$$T(\alpha)e^{\lambda}_{bg} = \sum_{c=1}^{t_\lambda} e^{\lambda}_{cg} T^{\lambda}_{cb}(\alpha), \ \forall \ \alpha \in U_m, \tag{7}$$

where $d_\lambda(t_\lambda)$ is the dimension of the irreducible representation $D^\lambda(T^\lambda)$ of $S_m(U_m)$. Using the definition (3) of form f, one easily proves that f is invariant under $T(\alpha)(\alpha \in U_m)$:

$$\langle T(\alpha)j_1\cdots j_m|T(\alpha)k_1\cdots k_m\rangle = \langle j_1\cdots j_m|k_1\cdots k_m\rangle.$$

Thus we have

$$\langle T(\alpha)e^{\lambda}_{bg}|T(\alpha)e^{\mu}_{ch}\rangle = \langle e^{\lambda}_{bg}|e^{\mu}_{ch}\rangle, \ \forall \ \alpha \in U_m, \tag{8}$$

that is,

$$\overline{T^{\lambda}_{b'b}(\alpha)}\langle e^{\lambda}_{b'g}|e^{\mu}_{c'h}\rangle T^{\mu}_{c'c}(\alpha) = \langle e^{\lambda}_{bg}|e^{\mu}_{ch}\rangle, \ \forall \ \alpha \in U_m,$$

or

$$\langle e^{\lambda}_{b'g}|e^{\mu}_{c'h}\rangle T^{\mu}_{c'c}(\alpha) = T^{\lambda}_{b'b}(\alpha)\langle e^{\lambda}_{bg}|e^{\mu}_{ch}\rangle, \ \forall \ \alpha \in U_m.$$

Using Schur's lemma one obtains

$$\langle e^{\lambda}_{bg}|e^{\mu}_{ch}\rangle = \delta_{\lambda\mu}\delta_{bc}M^{\lambda}_{gh}(q). \tag{9}$$

Thus we see that the Gram matrix of the restricted form $f|_{F^J\times F^J}$ under the new basis is the direct sum of some $M^\lambda(q)$ to which the corresponding irreducible representations D^λ occur in F^J. Denote by Q_m the subspace spanned by all the old base vectors equivalent to $|1\ 2\cdots m\rangle$. It carries the regular representation of S_m. Thus the Gram matrix of the restricted form $f|_{Q_m\times Q_m}$ under the new basis is a direct sum containing $M^\lambda(q)$ d_λ times. Now, to prove the positive definiteness of the form f, it suffices to show that $f|_{Q_m\times Q_m}$ is nondegenerate for $-1<q<1$. Its $m!\times m!$ Gram matrix under the old basis has the following (P,Q)-elements:

$$\langle D(P)1\ 2\cdots m|D(Q)1\ 2\cdots m\rangle = q^{\nu(PQ^{-1})}. \tag{10}$$

Zagier[4] has proven that the Gram determinant is

$$\prod_{t=1}^{m-1}(1-q^{t(t+1)})^{(m-t)m!/t(t+1)}. \tag{11}$$

The only zeroes in (11) are roots of unity. We conclude that the form f can be taken as the inner product of F for $-1<q<1$.

Define a set of linear operator in F, $\{x_1, x_2, \cdots\}$, such that

$$x_j|j_1\cdots j_m\rangle = |j\ j_1\cdots j_m\rangle. \tag{12}$$

Their Hermitian conjugate operators $x_j^\dagger$'s are completely determined by Eq. (12). Using Eq. (3) we can prove

$$x_j^\dagger x_k - q x_k x_j^\dagger = \delta_{jk}.$$

Let $a_j(q)=x_j^\dagger$. We obtain the required operators (1). Thus we complete the proof for the case with countably infinite annihilation operators. Consider the subspace spanned by $\{|\,\rangle\}\cup\{|j_1\cdots j_m\rangle\,|m\geqslant 1 \ \&\ 1\leqslant j_i\leqslant N\}$. It provides the Fock-like space with N annihilation operators. Thus we have proven the existence and uniqueness theorem of the Fock-like space for quons.

Appendix

To prove the positive definiteness of the Hermitian form (2), Fivel's strategy was to show that "if there is one isotropic vector then there are a lot , indeed too many of them". He argued, "Decompose R_m into the direct sum of subspaces R_m^k, $k=1,2,\cdots$, which are representation spaces for irreducible representation V_k of U_m. Let P_k be the projector onto R_m^k. If (the isotropic vector) $|J\rangle\neq 0$, there is at least one k such that $|J_k\rangle\equiv P_k|J\rangle\neq 0$. Evidently $|J_k\rangle$ is isotropic, as are its images under h for all h in H. (To see this note that the projectors onto irreducible subspaces can be expressed as linear combinations of the elements h.) Since R_m^k is irreducible, by the above argument R_m^k is an isotropic subspace. Let $P'_k\equiv hP_kh^{-1}$ for some h and let $|J'\rangle\equiv h|j\rangle$. $|J'\rangle$ is isotropic and $P'_k|J'\rangle\neq 0$. It follows that the subspace projected by P'_k is also isotropic. Thus the Union of the subspaces belonging to the same class μ of irreducible representations is isotropic for at least one μ."

First, for $m\geqslant 3$, there are irreducible representations of S_m with more than one dimensions. Let D^λ be an irreducible representation of S_m with dimension $d_\lambda(\geqslant 2)$. The corresponding irreducible representation T^λ of U_m with dimension t_λ occurs in R_m d_λ times. If the restricted representation to irreducible subspace R_m^k, V_k, occurs in R_m more than once, then the projector P_k can not be expressed as a linear combination of the elements h. In fact, eq. (7.14) on p. 95 in Ref. 5, cited by Fivel as Ref. 7, does not provide the desired projector P_k, which projects the whole space R_m onto irreducible subspace R_m^k, for any linear combination of $h(T(a)$, $a\in U_m)$ keeps the quasi-diagonal structure of $T(a)$, expressed by eq. (7). In spite of this mistake, one can still say that if there is one nonvanishing isotropic vector $|J\rangle$ in R_m, then there must be (at least) one isotropic irreducible subspace, spanned by $\{P_j^k|J\rangle\,|\,1\leqslant j\leqslant t_\lambda\}$ for some Young diagram λ and k. Secondly, if P_k projects R_m onto R_m^k, then hP_kh^{-1} does too. To see this note that R_m^k is invariant under h. This conflicts with Fivel's claiming, and makes the rest of his proof useless.

References

1 F. Wilczek, Phys. Rev. Lett. 48, 1144(1982); 49, 957(1982).

2 O. W. Greenberg, Phys. Rev. Lett. 64, 705(1990); Phys. Rev. D43, 4111 (1991).

3 D. I. Fivel, Phys. Rev. Lett. 65, 3361(1990).

4 D. B. Zagier, unpublished, quoted in Ref. 2.

5 W. Miller, Symmetry Groups and Their Applications (Academic, New York, 1972), Chapter 4.

Conformal Theory of the Statistical Mechanical Models

Bowei Xu and Haiyan Zhang

Department of Physics, Shanghai Jiaotong University, Shanghai 200030, China

In this paper we first make simple calculation for Z_2 symmetric quantum chain, then we discuss an NNN interaction model by numerical calculation. The conformal field theories of both models are constructed.

Two-dimensional conformal field is related to two commuting Virasoro algebra [1,2,3]. The unitarity restricts the values of the central charge c, for $0 < c < 1$, to [4]

$$c = 1 - \frac{6}{(m+2)(m+3)}\,, \quad m = 1, 2, \ldots \tag{1}$$

The corresponding anomalous scaling dimensions Δ_p, q ($\bar{\Delta}_p, q$) are

$$\Delta_p, q = \bar{\Delta}_p, q = \frac{[p(m+3) - q(m+2)]^2 - 1}{4(m+2)(m+3)}, 1 \le p \le m+1, 1 \le q \le m+2 \tag{2}$$

In statistical mechanical models, often one-dimensional quantum chains with N lattices are discussed.[5,6] In these cases the values of Δ are related to the energy gaps between the ground state and the excited states as

$$E_\Delta(n, \bar{n}) - E_0 = \frac{2\pi\xi}{N}(\Delta + \bar{\Delta} + n + \bar{n}) = \frac{2\pi\xi}{N}(\chi + n + \bar{n}), n, \bar{n} = 0, 1, 2, \ldots \tag{3}$$

where $\chi = \eta/2$ is the scaling dimension of the operator concerned and ξ is an unknown constant since the Hamiltonian may in principle be multiplied by an overall factor, and E_0 is the ground state energy

$$E_0 = -\frac{c\pi\xi}{6N} + K \tag{4}$$

here K is the integral constant. Therefore, from the energy spectrum and with the aid of eq.s (3) and (4), one can obtain the value of c and Δ for a given model.

In this paper we first make simple calculation for an integrable statistical system, the Z_2 symmetric quantum chain. Then we discuss a one-dimensional quantum model with next-nearest-neighbour interaction. Only numerical calculation has been made.

The Hamiltonian of Z_2 symmetric quantum chain is

$$H_F = g\sum_{n=1}^{N}\sigma_n^z - \sum_{n=1}^{N-1}(\frac{1+\gamma}{2}\sigma_n^x\sigma_{n+1}^x + \frac{1-\gamma}{2}\sigma_n^y\sigma_{n+1}^y) \tag{5}$$

Using Jordan-Wigner transformation [7], eq.(5) can be written as

$$H_F = 2g\sum_{n=1}^{N}c_n^+c_n - \sum_{n=1}^{N-1}(\gamma c_n^+c_{n+1}^+ + c_n^+c_{n+1} + H.c.) \tag{6}$$

where c_n and c_n^+ are Fermi operators. The field equations of the systems are

$$\frac{dc_n}{dx} = \frac{1}{\xi}[H_F, c_n], \frac{dc_n^+}{dx} = \frac{1}{\xi}[H_F, c_n^+] \tag{7}$$

From eq.s (6) and (7), one obtains

$$\frac{dc_n}{dx} = \frac{2}{\xi}(1-g)c_n + \frac{2\gamma}{\xi}\frac{dc_n^+}{dy}, \frac{dc_n^+}{dx} = -\frac{2}{\xi}(1-g)c_n^+ - \frac{2\gamma}{\xi}\frac{dc_n}{dy}. \tag{8}$$

The conformal covariance demands

$$g = 1, \xi = 2\gamma. \tag{9}$$

Considering the periodic boundary condition, the conformally covariant Hamiltonian H_c is [8]

$$H_c = \frac{1}{\gamma}\sum_{n=1}^{N}c_n^+c_n - \frac{1}{2\gamma}\sum_{n=1}^{N-1}(\gamma c_n^+c_{n+1}^+ + c_n^+c_{n+1} + H.c.) + \frac{1}{2\gamma}(\gamma c_N^+c_1^+ + c_N^+c_1 + H.c.)e^{i\pi\mathcal{N}} \tag{10}$$

$$\mathcal{N} = \sum_{n=1}^{N}c_n^+c_n. \tag{11}$$

With the aid of a canonical transformation one can diagonize Hamiltonian H_c and obtain the eigenvalues of H_c [9,10]

$$\Lambda^+(n) = \frac{1}{\gamma}[(1-cos\frac{2\pi(n+\frac{1}{2})}{N})^2 + \gamma^2 sin^2\frac{2\pi(n+\frac{1}{2})}{N}]^{\frac{1}{2}}, for \mathcal{N} even; \tag{12}$$

$$\Lambda^-(n) = \frac{1}{\gamma}[(1-cos\frac{2\pi n}{N})^2 + \gamma^2 sin^2\frac{2\pi n}{N}]^{\frac{1}{2}}, for \mathcal{N} odd. \tag{13}$$

Obviously, the ground state energy is

$$E_0 = -\frac{1}{2}\sum_{n=0}^{N-1}\Lambda^+(n) = -cosec(\frac{\pi}{2N}) \approx -(\frac{2N}{\pi} + \frac{\pi}{12N} + \ldots) \tag{14}$$

Comparing with eq.(4), one has

$$c = \frac{1}{2} \tag{15}$$

The lowest excited state energy for $\mathcal{N}$ even is

$$E_0^+ = \Lambda^+(0) + \Lambda^+(N-1) + E_0 = 4sin(\frac{\pi}{2N}) + E_0 \approx \frac{2\pi}{N} + E_0 \tag{16}$$

Comparing with eq.(3), one has

$$\Delta_\epsilon = \bar{\Delta}_\epsilon = \frac{1}{2} \tag{17}$$

The lowest excited state energy for $\mathcal{N}$ odd is

$$E_0^- = -\frac{1}{2}\sum_{n=0}^{N-1} \Lambda^-(n) = tan(\frac{\pi}{4N}) + E_0 \approx \frac{\pi}{4N} + E_0 \tag{18}$$

Comparing with eq.(3), one has

$$\Delta_\sigma = \bar{\Delta}_\sigma = \frac{1}{16} \tag{19}$$

The next example is the NNN model with Hamiltonian

$$H = g\sum_n \sigma_n^z - \sum_n \sigma_n^x \sigma_{n+1}^x - \lambda\sum_n(\frac{1+\beta}{2}\sigma_n^x\sigma_{n+2}^x + \frac{1-\beta}{2}\sigma_n^y\sigma_{n+2}^y) \tag{20}$$

If a model has critical phenomenon, there must be a critical point at which Δ ($\bar{\Delta}$) and c will approach the fixed values as $N \to \infty$.

We use Lanczos method [11] to calculate the energy gap between the ground state and the first-excited state and obtain

$$g = 1 + \beta + \beta\lambda f \tag{21}$$

as the critical point condition, where f varies slowly with β and λ. We also calculate the values of c/η for different β and λ, which all approach 2, and thus verifies that this model belongs to Ising universality. From eq.s (3), (4), and with $c = 1/2$, we obtain the values of $\chi_\sigma(n = \bar{n} = 0)$. Obviously,

$$\Delta_\sigma = \bar{\Delta}_\sigma \approx \frac{1}{16} \tag{22}$$

From the energy gaps between the ground state and the upper-excited states, one can calculate the values of other anomalous scaling dimensions. The results of $\chi_\epsilon(n = \bar{n} = 0)$ are acquired. Also clearly

$$\Delta_\epsilon = \bar{\Delta}_\epsilon \approx \frac{1}{2} \tag{23}$$

The results are accurate to 10^{-2}. Likewise, we can calculate the values of

$$\Lambda_\Delta(n,\bar{n}) = \frac{N\cdot[E_\Delta(n,\bar{n})-E_0]}{2\pi\xi} \tag{24}$$

From our calculation of $n=\bar{n}=1$, the following relations are acquired

$$\Lambda_\sigma(1,1)\approx\chi_\sigma+1+1, \Lambda_\epsilon(1,1)\approx\chi_\epsilon+1+1 \tag{25}$$

These results are compatible with the prediction of conformal field theory that with scaling dimension χ_Δ, it will appear a tower of states with the corresponding mass gap amplitudes converging to $\chi_\Delta+n+\bar{n}(n,\bar{n}=0,1,2,...)$. The accuracies of $(\Lambda_\sigma(1,1)-\chi_\sigma)$ and $(\Lambda_\epsilon(1,1)-\chi_\epsilon)$ for different β and λ are approximately 97% and 93% respectively.

The calculated results of c and Δ of both models are consistent with the predictions of eq.s (1) and (2) with $m=1$, which are

$$c=\frac{1}{2}, \Delta_{11}=\Delta_{23}=0, \Delta_{12}=\Delta_{22}=\frac{1}{16}, \Delta_{13}=\Delta_{21}=\frac{1}{2} \tag{26}$$

Conclusively, we have set up the conformal field theory of both models.

REFERENCES

[1] M.A.Virasoro, Phys. Rev. D 1, 2933 (1970).

[2] A.A.Belavin,A.M.Polyakov,A.B.Zamolodchikov, Nucl. Phys. B **241**, 333 (1984).

[3] S.Fubini,A.J.Hanson,R.Jackiw, Phys. Rev. D **7**, 1732 (1973).

[4] D.Friedan,Z.Qiu,S.Shenker, Phys. Rev. Lett. **52**, 1575 (1984).

[5] G v Gehlen,V.Rittenberg,H.Ruegg, J. Phys. A: Math. Gen. **19**, 107 (1985)

[6] J.L.Cardy, Nucl. Phys. B **270**, 186 (1986).

[7] T.D.Schultz,D.C.Mattis,E.H.Lieb, Rev. Mod. Phys. **36**, 856 (1964).

[8] B.W.Xu, ACTA Physica Sinica. 40, 681 (1991).

[9] E.H.Lieb,T.D.Schultz,D.C.Mattis, Ann. Phys. **16**, 407 (1961).

[10] S.Katsura, Phys. Rev. **127**, 1508 (1962).

[11] H.H.Roomany,H.W.Wyld,L.E.Holloway, Phys. Rev. D **21**, 1557 (1980).

Parallel Sessions 5

Unified Theories and Gravitation

Anyons and Coherent States for Gravitons

George Chapline
Lawrence Livermore National Laboratory
Livermore, California 94550

Abstract
The remarkable connection between 4-d quantum gravity and knot theory suggests that there is some deep connection between quantum gravity, quantum groups, and fractional statistics. In this paper a 3-d generalization of an exactly solvable model for interacting anyons is described whose ground-state wavefunctions satisfy a Wheeler-DeWitt-like equation. Interpreting the quasi-particles in this model as non-linear gravitons provides an attractive picture of how classical spacetime may emerge from quantum gravity. One can also exhibit a ground state corresponding to flat spacetime with ODLO, which may provide a clue as to why the observed universe has such low entropy.

One of the most important outstanding problems of quantum gravity is to find a solution to the Wheeler-DeWitt equation that could represent the macroscopic space-time of general relativity in much the same way that coherent states for photons are quantum representations for classical electromagnetic fields. In this talk I will describe a new approach[1] to this problem based on the recent discovery[2] of a new class of solvable models in 3 dimensions. The relation of this approach to a fundamental quantum theory of gravity, e.g., one based on string theory, is as yet unclear. Nevertheless, it does appear that a framework is now available for understanding at least qualitatively how classical space-time can emerge from a quantum theory of gravity. It is perhaps worth noting at the outset that the existence of this framework, independent of the explicit introduction of a fundamental quantum theory of gravity, seems to be intimately related to the discovery by Jones of a connection between knot invariants and soluble statistical models. Hence this work may perhaps be viewed as a "practical application" of the conference theme.

The new approach to finding a quantum representation of classical spacetime was derived in Ref. 1 as a 3-dimensional generalization of a model[3] for anyonic superconductivity. In particular, a candidate wavefunction for flat spacetime was constructed by considering static, self-dual solutions of the gauged non-linear Schrödinger equation[4]

$$i\frac{\partial\Psi}{\partial t} = D^2\Psi + [[\Psi^+,\Psi],\Psi] \tag{1}$$

in $2+1$ dimensions, where $D_\alpha\Psi \equiv -i\partial_\alpha\Psi + [A_\alpha,\Psi]$. The gauge fields A_α in this model are SU(N) Chern-Simons gauge fields where $N \to \infty$. It can be shown[5] that for arbitrary N the solution Ψ of the self-dual non-linear Schrödinger equation has the form $e^\phi\ \hat{e}_\alpha$, where $\hat{e}_\alpha$ is a raising or lowering operator in the Lie algebra of SU(N). In the limit $N \to \infty$ it follows that $\Psi = e^{\phi(x,y,r)}$, where (x,y) are the 2-dimensional coordinates, r is a continuous variable replacing the index α, and $\phi(x,y,r) \equiv \Delta\phi_\alpha$.

The self-dual static solutions of Ref. 5 are defined by the requirement that the field Ψ satisfies the massless 2-dimensional Dirac equation:

$$(D_x - iD_y)\Psi = 0 \quad , \tag{2}$$

where $D_a \equiv -i\partial_a + [A_a$. The gauge field A_a is not dynamical in this model but instead is an auxiliary field describing the interaction of quasi-particles. In the static limit it is determined by the "self-duality" condition

$$B \equiv F_{12} = -[\Psi^*,\Psi], \tag{3}$$

where the brackets represent an SU(N) commutator.

The existence of solutions to Eq. (2) for arbitrary magnetic field configurations is closely related to the Atiyah-Singer index theorem and supersymmetry.[6] This last observation provides a clue as to why Eqs. (2) and (3) might be used to construct a model of quantum gravity. In a supersymmetric quantum field theory the ground states Ψ_i with unbroken supersymmetry have zero energy; i.e. $H\Psi_i = 0$. In our model the quantum fields will be represented by the moduli for the zero energy solutions of Eqs. (2) and (3).

Invoking the relation between supersymmetry and the generator of time translations one finds that in the limit $N \to \infty$ the hamiltonian operator H for such a quantum field theory can be represented in the form

$$H = \sum_j Q_j^\dagger Q_j \tag{4}$$

where (x_j, y_j) is the position of the jth quasi-particle and $Q_j \equiv D_{x_j} - iD_{y_j}$ is the supersymmetry operator associated with (2). The reason the limit $N \to \infty$ is necessary to obtain a hamiltonian in the form (4) is that for general N the gauge potential A_a in Eq. (2) cannot be written as a sum of contributions from each quasi-particle.

In Ref. 5 it is shown that Eqs. (2) and (3) can be solved exactly for arbitrary N by setting $\Psi = \sum_\alpha e_{\pm\alpha} e^{\phi_\alpha}$ and $A_i = \sum_\alpha a_i^\alpha h_\alpha$, where h_α and $e_{\pm\alpha}$ are the usual generators for an SU(N) Lie algebra: $[h_\alpha, h_\beta] = 0, [e_\alpha, e_\beta] = \delta_{\alpha+\beta,0} h_\beta$, and $[h_\alpha, e_{\pm\beta}] = \pm K_{\alpha\beta} e_{\pm\beta}$. With this ansatz Eqs. (2) and (3) can be combined to give

$$\partial_z \partial_{\bar{z}} \phi_\alpha + \sum_\beta K_{\alpha\beta} \exp(\phi_\alpha) = 0 \tag{5}$$

Equation (5) is the Toda molecule equation, and it is exactly solvable for any N. In the limit $N \to \infty$ Eq. (5) becomes[7]

$$\partial_z \partial_{\bar{z}} \phi + \frac{\partial^2}{\partial r^2} e^\phi = 0 \tag{6}$$

This equation occurs in the theory of self-dual Einstein spaces,[8] and has a well-known solution: the Eguchi-Hanson solution[9]

$$\phi = \ell n \frac{r^2 - a^2}{(1 + z\bar{z})^2} \quad , r > a \tag{7}$$

By a change of variables[10] Eq. (6) can be transformed into a 3-dimensional Laplace equation for a certain function S(ϕ). The solution corresponding to (7) is

$$S = \frac{1}{2} \sum_{j=1}^{2} \ell n \frac{R_i + U - U_i}{R_i - U + U_i} \tag{8}$$

where $-\infty < U < \infty$ is now an unrestricted spatial coordinate, $R_i^2 \equiv (U - U_i)^2 + 4(z - z_i)(\bar{z} - \bar{z}_i)$, $z \equiv x + iy$, and (x_i, y_i, U_i) are the locations of the quasi-particles. Equation (8) describes the 2-center gravitational instanton of Gibbons and Hawking.[11] The multi-instanton solution corresponds to a sum of terms of the same form as in Eq. (8). It is not hard to show that even in the multi-center case e^S satisfies Eq. (2). The essential point is that in the limit $N \to \infty$ the operator Q_j has the form $-i\partial_{z_j} + A_{z_j}$, where the effective gauge field A_{z_j} is a sum of terms coming from each quasi-particle. In particular, the effective magnetic field B associated with A_z (more precisely the "U" component of the effective magnetic field) is given by

$$B(X) = \sum_k \partial_U \left(\frac{1}{|X - X_k|} \right) \tag{9}$$

where $\vec{X} = (x, y, U)$. It follows from Eqs. (8) and (9) that

$$B = \partial_U^2 S \tag{10}$$

However, it can be shown[8] that a self-dual Euclidean Einstein space with a translational Killing vector is described by the equation

$$(\partial_z \partial_{\bar{z}} + \partial_U^2)\, S = 0 \tag{11}$$

so that

$$B = -\partial_z \partial_{\bar{z}} S \tag{12}$$

Remarkably Eq. (12) is just the relation between B and S required for e^S to satisfy Eq. (2). Thus e^S, where S is a sum of terms of the form given in Eq. (8) will be annihilated by the hamiltonian (4).

The multi-center instanton solutions of Gibbons and Hawking asymptotically approach flat Euclidean space modulo a Z_p identification. Performing a Wick rotation this asymptotic space can be identified with a topologically non-trivial spacetime which forms a background for the instanton fluctuations. Our proposal, then, for a quantum version of macroscopic space time is simply to take the large p limit of the condensate wave function constructed in Ref. 8; viz.

$$\Psi = f(\bar{z}) \prod_{k>j}^{\infty} \left[\frac{R_{jk} + U_{jk}}{R_{jk} - U_{jk}} \right]^{1/2} \tag{13}$$

where $f(\bar{z})$ is an entire function of the $\bar{z}_j$, $R_{jk}^2 \equiv U_{jk}^2 + 4(z_j - z_k)(\bar{z}_j - \bar{z}_k)$, and $U_{jk} = U_j - U_k$. More generally, one could consider wave functions that mix self-dual and anti-self-dual gravitons. Indeed, by analogy with the anyonic superfluid[4] one would want to introduce pairing between self-dual and anti-self-dual gravitons.

In general, it is not possible to find exact solutions of classical field equations that mix instantons and anti-instantons. However, in the case of general relativity the full Riemann curvature of vacuum spacetime can be exactly decomposed[12] into a self-dual and anti-self-dual part. This suggests that a realistic quantum theory of macroscopic space-time should include the possibility of mixing self-dual and anti-self-dual gravitons. In fact, our approach to constructing a condensate wavefunction allows this possibility. Indeed, the Atiyah-Singer theorem guarantees[6] that zero-energy solutions to Eq. (2) will exist for arbitrary magnetic fields; in particular since the effective A_z in Eq. (2) comes from a sum of contributions from each quasi-particle, reversing the sign of the contribution from a particular quasi-particle (corresponding to reversing the sign on the r.h.s. of Eq. (3)) will not alter the existence of a solution to Eq. (2).

Let us partition the set of quasi-particles in Eq. (13) into 2 sets with indices $\{j\}$ and $\{[j]\}$. Then we can define a "pairing" wavefunction

$$\begin{aligned} \bar{\Psi} = \sum_{(\chi)\text{fixed}} & \prod_{i<j} (\bar{z}_i - \bar{z}_j)(\bar{z}_{[i]} - \bar{z}_{[j]}) \\ & \bullet \prod_{i<j} \frac{R_{ij} + U_{ij}}{R_{ij} - U_{ij}} \frac{R_{[i][j]} + U_{[i][j]}}{R_{[i][j]} - U_{[i][j]}} \\ & \bullet \prod_{i,j} \frac{R_{i[j]} + U_{i[j]}}{R_{i[j]} - U_{i[j]}} \end{aligned} \tag{14}$$

where $\sum_{(\chi)\text{fixed}}$ is a sum over all partitions of the set of quasi-particles such that the overall Euler characteristic χ is fixed. In a fashion similar to the pairing of anyons with opposite weights in an anyonic superfluid,[3] the wavefunction (14) favors pairing of self-dual and anti-self dual gravitons. To see this we follow Girvin and MacDonald[11] and introduce the singular gauge change transformation

$$\Psi_{GR} = \bar{\Psi} \prod_{ij} \frac{z_i - z_{[j]}}{| z_i - z_{[j]} |} \quad . \tag{15}$$

The $\{[j]\}$ here represent anti-self-dual gravitons. In the case of equal numbers of self-dual and anti-self-dual gravitons the overall Euler characteristic vanishes and so the wavefunction (15) might be considered as a candidate for a quantum version of flat Minkowski space.

The multi-instanton solutions of Gibbons and Hawking are asymptotically locally flat. However, the global topology of these solutions is non-trivial, and in a neighborhood of spatial infinity they resemble R^4/Γ_p, where Γ_p is the cyclic group of order p. The non-trivial global topology of these solutions is a consequence of the fact that the vector potential associated with the effective magnetic field ($B(x)$ of Eq. 9) has Dirac string type singularities attached to each point X_k. In order to remove these singularities one must surround each point with two coordinate patches, (τ_k, X) and (τ_{K+1}, X), where τ_{K+1} is identified with τ_k plus twice the aximuthal angle of the point X relative to the line joining X_k and X_{K+1}. The non-trivial global topology of these solutions is reflected in the fact that they have Euler number $\chi = p$.

In contrast to the classical multi-instanton solutions the condensate wavefunction (15) allows mixing self-dual and anti-self-dual instantons. Since the Dirac string singularities can now join self-dual and anti-self-instantous the global topology can be trivial. Indeed, in the case where these are equal numbers of self-dual and anti-self-dual gravitons it is reasonable to expect that over large distances one is representing (via a Wick rotation) flat spacetime. Locally the geometry and topology of spacetime will fluctuate. In particular the finite separation of self-dual and anti-self-dual gravitons will give rise to variations in the local Pontriagen density and topological signature. It should be noted though that in contrast with previous speculations concerning spacetime "foam" these local fluctuations in topology are completely consistent with the existence of quantum coherence. Indeed, as noted above the wavefunction (15) posses long range off-diagonal order, which would imply that a universe described by this wavefunction should manifest quantum effects on a macroscopic scale.

Because the condensate wavefunction (14) represents a superfluid there should be large fluctuations in the density of gravitons, which in turn should be associated with fluctuations in the curvature of spacetime. Indeed, our model for flat space-time provides a natural explanation for the type of curvature fluctuations recently observed in the cosmic microwave background. Further, since according to our model the universe is in pure quantum state the entropy of the universe should be extremely small, which is in accordance with observation.

We would also of course like to make contact with the fashionable view that string theory provides a fundamental theory of gravity. Unfortunately, it is not clear at the present time how to make contact with conventional string theory. It is conceivable that our model for spacetime is somehow related to the "dual" formulation of superstring theory in terms of 5-brane solutions of the superstring field equations.[14] Alternatively, one could take the point of view[15] that quantum gravity is more fundamental than superstring theory, with the quantum deformation parameter of knot theory specifying the number of extra dimensions. Unfortunately, the nonlinear gravitons in our model, being extended objects don't resemble the groundstate of a conventional superstring. Furthermore, the presence of a long-range gauge field, which is a reflection in the present model of the fractional statistics of anyons, means that our nonlinear gravitons are neither bosons nor fermions. One might imagine that the nonlinear gravitons in our model can be identified with excited states of a fractional superstring,[16] but this would require a theory of the internal quantum dynamics of nonlinear gravitons, which is only beginning to be understood.[17]

The author is very grateful for discussions with Steve Libby and Kengo Yamagishi. Work performed under the auspices of the U.S. Department of Energy by Lawrence Livermore National Laboratory under contract W-7405-Eng-48.

References

1. G. Chapline, *Mod. Phys. Lett.* **A**, (to be published).
2. G. Chapline and K. Yamagishi, *Phys. Rev. Lett.* **66** , 3064 (1991).
3. S. Girvin, A. MacDonald, M. Fisher, S. Rey, and J. Sethna, *Phys. Rev. Lett.* **65** , 1671 (1990).
4. R. Jackiw and S.Y. Pi, *Phys. Rev. Lett.* **64** , 2969 (1990).
5. B. Grossman, *Phys. Rev. Lett.* **65** , 3230 (1990).
6. G.W. Semenoff and P. Sodana, *Phys. Rev. Lett.* **57** , 1195 (1986).
7. Q.-H. Park, *Phys. Lett.* **B236** , 429 (1990).
8. C.P. Boyer and J.D. Finley, *M. Math. Phys.* **23** , 1126 (1982).
9. T. Eguchi and A. Hanson, *Phys. Lett.* **74B** , 249 (1978).
10. M.K. Prasad, *Phys. Lett.* **83B** , 310 (1979).
11. G.W. Gibbons and S.W. Hawking, *Phys. Lett.* **78B** , 430 (1978).
12. R. Penrose, *General Relativity and Gravitation* **7** , 31 (1976).
13. S.M. Girvin and A.H. MacDonald, *Phys. Rev. Lett.* **58** , 1252 (1987).
14. A. Strominger, *Nucl. Phys.* **B343** , 167 (1990).
15. G. Chapline, *Mod. Phys. Lett.* **A5** , 2165 (1990).
16. P.C. Argyres and S.-H. Henry Tye, *Phys. Rev. Lett.* **67** , 3339 (1991).
17. K. Yamagishi, and G. Chapline, *Class. Quant. Grav.* **8**, 427 (1991).

CLIFFORD VIEW OF THE K=1 INSTANTON

J.S.R.Chisholm,
Institute of Mathematics and Statistics,
University of Kent, Canterbury, ENGLAND.

Abstract: The algebraic structure of the k=1 instanton is defined entirely within the Clifford Algebra of Euclidean 4-space.

1.Introduction

My colleague Dr. Ruth Farwell and I have studied Spin Gauge Theories based on Clifford Algebras, and have begun to use these algebras to develop a simple description of metric manifolds [1]. We have a prejudice that many areas of mathematics can be linked through the Clifford formalism, and that this formalism is often simpler than standard descriptions. Here I am presenting the small example of the conversion to the Clifford formalism of the familiar k=1 instanton. I use as comparison the account in Chapter 10 of Nash and Sen [2], especially section 10.9; here the instanton is described by a mixture of differential geometry and quaternion algebra. My account will be based upon a single Clifford algebra, that appropriate to the flat vector space of the instanton. The quaternions can then be seen as a subalgebra of the larger algebra, linking them to the vector space. This approach underlines the fact that this space is four-dimensional Euclidean space, and cannot be space-time. I make no attempt to discuss topological properties of instantons , or to deal with solutions with higher values of k [3].

A distant motive for this study is the belief that there is a close association between Clifford Analysis and Cohomology. A relationship between Clifford Algebra and Topology has long been known [4]. Since that time, Clifford Analysis has been developed as a wide generalisation of Complex Analysis [5,6,7], and recent work [8,9] links it to topological concepts.

2. The Instanton

The vector space is taken to be $R^{0,4}$ ($R^{4,0}$ is equivalent), with basis vectors $\{e_\mu;\ \mu=0,1,2,3\}$ and coordinates $\{x^\mu\}$. The partial derivatives $\{\partial_\mu\}$ mutually commute, and $\{e_\mu\}$ satisfy the relations

$$\{e_\mu, e_\nu\} = -2I\delta_{\mu\nu}, \tag{2.1}$$

where I is the unit of the algebra $R_{0,4}$. The pseudoscalar

$$\eta = e_1 e_2 e_3 e_4 \tag{2.2}$$

satisfies $\eta^2 = I$, so that

$$P_\pm = (I \pm \eta)/2 \tag{2.3}$$

are idempotents of the algebra satisfying

$$(P_+)^2 = P_+,\ (P_-)^2 = P_-,\ P_+ + P_- = I,\ P_+ P_- = 0. \tag{2.4}$$

Within the algebra, the covariant derivative is taken to be

$$D_\mu = I\partial_\mu + A_\mu(x), \tag{2.5}$$

with

$$A_\mu(x) = (e_\mu \wedge x)\, P_-/(\lambda^2 + x^2), \tag{2.6}$$

λ constant, $x = x^\mu e_\mu$, $x^2 = -\Sigma(x^\mu)^2 I = -x^2 I$, and

$$e_\mu \wedge x = (e_\mu x - x e_\mu). \tag{2.7}$$

We note that in (2.6), $(e_\mu \wedge x)$ and P_- commute; the bivector $A_\mu(x)$ is equivalent to (10.106) of Nash and Sen [2].

By direct calculation, the Yang-Mills field

$$F_{\mu\nu} = \partial_\mu A_\nu - \partial_\nu A_\mu + [A_\mu, A_\nu]$$

is equal to

$$F_{\mu\nu} = -2\lambda^2 e_{\mu\nu} P_-/(\lambda^2 + x^2)^2, \tag{2.8}$$

where $e_{\mu\nu} = e_\mu e_\nu$, corresponding to (10.109) of Nash and Sen. The factor P_- ensures the self-duality of $F_{\mu\nu}$. Further direct calculation shows that the non-linear equation

$$[I\partial_\mu + A_\mu, F_{\mu\nu}] = 0 \tag{2.9}$$

is satisfied.

Dr. Farwell has pointed out [10] that adding the scalar part $e_\mu . x = (e_\mu x + x e_\mu)$ to $e_\mu \wedge x$ in (2.6) has no effect on either (2.8) or (2.9), so that (2.6) can be replaced by

$$A_\mu(x) = (e_\mu x)\, P_-/(\lambda^2 + x^2). \tag{2.10}$$

3. Equivalence to the quaternionic description

The potential or connection (2.6) and the field (2.8) are in the even part of the algebra $R_{0,4}$, which is isomorphic to $R_{0,3}$. We can take as basis of $R_{0,3}$ the elements

$$f_r = e_0 e_r = e_{0r} \quad (r=1,2,3); \tag{3.1}$$

then the bivectors and pseudoscalar of $R_{0,3}$ are

$$f_{23} = f_2f_3 = e_{23},\ f_{31} = e_{31},\ f_{12} = e_{12}, \qquad (3.2)$$

and $\phi = f_1f_2f_3 = \eta$. The idempotents are therefore those of $R_{0,4}$,

$$P_\pm = (I \pm \phi)/2; \qquad (3.3)$$

these commute with all elements of $R_{0,3}$ and divide $R_{0,3}$ into two mutually annihilating quaternion algebras:

$$R_{0,3} = P_+R_{0,3} \oplus P_-R_{0,3} = R_{0,2} \oplus R_{0,2}. \qquad (3.4)$$

Within the subalgebra $P_-R_{0,3}$, we can identify the quaternion elements $\mathbf{i}$, $\mathbf{j}$ and $\mathbf{k}=\mathbf{ij}$ by

$$\mathbf{i}=-f_{23}P_-=-f_1P_-\,,\ \mathbf{j}=-f_{31}P_-=-f_2P_-\,,\ \mathbf{k}=-f_{12}P_-=-f_3P_- \qquad (3.5)$$

Using the identifications (3.1)-(3.5), the components of the potential (2.6) become

$$A_0 = (x_1\mathbf{i} + x_2\mathbf{j} + x_3\mathbf{k})/(\lambda^2 + x^2), \qquad (3.6a)$$

$$A_1 = (-x_0\mathbf{i} - x_3\mathbf{j} + x_2\mathbf{k})/(\lambda^2 + x^2), \qquad (3.6b)$$

$$A_2 = (x_3\mathbf{i} - x_0\mathbf{j} - x_1\mathbf{k})/(\lambda^2 + x^2), \qquad (3.6c)$$

$$A_3 = (-x_2\mathbf{i} + x_1\mathbf{j} - x_0\mathbf{k})/(\lambda^2 + x^2); \qquad (3.6d)$$

these are the expressions given by (10.110) of Nash and Sen. The subalgebra $P_+R_{0,3}$ likewise defines the anti-self-dual instanton.

4. Comments and Acknowledgments

The instanton potential and field have been defined in terms of the natural algebra $R_{0,4}$ of the space $R^{0,4}$ of coordinates $\{x^\mu\}$. Within this algebra, the quaternion elements have been identified in (3.5) as a fixed combination of 'magnetic' and 'electric' bivectors. The work has been carried out entirely within the Clifford algebra formalism, and the self-duality of the potential is explicit in the factor P_-. Within this formalism, it is clear that the instanton cannot be 'analytically continued' to either of the space-time algebras $R_{1,3}$ or $R_{3,1}$. The even algebra of both of these algebras is the Pauli algebra, whose pseudoscalar ε satisfies $\varepsilon^2 = -I$. So real idempotents $P_\pm$ do not exist within these algebras, and the identification (3.4) and (3.5) of the quaternion algebra is invalid. It is therefore very difficult to accept the instanton as a relevant physical concept.

The potential A_μ and the field $F_{\mu\nu}$ are bivectors of the algebra $R_{0,4}$. The bivectors of any Clifford algebra form a Lie algebra; the six bivectors of $R_{0,4}$ generate rotations in four-space. This group is divided into two SU(2) groups by the split (3.4). The generators of the SU(2) group relevant to the instanton defined by (2.6) are

$$(e_{01}P_-, e_{02}P_-, e_{03}P_-)/2. \qquad (4.1)$$

Each of these elements generates two simultaneous rotations in different planes in four-space; for example, the first generator in (4.1) is equal to $(e_{01} + e_{23})/4$.

I am very grateful to Dr. Ruth Farwell for her comments on this work, and for our long collaboration during which we have learnt so much together, and I am pleased to thank colleagues in the Universities of Kent and Gent for their interest in our work.

References

1. J.S.R.Chisholm and R.S.Farwell, *'Clifford Approach to Metric Manifolds'*, Supp. ai Rec. del Circ. Mat. di Palermo, II, 26 (1991).
2. C.Nash and S.Sen, *'Topology and Geometry for Physicists'*, (Academic Press, 1983).
3. S.K.Donaldson, *'The Interface of Mathematics and Particle Physics'*, eds. D.G.Quillen, G.B.Segal and Tsou.S.T. (Clarendon,1990).
4. M.F.Atiyah, R.Bott and R.Shapiro, *'Clifford modules'*, Topology 3, 3 (1964).
5. D.Hestenes, *'Multivector Functions'*, J. Math, Analysis Appl. 24, 467 (1968).
6. R.Delanghe, *'On Regular-analytic Functions with Values in a Clifford Algebra'*, Math.Ann. 185, 91 (1970).
7. F.Brackx, R.Delanghe and F.Sommen, *'Clifford Analysis'*, Research Notes in Mathematics (Pitman, 1982).
8. F.Sommen and V.Soucek, *"Monogenic Differential Forms'*, Seminar of Algebra and Functional Analysis, University of Gent (1992).
9. R.Delanghe, F.Sommen and V.Soucek, *'Residues in Clifford Analysis'*, Seminar of Algebra and Functional Analysis, University of Gent (1992).
10. R.S.Farwell, *private communication* (1992).

FOUR-FORCE UNIFICATION USING CLIFFORD ALGEBRAS

J.S.R.Chisholm,
Institute of Mathematics and Statistics,
University of Kent, Canterbury, ENGLAND

R.S.Farwell,
Department of Mathematical Sciences,
University of Brighton, Brighton, ENGLAND.

1. Abstract

The principles of 'Spin Gauge Theories' originally were laid down independently by several groups [1,2,3,4]. More recently we have applied these principles in developing the basis of a unified theory of the four fundamental interactions [5,6,7,8,9,10]. In this brief talk, we summarise the basic assumptions of our model, and state the consequences of these assumptions. We have a number of serious problems still to solve, in particular establishing the renormalisability of the theory.

2. Basic assumptions

The model is based upon the Clifford algebra $R_{3,8}$, with basis vectors spanning an 11-dimensional space. In this space, the dimensions 1-4 constitute a curved space-time manifold. Dimensions 5-7 have negative signature, and constitute the flat 'electroweak space', since the helicity-dependent electroweak interactions arise from gauge transformations in this space. The remaining dimensions also form a flat subspace, with signature (+,+,-,-,), and the SU(4) gauge group in this space breaks to the SU(3) of QCD and the helicity-independent electroweak U(1) transformation.

The minimal representation of elements of the algebra $R_{3,8}$ is by 32x32 matrices. The states of the leptons and quarks of three colours, of one family, are represented by 8x8 blocks, each a left ideal within the algebra and we symmetrise over the quark colours. This structure accords with the breaking of the SU(4) group. The groups of gauge transformations associated with the subspaces of the 11-dimensional space, and their metrics, are set out in the first three lines of Table 1.

It is a characteristic of spin gauge theories that both the states and the operators transform under x-dependent gauge transformations. Thus the operators of the theory, including the representation $\{\Gamma_\mu(x); \mu=1,2,3,4\}$ of the Dirac matrices in space-time, are x-dependent in a general gauge; they also have an x-dependence as a result of the curvature [6]. These four basis vectors therefore form a field in space-time; we call it the 'Frame Field', since these basis vectors define the coordinate frame at all points. We regard the frame field as a *physical* field; we illustrate the reason by considering the Dirac equation for the electron. Then the frame field $\{\gamma_\mu(x)\}$ and the unit matrix U satisfy the identity

$$\gamma^\mu(x)\gamma_\mu(x) = 4U, \tag{2.1}$$

which is true on a curved manifold and in any gauge. Using this, we can write the Dirac equation in the form

$$\gamma^\mu[\partial_\mu - ieA_\mu(x) + im\gamma_\mu(x)/4]\,\psi(x) = 0. \tag{2.2}$$

Comparing the electromagnetic and frame field terms, we see that m/4 is the strength of interaction of the electron with the frame field, reducing the speed of the electron below that of light. Adding the frame field term to the usual covariant derivative gives the 'extended covariant derivative'

$$\Delta_\mu = U[\partial_\mu - ieA_\mu(x)] + im\gamma_\mu(x)/4. \tag{2.3}$$

In the full model, Δ_μ is a 32x32 matrix operator containing the SL(2,C), left-handed SU(2), right-handed U(1), and SU(3)xU(1) interactions. Since $\{\Gamma_\mu(x)\}$ does not commute with the SL(2,C) operators or with the helicity operators, it produces extra terms when boson Lagrangian terms are calculated from the usual expression

$$\mathrm{Tr}\{[\Delta_\mu,\Delta_\nu][\Delta_\rho,\Delta_\sigma]g^{\mu\rho}g^{\nu\sigma}\}. \tag{2.4}$$

The terms produced from the standard covariant derivative, and the extra terms produced by including the frame field term are listed in Table 1. It is interesting to note that, for a fermion of non-zero mass, the generators of the frame field interactions extend the Lorentz gauge group SL(2,C) to the Poincaré group.

3. Principal Results

(a) The standard electroweak interactions are generated by the helicity-dependent interactions and by the U(1) arising from the broken SU(4) interaction. The residual SU(3) symmetry generates the standard QCD interactions, automatically zero for leptons. The Einstein-Hilbert scalar curvature Lagrangian and a term quadratic in curvature arise from the Poincaré group.

(b) The frame field term gives rise to a boson mass matrix. This is diagonalised by a rotation through the 'Weinberg angle'. Then the photon and gluon masses are zero, and the W/Z mass ratio is correct.

(c) The boson masses are related to the fermion masses through the mass formula

$$\Sigma \, m_f^2 = 32 M_W^2/3, \qquad (3.1)$$

summed over all fermion masses, with multiplicity 3 for quarks. Assuming three families, we obtain from this formula the predicted value for the mass m_{top} of the top quark:

$$m_{top} = 151.7 \text{ Gev}. \qquad (3.2)$$

family mixing would give a correction of order 0.1 Gev.

References

1. L.Halpern, Proc. First Marcel Grossmann Meeting, ed. R.Ruffini, 113 (North-Holland, 1977).
2. J.S.R.Chisholm and R.S.Farwell, Lecture Notes in Physics **116**, 305 (1980) *and* Proc. Roy.Soc. London, **A337**,1 (1981).
3. Z.Dongpei, Phys. Rev. **D22**, 2027 (1980).
4. W.J.Wilson, Phys. Lett. **A75**, 172 (1980).
5. J.S.R.Chisholm and R.S.Farwell, J. Phys. **A20**, 6561 (1987).
6. J.S.R.Chisholm and R.S.Farwell, Gen. Rel. Grav. **20**, 371 (1988)
6. J.S.R.Chisholm and R.S.Farwell, J. Phys. **A22**, 1059 (1989).
7. J.S.R.Chisholm and R.S.Farwell, 'The Interface of Mathematics and Particle Physics', eds. D.G.Quillen, G.B.Segal and Tsou S.T.,193 (1990).
8. J.S.R.Chisholm and R.S.Farwell, Supp. ai Rec. del Circ. Mat. di Palermo, II, **26**, 123 (1991).
10. J.S.R.Chisholm and R.S.Farwell, J. Phys. **G18**, L117 (1992).

TABLE 1

Lagrangian Terms

Dimensions	1-4 (curved)	5-7	5-6	8-11
Metric	(- - - +)	(- - -)		(+ + - -)
Gauge group	SL(2,C) (Lorentz)	SU(2)h_+	U(1)h_-	SU(4)→SU(3)xU(1)
Standard Covariant Derivative Terms	Spin Gravity ($R_{\mu\nu\rho\sigma}R^{\mu\nu\rho\sigma}$).	Helicity-dependent electroweak. Kinetic (A,W,Z)		QCD (zero for leptons) Helicity-independent electroweak. Kinetic (gluons).
Frame Field Terms	E-H scalar curvature R. Mass of frame field quantum=0.	Correct W/Z mass ratio. Photon mass=0. Mass matrix diagonalised by θ_W rotation.		Gluon mass=0.

STRINGY ON-SHELL WARD IDENTITY

Jen-Chi Lee

Department of Electrophysics
National Chiao-Tung University
Hsin Chu, Taiwan, R.O.C.

ABSTRACT

An infinite sets of string-tree level on-shell Ward identities, which are valid to all σ-model loop orders, can be systematically constructed without refering to string field theory. Many interesting implications, including inter-particle Ward identities and triviality of scattering amplitudes for degenerate massive states, are presented and discussed.

One of the main difficulties of the theory of string today is the failure to identify the unique 4-dimensional vacuum from millions of possible candidates which have been discovered by various techniques during the last few years. The current strategy for this important issue is trying to understand both nonperturbative regimes (strong coupling and high energy regimes) of the theory. This is strongly suggested from the development of field theory in the late seventies, e.g. nonperturbative instanton physics which turned out to dramastically modify the structure of perturbative QCD vacuum. The current interest on matrix model and stringy solitons, including black holes etc., are along this line of thoughts.

In this talk we will derive some explicit form of stringy massive on-shell Ward identities by using zero-norm states in the bosonic string spectrum.This was suggested by the existence of stringy symmetries presented in ref [1]. Previous works on this direction can be found in [2]. The advantages of our approach are that off-shell ambiguities are avoided, and one can easily write down an infinite sets of string-tree level on-shell Ward identities after a systematic construction of zero-norm states in the spectrum. These Ward identities, which are valid to *all* σ-model loop (α') orders, indicate that quantum string theories do possess an infinite number of high energy symmetry structures as was conjectured in ref [3]. We have calculated the massive string-tree level scattering amplitudes up to the second massive states in the most general gauge choice, then use zero-norm states to derive the corresponding Ward identities. Of particular interest, the Ward identity corresponding to the $D_2 \otimes D_2'$ zero-norm state [1] explicitly relates amplitudes of *four different spin states* at the second massive level will be presented. The corresponding symmetry transformation law (in the first order weak field approximation) of the four background fields can then be constructed. This result justifies our previous consideration on higher inter-spin symmetries from the generalized σ-model point of view, and is a general feature for higher massive levels. Although the decoupling of zero-norm states from the string amplitudes has been proved for a long time by the so called "canceled propagator argument" in the context of "old fashioned" operator method[4], we find its implication on the physical amplitudes, as will be present in this talk, was always ignored and not clear so far. Moreover, through the uses of explicit form of zero-norm states which can be calculated to any higher massive level, one can easily write down infinite relations among string scattering amplitudes.

There are four physical propagating states at the second massive level of the closed bosonic string spectrum,

$$/\phi> = /\phi>_R \otimes /\phi>_L , \tag{1}$$

where $/\phi>_{R,L}$ are either of the following two states

$$\{\varepsilon_{\mu\nu\lambda}\alpha_{-1}{}^{\mu}\alpha_{-1}{}^{\nu}\alpha_{-1}{}^{\lambda}+\varepsilon_{(\mu\nu)}\alpha_{-1}{}^{(\mu}\alpha_{-2}{}^{\nu)}+\varepsilon_{\mu}\alpha_{-3}{}^{\mu}\} \mid 0,k> , \quad \varepsilon_{\mu\nu\lambda}=\varepsilon_{(\mu\nu\lambda)} \tag{2}$$

with gauge conditions

$$\varepsilon_{(\mu\nu)} = -\,{}^{3}/{}_{2}k^{\lambda}\varepsilon_{\mu\nu\lambda} , \ \varepsilon_{\mu} = {}^{1}/{}_{4}k^{\nu}k^{\lambda}\varepsilon_{\mu\nu\lambda} , \ -k^2=4 ,$$

$$2\varepsilon^{\mu}{}_{\mu\lambda}+k^{\mu}k^{\nu}\varepsilon_{\mu\nu\lambda}= 0 \tag{3}$$

and

$$\varepsilon_{[\mu\nu]}\alpha_{-1}{}^{[\mu}\alpha_{-2}{}^{\nu]} \mid 0, k > \tag{4}$$

with gauge conditions

$$k^{\mu}\varepsilon_{[\mu\nu]}=0 , \quad -k^2 = 4 . \tag{5}$$

The polization triplet $\{\varepsilon_{(\mu\nu\lambda)},\varepsilon_{(\mu\nu)},\varepsilon_{\mu}\}$ in eq (2) is the most general gauge choice for the first state. There are two types of zero-norm states in the bosonic open string spectrum [4]. They can be obtained either by standard spectrum analysis or generated in the following way

$$\text{Type I} \quad L_{-1}|\chi> \quad \text{where } L_m|\chi>= 0, \ m \geq 1 , \ L_0 |\chi>= 0, \tag{6}$$

$$\text{Type II} \quad (L_{-2} + 3/2\, L^2_{-1})|\chi> \quad \text{where } L_m|\chi>= 0, \ m \geq 1 , \ (L_0 + 1)|\chi>= 0. \tag{7}$$

Type I states have zero norm at any space time dimension, whereas type II states have zero norm only at D=26. A systematic construction of these zero-norm states has been given in ref [1]. One class of solution of eq (6), for example, is (omit all spin indices)

$$\theta\, (\alpha_{-1})^{n+1}(\alpha_{-1})^{n-1}\alpha_{-2} \mid 0,k > + k\, \theta\, (\alpha_{-1})^{n+1}(\alpha_{-1})^{n+1} \mid 0,k > \tag{8}$$

where θ is a $2n + 1$ spin index parameter that is symmetric on the first $n+1$ and last n indices. It is orthogonal to k_μ on each index, and traceless on any pair of the first $n+1$, or any pair of the last n indices. In the second term, the index of k_μ is symmetrized with the last n indices of θ. The mass of the state is $M^2 = 2n$. Another particular mixed type zero norm state at the second massive level which will generate inter-particle Ward identities is givern by

$$D_2, \qquad \{(^1/_2 k_\mu k_\nu \theta_\lambda + 2\eta_{\mu\nu}\theta_\lambda)\alpha_{-1}{}^\nu \alpha_{-1}{}^\mu \alpha_{-1}{}^\lambda + 9k_{[\mu}\theta_{\nu]}\alpha_{-2}{}^{[\mu}\alpha_{-1}{}^{\nu]} + 12\theta_\mu \alpha_{-3}{}^\mu\} \,|0,k>, \quad k\cdot\theta = 0 . \tag{9}$$

The Ward identity corresponding to $D_2 \otimes D_2$'zero-norm state is then given by

$$(1/4 k_\mu k_\nu k_\alpha k_\beta \theta_\lambda \theta_\gamma + \eta_{\mu\nu} k_\alpha k_\beta \theta_\lambda \theta_\gamma + \eta_{\alpha\beta} k_\mu k_\nu \theta_\lambda \theta_\gamma + 4\eta_{\mu\nu}\eta_{\alpha\beta}\theta_\lambda\theta_\gamma) T_2^{(\mu\nu\lambda)(\alpha\beta\gamma)} + (9/2\, k_\mu k_\nu k_\alpha \theta_\lambda \theta_\beta + 18\eta_{\mu\nu} k_\alpha \theta_\lambda \theta_\beta) T_2^{(\mu\nu\lambda)[\alpha\beta]} + (9/2 k_\alpha k_\beta k_\mu \theta_\gamma \theta_\nu + 18\eta_{\alpha\beta} k_\mu \theta_\gamma \theta_\nu) T_2^{[\mu\nu](\alpha\beta\gamma)} + 81 k_\mu k_\alpha \theta_\nu \theta_\beta T_2^{[\mu\nu][\alpha\beta]} + (6k_\mu k_\nu \theta_\lambda \theta_\alpha + 24\eta_{\mu\nu}\theta_\lambda\theta_\alpha) T_2^{(\mu\nu\lambda)\alpha} + (6k_\alpha k_\beta \theta_\gamma \theta_\lambda + 24\eta_{\alpha\beta}\, \theta_\gamma \theta_\lambda) T_2^{\lambda(\alpha\beta\gamma)} + 108 k_\mu \theta_\nu \theta_\alpha T_2^{[\mu\nu]\alpha} + 108 k_\alpha \theta_\beta \theta_\mu T_2^{\mu[\alpha\beta]} + 144\theta_\mu \theta_\lambda T_2^{\mu\lambda} = 0, \tag{10}$$

where $\theta_{\mu\nu}$ is a constant tensor with $k^\mu \theta_{\mu\nu} = k^\nu \theta_{\mu\nu} = 0$. In equation (10) the definition of, e.g. $T_2^{(\mu\nu\lambda)[\alpha\beta]}$ is

$$T_2^{(\mu\nu\lambda)[\alpha\beta]} = \int \prod_{i=1}^{4} d^2 z_i < e^{ik_1 X} \partial X^{(\mu} \partial X^\nu \partial X^{\lambda)} \partial^2 X^{[\alpha} \partial X^{\beta]} e^{ik_2 X} e^{ik_3 X} e^{ik_4 X} >, \tag{11}$$

other scattering amplitudes can be similarly written down. One can also write down the symmetry transformation in the first order weak field approximation for equation (10). Let $M_{\mu\nu\lambda,\alpha\beta\gamma}(X)$, $E_{\mu\nu,\alpha\beta\gamma}(X)$, $G_{\mu\nu\lambda,\alpha\beta}(X)$, and $A_{\mu\nu,\alpha\beta}(X)$ be the four physical propagating background fields of the second massive level ($\mu\nu\lambda = (\mu\nu\lambda)$ and $\mu\nu = [\mu\nu]$) , the inter-particle gauge transformation is

$$\delta M_{\mu\nu\lambda,\alpha\beta\gamma} = 1/4 \partial_{(\mu}\partial_\nu \partial_{(\alpha}\partial_\beta \theta_{\lambda)\gamma)} - \eta_{(\mu\nu}\partial_{(\alpha}\partial_\beta \theta_{\lambda)\gamma)} - \eta_{(\alpha\beta}\partial_{(\mu}\partial_\nu \theta_{\lambda)\gamma} + 4\eta_{(\mu\nu}\eta_{(\alpha\beta}\theta_{\lambda)\gamma)} ,$$

$$\delta E_{\mu\nu,\alpha\beta\gamma} = 9/2 \partial_{(\alpha}\partial_\beta \partial_{[\mu}\theta_{\gamma)\nu]} - 18\eta_{(\alpha\beta}\partial_{[\mu}\theta_{\gamma)\nu]} ,$$

$$\delta G_{\mu\nu\lambda,\alpha\beta} = 9/2 \partial_{(\mu}\partial_\nu \partial_{[\alpha}\theta_{\lambda)\beta]} - 18\eta_{(\mu\nu}\partial_{[\alpha}\theta_{\lambda)\beta]} ,$$

$$\delta A_{\mu\nu,\alpha\beta} = 81 \partial_{[\mu}\partial_{[\alpha}\theta_{\nu]\beta]} , \tag{12}$$

where $\partial^\mu \theta_{\mu\nu}(X) = \partial^\nu \theta_{\mu\nu}(X) = 0$ and $(\Box - 4)\, \theta_{\mu\nu}(X) = 0$, the other five background fields are just gauge artifacts according to equation (3). Equation (12) is consistent with the result obtained from generalized σ-model point of view in ref[1]. However, the method used here can be easily generalized to superstring massive cases. One can easily write down the symmetry transformation laws of the massive background fields after explicitely deriving the Ward identities. These transformation laws turn out to be too messy to obtain from the worldsheet σ-model approach [5].

Another interesting implication of the above infinite Ward identities, which we will just state briefly below, is the triviality of scattering amplitudes for degenerate massive states. It was known that there are ambiguity in defining the positive norm states due to the existence of zero-norm states in the same Young representation [1]. At the third massive level of the open bosonic spectrum, for example, there are degeneracies in the symmetric spin-two states. The Ward identities, generated by the spin-two zero norm states, can then be used to expressed the scattering amplitudes of the spin-two positive norm state in terms of the amplitudes of the other higher spin states. This means that the physical propagating spin-two state has no independent dynamics and are mere gauge artifacts of the theory. This result can be demonstrated to be valid to any orders in nonperturbative (all orders in α') weak field approximation [6].

Acknowledgement :

I would like to thank the government of Taiwan for financial support, which makes my visit to mainland China possible.

References :

1. J.C.Lee, Phys. Lett. **B241** (1990) 336; Phys. Rev. Lett. **64** (1990) 1636; Z. Phys. **C54** (1992) 283. J. C. Lee and B. A. Ovrut, Nucl. Phys. **B336** (1990) 222.

2. G.Veneziano, Phys. Lett. **B167** (1986) 388; J.Maharana and G.Veneziano, Phys. Lett. **B169** (1986) 177; Nucl. Phys. **B283** (1987) 126. T.Kubota, "*Stringy Symmetries and off-Shell Ward identities*", talk on B.Sakita's sixtieth birthday (1989).

3. D. J. Gross, Phys. Rev. Lett. **60** (1988) 129; D. J. Gross and P. Mende,Phys. Lett. **B197** (1987) 129; Nucl. Phys. **B303** (1988) 407. J.J. Atick and E. Witten, Nucl. Phys. **B310** (1988), 291; R. Hagedorn, Nuovo Cim **56A**, (1968), 1027.

4. Green, Schwarz and Witten." *Superstring Theorey* ", Vol. I.

5. J. C. Lee, Z. phys. **C 54** (1992) 283.

6. J.C. Lee, "*Triviality of stringy scattering amplitudes for degenerate massive states*" CTUP-105 (1992), in preparation.

A REAL STRUCTURE OF THE q-EUCLIDEAN SPACE

LI LIAO and XING-CHANG SONG

Department of Physics, Peking University, Beijing 100871, China

ABSTRACT

By introducing the left as well as the right derivatives, a real structure for the covariant differential calculus on the N-dimensional quantum Euclidean space is given within the framework of C* algebra. The method is also applicable to the quantum Minkowski space.

The non-commutative differential calculus on a quantum hyperplane was established by Wess and Zumino[1], and was then generalized to discuss the q-deformation of Schrödinger equation on the quantum Euclidean space[2,3] and Dirac equation on the quantum Minkowski space[4]. Up to now, however, physically required reality conditions have not been well established yet with the latter two cases. And we here present a kind of real structure for the quantum Euclidean space.

The covariant differential calculus on the quantum Euclidean space is defined as[2,3]

$$Q^{(1)}{}^{ij}{}_{kl}\, x^k\, x^l = 0\ , \tag{1}$$

$$\partial^i\, x^j = g^{ij} - \lambda_1\, \check{R}^{ij}{}_{kl}\, x^k\, \partial^l\ , \tag{2}$$

$$Q^{(1)}{}^{ij}{}_{kl}\, \partial^k \partial^l = 0\ . \tag{3}$$

Here $\check{R}^{ij}{}_{kl}$ is the $\check{R}$ matrix associated with $SO_q(N)$ and has three different eigenvalues λ_α, $\alpha = 0, 1, 2$; g^{ij}, the singlet eigenvector of the $\check{R}$ corresponding to λ_0, plays the role of a metric on the quantum Euclidean space; $Q^{(1)}{}^{ij}{}_{kl}$ is the projection operator onto the q-anti-symmetric subspace corresponding to λ_1.

In discussing the real structure we require the deformation parameter q real, i.e., $q^* = q$. Hoping the relation (2) still to make sense under the operation * which is an antimorphism, i.e., $(ab)^* = b^* a^*$, we introduce the right derivatives $\overleftarrow{\partial}$ which, joining x and ∂, enlarge the algebra by the following relations

$$x^i\ \overleftarrow{\partial}^j = g^{ij} - \lambda_1\, \overleftarrow{\partial}^k\ x^l\ \check{R}^{ij}{}_{kl}\ , \tag{4}$$

$$Q^{(1)ij}{}_{kl}\ \overleftarrow{\partial}^{k}\overleftarrow{\partial}^{l}=0\ , \tag{5}$$

$$\partial^{i}\ \overleftarrow{\partial}^{j}=-\lambda_{1}^{-1}\ \check{R}^{ij}{}_{kl}\ \overleftarrow{\partial}^{k}\ \partial^{l}\ . \tag{6}$$

The consistency conditions for the enlarged algebra can be obtained from a similar consideration of deriving (1-3), i.e., by requiring the associativity to hold in any triple product of the algebraic objects. It is easy to show that relations (1-6) do consititute a consistent differential calculus on the q-Euclidean space.

Due to the fact that x^i are q-deformed "spherical harmonical coordinates", the reality conditions are taken as in the classical case to be

$$(x^{i})^{*}=x^{j}\ g_{ji}\quad ,(\partial^{i})^{*}=\overleftarrow{\partial}^{j}\ g_{ji},\quad (\overleftarrow{\partial}^{i})^{*}=\partial^{j}\ g_{ji} \tag{7}$$

with attention paid to exchanging of the acting direction as required by the anti-morphism of the operation *.

It is easy to check that the complex conjugation described above is an involution, and under the complex conjugation (2) $\leftrightarrow$ (4), (3) $\leftrightarrow$ (5) while (1)and (6) go into themselves respectively. This means that the whole enlarged algebra generated by the coordinates, the left and right derivatives is compatible with the operation *.

All the relations of the differential calculus on the quantum Euclidean space developed above are covariant under the action of $SO_q(N)$ transformation

$$x^{i}\longrightarrow M^{i}{}_{j}\ x^{j}\ ,\quad \partial^{i}\longrightarrow M^{i}{}_{j}\ \partial^{j},\quad \tilde{\partial}^{i}\longrightarrow M^{i}{}_{j}\ (\tilde{\partial}^{j})\ . \tag{8}$$

For the real quantum Euclidean space the matrix elements $M^{i}{}_{j}$ must of course be real in the sense that

$$(M^{i}{}_{j})^{*}\equiv \bar{M}_{i}{}^{j}=g^{jk}\ M^{l}{}_{k}\ g_{li}\ . \tag{9}$$

Equipped with this kind of real structure, the q-Euclidean space will be more justified, and consequently the q-Schrödinger equation with the differential calculus presented above, along with its solutions such as the q-oscillator[2] and the q-hydrogen like atoms[3] will be more closely related to its classical counterpart. It is easy to see that the entire scheme goes back to the ordinary calculus in the limit $q\longrightarrow 1$. The discussion given here on the real structure of the quantum Euclidean space can also be applicable to the quantum Minkowski space and the q-Dirac equation[4].

This work is supported in part by the National Natural Science Foundation of China and the Doctoral Programme Foundation of Institution of High Education.

References

[1] J. Wess and B. Zumino, *Nucl. Phys. B(Proc. Suppl.)*18B (1990)302; W. Pusz and Woronowicz, *Reports in Math. Phys.* 27(1990)231.

[2] U.Carow-Watamura, M, Schlieker and S. Watamura, *Z. Phys.* **C49**(1991)439.

[3] X.C. Song & L. Liao, *J. Phys.* **A25**(1992)623.

[4] X.C. Song, Covariant Differential Calculus on The Quantum Minkowski Space and The q-Analogue of Dirac Equation, Preprint CCAST-91-63, to appear in *Z. Phys.*C.

A Noncommutative Differential Calculus and its Relation to Gauge Theory and Gravitation

A. Dimakis and **F. Müller-Hoissen**

Institut für Theoretische Physik
Bunsenstr. 9, D-3400 Göttingen, Germany

ABSTRACT

We consider a deformation of the standard differential calculus on the commutative algebra of functions on a manifold. The calculus is noncommutative in the sense that functions and differentials satisfy nontrivial (generally covariant) commutation relations involving the space-time metric and a "proper time" differential. There is a natural way in which field equations for physical fields emerge from this differential calculus.

1. *Prologue.* In 1941 Stueckelberg suggested to describe the relativistic motion of a point particle in a gravitational and an electromagnetic field by

$$\frac{d^2x^\mu}{dT^2} = -\Gamma^\mu{}_{\kappa\lambda}\,\frac{dx^\kappa}{dT}\,\frac{dx^\lambda}{dT} + e\,F^\mu{}_\nu\,\frac{dx^\nu}{dT}$$

in terms of a parameter T without reference to the mass. As a consequence of this equation, .

$$m^2 := -g_{\mu\nu}\,\frac{dx^\mu}{dT}\,\frac{dx^\nu}{dT}$$

is a constant of motion. If $m^2 > 0$, this expression relates T to the proper time τ by $m^2 = (d\tau/dT)^2$. Eliminating T in favour of τ in Stueckelberg's equation yields the more familiar relativistic equation of motion. The interesting point is that in terms of the "evolution parameter" T the dynamical law is "universalized", i.e. independent of the (mass of the) particle.

Stueckelberg also discussed the "universal" five-dimensional Schrödinger equation

$$i\hbar\,\frac{\partial}{\partial T}\,\phi = -\frac{\hbar^2}{2}\,\Box\,\phi\,.$$

A transformation $\phi \mapsto \exp(-i\,m^2T/2\hbar)\,\phi$ generates a mass term on the r.h.s. and we recover the Klein-Gordon equation for stationary states (with respect to T). However, Stueckelberg's theory a priori does not make reference to the notion of mass. Hence, there is no place for such a transformation and we should look for a different origin of mass. We may regard the above transformation as a special gauge transformation $\phi \mapsto \exp(i\alpha(T))\phi$ and consider the gauge invariant version of Stueckelberg's equation:

$$i\hbar\,D_T\,\phi = -\frac{\hbar^2}{2}\,\Box\,\phi$$

where $D_T := \partial/\partial T + i\,M/2\hbar$ with a gauge potential M transforms as a covariant derivative. If we break the gauge invariance (for example, by restricting ϕ to be T-independent), then M becomes a scalar. In the special case where this scalar is constant, it becomes the ordinary mass term when shifted to the r.h.s. of the Stueckelberg equation.

The model which we describe in the following shares many features with the above example. The central idea is that there is an evolution parameter T in terms of which dynamics is "universal". The field equations for the various physical fields should then emerge from the universal law (via breaking of T-gauge invariance). A "universal dynamical law" has to be formulated at a level which is common to all dynamical laws, i.e. it has to be encoded in the *differential calculus*.

2. *The model.*[1] We consider a differential calculus on the (commutative) algebra of functions on a manifold which deviates from the classical differential calculus only by

$$[x^\mu, \hat{d}x^\nu] = i\,\hbar\,\hat{d}T\,g^{\mu\nu}$$

where x^μ are coordinates. There is a connection with the canonical commutation relations of quantum mechanics which explains the appearance of $\hbar$.[1] The universality of the gravitational interaction justifies that the metric g enters the commutation relations. We are led to a consistent differential calculus with the deformed $\wedge$ product

$$\hat{d}x^\mu\,\hat{d}x^\nu + \hat{d}x^\nu\,\hat{d}x^\mu = -i\hbar\,\hat{d}T\,dx^\rho\,\partial_\rho g^{\mu\nu} \,.$$

($\hat{d}$ has all the properties of an exterior derivative. In particular, it satisfies $\hat{d}^2 = 0$ and the graded Leibniz rule.) For $\hbar \to 0$ we recover the ordinary differential calculus. The commutation relation postulated above is generally covariant. Furthermore, since all deviations from the ordinary calculus are proportional to $\hat{d}T$ we may drop the hats in expressions whenever they are multiplied by $\hat{d}T$.

One can show that

$$\hat{d}\phi = \hat{d}T\left(\partial_T\phi + \frac{i\hbar}{2}g^{\mu\nu}\,\partial_\mu\partial_\nu\phi\right) + \hat{d}x^\mu\,\partial_\mu\phi$$

i.e., the differential of a function ϕ involves a *second order* differential operator. Now

$$\hat{d}x'^\mu = \frac{i\,\hbar}{2}\,\hat{d}T\,\partial^2 x'^\mu + \hat{d}x^\nu\,\partial_\nu x'^\mu$$

shows that $\hat{d}x^\mu$ is not covariant. But

$$\hat{D}x^\mu := \hat{d}x^\mu + \frac{i\,\hbar}{2}\,\hat{d}T\,\Gamma^\mu \qquad \Gamma^\mu := \Gamma^\mu_{\kappa\lambda}\,g^{\kappa\lambda}$$

is *right-covariant* (Γ is the Levi-Civita connection of g). Then

$$\hat{d}\phi = \hat{d}T\,\frac{i\hbar}{2}\,\Box\phi + \hat{D}x^\mu\,\partial_\mu\phi$$

if $\partial_T \phi = 0$. The vanishing of the "evolution part" leads to the (covariant) wave equation $\Box\phi = 0$. If we drop the constraint $\partial_T \phi = 0$ we are led to Stueckelberg's five-dimensional Schrödinger equation (with T replaced by $-T$).

3. *Gauge fields.* We restrict our considerations to the flat metric $g = \eta$. Let ψ be a representation of a Lie group G. In the following all quantities are assumed not to depend on T. We introduce a covariant derivative

$$\hat{\mathcal{D}}\psi := \hat{d}\psi + \hat{A}\,\psi\,.$$

The gauge transformation law $\hat{A}' = U\hat{A}U^{-1} - \hat{d}U\,U^{-1}$ for the (generalized) connection

$$\hat{A} = \hat{d}T\,\frac{i\hbar}{2}\,A_T + \hat{d}x^\mu\,A_\mu$$

(with an ordinary gauge potential A_μ) requires that

$$A_T = \eta^{\mu\nu}(\partial_\mu A_\nu - A_\mu A_\nu) + M$$

with an undetermined tensorial part M. Then

$$\hat{\mathcal{D}}\psi = \hat{d}T\,\frac{i\hbar}{2}\,(\nabla^2\psi + M\,\psi) + \hat{\mathcal{D}}x^\mu\,\nabla_\mu\psi$$

where ∇_μ denotes the covariant derivative with respect to A_μ. Here we have introduced a covariant differential $\hat{\mathcal{D}}x^\mu := \hat{d}x^\mu - i\hbar\,\hat{d}T\,A^\mu$ since $U\hat{d}x^\mu U^{-1} \neq \hat{d}x^\mu$. The field strength of $\hat{A}$ is

$$\hat{F} = \hat{d}\hat{A} + \hat{A}^2 = \hat{d}T\,\frac{i\hbar}{2}\,(\mathcal{D}^*F - \mathcal{D}M) + \frac{1}{2}\,\hat{\mathcal{D}}x^\mu\,\hat{\mathcal{D}}x^\nu\,F_{\mu\nu}$$

where F is the field strength of A. In the "evolution part" we recover the Yang-Mills operator.

4. *Gravity.* We introduce a covariant derivative

$$\hat{D}Y^\mu := \hat{d}Y^\mu + Y^\nu\,{}_\nu\hat{\Gamma}^\mu$$

with the generalized linear connection

$${}_\nu\hat{\Gamma}^\mu \;:=\; \hat{d}T\,\frac{i\hbar}{2}\,\left[g^{\rho\sigma}\,(\partial_\rho\Gamma^\mu_{\nu\sigma} + \Gamma^\mu_{\lambda\rho}\,\Gamma^\lambda_{\nu\sigma}) + M^\mu{}_\nu\right] + \hat{d}x^\rho\,\Gamma^\mu_{\nu\rho}$$

where $M^\mu{}_\nu$ is an undetermined tensorial part and $\Gamma^\mu_{\nu\rho}$ an ordinary linear connection. Then

$$\hat{D}Y^\mu = \hat{d}T\,\frac{i\hbar}{2}\,(\nabla^2 Y^\mu + M^\mu{}_\nu Y^\nu) + \hat{D}x^\mu\,\nabla_\rho Y^\mu\,.$$

The product of covariant differentials $\hat{D}x^\mu$ is neither antisymmetric nor covariant. But

$$\hat{D}x^\mu \wedge \hat{D}x^\nu := \hat{D}x^\mu\,\hat{D}x^\nu - i\,\hbar\,\hat{d}T\,dx^\rho\,g^{\sigma\nu}\,\Gamma^\mu_{\sigma\rho}$$

(where the linear connection Γ may carry torsion) is antisymmetric and right-covariant. Then

$$\hat{D}\hat{D}x^\mu = \hat{\Theta}^\mu - \hat{d}T \frac{i\hbar}{2} dx^\nu \left(R^\mu{}_\nu - M^\mu{}_\nu - \nabla^\rho Q^\mu{}_{\rho\sigma} + Q^{\mu\rho\sigma} Q_{\rho\sigma\nu}\right)$$

with the torsion 2-form

$$\hat{\Theta}^\mu := \frac{1}{2} \hat{D}x^\kappa \wedge \hat{D}x^\lambda Q^\mu{}_{\kappa\lambda} .$$

The previous equation may be regarded as a *generalized torsion 2-form.* Its vanishing implies vanishing of the ordinary torsion and an equation $R_{\mu\nu} = M_{\mu\nu}$ of the form of the Einstein equations (if we relate $M^\mu{}_\nu$ to the energy-momentum tensor).

5. *Epilogue.* So far applications of noncommutative differential calculus concentrated on noncommutative algebras. The formalisms developed by Connes[3] and Dubois-Violette[4] reduce to the ordinary differential form calculus when applied to the commutative algebra of C^∞-functions on a manifold. In contrast, the formalism which we applied is more general and allows non-standard differential calculi on commutative algebras (see also Ref. 5 for applications to quantum groups). Our work demonstrates that deviations from the standard differential calculus on the algebra of functions on a manifold indeed have physical significance (see also Ref. 2 for another striking example). We have shown how field equations for matter fields, gauge fields and gravity emerge from the universal dynamics encoded in our noncommutative differential calculus. For further results and discussion we refer to Ref. 1. Some problems which have to be addressed next are

- Which mechanism relates the M's to masses / energy-momentum tensor ? To deal with this problem the restriction to quantities not depending on T has to be relaxed. The "mass terms" should then arise via breaking of T-gauge invariance.
- How to obtain the (first order) Dirac equation ?

References

1. A. Dimakis and F. Müller-Hoissen, "Noncommutative differential calculus, gauge theory and gravitation" Göttingen preprint GOET-TP 33/92.
2. A. Dimakis, F. Müller-Hoissen and T. Striker, "Noncommutative differential calculus and lattice gauge theory" Göttingen preprint GOET-TP 44/92.
3. A. Connes, "Non-commutative differential geometry" *Publ. I.H.E.S.* **62** (1986) 257.
4. M. Dubois-Violette, "Dérivations et calcul différentiel non commutatif", *C. R. Acad. Sci. Paris* **307** (1988) 403.
5. S. L. Woronowicz, "Differential calculus on compact matrix pseudogroups (quantum groups)", *Commun. Math. Phys.* **122** (1989) 125.

A Note on Representations of the w_∞ Algebra

C.N. Pope* and X.J. Wang

Center for Theoretical Physics, Texas A&M University,
College Station, TX 77843-4242, USA.

ABSTRACT

We explicitly demonstrate that the unitary representations of the w_∞ algebra and its truncations are just the unitary representations of the Virasoro algebra.

The representation theory of the Virasoro algebra plays a crucial rôle in our understanding of strings and conformal field theories. An important feature of the Virasoro algebra is that it admits a central extension; it is only for positive values of the central charge c that non-trivial unitary representations can occur.

Similar questions about the existence of unitary representations arise for the various extended conformal algebras, known generically as W algebras. These are generated by higher-spin conformal currents in addition to the spin-2 current, the energy-momentum tensor, which generates the Virasoro algebra. For example, the W_N algebras have currents of spins $2, 3, \ldots, N$, for each of which there is a non-trivial central extension [1,2]. These algebras are non-linear, but there is an $N \rightarrow \infty$ limit, known as W_∞, in which linearity is regained [3]. However the W_∞ algebra still retains the other feature of the W_N algebras mentioned above, namely that it has non-trivial central extensions in all the higher-spin sectors. There is a classical limit of W_∞, known as w_∞ [4], in which all the central extensions in the higher-spin sectors are lost, leaving only the usual one in the Virasoro subalgebra. The w_∞ algebra can be truncated to what one may call w_N, by setting all generators with spins greater than N to zero. Among the extended conformal algebras, the w_N and w_∞ algebras are the simplest in structure.

One might expect that the absence of central terms for higher spins in the w_N and w_∞ algebras should result in their unitary representations being equivalent to the usual Virasoro

* Supported in part by the U.S. Department of Energy, under grant DE-FG05-91ER40633.

unitary representations. Since a simple and explicit proof of this can be given, it seems worthwhile to present it here.

We shall consider the w_N algebra first. It has conformal currents of spins $2, 3, \ldots, N$. Let w_m^i denote the m'th Fourier mode of the spin $i+2$ current, where $i = 0, 1, 2, \ldots, N-2$. They satisfy

$$[w_m^i,\ w_n^j] = \left[(j+1)m - (i+1)n\right] w_{m+n}^{i+j} + \frac{c}{12}(m^3 - m)\delta^{i,0}\delta^{j,0}\delta_{m+n,0} \ , \tag{1}$$

where on the right-hand side w_{m+n}^{i+j} is defined to be zero if $i + j > N - 2$. Note that the central extension is non-trivial only for the Virasoro sector.

The Hilbert space of physical states consists of highest-weight states and their descendants. For the w_N algebra, we can denote a highest-weight state by $|\vec{h}\rangle$, where $\vec{h} = (h_0, h_1, h_2, \ldots, h_{N-2})$, satisfying

$$\begin{aligned} w_0^i|\vec{h}\rangle &= h_i|\vec{h}\rangle \ , \\ w_m^i|\vec{h}\rangle &= 0, \qquad m \geq 1 \ . \end{aligned} \tag{2}$$

Descendant states are those created by acting on $|\vec{h}\rangle$ with a string of w_{-m}^i's, with $m \geq 1$. Unitarity of the Hilbert space requires the matrix of inner products of any set of physical states to be non-negative. A particular case is the matrix of inner products of states $w_{-m}^i|\vec{h}\rangle$, with $m \geq 1$ and $i = 0, 1, 2, \ldots, N-2$;

$$M_{ij} = \langle\vec{h}|w_m^i w_{-m}^j|\vec{h}\rangle \ . \tag{3}$$

To see what we can learn about unitarity from this matrix, we look at the w_3 and w_4 algebras first, which are the first two w_N extensions of the Virasoro algebra.

The w_3 algebra takes the form

$$\begin{aligned} [L_m, L_n] &= (m-n)L_{m+n} + \frac{c}{12}(m^3 - m)\delta_{m+n,0} \ , \\ [L_m, W_n] &= (2m-n)W_{m+n} \ , \\ [W_m, W_n] &= 0 \ , \end{aligned} \tag{4}$$

where L_m and W_m are the spin 2 and 3 generators. The matrix M_{ij} is given by

$$(M_{ij}) = \begin{pmatrix} 2mh_0 + \frac{c}{12}(m^3 - m) & 3mh_1 \\ 3mh_1 & 0 \end{pmatrix} \ . \tag{5}$$

One can easily see that for this matrix to be non-negative, it must be that $h_1 = 0$. Since there is no central term for W_m, and, as we have just seen, the weight under W_0 is zero, it follows that all descendant states involving W_{-m}'s are spurious. The remaining descendant states are those of the Virasoro Verma module, and the conditions for their unitarity are standard, with no modification due to the higher-spin generators. Thus the unitary representations of w_3 are the same as the unitary Virasoro representations.

The w_4 algebra takes the form

$$\begin{aligned}
[L_m, L_n] &= (m-n)L_{m+n} + \frac{c}{12}(m^3 - m)\delta_{m+n,0} \,, \\
[L_m, W_n] &= (2m-n)W_{m+n} \,, \\
[W_m, W_n] &= 2(m-n)V_{m+n} \,, \\
[L_m, V_n] &= (3m-n)V_{m+n} \,, \\
[W_m, V_n] &= 0 = [V_m, V_n] \,,
\end{aligned} \tag{6}$$

where L_m, W_m and V_m are the spin 2, 3 and 4 generators. The corresponding matrix M_{ij} is

$$(M_{ij}) = \begin{pmatrix} 2mh_0 + \frac{c}{12}(m^3 - m) & 3mh_1 & 4mh_2 \\ 3mh_1 & 4mh_2 & 0 \\ 4mh_2 & 0 & 0 \end{pmatrix} . \tag{7}$$

For the outer 2×2 submatrix to be non-negative, we must have $h_2 = 0$. Now consider the upper diagonal 2×2 block. Having found that $h_2 = 0$, the requirement of non-negativity implies that $h_1 = 0$. So we have

$$h_1 = h_2 = 0 \,. \tag{9}$$

Thus unitary w_4 representations are the same as unitary Virasoro representations.

The pattern that we have seen in the two examples above continues for all the finite-N w_N algebras. The matrix M of inner products (3) for the w_N algebra is

$$(M_{ij}) = \begin{pmatrix} 2mh_0 + \frac{c}{12}(m^3 - m) & 3mh_1 & 4mh_2 & \cdots & Nmh_{N-2} \\ 3mh_1 & 4mh_2 & 5mh_3 & \cdots & 0 \\ 4mh_2 & 5mh_3 & 6mh_4 & \cdots & 0 \\ \vdots & \vdots & \vdots & \ddots & \vdots \\ Nmh_{N-2} & 0 & 0 & \cdots & 0 \end{pmatrix} . \tag{10}$$

Following the same method as we used above, we get, for the matrix to be non-negative, that

$$h_1 = h_2 = \cdots = h_{N-3} = h_{N-2} = 0 \,, \tag{11}$$

showing that the unitary representations of w_N are the same as those of the Virasoro algebra.

The w_∞ algebra is somewhat different from the finite-N w_N algebras, since it has infinitely-many conformal currents. The algebra takes the same form as (1);

$$[w^i_m,\, w^j_n] = [(j+1)m - (i+1)n]\, w^{i+j}_{m+n} + \frac{c}{12}(m^3 - m)\delta^{i,0}\delta^{j,0}\delta_{m+n,0} \,, \tag{12}$$

except that now we have $i = 0, 1, 2, \ldots, \infty$. A highest-weight state $|\vec{h}\rangle$ satisfies

$$\begin{aligned} w_0^i|\vec{h}\rangle &= h_i|\vec{h}\rangle\ , \\ w_m^i|\vec{h}\rangle &= 0, \qquad m \geq 1\ . \end{aligned} \tag{13}$$

The M matrix is now infinite dimensional, given by

$$(M_{ij}) = \begin{pmatrix} 2mh_0 + \frac{c}{12}(m^3 - m) & 3mh_1 & 4mh_2 & \cdots \\ 3mh_1 & 4mh_2 & 5mh_3 & \cdots \\ 4mh_2 & 5mh_3 & 6mh_4 & \cdots \\ \vdots & \vdots & \vdots & \ddots \end{pmatrix}\ . \tag{14}$$

The method we used for the w_N algebras, starting by considering the outer 2×2 submatrix, is not appropriate in this infinite-dimensional case. Instead, we first consider another subspace of physical states, consisting of

$$|\alpha\rangle = w_{-2m}^{2i}|\vec{h}\rangle\ , \qquad |\beta\rangle = (w_{-m}^i)^2|\vec{h}\rangle\ , \quad i \geq 1,\ m \geq 1\ . \tag{15}$$

The inner products in this two-dimensional subspace are

$$\begin{aligned} \langle\alpha|\alpha\rangle &= \langle\vec{h}|w_{2m}^{2i}w_{-2m}^{2i}|\vec{h}\rangle = 4m(2i+1)h_{4i}\ , \\ \langle\beta|\beta\rangle &= \langle\vec{h}|(w_m^i)^2(w_{-m}^i)^2|\vec{h}\rangle = 8m^2(i+1)^2h_{2i}^2 + 4m^3(i+1)(2i+1)^2h_{4i}\ , \\ \langle\beta|\alpha\rangle &= \langle\vec{h}|(w_m^i)^2w_{-2m}^{2i}|\vec{h}\rangle = 2m^2(4i+3)(2i+1)h_{4i}\ , \end{aligned} \tag{16}$$

where we have used the conditions (13) for highest-weight states. The matrix of inner products in this subspace is

$$(N_{ij}) = \begin{pmatrix} \langle\alpha|\alpha\rangle & \langle\alpha|\beta\rangle \\ \langle\beta|\alpha\rangle & \langle\beta|\beta\rangle \end{pmatrix} \tag{17}$$

The unitarity of this space requires $\mathrm{Det}(N_{ij}) \geq 0$, *i.e.*

$$\langle\alpha|\alpha\rangle\langle\beta|\beta\rangle - \langle\alpha|\beta\rangle\langle\beta|\alpha\rangle \geq 0\ . \tag{18}$$

Substituting (16), we find

$$\mathrm{Det}(N_{ij}) = 32m^3(2i+1)(i+1)^2h_{4i}h_{2i}^2 - 4m^4(2i+1)^2(8i^2+12i+5)h_{4i}^2\ . \tag{19}$$

Since $i > 0$, it follows that for sufficiently large m, $\mathrm{Det}(N_{ij})$ is negative, unless $h_{4i} = 0$ when $i \geq 1$. In the above calculations, we see that it is crucial that there are no central extensions in the $i \geq 1$ sectors.

Having established that $h_{4i} = 0$ for $i \geq 1$, we may now apply the previous argument that we used for the w_N algebras, where we choose 2×2 submatrices of the inner products

(M_{ij}) given in (14), with h_{4i} as a diagonal entry. This immediately shows that all the h_i's except h_0 are zero. For example, knowing $h_4 = 0$, the submatrix

$$\begin{pmatrix} 4mh_2 & 5mh_3 \\ 5mh_3 & 6mh_4 \end{pmatrix} \tag{20}$$

of (14) tells us that $h_3 = 0$. The submatrix

$$\begin{pmatrix} 2mh_0 + \frac{c}{12}(m^3 - m) & 4mh_2 \\ 4mh_2 & 6mh_4 \end{pmatrix} \tag{21}$$

then tells us that $h_2 = 0$, and so on. Thus we arrive at the conclusion that unitary representations of the w_∞ algebra coincide with Virasoro unitary representations.

There are other interesting examples in the w algebra family, namely the w_{1+N} algebras (including the $N \to \infty$ limit). They have the spin-1 current w^{-1} besides all the currents in the w_N algebra. They take the same form as (1), but with i and j now allowed to range from -1 to $N-2$. They allow a central extension also in the spin-1 sector, with a central charge $\tilde{c}$ that is independent of the spin-2 central charge c:

$$[w_m^{-1}, w_n^{-1}] = \tilde{c} m \delta_{m+n,0} \,. \tag{22}$$

Denoting by h_{-1} the w_0^{-1} weight of a highest-weight state, the M matrix (3) is now given by

$$(M_{ij}) = \begin{pmatrix} \tilde{c}m & mh_{-1} & 2mh_0 & \cdots \\ mh_{-1} & 2mh_0 + \frac{c}{12}(m^3 - m) & 3mh_1 & \cdots \\ 2mh_0 & 3mh_1 & 4mh_2 & \cdots \\ \vdots & \vdots & \vdots & \ddots \end{pmatrix} . \tag{23}$$

The arguments that we used previously for w_N and w_∞ can be applied *mutatis mutandis* to w_{1+N} and $w_{1+\infty}$. One immediately concludes that here too, $0 = h_1 = h_2 = h_3 = \cdots$. The new feature here is that one must also demand the non-negativity of 2×2 submatrices in (23) that have $\tilde{c}m$ as the upper-left entry. By taking the case where the lower-right entry is $4mh_2$, one can see that h_0 must be zero. Then, taking the upper-left 2×2 submatrix in (23), with $m = 1$, one sees that h_{-1} must also vanish. Thus we have

$$0 = h_{-1} = h_0 = h_1 = h_2 = \cdots \,. \tag{24}$$

So for any w_{1+N} algebra, including $w_{1+\infty}$, its unitary representations are totally trivial. The only state in its Hilbert space is the $SL(2,R)$-invariant vacuum $|0\rangle$.

References

[1] A.B. Zamolodchikov, *Teo. Mat. Fiz.* **65** (1985) 347.

[2] V.A. Fateev and S. Lukyanov, *Int. J. Mod. Phys.* **A3** (1988) 507.

[3] C.N. Pope, L.J. Romans and X. Shen, *Phys. Lett.* **236B** (1990) 173; *Nucl. Phys.* **B339** (1990) 191.

[4] I. Bakas, *Phys. Lett.* **228B** (1989) 57.

W-ALGEBRAS FROM MAPPINGS ON S^1*

Chao-Zheng Zha
CCAST (World Laboratory)
P. O. Box 8730, Beijing, 100080
and
Department of Physics, Xinjiang University**
Wulumuqi, Xinjiang 830046
PR China

Abstract

The operators of the W-algebras can be obtained by the mappings on S^1. The commutators for the generators of unequal spins admit central terms which are necessary to constructing primary fields out of the quasi-primary fields corresponding to the algebra.

1. Introduction

In two dimensional conformal field theories the Virasoro algebra of the spin-2 generators L_m plays a central role. While the W_N algebras containing generators of spin $2, 3, \cdots$, N for finite N are the natural extensions of the Virasoro algebra [1-5], and the W_∞ algebra is the $N \to \infty$ limit of finite N. We start directly from the diffeomorphisms on a circle and come to a set of simple high order generators for W_∞ which faciliates considerably the manipulations afterwards, thus the commutators are easily obtained. Solving algebraically the Jacobi identities, it is found that all the commutators admit central terms, and the operator product expansions can be given. But the operators are quasi-primary fields. In constructing the primary fields out of the quasi-primary ones, the full set of central terms are necessary.

2. Commutators and Central Terms

Consider the group G of smooth 1-1 mapping $S^1 \to S^1$ with the multiplication rule as follows, [6]

$$\xi_1 \cdot \xi_2(z) = \xi_1(\xi_2(z)). \tag{2.1}$$

Its faithful representation can be defined by its action on functions $f: S^1 \to V$, where V is a vector space,

$$D_\xi f(z) = f(\xi^{-1}(z)). \tag{2.2}$$

---

* This work is supported by NSFC.

** Mailing address

In the neighbourhood of identity, an element $\xi \in G$ is represented as

$$\xi(z)=ze^{-i\varepsilon(z)}, \tag{2.3}$$

thus

$$\begin{aligned}\xi^{-1}(z)&=ze^{i\varepsilon(z)}\\&=z\left(1+i\varepsilon(z)+\frac{i^2}{2!}\varepsilon^2(z)+\frac{i^3}{3!}\varepsilon^3(z)+\cdots+\frac{i^n}{n!}\varepsilon^n(z)+\cdots\right).\end{aligned} \tag{2.4}$$

The Laurant expansion of $\varepsilon(z)$ is as follows,

$$\varepsilon(z)=\sum_{n=-\infty}^{\infty}\varepsilon_{-n}^{(1)}z^n. \tag{2.5}$$

We then have

$$\varepsilon^k(z)=\sum_{n=-\infty}^{\infty}\varepsilon_{-n}^{(k)}z^n, \tag{2.6}$$

where

$$\varepsilon^{(k)}(z)=\sum_{\substack{n_1+n_2+\cdots+n_k=n\\ n_1,n_2,\cdots,n_k=-\infty}}^{\infty}\varepsilon_{-n_1}^{(1)}\varepsilon_{-n_2}^{(1)}\cdots\varepsilon_{-n_k}^{(1)}. \tag{2.7}$$

Inserting eqs. (2.4) and (2.5) into eq. (2.2), we obtain

$$\begin{aligned}D_\xi f(z)=&f(z)+i\sum_{-\infty}^{\infty}\varepsilon_n^{(1)}z^{n+1}\frac{d}{dz}f(z)+\frac{i}{2!}\sum_{-\infty}^{\infty}\varepsilon_{-n}^{(2)}\left(z^{n+2}\frac{d^2}{dz^2}\right)f(z)\\&+\frac{i^3}{3!}\sum_{-\infty}^{\infty}\varepsilon_{-n}^{(3)}\left(z^{n+1}\frac{d}{dz}+3z^{n+2}\frac{d^2}{dz^2}+z^{n+3}\frac{d^3}{dz^3}\right)f(z)+\cdots\\&+\frac{i}{k!}\sum_{-\infty}^{\infty}\varepsilon_{-n}^{(k)}\left(z^{n+1}\frac{d}{dz}+\frac{1}{2!}\sum_{\substack{-\infty\\p_1+p_2=k}}^{\infty}\frac{k!}{p_1!\ p_2!}z^{n+2}\frac{d^2}{dz^2}+\cdots\right.\\&+\frac{1}{r!}\sum_{\substack{-\infty\\ \sum_1^r p_i=k}}^{\infty}\frac{k!}{p_1!\ p_2!\ \cdots p_r!}z^{n+r}\frac{d^r}{dz^r}+\cdots\\&\left.+\frac{1}{k!}\sum_{\substack{-\infty\\ \sum_1^r p_i=k}}^{\infty}\frac{k!}{p_1!\ p_2!\ \cdots p_k!}z^{n+k}\frac{d^k}{dz^k}\right)f(z)+\cdots\end{aligned} \tag{2.8}$$

Therefore the k-th order differential neighbourhood of the holomorphic mapping is generated by the k-th order Virasoro generators L_m^k,

$$L_m^k=(-1)^k z^{m+k}\frac{d^k}{dz^k}. \qquad (k=1,2,3,\cdots;\ m=0,\pm1,\pm2,\pm3,\cdots) \tag{2.9}$$

The commutators of the high order Virasoro generators can be constructed through direct manipulation,

$$\begin{aligned}[L_m^k,L_n^{k+r}]=&\sum_{p=1}^{k}(-1)^p\left(C_k^pB_p^{n+k+r}-C_{k+r}^pB_p^{m+k}\right)L_{m+n}^{2k+r-p}\\&-\sum_{p+k+1}^{k+r}(-1)^pC_{k+r}^pB_p^{m+k}L_{m+n}^{2k+r-p}, \qquad (r\geqslant 0)\end{aligned} \tag{2.10}$$

where

$$C_k^p = \frac{k!}{(k-p)!\,p!}, \tag{2.11}$$

$$B_p^n = \begin{cases} 0, & (p>n) \\ \dfrac{n!}{(n-p)!} & (p\leqslant n) \end{cases} \tag{2.12}$$

Choosing appropriate linear dependence of the central term $c_{m,n}^{k,k+r}$ on m, we find

$$c_{m,n}^{k,k+r} = (-1)^{k+r+1}\frac{1}{2}\frac{(k+r)!\,k!}{(2k+r+1)!}\frac{(m+k)!\,c}{(m-k-r-1)!}\delta_{m+n,0}. \tag{2.13}$$

Note that $c_{m,n}^{k,k+r}$ with $r\neq 0$, i. e. , for different spins are nonvanishing contrary to the ansatz in ref. 2.

3. Operator Product Expansions (OPE)

Corresponding to L_m^k, the local field U_k is defined by

$$L_m^k = \oint \frac{dz}{2\pi i} z^{m+k} U_k(z). \tag{3.1}$$

Then the terms in OPE corresponding to the second term in the right hand side of eq. (2. 10) and the central term can be obtained easily through direct observation, while those corresponding to the first term in eq. (2. 10) must be

$$+\sum_{p=1}^{k}(-1)^p\sum_{q=0}^{p}\frac{(p-q)!\,a_{p;q}^{k;r}}{(z-w)^{p-q+1}}\partial^q U_{2k+r-p}, \tag{3.2}$$

where $a_{p;q}^{k;r}$ satisfies the following equation,

$$\sum_{q=0}^{k}(-1)^q a_{p;q}^{k;r} B_{p-q}^{m+k} B_q^{m+n+2k+r-p} = C_k^p B_p^{n+k+r} - C_{k+r}^p B_p^{m+k}. \tag{3.3}$$

Solving eq. (3. 3) for $a_{p;q}^{k;r}$, we have the OPE as follows,

$$\begin{aligned} U_k(z)U_{k+r}(w) = {} & (-1)^{k+r+1}1/2\frac{k!\,(k+r)!\,c}{(z-w)^{2k+r+2}} - \sum_{p=k+1}^{k+r}(-1)^p C_{k+r}^p\frac{p!}{(z-w)^{p+1}}U_{2k+r-p}(w) \\ & + \sum_{p=1}^{k}\Big\{\frac{p!\,(C_k^p-(-1)^p C_{k+r}^p)}{(z-w)^{p+1}}U_{2k+r-p}(w) \\ & + \sum_{q=1}^{p}\frac{(p-q)!\,C_k^q C_k^{p-q}}{(z-w)^{p-q+1}}\partial^q U_{2k+r-p}(w)\Big\}. \end{aligned} \tag{3.4}$$

4. Primary Fields

It is obvius that U_1 and U_2 are just the operators T and $\hat{W}_3$ is the W-slgebras respectively, where $\hat{W}_3$ is the quasi-primary field of conformal weight 3 [5]. In general W_n has a conformal weight n, and can be expressed in terms of $U_k(k\leqslant n-1)$ and their derivatives of conformal weight k+1. These expressions can be manipulated using the following equation,

$$[L_m^1, W_n(z)] = (z^{m+1}\frac{d}{dz} + n(m+1)z^m)W_n(z). \tag{4.1}$$

For example,

$$W_3(z) \sim 2U_2(z)+U_1(z), \tag{4.2}$$

$$W_4(z) \sim 5U_3(z)+5\partial U_2(z)+(\partial^2-\frac{12}{c}U_1(z))U_1(z), \tag{4.3}$$

$$W_5(z) \sim 14U_4(z)+21\partial U_3(z)+9(\partial^2-\frac{12}{c}U_1(z))U_2(z)$$
$$+(\partial^3-\frac{120}{c}U_1(z)-\frac{132}{c}U_2(z))U_1(z), \tag{4.4}$$

$$W_6(z) \sim 42U_5(z)+84\partial U_4(z)+56(\partial^2-\frac{12}{c}U_1(z))U_3(z)$$
$$+14(\partial^3-\frac{120}{c}U_1(z)\partial-\frac{60}{7c}U_2(z))U_2(z)$$
$$+(\partial^4-\frac{360}{c}U_1(z)\partial^2-\frac{120}{c}U_2(z)\partial-\frac{1008}{c}U_3(z)+\frac{2880}{c}U_1^2(z))U_1(z). \tag{4.5}$$

It is obvious that the terms containing central charge on the left hand side of eq. (4. 1) should be cancelled, thus the central terms in the commutators of L^k_m are crucial for the consistent expressions for the primary fields W_n in terms of the quasi-primary U_k.

References

1. V. A. Fateev and A. B. Zamolodchikov, Nucl. Phys. B280[FS18](1987)644.
2. C. N. Pope, L. J. Romans and X. Shen, Phys. Lett. B236 (1990) 173; B242 (1990) 401
3. I. Bakas, Phys. Lett. B228 (1989) 57.
4. C. N. Pope, L. J. Romans and X. Shen, Phys. Lett . B245 (1990) 72.
   E. Bergshoeff, M. Vasilive and B. de Wit, Phys. Lett. B256 (1990) 199.
5. A. Bilal, V, V. Fock and I. I. Kogen, Nucl. Phys. B359 (1991) 635.
6. P. Goddard and D. Olive, Int. J. Mod. Phys. A1 (1986) 303.

Parallel Sessions 6
Differential Equations and Differential Geometry

NEW REDUCTIONS FOR SOLITON EQUATIONS IN 2+1 DIMENSIONS

YI CHENG
Department of Mathematics, University of Science and Technology of China, Hefei 230026, Anhui, China.

ABSTRACT

A new reduction procedure is shown, which leads many soliton equations in 2+1 dimensions to integrable systems in 1+1 dimensions, such that solutions of the reduced systems also solve explicitly the original equations in 2+1 dimensions.

This is a brief survey of some results about the new reductions (or called constraints) on integrable soliton equations in two spatial and one temporal (2+1) dimensions, which were developed recently by Konopelchenko, Li, Oevel, Strampp and myself. In this survey I only emphasize the basic ideas of this new reduction technique through some examples. Further details can be found in the sited papers.

The first example is for the following Kadomtsev–Petviashvili (KP) equation

$$4u_t - 12uu_x - u_{xxx} - 3\partial_x^{-1}u_{yy} = 0 \tag{1}$$

which is the compatibility condition of the following linear systems

$$\psi_y = \psi_{xx} + 2u\psi, \qquad \psi_t = \psi_{xxx} + 3u\psi_x + (3/2)(u_x + \partial_x^{-1}u_y)\psi. \tag{2,3}$$

It is known the quantity $(\psi\psi^*)_x$ solves the linearized KP equation, where ψ^* satisfies the adjoint linear systems of (2) and (3)

$$\psi_y^* = -\psi_{xx}^* - 2u\psi^*, \qquad \psi_t^* = \psi_{xxx}^* + 3u\psi_x^* + (3/2)(u_x - \partial_x^{-1}u_y)\psi^*. \tag{4,5}$$

Thus $(\psi\psi^*)_x$ is the symmetry generator, or $\psi\psi^*$ the conserved covariant generator of the KP equation.

It was shown in [1-4] that if we identify the first covariant u of the KP equation to its covariant generator $\psi\psi^*$, $u = \psi\psi^*$, where $q = \psi, r = \psi^*$ is used for the simplicity, then (2) and (4) become

$$q_y = q_{xx} + 2q^2r, \qquad r_y = -r_{xx} - 2qr^2, \tag{6}$$

while (3) and (5) are reduced to

$$q_t = q_{xxx} + 6qrq_x, \qquad r_t = r_{xxx} + 6qrr_x \tag{7}$$

where in the derivation of (7) we have used $\partial_x^{-1}(qr)_y = q_x r - qr_x$ which is from (6). Eqs (6) and (7) is the first two nontrivial flows in the AKNS hierarchy. The above procedure is what we call the new reduction. It reduces the KP equation to the system consisting of (6) and (7) both in 1+1 dimensions. Originally the reduction of (1) means the requirement of the stationarity with respect to some variables. For example if $u_y = 0$, (1) is reduced to the KdV equation $4u_t = 12uu_x + u_{xxx}$ while $u_t = 0$ leads (1) to the Boussinesq equation. Although both reductions leads the KP in 2+1 dimensions to equations in 1+1 dimensions, only the solutions of the new reduced systems provide an explicit submanifold solutions for the KP equation itself.

There are two generalizations of the above new reduction. the first is to generalize $u = qr$ to the higher-order KP equations and make sure the reduced system coincides with the AKNS hierarchy. This has been done in [4] by using the mastersymmetry construction of [5]. The second comes from the identification of other symmetries of the KP to $(\psi\psi^*)_x$ (for examples, $u_y = (\psi\psi^*)_x$, $u_t = (\psi\psi^*)_x$, etc.) and some interesting integrable equations were reduced [1,2].

It was shown in [6] that all constraints for the KP equation have a simple description by using of Sato's construction of [7] for the KP hierarchy, there the KP hierarchy is given by the compatibility condition of

$$L\psi = \lambda\psi, \qquad \psi_{t_n} = B_n\psi, \tag{8}$$

where $L = \partial + u_2\partial^{-1} + u_3\partial^{-2} + \cdots$ is a microdifferential operator, $\partial = \partial/\partial x$, $u_n, n = 2, 3, \cdots$ are functions of $t = (t_1, t_2, t_3, \cdots)$ with $t_1 = x$ and $B_n = (L^n)_+$ denotes the differential part of L^n. The new reduction is then given by the condition [6,8]

$$L^k = B_k + q\partial^{-1}r \tag{9}$$

on coefficients of L and the reduced systems are given by

$$(L^k)_{t_n} = [B_n, \ L^k], \qquad q_{t_n} = B_n q, \qquad r_{t_n} = -B_n^* r, \tag{10}$$

for dynamical coordinates represented by $u_2, u_3, \cdots, u_k$ (or coefficients of B_k), and q and r, where k is a fixed and arbitrary integer, B_n^* means the adjoint of B_n. An important fact is that the Lax pairs of the reduced hierarchy (10) can be derived also from Sato's theory [6], they are

$$L^k\psi + q\phi = \lambda^k\psi, \qquad \phi_x = r\psi \tag{11}$$

$$\psi_{t_n} = B_n\psi, \qquad \phi_{t_n} = A_n\psi, \tag{12}$$

where ϕ is defined by the second equation in (11). A_n is an operator given by $\partial^{-1}A_n = rB_n - (B_n^* r)$. We notice that 1) if $q = r = 0$, (9) is reduced to $L^k = B_k$ which is nothing but the condition of the usual reduction in Sato's theory. 2) when $k = 1, 2, 3$ the first two flows of the reduced systems in (10) coincide with the previous discussion in [1,2,3]. With the help of their Lax pairs one is able to derive the bi-Haniltonian structures and the conserved quantities

We have also shown in [9] the bilinear equations for the reduced systems of (10), thus providing a understanding of new reductions from the view-point of the bilinear construction.

In the following we briefly give new reductions for some other soliton equations

(1) *The Matrix KP Equation* [10]

$$U_t = (1/4)(3\partial_x^{-1}U_{yy} + U_{xxx} + 3(UU_x + U_xU) + 3[\partial_x^{-1}U_y,\ U]) \tag{13}$$

The new reduction is given by $U = -2QR$ where Q and R are the $n \times m$ and $m \times n$ matrices and solve the linear systems and their adjoint. The reduced systems read

$$\pm Q_y^\pm = Q_{xx}^\pm - 2Q^\pm Q^\mp Q^\pm; \qquad Q_t^\pm = Q_{xxx}^\pm - 3(Q_x^\pm Q^\mp Q^\pm + Q^\pm Q^\mp Q_x^\pm) \tag{14}$$

where

$$Q^+ = \begin{pmatrix} 0 & Q \\ 0 & 0 \end{pmatrix}, \quad Q^- = \begin{pmatrix} 0 & 0 \\ R & 0 \end{pmatrix} \tag{15}$$

Eqs. (14) and (15) are associated with the symmetric spaces in terms of the construction in [11], and the related symmetric spaces are labelled by the Cartan classification: AIII, CI and DIII except BDI.

(2) *The Modified KP Equation* [4]

$$v_t = (1/4)(v_{xxx} - 6v^2v_x - 6v_x\partial_x^{-1}v_{yy}). \tag{16}$$

By identifying $v = qr$, where q and r solve the linear systems and the adjoints of (16) then the reduced equations are

$$q_y = -q_{xx} - 2qrq_x, \qquad r_y = r_{xx} - 2qrr_x, \tag{17}$$

$$q_t = q_{xxx} + 3qrq_{xx} + 3(q^2r^2 + q_xr)q_x, \quad r_t = r_{xxx} - 3qrr_{xx} + 3(q^2r^2 - qr_x)r_x, \tag{18}$$

inparticular (17) was derived in [12] and (18) is the higher-order one of (17).

(3) *The 2+1 Dimensional KS Equation* [4]

$$9u_t = -u_{xxxxx} + 5(\partial_x^{-1}u_y - u_x u_{xx} - uu_{xxx} - u^2 u_x - u_{xxy} - uu_y) \tag{19}$$

If we identify u to q which solves the linear system of (19) then we have

$$q_y + q_{xxx} + 6qq_x = 0; \qquad q_t = q_{xxxxx} + 10qq_{xxx} + 20q_x q_{xx} + 30q^2 q_x \tag{20}$$

which are the first two flows in the KdV hierarchy. This identification is also found to be a simple description in terms of Sato's theory [6].

To end the paper we emphasize that the new reductions should, at least, follow the fact that solutions of the reduced systems also solve the original equations in 2+1 dimensions. Keep this in mind we find that the type of new reduction is not unique. For example in the case of the KP equation (1), the identification $u = 2q_x$ with q satisfying (2) and (3), leads to

$$q_y + q_{xx} + 2qq_x = 0; \qquad q_t = q_{xxx} + 3qq_{xx} + 3q^2 q_x, \tag{21}$$

they are in the hierarchy of the Burgers equation. Indeed, when q solves Eq.(21), $u = 2q_x$ solves the KP equation (1) (see [4]).

Acknownledgement.

The work was supported by the National Natural Science Fund and National Basic Research Project for "Nonlinear Science".

References.

1. B.G. Konopelchenko & W. Strampp, *Inverse Problem* 7(1991), L17; ibid L37.
2. B.G. Konopelchenko, J. Sidorenko & W. Strampp, *Phys. Lett. A* **157**(1991),17.
3. Y. Cheng & Y. S. Li, *Phys. Lett. A* **157**(1991), 22.
4. Y. Cheng & Y. S. Li, *J. Phys. A: Math. Gen.* **25**(1992), 419.
5. B. Fuchssteiner, *Progr. Theor. Phys.* **70**(1983), 1508.
6. Y. Cheng, *J. Math. Phys.* **33**(1992). in press.
7. M. Jimbo & T. Miwa, *Publ. RIMS. Kyoto Univ.* **19**(1983), 943.
8. B. Xu, *Inverse Problem* **8**(1992), L7.
9. Y. Cheng & Y. J. Zhang, "*Bilinear equations for the constrained KP hierarchy,* Preprint,(1992) USTC, Hefei, China.
10. Y. Cheng, *Phys. Lett. A* **166** (1992), 217.
11. A.P. Fordy & P.P. Kulish, *Commun. Math. Phys.* **89**(1983), 427.
12. H.H. Chen, Y.C. Lee & C.S. Liu, *Phys. Scr.* **20**(1979), 490.

SYMMETRIES AND INVARIANT GROUPS OF DIFFERENTIAL EQUATIONS

TIAN CHOU*

Department of Mathematics, University of Science and Technology of China, Hefei, Anhui, P.R.China

ABSTRACT

The relationship between the symmetry and the group–invariant solution of an evolution equation is discussed. New invariant groups are presented for the MKdV equation and the CDF equation.

1. Invariant group

We consider

$$M = \{u(x,t) | u \in C^{\infty}\}$$

and the evolution equation

$$u_t = K(x,t,u,u_x,u_{xx},\cdots) \tag{1.1}$$

which is written as in brief,

$$u_t = K(x,t,u) \text{or } u_t = K(u).$$

Assume

$$N = \{u \in M | u_t = K(u)\},$$

where N is the set of the solutions of (1.1). $G=\{g\}$ is a Lie group which acts on M:

$$g: M \to M.$$
$$u \to \bar{u} = gu$$

Definition. G is called an invariant group of (1.1), if $gN \subset N$, for any $g \in G$; i.e., if u is a solution of (1.1), then $\bar{u} = gu$ is a solution of (1.1) as well.

As is known, it is meanful to find the invariant group for the differential equations.

2. One–parameter invariant groups, symmetries and group–invariant solutions.

The implest and most important case is the one–parameter invariant group $G=\{g_\varepsilon | \varepsilon \in R\}$.

$$g_\varepsilon : u \to \bar{u} = \bar{u}(u,\varepsilon),$$
$$g_0 u = \bar{u}(u,o) = u.$$

* Work supported by the National Scientific Fund of China

Its infinitesimal transformation is

$$\sigma = \frac{d\bar{u}}{d\varepsilon}\Big|_{\varepsilon=0}$$

It is not difficult to find the condition that $\sigma(u)$ should be satisfied:

$$\sigma_t = K'(u)\sigma, \tag{2.1}$$

where σ_t is the total derivitive of u to t,$K'(u)$ is the linerized operator of K to u, and u satisfies (1.1). Since

$$\sigma_t = \frac{\partial\sigma}{\partial t} + \sigma' K,$$

condition (2.1) can be written as

$$\frac{\partial\sigma}{\partial t} = [K,\sigma] \tag{2.2}$$

where $[K,\sigma] = K'\sigma - \sigma' K$. We can suppose that σ doesn't include u_t (can be replaced by $K(u)$), therefore (2.2) should be an indentical relation for u.

Definition. $\sigma(x,t,u)(\equiv\sigma(x,t,u,u_x,u_{xx},\cdots))$ is called a symmetry of (1.1), if

$$\frac{\partial\sigma}{\partial t} = [K,\sigma]$$

for any $u \in M$.

Inversely, does a symmetry correspond to a one—parameter invariant group?

Theorem 2.1. $\sigma(x,t,u)$ is a symmetry of (1.1), $\bar{u}(u,\varepsilon)$ satisfies

$$\frac{d\bar{u}}{d\varepsilon} = \sigma(x,t,\bar{u})$$

$$\bar{u}|_{\varepsilon=0} = u,$$

then

$$g_\varepsilon : u \to \bar{u}$$

is a one—parameter invariant group of (1.1).

Theorem 2.2. If $g_\varepsilon(\varepsilon \in R)$ is a one—parameter invariant group of (1.1) corresponding to the symmetry σ, then u is a g_ε invariant solution if and only if u satisfies the equations:

$$\begin{cases} \sigma(u) = 0, \\ u_t = K(u), \end{cases} \tag{2.3}$$

and (2.3) can be reduced to an ordinary differential equation.

Example 2.1. For the KdV equation

$$u_t = u_{xxx} + 6uu_x, \tag{2.4}$$

$K_0 = u_x$, $K_1 = u_{xxx} + 6uu_x = u_t$ are symmetries of (2.4). We take the symmetry $\sigma = u_t + cu_x$(c is an arbitrary constant) which corresponds to the invariant group G_σ. To find the G_σ—invariant solutions, we only need to solved the equations:

$$\begin{cases} u_t + cu_x = 0, \\ u_t = u_{xxx} + \sigma uu_x. \end{cases}$$

They can be reduced to the ordinary differential equation

$$f''' + 6ff' + cf' = 0,$$

where $u=f(\xi)$, $\xi=x-ct$. In fact it is the travelling solution of (2.4).

Example 2.2. KdV equation (2.4) has the following Lax pair:

$$\begin{cases}\varphi_{xx}=(\lambda-u)\varphi\\ \varphi_t=u_x\varphi+(4\lambda+2u)\varphi_x\end{cases} \tag{2.5}$$

(λ is an arbitrary constant), (2.4) and (2.5) are compatable. As is known, $(\varphi^2)_x$ and $\tau_0=3tu_x+\frac{1}{2}$ are symmetries of (2.4) as well. We take the symmetry $\sigma=\tau_0-(\varphi^2)_x$. To find the corresponding G_σ–invariant solutions, we only need to solve the equations:

$$\begin{cases}\varphi_{xx}=(\lambda-u)\varphi,\\ \varphi_t=-u_x\varphi+(4\lambda+2u)\varphi_x,\\ 3tu_x+\frac{1}{2}-(\varphi^2)_x=0.\end{cases}$$

They can be reduced to the following ordinary differentail equation:

$$9f''+\frac{1}{18}\xi f+\frac{1}{3}f^3=0.$$

The G_σ–invariant solution is

$$u=\frac{1}{2}t^{-3/2}f^2(\xi)-\frac{x}{6t},\xi=-18t^{2/3}\left(\lambda+\frac{x}{6t}\right)+\beta$$

(λ and β are arbitrary contants).

3. New invariant groups of the MKdV equation and CDF equation.

In the following, $\int f dx$ (or $\int f dt$) means an arbitrary primative function of f and it is taken definitely.

Lemma 3.1. If q is a solution of the MKdV equation

$$q_t=q_{xxx}-6q^2q_x, \tag{3.1}$$

then

$$c=(\int q dx)_x-(q_{xx}-2q^3),$$

$$h=-(\int e^{-2\int q dx}dx)_t-2(q_x+q^2)e^{-2\int q dx}-2c\int e^{-2\int q dx}dx$$

are the functions of t only (i.e. $c_x=o=h_x$).

Theorem 3.1. If q is a solution of the MKdV equation

$$q_t=q_{xxx}-6q^2q_x,$$

then

$$\bar{q}=q+r$$

is a solution of (3.1) as well, where

$$r=e^{-2\int q dx}/(\int e^{-2\int q dx}dx+s(t)),$$

$$s(t)=e^{-2\int c(t)dt}(\int h e^{2\int c(t)dt}dt+\alpha),$$

α is an arbitrary constant.

Lemma 3.2. If ϕ is a solution of the CDF equation

$$\phi_t = \phi_{xxx} + 2\phi_x^3 + 6\eta^2\phi_x \sin 2\phi, \tag{3.2}$$

where η is an arbitrary constant, then

$$c = -\frac{1}{2}\left(\int \cos 2\phi dx\right)_t - \phi_{xx}\sin 2\phi + \phi_x^2 \cos 2\phi + 3\eta^2 \cos 2\phi - \eta^2 \cos^3 2\phi,$$

$$h = 2\eta\left(\int \sin 2\phi e^{-2\eta\int \cos 2\phi dx} dx\right)_t - 8\eta c \int \sin 2\phi e^{-2\eta\int \cos 2\phi dx} dx$$

$$- 4\eta(\phi_{xx}\cos 2\phi + \phi_x^2 \sin 2\phi + 2\eta\phi_x + 2\eta^2 \sin 2\phi + \eta^2 \sin^3 2\phi)e^{-2\eta\int \cos 2\phi dx}$$

are the fuctions of t only (i.e. $c_x = o = h_x$).

Theorem 3.2. If ϕ is a solution of (3.2), then

$$\overline{\phi} = arctg(tg\phi + y)$$

is a solution of (3.2) as well, where

$$y = sec^2\phi e^{-2\eta\int \cos 2\phi dx} / Q,$$

$$Q = -tg\phi e^{-2\eta\int \cos 2\phi dx} - 2\eta\int \sin 2\phi e^{-2\eta\int \cos 2\phi dx} dx + \varepsilon,$$

$$\varepsilon = e^{4\eta\int c(t)dt}\left(\int h(t) e^{-4\eta\int c(t)dt} dt + \alpha,\right.$$

α is an arbitrary constant.

According to the theorem 3.1 and theorem 3.2, we can obtain the new solution $\overline{q}$(or $\overline{\phi}$) from a known solution $q(or\ \phi)$ by quadrature, and this process can be contiuned no stop by using the symmetrical property for q(MKdV) and η(CDF).

Example 3.1. For the MKdV equation (3.1), by using the theorem 3.1,

1). we take $q = -\frac{1}{x}$, and

$$\int q dx = -\ln x, \int e^{-\int q dx} dx = x^3 / 3,$$

then we have $c=0$, $h=-4$ and the solution of (3.1):

$$\overline{q} = (2x^3 - 3\varepsilon) / x(x^3 + 3\varepsilon),$$

where $\varepsilon = -4(t+\alpha)$, α is an arbitrary constant.

2). we take $q = -(2x^3 - 3\varepsilon) / x(x^3 + 3\varepsilon)$, $\varepsilon = -4(t+\alpha)$, α is an arbitrary constant, and

$$\int q dx = \ln x - \ln(x^3 + 3\varepsilon),$$

$$\int e^{-2\int q dx} dx = x^5 / 5 + 3x^2\varepsilon - 9\varepsilon^2 / x,$$

then we have $c = o = h$, and the solution of (3.1):

$$\overline{q} = (3x^8 + 18\varepsilon x^5 - 10\delta x^3 + 270\varepsilon^2 x^2 + 15\varepsilon\delta) / (x^3 + 3\varepsilon)(x^6 + 15\varepsilon x^3 + 5\delta x - 45\varepsilon^2),$$

where $\varepsilon = -4(t+\alpha)$, α and δ are arbitrary constants.

REFERENCES

[1] P.J.Olver, Application of Lie group to differential equations, 1986(Springer—Verlag).

[2] Tian C,Symmetries, Solution theory and its application, Edited by Gu Chao hao, Zhejiang Publishing House of Science and Technology, 1990, 217–267.

[3] Tian C, An invariant group of MKdC equation, Preprint, USTC, 1990.

[4] Tian C, An invariance of CDF equation, Preprint, ICTP, 1991, IC / 91 / 73.

Ill-posed Problems and Chaos

D.Y. Hsieh
Department of Mathematics
Hong Kong University of Science and Technology

Abstract

The switching back and forth between elliptic and hyperbolic regimes of ill-posed partial differential equations is explored in relation to the study of chaos. The study of some model linear problems illustrates the feasibility of the idea that chaos may be related to the ill-posedness of partial differential equations.

1. Ill-posed Problems-Hadamard's example

Consider the hyperbolic problem, i.e. the wave equation:

$$\frac{\partial^2 u}{\partial x^2} - \frac{\partial^2 u}{\partial t^2} = 0, \qquad -\infty < x < \infty, \qquad t > 0, \tag{1}$$

with the initial conditions:

$$u(x, 0) = \frac{1}{k^2} \sin kx, \qquad \frac{\partial u}{\partial t}(x, 0) = \frac{1}{k} \cos kx. \tag{2}$$

The solution is easily found to be

$$u(x, t) = \frac{1}{k^2} \sin kt \sin kx + \frac{1}{k^2} \cos kt \sin kx \tag{3}$$

As $k \to \infty$, we have $u(x, 0) \to 0$, and $\frac{\partial u}{\partial t}(x, 0) \to 0$. We also have $u(x, t) \to 0$, for all x and t. $u(x, t)$ is also bounded for $k > 0$.

Now let us consider the elliptic problem, i.e. the Laplace equation:

$$\frac{\partial^2 u}{\partial x^2} + \frac{\partial^2 u}{\partial t^2} = 0, \qquad -\infty < x < \infty, \qquad t > 0, \tag{4}$$

with the same initial conditions (2), the solution is

$$u(x, t) = \frac{1}{k^2} e^{kt} \sin kx \tag{5}$$

As $k \to \infty$, although the initial conditions are vanishingly small, the solution $u(x, t)$ becomes unbounded and rapidly oscillating. The system become unstable. We call the problem ill-posed. In general, initial value problems for elliptic equations are ill-posed. $u(x, t)$ is unbounded also for all $k > 0$, as $t \to \infty$.

2. Equations of mixed type - Linear Examples

Consider

$$F(t)\frac{\partial^2 u}{\partial x^2}+\frac{\partial^2 u}{\partial t^2}+\sigma\frac{\partial u}{\partial t}=0, \qquad -\infty<x<\infty, \qquad t>0, \tag{6}$$

where

$$F(t)=\begin{cases} 1 \text{ for t in } (0, 1), (2, 3), (4, 5), \ldots\ldots \\ -1 \text{ for t in } (1, 2), (3, 4), (5, 6), \ldots\ldots \end{cases} \tag{7}$$

and $\sigma > 0$.

As t changes, equation (6) is alternatively elliptic and hyperbolic. Due to the damping term associated with σ, it may be shown that with the initial conditions (2), u(x, t) will in general be bounded as $t \to \infty$ if and only if

$$\frac{k}{\sigma}\le\frac{\sqrt{3}}{2}. \tag{8}$$

To remove the short wavelength, i.e. large k, instability, we may add a diffusive term and consider the following equation:

$$F(t)\frac{\partial^2 u}{\partial x^2}+\frac{\partial^2 u}{\partial t^2}+\sigma\frac{\partial u}{\partial t}-\gamma\frac{\partial^3 u}{\partial^2 x\partial t}=0, \qquad (\gamma>0). \tag{9}$$

Then the stability condition becomes

$$\sqrt{3}\,(\sigma+\gamma k^2)>2k. \tag{10}$$

Thus the system is always stable if

$$3\sigma\gamma>1. \tag{11}$$

For $\sigma\gamma<\frac{1}{3}$, the system is unstable for

$$k_1<k<k_2, \tag{12}$$

where

$$k_{1,2}=\frac{\sqrt{3}}{\gamma}[1\mp(1-3\sigma\gamma)^{1/2}], \tag{13}$$

3. Partial Differential Equations Related to Chaos

Consider the system

$$\frac{\partial\psi}{\partial t}=-\sigma\psi-\sigma\frac{\partial\theta}{\partial x}, \tag{14}$$

$$\frac{\partial\theta}{\partial t}=-(1-\beta)\theta+\beta\frac{\partial^2\theta}{\partial x^2}+r\frac{\partial\psi}{\partial x}+\frac{\partial}{\partial x}(\theta\psi), \tag{15}$$

where σ and r are positive parameters, and $0\le\beta<1$.
If we take

$$\psi=\sqrt{2}\,X(t)\sin x, \tag{16}$$

$$\theta=\sqrt{2}\,Y(t)\cos x+2Z(t)\cos 2x, \tag{17}$$

and substitute (16) and (17) into the system (14) and (15), and retain only the coefficients of sin x, cos x and cos 2x, the differential equations governing X(t),

Y(t) and Z(t) are the Lorenz equations.

It may be remarked that in connection with the physical Rayleigh-Benard problem of thermal convection, σ is the Prandtl number and r is the Rayleigh number. The equations (14) and (15) can also be interpreted as those governing a gas with internal damping and diffusion but with imaginary sound speed which will make the initial problem ill-posed.

4. Ellipticity and Hyperbolicity

In the system (14) and (15), the term $\beta\frac{\partial^2\theta}{\partial x^2}$ is "good" term, whose presence can smooth out irregularities that may arise otherwise. If we neglect this diffusive term, i.e. set $\beta = 0$, then the characteristics of the system, which we may denote as (K_0), are given by

$$\frac{dx}{dt} = \frac{1}{2}[-\psi\pm(\psi^2 - 4\sigma\theta - 4\sigma r)^{1/2}]. \tag{18}$$

Thus the system is elliptic when $\psi^2 < 4\sigma(r+\theta)$, and hyperbolic when $\psi^2 > 4\sigma(r+\theta)$. Roughly speaking, the system is elliptic when ψ and θ are small, and hyperbolic when ψ is large enough. Ellipticity is associated with instability of the sytem. More precisely, for initial value problems, solutions of the elliptic equations tend to depend sensitively on initial data. Or in other words, the initial value problem is ill-posed. The instability due to ellipticity will make ψ^2 grow with t. But when ψ^2 exceeds the value of $4\sigma(r+\theta)$, the system (K_0) becomes hyperbolic. Hyperbolity implies stability of the system. Therefore the initial unstable growth will be arrested. The damping terms $(-\sigma\psi)$ and $(-\theta)$ in (K_0) will diminish the magnitude of ψ^2 and drive the system back to the elliptic regime. As the system becomes elliptic, the instability mechanism will operate again and raise the magnitude of ψ^2 over $4\sigma(r+\theta)$, thus push the system to the hyperbolic regime. This type of switching back and forth from ellipticity to hyperbolicity may be what corresponds to the chaotic behaviour of the Lorenz system.

5. Numerical Program

Preliminary numerical compution using the Essentially Non-oscillatory Scheme has been carried out. With a stable set of data, the strange attractor associated with the system will be constructed. Other parameters such as fractal dimension and Lyapunov exponent can be determined. Thus a direct link may be established between these properties of dynamical systems to partial differential equations.

6. Other Ill-posed Problems

For van der Waal fluids, the relation between pressure p are specific volume v is given by

$$p(v) = \frac{RT}{v-b} - \frac{a}{v^2}\,. \tag{18}$$

In part of the phase transition region, $p'(v) > 0$. Thus the sound speed is imaginary in that part of the phase transition region. In other words, the governing equation for that part is elliptic, and the corresponding initial value problem is ill-posed.

Ill-posed problems are also encountered in the study of granular flow. Typically, for stress-strain relation $\sigma(\varepsilon)$, the wave speed is proportional to $[\sigma'(\varepsilon)]^{1/2}$. For strains larger than the yielding point, $\sigma'(\varepsilon)$ becomes negative, and again the problem is ill-posed.

Mathematically, ill-posed problem is by definition a taboo. Physically, the problem implies instability or noises. That is why these problems have not been touched before. Now perhaps we are ready to probe into this area.

THE SOLITON BEHAVIOR OF THE PRINCIPAL CHIRAL FIELDS

GU CHAOHAO (C. H. GU) HU HESHENG (H. S. HU)

Dedicated to Prof. C. N. Yang on his 70th birthday.

ABSTRACT. The asymptotic behavior of the multi-soliton solution to the principal chiral fields of unitary group is presented. The solutions are defined as k-th Darboux transformation of the trivial solutions. It is found that these solutions are splitting to k single soliton solutions as $t \to \pm\infty$ and the collision of these solitons is elastic if we consider their amplitudes only.

§ 1 INTRODUCTION

In the present paper we use the Darboux matrix method[1] to study the multi-soliton solutions of principal chiral fields of the special unitary group $SU(2)$[2]. The single soliton solution is obtained by using the Darboux transformation to the trivial solution. Its behaviour can be easily seen from its explicit expression. Moreover, if we consider the principal chiral field as the harmonic maps from the 2-dimensional Minkowski space-time R^{1+1} to the sphere S^3[3], then for a single soliton the images of the line t = const. and the images of the time orbit x = const are all spiral curves on the sphere S^3. It is found that the k-th Darboux transformation of the trivial solution behaves as a k multi-soliton solution. It splits to k-single solitons when t approaches to $+\infty$ or $-\infty$. Besides, the collision of these solitons is elastic if we consider their amplitudes only. However, the change of the phase angles is rather complicated.

In § 2, we recall briefly the Darboux transformation method for $U(n)$ principal chiral fields. we construct the explicit formula for single solitons and the geometrical interpretation is presented in § 3. § 4 devoted to the behavior of the multi-soliton solutions. The main theorem is proved there. In § 5 we give a proof for the theorem of permutability without using any boundary conditions. This theorem is needed in the proof of the main theorem in § 4.

The result obtained in the present paper can be applied to the case of group $SU(N)$ ($N > 2$), but the calculations are much more complicated. It is noted that a certain kind of "multi-soliton solutions" were defined by the dressing method with explicit formulas in [4, 5]. However, the soliton behavior has not been discussed there.

This work is supported by Chinese National Program for Fundamental Research "Nonlinear Sciences" and Tianyuan Fund for mathematics .

§ 2 Darboux transformations

We give a review on the Darboux transformations for $GL(N)$ and $SU(N)$ principal chiral fields.

A $GL(N)$ principal chiral field over Minkowski space-time R^{1+1} is a map $g : R^{1+1} \to GL(N)$ such that the Euler-Lagrange equation of the action integral

$$S[g] = -\int \operatorname{tr}(g_\xi g^{-1} g_\eta g^{-1}) d\xi d\eta \tag{2.1}$$

is satisfied. Here g_ξ and g_η are the derivatives of g with respect to the light-cone coordinates:

$$\xi = x + t, \quad \eta = x - t \tag{2.2}$$

Let

$$A = g_\eta g^{-1}, \quad B = g_\xi g^{-1}. \tag{2.3}$$

As is well-known, A, B should satisfy the curvatureless equation

$$A_\xi - B_\eta + [A, B] = 0, \tag{2.4}$$

and the Euler-Lagrange equations of (2.1) can be written in the form

$$A_\xi + B_\eta = 0. \tag{2.5}$$

Thus, the principal chiral field is also defined by (2.4) and (2.5). The map g can be determined from (2.3).

The Lax pair of (2.4) and (2.5) can be written in the form

$$\Phi_\eta = \lambda A \Phi, \quad \Phi_\xi = \lambda(2\lambda - 1)^{-1} B \Phi \tag{2.6}$$

This means that the condition of integrability of the linear system (2.6) with a spectral parameter λ is just (2.4) and (2.5). Here Φ is an unknown matrix-valued function.

Moreover, the map g is

$$g = \Phi(1) \quad \text{or} \quad \Phi(1)K \tag{2.7}$$

where K is a constant nondegenerate matrix in $GL(N)$, since (2.6) with $\lambda = 1$ is just (2.3).

Let

$$S = I + \lambda\alpha \quad (I \text{ is the } N \times N \text{ unit matrix}) \tag{2.8}$$

be a matrix depending on λ linearly and the $N \times N$ matrix α will be determined later. If the matrix-valued function

$$\Phi_1(\lambda) = S(\lambda)\Phi(\lambda) \tag{2.9}$$

satisfies the Lax pair

$$\Phi_{1\eta} = \lambda A_1 \Phi_1, \quad \Phi_{1\xi} = \lambda(2\lambda - 1)^{-1} B_1 \Phi_1 \tag{2.10}$$

where A_1 and B_1 are matrix-valued function of t and x, then S is called a Darboux matrix of first order. From $\Phi_1(\lambda)$ we can use (2.9) and (2.7) to obtain g_1.

The problem is to find the $N \times N$ matrix-valued function α so that (2.10) holds.

Substituting (2.9) in (2.10) we obtain

$$A_1 = A + \alpha_\eta \tag{2.11}$$

$$B_1 = B - \alpha_\xi \tag{2.12}$$

and

$$\alpha_\eta \alpha = \alpha A - A\alpha \tag{2.13}$$

$$\alpha_\xi \alpha + 2\alpha_\xi = B\alpha - \alpha B \tag{2.14}$$

The following construction gives an explicit formula for the matrix α whose Jordan form is diagonal.

Let λ_1, λ_2, $\cdots$, λ_N be a set of constants (each $\lambda_i \neq 0, \frac{1}{2}, 1$) and at least two of them are unequal. Define

$$H = [\Phi(\lambda_1)l_1, \Phi(\lambda_2)l_2, \cdots, \Phi(\lambda_N)l_N] \tag{2.15}$$

where $l_1, \cdots, l_N$ are arbitrary constant column vectors such that $\det H \neq 0$. Then

$$\alpha = -H\Lambda^{-1}H^{-1} \tag{2.16}$$

with

$$\Lambda = \begin{bmatrix} \lambda_1 & & \\ & \ddots & \\ & & \lambda_N \end{bmatrix} \tag{2.17}$$

is a solution to (2.13) and (2.14) and hence $g_1 = \Phi_1(1) = (1+\alpha)g$ (or $\Phi_1(1)K$), A_1 and B_1 are new solutions of $GL(N)$ principal chiral field. Thus from a given solution g, A, B, we can construct a new solution g_1, A_1, B_1 by using the solution $\Phi(\lambda)$ of the Lax pair. The transformation $(A, B, g, \Phi) \rightarrow (A_1, B_1, g_1, \Phi_1)$ is called Darboux transformation. The advantage of the Darboux transformation method is that the algorithm from $(A, B, g, \Phi) \rightarrow (A_1, B_1, g_1, \Phi_1)$ is purely algebraic and universal. Hence we can use the transformations successfully to obtain an infinite number of new solutions with the same algorithm.

For the cases of $U(N)$ and $SU(N)$ principal chiral fields we can use the formulas (2.8), (2.9), (2.16), (2.11) and (2.12) too. But the constants $\lambda_1, \cdots, \lambda_N$ and the column vectors $l_1, l_2, \cdots, l_N$ should be specified. We set

$$\lambda_i = \lambda_0 \quad \text{or} \quad \overline{\lambda}_0 \tag{2.18}$$

where λ_0 is a non-real complex number and it has been shown that[2] we can choose l_i $(i = 1, 2, \cdots, N)$ such that

$$h_i^+ h_j = 0, \quad \text{if } \lambda_i \neq \lambda_j \tag{2.19}$$

Here h_i is the ith column in H, i.e.

$$h_i = \Phi(\lambda_i)l_i \tag{2.20}$$

In this case we have

$$\alpha^+ + \alpha = (-\lambda_0^{-1} + \overline{\lambda}_0^{-1})I$$

$$\alpha^+ \alpha = |\lambda_0|^{-2} I$$

$$(I+\alpha)^+(I+\alpha) = |1-\lambda_0^{-1}|^2 I \tag{2.21}$$

Hence

$$A_1^+ = -A_1, \quad B_1^+ = -B_1 \tag{2.22}$$

provided that

$$A^+ = -A, \quad B^+ = -B.$$

Moreover

$$trA_1 = 0, \quad trB_1 = 0 \tag{2.23}$$

provided that

$$trA = trB = 0.$$

the $U(N)$ or $SU(N)$-valued matrix function g_1 has the explicit expression

$$g_1 = (I+\alpha)g_0\Lambda(\Lambda-1)^{-1} \tag{2.24}$$

up to a right-multiplication of a constant matrix in $U(N)$ or $SU(N)$.

§ 3. Single soliton

For simplicity we consider the group $SU(2)$ only. The elements of $SU(2)$ are matrices

$$\begin{bmatrix} \gamma & \beta \\ -\overline{\beta} & \overline{\gamma} \end{bmatrix} \tag{3.1}$$

with $\gamma\overline{\gamma}+\beta\overline{\beta}=1$. The seed solution is

$$g_0 = \begin{bmatrix} e^{i(\phi\eta+\psi\xi)} & 0 \\ 0 & e^{-i(\phi\eta+\psi\xi)} \end{bmatrix} \tag{3.2}$$

Here ϕ, ψ are real constants, not all zero. Then

$$A = \begin{bmatrix} i\phi & 0 \\ 0 & -i\phi \end{bmatrix}, \quad B = \begin{bmatrix} i\psi & 0 \\ 0 & -i\psi \end{bmatrix} \tag{3.3}$$

and

$$\Phi_0 = \begin{bmatrix} l(\lambda) & 0 \\ 0 & l^{-1}(\lambda) \end{bmatrix} \tag{3.4}$$

where

$$l(\lambda) = \exp(i\lambda\phi\eta + \frac{i\lambda}{2\lambda-1}\psi\xi) \tag{3.5}$$

Take $\lambda_1 = \lambda_0$, $\lambda_2 = \overline{\lambda}_0$ and

$$H = \begin{bmatrix} l(\lambda_0) & bl(\overline{\lambda}_0) \\ al^{-1}(\lambda_0) & l^{-1}(\overline{\lambda}_0) \end{bmatrix} \tag{3.6}$$

The condition (2.19) implies $b = -\overline{a}$.

By calculations we have

$$\det H = e^p + |a|^2 e^{-p} > 0 \tag{3.7}$$

$$\alpha = -H\Lambda^{-1}H^{-1} = \frac{-1}{e^p+|a|^2e^{-p}} \begin{bmatrix} \frac{e^p}{\lambda_0} + \frac{e^{-p}}{\overline{\lambda}_0}|a|^2 & \left(\frac{1}{\overline{\lambda}_0} - \frac{1}{\lambda_0}\right)\overline{a}e^{iq} \\ \left(\frac{1}{\overline{\lambda}_0} - \frac{1}{\lambda_0}\right)\overline{a}e^{-iq} & \frac{e^p}{\overline{\lambda}_0} + \frac{e^{-p}}{\lambda_0}|a|^2 \end{bmatrix} \tag{3.8}$$

Here

$$p = i(\lambda_0 - \overline{\lambda}_0)\phi\eta + i\left(\frac{\lambda_0}{2\lambda_0 - 1} - \frac{\overline{\lambda}_0}{2\overline{\lambda}_0 - 1}\right)\psi\xi$$
$$q = (\lambda_0 + \overline{\lambda}_0)\phi\eta + \left(\frac{\lambda_0}{2\lambda_0 - 1} + \frac{\overline{\lambda}_0}{2\overline{\lambda}_0 - 1}\right)\psi\xi \tag{3.9}$$

are real linear functions. Moreover

$$g_1 = (I + \alpha)g_0\Lambda(\Lambda - 1)^{-1} = \frac{1}{e^p + |a|^2 e^{-p}} \cdot \begin{bmatrix} a_{11} & a_{12} \\ -\overline{a}_{12} & \overline{a}_{11} \end{bmatrix} \tag{3.10}$$

where

$$a_{11} = \left[\left(\frac{\lambda_0 - 1}{\lambda_0}\right)e^p + |a|^2\left(\frac{\overline{\lambda}_0 - 1}{\overline{\lambda}_0}\right)e^{-p}\right]\left(\frac{\lambda_0}{\lambda_0 - 1}\right)e^{i(\phi\eta + \psi\xi)}$$
$$a_{12} = \left(\frac{1}{\lambda_0} - \frac{1}{\overline{\lambda}_0}\right)\overline{a}e^{i(q + \phi\eta + \psi\xi)}\left(\frac{\lambda_0}{\lambda_0 - 1}\right)$$

Comparing with (3.1) we have

$$\beta = \rho_1 e^{i\theta_1}, \quad \gamma = \rho_2 e^{i\theta_2} \tag{3.11}$$

which can be considered as two complex-valued wave functions. We call them the β-wave and γ-wave. Their amplitudes (or the density distributions) are

$$\rho_1 = \frac{|a|}{e^p + |a|^2 e^{-p}}\left|\frac{\lambda_0 - \overline{\lambda}_0}{\lambda_0(\lambda_0 - 1)}\right|$$
$$\rho_2 = (1 - \rho_1^2)^{1/2} \tag{3.12}$$

They are travelling waves since p is linear with respect to t and x. From (3.12) we see

$$\lim_{p \to \pm\infty} \rho_1 = 0, \quad \lim_{p \to \pm\infty} \rho_2 = 1 \tag{3.13}$$

The soliton behavior is clear. The phase angle θ_1 is a linear function of x and t too. In general the phase speed and the amplitude speed are different.

We turn to the geometry of the soliton solution. As is well known, a principal chiral field over R^{1+1} is a harmonic map from R^{1+1} to the group manifold with respect to the invariant Riemannian (or pseudo Riemannian) metric. For the case of group $SU(2)$, the group manifold is isometric to the standard sphere S^3:

$$\gamma = x_1 + ix_2, \quad \beta = x_3 + ix_4 \tag{3.14}$$

with

$$|\gamma|^2 + |\beta|^2 = x_1^2 + x_2^2 + x_3^2 + x_4^2 = 1. \tag{3.15}$$

The seed solution (3.2) is the geodesic

$$x_3 = x_4 = 0, \quad x_1 + ix_2 = e^{i(\phi\eta + \psi\xi)}$$

(a great circle $C \subset S^3$). In the generic case that $\phi\eta + \psi\xi$ essentially depends on both t and x, the images of the time section $t = t_0$ and the images of the world line $x = x_0$ are the circle C.

For the single soliton solution (3.10), if p depends on t, the image of the time orbit $x = x_0$ is a two-side spiral on S^3 which approaches to the circle C as $t \to \infty$ and $t \to -\infty$. If p depends on x, the image of the time section $t = t_0$ approaches to the circle C in a spiral like way too.

§ 4. Multi-soliton solutions

By applying the Darboux transformations successively, we obtain multi-soliton solutions.

Let $\lambda_0, \lambda_1, \cdots, \lambda_k$ ($\neq 0, 1, 1/2$) be $k+1$ complex numbers which are not real. Assume that

(1) $|2\lambda_l - 1|$ ($l = 0, 1, \cdots, k$) are different.
(2) $(\lambda_l - \overline{\lambda}_l)\phi \pm (\lambda_l/(2\lambda_l - 1) - \overline{\lambda}_l/(2\overline{\lambda}_l - 1))\,\psi \neq 0$.

Introduce notations

$$p_l = i\left\{(\lambda_l - \overline{\lambda}_l)\phi\eta + \left(\frac{\lambda_l}{2\lambda_l - 1} - \frac{\overline{\lambda}_l}{2\overline{\lambda}_l - 1}\right)\psi\xi\right\} \tag{4.1}$$

$$q_l = (\lambda_l + \overline{\lambda}_l)\phi\eta + \left(\frac{\lambda_l}{2\lambda_l - 1} + \frac{\overline{\lambda}_l}{2\overline{\lambda}_l - 1}\right)\psi\xi \tag{4.2}$$

The assumption (1) implies that p_i/p_j ($i \neq j$) are not constant and (2) means that p_l are dependent on x and t both.

The seed solution is

$$g_0 = \begin{bmatrix} e^{i(\phi\eta + \psi\xi)} & 0 \\ 0 & e^{-i(\phi\eta+\psi\xi)} \end{bmatrix} \tag{4.3}$$

with

$$\Phi_0(\lambda) = \begin{bmatrix} l(\lambda) & 0 \\ 0 & l^{-1}(\lambda) \end{bmatrix} \tag{4.4}$$

Recall that

$$l(\lambda) = \exp i(\lambda\phi\eta + \frac{\lambda}{2\lambda - 1}\psi\xi) \tag{4.5}$$

By using the Darboux transformations we obtain

$$\Phi_l(\lambda) = (I + \lambda\alpha_{l-1})\cdots(I + \lambda\alpha_0)\Phi_0(\lambda) \tag{4.6}$$

$$g_l = (I + \alpha_{l-1})\cdots(I + \alpha_0)g_0\Lambda_0(\Lambda_0 - 1)^{-1}\cdots\Lambda_{l-1}(\Lambda_{l-1} - 1)^{-1} \tag{4.7}$$

for $l = 1, 2, \cdots, k$. Here α_0, Φ_1, g_1 has been constructed before. Note that α_h is constructed from $\Phi_h(\lambda)$ by the general formula (2.16) in§ 2.

From the expression of α_0, we see that

$$\lim_{\substack{t\to\pm\infty \\ p_0\to\pm\infty}} \alpha_0 \to \begin{bmatrix} -\frac{1}{\lambda_0} & 0 \\ 0 & -\frac{1}{\overline{\lambda}_0} \end{bmatrix} \quad \text{or} \quad \begin{bmatrix} -\frac{1}{\overline{\lambda}_0} & 0 \\ 0 & -\frac{1}{\lambda_0} \end{bmatrix} \tag{4.8}$$

we will show that

$$\lim_{\substack{t\to\pm\infty \\ p_0, p_1, \cdots, p_i\to\pm\infty}} \alpha_i \to \begin{bmatrix} -\mu_i & 0 \\ 0 & -\overline{\mu}_i \end{bmatrix} \tag{4.9}$$

for all $i = 0, 1, \cdots k$. Here

$$\mu_i = \frac{1}{\lambda_i} \quad \text{or} \quad \frac{1}{\overline{\lambda}_i} \tag{4.10}$$

We assume that (4.9) holds for $i = 1, 2, \cdots, l-1$ and consider the asymptotic behavior of α_l.

In this case, we see that as $p_i \to \pm\infty$ $(i = 0, 1, \cdots, l-1)$

$$\Phi_{l-1}(\lambda) \sim \begin{bmatrix} A(\lambda)l(\lambda) & 0 \\ 0 & \tilde{A}(\lambda)l^{-1}(\lambda) \end{bmatrix} \tag{4.11}$$

Here

$$A(\lambda) = \prod_{i=0}^{l-1}(1 - \lambda\mu_i) \quad \tilde{A}(\lambda) = \prod_{i=0}^{l-1}(1 - \lambda\overline{\mu}_i) \tag{4.12}$$

in which μ_i are defined by (4.10). From (4.12) we have

$$\overline{A(\lambda)} = \tilde{A}(\overline{\lambda}). \tag{4.13}$$

Consequently, the asymptotic behavior of the matrix H_l is

$$H_l \sim \begin{bmatrix} A(\lambda_l)l(\lambda_l) & b_l A(\overline{\lambda}_l)l(\overline{\lambda}_l) \\ a_l\tilde{A}(\lambda_l)l^{-1}(\lambda_l) & \tilde{A}(\overline{\lambda}_l)l^{-1}(\overline{\lambda}_l) \end{bmatrix} \tag{4.14}$$

As before, we have

$$b_l = -\overline{a}_l. \tag{4.15}$$

By calculations, we obtain

$$\alpha_l = -H_l\Lambda_l^{-1}H_l^{-1} \sim \frac{-1}{|A(\lambda_l)|^2 e^{p_l} + |A(\overline{\lambda}_l)|^2|a_l|^2 e^{-p_l}} \cdot$$
$$\cdot \begin{bmatrix} \frac{|A(\lambda_l)|^2}{\lambda_l}e^{p_l} + \frac{|A(\overline{\lambda}_l)|^2|a_l|^2}{\overline{\lambda}_l}e^{-p_l} & \left(\frac{1}{\lambda_l} - \frac{1}{\overline{\lambda}_l}\right)\overline{a}_l\overline{A}(\lambda_l)A(\overline{\lambda}_l)e^{iq_l} \\ \left(\frac{1}{\lambda_l} - \frac{1}{\overline{\lambda}_l}\right)a_l A(\lambda_l)\overline{A}(\overline{\lambda}_l)e^{-iq_l} & \frac{|A(\lambda_l)|^2}{\overline{\lambda}_l}e^{p_l} + \frac{|A(\overline{\lambda}_l)|^2|a_l|^2}{\lambda_l}e^{-p_l} \end{bmatrix} \tag{4.16}$$

From (4.16) it is not difficult to see that (4.9) is true for $i = l$. Thus, by mathematical induction, we obtain

Lemma. *As $t \to \pm\infty$, and $p_0, p_1, \cdots, p_l \to \pm\infty$, we have*

$$\alpha_l \to -\begin{bmatrix} 1/\lambda_l & 0 \\ 0 & 1/\overline{\lambda}_l \end{bmatrix} \quad \text{or} \quad -\begin{bmatrix} 1/\overline{\lambda}_l & 0 \\ 0 & 1/\lambda_l \end{bmatrix} \tag{4.9'}$$

Let t approach $\pm\infty$ and p_i remains finite for a fixed i, from the assumption (1) we see that $p_j \to \pm\infty$ or $\mp\infty$, for all $j \neq i$.

Consider the asymptotic behavior of the k-soliton solution. Let $t \to \pm\infty$ and p_{k-1} remains finite. Then $p_0, p_1, \cdots, p_{k-2} \to +\infty$ or $-\infty$.

From the lemma, we have

$$g_k \sim (I + \alpha_{k-1})\begin{bmatrix} 1-\mu_{k-2} & 0 \\ 0 & 1-\overline{\mu}_{k-2} \end{bmatrix} \cdots \begin{bmatrix} 1-\mu_0 & 0 \\ 0 & 1-\overline{\mu}_0 \end{bmatrix} \begin{bmatrix} e^{i(\phi\eta+\psi\xi)} & 0 \\ 0 & e^{-i(\phi\eta+\psi\xi)} \end{bmatrix} M_{k-1} \tag{4.17}$$

Here

$$M_{k-1} = \Lambda_0(\Lambda_0 - 1)^{-1} \cdots \Lambda_{k-1}(\Lambda_{k-1} - 1)^{-1} \tag{4.18}$$

From (4.17) and (4.18), it is readily seen that when $t \to \pm\infty$ and p_{k-1} keeps finite, then the solution g_k behave as the single soliton g_1 expressed as (3.10), and the speed of propagation is determined by $p_{k-1} = \text{const}$. If we write

$$g_k = \begin{bmatrix} \gamma_k & \beta_k \\ -\overline{\beta}_k & \overline{\gamma}_k \end{bmatrix}$$

we can see that $|\beta_k|$ and $|\gamma_k|$ have the same shape for $t \to +\infty$ and $t \to -\infty$. However, the alternative choice of the asymptotic values of α_l in (4.9') are different as $t \to +\infty$ and $t \to -\infty$, and hence there will be some "phase shift" after the scattering.

If we consider the case $t \to \pm\infty$ and p_i = finite ($i < k-1$), by using the theorem of permutability (see § 5), we complete the proof of the following.

Main theorem. *The solution g_k of the $U(n)$ (or $SU(n)$) principal chiral field obtained by using Darboux transformations successively from the trivial solution (3.2) for k times behaves as a k multi-soliton solution i.e. (1) When $t \to +\infty$ (or $-\infty$) the asymptotic behavior of the solution g_k is splitting to a train of k solitons. (2) After the scattering of solitons, the speeds and the amplitudes of the solitons are unchanged.*

§ 5. Theorem of Permutability

In the present section we state and prove the theorem of permutability.

Theorem of permutability. *Let $\Phi_1(\lambda)$ be the Darboux transform of $\Phi(\lambda)$ with the parameters Λ_1 and l_i, $\Phi_2(\lambda)$ the Darboux transform of $\Phi_1(\lambda)$ with the parameters Λ_2 and m_i, $\tilde{\Phi}_1(\lambda)$ be the Darboux transform of $\Phi(\lambda)$ with the parameters Λ_2 and m_i, and $\tilde{\Phi}_2(\lambda)$ the Darboux transformation of $\tilde{\Phi}_1(\lambda)$ with the parameter Λ_1 and l_i. Then $\Phi_2(\lambda) = \tilde{\Phi}_2(\lambda)$.*

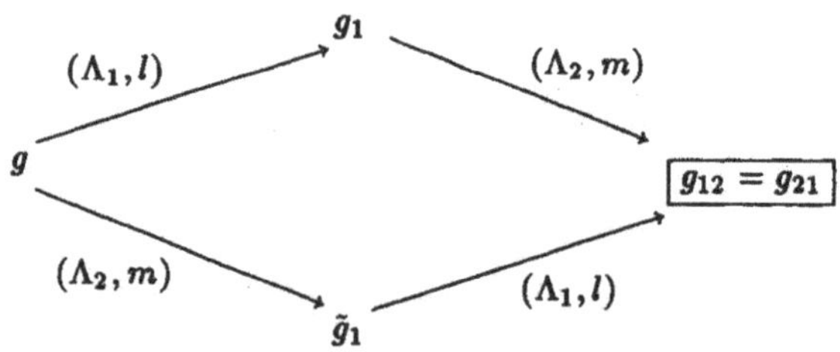

Proof. From the definition

$$\Phi_1(\lambda) = (I + \lambda\alpha_1)\Phi(\lambda), \quad \Phi_2(\lambda) = (I + \lambda\alpha_2)\Phi_1(\lambda) = (I + \lambda\alpha_2)(I + \lambda\alpha_1)\Phi(\lambda). \tag{5.1}$$

Here

$$\alpha_1 = -H_1\Lambda_1^{-1}H_1^{-1}, \quad \Lambda_1 = \begin{bmatrix} \underset{1}{\lambda_1} & & \\ & \ddots & \\ & & \underset{1}{\lambda_N} \end{bmatrix}, \quad H_1 = \left[\Phi(\underset{1}{\lambda_1})l_1, \cdots, \Phi(\underset{1}{\lambda_N})l_N\right] \tag{5.2}$$

$$\alpha_2 = -H_2\Lambda_2^{-1}H_2^{-1}, \quad \Lambda_2 = \begin{bmatrix} \underset{2}{\lambda_1} & & \\ & \ddots & \\ & & \underset{2}{\lambda_N} \end{bmatrix}, \quad H_2 = \left[\Phi_1(\underset{2}{\lambda_1})m_1, \cdots, \Phi_1(\underset{2}{\lambda_N})m_N\right] \tag{5.3}$$

Similarly

$$\tilde{\Phi}_1 = (I + \lambda\tilde{\alpha}_1)\Phi(\lambda), \quad \tilde{\Phi}_2(\lambda) = (I + \lambda\tilde{\alpha}_2)(I + \lambda\tilde{\alpha}_1)\Phi(\lambda) \tag{5.4}$$

with

$$\tilde{\alpha}_1 = -\tilde{H}_1\Lambda_2^{-1}\tilde{H}_1^{-1} \tag{5.5}$$

$$\tilde{\alpha}_2 = -\tilde{H}_2\Lambda_1^{-1}\tilde{H}_2^{-1} \tag{5.6}$$

and

$$\tilde{H}_1 = \left[\Phi(\underset{2}{\lambda_1})m_1, \cdots, \Phi(\underset{2}{\lambda_N})m_N\right] \tag{5.7}$$

$$\tilde{H}_2 = \left[\tilde{\Phi}_1(\underset{1}{\lambda_1})l_1, \cdots, \tilde{\Phi}(\underset{1}{\lambda_N})l_N\right] \tag{5.8}$$

we are going to prove

$$(I + \lambda\alpha_2)(I + \lambda\alpha_1)\Phi(\lambda) = (I + \lambda\tilde{\alpha}_2)(I + \lambda\tilde{\alpha}_1)\Phi(\lambda) \tag{5.9}$$

From $\alpha_1 H_1\Lambda_1 + H_1 = 0$, we have

$$\left(\Phi(\underset{1}{\lambda_i}) + \underset{1}{\lambda_i}\alpha_1\Phi(\underset{1}{\lambda_i})\right) l_i = 0, \quad i = 1, 2, \cdots, N. \tag{5.10}$$

Moreover, we have

$$\left(\Phi_1(\underset{2}{\lambda_i}) + \underset{2}{\lambda_i}\alpha_2\Phi_1(\underset{2}{\lambda_i})\right) m_i = 0 \tag{5.11}$$

Hence, we have

$$\begin{aligned} &(I + \underset{1}{\lambda_i}\alpha_2)(I + \underset{1}{\lambda_i}\alpha_1)\Phi(\underset{1}{\lambda_i})l_i = 0, \\ &(I + \underset{2}{\lambda_i}\alpha_2)(I + \underset{2}{\lambda_i}\alpha_1)\Phi(\underset{2}{\lambda_i})m_i = 0. \end{aligned} \tag{5.12}$$

Similarly, we have

$$(I + \underset{2}{\lambda_i}\tilde{\alpha}_2)(I + \underset{2}{\lambda_i}\tilde{\alpha}_1)\Phi(\underset{2}{\lambda_i})m_i = 0, \quad (I + \underset{1}{\lambda_i}\tilde{\alpha}_2)(I + \underset{1}{\lambda_i}\tilde{\alpha}_1)\Phi(\underset{1}{\lambda_i})l_i = 0. \tag{5.13}$$

(5.12) and (5.13) are both of the following form

$$\begin{aligned} &(I + A\underset{1}{\lambda_i} + B\underset{1}{\lambda_i^2})\Phi(\underset{1}{\lambda_i})l_i = 0 \\ &(I + A\underset{2}{\lambda_i} + B\underset{2}{\lambda_i^2})\Phi(\underset{2}{\lambda_i})m_i = 0 \end{aligned} \tag{5.14}$$

Consider (5.14) as a system of linear equations with A, B as unknowns. From the definition of H_1, $\tilde{H}_1$, Λ_1 and Λ_2 we see that (5.14) can be written in the form

$$H_1 + AH_1\Lambda_1 + BH_1\Lambda_1^2 = 0, \quad \tilde{H}_1 + A\tilde{H}_1\Lambda_2 + B\tilde{H}_1\Lambda_2^2 = 0 \tag{5.15}$$

From (5.15) we have

$$H_1\Lambda_1^{-1}H_1^{-1} + A + BH_1\Lambda_1H_1^{-1} = 0, \quad \tilde{H}_1\Lambda_2^{-1}\tilde{H}_1^{-1} + A + B\tilde{H}_1\Lambda_2\tilde{H}_1^{-1} = 0. \tag{5.16}$$

hence

$$B(H_1\Lambda_1H_1^{-1} - \tilde{H}_1\Lambda_2\tilde{H}_1^{-1}) + (H_1\Lambda_1^{-1}H_1^{-1} - \tilde{H}_1\Lambda_2^{-1}\tilde{H}_1^{-1}) = 0. \tag{5.17}$$

In the generic case

$$\det\left(H_1\Lambda_1H_1^{-1} - \tilde{H}_1\Lambda_2\tilde{H}_1^{-1}\right) \neq 0. \tag{5.18}$$

The solution A, B is unique. Comparing (5.14) with (5.12) and (5.13), we have

$$A = \alpha_1 + \alpha_2 = \tilde{\alpha}_1 + \tilde{\alpha}_2, \quad B = \alpha_2\alpha_1 = \tilde{\alpha}_2\tilde{\alpha}_1 \tag{5.19}$$

and hence

$$\Phi_2(\lambda) = \tilde{\Phi}_2(\lambda) \tag{5.20}$$

The theorem of permutability holds. If (5.18) is not true we can find a sequence of parameters $l_i^{(n)}$, $m_i^{(n)}$, $\Lambda_1^{(n)}$, $\Lambda_2^{(n)}$ such that $l_i^{(n)} \to l_i$, $m_i^{(n)} \to m_i$, $\Lambda_1^{(n)} \to \Lambda_1$, $\Lambda_2^{(n)} \to \Lambda_2$ and

$$\det\left(H_1^{(n)}\Lambda_1^{(n)}H_1^{(n)-1} - \tilde{H}_1^{(n)}\Lambda_2^{(n)}\tilde{H}_1^{(n)-1}\right) \neq 0 \tag{5.21}$$

Hence,

$$\alpha_1^{(n)} + \alpha_2^{(n)} = \tilde{\alpha}_1^{(n)} + \tilde{\alpha}_2^{(n)}, \quad \alpha_2^{(n)}\alpha_1^{(n)} = \tilde{\alpha}_2^{(n)}\tilde{\alpha}_1^{(n)} \tag{5.22}$$

When $n \to \infty$, we still have

$$\alpha_1 + \alpha_2 = \tilde{\alpha}_1 + \tilde{\alpha}_2, \quad \alpha_2\alpha_1 = \tilde{\alpha}_2\tilde{\alpha}_1 \tag{5.23}$$

and the theorem of permutability holds too. The proof is completed.

References

1. Gu, C. H., Nankai Lectures on Mathematical Physics, World scientific publishing Company, Singapore, 1989, pp. 162-168.
2. Gu, C. H. & Zhou, Z., Nonlinear evolution equations: integrability and spectral methods, Manchester Univ. Press, 1990, pp. 115-123.
3. Gu Chaohao, Commun. Pure and Appl. Math. **33** (1980), 727-737.
4. Harnad, I, Saint-Aubin, Y. and Shnider, S., Commun. Math. Phys. **92** (1984), 329-367.
5. Zakharov, V. E. & Mikhailov, Commun. Math. Phys. **74** (1980), 21-40.

Gu Chaohao (C. H. Gu)
Univ. of Science and Technology of China, Hefei, Anhui, China
Institute of Math., Fudan Univ. Shanghai, China

Hu Hesheng (H. S. Hu)
Institute of Math., Fudan Univ. Shanghai, China

UNIFIED MATHEMATICAL APPROACH TO SPINORS AND MULTIVECTORS, GEOMETRIC SUPERALGEBRA

JAIME KELLER
División de Ciencias Básicas, F.Q., and Quantum Theory of Matter Project
Facultad de Estudios Superiores-Cuautitlán, Universidad Nacional Autónoma de México
A. Postal 70-528, 04510 México, D. F.

Abstract.
A unified mathematical approach to spinors and multivectors (here called geometric superalgebra) is constructed in such a form useful to study the mathematical description of spacetime, matter and its interaction fields.

I. Introduction.

It is convenient, Keller and Rodríguez (1991)[1], to show that we can regard the mathematical formalism of the "spacetime" as a joint system of vectors, multivectors and spinors, in such a way that the multivector geometry of spacetime turns out to be explicitly expressable as an algebraic structure in terms of spinors and their outer products. Multivector being in turn generated by the outer product (fully antisymmetric product) of a vector basis set or directly as are vectors. The development of such formalism, which we have called *geometric superalgebra*, allows a simple and straightforward statement of many of the properties and relations of the corresponding spaces.
As a result a new type of algebra, with an interesting structure of internal mappings, is then developed: it embeddes the spacetime Clifford algebra and the spinor algebra, (see in particular references 2-39 in ref. 1 and Crumeyrolle (1990)[2]) both are widely used in physics[3]. We have also shown, Keller (1991)[4], that considering the full spacetime complex Clifford algebra as the tangent space manifold a unified theory of matter and interaction fields is obtained suitable to recover all the features of the standard electroweak-color theory of elementary particle physics.
Here we consider that for a base space of $n = p+q$ dimensions corresponds to a system with $(2^d+1)^2$ complex degrees of freedom (d = integer $(\frac{p+q}{2})$).

II. Naive Construction of a Simple, Meaningful Example.

As in Keller and Rodríguez[1] let us begin with the example of the most simple space where all basic features are already present: $\mathcal{R}^{2,0}$, spanned by two mutually perpendicular unit vectors e_1 and e_2, to which we assign correspondingly the real Clifford algebra $\mathcal{C}^{2,0}$, spanned by the basis elements $\{1, e_1, e_2, e_{12}\}$. We denote by $\{\xi_1, \xi_2\}$ and by $\{\xi_1^T, \xi_2^T\}$ the basis for the spinor space $\mathcal{L}$ and for the corresponding dual spinor space $\mathcal{L}_\mathcal{D}$. The algebra is completed introducing the complex scalar field $\hat{\mathcal{S}}$.

To be specific let us introduce a conventional 2×2-matrix representation for the Clifford algebra elements of $\mathcal{C}^{2,0}$:

$$\mathbf{1} = \begin{pmatrix} 1 & 0 \\ 0 & 1 \end{pmatrix}, \quad e_1 = \begin{pmatrix} 1 & 0 \\ 0 & -1 \end{pmatrix}, \quad e_2 = \begin{pmatrix} 0 & 1 \\ 1 & 0 \end{pmatrix}, \quad e_{12} = e_1 e_2 = \begin{pmatrix} 0 & 1 \\ -1 & 0 \end{pmatrix}, \qquad [1]$$

We correspondingly represent the spinor basis elements by

$$\xi_1 = \begin{pmatrix} 1 \\ 0 \end{pmatrix}, \quad \xi_2 = \begin{pmatrix} 0 \\ 1 \end{pmatrix}, \quad \xi_1^T = (1 \quad 0), \quad \xi_2^T = (0 \quad 1). \qquad [2]$$

We can easily verify that $1 = \xi_1\xi_1^T + \xi_2\xi_2^T$, $e_1 = \xi_1\xi_1^T - \xi_2\xi_2^T$, etc. The metric of the spinors is seen to be $\xi_a^T\xi_b = \delta_{ab}$.
It is important to keep in mind the closure property inherent to the matrix structure of any representation of a Clifford algebra and its modules: Being $\mathcal{M}^{p,q} \in \mathcal{C}^{p,q}$, we are allowed to write formally, for the left module $\mathcal{M}^{p,q}\mathcal{L}^{p,q} = \mathcal{L}^{p,q}$, products of the form $\mathcal{M}^{p,q}\mathcal{L}_D^{p,q}$ are not defined, for them the closure property is to be written as corresponds to a right module $\mathcal{L}_D^{p,q}\mathcal{M}^{p,q} = \mathcal{L}_D^{p,q}$. For spinors $\phi \in \mathcal{L}^{p,q}$ and $\phi'^T \in \mathcal{L}_{\mathcal{D}}{}^{p,q}$ we would identify outer products of the form $\phi\,\phi'^T$ as members of the field $\mathcal{M}^{p,q}$ and products of the form $\phi'^T\,\phi$ as members of the complex scalars field $\mathcal{S}$.
So, we first consider: how could we construct a system in which both closure properties of field and ring are kept together inside a unique product operation? For this purpose let us now consider a matrix algebra whose elements possess the general block structure

$$\hat{K} = \begin{pmatrix} M & \chi \\ \phi^T & s \end{pmatrix}, \quad \text{and} \quad \hat{K}\,\hat{K}' = \begin{pmatrix} M\,M' + \chi\phi'^T & M\,\chi' + \chi s' \\ \phi^T\,M' + s\,\chi'^T & \phi^T\chi' + s\,s' \end{pmatrix}. \qquad [3]$$

with $M \in \mathcal{M}^{p,q}$, $\chi \in \mathcal{L}^{p,q}$, $\phi^T \in \mathcal{L}_{\mathcal{D}}{}^{p,q}$ and $s \in \mathcal{S}$. It turns out that under matrix multiplication, this block structure of multivector, spinor and scalar is being properly held.
In particular, for our example, we are in position to construct the algebra $\hat{\mathcal{K}}^{2,0}$ whose generators $\{\hat{1}, \hat{e}_1, \hat{e}_2, \hat{e}_{12}, \hat{\chi}_1, \hat{\chi}_2, \hat{\chi}_1^T, \hat{\chi}_2^T, \hat{1}\}$ would appear in a 3 x 3–matrix representation:

$$\begin{gathered}
\hat{1} = \begin{pmatrix} 1 & 0 & 0 \\ 0 & 1 & 0 \\ 0 & 0 & 0 \end{pmatrix}, \quad \hat{e}_1 = \begin{pmatrix} 1 & 0 & 0 \\ 0 & -1 & 0 \\ 0 & 0 & 0 \end{pmatrix}, \quad \hat{e}_2 = \begin{pmatrix} 0 & 1 & 0 \\ 1 & 0 & 0 \\ 0 & 0 & 0 \end{pmatrix}, \\
\hat{e}_{12} = \begin{pmatrix} 0 & 1 & 0 \\ -1 & 0 & 0 \\ 0 & 0 & 0 \end{pmatrix}, \quad \hat{\chi}_1 = \begin{pmatrix} 0 & 0 & 1 \\ 0 & 0 & 0 \\ 0 & 0 & 0 \end{pmatrix}, \quad \hat{\chi}_2 = \begin{pmatrix} 0 & 0 & 0 \\ 0 & 0 & 1 \\ 0 & 0 & 0 \end{pmatrix}, \\
\hat{\chi}_1^T = \begin{pmatrix} 0 & 0 & 0 \\ 0 & 0 & 0 \\ 1 & 0 & 0 \end{pmatrix}, \quad \hat{\chi}_2^T = \begin{pmatrix} 0 & 0 & 0 \\ 0 & 0 & 0 \\ 0 & 1 & 0 \end{pmatrix}, \quad \hat{1} = \begin{pmatrix} 0 & 0 & 0 \\ 0 & 0 & 0 \\ 0 & 0 & 1 \end{pmatrix},
\end{gathered} \qquad [4]$$

notice that this corresponds to an outer sum of [1] and [2]. The representation can be changed by similarity transformations. The closure of the algebra is inmediate and we have *one* product only to consider, here matrix product.

III. Formal Construction

Now we are in position to discuss the more general euclidian space $\mathcal{R}^{p,q}$ with $n = p + q$ basis vectors such that in that basis for a general vector u with components u_i, $\{u\} = \{u \mid \|u\|^2 = \sum^p u_k^2 - \sum^q u_{p+k}^2\}$. They generate the Clifford ring $\hat{\mathcal{C}}^{p,q}$.
Our complex superalgebra $\hat{\mathcal{K}}^{p,q}$ is defined to have the structure: $\hat{\mathcal{K}}^{p,q} = \{\hat{\mathcal{C}}^{p,q}, \hat{\mathcal{L}}_{pq}, \hat{\mathcal{L}}_D^{p,q}, \hat{\mathcal{S}}\}$, such that formally $\hat{\mathcal{C}}^{p,q}\hat{\mathcal{L}}^{p,q} = \hat{\mathcal{L}}^{p,q}, \hat{\mathcal{L}}_D^{p,q}\hat{\mathcal{C}}^{p,q} = \hat{\mathcal{L}}_D^{p,q}, \hat{\mathcal{L}}^{p,q}\hat{\mathcal{L}}_D^{p,q} = \hat{\mathcal{C}}^{p,q}$ and $\hat{\mathcal{L}}_D^{p,q}\hat{\mathcal{L}}^{p,q} = \hat{\mathcal{S}}$.
The product of any two terms is allowed; the algebra is noncommutative but associative.
The number of generators of $\hat{\mathcal{K}}^{p,q}$ is given as the sum of the following constituents:

Number of elements:	Type:
2^{p+q}	multivectors
$2^d + 2^d$	spinors
1	scalar

where $d = integer(\frac{p+q}{2})$.

For example, in the case above $\mathcal{C}^{2,0}$, where the parameter $d = 1$, we should use for the standard matrix representation of multivectors and spinors: $2^1 \times 2^1$–matrices for the multivectors, 1×2^1–matrices for the spinors and 1×1–matrices for the complex scalars, and we would have 2^2 multivectors, $2^1 + 2^1$

spinors and 1 scalar. Whereas for the $\hat{\mathcal{K}}^{2,0}$–geometric superalgebra the total minimal representation would be in terms of matrices with $(2^d+1)\times(2^d+1)$ entries, in this case 3×3–matrices.
This allows us to construct a noncommutative, associative algebra, such that, if $\mathbf{K},\mathbf{K}',\mathbf{K}''\in\hat{\mathcal{K}}^{p,q}$, then $\mathbf{K}+\mathbf{K}'=\mathbf{K}''$, $a(\mathbf{K}+\mathbf{K}')=a(\mathbf{K})+a(\mathbf{K}')$, $\mathbf{K}\,\mathbf{K}'=\mathbf{K}'''$, and $\mathbf{K}(\mathbf{K}'\mathbf{K}'')=(\mathbf{K}\,\mathbf{K}')\,\mathbf{K}''$. All of these from the basic definition with representation [3].
But the main usefulness of the algebra araises from a further algebraic structure which is already clearly shown in the representation [3]. In fact, we have

a) The part $\hat{\mathcal{C}}^{p,q}$ of $\hat{\mathcal{K}}^{p,q}$ constitutes a normed ring generated by $n=p+q$–anticommuting elements generating the entire ring.
b) The part $\hat{\mathcal{L}}^{p,q}$ of $\hat{\mathcal{K}}^{p,q}$ is dual to the part $\hat{\mathcal{L}_{\mathcal{D}}}^{p,q}$ of $\hat{\mathcal{K}}^{p,q}$. They can be obtained from each other through the *transpose* operation.
c) The part $\hat{s}$ of $\hat{\mathcal{K}}^{p,q}$ can be obtained from the elements of $\hat{\mathcal{L}}^{p,q}$ and $\hat{\mathcal{L}_{\mathcal{D}}}^{p,q}$ through the *spinor metric product* operation.
d) Finally and most important, the members of the $\hat{\mathcal{C}}^{p,q}$–ring can be generated from the elements of $\hat{\mathcal{L}}^{p,q}$ and $\hat{\mathcal{L}_{\mathcal{D}}}^{p,q}$ through the *spinor outer product* operation.

Remark: The products of the vectors and multivectors ring, the spinor metric product, and the spinor outer product, as well as the products of multivectors onto spinors (defining the spinors as the left- or right-module) are standard products within the $\hat{\mathcal{K}}^{p,q}$– algebra. This can be more clearly seen in the representation [3] of the algebra, where any multiplication is standard matrix multiplication.

IV. Example: Spacetime (Complex) Geometric Superalgebra.

To construct the geometric superalgebra of spacetime, we will start from the spinors spaces as, as mentioned in item d) above, the multivectors can be constructed from them and, as mentioned in item c) above, the complex scalars are also obtainable from them. We will construct the $\hat{\mathcal{K}}^{1,3}$–space–time–complex algebra and its embedding in $\hat{\mathcal{K}}^{0,5}\approx\hat{\mathcal{K}}^{2,3}\approx\hat{\mathcal{K}}^{4,1}$.

We start then by considering two subspaces of $\hat{K}$: the vectorial space $\hat{\mathcal{L}}^{1,3}$, corresponding in spacetime to the Dirac–spinors with basis elements $\chi_a\in\hat{\mathcal{L}}^{1,3}$, $a=1,2,3,4$ (here $d=integer(\frac{3+1}{2})=2$, and the dimension r of the spinor space $r=2^d=4$), and its dual space $\hat{\mathcal{L}}_{\mathcal{D}}13$, with basis elements $\chi^\dagger_a\in\hat{\mathcal{L}}_{\mathcal{D}}13$, taken as the primitive set of elements of $\hat{\mathcal{K}}^{1,3}$. A general element is $\psi=\sum_a\psi^a\chi_a\in\hat{\mathcal{L}}^{1,3}$, or $\psi^\dagger=\sum_a{\psi^\dagger}^a\chi^\dagger_a\in\hat{\mathcal{L}_{\mathcal{D}}}^{1,3}$. Here $\psi^a\in C^1$ and $\psi^{\dagger a}\in C^1$.
A matrix representation of the geometric superalgebra should have dimension $(r+1)$ x $(r+1)$. The ψ and $\psi^\dagger$ correspond to antidiagonal blocks of dimension 1 x r and r x 1 respectively.
The spinor metric is defined by the noncommutative product:

$$\chi^\dagger_a\mathcal{M}\chi_b=\delta^a_b\,\hat{1}\in\hat{\mathcal{S}}, \qquad [5]$$

(Remark: The unit 1 of the $\hat{K}$–geometric superalgebra corresponds to the sum of the multivectors unit scalar $\hat{\mathbf{1}}$ and the spinor metric unit $\hat{1}$, that is $1=\hat{\mathbf{1}}+\hat{1}$. Here we will use a symmetric metric with $\mathcal{M}=\hat{1}$. The matrix representation of $\hat{1}$ will correspond to a (lower) diagonal 1 x 1 block). A multivector $\vartheta^b_a\in\hat{M}^{1,3}$ is obtained by the noncommutative product

$$\vartheta^a_b=\chi_b\,\chi^\dagger_a. \qquad [6]$$

The set $\{\vartheta^b_a\in\hat{M}^{1,3}\}$ corresponds to a basis for the multivectors ($\hat{M}^{1,3}=$ complexification of $\hat{C}^{1,3}$) considered as generated by the "outer" product [6] of the spinors.
Given the previous operators there are thirteen others, they are all step operators that map the set $\{\chi_a\}$ into itself: each $\{\chi_a\}\rightarrow\{\chi_b\neq\chi_a\}$. Together they correspond to a new arrangement of the 16 elements ϑ^b_a, which have a direct meaning as operators in spinor space. As each of them acts non trivially only on one spinor χ_b (or $\chi^\dagger_a$ on the right) it is more meaningful to show explicitly the linear combinations which have a nontrivial action on the complete set of four χ_b (or $\chi^\dagger_a$), mapping each and all spinors χ_b in a corresponding spinor χ_a

$$\mathbf{M_A} = \mathbf{A}_1\chi_{a_1}\chi_1^{\dagger} + \mathbf{A}_2\chi_{a_2}\chi_2^{\dagger} + \mathbf{A}_3\chi_{a_3}\chi_3^{\dagger} + \mathbf{A}_4\chi_{a_4}\chi_4^{\dagger}, \quad [7]$$

$a_i \neq i$, with the particular, normalized, choice of the complex coefficients

$$\mathbf{A}_a \neq 0 \text{ and } \sum_a |\mathbf{A}_a|^2 = 4 \quad [8]$$

the set $\mathbf{M_A}$ constitutes a Clifford ring $\hat{\mathcal{M}}^{p,q} = R_{p,q} + iR_{p,q}$ (of 16 or 32 elements in our case depending on the $\mathbf{A}_a$ being complex) and generates a Grassmann-Clifford algebra (invariant under a similitude transformation). $R_{1,3}$ has 16 elements.
The explicit forms of e_α are (up to a complex phase factor)

$$e_0 = \vartheta_3^1 + \vartheta_4^2 + \vartheta_1^3 + \vartheta_2^4,\ e_1 = \vartheta_4^1 + \vartheta_3^2 - \vartheta_2^3 - \vartheta_1^4\ e_2 = i(-\vartheta_4^1 + \vartheta_3^2 + \vartheta_2^3 - \vartheta_1^4),\ e_3 = \vartheta_3^1 - \vartheta_4^2 - \vartheta_1^3 + \vartheta_2^4. \quad [9]$$

The particular choice of the four e_α was made to fulfill the condition for the chirality operator $\mathbf{M}_{ch} = ie_0e_1e_2e_3$, ensuring at the same time the condition $e_\alpha \mathbf{M}_{ch} = -\mathbf{M}_{ch} e_\alpha$ given above (the factor i makes $\mathbf{M}_{ch}$ linearly independent from e_0, e_1, e_2, and e_3). The set of e_μ will be used to give the spacetime its best known representation with signature $\mathbf{R}^{1,3}$.
For a further discussion of this construction we refer again the reader to reference 1.

V. Symmetry and Supersymmetry Transformations.
The K-algebra allows a large group of transformations: those corresponding to the usual transformations $\hat{L}$ for multivectors and spinors (in spacetime the Lorentz transformations are a well known special case) and a new set of transformations $\hat{S}$ corresponding to a "mixing" between multivectors and spinors. The general form of the transformations is $SKS^{-1} = K'$ where now the symmetries S will correspond in matrix representation to the following two types and their combinations

$$S = \left\{ \hat{L} = \begin{pmatrix} L & 0 \\ 0 & 1 \end{pmatrix};\ \hat{L}^{-1} = \begin{pmatrix} L^{-1} & 0 \\ 0 & 1 \end{pmatrix};\ \hat{S} = \begin{pmatrix} 0 & a \\ b^1 & 0 \end{pmatrix};\ \hat{S}^{-1} = \begin{pmatrix} 0 & b^{-1} \\ a^{\dagger -1} & 0 \end{pmatrix} \right\} \quad [10]$$

The transformations of the type L will have the following actions

$$\hat{L}K\hat{L}^{-1} = \begin{pmatrix} L & 0 \\ 0 & 1 \end{pmatrix} \begin{pmatrix} M & \psi \\ \phi^{\dagger} & s \end{pmatrix} \begin{pmatrix} L^{-1} & 0 \\ 0 & 1 \end{pmatrix} = \begin{pmatrix} LML^{-1} & L\psi \\ \phi^{\dagger}L^{-1} & s \end{pmatrix} = K' \quad [11]$$

which in fach justify the names given as scalars, spinors and multivectors.
The transformations of the type $\hat{S}$ will have the action:

$$\hat{S}K\hat{S}^{-1} = \begin{pmatrix} 0 & a \\ b^{\dagger} & 0 \end{pmatrix} \begin{pmatrix} M & \psi \\ \phi^{\dagger} & s \end{pmatrix} \begin{pmatrix} 0 & b^{-1} \\ a^{\dagger -1} & 0 \end{pmatrix} = \begin{pmatrix} asa^{\dagger -1} & a\phi^{\dagger}b^{-1} \\ b^{\dagger}\psi a^{\dagger -1} & b^{\dagger}Mb^{-1} \end{pmatrix} = K'' \quad [12]$$

Acknowledgements
The technical assistance of Mrs. Irma Aragón is greatly appreciated.

References.

1. J. Keller, A. Rodríguez, *J. Math. Phys.* **33** (1), 161-170 (1992).
2. Crumeyrolle A., *Orthogonal and Symplectic Clifford Algebras. Spinor Structures.* (Kluwer A.P. Dordrecht 1990).
3. J.S.R. Chisholm and A.K. Common, eds., *Clifford Algebras and Their Applications in Mathematical Physics* (Reidel, Dordrecht, 1986), NATO ASI C-183.
4. Keller J., *Int. J. Theor. Phys.* **30**, 137 (1991).
5. Keller J., *Advances in Applied Clifford Algebras* **1** (1), 31-50 1991, J. Keller, ed.

ON THE COMPLETE INTEGRABILITY OF DYNAMICAL SYSTEMS WITH FERMIONIC VARIABLES

G. Landi

Dipartimento di Scienze Matematiche - Università di Trieste – Italy.
Istituto Nazionale di Fisica Nucleare - Sezione di Napoli – Italy.

1. Introduction.

In the analysis of complete integrability of dynamical systems, a crucial rôle is played by the so called recursion operator which appears as a $(1,1)$ tensor field (and gives rise to an endomorphism of vector fields or 1-forms) on the carrier space. Such an operator, which is used to generate sequences of conserved quantities or sequences of symmetries, has to satisfies few requirements, the most important being that its Nijenhuis torsion vanishes.

We study the rôle of $(1,1)$ graded tensor field R in the analysis of complete integrability of dynamical systems with fermionic variables [LMV]. We find that such a tensor R can be a recursion operator, if and only if T is even as a graded map.

We show that such odd tensors do exists by giving examples of dynamical systems admitting a bi-Hamiltonian description by means of Hamiltonian functions of opposite parity. The corresponding Poisson structures allows to construct odd $(1,1)$ graded tensor fields R which could play the rôle of recursion operator. We explicitly show that this is not possible.

2. Graded dynamical Hamiltonian systems.

We sketch some basic notions on supermanifolds while refering to the literature for details (see for instance [BBH]). The basic algebraic object is a real exterior algebra $B_L = (B_L)_0 \oplus (B_L)_1$ with L generators. An (m,n) dimensional supermanifold is a topological manifold S modelled over the "vector superspace" $B_L^{m,n} = (B_L)_0^m \times (B_L)_1^n$, by means of an atlas whose transition functions fulfil a suitable "supersmoothness" condition. A supersmooth function $f : U \subset B_L^{m,n} \to B_L$ has the usual superfield expansion

$$f(x^1 \dots x^m, \theta^1 \dots \theta^n) = f_0(x) + \sum_{\alpha=1}^{n} f_\alpha(x)\theta^\alpha + \dots + f_{1\dots n}(x)\theta^1 \dots \theta^n \tag{1}$$

where the x's are the even (Grassmann) coordinates, the θ's are the odd ones, and the dependence of the coefficient functions $f_{\dots}(x)$ on the even variables is fixed by their values for real arguments.

The graded tangent space TS is constructed in the usual algebraic manner. If $\mathcal{G}(S)$ denotes the graded ring of supersmooth B_L-valued functions on S , then the sections $\mathcal{X}(S)$ of TS are a graded $\mathcal{G}(S)$-module and are identified with the graded Lie algebra $Der\mathcal{G}(S)$ of derivations of $\mathcal{G}(S)$. Derivations (or vector fields) are said to be even (or odd) if they are even (or odd) as maps (satisfying in addition a graded Leibnitz rule) from $\mathcal{G}(S) \to \mathcal{G}(S)$. A local basis is given by $\frac{\partial}{\partial x^1}, \dots, \frac{\partial}{\partial x^m}, \frac{\partial}{\partial \theta^1}, \dots, \frac{\partial}{\partial \theta^n}$. A crucial difference between even and odd vector fields on a supermanifold is that the latter do not admit local flows.

In a similar way one defines the cotangent bundle T^*S. The sections $\mathcal{X}(S)^*$ of T^*S are a graded $\mathcal{G}(S)$-module and are identified with the graded $\mathcal{G}(S)$-linear maps from $Der\mathcal{G}(S) \to \mathcal{G}(S)$. They are the 1-forms on S.

On the cotangent bundle T^*S there exists a canonical symplectic structure. More generally, one can introduce super Poisson structures on a supermanifold S which, when invertible, give symplectic structures for which there is a Darboux theorem (for details see [Be] and [Le]). Let $z^i = (x^j, \theta^k)$, $i \in \{1, \ldots, m+n\}$ be local coordinates on S. Suppose we have a point-dependent $(m+n) \times (m+n)$ matrix $\|\omega^{ij}\|$ with the properties: 1. the elements ω^{ij} are homogeneous with parity $p(\omega^{ij}) = p(z^i) + p(z^j) + p(\omega)$ and $p(\omega)$ not depending on the indices i and j ; 2. $\omega^{ji} = -(-1)^{[p(z^i)+p(\omega)][p(z^j)+p(\omega)]}\omega^{ij}$; 3. an additional property ensuring the Jacobi identity. Then, with F and G any two superfunctions, the following bracket

$$\{F, G\} =: F\frac{\overleftarrow{\partial}}{\partial z^i}\omega^{ij}\frac{\overrightarrow{\partial}}{\partial z^j}G \tag{2}$$

is a Poisson structure on S. We have two different kind of structures according to the fact that $p(\omega) = 0$ (even Poisson structure) or $p(\omega) = 1$ (odd Poisson structure). From (2), if H is the hamiltonian, the corresponding Hamilton's equations are

$$\dot{z}^i = \omega^{ij}\frac{\overrightarrow{\partial}}{\partial z^j}H. \tag{3}$$

To maintain the possibility of explicitly constructing the flow of (3) one needs the dynamical evolution to be an even vector field. In turn this implies that the Poisson structure and the Hamiltonian function should have the same parity.

3. Examples.

Here we consider two simple examples.

Example 3.1. A graded harmonic oscillator.

On $T^*B_L^{1,1}$ and with coordinates (q, p, θ, ξ) the mixed bosonic-fermionic harmonic oscillator has the following equations of motion

$$\frac{dq}{dt} = p\ , \quad \frac{dp}{dt} = -q\ , \qquad \frac{d\theta}{dt} = \xi\ , \quad \frac{d\xi}{dt} = -\theta\ . \tag{4}$$

Equations (4) can be given two Hamiltonian descriptions with an even and an odd Hamiltonian given respectively by

$$H = \frac{1}{2}(p^2 + q^2) + i\xi\theta\ , \qquad K = p\xi + q\theta\ , \tag{5}$$

while the two symplectic structures are respectively

$$\omega_H = dq \wedge dp - \frac{i}{2}d\theta \wedge d\theta\frac{i}{2}d\xi \wedge d\xi\ , \qquad \omega_K = dq \wedge d\xi - dp \wedge d\theta\ . \tag{6}$$

We can construct an endomorphism of 1-forms R by

$$R =: \omega_H \circ (\omega_K)^{-1} = \begin{pmatrix} 0 & 0 & 1 & 0 \\ 0 & 0 & 0 & 1 \\ 0 & -i & 0 & 0 \\ i & 0 & 0 & 0 \end{pmatrix} \tag{7}$$

which is numerical and of odd parity. A surprising consequence of the latter fact is that R, although numerical is not a recursion operator.

$$RdK = dH \ , \qquad RdH = -i(dq)\xi + i(dp)\theta - i(d\theta)p + i(d\xi)q \ , \quad d(RdH) \neq 0 \ . \qquad (8)$$

Example 3.2. A supersymmetric Toda chain.

It has been shown [So] that the dynamics of Witten's Hamilton systems [Wi] can be given a bi-Hamiltonian description with Hamiltonian functions of opposite parity. Here we consider the simple case of a supersymmetric Toda chain on $T^*B_L^{1,1}$ with coordinates (q, p, θ, ξ).

The even Hamiltonian is given by

$$H = \frac{1}{2}(p^2 + e^q) + \frac{1}{2}i\xi\theta e^{\frac{q}{2}} \ . \qquad (9)$$

With the even Poisson structure

$$\omega_H = dq \wedge dp - \frac{i}{2}d\theta \wedge d\theta \frac{i}{2} d\xi \wedge d\xi \ , \qquad (10)$$

the equations of motion read

$$\begin{aligned} \dot{q} &= p \ , & \dot{p} &= -\frac{1}{2}e^q \ - \frac{1}{4}i\xi\theta e^{\frac{q}{2}}, \\ \dot{\theta} &= \frac{1}{2}\xi e^{\frac{q}{2}} \ , & \dot{\xi} &= -\frac{1}{2}\theta e^{\frac{q}{2}} \ . \end{aligned} \qquad (11)$$

Then the following functions are constants of the motion

$$K = p\xi + e^{\frac{q}{2}}\theta \ , \qquad L = p\theta - e^{\frac{q}{2}}\xi \ , \qquad F = i\xi\theta \ . \qquad (12)$$

We can use K in (12) (or L) as an alternative Hamiltonian. The corresponding symplectic structure can be written as

$$\begin{aligned} \omega_K &= dq \wedge d\xi + dp \wedge dq(e^{-\frac{q}{2}}\theta) + dp \wedge d\theta(-2e^{-\frac{q}{2}}) + df \wedge dH \\ &= d\{dq(-\xi) + dp(2e^{-\frac{q}{2}}\theta) + f dH\} \ , \end{aligned} \qquad (13)$$

where $f(q, p, \theta, \xi)$ is a function explicitly given by

$$\begin{aligned} f &= A\xi + B\theta \\ A(q,p) &= \frac{1}{p^2 + e^q}\left(\frac{2p}{\sqrt{p^2 + e^q}} log(\frac{e^{\frac{q}{2}}}{p + \sqrt{p^2 + e^q}}) + \frac{2e^{\frac{q}{2}}}{\sqrt{p^2 + e^q}} - 2\right) \\ B(q,p) &= \frac{1}{p^2 + e^q}\left(\frac{2e^{\frac{q}{2}}}{\sqrt{p^2 + e^q}} log(\frac{e^{\frac{q}{2}}}{p + \sqrt{p^2 + e^q}}) - \frac{2p}{\sqrt{p^2 + e^q}} - 2pe^{-\frac{q}{2}}\right) \ . \end{aligned} \qquad (14)$$

It take some algebra to check that the endomorphism $R =: \omega_H \circ (\omega_K)^{-1}$ is not a recursion operator. Indeed, one finds that

$$RdH = dK \ , \qquad d(R^2 dH) \neq 0 \ . \qquad (15)$$

4. Super Nijenhuis torsion.

One of the most relevant consequences deriving from a (not graded) (1-1) tensor field R with vanishing Nijenhuis torsion $\mathcal{N}_R$ is the possibility to generate sequences of exact 1-forms. We first recall that the tensor field $\mathcal{N}_R$ is defined by

$$\mathcal{N}_R(X,Y) =: R^2[X,Y] + [RX,RY] - R[RX,Y] - R[X,RY] \,. \tag{16}$$

Proposition.

$$\mathcal{N}_R = 0, \quad d(RdF) = 0 \Longrightarrow d(R^k dF) = 0 \,. \tag{17}$$

Proof. Let α be any 1-form. It takes some easy algebra to find that

$$X \wedge Y \lrcorner\, d(R^2\alpha) = \{X \wedge RY + RX \wedge Y\} \lrcorner\, d(R\alpha) - \{RX \wedge RY\} \lrcorner\, d\alpha - \mathcal{N}_R(X,Y) \lrcorner\, \alpha \,. \tag{18}$$

Assume now that both α and $R\alpha$ are closed. From (19) we see that $R^2\alpha$ is closed if and only if $\mathcal{N}_R = 0$, namely if and only if the Nijenhuis torsion of R vanishes. One gets the proposition with a recursive application of R. With extra condition, for instance the vanishing of the first cohomology group, the forms $R^k dF$ will be exact.

Let us analize now the graded situation. Suppose R is a graded $(1,1)$ tensor field which is homogeneous of parity $p(R)$. Then, if α is any 1-form, after some (graded) algebra, the analogue of (19) reads

$$\begin{aligned} X \wedge Y \lrcorner\, d(R^2\alpha) = {} & \{(-1)^{p(R)p(Y)} X \wedge RY + (-1)^{p(R)[p(X)+p(Y)]} RX \wedge Y\} \lrcorner\, d(R\alpha) \\ & -(-1)^{p(R)[p(X)+p(R)]} RX \wedge RY \lrcorner\, d\alpha - (-1)^{p(R)}\, {}^G\mathcal{N}_R(X,Y) \lrcorner\, \alpha \\ & +(-1)^{p(R)p(X)}[1 - (-1)^{p(R)}] L_{RX}(RY \lrcorner\, \alpha) \,. \end{aligned} \tag{19}$$

Where ${}^G\mathcal{N}_R$ is defined by

$${}^G\mathcal{N}_R(X,Y) = R^2[X,Y] + (-1)^{p(R)p(X)}[RX,RY] - R[RX,Y] - (-1)^{p(R)p(X)} R[X,RY] \,. \tag{20}$$

From (20) it is clear that for a (1,1) odd tensor a conclusion like (18) can be obtained only when $p(R) = 0$. This is a consequence of the fact that a Nijenhuis supertorsion) can be defined only when $p(R) = 0$, which in turn reflects the fact that the map ${}^G\mathcal{N}_R$ in (21) defines a tensor fields if and only if $p(R) = 0$.

Acknowledgements.

The work reported here has been done in collaboration with G. Marmo and G. Vilasi. Supported in part by the Italian Ministero dell' Università e della Ricerca Scientifica e Tecnologica.

References.

BBH C. Bartocci, U. Bruzzo, D. Hernández-Ruipérez, The Geometry of Supermanifolds (Kluwer, Dordrecht, 1991).

Be F. A. Berezin Introduction to superanalysis (Kirillov ed.) (D. Reidel Publishing Company, Dordrecht, 1987).

LMV G. Landi, G. Marmo, G. Vilasi, J. Phys. A, Math. Gen., **25** (1992) 4413.

So V. A. Soroka, Lett. Math. Phys. **17** (1989) 201. D. V. Volkov, A. I. Pashnev, V. A. Soroka and V.I.Tkach, JETP Lett. **44** (1986) 70.

Wi E. Witten, Nucl. Phys. **B188** (1981) 513.

A $N \times N$ ZAKHAROV-SHABAT SYSTEM WITH A POLYNOMIAL SPECTRAL PARAMETER AND THE ASSOCIATED EVOLUTION EQUATIONS

JYH-HAO LEE
INSTITUTE OF MATHEMATICS, ACADEMIA SINICA
TAIPEI 11529, TAIWAN

ABSTRACT. Here we review some results of J.-H. Lee of the $N \times N$ Zakharov-Shabat system with a polynomial spectral parameter. We define a scattering transform following the set-up of Beals-Coifman. We also derive the associated evolution equations. Some examples are given here.

1. Forward Problem

Let $J = diag(id_1, id_2, ..., id_N), d_1 < d_2 < \cdots < d_N$, $i = \sqrt{-1}$, $\sum = \{z : Im(z^n) = 0\}$. Let q_i, $i = 1, 2, ..., n$ be $M_N(C)$-valued functions and q_i, dq_i/dx belong to L^1 for all i. We consider the following system: $dM/dx = z^n[J, M] + (z^{n-1}q_1 + z^{n-2}q_2 + \cdots + q_n)M$, $z \notin \sum$, $M(\cdot, z)$ is bounded and absolutely continous, $M(x, z) \to I$ as $x \to -\infty$. We also have the following constraints on the potentials: the diagonal part of q_i must be expressible by the off-diagonal parts of the preceding potentials [13]. These constraints enable us to have $M(x, z)$ normalized at $z = \infty$, which is convenient to pose the inverse problem properly. Our procedure is similar to that in [2], but with the following new feature: for the forward problem, we have to control the z-dependence in the system above when $z \to \infty$. Following the set-up of Beals-Coifman, we may solve M via Fredholm integral equation or we may also use Volterra equation to solve the wedge product of the columns of M [4].

Let $\Omega_+ = \{z : Im(z^n) > 0\}$, $\Omega_- = \{z : Im(z^n) < 0\}$. For the generic potential $(q_1, q_2, ..., q_n)$, M has the following properties: (i) For any $z \in \sum \backslash \{0\}$, there is a unique $v(z)$ such that for all $M^+(x, z) = M^-(x, z)\exp(xz^nJ)v(z)\exp(-xz^nJ)$, where $M^\pm$ = limit of M on $\sum$ from $\Omega_\pm$; (ii) $M(x, \cdot)$ has finite number of poles at $D = \{z_1, ..., z_L\}$. For any z, there is a matrix $v(z_j)$ such that $Res(M(x, \cdot), z_j) =$

$N(x,z_j)\exp(xz_j^nJ)\cdot v(z_j)\exp(-xz_j^nJ)$, where $N(x,z)$ is the regular part of $M(x,z)$ near $z=z_j$. (iii) The map $(q_1,q_2,...,q_n)\to(v(z);z_1,...,z_n;v(z_1),v(z_2),v(z_n))$ is injective. If the potential has the symmetric conditions $q_j^*=-q_j$, by uniqueness $M^*(x,\bar{z})=M(x,z)^{-1}$. It follows that $v^*(\bar{z})=v(z)$, $z\in\sum$, $v^*(\bar{z}_j)=-v(z_j)$ for $j=1,2,...,L$.

2. Inverse Problem

Given scattering data v satisfying the constraints as desccribed in [20, p.328], the inverse problem amounts to solving an analytic factorization problem (Riemann-Hilbert) problem. We may use the argument of Beals-Coifman [2] and the property of weighted-boundedness of Hilbert transform to show that the inverse problem is solvable if $v(z)-I$ is small. The argument in Beals-Coifman establishes a Fredholm property [3]. We may use this to show that if the scattering data satisfies the symmetric conditions $v^*(\bar{z})=v(z)$, $z\in\sum$, $v^*(\bar{z}_j)=-v(z_j)$ for $j=1,2,...,L$, then the inverse problem is solvable.

Remarks. The analytic results of the scattering and the inverse scattering for the generalized Zakharov-Shabat system with rational spectral parameter have been worked out by X. Zhou [27].

3. Evolution Equations

Suppose the evolution of the scattering data is given by $dv(z,t)/dt=z^k[J,v(z,t)]$; $dv(z_j,t)/dt=z_j^k[J,v(z_j,t)]$, z_j is fixed for all $j, j=1,2,...,L$. Then the evolution equation for q_j is as follows:

$$dq_k/dt=-([q_{k-1},G_1]+[q_{k-2},G_2]+\cdots+[q_1,G_n]+[J,G_{n+1}]),\quad k=1,2,...,n,$$

where G_j is the coefficient of $1/z^j$ in the asymptotic expansion of MJM^{-1} as $z\to\infty$ [13].

Remarks. (i) The wedge product method to construct M enables us to control the L^∞-norm of M by the L^2-norm of q_i in the following case: $dM/dx=z^n[J,M]+(z^kq_{n-k}+z^{k-1}q_{n-k-1}+\cdots+q_n)M$, $k\le\frac{n}{2}$. By this result, we have the global existence of the dissipative evolution equation associated with this system [3]. [7]. (ii) We note that the symmetric conditions of the scattering data described in Section 2 are invariant under the evolution described above. Hence we have the global existence of the associated evolution equation with $q_j^*=-q_j$.

4. Examples

By some modification the augument in Section 1, 2 works for the case $dM/dx = z^2[J,M] + zQM$ and $dM/dx = (z^2 - 1/z^2)[J,M] + (zQ + P + R/z)M$.

(I) Let $Q = \begin{pmatrix} 0 & q \\ r & 0 \end{pmatrix}$, $r = -q^*$, $P = Q(adJ)^{-1}Q$. An associated evolution equation for the system $dM/dx = z^2[J,M]+(zQ+P)M$ is $q_t = (i/2)q_{xx}+(i/2)q^2q_x+q \mid q \mid^4$. The global solvability in t was solved in [16]. By the argument used by Zakharov-Shabat in [26] for NLS equation, we know the long-time behavior of the N-soliton solutions. Let $u = q\exp(\int_{-\infty}^{x} -iqq^*)$, then u satisfies the derivative nonlinear Schrodinger equation (DNLS): $u_t = iu_{xx} - (u^2u^*)_x$, which was considered by Kaup-Newell [10], Gerzhikov [9].

(I) Let $J = \begin{pmatrix} i\alpha I_K & \\ & i\beta I_N \end{pmatrix}$, $Q = \begin{pmatrix} 0 & q \\ r & 0 \end{pmatrix}$, $r^* = -q, I_K$ is the $K \times K$ identity matrix and I_N is the $N \times N$ identity matrix. An interesting associated evolution equation is $q_t = icq_{xx} - 2c^2(qq^*q)_x$, $c = 1/(\beta - \alpha)$. This is a generalized vector derivative nonlinear Schrodinger equation, which is related to Hermitian symmetric spaces, see Fordy [8]. In the case $K = 1$ this equation is N-component derivative nonlinear [25]. R. Sasaki gave some geometrc interpretation to this equation [25]. Specially the case $K = 1, N = 2$ was considered by Morris and Dodd [23].

(III) $J = \begin{pmatrix} -i & 0 \\ 0 & i \end{pmatrix}$, $Q = \begin{pmatrix} 0 & q \\ r & 0 \end{pmatrix}$, $R = \begin{pmatrix} 0 & p \\ u & 0 \end{pmatrix}$, $P = Q(adJ)^{-1}Q$, $r = -q^*$, $u = -p^*$. An interesting associated evolution equation for the system $dM/dx = (z^2 - 1/z^2)[J,M] + (zQ + P + R/z)$ is of the form $Q_t - Q_x - 2[J,R] = 0$; $R_t + R_x + 2[R,P] - 2[J,Q] = 0$. This equation can be transformed into the massive Thirring model (MTM) in laboratory coordinates considered by Kuznetsov and Mikhailov in [14].

Acknowledgement This work was partially supported by the National Science Council at Taipei.

References

1. R. Beals and R. R. Coifman, *Scattering, transformation spectrales, et equations d'evolution non lineaires,*, Seminaire Goulaouic-Meyer-Schwartz (1980-1981, exp. 22, Ecole Polytechnique, Palaiseau).
2. R. Beals and R. R. Coifman, *Scattering and inverse scattering for first order system*, Comm. Pure Appl. Math. 37 (1984), 39-90.

3. R. Beals and R. R. Coifman, *Inverse scattering and evolution equations*, Comm. Pure Appl. Math. **38** (1985), 29-42.
4. R. Beals and R. R. Coifman, *Scattering and inverse scattering for first-order system: II*, Inverse Problem **3** (1987), 577-593.
5. R. Beals and R. R. Coifman, *Direct and Inverse Scattering on the line*, Math. Surveys and Monographs **28** (AMS), AMS.
6. R. Beals and R. R. Coifman, *Linear spectral Problems, non-linear equations and D-BAR-method*, Inverse Problems **5** (1989), 87-130.
7. R. K. Bullough, P. J. Caudrey(eds), *Solitons, Topics in Current Physics*, No. 17, Apringer-Verlag, 1980.
8. A. P. Fordy, *Derivative nonlinear Schrödinger equations and Hermitian symmetric spaces*, J. Phys. A: Math. Gen. **17** (1984), 1235-45.
9. V S Gerzhikov et al, *Quadratic Bundle and Nonlinear Equations*, Theoret. and Math. Phys. **44, No.3** (1980), 784-795.
10. D. J. Kaup and A. C. Newell, *An exact solution for a derivative nonlinear Schrodinger equation*, J. Math. Phys. 19(4), April 1978, 789-801.
11. E A, Kuznetsov and A B, Mikhailov, *On the complete integrability of the two-dimensional cclassical Thirring Model*, Thet. Math. Physics **30, No. 3** (1977), 193-200.
12. J.H. Lee, *Analytic properties of Zakharov-Shabat inverse scattering problem with polynomial spectral dependence of degree 1 in the potential*, Ph.D. dissertation, Yale University, May 1983.
13. J.H. Lee, *A Zakharov-Shabat inverse scattering problem and the associated evolution equations*, Chinese J. of Mathematics **12, No.4** (1984), 223-233.
14. J.H. Lee, *Analytic properties of a Zakharov-Shabat inverse scattering problem (I)*, Chinese J. of Mathematics **14, No.4 Dec. 1986**, 225-248.
15. J. H. Lee, *Analytic properties of a Zakharov-Shabat inverse scattering problem (II)*, Chinese J. of Mathematics **16, No.2**, 81-110.
16. J.H. Lee, *Global solvability of the derivative nonlinear Schrödinger equation*, to appear in the Transactions of the Americal Mathematical Society **314, No. 1** (1989), 107-118.
17. J.H. Lee, *On the dissipative evolution equations associated with the Zakharov-Shabat system with a quadratic spectral parameter,*, Transactions of the A.M.S. **316, No. 1** (1989), 327-336.
18. J.H. Lee, *A $N \times N$ Zakharov-Shabat System with a quadratic spectral parameter*, Proc. of the 5th workshop on nonlinear evolution equations and dynamical systems, Kolymbari, Crete, Greece, July 2-16, Springer-Verlag.
19. J.H. Lee, *A $N \times N$ Zakharov-Shabat system with a quadratic spectral parameter and Skew-Hermitian potentials*, Preprint.
20. J.H. Lee, *Inverse scattering transform and D-Bar method*, NSC Report, Taipei (1989-90).
21. W. Mio et al, *Modified nonlinear Schrodinge equation for Alfven waves propagating along the magnetic field in cold plasmas*, J. Phys Soc. Japan **41** (1976), 265-271.
22. E. MJϕLHUS, *On the modulational instability of hydromagnetic waves parallel to the magnetic field*, J. Plasma Physics **16, part 3** (1976), 321-334.
23. H. C. Morris and R. K. Dodd, *The two component derivative nonlinear Schrodinger equation*, Physica Scripta **20** (1979), 505-508.
24. R. Sasaki,, *Nucl*, Phys **B 154**, 343-.
25. R. Sasaki, *Canonical structure of soliton equations, II. The Kaup-Newell syem*, Physica **5D**, 66-74.
26. V. E. Zakharov and A. B. Shabat, *A refined theory of two-dimensional self-focussing and one-dimensional self-modulation of waves in non-linear media*, Soviet Physics JETP **34** (1972), 62-69.
27. X. Zhou, *Riemann Hilbert problem and inverse scattering*, SIAM J. Math. Anal. **20** (1989), 966-986.

REMARKS ON THE BREAKING SOLITON EQUATIONS

Li Yi–shen
Department of mathematics, University of Science and Technology of China, Hefei, 230026 P.R.C.

1. Introduction

O.I. Bogoyavlenskii[1] has published a series papers on the breaking soliton equations. which is used to discribe the 2+1 dimensional interaction of Riemann wave propagation along the y–axis with long wave propagation along x–axis, In § 2 of this paper, we deduce two typical breaking soliton equations from the nonisospectral zero curvature equation. In § 3. a geometrical approach is used to prove that these equations possess bi–Hamiltonian systems, two sets of infinite constants of motion in involution, four sets of symmetries with infinite dimensional Lie algebraic structure. In § 4. we prove that the solutions of these equations can be obtained by Darboux transformation.

2. Two typical equations

Consider two linear spectral problems

$$-\varphi_{xx}+u_x(x,y,t)\varphi=\xi^2\varphi \tag{2.1}$$

$$\varphi_t=4\xi^2\varphi_y-u_{xy}\varphi+2u_y\varphi_x \tag{2.2}$$

and assume that the spectral parameter ξ satisfies the Riemann equation

$$\xi_t=4\xi^2\xi_y \tag{2.3}$$

From the compatibility condition $\varphi_{xxt}=\varphi_{txx}$. it yields that if φ satisfies both equations (2.1) and (2.2). then the function u satifies the equation

$$u_{xt}=4u_xu_{xy}+2u_yu_{xx}-u_{xxxy} \tag{2.4}$$

This type of equation is first considered in ref[2]

Similary, we consider two linear spectral problems

$$\varphi_x=M\varphi \qquad \varphi=\begin{bmatrix}\varphi_1\\ \varphi_2\end{bmatrix} \qquad M=\begin{bmatrix}-i\xi & q(x,y,t)\\ r(x,y,t) & i\xi\end{bmatrix} \tag{2.5}$$

$$\varphi_t=2\xi\varphi_y+N\varphi \qquad N=\begin{bmatrix}-i\partial_x^{-1}(qr)_y & iq_y\\ -ir_y & i\partial_x^{-1}(qr)_y\end{bmatrix} \tag{2.6}$$

and assume that ξ satisfies the equation

$$\xi_t=2\xi\xi_y \tag{2.7}$$

From the compatibility condition, it yields that the functions q,r satisfy the equation

$$\begin{cases} q_t = -iq_{xy} + 2i\partial_x^{-1}(qr)_y \\ r_t = ir_{xy} - 2i\partial_x^{-1}(qr)_y \end{cases} \tag{2.8}$$

we note that if one takes x=y, then two typical equations (2.4) and (2.8) are reduced to the kdv and generalized nonlinear schrodinger equation respectively.

3. The algebraic properties of these equations

In this section, we use the notation and some results list in ref[3], we assume that M is a Schwartze space, S is a tangent bundle of M, S_ω is a tangent space of $\omega \in M$. K is a field, it maps M to S, $K(\omega) \in S_\omega$. $\omega \in M$, we define it Cateax derivative in direction $\sigma(\omega) \in S_\omega$. as

$$K'[\sigma] = \lim_{\varepsilon \to 0} \frac{\partial}{\partial \varepsilon} K(\omega + \varepsilon\sigma) \tag{3.1}$$

and define the commutator

$$[A.B] = A'[B] - B'[A] \qquad A,B \in S_\omega \tag{3.2}$$

Now, we set $u_x = \omega$, The equation (2.4) can be written as follows

$$\omega_t = 4\omega\omega_y + 2(\partial_x^{-1}\omega_y)\omega_x - \omega_{xxy} \equiv K(\omega) \tag{3.3}$$

The equation (3.3). can be written in the Bi–Hamiltonian system

$$\omega_t = \Phi_1 \frac{\delta H_1}{\delta\omega} = \Phi_2 \frac{\delta H_2}{\delta\omega} \tag{3.4}$$

where the Hamiltonian operators Φ_1 Φ_2 read

$$\Phi_1 = \partial_x, \qquad \Phi_2 = -\partial_x^3 + 4\omega\partial_x + 2\omega_x \tag{3.5}$$

and

$$H_1 = \iint_{-\infty}^{\infty} \left[\left(\omega^2 - \frac{1}{2}\omega_{xx} \right) \omega\partial_x^{-1}\omega_y \right] dxdy, \ H_2 = \iint_{-\infty}^{\infty} \left[\frac{1}{2}\omega\partial_x^{-1}\omega_y \right] dxdy \tag{3.6}$$

The linearlized equation (3.3), reads

$$\sigma_t = K'[\sigma] = 4\omega_y\sigma + 4\omega\sigma_y + 2(\partial_x^{-1}\omega_y)\sigma_x + 2(\partial_x^{-1}\sigma_y)\omega_x - \sigma_{xxy} \tag{3.7}$$

If ω is a solution of (3.3), $\sigma = \sigma(\omega)$ is a solution of (3.7), then $\sigma(\omega)$ is a symmetry of the equation (3.3). It is easy to verify that the following functions are symmetries of the equation (3.3).

$$K_0 = \omega_x, \quad K_0^2 = \omega_y, \quad \tau_0^1 = 2t\omega_t + \frac{1}{2}, \quad \tau_0^2 = t\omega_t + y\omega_y \tag{3.8}$$

It is well known that the operator

$$\Phi = -\partial_x^2 + 4\omega + 2\omega_x\partial_x^{-1} \tag{3.9}$$

is hereditary symmetry and the operator Φ is strong symmetng for the equation (3.3). i.e.

$$\Phi'[K] + \Phi K' - K'\Phi = 0 \tag{3.1}$$

using the lemma in ref[3]. one can prove the propositions.

Proposition 3.1. There are four sets of symmetries of the equation (3.3).

$$K_n^1 = \Phi^n K_0^1, \quad K_n^2 = \Phi^n K_0^2, \quad \tau_n^1 = \Phi^n \tau_0^1, \quad \tau_n^2 = \Phi^n \tau_0^2, \tag{3.11}$$

They form the algebraic structure

$$\left[K^1_m,K^1_n\right]=[K^1_m,K^2_n]=[K^1_m,\tau^2_n]=[K^2_m,K^2_n]=[\tau^2_m,\tau^2_n]=0,\ [K^2_m,\tau^1_n]=2mK^2_{m+n-1},[K^2_m,\tau^2_n]=K^2_{m+n}$$

$$[K^1_m,\tau^1_n]=(2m+1)K^1_{m+n-1},[\tau^1_m,\tau^1_n]=2(m-n)\tau^1_{m+n-1},\ [\tau^1_m,\tau^2_n]=-2n\tau^2_{m+n-2}. \tag{3.12}$$

Two sets of constants of motion read

$$I^1_n=\int_0^1<\partial_x^{-1}k^1_n(\lambda\omega),\ \omega>d\lambda,\ I^2_n=\int_0^1<\partial_x^{-1}k^2_n(\lambda\omega),\omega>d\lambda,<A,B>=\iint_{-\infty}^{\infty}A.Bdxdy \tag{3.13}$$

They are in involution

$$\left\{I^i_n,i^j_m\right\}+\iint_{-\infty}^{\infty}\frac{\delta I^i_n}{\delta\omega}\partial_x\frac{\delta I^j_m}{\delta\omega}dxdy=0\quad i,j=1,2. \tag{3.14}$$

The equation (2.8) can be writtens in vector's form,

$$\omega_t=K(\omega),\quad \omega=\begin{bmatrix}q\\ r\end{bmatrix},\quad K(\omega)=\begin{bmatrix}iq_{xy}-2iq\partial_x^{-1}(qr)_y\\ -ir_{xy}+2ir\partial_x^{-1}(qr)_y\end{bmatrix} \tag{3.15}$$

The linearlized equation reads,

$$\begin{cases}\sigma_{1t}=i\sigma_{1xy}-2i\sigma_1\partial_x^{-1}(qr)_y-2iq\partial_x^{-1}(\sigma_1r)_y-2iq\partial_x^{-1}(q\sigma_2)_y\\ \sigma_{2t}=-i\sigma_{2xy}+2i\sigma_2\partial_x^{-1}(qr)_y+2ir\partial_x^{-1}(\sigma_1r)_y+2ir\partial_x^{-1}(q\sigma_2)_y\end{cases} \tag{3.16}$$

(3.15) is a Bi–Hamiltonian systems

$$\omega_t=\Phi_1\frac{\delta H_1}{\delta\omega}=\Phi_2\frac{\delta H_2}{\delta\omega} \tag{3.17}$$

$$\Phi_1=\begin{bmatrix}0&-1\\1&0\end{bmatrix}\quad \Phi=\begin{bmatrix}\partial_x-2q\partial_x^{-1}r&-2q\partial_x^{-1}q\\ 2r\partial_x^{-1}r&-\partial_x+2r\partial_x^{-1}q\end{bmatrix},\quad \Phi_2=\Phi\Phi_1, \tag{3.18}$$

$$H_1=\frac{1}{2}\iint_{-\infty}^{\infty}\{-r_{xy}q-q_{xy}r+2qr\partial_x^{-1}(qr)_y\}dxdy,\quad H_2=\frac{1}{2}\iint_{-\infty}^{\infty}(qr_y-rq_y)dxdy \tag{3.19}$$

Proprosition 3.2. There are four sets symmetries of the equation (3.15)

$$K^1_n=\Phi^nK^1_0,\quad K^2_n=\Phi^nK^2_0,\quad \tau^1_n=\Phi^n\tau^1_0,\quad \tau^2_n=\Phi^n\tau^2_0,$$

$$K^1_0=\begin{bmatrix}-q\\ r\end{bmatrix},\ K^2_0=\begin{bmatrix}q_y\\ ry\end{bmatrix},\ K^1_0=itK^2_0+xK^1_0,\ \tau^2_0=t\begin{bmatrix}q_t\\ r_t\end{bmatrix}+yK^2_0 \tag{3.20}$$

They form the abgebraic structure

$$\left[K^1_m,K^1_n\right]=[K^1_m,K^2_n]=[K^1_m,\tau^2_n]=[K^2_m,K^2_n]=[\tau^2_m,\tau^2_n]=0,[K^1_m,\tau^1_n]=mK^1_{m+n-1}[K^2_m,\tau^2_n]=K^2_{m+n},$$

$$[K^2_m,\tau^1_n]=mK^1_{m+n-1},\ [\tau^1_m,\tau^1_n]=(m-n)\tau^1_{m+n-1},[\tau^1_m,\tau^2_n]=-n\tau^2_{m+n-1} \tag{3.21}$$

Two sets of constants of motion read,

$$I^1_n=\int_0^1<\Phi_1^{-1}K^1_n(\lambda\omega),\omega>d\lambda,\ I^2_n=\iint_{-\infty}^{\infty}<\Phi_1^{-1}K^2_n(\lambda\omega),\omega>d\lambda,\ <A.B>=\iint_{-\infty}^{\infty}A^T.Bdxdy \tag{3.22}$$

They are in involution

$$\left\{I^i_n,I^j_m\right\}=\iint_{-\infty}^{\infty}\frac{\delta I^i_n}{\delta\omega}\Phi_1\frac{\delta I^j_m}{\delta\omega}dxdy=0\qquad i.j=1.2 \tag{3.23}$$

4. Darboux transformation

We can prove the following two propositions by the same approach, inref[4][5].

Poroposition 4.1. If. u is a solution of the equation (2.4). φ, f, satisfy the equations (2.1). (2.2) with ξ and ξ_0 respectively, ξ. ξ_0 satisfy the equation (2.3). Then, by the Darboux transformation

$$\bar{\varphi}=\varphi_x-\frac{f_x}{f}\varphi,\quad \bar{u}_x=u_x-2(log f)_{xx} \tag{4.1}$$

The equation (2.1), (2.2) are invartiant. i.e.

$$-\bar{\varphi}_{xx}+\bar{u}_x\bar{\varphi}=\xi^2\bar{\varphi}<\quad \bar{\varphi}_t=4\xi^2\bar{\varphi}_y-\bar{u}_{xy}\bar{\varphi}-2\bar{u}_y\bar{\varphi}_x, \tag{4.2}$$

and $\bar{u}$ is a new solution of the equation(2.4)

Proposition 4.2. If. q,r is a solution of the equation (2.8), φ satisfies the equations (2.5), (2.6), and ξ satisfies (2.7). The nonsingular matrix K satisfies

$$K_x=-i\begin{bmatrix}1&0\\0&1\end{bmatrix}K\wedge+\begin{bmatrix}0&q\\r&0\end{bmatrix}K,\ K_t-2K_y\wedge=\begin{bmatrix}-i\partial_x^{-1}(qr)_y & iq_y\\ -ir_y & i\partial_x^{-1}(qr)_y\end{bmatrix}K \tag{4.3}$$

where $\wedge=\begin{bmatrix}\xi_1&0\\0&\xi_2\end{bmatrix}$ $\xi_1\neq\xi_2$, and satisfy $\xi_{it}=2\xi_t\xi_{iy}$, $i=1,2$.

By the Darboux transformation

$$\bar{\varphi}=T\varphi,\quad T=\xi I-S\quad S=K\wedge K^{-1},\quad S=\begin{bmatrix}s_{11}&s_{12}\\s_{21}&s_{22}\end{bmatrix} \tag{4.4}$$

Then the equations (2.5), (2.6) are invariant. and

$$\bar{q}=q-2is_{12},\ \bar{r}=r+2is_{21} \tag{4.5}$$

is a new solution of the equation (2.8).

Acknowledgement

Part of this work was done when the author visits the Pudue University according to an agreement. This work is part supported by National Basic Research Project. "Nonlinear Science" and the Ministry of Education of PRC.

Reference

[1] O.I.Bogogavlenskii, YMH> T45, N04(1990)17.

[2] F.Calogero, A.Degasperis. Nuovo Cimento. 31B 2(1976) 201,39B 1(1977)54.

[3] Li Yishen, Cheng Yi, J.Partial Diff. Equ. Vol.3(1990)4.17.

[4] Li Yishen. International workshop on applied differential equations. World Scientific. (1986) 76.

[5] Zhou, Zixiang, Research Reports in Physics Physics, Nonlinear Springer, verlay, (1990)23.

ON THE HEINZ INEQUALITY FOR THE SPACE LIKE EXTREMAL SURFACES IN $R^{2,1}$

LIN JUNMIN

Inst. of Math, Fudan Univ., Shanghai 200433, China

ABSTRACT. In this paper, we show that following inequality holds true for space like extremal graph $X = (x, y, F(x, y)) : B_R(0) \to R^{2,1}$:

$$K(0) \leq \frac{2\pi^2}{R^2(1 - |\nabla F|^2(0))^2}$$

Here, $K(0)$ is the Gaussian curvature of the surface X at the origin. This result is an extension of Heinz inequality for minimal graph in R^3, and a quantitative version of the Bernstein type theorem for space like extremal surface in $R^{2,1}$ due to E. Calabi [8], and S. Y. Cheng and S. T. Yau [4]. The proof based on the comparison with a special space like extremal graph: $F(x, y) = \tan^{-1}\left(\frac{\tan x \tan y}{\sqrt{1+\tan^2 x+\tan^2 y}}\right)$.

§ 1. INTRODUCTION

In the theory of minimal surface in R^3, the famous Bernstein theorem states that the only minimal graph over R^2 is linear plane. This result has a quantitative version, which says that there exist a constant $c > 0$, such that for all minimal graph in $B_R(0) = \{ (x, y) \mid x^2 + y^2 \leq R^2\}$, the Gaussian curvature of the surface at the origin satisfies

$$|K(0)| \leq \frac{C}{R^2}$$

This kind of inequality is first obtained by E. Heinz [2], and improved to the above form by E. Hopf [3], Finn and Osserman [1].

In the theory of space like extremal surface in $R^{2,1}$, similar Bernstein theorem holds true, as shown by E. Calabi [5], S. Y. Cheng and S. T. Yau [4]. It is a natural problem whether or not it has a similar quantitative version, i. e. Heinz type inequality. In this paper, we give the following answer to this problem.

Key words and phrases. Extremal Surfaces, Heinz inequality, Maximum principle.
The present work is partially supported by NNSFC and SFECC .

Main Theorem. *Let $F : B_R(0) \subset R^2 \to R$ defines a space like extremal surface $X = (x, y, F(x, y)) \in R^{2,1}$. Then,*

$$(1 - |\nabla F|^2(0))^2 K(0) \leq \frac{2\pi^2}{R^2}.$$

Here, $K(0)$ is the Gaussian curvature of the surface X at $(0, 0, F(0, 0)) \in R^{2,1}$.

From this, we know that: Let $F : \Omega \to R$ be a smooth function which defines a space like extremal graph in $R^{2,1}$. Then, for any $p \in \Omega$, we have:

$$(1 - |\nabla F|^2(p))^2 K(p) \leq \frac{2\pi^2}{\operatorname{dist}^2(p, \partial\Omega)}.$$

Here, $\operatorname{dist}(p, \partial\Omega)$ is the Euclidean distance between p and $\partial\Omega$. Let $\Omega = R^2$, we give an easy proof of the Bernstein type theorem:

Theorem. *Let $F : R^2 \to R$ defines a space like extremal surface $X = (x, y, F(x, y)) \in R^{2,1}$. Then, F is a linear function.*

In fact, $\operatorname{dist}(p, \partial\Omega) = \infty$, we know the Gaussian curvature of the extremal surface X is zero. Hence, by a simple calculation, $F_{xx} = F_{yy} = F_{xy} = 0$, or F is a linear function. In this sense, our inequality is a quantitative version of Bernstein type theorem.

§ 2. Notation and Example

Let $R^{2,1} = \{(x, y, z) \mid x, y, z \in R\}$ with Lorentz metric $ds^2 = dx^2 + dy^2 - dz^2$. Let $F : \Omega \subset R^2 \to R$ be a smooth function. Then, F defines a surface $X = (x, y) \mapsto (x, y, F(x, y)) \in R^{2,1}$. Its first fundamental form is $ds^2 = (1 - F_x{}^2)dx^2 - 2F_xF_ydxdy + (1 - F_y{}^2)dy^2$. Hence, this surface is space like if $|\nabla F|^2 < 1$, and is light like if $|\nabla F|^2 = 1$. This surface is said to be extremal surface, if

$$(1 - F_y{}^2)F_{xx} + 2F_xF_yF_{xy} + (1 - F_y{}^2)F_{yy} = 0.$$

This equation is a second order elliptic partial differential equation when the surface is space like. This is known that the above equation is just the Euler-Lagrange equation for the area functional $A(S) = \int \sqrt{1 - F_x{}^2 - F_y{}^2}dxdy$. Or, the mean curvature of the graph of F is zero.

It can be directly verified that extremal surface equation has a special solution

$$G(x, y) = \tan^{-1}\left(\frac{\tan x \tan y}{\sqrt{1 + \tan^2 x + \tan^2 y}}\right).$$

Here, $(x,y) \in (-\pi/2,\pi/2) \times (-\pi/2,\pi/2)$. Furthermore, its Gaussian curvature is: $K(0) = 1$. And at the boundary of $\Omega = (-\pi/2,\pi/2) \times (-\pi/2,\pi/2)$, we have $G(\pi/2,y) = y$, $G(x,\pi/2) = x$, $G(-\pi/2,y) = -y$, $G(x,-\pi/2) = -x$. We also notice that at the boundary points the surface is light like and the out normal derivative $\partial G/\partial\nu = 0$ except at the four corner points $(\pm\pi/2,\pm\pi/2)$ of the boundary which is not smooth. This surface can also be extended to a extremal surface over R^2 via reflection. The extended surface is space like except at $\{(x,y)\,|\,x = (2k+1)\pi/2 \text{ or } y = (2l+1)\pi/2,\ k,l \in Z\}$ where surface is light like.

§ 3. THE MAIN IDEA OF THE PROOF.

Let $F(x,y)$ be a solution of space like extremal surface equation in $B_R(0)$, then, $F_R(x,y) = F(Rx,Ry)/R$ is a solution of extremal surface equation in $B_1(0)$, too. And, we have a relation of the Gaussian curvature: $K_{F_R}(0) = R^2 K_F(0)$. Hence, to prove the main theorem, we need only prove the theorem for the unit ball. To illustrate the idea of the prove, we give a prove of the simplest case.

Let F_1, F_2 be space like solutions of the extremal surface equation over a region $\Omega \subset R^2$, then the standard maximum principal for elliptic equation shows the supremum and infimum of $F_1 - F_2$ is achieved at the boundary of Ω. This can be improved to the following in our case.

Lemma 1. *Let F_1, F_2 be space like solutions of the extremal surface equation over a region Ω. Let $\partial\Omega = \Gamma_1 \bigcup \Gamma_2$, and Γ_1 is smooth on which F_1 is space like, while F_2 is light like. Furthermore, $\partial F_2/\partial\nu \geq 0$ holds true on Γ_1. Here ν is the unit outward normal of $\partial\Omega$. Then the supremum of $F_1 - F_2$ is achieved at Γ_2.*

Remark. if $\partial F_2/\partial\nu \leq 0$ holds true on Γ_1, then the infimum of $F_1 - F_2$ is achieved at Γ_2.

Lemma 2. *Let F_1, F_2 be space like solutions of the extremal surface equation in $\Omega \subset R^2$. Suppose that $J^r F_1(p) = J^r F_2(p)$, and $J^{r+1}F_1(p) \neq J^{r+1}F_2(p)$, $p \in \Omega$. Then, under certain coordinate transformation near p, $(x,y) \mapsto (u,v)$, we have*

$$(F_1 - F_2)(u,v) = \Re\left(\zeta^{r+1}\right) \quad \zeta = u + iv, \zeta(p) = 0.$$

Here, J^r is r-jet extension of a function.

Lemma 3. *Let F_1, F_2 be space like solutions of the extremal surface equation in $\Omega \subset R^2$. Suppose that $\nabla F_1(p) = \nabla F_2(p)$, $K_{F_1}(p) = K_{F_2}(p)$, and the eigen directions of F_1 and F_2 are coincide at p. Then, $J^2F_1(p) = J^2F_2(p)$ up to a sign.*

Proposition 1. *Let F be a space like solution of extremal surface equation in the unit ball around the origin, and $|\nabla F|^2(0)$=0. Then,*

$$K_F(0) \leq \frac{\pi^2}{2}.$$

Proof. If $K_F(0) > \pi^2/2$, then, there exist $R > \sqrt{2}/\pi$, such that $R^2K_F(0) = 1$. Let $F_R(x,y) = F(Rx, Ry)/R$. Then, F_R is a space like solution of the extremal surface equation, such that $K_{F_R}(0) = R^2K_F(0) = 1$, and $|\nabla F_R|^2(0) = 0$, in $B_{\sqrt{2}\pi/2}(0)$. Now, we rotate the coordinate such that the principle directions of F_R and G coincide. Then, F is space like in $\Omega = (-\pi/2, \pi/2) \times (-\pi/2, \pi/2)$.

From Lemma 3, we know that $J^rF_R = J^rG$ and $J^{r+1}F_R \neq J^{r+1}G$ for some $r \geq 2$. If not, $J^rF_R = J^rG$ for all $r > 0$, hence, $F_R = G$ which is contradict to the fact that G is light like at $\partial\Omega$. Hence, under certain coordinate transformation $(x,y) \mapsto (u,v)$,

$$\varphi(x,y) = (F_R - G)(x,y) = \Re\zeta^{r+1}, \qquad \zeta = u + iv.$$

Or, for some $\delta > 0$, $B_\delta(0) - \{(x,y) \,|\, \varphi(x,y) = 0\}$ has $2(r+1) \geq 6$ components, and φ change its sign as we count the components clockwise.

Now, consider the component decomposition of $\Omega - \{(x,y) \,|\, \varphi(x,y) = 0\} = \bigcup \Omega_i$. If one of the components satisfies $(0,0) \in \bar{\Omega}_i \subset \Omega$, suppose $\varphi(x,y) > 0$ for some (x,y), $x^2 + y^2 < \delta$ and $(x,y) \in \Omega_i$. Then, $\sup_{\Omega_i} \varphi > 0$, which is achieved in Ω_i, as $\varphi = 0$ at the boundary of Ω_i. But, this is contradict to the standard maximum principle. Hence, all the components extends to the boundary of Ω.

Now take one component Ω_i, suppose $\varphi(x,y) > 0$ at some point in Ω_i. Then, $\sup \varphi > 0$ is achieved at some point of the boundary of Ω. But, from Lemma 1, the $\sup \varphi$ can only be achieved at one of the four corner points. Hence, Ω_i contains, at least one of the four corner points. Similar discussion holds true for the components where $\varphi(x,y) < 0$. Hence, we can have more than four components.

Hence, at least two components of $B_\delta(0) - \{(x,y) \,|\, \varphi(x,y) = 0\}$ having the same sign, say positive, belong to a component Ω_i. Then, one of the component on which φ is negative has its closure in Ω, which is impossible.

We don't give the proof of the lemma and main theorem due to limit of space.

References

[1] R. Finn and R. Osserman, *On the Gauss curvature of non-parametric minimal surfaces*, J. Analyse Math. **12** (1964), 351-364.

[2] E. Heinz, *Über die Lösungen der Minimalflächengleichung*, Nachr. Akad. Wiss. Göttingen, Math-Phys. Kl. (1952), 51-56.

[3] E. Hopf, *On an inequality for minimal surfaces $z = z(x,y)$*, J. Rat. Mech. Anal. **2** (1953), 519-522.

[4] S. Y. Cheng and S. T. Yau, *Maximal space-like hypersurfaces in the Lorentz-Minkowski spaces*, Annals of Mathematics **104** (1976), 407-419.

[5] E. Calabi, *Examples of Berstein problems for some nonlinear equations*, Proc. Symps. Pure Appl. Math. **15** (1968), 223-230.

INST. OF MATH, FUDAN UNIV., SHANGHAI 200433, CHINA.

FACTORIZATION AND INVERSE OF THE RECURSION OPERATORS FOR SOME NONLINEAR INTEGRABLE MODELS

Sen-yue Lou
Institute of Modern Physics, Ningbo Normal College,
Ningbo 315211, P. R. China

ABSTRACT

With help of the factorization procedure, the inverses of the recursion operators for some integrable models are obtained explicitly and then various new sets of the symmetries can be obtained. At the same time, the hereditary properties of the recursion operators can be proved quite easy when the factorization procedure is finished.

It is well known that there exist two sets of symmetries for many integrable nonlinear evolution equations. They are so-called K symmeties and τ symmeties.[1]

The symmeties of an evolution equation

$$u_t=K(x,t,u) \tag{1}$$

are the solutions of the linearized equation of (1)

$$\sigma_t=K'(x,t,u)[\sigma]=K'\sigma \tag{2}$$

An effective method to find new symmeties of (1) is acting a strong symmetry on some known seed symmetries of (1). The strong symmetry is defined by

$$d\Phi/dt=K'\Phi-\Phi K'=[K',\ \Phi] \tag{3}$$

which is also called hereditary if the following relation

$$\Phi'[\Phi f]h-\Phi'[\Phi h]f=\Phi(\Phi'[f]h-\Phi'[h]f) \tag{4}$$

holds for arbitrary functions f and h.

But however to check an operator whether is hereditary or not is quite difficult especially when K is complicatied. For instance, the recursion operator of the Kawamoto-type equation

$$\rho_t=K_0(\rho)=10\rho^4\rho_{xx}\rho_{xxx}+5\rho^4\rho_x\rho_{xxxx}+\rho^5\rho_{xxxxx} \tag{5}$$

reads[2]

$$\Phi=\Theta(\rho)J(\rho) \tag{6}$$

where the operator

$$\Theta(\rho)=\rho^2 D\Theta_1(u,\ \rho)^{-1}\rho^2 D, \qquad (u=\rho\rho_{xx}-(1/2)(\rho_x)^2,\quad D=d/dx) \tag{7}$$

*Work supported by the Nature Science Foundation of Zhejiang Province in China and the NSF in China

is implectic and

$$J(\rho)=D^{-1}\rho^{-2}\Theta_1 J_1 \Theta_1 D^{-1}\rho^{-2} \qquad (D^{-1}D=DD^{-1}=1) \tag{8}$$

is symplectic with

$$\Theta_1(u,\ \rho)=(\rho D)^2+u\rho D+\rho Du \tag{9}$$

and

$$J_1(u,\ \rho)=(\rho D)^3+2[(\rho D)^2u(\rho D)^{-1}+(\rho D)^{-1}u(\rho D)^2]+8[u^2(\rho D)^{-1}+(\rho D)^{-1}u^2]$$

$$+3(u\rho D+\rho Du) \tag{10}$$

In principle, the hereditary property of Φ can be checked by substituting (6)-(10) into (4). However, after carrying out all differentiations one discovers that (as a result of the product rule) the variational derivative of Φ has more than 1 000 000 terms.[a] And then we have to check whether or not an integrodifferential operator of this size is equal to zero. This is not an easy task even by the computor algebra.[2,3]

But however, if we can take some factorization procedures such that Φ becomes a single term

$$\Phi=\prod_i f_i D g_i D^{-1} \tag{11}$$

where f_i and g_i are some functions of ρ then one can easily to finish the task even by hand. Fortunatally, such types of factorization procedures can be worked out quite easy for various integrable models. For instance, after finishing the factorization procedure, Φ shown by (6) with (7)-(10) can be rewritten as

$$\Phi=\rho^{-4}D^{3}\rho^{2}D^{-1}\rho D^{3}\rho D^{-1}\rho^{-2} \tag{12}$$

Here are some other examples:
(i). The recursion operator of the well known mKdV equation

$$u_t=u_{xxx}-6u^2u_x \tag{13}$$

can be factorized as

$$\Phi=DgDu^{-1}g^{-2}DgD^{-1}u \tag{14}$$

with the function g being given by

$$g=\exp 2D^{-1}u\,. \tag{15}$$

(ii). The factorized form of the recursion operator for the KdV equation

$$u_t=u_{xxx}+6uu_x \tag{16}$$

reads:

$$\Phi=\psi^{-2}D\psi^{2}D\psi^{2}D\psi^{-2}D^{-1} \tag{17}$$

where ψ is determined by the Schrodinger equation

$$\psi_{xxx}+u\psi=0 \tag{18}$$

(iii). Kaup–Kupershmidt equation

$$u_t=u_{xxxxx}+20u^2u_x+10uu_{xxx}+25u_xu_{xx} \tag{19}$$

possesses the factorized recursion operator

$$\Phi=\psi^{-2}D^{-1}(\psi^{2}D)^{2}\psi^{-2}D^{-1}\psi^{-3}(\psi^{2}D)^{3}\psi^{-4}D^{-1} \tag{20}$$

here ψ is also given by a Schrodinger equation

$$\psi_{xxx}+(1/2)u\psi=0. \tag{21}$$

Whence the factorization procedure is finished, the task to check the hereditariness of a strong symmetry is quite easy. For instance, after substituting (12) into (6), the remained terms are only 14 rather than 1 000 000.

In the conventional theory about symmetries and strong symmetries, one treat the inverse of the recursion operator only in the formal form. Now, when the factorization procedure is finished, the explicit form of the inverse of the strong symmetry follows immediately, e.g., the inverse strong symmetries of the Kawamoto-type, mKdV, KdV and Kaup–Kupershmidt equations have the forms

$$\Phi^{-1}=\rho^{2}D\rho^{-1}D^{-2}\rho^{-1}D\rho^{-2}D^{-3}\rho^{-4}, \tag{22}$$

$$\Phi^{-1}=u^{-1}Dg^{-1}D^{-1}ug^{2}D^{-1}g^{-1}D^{-1}, \tag{23}$$

$$\Phi^{-1}=D\psi^{2}D^{-1}\psi^{-2}D^{-1}\psi^{-2}D^{-1}\psi^{2} \tag{24}$$

and

$$\Phi^{-1}=D\psi^{4}(\psi^{2}D)^{-3}\psi^{3}D\psi^{2}(\psi^{2}D)^{-2}D\psi^{2} \tag{25}$$

respectively. Because the explicit form of the inverse operator of the recursion operator has been obtained then various new sets of symmetries can be obtained by acting the inverse recursion operator on the trivial symmeting and fixing the integral function directly from the definition equation of symmetry (3). For instance, for the mKdV equation we get six sets of symmetries[4]

$$K^{(0)}_{n}=\Phi^{n}u_{x},\ K^{(1)}_{-n}=\Phi^{-n}g^{-1},\ K^{(2)}_{-n}=\Phi^{-n}g,\ K^{(3)}_{-n}=\Phi^{-n}(gD^{-1}g^{-1}+g^{-1}D^{-1}g)/2,$$

$$\tau^{(0)}_{-n}=3tK^{(0)}_{n}+(1/2)\Phi^{n}gD^{-1}g^{-1},\ \tau^{(1)}_{-m}=\Phi^{-m}gD^{-1}g^{-1}, \tag{26}$$

where $n=0, 1, 2, \ldots$, while $m=1, 2, \ldots$. Only two sets of them, $K^{(0)}_{n}$ and $\tau^{(0)}_{-n}$, are known in literatures.Starting from these new sets of symmetries, some new integrable hierarchies can be obtained, for instance.

(i) Liouville I hierarchy:

$$u_t=K^{(2)}_{-n},\quad (n=0, 1, 2, \ldots) \tag{27}$$

The first one of (27) is

$$u_t=\exp D^{-1}2u,\ \text{i.e.,}\quad \phi_{xt}=\exp 2\phi \tag{28}$$

with $u=\phi_x$.

(ii) sinh–Gordon hierarchy:

$$u_t=K^{(2)}_{-n}-K^{(1)}_{-n},\quad (n=0,\ 1,\ 2,\ \ldots) \tag{29}$$

The first one of (29) reads

$$u_t=\exp D^{-1}2u-\exp D^{-1}(-2u),\ \text{i.e.,}\ \phi_{xt}=2\sinh 2\phi\ . \tag{30}$$

Simmilar to mKdV case, various new sets of symmetries can be obtained for other models. For example, twelve sets of symmetries for the Kawamoto-type equation read

$$K_n=\Phi^n\rho^4D^2\rho^2,\quad J_n=\Phi^n\rho^4D^2\rho^2D^{-1}\rho D^2\rho\ ,\quad T_{-n}=\Phi^{-n}\rho_x,\quad R_{-n}=\Phi^{-n}(x\rho_x-\rho),$$

$$N_{-n}=\Phi^{-n}(x^2\rho_x/2-x\rho),\quad P_{-n}=\Phi^{-n}\rho^2D\rho^{-1}D^{-2}\rho^{-1}D\rho^{-2},$$

$$Q_{-n}=\Phi^{-n}\rho^2D\rho^{-1}D^{-2}\rho^{-1}D\rho^{-2}x,\quad S_{-n}=\Phi^{-n}\rho^2D\rho^{-1}D^{-2}\rho^{-1}D\rho^{-2}x^2,$$

$$U_{-n}=\Phi^{-n}\rho^2D\rho^{-1}D^{-2}\rho^{-1}D\rho^{-2}x^3,\quad V_{-n}=\Phi^{-n}\rho^2D\rho^{-1}D^{-2}\rho^{-1}D\rho^{-2}x^4,$$

$$M_n=\Phi^n(5tK_0-x\rho_x),\quad W_{-m}=\Phi^{-m}x\rho_x, \tag{31}$$

where, $n=0,\ 1,\ 2,\ \ldots,\ m=1,\ 2,\ldots$ and Φ and Φ^{-1} are given by eqs.(12) and (22). Only T_0, K_n, R_0, N_0 and M_0 are known in literatures.[2]

It is also well known that K symmetries and τ symmetries for many integrable models constitute an infinite dimensional Lie algebra. One can also to prove that all the new symmetries for some integrable equations, say the mKdV equation, the Kawamoto-type equation etc, constitute also a Lie algebra. But here we do not discuss it further.

References

1. C. H. Gu et. al., *Soliton theory and its application*, (Zhejiang Publishing House of Science Technology 1990) pp216–267.
2. S. Carillo and B. Fuchssteiner, J. Math. Phys., 30(1989)1606.
3. B. Fuchssteiner, W. Oevel and W. Wiwianka, Comp. Phys. Commun. 44(1987)47.
4. S-y Lou, *Symmetries of the KdV equation and four hierarchies of the integro-differential KdV equation*, Preprint: NBN-IMP, 07/90, 1990; *Symmetries, Conservation laws and infinite dimentional Lie algebra of the modified KdV equation*, Preprint: NBN-IMP 10/92, 1992.

A METHOD OF ZERO CURVATURE REPRESENTATION FOR CONSTRUCTING SYMMETRY ALGEBRAS OF INTEGRABLE SYSTEMS

WEN-XIU MA
Institute of Mathematics, Fudan University
Shanghai 200433, P. R. of China

ABSTRACT

By establishing Lax operator algebras related to zero curvature (zc) representation, a general skeleton of constructing symmetry algebras of integrable systems is proposed along with an application to coupled KdV hierarchy. Moreover it is clearly explained why nonisospectral vector fields can become the first order common master symmetries of the hierarchy of isospectral flows for a given spectral problem.

1. Introduction

In the last two decades, a lot of progress in the investigation of symmetry algebras of integrable systems has been achieved.[1] The recursion operator, which was first proposed by Olver[2] to provide a mechanism for generating an infinite family of symmetries, plays an important role in the theory. Moreover this kind of operators usually possesses a fundamental algebraic-geometrical property called hereditary[3] or Nijenhuis[4,5] property, and has a close relation[4-10] to the multi-Hamiltonian formulation. However, by the recursion operator it can not be exposed why an integrable system possesses a symmetry algebra, particularly a τ-algebra[11] of symmetries which is a semi-product Lie algebra of a Kac-Moody algebra and a Virasoro algebra. Recently for integrable systems with Lax representations, through introducing Lax operator algebras we have given a clear explanation for the existence of τ-algebras of symmetries.[12] In the present paper for integrable systems with zero curvature (zc) representations, we want to display tersely a general skeleton for constructing symmetry algebras by establishing analogous Lax operator algebras, and thus also explain the same problem, that is, why nonisospectral vector fields can become the first order master symmetries of isospectral flows.

2. The Algebraic Structure Related to ZC Representation

Assume that the spectral operator $U = U(u,\lambda)$ has an injective Gateaux derivative operator U'. Let us discuss the spectral problem

$$\begin{cases} \phi_x = U\phi = U(u,\lambda)\phi, \ \lambda_t = f(\lambda), \ f \in C^{\infty}, \\ \phi_t = V\phi = V(u,\lambda)\phi, \ u = (u_1, u_2, \cdots, u_q)^T. \end{cases} \tag{2.1}$$

Noticing that $U_t = U'[u_t] + \lambda_t U_\lambda$, we can see that the equation $u_t = K = K(x,t,u)$ has a zc representation $U_t - V_x + [U,V] = 0$, which is the compatibility condition of (2.1), iff

$$U'[K] + f(\lambda)U_\lambda - V_x + [U,V] = 0. \tag{2.2}$$

This equality provides a connection between the equation $u_t = K$ and the spectral problem (2.1) and plays a crucial role in our deduction. For two groups of (V,K,f), (W,S,g) satisfying (2.2), we introduce

$$[f,g](\lambda) = f'(\lambda)g(\lambda) - f(\lambda)g'(\lambda), \tag{2.3}$$

$$[\![V,W]\!] = V'[S] - W'[K] + [V,W] + gV_\lambda - fW_\lambda. \tag{2.4}$$

We may verify the following by a direct computation.

Theorem 2.1. *We generally have the equality*

$$U'[[K,S]] + [f,g](\lambda)U_\lambda - [\![V,W]\!]_x + [U,[\![V,W]\!]] = 0, \tag{2.5}$$

namely, the equation $u_t = [K,S]$ is the compatibility condition of the spectral problem $\phi_x = U\phi$, $\lambda_t = [f,g](\lambda)$, and the auxiliary problem $\phi_t = [\![V,W]\!]\phi$ provided that (2.2) holds for two groups of (V,K,f), (W,S,g).

By the above theorem, we easily show an important fact that the eigenvector fields (K) corresponding to a given evolution function (f) are certain to possess exactly the same algebraic structure as their Lax operators (V) because of the injective property of U'. This may give rise to a way to construct symmetry algebras, which will be discussed in the following sects. 3 and 4.

3. ZC Representations of Hierarchies

Let Φ be a hereditary symmetry associated with the spectral problem $\phi_x = U\phi$.[13] For a given vector field X, we construct an operator equation of Ω:

$$[\Omega, U] + \Omega_x = U'[\Phi X] - \lambda U'[X] \tag{3.1}$$

and call it the characteristic equation of U at X. Assume that (3.1) has solution for any X and $\Omega = \Omega(X)$ is a solution at X.

Theorem 3.1. *Suppose $k \geq 0$ and there exist starting Lax operators A_0, B_0 and vector fields f_0, g_0 such that the equalities*

$$U'[f_0] - A_{0x} + [U, A_0] = 0, \quad U'[g_0] + \lambda^k U_\lambda - B_{0x} + [U, B_0] = 0 \tag{3.2}$$

hold. Let $K_m = \Phi^m f_0$, $\sigma_n = \Phi^n g_0$, $m,n \geq 0$, and further let

$$V_m = \sum_{i=0}^{m} \lambda^{m-i} A_i = \lambda^m A_0 + \sum_{i=1}^{m} \lambda^{m-i}\Omega(K_{i-1}), \tag{3.3a}$$

$$W_n = \sum_{j=0}^{n} \lambda^{n-j} B_j = \lambda^n B_0 + \sum_{j=1}^{n} \lambda^{n-j}\Omega(\sigma_{j-1}). \tag{3.3b}$$

Then V_m, W_n are respectively Lax operators associated with K_m, σ_n, i.e. we have

$$U'[K_m] - V_{mx} + [U, V_m] = 0, \quad m \geq 0, \tag{3.4a}$$

$$U'[\sigma_n] + \lambda^{n+k}U_\lambda - W_{nx} + [U, W_n] = 0, \quad n \geq 0. \tag{3.4b}$$

This theorem implies that two hierarchies of equations $u_t = K_m$, $m \geq 0$, and $u_t = \sigma_n$, $n \geq 0$, possess zc representations $U_t - V_{mx} + [U, V_m] = 0$, $m \geq 0$, and $U_t - W_{nx} + [U, W_n] = 0$, $n \geq 0$, respectively. It also tells us that a starting master symmetry g_0 may be worked out by solving an equation of g_0, B_0 in (3.2).

4. Lax Operator Algebras

The aim of this section is to explain why the vector fields K_m, σ_n, $m, n \geq 0$, defined in the previous section, possess a semi-product Lie algebraic structure.

Theorem 4.1. *Let $k' \geq 0$. Suppose that (1) $f_0|_{u=0} = 0$, $\sigma_{k'}|_{u=0} = 0$; (2) if $U'[T] - \Theta_x + [U, \Theta] = 0$ and $\Theta|_{u=0} = 0$, then $\Theta = 0$; (3) the equality $[V_0|_{u=0}, W_{k'}|_{u=0}] = \gamma V_{k'+k-1}|_{u=0}$, γ =const. holds if $k' > 0$ or the equality*

$$[V_0|_{u=0}, W_0|_{u=0}] = 0,\ [V_0|_{u=0}, W_1|_{u=0}] = \gamma V_k|_{u=0},\ [V_1|_{u=0}, W_0|_{u=0}] = (1+\gamma)V_k|_{u=0},$$

where γ is a constant, holds if $k' = 0$. Then we have a semi-product Lie algebra of Lax operators

$$[\![V_m, V_n]\!] = 0,\ m, n \geq 0, \tag{4.1a}$$

$$[\![V_m, W_n]\!] = (m+\gamma)V_{m+n+k-1},\ V_{-1} = 0,\ m \geq 0,\ n \geq k', \tag{4.1b}$$

$$[\![W_m, W_n]\!] = (m-n)W_{m+n+k-1},\ W_{-1} = 0,\ m, n \geq k', \tag{4.1c}$$

where $[\![\cdot,\cdot]\!]$ is determined by (2.4).

We refer to the above algebra (4.1) as a Lax operator algebra. From a Lax operator algebra (4.1), we can readily deduce the same semi-product algebraic structure of the vector fields K_m, σ_n, $m, n \geq 0$, by noticing the explanation in Sec. 2 or directly using (2.5), and thus may obtain the following at once.

Theorem 4.2 (τ-algebra). *Under the assumption of Theorem 4.1, every integrable systems $u_t = K_s\,(s \geq 0)$ possess the following symmetry algebra:*

$$[K_m, K_n] = 0,\ m, n \geq 0,$$

$$[K_m, \tau_n^{(s)}] = (m+\gamma)K_{m+n+k-1},\ K_{-1} = 0,\ m \geq 0,\ n \geq k',$$

$$[\tau_m^{(s)}, \tau_n^{(s)}] = (m-n)\tau_{m+n+k-1}^{(s)},\ \tau_{-1}^{(s)} = 0,\ m, n \geq k',$$

where τ-symmetries $\tau_n^{(s)}$ read as $\tau_n^{(s)} = t[K_s, \sigma_n] + \sigma_n$, $n \geq 0$.

Summarizing, we can state that symmetry algebras may be generated by computing Lax operator algebras, which is a kind of new trick.

5. Application: An Illustrative Example

In terms of the above theory, we consider the coupled KdV case[14] as a particular example. The spectral problem in this case can be rewritten as

$$\phi_x = U(u,\lambda)\phi = \begin{bmatrix} 0 & 1 \\ -Q & 0 \end{bmatrix}\phi,\ Q = Q(u,\lambda) = \lambda^{-l}\sum_{i=0}^{q-1} v_i\lambda^i - \lambda^{q-l},\ 0 \leq l \leq q-1$$

with $u = (v_0, v_1, \cdots, v_{q-1})^T$. The corresponding isospectral hierarchy is $u_t = K_m = \Phi^m u_x$, $m \geq 0$, in which Φ is the common hereditary recursion operator

$$\Phi = \begin{bmatrix} 0 & 0 & \cdots & 0 & R_0^* \\ 1 & 0 & \cdots & 0 & R_1^* \\ 0 & 1 & \cdots & 0 & R_2^* \\ \cdots & \cdots & \cdots & \cdots & \cdots \\ 0 & 0 & \cdots & 1 & R_{q-1}^* \end{bmatrix}, \quad R_i^* = \frac{1}{4}\delta_{il}\partial^2 + v_i + \frac{1}{2}v_{ix}\partial^{-1}, \, 0 \leq i \leq q-1.$$

The characteristic eq. (3.1) has a special solution for any $X = (X_1, \cdots, X_q)^T$:

$$\Omega = \Omega(X) = \begin{bmatrix} -\frac{1}{4}X_q & \frac{1}{2}\partial^{-1}X_q \\ -\frac{1}{2}Q\partial^{-1}X_q - \frac{1}{4}X_{qx} & \frac{1}{4}X_q \end{bmatrix}$$

and Eq. (3.2) with $k = 1$ has two pairs of solutions: $f_0 = u_x$, $A_0 = U$ and

$$g_0 = (\, qv_0 + \frac{1}{2}(q-l)xv_{0x}, \cdots, v_{q-1} + \frac{1}{2}(q-l)xv_{q-1,x}\,)^T,$$

$$B_0 = \begin{bmatrix} -\frac{1}{4}(q-l) & \frac{1}{2}(q-l)x \\ -\frac{1}{2}(q-l)xQ & \frac{1}{4}(q-l) \end{bmatrix}.$$

According to Theorem 4.1 with $k' = 0$, for coupled KdV hierarchy we can present a Lax operator algebra determined by (4.1) with $k = 1$ and $\gamma = \frac{1}{2}(q-l)$, and thus the same structure admitted by the symmetry algebra of the hierarchy can be formulated by Theorem 4.2. The effectiveness of our method may be also demonstrated by applications to the other typical integrable hierarchies, which will be reported elsewhere.

Acknowledgement: The author would like to express his sincere thanks to Profs. C. H. Gu and H. S. Hu for their guidance and support.

References

1. C. H. Gu *et al.*, *Soliton Theory and Its Application*, (Zhejiang Publishing House of Science and Technology, Hangzhou, China, 1990).
2. P. J. Olver, *J. Math. Phys.* **18** (1977), 1212–1215.
3. B. Fuchssteiner, *Nonlinear Anal. Theor. Math. Appl.* **3** (1979), 849–862.
4. F. Magri, *J. Math. Phys.* **19** (1978), 1156–1162.
5. ——, *in: Nonlinear Evolution Equations and Dynamical Systems edited by M. Boiti, F. Pempinelli and G. Soliani*, (Springer-Verlag, New York, 1980), pp. 233–263.
6. A. S. Fokas and B. Fuchssteiner, *Lett. Nuovo Cimento* **28** (1980), 299–303.
7. B. Fuchssteiner and A. S. Fokas, *Physica* **4D** (1981), 47–66.
8. I. M. Gel'fand and I. Y. Dorfman, *Funct. Anal. Appl.* **13** (1979), 248–262.
9. ——, *Funct. Anal. Appl.* **14** (1980), 223–226.
10. G. Magnano and F. Magri, *Rev. Math. Phys.* **3** (1991), 403–466.
11. W. X. Ma, *J. Phys. A: Math. Gen.* **25** (1992), to appear.
12. ——, *J. Math. Phys.* **33** no. 7 (1992).
13. A. S. Fokas and R. L. Anderson, *J. Math. Phys.* **23** (1982), 1066–1073.
14. M. Boiti, P. J. Caudrey and F. Pempinelli, *Il Nuovo Cimento* **83B** (1984), 71–87.

HAMILTONIAN FORMALISM FOR FIELD THEORY*

GIOVANNI GIACHETTA and LUIGI MANGIAROTTI
Department of Mathematics and Physics
University of Camerino, 62032 Camerino, Italy

OLEG ZAKHAROV
Department of Theoretical Physics
Moscow State University, 117234 Moscow, Russia

ABSTRACT

We propose a new formulation of the Hamiltonian (and Lagrangian) formalism in which connections and base-tangent valued symplectic forms play a key role.

1. Introduction.

In this paper we outline a geometric formulation of the Hamiltonian (and Lagrangian) formalism based on the notions of connections and base-tangent valued symplectic forms.[1] As a mathematical tool, we use the language of jet bundles.[2,3] Surprisingly enough, it turns out that this approach is similar to the conventional geometric approach for finite-dimensional systems.[4] As examples in developing this new point of view, we consider regular and hyperregular Lagrangian systems. The case of nonregular Lagrangian theories is currently under study.

2. Technical Preliminaries.

Let $E \to M$ be a fiber bundle (*configuration bundle*) with total space E and base M, and let $\dim M = m$, $\dim E = m + l$ (all manifolds and maps are smooth, i.e. C^∞).We use adapted coordinates (x^λ, ϕ^i) compatible with the fibration. We need the jet prolongations of E up to the second order, namely[2]

$$J^2E \to J^1E \to E \to M, \qquad J^2E \subset \widehat{J}^2E \subset J^1J^1E. \tag{1}$$

The coordinates on J^1E and J^2E are denoted by $(x^\lambda, \phi^i, \phi^i_\lambda)$ and $\left(x^\lambda, \phi^i, \phi^i_\lambda, \phi^i_{\lambda\mu}\right)$, respectively. Here $\widehat{J}^2E$ denotes the sesqui-holonomic prolongation while J^1J^1E is the repeated prolongation.

As is well known,[2] a *first order connection* on E is a section $\Gamma : E \to J^1E$. In an equivalent way, Γ can be seen as a projectable vector valued 1-form on E, namely

$$\Gamma : E \to J^1E, \qquad \Gamma = dx^\lambda \otimes \left(\partial_\lambda + \Gamma^i_\lambda \partial_i\right). \tag{2}$$

The connections parameters of Γ are just the local functions Γ^i_λ (defined on E).

*Work supported by the Ministero della Pubblica Istruzione, Italy (national and local funds).

A *second order connection* on E is a first order connection on J^1E, i.e. a section $\Gamma : J^1E \to J^1J^1E$ which takes its values in $J^2E \subset \widehat{J}^2E \subset J^1J^1E$.

The *momentum bundle* associated with the configuration bundle is the bundle $P = V^*E \otimes \overset{m-1}{\wedge} T^*M \to E \to M$ with induced coordinates $(x^\lambda, \phi^i, \pi_i^\lambda)$. Here $VE \subset TE$ denotes the vertical subspace.

The *(generalized) symplectic structure* on P is the base-tangent valued form defined by[1]

$$\Omega : P \to \overset{m+2}{\wedge} T^*P \otimes TM, \qquad \Omega = d\phi^i \wedge d\pi_i^\lambda \wedge \omega \otimes \partial_\lambda. \tag{3}$$

Indeed it is easily verified that Ω is a globally object defined on P.

Given a Lagrangian density

$$\mathcal{L} : J^1E \to \overset{m}{\wedge} T^*M, \quad \mathcal{L} = L\omega, \quad \omega = dx^1 \wedge \cdots \wedge dx^m, \tag{4}$$

where L is a local function on J^1E, its *Legendre map* is the morphism over E given by

$$\Pi : J^1E \to P, \qquad \pi_i^\lambda \circ \Pi = \partial_i^\lambda L \equiv \frac{\partial L}{\partial \phi_\lambda^i}. \tag{5}$$

The *symplectic structure associated with* $\mathcal{L}$ is the base-tangent valued form on J^1E given by

$$\Omega_{\mathcal{L}} = \Pi^*\Omega : J^1E \to \overset{m+2}{\wedge} T^*J^1E \otimes TM,$$
$$\Omega_{\mathcal{L}} = \left(\partial_j^\mu \partial_i^\lambda L d\phi^i \wedge d\phi_\mu^j + \partial_j \partial_i^\lambda L d\phi^i \wedge d\phi^j\right) \wedge \omega \otimes \partial_\lambda. \tag{6}$$

In the same way, given a *Hamiltonian form*[1]

$$\mathcal{H} : P \to \overset{m}{\wedge} T^*E, \qquad \mathcal{H} = \pi_i^\lambda d\phi^i \wedge \omega_\lambda - H\omega, \qquad \omega_\lambda = \partial_\lambda \rfloor \omega, \tag{7}$$

where H is a local function on P, its *Hamiltonian map* is the morphism over E given by

$$\Phi : P \to J^1E, \qquad \phi_\lambda^i \circ \Phi = \partial_\lambda^i H \equiv \frac{\partial H}{\partial \pi_i^\lambda}. \tag{8}$$

The *Poincaré-Cartan form associated with* $\mathcal{L}$ is the form on J^1E given by[5]

$$\mathcal{H}_{\mathcal{L}} : J^1E \to \overset{m}{\wedge} T^*E, \quad \mathcal{H}_{\mathcal{L}} = \partial_i^\lambda L d\phi^i \wedge \omega_\lambda - H_{\mathcal{L}}\, \omega, \tag{9}$$

where $H_{\mathcal{L}} = L - \phi_\lambda^i \partial_i^\lambda L$. Note that, in a sense, $\mathcal{H}_{\mathcal{L}}$ is a Hamiltonian form canonically associated with $\mathcal{L}$.

Finally, the *Euler-Lagrange operator associated with* $\mathcal{L}$ is the morphism over J^1E given by[5]

$$\mathcal{E}_{\mathcal{L}} : J^2E \to \overset{m}{\wedge} T^*M \otimes V^*E,$$
$$\mathcal{E}_{\mathcal{L}} = \left(\partial_i L - \partial_\lambda \partial_i^\lambda L - \phi_\lambda^j \partial_j \partial_i^\lambda L - \phi_{\lambda\mu}^j \partial_j^\mu \partial_i^\lambda L\right) d\phi^i \otimes \omega. \tag{10}$$

3. Hamiltonian Formalism.

We begin with the following basic concept.

DEFINITION 1. *A first order connection $\gamma : P \to T^*M \otimes TP$ is called a Hamiltonian connection if we have*

$$\gamma \rfloor \Omega = -d\mathcal{H} \tag{11}$$

for some Hamiltonian form $\mathcal{H}$.

REMARK 2. *Let us write*

$$\gamma = dx^\lambda \otimes \left(\partial_\lambda + \gamma_\lambda{}^i \partial_i + \gamma_{\lambda i}^{\ \mu} \partial^i_\mu\right), \tag{12}$$

where the parameters $\gamma_\lambda{}^i$ and $\gamma_{\lambda i}^{\ \mu}$ are local functions on P. Then using Eqs.3 and 7, we easily see that Eq.11 is equivalent to

$$\gamma_{\lambda i}^{\ \lambda} = -\frac{\partial H}{\partial \phi^i}, \qquad \gamma_\lambda{}^i = \frac{\partial H}{\partial \pi_i^\lambda}. \tag{13}$$

In other words, a Hamiltonian connection satisfies ***Hamilton's equations.***

The following proposition provides an intrinsic formulation of Hamilton's equations for a section $s : M \to P$ of the momentum bundle $P \to M$.

PROPOSITION 3. *Let $\mathcal{H}$ be a Hamiltonian form and let s be a section: $s(x) \equiv (\phi^i(x), \pi_i^\lambda(x))$, $x \in M$. Then the two following conditions are equivalent:*

(1)

$$s^*(u \rfloor d\mathcal{H}) = 0 \quad \textit{for each } u : P \to VP, \tag{14}$$

(2)

$$\partial_\lambda \pi_i^\lambda = -\frac{\partial H}{\partial \phi^i}, \quad \partial_\lambda \phi^i = \frac{\partial H}{\partial \pi_i^\lambda}, \tag{15}$$

where u denotes a vertical field over P. Further let γ be a Hamiltonian connection for $\mathcal{H}$ and let $s : M \to P$ be an integral section of γ, i.e. $j^1 s = \gamma \circ s$, where $j^1 s : M \to J^1 P$ is the first jet prolongation[2] of s. Then s satisfies Eq.14 or, equivalently, Eq.15.

Coming to the Lagrangian formalism, we begin with the following proposition.

PROPOSITION 4. *Let $\mathcal{L}$ be a Lagrangian density and let $\Gamma : J^1 E \to J^2 E$ be a second order connection. Then the two following conditions are equivalent:*

$$\Gamma \subset Ker\, \mathcal{E}_\mathcal{L}, \tag{16}$$

$$\Gamma \rfloor \Omega_\mathcal{L} = -d\mathcal{H}_\mathcal{L}. \tag{17}$$

Such a connection is called a ***Lagrangian connection.***

We recall that a Lagrangian density $\mathcal{L}$ is called *regular* if the intrinsic condition $\left\| \dfrac{\partial^2 L}{\partial \phi^i_\lambda \partial \phi^j_\mu} \right\| \neq 0$ holds.

About regular Lagrangians we have the following theorem.

THEOREM 5. *Let $\mathcal{L}$ be a regular Lagrangian density. Let $\Gamma : J^1E \to J^1J^1E$ be a first order connection on J^1E and let $R_\Gamma : J^1E \to \overset{2}{\wedge} T^*M \otimes VJ^1E$ be its curvature.*[2] *Then the two following conditions are equivalent:*

(1)

$$\Gamma \rfloor \Omega_{\mathcal{L}} = -d\mathcal{H}_{\mathcal{L}}, \qquad R_\Gamma = 0, \tag{18}$$

(2)

$$\Gamma : J^1E \to J^2E \subset \widehat{J}^2E \subset J^1J^1E, \text{ and } \Gamma \subset \operatorname{Ker} \mathcal{E}_{\mathcal{L}}, \qquad R_\Gamma = 0. \tag{19}$$

Note that, according to Eq.19, Γ becomes a second order connection.

We recall that a Lagrangian density $\mathcal{L}$ is called *hyperregular* if its associated Legendre map Π is a diffeomorphism.

About hyperregular Lagrangians we have the following theorem.

THEOREM 6. *Let $\mathcal{L}$ be a hyperregular Lagrangian density and let $\Gamma : J^1E \to J^1J^1E$ be a first order connection on J^1E. It follows that the composition*

$$P \xrightarrow{\Pi^{-1}} J^1E \xrightarrow{\Gamma} J^1J^1E \xrightarrow{J^1\Pi} J^1P \tag{20}$$

gives rise to a first order connection γ on P ($J^1\Pi$ denotes the jet prolongation[2] *of Π). Then the two following conditions are equivalent:*

$$\Gamma \rfloor \Omega_{\mathcal{L}} = -d\mathcal{H}_{\mathcal{L}}, \quad R_\Gamma = 0, \tag{21}$$

$$\gamma \rfloor \Omega = -d\left(\Pi^{-1}\right)^* \mathcal{H}_{\mathcal{L}}, \qquad R_\gamma = 0. \tag{22}$$

References.

1. O. Zakharov, *J. Math. Phys.* **33** (1992) 607.
2. L. Mangiarotti and M. Modugno, in *Proceedings of the International Meeting on Geometry and Physics*, ed. M. Modugno, (Pitagora Editrice, Bologna, 1983).
3. G. Sardanashvily and O. Zakharov, *Gauge Gravitation Theory*, (World Scien tific, Singapore, 1992).
4. R. Abraham and J. E. Marsden, *Foundations of mechanics*, (Benjamin Inc., London, 1978).
5. L. Mangiarotti and M. Modugno, *Ann. Inst. H. Poincaré* **31** (1983) 29.

AN OUTLINE OF A NEW GEOMETRICAL APPROACH TO GALILEI GENERAL RELATIVISTIC QUANTUM MECHANICS

Arkadiusz Jadczyk†, Marco Modugno††

†Institute of Theoretical Physics
pl. Maxa Borna 9, 50-204 Wrocław, Poland

††Department of Applied Mathematics "G. Sansone"
Via S. Marta 3, 50139 Florence, Italy

(Talk presented by M. Modugno)

Abstract

We outline a new general relativistic approach to the quantum mechanics of a charged particle subject to an external gravitational and electromagnetic field in a curved space-time with absolute time.

This approach has some ideas in common with the standard "geometrical quantization", but it contains several differences and new results as well. Our formulation is explicitly covariant. Moreover, a fundamental role is played by the fibring over time, connections and jets; conversely, the Hamiltonian and Lagrangian formalisms have a limited application.

More details can be found in the extended book in preparation [1].

The classical theory

We need an appropriate model for the classical space-time as necessary support for the quantum theory. Therefore, we present the following Galilei general relativistic classical space-time. The model below takes into accounts classical ideas due to E. Cartan, C. Duval, K. Kuchar, A. Trautman and several other authors and contains new ideas and results as well (for references see [1]).

Space-time is defined to be a 4-dimensional fibred manifold

$$t: \mathbf{E} \to \mathbf{T}$$

over a 1-dimensional oriented affine space $\mathbf{T}$. Thus, we have the absolute time $\mathbf{T}$ and the absolute time function t, but no absolute space of positions, as we do not assume any distinguished splitting of the space-time fibred manifold. We denote by $\mathbb{T}$ the vector space associated with $\mathbf{T}$.

We denote the space-time coordinates adapted to the fibring by (x^0, y^i); the associated basis of vector fields, 1-forms and vertical 1-forms are (∂_0, ∂_i), (d^0, d^i) and $(\check{d}^i)$. The Capital Greek indices Φ, Ψ, ... run from 0 to 3 and the Latin indices i,j,h,k, ... rum from 1 to 3. In general, the vertical restrictions will be denoted by "ˇ".

The first jet bundle $J_1\mathbf{E} \to \mathbf{E}$ plays an important role in the classical and quantum theory. We denote the induced coordinates of $J_1\mathbf{E}$ by (x^0, y^i, y^i_0). We shall be involved with the canonical maps

$$д: J_1\mathbf{E} \to \mathbb{T}^* \otimes T\mathbf{E} \qquad \vartheta: J_1\mathbf{E} \to T^*\mathbf{E} \otimes V\mathbf{E},$$

whose coordinate expressions are

$$д = u^0 \otimes д_0 \equiv u^0 \otimes (\partial_0 + y^i_0\, \partial_i) \qquad \vartheta = \vartheta^i \otimes \partial_i \equiv (d^i - y^i_0\, d^0) \otimes \partial_i.$$

A *motion* is defined to be a section $s: \mathbf{T} \to \mathbf{E}$ and its *absolute velocity* is described by the first jet prolongation $j_1 s: \mathbf{T} \to J_1\mathbf{E}$. An *observer* is defined to be a section $o: \mathbf{E} \to J_1\mathbf{E}$. Each observer determines a splitting of the tangent bundle of space-time into its observed time-like and space-like components. Accordingly, we obtain the corresponding observed space-like component of the velocity of a given motion. A *frame of reference* is defined to be a pair (u_0, o), where $u_0 \in \mathbb{T}^+$ is a time unit of measurement and o is an observer.

We assume space-time to be equipped with a Riemannian vertical metric

$$g: \mathbf{E} \to \mathbb{A} \otimes S^2 V^*\mathbf{E}.$$

Here, the "1-dimensional positive half-vector space" $\mathbb{A}$ represents the space of area units; we denote by $\mathbb{L}$ the 1-dimensional positive half-vector space obtained by taking the "square root" of $\mathbb{A}$. We can also regard g as a degenerate 4-dimensional metric of signature (0,3) by considering the associated contravariant tensor $\bar{g}: \mathbf{E} \to \mathbb{A} \otimes S^2 V\mathbf{E}$. The coordinate expression of g is $g = g_{ij}\, \check{d}^i \otimes \check{d}^j$, where $g_{ij}: \mathbf{E} \to \mathbb{A} \otimes \mathbb{R}$.

Thus, the differences between the Einstein and Galilei general relativistic space-times can be summarized as follows: in the Einstein case, we have a Lorentz metric and no fibring over absolute time; in the Galilei case, we have a "space-like" metric $\bar{g}$ and a "time-like" 1-form $dt: \mathbf{E} \to \mathbb{T} \otimes T^*\mathbf{E}$, which determines the fibring over absolute time. In brief words, we can say that the essential difference between the two theories consists in the replacement of the light cones with the vertical subspaces.

The metric and time-fibring yield a space-time volume form $\upsilon: \mathbf{E} \to (\mathbb{T} \otimes \mathbb{L}^3) \otimes \Lambda^4 T^*\mathbf{E}$, with coordinate expression $\upsilon = \sqrt{|g|}\; u_0 \otimes d^0 \wedge d^1 \wedge d^2 \wedge d^3$.

A (generic) *space-time connection* is defined to be, equivalently, a dt-preserving linear connection on the manifold $\mathbf{E}$, or an affine torsion free connection on the affine bundle $J_1\mathbf{E} \to \mathbf{E}$, respectively,

$$K: T\mathbf{E} \to T^*\mathbf{E} \otimes TT\mathbf{E} \qquad \text{or} \qquad \Gamma: J_1\mathbf{E} \to T^*\mathbf{E} \otimes TJ_1\mathbf{E},$$

with coordinate expression

$$\Gamma = d^{\Phi}\otimes\partial_{\Phi} + \left((\Gamma_{0h}{}^{i}\, y_0^h + \Gamma_{0\circ}{}^{i})\, d^0 + (\Gamma_{jh}{}^{i}\, y_0^h + \Gamma_{j\circ}{}^{i})\, d^j\right)\otimes\partial_i^0,$$

where $\Gamma_{\Phi}{}^{i}{}_{\Psi}: \mathbf{E}\rightarrow\mathbb{R}$ and $\Gamma_{\Phi}{}^{i}{}_{\Psi} = K_{\Phi}{}^{i}{}_{\Psi}$. The first viewpoint is more suitable for field theory, the second one for particle mechanics. The vertical valued 1-form associated with Γ is denoted by

$$\nu_\Gamma : \mathbf{E}\rightarrow\mathbb{T}^*\otimes T^*J_1\mathbf{E}\otimes V\mathbf{E}.$$

A connection Γ yields the connection on the fibred manifold $J_1\mathbf{E}\rightarrow\mathbf{T}$

$$\gamma \equiv \Delta \lrcorner \Gamma : J_1\mathbf{E} \rightarrow T^*\mathbf{T}\otimes TJ_1\mathbf{E}.$$

Moreover, g and a connection Γ yield the *contact 2-form*

$$\Omega \equiv \nu_\Gamma \bar{\wedge} \vartheta : J_1\mathbf{E} \rightarrow (\mathbf{A}\otimes\mathbb{T}^*)\otimes\Lambda^2 T^*J_1\mathbf{E},$$

We have

$$\gamma \lrcorner \Omega = 0.$$

Conversely, γ and Ω characterize Γ, through the above relations. Moreover, Ω yields the volume form on the manifold $J_1\mathbf{E}$

$$dt\wedge\Omega\wedge\Omega\wedge\Omega : J_1\mathbf{E}\rightarrow(\mathbf{A}^3\otimes\mathbb{T}^{*2})\otimes\Lambda^7 T^*J_1\mathbf{E}.$$

The coordinate expressions of γ and Ω are

$$\gamma = u^0\otimes\left(\partial_0 + y_0^i\,\partial_i + (\Gamma_{hk}{}^{i}\, y_0^h\, y_0^k + 2\,\Gamma_{h\circ}{}^{i}\, y_0^h + \Gamma_{0\circ}{}^{i})\,\partial_i^0\right)$$

$$\Omega = g_{ij}\, u^0\otimes(d_0^i - \gamma^i\, d^0 - \Gamma_h{}^i\, \vartheta^h)\wedge\vartheta^j.$$

We observe that the contact 2-form Ω restricts to the vertical symplectic form $\check{\Omega}$ associated with the vertical metric, but $\check{\Omega}$ has no dynamical role.

Given an observer o, let

$$\Phi : \mathbf{E}\rightarrow(\mathbb{T}^*\otimes\mathbf{A})\otimes\Lambda^2 T^*\mathbf{E},$$

be the form

$$\Phi = 2\, o^*\Omega,$$

whose coordinate expression in adapted coordinates is

$$\Phi \equiv u^0\otimes\left(2\,\Phi_{0j}\, d^0\wedge d^j + \Phi_{ij}\, d^i\wedge d^j\right) \qquad \text{with} \qquad \Phi_{0j} = -\,\Gamma_{0j\circ},\ \Phi_{ij} = -\,(\Gamma_{ij\circ} - \Gamma_{ji\circ}).$$

Next, we assume space-time to be equipped with a "gravitational" connection (which is a particular space-time connection) and an "electromagnetic" field

$$\Gamma^{\natural}: J_1\mathbf{E} \to T^*\mathbf{E} \otimes TJ_1\mathbf{E} \qquad\qquad F: \mathbf{E} \to \mathbb{B} \otimes \Lambda^2 T^*\mathbf{E},$$

where, $\mathbb{B}$ represents the space of electromagnetic units. We stress that, as the metric is degenerate, it will not determine fully the gravitational connection. Hence, we must assume that the gravitational field is described both by the metric g and the connection $\Gamma^{\natural}$. The coordinate expression of F is $F = 2\, F_{0j}\, d^0 \wedge d^j + F_{ij}\, d^i \wedge d^j$, where $F_{\phi j}: \mathbf{E} \to \mathbb{B} \otimes \mathbb{R}$.

The *mass* and the *charge* of a classical or quantum particle will be represented by elements $m \in \mathbb{M}$ and $q \in \mathbb{Q}$ of a 1-dimensional positive half-vector space $\mathbb{M}$ and a 1-dimensional vector space $\mathbb{Q}$. In view of the consistency of the equations of classical and quantum theories, we assume the following equalities $\mathbb{B} = \mathbb{M}^{1/2} \otimes \mathbb{L}^{1/2}$, $\mathbb{Q} = \mathbb{M}^{1/2} \otimes \mathbb{L}^{3/2} \otimes \mathbb{T}^*$. Moreover, given $u_0 \in \mathbb{T}$, we set $q \equiv \mathbf{q}(u_0) \in \mathbb{M}^{1/2} \otimes \mathbb{L}^{3/2}$. The mass and the charge will play the role of coupling constants for the metric and the electromagnetic field.

Given a *mass* $m \in \mathbb{M}$ and a *charge* $q \in \mathbb{Q}$, there is a natural way to deform the gravitational space-time connection $\Gamma^{\natural}$ and the related second order connection $\gamma^{\natural}$ and contact 2-form $\Omega^{\natural}$ into corresponding "dynamical" objects Γ, γ and Ω, through the electromagnetic field F.

In fact, it is natural to consider the deformed "dynamical" 2-form

$$\Omega \equiv \Omega^{\natural} + \Omega^{e} \equiv \Omega^{\natural} + \frac{1}{2}\frac{\mathbf{q}}{m} F,$$

obtained by adding the electromagnetic 2-form to the gravitational contact 2-form. Here, the charge and the mass are necessary for the equality of the left and right hand spaces (i.e. the correct dimensionality of the above formula). On the other hand, we could multiply F by any other non-zero scalar factor; 1/2 has been chosen just in order to obtain the standard normalizations in the following classical and quantum equations.

If we demand to maintain the equations (valid in the gravitational case)

$$\gamma \equiv \mathcal{A} \lrcorner \Gamma \qquad\qquad \Omega \equiv \nu_{\Gamma} \bar{\wedge} \vartheta \qquad\qquad \gamma \lrcorner \Omega = 0,$$

then we can prove that there is a unique consistent way to obtain the deformed dynamical objects Γ, γ. Namely, we obtain

$$\gamma = \gamma^{\natural} + \gamma^{e} \qquad\qquad \Gamma = \Gamma^{\natural} + \Gamma^{e},$$

where γ^e turns out to be the Lorentz force and Γ^e an intrinsic tensor with coordinate expressions

$$\gamma^{e} = -\frac{q}{m}\,(F_0{}^i + F_h{}^i\, y^h_0)\, u^0 \otimes \partial^0_i \qquad\qquad \Gamma^{e} = \frac{1}{2}\frac{q}{m}\,\big((F^i{}_h\, y^h_0 + 2\, F^i{}_0)\, d^0 + F^i{}_j\, d^j\big) \otimes \partial^0_i.$$

So, we have found a natural way to incorporate the electromagnetic field together with the gravitational field into the geometric structures of space-time. This coupling works well in classical and quantum theories.

Now, we want to introduce the gravitational and electromagnetic field equations. We consider two equations: the first one couples just the gravitational and electromagnetic fields, while the second one couples the gravitational and electromagnetic fields with the charged matter sources.

First, we assume that, for any charge and mass, the corresponding dynamical contact 2-form Ω is closed, i.e.

$$d\Omega = 0.$$

After long computations, we can interpret the equation $d\Omega = 0$ in several interesting ways.

First, the equation $d\Omega = 0$ is equivalent to the system

$$\nabla \bar{g} = 0 \qquad R^{i}{}_{\Phi}{}^{j}{}_{\Psi} = R^{j}{}_{\Psi}{}^{i}{}_{\Phi},$$

where R is the curvature of K. In other words, this equation says that the connection K is metrical and the standard algebraic Riemannian identities for the curvature of K hold; we stress that we cannot apply fully the standard procedures of Riemannian geometry in our context, because the metric is degenerate. On the other hand, in virtue of the arbitrariness of the mass and charge, the above equations can be split into their gravitational and electromagnetic components, respectively,

$$\nabla^{\natural}\bar{g} = 0 \qquad F^{ij} + F^{ji} = 0 \qquad R^{\natural i}{}_{\Phi}{}^{j}{}_{\Psi} - R^{\natural j}{}_{\Psi}{}^{i}{}_{\Phi} = 0 \qquad \partial_{\Phi} F_{\Psi\Theta} + \partial_{\Theta} F_{\Phi\Psi} + \partial_{\Psi} F_{\Theta\Phi} = 0.$$

Thus, we obtain the standard Riemannian identities involving the pure gravitational objects, the antisymmetry of the electromagnetic field and the first Maxwell equation.

Moreover, given an observer o, the equation $d\Omega = 0$ is equivalent to

$$K_{0ij} + K_{0ji} = -\partial_0 g_{ij} \qquad K_{hik} = -\frac{1}{2}(\partial_h g_{ik} + \partial_k g_{ih} - \partial_i g_{hk}) \qquad d\Phi = 0.$$

This system can be summarized by the equation

$$K_{\Phi i \Psi} = -\frac{1}{2}(\partial_{\Phi} g_{i\Psi} + \partial_{\Psi} g_{i\Phi} - \partial_i g_{\Phi\Psi}),$$

where we have set $g_{i0} \equiv g_{0i} \equiv a_i$, $g_{00} \equiv 2\, a_0$, being $a: \mathbf{E} \to (\mathbb{T}^* \otimes \mathbf{A}) \otimes T^*\mathbf{E}$ a local potential of the closed 2-form Φ, which involves both the gravitational and electromagnetic potentials.

Similarily, in virtue of the arbitrariness of m and $\mathbf{q}$, the equation $d\Omega = 0$ can be split into its gravitational and electromagnetic components

$$d\Omega^{\natural} = 0 \qquad dF = 0.$$

The first equation involving pure gravitational objects can be interpreted as above, while the second equation is nothing but the first Maxwell equation.

Thus, the equation $d\Omega = 0$ turns out to be a compact way to express several important conditions involving the classical fields. But it will be also an essential integrability condition in the quantum theory.

Secondly, just as an example os source, we consider an incoherent charged fluid described by a *velocity field* $\upsilon: \mathbf{E} \to J_1\mathbf{E} \subset \mathbb{T}^* \otimes T\mathbf{E}$, a *mass density* $\mu: \mathbf{E} \to \mathbb{M} \otimes \mathbb{L}^{*3}$ and a *charge density* $\rho: \mathbf{E} \to \mathbb{Q} \otimes \mathbb{L}^{*3}$; for the sake of simplicity, we assume $\rho = \varepsilon\,\mu$, with $\varepsilon \in \mathbb{Q} \otimes \mathbb{M}^*$. We assume the continuity equation $\mathrm{div}^{\natural}(\mu \otimes \upsilon) = 0$. Moreover, we can assume a dynamical equation for the charged fluid, in the considered gravitational and electromagnetic field, in agreement with the subsequent equation of motion of classical charged particles. Then, we define the *dynamical energy tensor* to be

$$\tau \equiv \tau^{\natural} + \tau^{e} \equiv \mathbf{k} \otimes \mu \otimes dt \otimes dt + \frac{1}{2}\rho \otimes (\langle \frac{\mathbf{q}}{m}, dt \rangle \otimes dt + dt \otimes \langle \frac{\mathbf{q}}{m}, dt \rangle),$$

where $\mathbf{k} \in \mathbb{M}^* \otimes \mathbb{L}^3 \otimes \mathbb{T}^2$ is the *gravitational coupling constant* .

Then, we postulate the *dynamical Einstein equation*

$$r = \tau,$$

where r is the Ricci tensor of the dynamical connection K. In virtue of the arbitrariness of the mass and charge, the equation $r = \tau$ can be split into its gravitational and electromagnetic components

$$r^{\natural} = \tau^{\natural} \qquad \mathrm{div}^{\natural} F = \rho \otimes dt.$$

The first equation is nothing but the pure gravitational Einstein equation and the second one the second Maxwell equation.

In particular, the vertical projections of the above equations say that: i) the vertical metric is Ricci flat and its scalar curvature vanishes (but this does not mean that the gravitational connection is Ricci flat); ii) the vertical divergence of the (observer independent) magnetic field vanishes.

The coordinate expression of the dynamical Einstein equation is

$$r^{\natural}{}_{00} = k \otimes \mu \qquad r^{\natural}{}_{0h} = r^{\natural}{}_{h0} = 0 \qquad r^{\natural}{}_{hk} = 0$$

$$g^{ij}\,\partial_i F_{j0} + K_i{}^h{}_j\, F_{h0} = \rho \qquad g^{ij}\,(\partial_i F_{jk} + K_i{}^h{}_j\, F_{hk}) = 0.$$

The standard integrability conditions related to the divergence operator turn out to be identities, because of the time-like character of the energy tensor.

We observe that, in the Galilei framework, the effects of the fluid velocity on the gravitational field and of the current on the magnetic field are lost. Because of the degeneracy of the metric, there is no way to write reasonable general relativistic equations involving our objects and accounting for the above effects. Moreover, we observe that we can define a contravariant mass and charge currents and energy tensors, but there is no way to couple them with the Ricci tensor and the electromagnetic field, in a reasonable way. Some authors describe the effects of the charge current on the magnetic field by considering also a contravariant electromagnetic field; but we think that it would be more reasonable to consider an additional 1-form, describing a "current of magnets", independent of the charge current. So, the essential differences between the classical electromagnetic theory in an Einstein or in a Galilei general relativistic framework consists in the replacement of the light cones with the vertical subspaces and the covariant charge current with two independent charge and magnet currents.

Now, let us consider the dynamics of classical charged particles in the given gravitational and electromagnetic field. The only observer independent approach to classical mechanics can be achieved in terms of the dynamical connection γ. Namely, given a particle, with mass $m \in \mathbf{M}$ and charge $\mathbf{q} \in \mathbf{Q}$, whose motion is $s: \mathbf{T} \to \mathbf{E}$, we postulate the equation

$$\nabla_{\gamma} s = 0,$$

which can be written as

$$\nabla_{\gamma^{\natural}} s = \gamma^{e} \circ j_1 s.$$

We observe that the above equation can be expressed by Hamiltonian and Lagrangian approaches, but these approaches depend essentially on the choice of a frame of reference, hence there is no way of achieving explicit general relativistic formulations for them.

The special relativistic Galilei classical fields and mechanics can be obtained as a particular case of the previous model.

We define the *special Galilei relativistic space-time* to be a general relativistic space-time with the further conditions that $\mathbf{E}$ is an affine space, $t: \mathbf{E} \to \mathbf{T}$ is an affine map, the vector bundle $\bar{t}: \bar{\mathbf{E}} \to \mathbf{T}$ associated with the affine bundle $t: \mathbf{E} \to \mathbf{T}$ is a product $\bar{\mathbf{E}} = \mathbf{T} \times \mathbf{S}$, where $\mathbf{S}$ is a three dimensional vector space, and the vertical metric is constant $g \in \mathbf{A} \otimes S^2 \mathbf{S}^*$. We observe that the hypothesis $\bar{\mathbf{E}} = \mathbf{T} \times \mathbf{S}$ is necessary in the special Galilei case (and not in the special Einstein case) in order to express the constancy of the degenerate metric g with respect to time. Thus, $\mathbf{S}$ is to be interpreted as the *absolute space* of space-like vectors; we observe that, even in the special Galilei case, an absolute space of space-like positions cannot be considered, because it would destroy the special relativistic character of the theory (space-like positions depend on the choice of an observer).

In the special relativistic Galilei case, the Einstein equations reduce exactly to the classical Newton law of gravitation (and the space-like components $\alpha^{\natural}{}_i$ of the gravitational potential $\alpha^{\natural}$ vanish).

The quantum theory

The quantum mechanical model is introduced in the following way.

The *quantum bundle* of a given particle of mass $m \in \mathbf{M}$ and charge $q \in \mathbf{Q}$ is defined to be a complex 1-dimensional bundle over the classical space-time

$$\pi : \mathbf{Q} \to \mathbf{E},$$

equipped with a Hermitean product

$$h : \mathbf{Q} \underset{\mathbf{E}}{\times} \mathbf{Q} \to \mathbb{C}.$$

The quantum histories are described by sections $\Psi : \mathbf{E} \to \mathbf{Q}$. We shall use normal complex charts for $\mathbf{Q}$ and write $\Psi = \psi\, b$, where $\psi : \mathbf{E} \to \mathbb{C}$ is the wave function and $b : \mathbf{E} \to \mathbf{Q}$ is a normal basis (a "gauge").

We observe that the quantum bundle lives on the space-time $\mathbf{E}$. However, $\mathbf{E}$ does not carry sufficient classical information; for instance, Γ, γ and Ω live on $J_1\mathbf{E}$. Therefore, we are led to consider the pull-back bundle $\mathbf{Q}^{\uparrow} \to J_1\mathbf{E}$ of the quantum bundle $\mathbf{Q} \to \mathbf{E}$ with respect to $J_1\mathbf{E} \to \mathbf{E}$. Here, this bundle $J_1\mathbf{E} \to \mathbf{E}$ has to be interpreted as the space of classical observers.

Our main assumption is the following. We postulate a Hermitean linear complex connection

$$\chi : \mathbf{Q}^{\uparrow} \to T^*J_1\mathbf{E} \otimes T\mathbf{Q}^{\uparrow},$$

which is "universal" (in the sense of [2]) and whose curvature is $i\,\frac{m}{\hbar}\,\Omega$, where $\hbar \in \mathbb{T}^{+*} \otimes \mathbf{A} \otimes \mathbf{M}$ is the Planck constant. The coordinate expression of such a connection turns out to be of the type

$$\chi_0 = -H/\hbar \qquad \chi_j = p_j/\hbar \qquad \chi^0_j = 0,$$

where $H \equiv m\,(\frac{1}{2} g_{hk}\, y^h_0\, y^k_0 + \alpha_0)$ and $p_j \equiv m\,(g_{jh}\, y^h_0 + \alpha_j)$ are the classical frame dependent Hamiltonian and momentum of the particle. We stress that, in the classical context, any potential α of the observer dependent form Φ is good; however, in the quantum context, for each chart, we have to choose just one local potential α of Φ, as χ must be a global object. Therefore, the above coordinate formula determines χ only up to the differential of an imaginary function of $\mathbf{E}$. We observe that the condition $\text{ч}^0_j = 0$ is just the coordinate expression of the universality of the connection (which is a frame independent condition). We stress that the requirement of universality is very important, because it solves the well known problem of the choice of polarizations, by yielding the only canonical polarization, namely the Schrödinger one.

Now, we want to derive all other quantum structures and equations from the above assumption. But the quantum connection lives on the frame dependent bundle $\mathbf{Q}^\uparrow$, while we require that the physically significant objects be frame independent. We shall solve this problem by means of a projectability criterion, which yields an effective heuristic procedure.

According to the above idea, let us start by finding a natural frame independent quantum 4-momentum associated with a quantum section Ψ. We observe that there is a natural way to define two fibred morphisms from $J_1\mathbf{E}$ to $\mathbb{T}^*\otimes T\mathbf{E}\otimes\mathbf{Q}$, namely

$$\mathring{p}_\Psi \equiv \text{д}\otimes\Psi : J_1\mathbf{E} \to \mathbb{T}^*\otimes T\mathbf{E}\otimes\mathbf{Q} \qquad\qquad \check{p}_\Psi \equiv \frac{\hbar}{m}\check{\nabla}^\#\Psi : J_1\mathbf{E} \to \mathbb{T}^*\otimes T\mathbf{E}\otimes\mathbf{Q},$$

where $\check{\nabla}^\#$ denotes the contravariant vertical quantum differential. Then, we can easily see that there is a unique coupling of these objects which projects onto a (observer independent) section over $\mathbf{E}$; namely, we obtain

$$p_\Psi \equiv \mathring{p}_\Psi - i\,\check{p}_\Psi : \mathbf{E} \to \mathbb{T}^*\otimes T\mathbf{E}\otimes\mathbf{Q},$$

with coordinate expression

$$p_\Psi = u^0\otimes\left(\psi\,\partial_0 - i\,\frac{\hbar}{m}g^{ij}\,\nabla^0{}_j\psi\,\partial_i\right)\otimes b,$$

where ∇^0 is the quantum covariant differential related to the observer associated with the chosen chart (hence, ∇^0 lives on $\mathbf{E}$).

Next, let us introduce the quantum dynamical equation, i.e. the generalized Schrödinger equation, which takes into account the given gravitational field in our curved space-time. We arrive to the same result by following the projectability criterion in two independent ways. Let us consider a given section $\Psi:\mathbf{E}\to\mathbf{Q}$.

First, we observe that there is a natural way to define two $\mathbb{T}^*$-valued functions on $J_1\mathbf{E}$, namely

$$\mathring{\mathcal{L}}_\Psi \equiv h(\Psi,\, i\,\mathring{\nabla}\Psi) + h(i\,\mathring{\nabla}\Psi,\, \Psi) : J_1\mathbf{E} \to \mathbb{T}^*$$

$$\check{\mathcal{L}}_\Psi \equiv \frac{\hbar}{m}(\bar{g}\otimes h)(\check{\nabla}\Psi,\, \check{\nabla}\Psi) : J_1\mathbf{E} \to \mathbb{T}^*.$$

Then, we can easily see that there is a unique coupling of these objects which projects onto a (observer independent) function on $\mathbf{E}$; namely, we obtain

$$\mathcal{L}_\Psi \equiv \mathring{\mathcal{L}}_\Psi - \check{\mathcal{L}}_\Psi : \mathbf{E} \to \mathbb{T}^*,$$

with coordinate expression

$$\mathcal{L}_\Psi = u^0 \Big(- \tfrac{1}{2} i \,(\partial_0\bar{\psi}\,\psi - \bar{\psi}\,\partial_0\psi) - \tfrac{\hbar}{2m} g^{ij}\,\partial_i\bar{\psi}\,\partial_j\psi +$$

$$+ \tfrac{i}{2}\, a^i\,(\partial_i\bar{\psi}\,\psi - \bar{\psi}\,\partial_i\psi) + \tfrac{m}{\hbar}\,(a_0 - \tfrac{1}{2} a^2)\,\psi\,\bar{\psi}\Big).$$

We assume the above function to be the *quantum Lagrangian* responsible of the quantum dynamical equation. So, we obtain a Euler-Lagrange equation, whose coordinate expression is

$$i\,\partial_0\psi + \tfrac{\hbar}{2m} g^{ij}\,\partial_{ij}\psi + \tfrac{m}{\hbar}\,(a_0 - \tfrac{1}{2} a^2)\,\psi - i\,a^i\,\partial_i\psi - i\,\tfrac{1}{2}\,\partial_i a^i\,\psi +$$

$$+ \tfrac{\hbar}{2m}\,\partial_i g^{ij}\,\partial_j\psi + i\,\tfrac{1}{2}\,\frac{\partial_0\sqrt{|g|}}{\sqrt{|g|}}\,\psi + \tfrac{1}{2}\,\frac{\partial_i\sqrt{|g|}}{\sqrt{|g|}}\,\frac{\hbar}{m}\, g^{ij}\,(\partial_j\psi - i\,\tfrac{m}{\hbar}\,a_j\,\psi) = 0.$$

We observe that the 4-momentum associated with the above Lagrangian coincides with the momentum $\mathfrak{p}_\Psi$, that we have found independently. The first line of the above equation is apparently equal to the standard Schrödinger equation, but we have to take into account that here a includes both the gravitational and electromagnetic potentials. On the other hand, in the special Galilei case, the above equation reduces exactly to the standard Schrödinger equation.

By considering the invariance of the Lagrangian under the action of the group U(1), the Nöther theorem yields the conserved current

$$\jmath_\Psi \equiv \tfrac{1}{2}\,\langle \upsilon,\, h(\Psi,\mathfrak{p}_\Psi) - h(\mathfrak{p}_\Psi,\Psi)\rangle : \mathbf{E} \to \mathbb{L}^3 \otimes \Lambda^3 T^*\mathbf{E},$$

with coordinate expression

$$\jmath_\Psi = \sqrt{|g|}\,\Big(\bar{\psi}\,\psi\, d^1\wedge d^2\wedge d^3 +$$

$$+ (-1)^h\,(- i\,\tfrac{\hbar}{2m} g^{hk}\,(\bar{\psi}\,\partial_k\psi - \partial_k\bar{\psi}\,\psi) - a^h\,\bar{\psi}\,\psi)\, d^0\wedge d^1 \ldots\wedge \hat{d}^h \ldots\wedge d^3\Big),$$

which will be interpreted as *probability current*.

There is a second geometrical approach to the Schrödinger equation. In fact, there is a natural way to define two fibred morphisms from $J_1\mathbf{E}$ to $\mathbb{T}^*\otimes\mathbf{Q}$, namely

$$\overset{\circ}{\nabla}\Psi \equiv \text{д} \lrcorner \nabla\Psi : J_1\mathbf{E} \to \mathbb{T}^* \otimes \mathbf{Q} \qquad \delta\, \mathbf{p}_\Psi \equiv \langle \bar{\upsilon}, d \langle \upsilon, \mathbf{p}_\Psi \rangle \rangle : J_1\mathbf{E} \to \mathbb{T}^* \otimes \mathbf{Q},$$

where $\bar{\upsilon}$ is the dual of the volume form υ. Then, we can easily see that there is a unique coupling of these objects which projects onto a (observer independent) section on $\mathbf{E}$; namely, we obtain

$$\overset{\circ}{\nabla}\Psi + \delta\, \mathbf{p}_\Psi : \mathbf{E} \to \mathbb{T}^* \otimes \mathbf{Q}.$$

Moreover, a comparison in coordinates shows that the generalized Schrödinger equation can be written as

$$\overset{\circ}{\nabla}\Psi + \delta\, \mathbf{p}_\Psi = 0.$$

The next step will be the introduction of the quantum operators.

For this purpose we need a few further additional results of classical mechanics.

The contact 2-form $\frac{m}{\hbar}\,\Omega$ maps each vector field of $J_1\mathbf{E}$ into a 1-form orthogonal to γ. We can easily see that, chosen any time scale function $\tau : J_1\mathbf{E} \to \mathbb{T}$, the 2-form $\frac{m}{\hbar}\,\Omega$ yields a linear isomorphism $\Omega^\flat_\tau : X \mapsto \frac{m}{\hbar}\, i_X\Omega$ between the space of vector fields of $J_1\mathbf{E}$, with time component τ, and the space of 1-forms orthogonal to γ. The inverse isomorphism is denoted by $\Omega^\#_\tau$.

Then, for each τ, we obtain the Hamiltonian lift

$$\varphi \mapsto \varphi^\#_\tau \equiv \Omega^\#_\tau (d\varphi - \gamma \lrcorner d\varphi),$$

which maps the functions of $J_1\mathbf{E}$ into vector fields of $J_1\mathbf{E}$. Its coordinate expression is

$$\varphi^\#_\tau = \tau\,(\partial_0 + y^i_0\, \partial_i + \gamma^i\, \partial^0_i) + \frac{\hbar}{m}\, g^{ij}\left(- \partial^0_j\varphi\, \partial_i + (\partial_j\varphi + (\Gamma_j{}^k - \Gamma^k{}_j)\, \partial^0_k\varphi)\, \partial^0_i\right).$$

By means of the Hamiltonian lift we can define the observer independent Poisson Lie bracket

$$\{\varphi', \varphi''\} \equiv \frac{m}{\hbar}\, \varphi'^\#_0 \lrcorner \varphi''^\#_0 \lrcorner \Omega.$$

On the other hand, we can prove that the vector field $\varphi^\#_\tau$ is projectable over $\mathbf{E}$ if and only if the function φ is, with respect to the fibres of $J_1\mathbf{E} \to \mathbf{E}$, a polynomial of degree 2, whose second derivative is of the form $\tau \frac{m}{\hbar} g : \mathbf{E} \to (\mathbb{T} \otimes \mathbb{T}) \otimes S^2 V^*\mathbf{E}$, where $\tau : \mathbf{E} \to \mathbb{T}$, i.e., if and only if the coordinate expression of φ is of the following type

$$\varphi = \tau \frac{m}{2\hbar} g_{ij}\, y^i_0\, y^j_0 + \varphi_i\, y^i_0 + \varphi_0 \qquad\qquad \tau, \varphi_0, \varphi_i : \mathbf{E} \to \mathbf{R}.$$

So, we define the *quantizable functions* to be the functions of the above type, for a reason which will be clear soon. The space-time coordinates, the momentum and the Hamiltonian associated with a given frame of reference are quantizable functions. We can prove that the quantizable functions constitute a Lie algebra with respect to the bracket

$$[\varphi', \varphi''] \equiv \{\varphi', \varphi''\} + \gamma(\tau').\varphi'' - \gamma(\tau'').\varphi'.$$

Now, we are ready to study the quantum operators.

We start by considering the vector fields $X^\uparrow : \mathbf{Q}^\uparrow \to T\mathbf{Q}^\uparrow$, which preserve the basic quantum structures, namely which project over a vector field $X^\uparrow$ of $J_1\mathbf{E}$, are Hermitean complex linear over $X^\uparrow$ and fulfill the condition $d_ч X^\uparrow = \gamma \lrcorner d_ч X^\uparrow$, where $d_ч X^\uparrow$ is the covariant exterior differential of $X^\uparrow$ defined in [2] (by means of the Frölicher-Nijenhuis algebra). We can prove that, chosen a time scale τ, there is a bijection between the functions $\varphi : J_1\mathbf{E} \to \mathbf{R}$ and the above vector fields, given by the formula

$$\varphi \mapsto X^\uparrow{}_{\varphi,\tau} \equiv ч(\varphi^{\#}_{\tau}) + i\,\varphi$$

(here i stands for the imaginary vertical Liouville vector field of $\mathbf{Q}^\uparrow$).

This formula resembles a well known idea of geometrical quantization and is nice, but we have got two problems. In fact, as the quantum connection lives on $\mathbf{Q}^\uparrow$, we have looked for distinguished vector fields on $\mathbf{Q}^\uparrow$. But, if we apply them to sections of the quantum bundle, then they map sections of $\mathbf{Q}$ into sections of $\mathbf{Q}^\uparrow$. This problem would be solved if, additionally, the above vector fields are projectable over $\mathbf{E}$. Moreover, the above isomorphism depends on the choice of a time scale and we need a reasonable criterion to make this choice.

Luckily, the two problems can be solved together. In fact, we can prove that the vector field $X^\uparrow{}_{\varphi,\tau}$ of $\mathbf{Q}^\uparrow$ is projectable over a vector field X_φ of $\mathbf{Q}$ if and only if φ is a quantizable function and the time scale τ is just the one appearing in φ. So, we define the *quantum vector fields* to be the above vector fields X_φ of $\mathbf{Q}$, associated with quantizable functions φ. The coordinate expression of X_φ is

$$X_\varphi = \tau\, \partial_0 - \frac{\hbar}{m} \varphi^i\, \partial_i + i\,(\varphi_0 + \tau \frac{m}{\hbar} a_0 - \varphi_i\, a^i).$$

The map $\varphi \mapsto X_\varphi$ is a Lie algebra isomorphism; thus, we have

$$[X_{\varphi'}, X_{\varphi''}] = X_{[\varphi', \varphi'']}.$$

The quantum vector fields act naturally on the sections of $\mathbf{Q}$ by means of a Lie derivative. Then, we define the *quantum Lie operators* to be the maps of the type

$$\Xi_{\varphi} : \Psi \mapsto i\,[X_{\varphi}, \Psi],$$

where the section $\Psi: \mathbf{E} \to \mathbf{Q}$ is regarded as a vertical vector field $\Psi: \mathbf{Q} \to V\mathbf{Q}$. Moreover, we obtain the following commutator

$$[\Xi_{\varphi'}, \Xi_{\varphi''}] = i\,\Xi_{[\varphi', \varphi'']}.$$

As a particular case, we obtain the following operators

$$\Xi_{x^0} = x^0 \qquad \Xi_{y^i} = y^i \qquad \Xi_{p_i/\hbar} = -\,i\,\partial_i \qquad \Xi_{H/\hbar} = i\,\partial_0$$

and commutators

$$[\Xi_{x^0}, \Xi_{\varphi}] = -\,i\,\tau \qquad [\Xi_{y^i}, \Xi_{\varphi}] = i\,\frac{\hbar}{m}\,\varphi^i \qquad [\Xi_{p_i/\hbar}, \Xi_{p_j/\hbar}] = [\Xi_{H/\hbar}, \Xi_{p_i/\hbar}] = 0$$

hence, in particular,

$$[\Xi_{y^i}, \Xi_{p_j/\hbar}] = i\,\delta^i{}_j \qquad [\Xi_{y^i}, \Xi_{H/\hbar}] = 0.$$

The above formulas generalize the standard quantum operators to our curved framework.

To conclude our scheme, we just briefly sketch how to introduce the Hilbert stuff.

Of course, we want to preserve the general relativistic character of the theory, which is based on the fact the the classical space-time cannot be split into a product in a distinguished way. Then, according to the standard procedures we can construct, for each time $\tau \in \mathbf{T}$, the space $\mathbf{S}_{\tau} \equiv \{\Psi_{\tau}\}$ consisting of the smooth sections $\Psi_{\tau}: \mathbf{E}_{\tau} \to \mathbf{Q}_{\tau}$. In this way, we obtain the (infinite dimensional) fibred set

$$\mathbf{S} \equiv \coprod_{\tau \in \mathbf{T}} \mathbf{S}_{\tau} \to \mathbf{T},$$

whose fibres are vector spaces. Thus, each section $\Psi: \mathbf{E} \to \mathbf{Q}$ can be regarded as a section $\hat{\Psi}: \mathbf{T} \to \mathbf{S}$, and conversely. We do not need to consider any explicit differential structure on $\mathbf{S}$; in fact, the construction of $\mathbf{S}$ through smooth fibre-wise sections of $\mathbf{Q}$ allows us to define directly the smoothness of sections $\hat{\Psi}: \mathbf{T} \to \mathbf{S}$, the tangent space $T\mathbf{S}$ and so on.

Again, the space-time structures do not provide any distinguished splitting of the bundle $\mathbf{S} \to \mathbf{T}$. On the other hand, we can exhibit a distinguished connection $\mathfrak{J}$ on the bundle $\mathbf{S} \to \mathbf{T}$. In fact, we can prove that the Schrödinger equation can be regarded as the vanishing of a covariant derivative

$$\nabla_{\mathfrak{J}} \hat{\Psi} = 0,$$

hence it defines a connection $\mathfrak{J}$, which will be called *Schrödinger connection*.

Then, starting from $\mathbf{S} \to \mathbf{T}$, we can build a Hilbert bundle $\mathbf{H} \to \mathbf{T}$, by means of the standard procedures applied, fibre by fibre, to the double fibred manifold $\mathbf{Q} \to \mathbf{E} \to \mathbf{T}$. Moreover, the above Schrödinger connection $\mathfrak{J}$ is sufficient to adapt the standard results concerning the standard quantum Hilbert space to our bundle framework.

Our research will continue with the aims of extending the above approach to particles with spinand to interaction of particles, and of replacing the Galilei with the Einstein general relativistic framework.

Acknowledgements . This work has been supported by GNFM of Consiglio Nazionale delle Ricerche (Italy), by national and local funds of MURST (Italy) and by Committee of Scientific Research (Poland).

One of the two authors (M. Modugno) would like to thank the organizers of the XXI Meeting on Differential Geometric Methods in Theoretical Physics for their very kind hospitality; a special gratitude is due to Prof. Hu Hesheng.

References

[1] A. JADCZYK, M. MODUGNO: *A new geometrical approach to general relativistic Galilei quantum mechanic*, book in preparation.

[2] M. MODUGNO: *Systems of vector valued forms on a fibred manifold and applications to gauge theories*, Lect. Notes in Math. **1251**, Springer-Verlag, 1987.

DIFFERENTIAL EQUATIONS ASSOCIATED TO MANIFOLDS OF CONSTANT SECTIONAL CURVATURE

KETI TENENBLAT [1]

Departamento de Matemática, Universidade de Brasília
70910 Brasília, D.F. Brazil

ABSTRACT

Backlund transformations and superposition formulae are given for a class of partial differential equations for matrix valued functions of n independent variables. These equations are geometric generalizations of the homogeneous wave equation, the sine-Gordon equation, the Laplace equation and the elliptic sinh-Gordon equation.

We consider the system of differential equations for a matrix valued function $a : \Omega \subset \mathbf{R}^n \longrightarrow O(p, q)$ where $p + q = n$ and $p \geq 1$ given by

$$\begin{aligned} aJa^t &= J, \\ \frac{\partial a_{li}}{\partial x_j} &= a_{lj}h_{ji} \qquad i \neq j, \quad h_{ii} = 0, \\ \frac{\partial h_{ij}}{\partial x_i} &+ \frac{\partial h_{ji}}{\partial x_j} + \sum_{s\neq i, s\neq j} h_{si}h_{sj} = -Kv_iv_j, \qquad i \neq j, \\ \frac{\partial h_{ij}}{\partial x_s} &= h_{is}h_{sj} \qquad i, j, s \text{ distinct}, \end{aligned} \tag{1}$$

where $J = \operatorname{diag}\,(\overbrace{1, \ldots, 1}^{p\ times}, \overbrace{-1, \ldots, -1}^{q\ times})$ and the off-diagonal matrix fuunction $h = (h_{ij})$ is defined in the second equation.

For $p = n$ this is called a *Generalized Wave Equation* when $K = 0$ and a *Generalized Sine-Gordon Equation* when $K \neq 0$. This is due to the fact that when $n = 2$, the system of equations (1) reduces to the classical equation $u_{x_1x_1} - u_{x_2x_2} = -K \sin u$. Similarly, for $1 \leq p < n$ the above system is called a *Generalized Laplace Equation* when $K = 0$ and a *Generalized Elliptic Sinh-Gordon Equation* when $K \neq 0$. In this case, when $n = 2$, the system (1) reduces to the classical equation $u_{x_1x_1} + u_{x_2x_2} = -K\sinh u$.

[1] Partially supported by CAPES and CNPq, Brasil.

We will describe the geometrical motivation to consider the above generalization. Let $M^n(K)$ be an n-dimensional manifold of constant sectional curvature K isometrically immersed in a space form $\bar{M}^{2n-1}(\bar{K})$ of curvature $\bar{K}$ such that $K \neq \bar{K}$. We can normalize so that $|K - \bar{K}| = 1$. Suppose that $M^n(K)$ is free from weak umbilic points when $K > \bar{K}$. Then there exist local coordinates $(x_1, \ldots, x_n)$ and an orthonormal frame e_{n+j-1}, $2 \leq j \leq n$, normal to M such that the first and second fundamental forms are given by[6,8,9]

$$I = \sum_{i=1}^{n} a_{1i}^2 \, dx_i^2 \qquad II = \sum_{i=1,j=2}^{n} a_{ji} a_{1i} \, dx_i^2 e_{n+j-1}. \tag{2}$$

Moreover, one can show that the $n \times n$ matrix function $a = (a_{ij})$ satisfies the system of equations (1) and $p = n$ if $K < \bar{K}$ and $p = 1$ if $K > \bar{K}$.

Conversely, given a matrix function a, which satisfies the equations (1), with $p = n$ or $p = 1$, defined on a simply connected open subset $\Omega \subset \mathbf{R}^n$, such that a_{1i} do not vanish, there exists an isometric immersion $X : \Omega \subset \mathbf{R}^n \longrightarrow \bar{M}^{2n-1}(\bar{K})$, $|\bar{K} - K| = 1$, whose first and second fundamental forms are given by (2).[2,4,10,12,14]

The above equations are equivalent to the Gauss and Codazzi equations, when $p = 1$ and $p = n$. We observe that the matrix function h reduces the system of second order differential equations to a first order one.

Solutions for the system of differential equations (1) can be obtained by Backlund transformations, superposition formulae and also by the Inverse Scattering Method[1] for $p = n$. In order to state the Backlund transformation we rewrite (1) in matrix notation as

$$\begin{aligned} aJa^t &= J \\ da \wedge E &= aE \wedge C \\ dC &= C \wedge C - \frac{1}{2} KEJ \wedge QE \end{aligned} \tag{3}$$

where

$$E = \text{diag}\,(dx_1, \ldots, dx_n) \quad C = hE - Eh^t \quad Q = 2Jv^t v - I \quad v = (a_{11}, \ldots, a_{1n}) \tag{4}$$

and I denotes the $n \times n$ identity matrix.

Theorem 1. *Let $a: \Omega \subset \mathbf{R}^n \to O(p,q)$, be a solution of (3), where Ω is simply connected. For each z ($\neq 0$ if $K \neq 0$) and a given initial condition $X_0 = X(x_0)$, $x_0 \in \Omega$, there exists a unique solution X for the system*

$$dX + XJ^{-\frac{1}{2}} A_z^t JX = A_z J^{\frac{1}{2}} - XJ^{-\frac{1}{2}} CJ^{\frac{1}{2}}, \tag{BT(z)}$$

where

$$A_z = JM_z aEJ^{-\frac{1}{2}} \qquad M_z = \frac{1}{2}(zI - Kz^{-1}\tilde{I}) \qquad \tilde{I} = \text{diag}(1, -1, \ldots, -1), \tag{5}$$

Moreover, if $X_0 \in O(p,q)$, then $X(x) \in O(p,q)$ and it satisfies (3).

Theorem 1 provides a method of obtaining a new solution X for (3) from a given solution a, by solving BT(z). We say that X is associated to a by BT(z). We observe that one can always start with a trivial solution, $a = I$ and $h = 0$.

The superposition formula provides an algebraic procedure of obtaining more solutions for the system (3) and it is given in the following result.

Theorem 2. *Let a be a solution of (3). If a_i, $i = 1, 2$, are solutions associated to a by $BT(z_i)$, $z_1 \neq z_2$, then there exists a solution a^* associated to a_1 and a_2 by $BT(z_2)$ and $BT(z_1)$ respectively. Moreover, a^* is algebraically given by*

$$a^* J a^t = J(a_1 J a_2^t M_{z_1} - J M_{z_2})^{-1}(J M_{z_1} - a_1 J a_2^t M_{z_2}).$$

The above theorems generalize the results initially obtained [12,13] for $p = n$ and $K = -1$. Later[10], the case $p = n$ and $K = 0$ was given. More recently, the results for $p = 1$ were obtained[5]. Moreover, we observe that whenever $n = 2$ and $K \neq 0$, the above theorems reduce to the classical Backlund transformations and superposition formulae for the sine-Gordon ($p = 2$) or elliptic sinh-Gordon ($p = 1$) equations[4].

Whenever $p \neq n$ the transformation (BT(z)) may provide a complex solution for (3). Using conviniently the superposition formula one gets a real solution from a given real one. More precisely:

Corollary *Let a be a real solution of (3) and let a_1 be associated to a by $BT(z)$, $z \in \mathbf{C}$. Then a^* defined by*

$$a^* J a^t = (a_1 \overline{a_1}^t M_z - J\overline{M_z})^{-1}(J M_z - a_1 \overline{a_1}^t \overline{M_z}).$$

is another real solution of (3).

Another generalization of the classical equations is given by the following system for a vector valued function $v : \Omega \subset \mathbf{R}^n \longrightarrow \mathbf{R}^n$ and an off diagonal $n \times n$ matrix function h defined on Ω.

$$\begin{aligned} vJv^t &= 1 \\ dv &= -vBJ \\ dC &= C \wedge C - \frac{1}{2} KEJ \wedge QE \\ dB &= BJB \end{aligned} \tag{6}$$

where h, C, Q, J are defined as before and $B = Jh^t E - EhJ$.

This system of equations can be considered as a subset of (3). In fact there is a correspondence[2,3,14] between solutions of (3) and those of (6). As in the previous case, when $n = 2$, the system (6) reduces to the sine-Gordon or the wave equations when $p = 2$ and it reduces to the Laplace or Elliptic Sinh-Gordon equations when $p = 1$.

The name Intrinsic Generalized Equation given to this system, is due to the fact that each solution v, whose coordinate functions v_i do not vanish, defines a metric of constant sectional curvature K on Ω. In this direction, there is a theory[7] of differential equations, for a function of two independent variables, which describe pseudo-spherical surfaces.

One can show that there is a Backlund Transformation for (6), which generalizes the results obtained previously[3,5] for $p = n$ and $p = 1$. Examples of solutions (1-solitons) for (6) obtained by such a transformation for $p = n = 3$ were given in[11]. Moreover, it was shown[3] that the Inverse Scattering Method can be applied when $p = n$.

Complete proofs of the theorems announced in this note and further results will appear in a joint article with P.T. Campos.

References

1. M. Ablowitz, R. Beals, K.Tenenblat, *Stud. Appl. Math.* **74** (1986), 177-203.
2. J. A. Aminov, *Soviet Math. Dokl.* **18** (1977), 1210-1213.
3. R. Beals and K. Tenenblat, *Differential Geometry,* ed. B. Lawson and K. Tenenblat, Pitman Monographs and Surveys in Pure and Appl. Math. **52** 1991, 25-46.
4. L. Bianchi, *Lezioni di Geometria Differenziale*, N. Zanichelli Ed., 1927.
5. P. T. Campos, *PhD. thesis*, Univ. de Brasilia, 1992.
6. E. Cartan, *Bull. Soc. Math. France* **47** (1919), 125-160; **48** (1920), 132-208.
7. S.S. Chern and K. Tenenblat, *Stud. Appl. Math.* **74** (1986), 55-83.
8. J. D. Moore, *Pac. J. of Math.* **40** (1972), 157-166.
9. ——, *Duke Math. J.* **44** (1977), 449-484.
10. K. Tenenblat, *Bol.Soc. Bras. Mat.* **16** (1985), 69-94.
11. K. Tenenblat, *J. Math. Anal. and Appl.* **166** (1992), 288-301.
12. K. Tenenblat and C. L. Terng, *Ann.of Math.* **111** (1980), 477–490.
13. C. L. Terng, *Ann.of Math.* **111** (1980), 491–510.
14. R. Tojeiro, *PhD. thesis*, IMPA, 1991.

THE DECOMPOSITIONS OF ZERO–CURVATURE EQUATIONS VIA HIGHER–ORDER CONSTRAINTS *

YUNBO ZENG
Department of Mathematics
University of Science and Technology of China, Hefei 230026, China

ABSTRACT

The general scheme and uniform formulas for the higher–order decompositions of each equation in a hierarchy of zero–curvature equations associated with loop algebra $\widetilde{sl}(2)$ are presented.

Consider the isospectral problem[1,2]

$$\begin{bmatrix}\varphi_1\\ \varphi_2\end{bmatrix}_x = U(u,\lambda)\begin{bmatrix}\varphi_1\\ \varphi_2\end{bmatrix} \equiv \sum_{i=0}^{s} u_i e_i \lambda^{\varepsilon_i}, \quad e_i = \begin{bmatrix} e_{i1} & e_{i2}\\ e_{i3} & -e_{i1}\end{bmatrix} \tag{1}$$

where $u_0 = 1$, $u=(u_1,\cdots,u_s)^T$, $e_i \in sl(2)$, $\varepsilon_i \in Z$, $\varepsilon_0 > \varepsilon_i \geq 0$ and $\min\{\varepsilon_1,\cdots,\varepsilon_s\} = 0$. Let

$$V = \begin{bmatrix} a & b\\ c & -a\end{bmatrix} = \sum_{m=0}^{\infty} V_m \lambda^{-m}, \quad V_m(u) = \begin{bmatrix} a_m & b_m\\ c_m & -a_m\end{bmatrix}$$

satisfy

$$V_x = [U,V] \equiv UV - VU. \tag{2}$$

There exist $\Delta_n \in \widetilde{sl}(2)$ so that[2] the t_n evolution of $(\varphi_1,\varphi_2)^T$ is given by

$$\begin{bmatrix}\varphi_1\\ \varphi_2\end{bmatrix}_{t_n} = V^{(n)}(u,\lambda)\begin{bmatrix}\varphi_1\\ \varphi_2\end{bmatrix} = \left[\sum_{m=0}^{n} V_m \lambda^{n-m} + \Delta_n(u,\lambda)\right]\begin{bmatrix}\varphi_1\\ \varphi_2\end{bmatrix}. \tag{3}$$

The zero–curvature equations $U_{t_n} - V_x^{(n)} + [U,V^{(n)}] = 0$ can be cast in Hamiltonian form[2]

$$u_{t_n} = J\frac{\delta H_{n+n_2}}{\delta u}, \quad n = 1,2,\cdots, \tag{4}$$

$$\frac{\delta H_n}{\delta u} = L\frac{\delta H_{n-1}}{\delta u} = L^{n-n_0}\frac{\delta H_{n_0}}{\delta u}, \quad \frac{\delta H_{n_0}}{\delta u} \neq 0, \quad \frac{\delta H_k}{\delta u} = 0 \ \ for\ k < n_0. \tag{5}$$

V also satifies[1]

$$V_{t_n} = [V^{(n)},V], \quad n = 1,2,\cdots. \tag{6}$$

For N distinct λ_j, $j = 1,\cdots,N$, Eq.(1) and (3) give

$$\begin{bmatrix}\varphi_{1j}\\ \varphi_{2j}\end{bmatrix}_x = \sum_{i=0}^{s} u_i e_i \lambda_j^{\varepsilon_i}\begin{bmatrix}\varphi_{1j}\\ \varphi_{2j}\end{bmatrix}, \quad j = 1,\cdots,N, \tag{7}$$

* Work supported by the Chinese National Basic Research project "Nonlinear Science".

$$\begin{bmatrix}\varphi_{1j}\\ \varphi_{2j}\end{bmatrix}_{t_n}=\left[\sum_{m=0}^{n}V_m(u)\lambda_j^{n-m}+\Delta_n(u,\lambda_j)\right]\begin{bmatrix}\varphi_{1j}\\ \varphi_{2j}\end{bmatrix},\quad j=1,\cdots,N. \tag{8}$$

For an arbitrary $k_0>n_0$, the following equation[3-5]

$$\frac{\delta H_{k_0}}{\delta u_i}=\sum_{j=1}^{N}\frac{\delta\lambda_j}{\delta u_i}=e_{i1}<\Lambda^{s_i}\Phi_1,\Phi_2>$$
$$+\frac{1}{2}e_{i2}<\Lambda^{s_i}\Phi_2,\Phi_2>-\frac{1}{2}e_{i3}<\Lambda^{s_i}\Phi_1,\Phi_1>,\qquad i=1,\cdots,s, \tag{9}$$

where $\Lambda=\text{diag}(\lambda_1,\cdots,\lambda_N)$, $\Phi_i=(\varphi_{i1},\cdots,\varphi_{iN})^T$, $<\cdot,\cdot>$ is the inner product in R^N,determines a finite–dimensional invariant set of the flows (4). This property guarantees that Eq.(4), (7),(8) and (9) are consistent, and enables us to factor the x–and t_n–dependence of (4) by two commuting and integrable finite–dimensional Hamiltonian systems (FDHS) constructed from Eq.(7) ,(8) and (9). This provides a method of seperation of variables to solve Eq.(4). For the explicit constraint, the systems (7) and (8) were first considered by Cao[6]. For the higher–order constraint (9), which cannot give an explicit expression of u in terms of Φ_i, if $H_{k_0}=H_{k_0}(u,u_x,\cdots,u^{(k_1)})$, where $u^{(k)}=\frac{\partial^k u}{\partial x^k}$, is regular[7], then Jacobi–Ostrogradsky coordinates can be introduced as follows[7]

$$q_{js+i}=u_i^{(j)},\qquad i=1,\dots,s,\qquad j=0,1,\dots k_1-1,$$
$$p_{js+i}=\frac{\delta H_{k_0}}{\delta u_i^{(j+1)}}=\sum_{l\geq 0}(-\partial)^l\frac{\partial H_{k_0}}{\partial u_i^{(j+1+l)}}. \tag{10}$$

Set $Q=(\varphi_{11},\dots,\varphi_{1N},q_1,\dots,q_{k_1s})^T,P=(\varphi_{21},\dots,\varphi_{2N},p_1,\dots,p_{k_1s})^T$, then the system consisting of Eq.(7) and (9) can be written as a FDHS (take $q_0=1$)

$$Q_x=\frac{\partial\hat{H}_0}{\partial P},\quad P_x=-\frac{\partial\hat{H}_0}{\partial Q}, \tag{11}$$

$$\hat{H}_0=\sum_{l=0}^{k_1s}q_{lx}p_l-H_{k_0}$$
$$+\sum_{i=0}^{s}q_i(e_{i1}\langle\Lambda^{s_i}\Phi_1,\Phi_2\rangle+\frac{1}{2}e_{i2}\langle\Lambda^{s_i}\Phi_2,\Phi_2\rangle-\frac{1}{2}e_{i3}\langle\Lambda^{s_i}\Phi_1,\Phi_1\rangle)$$

To show the integrability of Eq.(11), for the moment, we assume that λ_j are discrete eigenvalues, u and its derivatives and $(\varphi_{1j},\varphi_{2j})^T$tend to zero as $|x|\to\infty$, (called as (A)), and

$$L\frac{\delta\lambda_j}{\delta u}=\lambda_j\frac{\delta\lambda_j}{\delta u}. \tag{12}$$

We denote the expressions of V_m obtained from (2) under (11) and (12) by $\tilde{V}_m$. We can obtain[4,5]

$$\tilde{V}_m=\tilde{V}_m(P,Q)\qquad m<k_0, \tag{13a}$$

$$\tilde{V}_m = \frac{1}{2}\begin{bmatrix} \langle\Lambda^{m-k_0}\Phi_1,\Phi_2\rangle & -\langle\Lambda^{m-k_0}\Phi_1,\Phi_1\rangle \\ \langle\Lambda^{m-k_0}\Phi_2,\Phi_2\rangle & -\langle\Lambda^{m-k_0}\Phi_1,\Phi_2\rangle \end{bmatrix}, \quad m \geqslant k_0. \tag{13b}$$

Eq.(2) and (6) give rise to

$$\frac{d}{dx}(a^2+bc) = \frac{1}{2}Tr(V^2)_x = \frac{1}{2}Tr[U,V^2] = 0, \quad a^2+bc = \tilde{r}_0\lambda^{-n_1}, \tag{14}$$

$$\frac{d}{dt_n}(a^2+bc) = \frac{1}{2}Tr(V^2)_{t_n} = \frac{1}{2}T_r[V^{(n)},V^2] = 0. \tag{15}$$

We obtain from (13) and (14) that

$$F = (\tilde{a}^2+\tilde{b}\tilde{c}) = \sum_{m=0}^{\infty} F_m\lambda^{-m}, \qquad F_m = \sum_{l=0}^{m}(\tilde{a}_l\tilde{a}_{m-l}+\tilde{b}_l\tilde{c}_{m-l}), \tag{16a}$$

$$F_m = G_{m,m-k_0} + \sum_{l=m-k_0+1}^{k_0-1}[\tilde{a}_l\tilde{a}_{m-l}+\tilde{b}_l\tilde{c}_{m-l}], \qquad k_0 \leqslant m \leqslant 2k_0-1, \tag{16b}$$

$$F_m = G_{m,k_0-1} + \frac{1}{4}\sum_{l=0}^{m-2k_0}[\langle\Lambda^l\Phi_1,\Phi_2\rangle\langle\Lambda^{m-2k_0-l}\Phi_1,\Phi_2\rangle$$
$$-\langle\Lambda^l\Phi_1,\Phi_1\rangle\langle\Lambda^{m-2k_0-l}\Phi_2,\Phi_2\rangle], \qquad m \geqslant 2k_0-1, \tag{16c}$$

$$G_{m,k} = \sum_{l=0}^{k}[\tilde{a}_l\langle\Lambda^{m-k_0-l}\Phi_1,\Phi_2\rangle + \frac{1}{2}\tilde{b}_l\langle\Lambda^{m-k_0-l}\Phi_2,\Phi_2\rangle$$
$$-\frac{1}{2}\tilde{c}_l\langle\Lambda^{m-k_0-l}\Phi_1,\Phi_1\rangle],$$

$$F_{n_1} = \tilde{r}_0, \qquad F_m = 0 \textit{ for } m \neq n_1 \textit{ and } m < max\{k_0,n_1\}. \tag{16d}$$

Eq.(14) implies that if (P,Q) satisfies Eq.(11), then

$$\frac{dF_m(P,Q)}{dx} = 0, \ m = 0,1,.... \tag{17}$$

From Eq.(4), (10) and (11), it can be found that[5]

$$q_{l,t_n} = \tilde{f}_{l,n}(P,Q), \quad p_{l,tn} = \tilde{g}_{l,n}(P,Q), \quad l = 1,...,k_1 s, \tag{18}$$

For specific (1), it is easy to verify that the t_n–evolution equation of P and Q consisting of Eq.(8) and (18) can be written as a FDHS

$$Q_{t_n} = \frac{\partial F_{n+k_0}}{\partial P}, \qquad P_{t_n} = -\frac{\partial F_{n+k_0}}{\partial Q}, \qquad n = 1,2,.... \tag{19}$$

Eq.(15) shows that if (P,Q) satisfies Eq.(19), then

$$\frac{dF_m(P,Q)}{dt_n} = 0, \quad m,n = 1,2,..., \tag{20}$$

which together with Eq.(17) and (19) leads to

$$\{F_m, F_n\} = 0, \quad \{F_m, H\} = 0, \quad m, \ n = 1, \ 2, \ ..., \tag{21}$$

where { • , • } stands for the ordinary Poisson bracket.

From now on, we remove (A), it is obvious that Eq.(17) and (21) remain hold due to the algebraic structure of F_m and Eq.(11). After having removed (A), we have instead of Eq.(12)

$$L\frac{\delta\lambda_j}{\delta u}=\lambda_j\frac{\delta\lambda_j}{\delta u}+\tilde{E}_j\frac{\delta H_{n_0}}{\delta u}. \tag{22}$$

Denoting the expressions of V_m under Eq.(11) and (22) by $\overline{V}_m$, we can obtain

$$\overline{V}_m=\sum_{l=0}^{m}E_l\tilde{V}_{m-l},\qquad E_0=1,\qquad E_1=\ldots=E_{k_0-n_0}=0, \tag{23}$$

where E_l are x−independent. Substituting Eq. (23) into (14) and using (16a) gives rise to[4,5]

$$E_m=\sum_{j=1}^{m}\beta_j\gamma_0^j\sum_{m_1+\ldots+m_j=m}F_{m_1+n_1}\ldots F_{m_j+n_1}, \tag{24}$$

where $m_l\geqslant 1$, $\gamma_0=-\frac{1}{2F_{n_1}}$, $\beta_0=\beta_1=1$, $\beta_2=\frac{3}{2}$,

$$\beta_j=\sum_{l=1}^{j-2}\beta_l\beta_{j-1-l}+2\beta_{j-1}-\frac{1}{2}\sum_{l=1}^{j-1}\beta_l\beta_{j-l},\qquad j\geqslant 3. \tag{25}$$

Notice Eq.(19), using Eq(8), (18) and (23), we find

$$Q_{t_n}=\sum_{l=0}^{n}E_l\frac{\partial F_{n+k_0-l}}{\partial P},\qquad P_{t_n}=-\sum_{l=0}^{n}E_l\frac{\partial F_{n+k_0-l}}{\partial Q},\qquad n=1,2,\ldots, \tag{26}$$

which by means of (24) can be reduced to a FDHS

$$Q_{t_n}=\frac{\partial\hat{H}_n}{\partial P},\qquad P_{t_n}=-\frac{\partial\hat{H}_n}{\partial Q},\qquad n=1,2,\ldots, \tag{27}$$

$$\hat{H}_n=\sum_{j=0}^{n+k_0-n_1-1}\frac{\beta_j\gamma_0^j}{j+1}\sum_{m_1+\ldots+m_{j+1}=n+k_0-n_1}F_{m_1+n_1}\ldots F_{m_{j+1}+n_1}.$$

Theorem 1. If Eq.(19) holds and grad F_m, $m=k_0+\varepsilon_0,\ldots,k_0+\varepsilon_0+N+k_1s-1$ are linearly independent, then the FDHS's (11) and (27) (n=1,2,...) are completely integrable and commute with each other. For a fixed n, if(P,Q) satisfies Eq.(11) and (27) simultaneously, then $u=(q_1,\ldots,q_s)^T$salves Eq.(4).

Theorem 2. If (P,Q) is a solution of Eq.(11), then $u=(q_1,\ldots,q_s)^T$ satisfies

$$\sum_{m=0}^{N+k_0}d_mJ\frac{\delta H_m}{\delta u}=0, \tag{28}$$

where d_m are x−independent.

References

[1] A.C. Newell, Soliton in Mathematics and Physics, (SIAM, Philadelphia, 1985).
[2] Guizhang Tu, J. Phys. A, 22 (1989) 2375.
[3] Yunbo Zeng and Yishen Li, J. Math. Phys., 30 (1989) 1679.
[4] Yun bo Zeng and Yishen Li, J. Phys. A, 23(1990) L89.
[5] Yun bo Zeng, Phys. Lett. A, 160(1991) 541.
[6] Cewen Cao, Chinese Quartenly J. of Math., 3(1988) 90.
[7] M. Deleon and P.R. Rodrigues, Generalized Classical Mechanics and Field Theory, (North−Holland, Amsterdam, 1985).

A Method to Obtain Explicit Solutions of 1+2 Dimensional AKNS System[1]

Zhou Zixiang

Institute of Mathematics, Fudan University, Shanghai 200433, China

1.Introduction.

The 1+2 dimensional AKNS system is a generalization of usual AKNS system in 1+1 dimensions. It includes some of the important equations such as Davey-Stewartson (DS) equation, N-wave equation etc. It takes the form

$$\Psi_y = J\Psi_x + U\Psi, \qquad \Psi_t = \sum_{j=0}^{n} V_j \partial^{n-j}\Psi \tag{1}$$

where $\partial = \partial/\partial x$, J is a constant diagonal matrix with mutually different diagonal entries, $U(x, y, t)$ is off-diagonal (i.e. all the diagonal entries of U are zero), V_j's are $r \times r$ matrices.

The integrability conditions of (1) are

$$[J, V_{j+1}^A] = V_{j,y}^A - JV_{j,x}^A - [U, V_j]^A + \sum_{k=0}^{j-1} C_{n-k}^{n-j}(V_k \partial^{j-k} U)^A \tag{2}$$

$$V_{j,y}^D - JV_{j,x}^D = [U, V_j^A]^D - \sum_{k=0}^{j-1} C_{n-k}^{n-j}(V_k \partial^{j-k} U)^D \tag{3}$$

$$U_t = V_{n,y}^A - JV_{n,x}^A - [U, V_n]^A + \sum_{k=0}^{n-1}(V_k \partial^{n-k} U)^A \tag{4}$$

where the superscript D and A refer to the diagonal and off-diagonal parts of a matrix.

We look (3) and (4) as a system of partial differential equations of unknowns (U, V_j^D) $(j = 1, \cdots, n)$, in which V_j^A's are defined inductively by (2). So (1) is the Lax pair of (3) and (4).

The Lax pair (1) has been studied by various methods. In this paper, we shall study the relation between the 1+2 dimensional AKNS system and a linear system with spectral parameter, and apply it to DS equation. This gives a way to get explicit solutions of AKNS equation and especially, DS equation. This kind of relations have been studied in [1,2,4,6] for KP hierarchy and N-wave equation.

2. A Lax pair with spectral parameter connecting with the 1+2 dimensional AKNS system.

We introduce the following linear system

$$\begin{gathered}\Phi_x = \begin{pmatrix} \lambda I & p \\ q & 0 \end{pmatrix}\Phi, \qquad \Phi_y = \begin{pmatrix} \lambda J + U & Jp \\ qJ & 0 \end{pmatrix}\Phi \\ \Phi_t = \begin{pmatrix} W & X \\ Y & Z \end{pmatrix}\Phi = \sum_{j=0}^{n}\begin{pmatrix} W_j & X_j \\ Y_j & Z_j \end{pmatrix}\lambda^{n-j}\Phi \end{gathered} \tag{5}$$

[1]Supported by National Basic Research Project "Nonlinear Science", by FEYUT-SEDC-CHINA and by Fok Ying-Tung Education Foundation of China

where λ is a complex parameter, and J, U are as before. W_j's are $r \times r$ matrices, p, X_j are $r \times 1$ matrices, q, Y_j are $1 \times r$ matrices, and Z_j are functions ($j = 0, 1, \cdots, n$). All of them are functions of x, y, t, independent of the parameter λ. From the integrability conditions of (5), we have

$$p_y = Jp_x + Up, \qquad q_y = q_x J - qU \tag{6}$$
$$p_t = X_{n,x} + (W_n - Z_n)p, \qquad q_t = Y_{n,x} - q(W_n - Z_n) \tag{7}$$
$$U_x = [pq, J], \qquad W_{j,x} = pY_j - X_j q, \qquad Z_{j,x} = qX_j - Y_j p \tag{8}$$
$$W_{j,y}^D = [U, W_j]^D + J(pY_j)^D - (X_j q)^D J, \qquad Z_{j,y} = qJX_j - Y_j Jp \tag{9}$$

$$\begin{aligned} &X_0 = Y_0 = 0, \quad [J, W_0] = 0 \\ &X_{j+1} = X_{j,x} + (W_j - Z_j)p, \qquad Y_{j+1} = -Y_{j,x} + q(W_j - Z_j) \\ &[J, W_{j+1}^A] = W_{j,y}^A - [U, W_j]^A - J(pY_j)^A + (X_j q)^A J - U_t \delta_{jn} \end{aligned} \tag{10}$$

According to [5], all the W_j, X_j, Y_j, Z_j are functions of the entries of p, q, U and their x, y-derivatives.

From (7), (8), (9) and (10),

$$p_t = \sum_{j=0}^{n} M_{n-j}^{(n)} \partial^j p, \qquad q_t = \sum_{j=0}^{n} (\partial^j q) N_{n-j}^{(n)} \tag{11}$$

where $M_j^{(n)}$, $N_j^{(n)}$ are given inductively by

$$M_0^{(0)} = W_0 - Z_0, \quad M_j^{(k+1)} = M_j^{(k)} + M_{j-1,x}^{(k)} + (W_{k+1} - Z_{k+1})\delta_{j,k+1} \tag{12}$$
$$N_0^{(0)} = W_0 - Z_0, \quad N_j^{(k+1)} = -N_j^{(k)} - N_{j-1,x}^{(k)} - (W_{k+1} - Z_{k+1})\delta_{j,k+1} \tag{13}$$

with $M_j^{(k)} = 0$, $N_j^{(k)} = 0$ if $j > k$ or $j < 0$.

Hence p satisfies a similar equation as Ψ in (1). Since V_j^A's are uniquely determined by the integrability condition of (1) if (U, V_j^D) are looked as unknowns, we have

Theorem 1. *If (5) is completely integrable, then $(U, M_j^{(n)D})$ gives a solution of (3),(4).*

It seems that the equations (6)-(10) are quite complicated. However, since they have a Lax pair (5), we can use various methods in 1+1 dimensions to obtain explicit solutions of (6)-(10). Here we use the Darboux transformation method. By the general method of constructing Darboux matrices[3], we have

Theorem 2. *For given solution (p, q, U) of (6)-(10), let $\Lambda = \mathrm{diag}(\lambda_1, \cdots, \lambda_{r+1})$ be a constant diagonal matrix, let h_i be a vector solution of (5) with $\lambda = \lambda_i$, $H = (h_1, \cdots, h_{r+1})$. If $\det H \neq 0$, let $S = H\Lambda H^{-1}$ and denote $S = \begin{pmatrix} T & \xi \\ \eta & \zeta \end{pmatrix}$ where T, ξ, η, ζ are $r \times r$, $r \times 1$, $1 \times r$, 1×1 matrices respectively. Then, for any solution Φ of (5), $\Phi' = (\lambda - S)\Phi$*

satisfies (5) with p, q, U, W_j, X_j, Y_j, Z_j, Φ replaced by p', q', U', W_j', X_j', Y_j', Z_j', Φ' respectively, where

$$\begin{aligned}
&p' = p + \xi, \quad q' = q - \eta, \quad U' = U + [J, T] \\
&W_{j+1}' = W_{j+1} + W_j' T - T W_j + X_j' \eta - \xi Y_j \\
&Z_{j+1}' = Z_{j+1} + Y_j' \xi - \eta X_j + \zeta(Z_j' - Z_j) \\
&X_{j+1}' = X_{j+1} + W_j' \xi - T X_j + \zeta X_j' - Z_j \xi \\
&Y_{j+1}' = Y_{j+1} + Y_j' T - \eta W_j + Z_j' \eta - \zeta Y_j .
\end{aligned} \tag{14}$$

The proof is direct. The matrix $\lambda - S$ here is called a Darboux matrix.

Therefore, for any solution of (6)-(10), if we can solve the corresponding Lax pair (5), then (14) gives a new solution (p', q', U') of (6)-(10), and the corresponding solution of the Lax pair is $\Phi' = (\lambda - S)\Phi$. Moreover, by considering Theorem 1 and [3], we know that a series of solutions of (3),(4) are obtained provided one solution of (6)-(10) and the corresponding solution of (5) are known. Especially, we can obtain infinite number of nontrivial solutions of (3),(4) from the zero solution of (6)-(10).

3. Solutions of DS equation.

The DS equation

$$\begin{aligned}
&i u_t = u_{xx} + u_{yy} - u(A - D) \\
&A_x - A_y = (\partial_x + \partial_y)|u|^2 \\
&D_x + D_y = -(\partial_x - \partial_y)|u|^2 .
\end{aligned} \tag{15}$$

has a Lax pair

$$\begin{aligned}
\Psi_y &= J\Psi_x + U\Psi = \begin{pmatrix} 1 & 0 \\ 0 & -1 \end{pmatrix} \Psi_x + \begin{pmatrix} 0 & u \\ -u^* & 0 \end{pmatrix} \Psi \\
\Psi_t &= -2iJ\Psi_{xx} - 2iU\Psi_x + Q\Psi \\
&= -2i \begin{pmatrix} 1 & 0 \\ 0 & -1 \end{pmatrix} \Psi_{xx} - 2i \begin{pmatrix} 0 & u \\ -u^* & 0 \end{pmatrix} \Psi_x + \begin{pmatrix} iA & -i(u_x + u_y) \\ i(u_x^* - u_y^*) & iD \end{pmatrix} \Psi .
\end{aligned} \tag{16}$$

We can take the system (5) as

$$\begin{aligned}
\Phi_x &= \begin{pmatrix} \lambda I & p \\ -p^* & 0 \end{pmatrix} \Phi \\
\Phi_y &= \begin{pmatrix} \lambda J + U & Jp \\ -p^* J & 0 \end{pmatrix} \Phi \\
\Phi_t &= \begin{pmatrix} -2iJ\lambda^2 - 2iU\lambda + W_2 & -2iJp\lambda - 2iJp_x - 2iUp \\ 2ip^* J\lambda - 2ip_x^* J + 2ip^* U & 2ip^* Jp \end{pmatrix} \Phi ,
\end{aligned} \tag{17}$$

whose integrability conditions are

$$\begin{aligned}
&p_y = Jp_x + Up, \quad p_t = -2iJp_{xx} - 2iUp_x + (W_2 - \alpha_2 - 2iJpp^*)p \\
&U_x = [J, pp^*], \quad U_y = \frac{i}{2}[J, W_2] \\
&U_t = V_{2,y} - [U, W_2] + 2iJ(pp^*)_x J - 2iJpp^* U + 2iUpp^* J \\
&W_{2,x} = -2i(pp_x^* J + Jp_x p^*) + 2i[pp^*, U].
\end{aligned} \tag{18}$$

The Darboux matrix for (17) can be constructed as in Theorem 1, provided that we take $\lambda_1 = \lambda_0$, $\lambda_2 = \lambda_3 = -\lambda_0^*$ for any given $\lambda_0 \in \mathbf{C}$, and $h_i^* h_j = 0$ if $\lambda_i + \lambda_j^* = 0$. (This is possible since $h_i^* h_j$ are constants by the Lax pair (17).) Let $h_1 = (\phi_1, \phi_2, \phi_3)^T$, direct calculation shows

$$S_{ij} = -\lambda_0^* \delta_{ij} + (\lambda_0 + \lambda_0^*) \frac{\phi_i \phi_j^*}{|\phi_1|^2 + |\phi_2|^2 + |\phi_3|^2}. \tag{19}$$

As in (16), denote iA, iD to be the diagonal entries of $Q = W_2 - \alpha_2 - 2iJpp^*$, then, from Theorem 2, we have

Theorem 3. *Suppose (p, U) satisfies (18), then (u, A, D) gives a solution of (15). Moreover, for given $\lambda_0 \in \mathbf{C}$, let $(\phi_1, \phi_2, \phi_3)^T$ be a solution of (17) with $\lambda = \lambda_0$, then*

$$\begin{aligned} u' &= u + 2\kappa\phi_1\phi_2^* \\ A' &= A - 2\kappa(u\phi_1^*\phi_2 + u^*\phi_1\phi_2^*) - 2\kappa(p_1\phi_1^*\phi_3 + p_1^*\phi_1\phi_3^*) - 4\kappa^2|\phi_1|^2(|\phi_2|^2 + |\phi_3|^2) \\ D' &= D + 2\kappa(u\phi_1^*\phi_2 + u^*\phi_1\phi_2^*) + 2\kappa(p_2\phi_2^*\phi_3 + p_2^*\phi_2\phi_3^*) + 4\kappa^2|\phi_2|^2(|\phi_1|^2 + |\phi_3|^2) \end{aligned} \tag{20}$$

where

$$\kappa = \frac{\lambda_0 + \lambda_0^*}{|\phi_1|^2 + |\phi_2|^2 + |\phi_3|^2} \tag{21}$$

give a new solution of (15). The solution p' of the corresponding Lax pair (16) is given by

$$p' = p + \kappa\phi_3^* \begin{pmatrix} \phi_1 \\ \phi_2 \end{pmatrix}. \tag{22}$$

From this theorem, we can obtain a series of solutions of DS equation from the zero solutions of (18).

Acknowledgements

The author would like to thank Prof. Gu Chaohao and Prof. Hu Hesheng for the helpful advice and encouragement. He also would like to thank Prof. Li Yishen who told the author that Prof. Cheng Yi has another approach to get the relation of equations (1) and (5).

References

1. Y.Cheng and Y.S.Li, *The constraint of the Kadomtsev-Petviashvili equation and its special solutions*, Phys. Lett. A157 (1991), 22.
2. Y.Cheng and Y.S.Li, *Constraints of the 2+1 dimensional integrable soliton systems*, preprint.
3. C.H.Gu, *On the Darboux form of Bäcklund transformations*, in Integrable System: Nankai Lectures on Mathematical Physics (1989), 162, World Scientific Publishing Company, Singapore.
4. B.Konopelchenko, J.Sidorenko and W.Strampp, *(1+1)-dimensional integrable systems as symmetry constraints of (2+1)-dimensional systems*, Phys. Lett. A157 (1991), 17.
5. G.Wilson, *Commuting flows and conservation laws for Lax equations*, Math. Proc. Camb. Phil. Soc., 86 (1979), 131.
6. Z.X.Zhou, *Explicit Solutions of N-Wave Equation in 1+2 Dimensions*, Phys. Lett. A168 (1992), 370.

Parallel Sessions 7
Yang-Mills Fields

N=2 EUCLIDEAN INSTANTON SOLUTIONS IN COVARIANT SUPERSPACE WITH UNCONSTRAINED FIELDS

Sultan Catto

Physics Department, The Rockefeller University
1230 York Avenue
New York, NY 10021-6399
and
Baruch College and The Graduate School and University Center
City University of New York, New York, NY 10010

Abstract

Four dimensional supersymmetric field equations depending on a chiral supermultiplet we have recently developed are shown to admit Euclidean instanton solutions in N=2.

Introduction

We show that certain four dimensional supersymmetric field equations depending on a chiral scalar supermultiplet admit instanton solutions in $N = 2$ theories when transition is made to Euclidean space. Instanton-like solutions were recently found [1] in $N = 1$ case, where we obtained monopole solutions but not supersymmetric instantons. In that case we could obtain Euclidean solutions but at the expense of losing supersymmetry. To have both supersymmetry and Euclidean solutions by analytic continuation we go into $N = 2$ theory and repeat our earlier procedure, which will be shown below. We had also obtained partial results [2] using methods developed previously [3,4] in connection with two and three dimensional supersymmetric equations discussed by Witten [5], Fronsdal [6], and Affleck, *et. al.* [7], in the conformally invariant case.

Let Φ be a chiral scalar supermultiplet. It has a standard form [8]

$$\Phi = A(y) + \sqrt{2}\theta\psi(y) + \theta\theta F \tag{1}$$

with $y^\mu = x^\mu + i\theta^\dagger\sigma^\mu\theta$, σ^k (k=1,2,3) are the usual Pauli matrices, and $\theta_\alpha(\alpha = 1,2)$ the complex Grassmann numbers (for their transformation properties see [1]).

Expanding Φ, we have

$$\begin{aligned}\Phi &= A(x) + i(\theta^\dagger\sigma^\mu\theta)\partial_\mu A + \sqrt{2}\theta\psi + i\sqrt{2}(\theta^\dagger\sigma^\mu\theta)\theta\partial_\mu\psi \\ &\quad + \theta\theta F - \frac{1}{4}\theta\theta\bar{\theta}\bar{\theta}\Box A.\end{aligned} \tag{2}$$

Powers of Φ are also scalar so that if $f(\Phi)$ is a power series in Φ we obtain

$$f(\Phi) = f(A) + \sqrt{2}f'(A)\theta\psi + \theta\theta I f'(A)F - \frac{1}{2}f''(A)\psi\psi \tag{3}$$

which has the chiral scalar multiplet structure.

The kinetic part of the Lagrangian, $\mathcal{L}_k$, is the $\theta\theta\bar{\theta}\bar{\theta}$ component of the product $\Phi^\dagger\Phi$ so that, up to a total divergence

$$\begin{aligned}\mathcal{L}_k &= -\frac{1}{2}A^*\Box A - \frac{1}{2}A\Box A^* + FF^* - \frac{i}{2}(\psi^\dagger\bar{\sigma}^\mu\partial_\mu\psi) \\ &\quad + i(\partial_\mu\psi^\dagger\bar{\sigma}^\mu\psi).\end{aligned} \tag{4}$$

The part of the Lagrangian which includes the interaction is

$$\begin{aligned}\mathcal{L}' &= \int d^2\theta f(\Phi) + \int d^2\bar{\theta} f(\Phi)^\dagger \\ &= Ff'(A) - \frac{1}{2}f''(A)\psi\psi + F^*f'(A)^* - \frac{1}{2}f''(A)^*\bar{\psi}\bar{\psi}\end{aligned} \tag{5}$$

with the total Lagrangian given by$\mathcal{L} = \mathcal{L}_k + \mathcal{L}'$. The nonpropagating field F can be eliminated as usual by varying F and F^* so that

$$-F^* + f'(A) = 0, \tag{6}$$

$$-\Box A + F^*f''(A)^* - \frac{1}{2}f'''(A)^*\bar{\psi}\bar{\psi} = 0, \tag{7}$$

and

$$i\bar{\sigma}^\mu\partial_\mu\psi + f''(A)^*\hat{\psi} = 0. \tag{8}$$

Equations (6) and (7) then yield

$$\Box A = f'(A)f''(A)^* - \frac{1}{2}f'''(A)^*\psi^\dagger\hat{\psi}. \tag{9}$$

The Lagrangian is conformally invariant if we choose

$$f(\Phi) = -\frac{\lambda}{3}\Phi^3, \quad f'(\Phi) = -\lambda\Phi^2 \tag{10}$$

where λ is dimensionless. Then the equation of motion for the propagating field ψ and A are

$$\Box A = 2\lambda^2A^2A^* + \lambda\psi^\dagger\hat{\psi}, \tag{11}$$

and

$$i\bar{\sigma}^\mu\partial_\mu\psi = 2\lambda A^*\hat{\psi}. \tag{12}$$

These equations can be written in Dirac form if we introduce the Majorana field represented by the 4×1 column ξ, with

$$\xi = \begin{pmatrix}\psi \\ -\hat{\psi}\end{pmatrix}, \quad \gamma_\mu = \begin{pmatrix}0 & -i\sigma^\mu \\ -i\bar{\sigma}^\mu & 0\end{pmatrix}. \tag{13}$$

Taking the complex conjugate of Eq.(12) and multiplying from the left by $-i\sigma_2$ we obtain

$$i\sigma^\mu\partial_\mu\hat{\psi} = 2\lambda A\psi. \tag{14}$$

Let

$$A = \phi + i\chi \tag{15}$$

where ϕ is a real scalar and χ a real pseudoscalar field. Then the Dirac equation for Majorana spinors

$$\gamma^\mu \partial_\mu \xi = 2\lambda(\phi + i\gamma_5\chi)\xi \tag{16}$$

is equivalent to the pair of two-component equations (12) and (14). Note that in the Majorana representation for the γ-matrices $\gamma^\mu\partial_\mu$ and $i\gamma_5$ are real as well as all the four components of ξ. Equation (11) can now be rewritten as

$$\Box A = 2\lambda^2(A^*A)A + \lambda\bar{\xi}\frac{1+\gamma_5}{2}\xi, \tag{17}$$

where

$$\bar{\psi} = \psi^\dagger\gamma_4 = i\psi^\dagger\gamma^0, \tag{18}$$

together with its complex conjugate equation

$$\Box A^* = 2\lambda^2(A^*A)A^* + \lambda\bar{\xi}\frac{1-\gamma_5}{2}\xi \tag{19}$$

From equations (17) and (19) we obtain

$$\Box\phi = 2\lambda^2(\phi^2 + \chi^2)\phi + \frac{1}{2}\lambda\bar{\xi}\xi, \tag{20}$$

and

$$\Box\chi = 2\lambda^2(\phi^2 + \chi^2)\chi - \frac{1}{2}i\lambda\bar{\xi}\gamma_5\xi, \tag{21}$$

which are the equations of motion together with equation (14).

Instead of working directly in the Euclidean version of supersymmetric equations we would like to find the solutions in the Minkowski space and work out the analytic continuation of the solutions for imaginary time. The four-spinor in this case still splits into two component left handed spinor ψ_L and a right handed ψ_R. But the transformation properties are drastically altered since the Lorentz group $SL(2,C)$ is replaced by $SU(2) \times SU(2) \sim O(4)$. For the transformation properties of the fields and their derivatives we refer the reader to our paper [1]. Writing Eq.(1) in terms of the left and right handed fields, and again taking the $\theta_L\theta_L\bar{\theta}_R\bar{\theta}_R$ component of the product $\Phi_R^\dagger\Phi_L$, namely

$$(\Phi_R^\dagger\Phi_L)_{\theta\theta\bar{\theta}\bar{\theta}} = \psi_R^\dagger\mathcal{D}^\dagger\psi_L + F_R^*F_L + A_R^*\Box A_L + A_L\Box A_R^* \tag{22}$$

we can write (up to a total divergence) the kinetic part of the Lagrangian as

$$\mathcal{L}_k = \int \Phi_R^\dagger\Phi_L d^2\theta_L d^2\bar{\theta}_R + \int \Phi_L^\dagger\Phi_R d^2\theta_R d^2\bar{\theta}_L. \tag{23}$$

Similarly, the part which includes the interaction may be written as

$$\mathcal{L}_1' = \frac{\lambda}{3}\int(\Phi_L^3 d^2\theta_L + \Phi_L^{\dagger 3} d^2\bar{\theta}_L), \tag{24}$$

and

$$\mathcal{L}_2' = \frac{\lambda'}{3}\int(\Phi_R^3 d^2\theta_R + \Phi_R^{\dagger 3} d^2\bar{\theta}_R). \tag{25}$$

Therefore, the general conformal invariant Lagrangian is given by

$$\mathcal{L} = \mathcal{L}_k + \mathcal{L}'_1 + \mathcal{L}'_2. \tag{26}$$

If we let $\lambda = \lambda'$ we also have parity invariance. If we keep λ and set $\lambda' = 0$, or vice versa, we recover the results of an earlier work [2]. The transformation to Euclidean case is discussed in [1]. The extended Euclidean Lagrangian we obtain is

$$\begin{aligned} \mathcal{L} &= \frac{1}{2}(\frac{i}{2}\psi^\dagger\gamma^\mu\partial_\mu\psi - \lambda\psi^\dagger(\phi + i\gamma_5\chi)\psi - \partial_\mu\phi\partial^\mu\phi \\ &\quad - \partial_\mu\chi\partial^\mu\chi - \lambda^2(\phi^2 + \chi^2)^2 + h.c.) \end{aligned} \tag{27}$$

which leads to the Euclidean equations of motion

$$i\gamma^\mu\partial_\mu\psi = 2\lambda(\phi + i\gamma_5\chi)\psi, \tag{28}$$

$$\Box\phi = 2\lambda^2(\phi^2 + \chi^2)\phi + \frac{\lambda}{2}\psi^\dagger\psi, \tag{29}$$

and

$$\Box\chi = 2\lambda^2(\phi^2 + \chi^2)\chi - i\frac{\lambda}{2}\psi^\dagger\gamma_5\psi. \tag{30}$$

To find the Euclidean instanton solutions in $N = 2$ case, corresponding to the left handed superfield Φ_L, we write Φ_R such that

$$\Phi_R = B_R(y) + \sqrt{2}\tau_R\psi_R(y) + \tau_R\tau_R G \tag{31}$$

where B_R, ψ_R and G_R are the right handed analogues of A_L, ψ_L and F_L, respectively, and $\tau_R^\alpha(\alpha = 1,2)$ the right handed Grassmann numbers whose transformation properties are similar to those of θ_L shown in [1]. We can than show that the kinetic part of the Lagrangian upto a total divergence is the $\theta_L\theta_L\bar{\tau_R}\bar{\tau_R}$ component of the product $\Phi_R^\dagger\Phi_L$. The equations of motion we obtain turn out to be exact left handed and right handed analogues of the above ones with mentioned replacements. Therefore, the instanton solutions we show below hold for the N=2 case as well. For details we refer the reader to our lengthy discussion in [9].

Instanton Solutions

The form of the fermionic solution is

$$\psi = \frac{8}{\lambda}e^{-i\gamma_5\frac{\alpha}{2}}\frac{\sqrt{k}}{1 + ik\gamma\cdot x}\frac{k}{1 + k^2x^2}\eta \tag{32}$$

where k has dimensions of inverse length, and the fermionic spinor η obeys $\eta^\dagger\eta = -cos\alpha$, and $\eta^\dagger\gamma_5\eta = isin\alpha$. The bosonic fields are given by

$$\phi = \frac{2}{\lambda}hcos\alpha, \tag{33}$$

and

$$\chi = \frac{2}{\lambda}hsin\alpha, \tag{34}$$

such that

$$h = \frac{k}{1 + k^2x^2}. \tag{35}$$

After a lengthy calculation (see[1,2 and9]) it is easy to show that the equations of motion are exactly satisfied.

The four dimensional equations we developed could arise as chiral fermion theories involving left-handed spinors only, which might be derived from effective Lagrangian approximating supersymmetric grand unified theories or weak interaction theories. Applications to high temperature limits of four dimensional supersymmetry are also possible. Equations (28 – 30) give rise to conformal invariant Lagrangians which produce the generalized supersymmetric nonlinear σ-models. We shall explore these aspects as well as spontaneous symmetry breaking in subsequent publications.

In the study of chiral models, and to a large extent of superstring theory, not only the methods of Lie group theory usually used in theoretical physics is important, but also methods of algebraic and differential geometry and topology. We are also investigating Kähler and hyper-Kähler supersymmetric chiral models because they admit $N = 4$ supersymmetric generalizations of the ones we described above.

Acknowledgements

I would like to thank Professor M.L. Ge for inviting me to this most stimulating conference and for the superb hospitality he and the other organizers have shown. Besides my collaborator Y. Gürsey, I also would like to thank professors M.L. Ge, H.H.K. Tang, H.-C. Tze, Julius Wess and C.N. Yang for stimulating discussions, inspiration and criticism. This work was supported in part by the grant under the Department of Energy Contract Grant No. DE-AC0287ER40325.B, and PSC-CUNY Research Award Nos. 6-6(1389,2363,7354,8369).

References

1. S. Catto, *J. Math. Phys.* *33, October 1992.*
2. S. Catto and Y. Gürsey, *Nucl. Phys.* *B6, (Proc. Supp.) 148 (1989).*
3. F. Gürsey and H. C. Tze, *Phys. Lett.* *129B, 205 (1983).*
4. S. Catto and F. Gürsey, *Phys. Rev.* *D29, 653 (1984).*
5. E. Witten, *Nucl. Phys.* *B185, 513 (1981).*
6. C. Fronsdal, *Phys. Rev.* *D26, 1988 (1982).*
7. I. Affleck, J. Harvey and E. Witten, *Nucl. Phys.* *B206, 413 (1982).*
8. See for instance, J. Wess and J. Bagger, "Supersymmetry and Supergravity." (Princeton University Press, 1983).
9. S. Catto, *Rockefeller University preprint, to be published.*

CRITICAL NON–LINEAR O(N) σ–MODELS AT 2 < d < 4: THE QUASI–PRIMARY FIELDS AND THEIR FUSION

K. LANG and W. RÜHL

Fachbereich Physik
Universität Kaiserslautern
Kaiserslautern, Germany

The conformal group at $d > 2$ has only a finite number of generators. So far it has not been shown that the diffeomorphism group plays any role in these critical models. Nor has the solvability of these models been established. We use a $\frac{1}{N}$ expansion to explore their field content, their fusion coefficients and field dimensions and hope to obtain this way a deeper insight into the algebraic structures of these models.

The σ–models are defined by a Euclidean path integral for an O(N) vector field $\vec{S}(x)$ and an auxiliary field $\alpha(x)$ which are both O(d) scalar

$$Z = \int D[\vec{S}]\, D[\alpha]\, \exp\left\{-\int dx \left[\frac{1}{2}\,\partial_\mu \vec{S}\,\partial_\mu \vec{S}(x) + z^{\frac{1}{2}}\,\alpha(x)\,(\vec{S}(x)^2-1)\right]\right\} \tag{1}$$

At all $d > 2$ they possess a paramagnetic–ferromagnetic phase transition which defines the critical point of interest. The $\frac{1}{N}$ expansion is a saddle point expansion of (1). This expansion is extended to all space dimensions $2 < d < 4$.

Operator product expansions of the basic fields $\vec{S}$ and α define all fields in these models and an associative algebraic structure on them. Properly conformally covariant fields are denoted "quasi–primary" (qp–fields), together with derived fields they exhaust the whole field content. Information on the qp–fields comes from the calculation of 2n–point functions by a $\frac{1}{N}$–expansion of (1).

The qp–fields form classes. A class is labelled by an irreducible O(N) representation (indicated by a Young frame Y) and a natural number p. At $N = \infty$ these models degenerate partly into a free field theory of massless bosonic vector fields. The auxiliary field has no representative at $N = \infty$. The label p is the polynomial degree of the normal product of free vector fields to which a qp–field tends at $N = \infty$. The label p is defined only in a neighborhood of $N = \infty$ and is set zero for the auxiliary field.

Each field in a class (Y,p) has a dimension δ and, as a symmetric traceless O(d) tensor, it has a rank ℓ. At $d = 3$ only symmetric tensors exist and it seems that they suffice also after the analytic extension to $2 < d < 4$. For technical reasons it is useful to label a qp–field ϕ first by its normal dimension $[\delta_\phi]$ which is independent of N

$$\delta_\phi = [\delta_\phi] + \eta(\phi) \tag{2}$$

where $\eta(\phi)$ is the anomalous dimension

$$\eta(\phi) = O\left[\frac{1}{N}\right] \tag{3}$$

The normal dimension can be decomposed as

$$[\delta_\phi] = p(\mu-1)+q, \quad \mu = \frac{1}{2}d \tag{4}$$

where q is the number of derivatives involved in the normal product at $N = \infty$. For the auxiliary field α we have by definition

$$p = 0, \quad q = 2 \tag{5}$$

and for the vector field S

$$p = 1, \quad q = 0 \tag{6}$$

Thus we end up with the labels $(Y,p;\ [\delta],\ell)$ of any qp–field. We shall see, however, that a degeneracy with respect to this labelling is left.

The field content of a generic class is plotted below. There are two exceptional classes $(\emptyset,0)$ ($\emptyset$ stands for the empty set) and $(\square,1)$ which have qp–fields

$(\emptyset,0;2+\ell,\ell)$ resp. $(\square,1;\ \mu-1+\ell,\ell)$
only for $\ell = 0$.

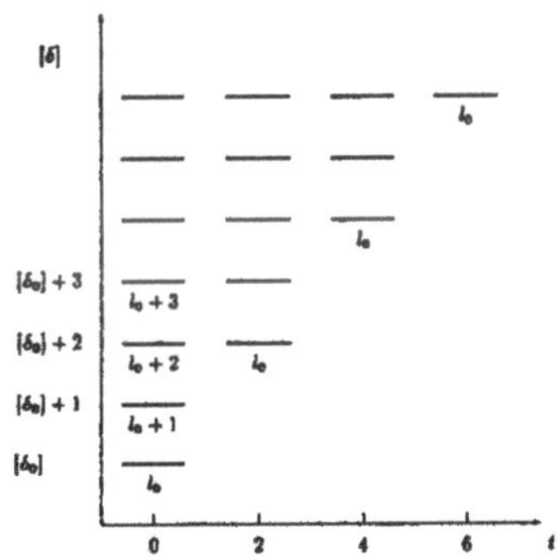

The classes can also be considered as vector spaces spanned by their qp–fields and the corresponding derived fields. Consider the classes of all O(N) scalar fields. There is the class $(\emptyset,0)$ containing the field α and the class $(\emptyset,2)$ containing the energy–momentum field. These fields are "primary" in the sense that they belong to minimal dimension $[\delta_0]$ and minimal tensor rank $[\ell_0]$ in their class. There are classes $(\emptyset,2n)$, $n > 1$, whose primary fields are the W–currents. These are symmetric traceless O(d) tensors of rank 2n, which are conserved and have the full dimension nd, i.e.

$$p = q = 2n, \qquad \eta = 0 \tag{7}$$

The direct sum

$$\bigoplus_{n=0}^{\infty} (\emptyset,2n) \tag{8}$$

is the minimal subalgebra of the field algebra. In fact, in [5] we showed that

$$(\emptyset,0) \circledast (\emptyset,0) \longrightarrow (\emptyset,2) \tag{9}$$

at order $\frac{1}{N}$ and the classes $(\emptyset,2n)$ are expected to appear at order $\left[\frac{1}{N}\right]^n$.

For an irreducible representation Y of O(N) let #Y denote the number of blocks and

$$p = \#Y + 2n \tag{10}$$

where n is the number of O(N) contractions. Then

$$\bigoplus_{n=0}^{\infty} (Y,\#Y + 2n) \tag{11}$$

serves as a module for the operation of the subalgebra (8).

Fusion of three qp–fields may occur at any order of $\frac{1}{N}$ but is most important at O(1). Basic examples are

$$(\square\!\square\!\square_n,p;\ p(\mu-1),0) \circledast (\square,1;\ \mu-1,0) \longrightarrow (\square\!\square\!\square_{n+1},p+1;\ (p+1)(\mu-1)+\ell,\ell) \tag{12}$$

$$(\emptyset,0;\ 2n,0) \circledast (\emptyset,0;\ 2,0) \longrightarrow (\emptyset,0;\ 2n+2+\ell,\ell) \tag{13}$$

At $O\left[\frac{1}{N}\right]$ it becomes possible to raise or lower p, e.g.

$$(\square,1;\ \mu-1,0) \circledast (\square,1;\ \mu-1,0) \longrightarrow (\emptyset,0;\ 2,0) \tag{14}$$

$$(\emptyset,0;\ 2,0) \circledast (\emptyset,0;\ 2,0) \longrightarrow (\emptyset,2;\ 2\mu+\ell,\ell), \quad (\ell \geq 2,\ \text{even}) \tag{15}$$

Some results on the anomalous dimensions are the following ones.

(1) For the basic field S and α the anomalous dimensions are known to $O\left[\frac{1}{N^3}\right]$, $O\left[\frac{1}{N^2}\right]$, respectively [2].

(2) For the scalar primary fields

$$C_0^{(n_1,n_2)} = (Y(n_1 n_2),n;\ n(\mu-1),0) \tag{16}$$

with

$$n_1 \geq n_2, \quad n_1 + n_2 = n \tag{17}$$

we have [1]

$$\eta(C_0^{(n_1,n_2)}) = \left[2(n_2-1)\,(n_1 + n_2\,\frac{\mu}{\mu+1}) - n(n-2)\mu\right] \cdot \frac{\eta(S)}{\mu-2} + O\left[\frac{1}{N^2}\right] \tag{18}$$

(3) For the scalar qp–fields

$$\alpha_{\circledast}^{n} = (\emptyset,0;\ 2n,0) \tag{19}$$

we have [1]

$$\eta(\alpha_{\circledast}^{n}) = -4\,(2\mu-1)\left[\frac{\mu-1}{\mu-2}\,n(n-2) + (2\mu-1)\left[\frac{n}{2}\right]\right]\eta(S) + O\left[\frac{1}{N^2}\right] \tag{20}$$

(4) For towers of qp–fields

$$C_\ell = (Y,p;\ [\delta]+\ell,\ell) \tag{21}$$

the limit of $\eta(C_\ell)$ for $Y,p,[\delta]$ fixed seems to exist (a proof is known in many but not all cases). Moreover, if there is no degeneracy (see below) in such tower and

$$A \circledast B \longrightarrow C_\ell \qquad (A,\ B\ \text{scalar}) \tag{22}$$

then [1]

$$\lim_{\ell\to\infty} \eta(C_\ell) = \eta(A) + \eta(B) + O\left[\frac{1}{N^2}\right] \tag{23}$$

The degeneracy has turned up in the following way. From n fields α, $n = p+q$, we form a p–cluster and a q–cluster first by repeated application of (13)

$$((\emptyset,0;\ 2,0))_{\circledast}^{n} \longrightarrow (\emptyset,0;\ 2p,0) \circledast (\emptyset,0;\ 2q,0) \longrightarrow (\emptyset,0;\ 2n+\ell,\ell) \tag{24}$$

If $\ell \geq 4$ the final qp–field depends on the coupling mode. Namely, for

$$p' + q' = p + q, \qquad p' \notin \{p,q\} \tag{25}$$

the result is different. A similar thing happens for the vector field S. A secular matrix for the anomalous dimensions has not been diagonalized yet.

For many more details, in particular also the fusion coefficients and the technical apparatus used, we refer to our published papers [1].

References

1. K. Lang and W. Rühl, *Z. Phys.* **C50** (1991) 285, *Z Phys.* **C51** (1991) 127, *Nucl. Phys.* **B377** (1992) 371, *Phys. Letters* **B275** (1992) 93, Kaiserslautern University preprints KL–TH–92/7 and 92/15.
2. E. Brézin, D.J. Wallace, *Phys. Rev.* **B7** (1973) 1967; S.K. Ma, *Phys. Rev.* **A7** (1973) 2172; A.N. Vasilèv, Yu.M. Pismak, Yu.R. Khonkonen, *Teor. Mat. Fiz.* **46** (1981) 157, **47** (1981) 291, **50** (1982) 195; W. Bernreuther, F. Wegner, *Phys. Rev. Lett.* **57** (1986) 1383.

ASYMPTOTIC PROPERTIES OF YANG–MILLS FLOW

SHEN CHUN-LI*

Department of Mathematics, East China Normal University
Shanghai 200062, P.R.China

§1. Introduction

Let $P(M^4, G)$ be a principal bundle over a 4-dimensional compact Riemannian manifold M with a compact simple Lie group G as its structure group. Uhlenbeck [7] proved the following weakly compactness theorem for Yang-Mills connections. Let $\{A_i\}$ be a sequence of Yang-Mills connections of $P(M^4, G)$, then we can take a subsequence of $\{A_i\}$ suitably (denoted still by $\{A_i\}$), and find a sequence of gauge transformations σ_i, a finite set $\Sigma_\infty = \{p_1, \cdots, p_l\}$ of points in M, such that $\{\sigma_i^* A_i\}$ C^∞-converges to a Yang-Mills connection A_∞ on $M \setminus \{p_1, \cdots, p_l\}$. This limiting connection A_∞ can be extended on whole M, but the extended Yang-Mills connection $\tilde{A}_\infty$ may belong to another principal bundle Q which is an extension of $P|_{M\setminus\{p_1,\cdots,p_l\}}$.

Moreover, Sedlacek [5] obtained a similar theorem with Uhlenbeck's weakly compactness theorem, but he replaced the Yang-Mills condition of A_i by the assumption that $\{A_i\}$ is a sequence of minimizing connections of Yang-Mills action, and got the $L^2_{1;loc}$-convergence in stead of C^∞-convergence.

Taubes [6] proved a result similar to Sedlacek's result for a sequence of so-called 'good' connections $\{A_i\}$ on P.

In this paper, we shall discuss the similar problem for Yang-Mills flow. Let A_t be a Yang-Mills flow with an initial connection A_0, i.e.,

$$\begin{cases} \dfrac{\partial A_t}{\partial t} = -2 D^*_{A_t} F_{A_t} \\ A_t|_{t=0} = A_0 \end{cases},$$

where F_{A_t} are the curvatures of A_t, D_{A_t} the gauge-covariant differentiation, and $D^*_{A_t}$ the adjoint operator of D_{A_t} with respect to the Riemannian metric of M and

* This is a joint work with Prof. Chen Yunmei, Department of Mathematics, University of Florida, Gainesvill, FL 32611, USA. The author was partially supported by National Science Foundation of China..

Cartan-Killing inner product of G. We assume that the Yang-Mills flow A_t is a regular global flow for any time t, and our interest is only to discuss the asymptotic properties of A_t when $t \to \infty$. The main result is the following.

Main Theorem *Let $P(M^4, G)$ be a principal bundle over a 4-dimensional compact Riemannian manifold (M^4, g) with compact, simple Lie structure group G, $k(P)$ Pontrjagin index of P. Let A_t be a regular global Yang-Mills flow with an initial connection A_0 for any time $t \geq 0$. Then we can take a subsequence t_i $(t_i \to \infty)$suitably, and find a sequence of gauge transformations σ_i, a subset $\Sigma_\infty = \{p_1, \cdots, p_l\}$ of finite points in M, such that $\{\sigma_i^*(A_{t_i})\}$ C^∞-converges to a regular Yang-Mills connection A_∞ on $M \setminus \Sigma_\infty$, A_∞ can be extended to a regular Yang-Mills connection $\tilde{A}_\infty$ over whole M, but $\tilde{A}_\infty$ may belong to an another bundle Q which is an extension of $P|_{M\setminus\Sigma_\infty}$. Furthermore, for each singular point p_α, $(\alpha = 1, \cdots, l)$, there corresponds a principal G-bundle P_α over (S^4, can) with Pontrjagin index k_α (where 'can' means the standard metric on S^4), and a non-flat Yang-Mills connection A_α on P_α, such that*

$$k(Q) + \sum_{\alpha=1}^{l} k(P_\alpha) = k(P), \qquad \text{and} \qquad YM(\tilde{A}_\infty) + \sum_{\alpha=1}^{l} YM(A_\alpha) = YM_\infty,$$

where $YM(A)$ is the Yang-Mills action of connection A, and $YM_\infty = \lim_{i\to\infty} YM(A_i)$.

§2. The sketch of the proof of main theorem

1° For any $(x_0, t_0) \in M \times R_+$, where R_+ is the set of all nonnegative real numbers, let

$$T_R(x_0, t_0) = \{(x, t) \in M \times R_+ | -4R^2 < t - t_0 < -R^2\},$$

$$G_{(x_0,t_0)}(x, t) = (4\pi(t_0 - t))^{-2} \exp\left\{-\frac{|x_0 - x|^2}{4(t_0 - t)}\right\}, \qquad t < t_0,$$

where $|x_0 - x|$ means the Riemannian distance between points x_0 and x, and R_M a lower bound for the injectivity radius of M. For $0 < R < R_M$, we define the function $\Psi_{(x_0,t_0)}(R)$ for Yang-Mills flow A_t as follows :

$$\Psi_{(x_0,t_0)}(R) = R^2 \int_{T_R(x_0,t_0)} |F|^2 G_{(x_0,t_0)} \phi^2 \sqrt{g}\, dx\, dt,$$

where $\phi \in C_0^\infty(B_{R_M}(x_0))$ is a suitable cut-off function satisfying $0 \leq \phi \leq 1$, $\phi \equiv 1$ on $B_{R_M/2}(x_0)$ and $|\nabla\phi| \leq C/R_M$.

In [4] we have proved the following results:

ε-Regularity Theorem *Suppose A_t is a regular solution of Yang-Mills flow, $t \in [-T, T]$, and $T \le R_M^2$. There exists a sufficiently small constant $0 < \varepsilon_0 < R_M$, depending only on M and the initial Yang-Mills action YM_0, such that if for some $0 < R < \min(\varepsilon_0, \sqrt{T}/2)$ and $x_0 \in M$, the inequality $\Psi_{(x_0,0)}(R) \le \varepsilon_0$ is satisfied, then there holds*

$$\mathop{Sup}_{P_{\delta R}(x_0,0)} |F_{A_t}|^2 \le \frac{c}{(\delta R)^4}$$

with constant c depending only on M, YM_0, ε_0, constant $0 < \delta < 1/4$ depending only on M, YM_0, and $P_{\delta R}(x_0, 0) = \{(x,\ t) \in M \times R \mid\ |x - x_0| < \delta R, |t| < (\delta R)^2\}$.

2° From the energy inequality of Yang-Mills flow [4], we can choose a sequence $\{t_m\}$ satisfying the following two conditions as $m \to \infty$:

$$\int_M |J_{A_{t_m}}|^2 \sqrt{g} dx \longrightarrow 0 \qquad \text{and} \qquad \int_{t_m - 1}^{t_m} \int_M |J_{A_t}|^2 \sqrt{g} dx dt \longrightarrow 0.$$

3° Using Sedlacek's technique [5,Proposition 3.3], we can divide the whole manifold M into two parts : the 'good' part $M \setminus \Sigma_\infty$ and the 'bad' part Σ_∞, according to the properties of function Ψ. We can choose a countable many covering $\{B_{R_i}(x_i)\}$ of $M \setminus \Sigma_\infty$, such that $\Psi_{(x_i,t_m)}(R_i/\delta) \le \varepsilon_0$ holds for sequence $\{t_m\}$, (or its some subsequence, if necessary). Using the similar argument as [1,p.100], we can prove that Σ_∞ is a finite set of points in M, i.e., $\Sigma_\infty = \{p_1, \cdots, p_l\}$.

4° Using ε-regularity theorem for every ball $B_{R_i}(x_i)$ in the above covering of $M \setminus \Sigma_\infty$, there holds $|F_{A_t}| < c/R_i^2$ on $P_{R_i}(x_i, t_m)$ for any t_m. Namely, the C^0-norm of F_{A_t} are bounded on a rather small domain in $P_{R_i}(x_i, t_m)$ uniformly (with respect to t_m). In particular, for the sequence of connections $\{A_{t_m}\}$, the norms $|F_{A_{t_m}}|$ are bounded uniformly on B', where B' is a rather small ball in B.

5° We fix any one of the balls $B_{R_i}(x_i)$ in this covering, and denote it simply by $B = B_{R_i}(x_i)$. Using the Bochner type inequality [4], not only $|F_{A_t}|$ but also the norms of higher gauge-covariant derivatives $|\nabla_{A_t}^k F_{A_t}|$ are uniformly bounded for all $k = 1, 2, \cdots$, on a rather small domain P' in $P = P_{R_i}(x_i, t_m)$, i.e., $|\nabla_{A_t}^k F_{A_t}| < C_k$, for any k on P'. In particular, for the sequence of connections $\{A_{t_m}\}$, the norms $|\nabla_{A_{t_m}}^k F_{A_{t_m}}|$ are bounded uniformly on B', where B' is a rather small ball in B.

6° Now we have controlled the curvatures $F_{A_{t_m}}$ and their higher gauge-covariant derivatives. But it is not enough to get the C^∞-convergence for a sequence of connections, and we have to esimate the norms of connections and their higher derivatives. To this end, we use the Uhlenbeck's Coulomb gauge fixing arguments. We want to estimate the norms of the connections and their higher derivatives under the Coulomb gauge.

Using bootstrapping technique, under the Coulomb gauge σ_m, the Sobolev norms of connections $\tilde{A}_m = \sigma_m^*(A_{t_m})$ can be estimated by

$$\|\tilde{A}_m\|_{L^2_{l+1}(B')} \leq C_l(\|F_{A_{t_m}}\|_{C^0(B)} + \sum_{k=1}^{l} \|\nabla^k_{A_{t_m}} F_{A_{t_m}}\|_{L^2(B)}),$$

where B' is a rather small domain in B ([2, Lemma 2.3.11]). As the right side of this estimate is uniformly bounded, $\|\tilde{A}_m\|_{L^2_{l+1}(B')}$ is also uniformly bounded for every l. Then, from Sobolev imbedding theorem, the C^l-norms of $\{\tilde{A}_m\}$ are uniformly bounded for $l = 0, 1, 2, \cdots$. Using the Arzela-Ascoli theorem, we can choose a subsequence of $\{t_m\}$ such that $\{\tilde{A}_m\}$ C^∞-converges to a limiting connection A_∞ over B'. Since

$$\int_M |J_{A_{t_m}}|^2 \sqrt{g} dx \longrightarrow 0$$

as $m \to \infty$, so $J_{A_\infty} = 0$. Hence the limiting connection A_∞ is a Yang-Mills connection over B'.

7° Using Uhlenbeck's patching arguments [3], these individual limiting connections A_∞ on each ball can be glued together, and become a limiting connection over $M \setminus \Sigma_\infty$.

8° Up to now, we could choose a sequence $\{t_i\}$ such that, under suitable gauge transformations σ_i, $\sigma_i^*(A_{t_i})$ C^∞-converges to a limiting connection A_∞ on $M \setminus \{p_1, \cdots, p_l\}$. Using Uhlenbeck's removable singularity theorem, these singular points p_α could be removed. Obviously, the sequence $\{\sigma_i^*(A_{t_i})\}$ is a 'good' sequence in the sense of Taubes [6], so the proof of main theorem is completed from [6].

References

[1]. Y.M.Chen and M.Struwe, *Existence and partial regularity for heat flow for harmonic maps*, Math. Zeit. **201** (1989), 83–103.

[2]. S.K.Donaldson and P.B.Kronheimer, *The geometry of four-manifolds*, Clarendon Press, Oxford, 1990.

[3]. D.S.Freed and K.K.Uhlenbeck, *Instantons and four-manifolds*, Springer-Verlag, New York, 1984.

[4]. C.L.Shen, *Monotonicity formula and partial regularity results in Yang-Mills flow, Qualitative aspects and applications of nonlinear evolution equations* (T.T.Li and P. de Mottoni, ed.), World Scientific, Singapore, 1991, p. 248-251.

[5]. S.Sedlacek, *A direct method for minimizing the Yang-Mills functional over 4-manifolds*, Commun. Math. Phys. **86** (1982), 515–527.

[6]. C.H.Taubes, *Path-connected Yang-Mills moduli spaces*, Jour. Diff. Geom. 19 (1984), 337–392.

[7]. K.K.Uhlenbeck, *Connections with L^p bounds on curvature*, Commun. Math. Phys. **83** (1982), 31–42.

YANG-MILLS HIERARCHY

D.H. Tchrakian

CPT, Ecole Polytechnique, F 91128 Palaiseau Cedex, France
Math-Phys, St Patrick's College, Maynooth, Ireland
STP, Dublin Institute for Advanced Studies, Dublin, 4 Ireland

Contents :

1 . Motivation

The Yang-Mills (YM) model plays a central role in the dynamics of both Strong and Electroweak interactions. Concerning instanton[1] effects, it is desirable in both these theories to include a Higgs field. In the strong theory, because of problems[2] of arbitrary scale, while in the Electroweak simply because there *is* a Higgs sector. It is however well known that Derrick's theorem forbids a nontrivial Higgs field in d = 4. To circumvent this obstacle is the ultimate motivation of what follows.

For orientation, we note that the Monopole[3] and the Abelian Higgs Vortex[4] are employed as the *localised* instantons in d =2 and 3 respectively[2,5]. Both these systems can be derived[6] from the SU(2) YM system on $\mathbb{R}_d \times S^{4-d}$ (d =2,3) by dimensional reduction.

To find systems with *localised* instantons in d = 4, we first seek a hierarchy of YM models in d =2n (n>2), with instanton solutions, and then descend to d = 4.

2 . Generalisations of YM : The hierarchy in d = 2n

The requirement of topologically stable solutions restricts the higher-dimensional generalisations of YM to 2n dimensions, since it is only in even dimensions that Chern Pontryagin (C-P) densities are defined, namely

$$\mathcal{C}_n = \mathrm{tr}\, F\wedge F\wedge\ldots\wedge F \ , \quad \text{n-times} \tag{2.1 a}$$

$$\mathcal{C}_{r,s} = (\mathrm{tr}\, F\wedge\ldots\wedge F)\wedge(\mathrm{tr}\, F\wedge\ldots\wedge F) \quad \text{r-times and s-times} \tag{2.1 b}$$

Both (2.1 a, b) are divergences of the corresponding Chern-Simons (C-S) forms. The C-P densities (2.1) are the essential ingeredients of the topological inequalities giving the higher dimensional instantons to be introduced below, their stability.

Let us denote the p-fold totally antisymmetrised product of the curvature 2 form F(2), namely the curvature 2p-form, by $F(2p) = F_{\mu_1, \ldots \mu_{2p}}$

Then its Hodge dual on d = 2 (p+q)

$$*(F(2p))\,(2q) = \frac{1}{(2p)!}\, \varepsilon_{\mu_1 \cdots \mu_{2q} \nu_1 \cdots \nu_{2q}}\, F_{\nu_1 \cdots \nu_{2q}} \tag{2.2}$$

is a 2q-form. When q=p, both *F and F are 2p-forms.

The Lagrangian densities, for which instantons exist, are defined[7] from the following inequalities

$$\mathrm{tr}\,[\kappa^{2(q-p)}\,F(2p) - {}^*(F(2q))\,(2p)]^2 \geq 0 \tag{2.3a}$$

$$[\kappa^{2(q-p)}\,\mathrm{tr}\,F(2p) - \mathrm{tr}\,{}^*(F2q))\,(2p)]^2 \geq 0\,,\ q\geq p, \tag{2.3b}$$

by developing them. Here κ is a constant with the dimension of a length. It follows that

$$L_{p+q} = \mathrm{tr}\,[\kappa^{4(q-p)}\,F(2p)^2 + F(2q)^2] \geq \mho_{p+q} \tag{2.4a}$$

$$L_{p,q} = \kappa^{4(q-p)}\;(\mathrm{tr}\,F(2p))^2 + (\mathrm{tr}\,F(2q))^2 \geq \mho_{p,q}, \tag{2.4b}$$

where the Lagrangian densities L_{p+q} and $L_{p,q}$ defined by (2.4 a, b), are bounded from below by $\mho p{+}q$ and $\mho p,q$ defined by (2.1a,b) respectively. The integrals of these C-P densities are converted to surface-integrals, which, provided that the gauge-connection is asymptotically *pure-gauge*, yield the (finite, integer) topological charges of the instantons. We restrict henceforth to models typified by (2.4a) only, since those pertaning to (2.4b) have not yet been studied in sufficient detail.

Before proceeding to study the solutions, we dispose of an alternative generalisation of YM in d = 2n dimensions, defined[8,9] by

$$L'_{2n} = \mathrm{Tr}\,(\Sigma^{\mu\nu} \otimes F_{\mu\nu})^n \tag{2.5}$$

where $\Sigma_{\mu\nu}$ are chiral SO(2n) generators, and the trace runs over both the SO(2n) and the gauge group. These systems (2.5), even when they are positive definite with n even, are not endowed with topological inequalities like (2.4), and hence their solutions are not be expected to be stable. This instability was explicitly verified[10] for the spherically symmetric solution in d = 6.

3 . Solutions to the selfdual equation in d = 4p

While non-selfdual solutions to the Euler-Lagrange equations of the systems (2.4a) with $p\neq q$ exist, the most interesting case is p=q, i.e. d = 4p. In these cases, the Lagrangians (2.4a) become conformally invariant[11], and the inequalities (2.4a) can be saturated by the selfdual equations

$$F(2p) = {}^*F(2p). \tag{2.5}$$

Solutions to (2.5) for arbitrary p, and gauge group $SO_{\pm}(4p)$, are known[12, 13, 14]. Subject to *spherical* symmetry, these solutions are

$$A_\mu^{(\pm)} = \frac{2}{(r^2 + a^2)}\;\Sigma_{\mu\nu}^{(\pm)}\,x_\nu \tag{2.6}$$

where $\Sigma_{\mu\nu}^{(\pm)}$ are the chiral spinor representations of $so_{\pm}(4p)$. The p=1 case is the BPST solution[1] Subject *to axial* symmetry, (2.5) reduce to one complex and one real p.d.e. The complex equation can be reduced to the Cauchy-Riemann equation and hence integrated, while the real equation is[14]

$$\Delta\psi = \frac{(2p-1)}{r}\;(1 - e^{-2\psi})\;-\frac{2(p-1)\,e^{-2\psi}}{(1 - e^{-2\psi})}\;(\nabla\psi)^2 \tag{2.7}$$

For p = 1, (2.7) can be converted to the Liouville equation[15], and solved explicitly. For $p\geq2$, solutions to (2.7) can be shown[16] to exist, but cannot be explicitly evaluated.

Subject to symmetry less stringent than *axial*, equations (2.5) are over determined[17].This is because the number of equations (2.5) is $4p!/2(2p!)^2$, which is larger than the number (4p-1) of independent gauge fields, except for the case of p = 1, for the usual YM.

Concerning spinor zero modes in the selfdual background (2.5), these can be easily checked for the *usual* Dirac equation in the special background (2.6). The general demonstration however is given[18] for a *new* Dirac equation whose symbol is $\Gamma_{\mu_1 \cdots \mu_{2p+1}} F_{\mu_1 \cdots \mu_{2p}} D_{\mu_{2p+1}}$ instead of $\Gamma_\mu D_\mu$.

Finally, there is the question of saturating (2.3a) when $p \neq q$. Since the selfdual equation here

$$\kappa^{2(p-q)} F(2q) = * (F(2p))(2q) \tag{2.8}$$

involves a dimansional constant, this cannot be done on $\mathbb{R}_{2(q+p)}$. It has been verified[19, 20] however that (2.8) can be solved on S^{2n}, and on $\mathbb{C}\mathbb{P}^n$ respectively, and for the special case[21] with $q = 1$, also on $\mathbb{H}\mathbb{P}^n$, indicating that (2.8) can be solved on symmetric coset spaces.

4. Higgs models descended from $\mathbb{R}_d \times S^{4p-d}$

To obtain models on $\mathbb{R}_d$, whose dynamics supports finite action topologically stable 'instanton' solutions, a topological inequality like (24) is needed. This in turn can be acheived by subjecting both sides of (2.4) to dimensional reduction. The residual Lagrangian on the left hand side then defines the Lagrangian of the Higgs model on $\mathbb{R}_d$. The crucial question then is, whether the C-P density on the right hand side reduces to a total divergence ? If so, then the residual action will be bounded from below by a surface integral, which, subject to the Higgs type asymptotic condition

$$|\Phi|^2 \xrightarrow[r \to \infty]{} \eta^2 \tag{3.1}$$

takes on a *finite* value- the topological charge of the 'instanton'. We know this is true from the results of refs.22 - namely, subjecting the integral of the C-P density (2.1a) on $\mathbb{R}_d \times (G/H)^{4p-d}$ to dimensional reduction, one obtains a *surface integral* on $\mathbb{R}_d$. We note that for even d, the residual C-P density is *gauge-variant*, while for odd d, it is *gauge-invariant*, this property is relevant to the study of anomalies.

For brevity, we illustrate explicitly one case only, the physically relevant $d = 4$ model derived from the p=2 SO(8) GYM system on $\mathbb{R}_4 \times S^4$. In this case, the residual C-P density reads

$$\begin{aligned}\Omega_\mu \approx \varepsilon_{\mu\nu\rho\sigma} \operatorname{tr} \gamma_5 [\eta^4 A_\nu (F_{\rho\sigma} - \tfrac{3}{3} A_\rho A_\sigma) - \tfrac{1}{2}\eta^2 \Phi \{F_{\mu\nu}, D_\rho \Phi\} \\ + \tfrac{1}{6} \Phi (\{S, F_{\rho\sigma}\} + [D_\rho \Phi, D_\sigma \Phi]) D_\nu \Phi],\end{aligned} \tag{3.2}$$

while they residual SU (2) x SU(2) xU(1) Lagrangian is

$$\begin{aligned}L_{res} = \operatorname{tr} [F_{\mu\nu\rho\sigma}{}^2 + \lambda_1 \{F_{\mu\nu}, D_\rho \Phi\}^2 - \lambda_2 (\{S, F_{\mu\nu}\} + [D_\mu \Phi, D_\nu \Phi])^2 \\ - \lambda_3 \{S, D_\mu \Phi\}^2 + \lambda_3 \{S, D_\mu \Phi\}^2 + \lambda_4 S^4]\end{aligned} \tag{3.3}$$

where $S = \mu^2 + \Phi^2$, and any three of the four $\lambda_i > 0$ can be considered independent 'coupling constants. When $\lambda_i = 1$ for all i, the action of (3.3) is absolutely minimised by the Bogomol'nyi equations, which however have only the trivial solution, corresponding to the fact that the original selfdual equations (2.5) from which they are derived, are overdetermined[17]. To have nontrivial 'instanton', the choice $\lambda_i = 1$ for all i is excluded. The details are given in ref 23. It has also been verified that in addition to the 'instanton', (2.3) also admits a Sphaleron[24] in the static limit.

This example captures the qualitative features of all d, except for d=2 and d=7=(4p -1). In the latter cases respectively[25,26], the action derived from the p=2 SO(8) GYM systems on $\mathbb{R}_2 \times S^6$ and $\mathbb{R}_7 \times S^1$, which are respectively Abelian and SO(7) Higgs models, *can* be absolutely minimised and the respective Bogomol'nyi equations solved[25,26,27]. Both these cases correspond to the axial field configuration[14] for which (2.5) is *not* overdetermined[17]. These features persist for all p.

References

1 . A.A. Belavin, A.M. Polyakov, A.S. Schwartz and Yu. S. Tyupkin, Phys. Lett. **B59** (1975)_85.
2 . A.M. Polyakov, Nucl. Phys. **B120** (1977) 429.
3 . G.' t Hooft, Nucl. Phys. **B79** (1974) 276 ; A.M. Polyakov, JETP Lett. **20** (1974) 194.
4 . A. Jaffe and C.H. Taubes, Vortices and Monopoles, Birkhäuser, Basel (1980).
5 . C.G. Callan, R.F. Dashen and D. Gross, Phys. Rev. **D17** (1978) 2717.
6 . J. Burzlaff and D.H. Tchrakian, Lett. Nuovo Cimento **40** (1984) 129.
7 . D.H. Tchrakian, J. Math. Phys **21** (1980) 166.
8 . C. Saclioglu, Nucl. Phys. **B277** (1986) 487.
9 . K. Fujii, Lett. Math. Phys. **12** (1986) 363.
10.Zh-Q.Ma and D.H. Tchrakian, Lett. Math. Phys. **19** (1990) 237.
11. D. O'Sé and D.H. Tchrakian, Lett. Math. Phys **13** (1987) 211.
12. D.H. Tchrakian, Phys. Lett. **B150** (1985) 360.
13. B. Grossman, T.W Kephart and J.D. Stasheff, Commun. Math. Phys. **96** 1984 431 ; **100** (1985) 311.
14. A.Chakrabarti, T.N. Sherry and D.H. Tchrakian, Phys. Lett. **B162** (1985) 340.
15. E. Witten, Phys. Rev. Lett. **38** (1977) 121.
16. J. Burzlaff, A. Chakrabarti and D.H. Tchrakian, Axially symmetric instantons in generalized Yang-Mills theory in 4p dimensions, to appear in J. Math. Phys.
17. D.H. Tchrakian and A. Chakrabarti, J. Math. Phys. **32** (1991) 2532.
18 . O. Lechtenfeld, W. Nahm and D.H. Tchrakian, Phys. Lett. **B162** (1985) 143.
19.G.M. O'Brien and D.H. Tchrakian, J. Math. Phys **29** (1988) 1212.
20. Zh - Q. Ma and D.H. Tchrakian, J. Math. Phys. **31** (1990) 1506.
21. F.A. Bais and P. Batenburg, Nucl. Phys. **B269** (1986) 363.
22. T.N. Sherry and D.H. Tchrakian, Phys. Lett. **B147** (1984) 129 ; D. O'Sé, T.N. Sherry and D.H. Tchrakian, J. Math. Phys.**27** (1986) 325 ; Z.H-Q. Ma, G.M. O'Brien and D.H. Tchrakian, Phys. Rev. **D33** (1986) 1177 ; Zh-Q.Ma and D.H. Tchrakian, Phys Rev. **D38** (1988) 3827
23. G.M. O'Brien and D.H. Tchrakian, Mod. Phys. Lett. **A4** (1989) 1389
24. G.M. O'Brien and D.H. Tchrakian, Phys. Lett. **B282** (1992) 111.
25. J; Burzlaff, A. Chakrabarti and D.H. Tchrakian, Generalised Abelian Higgs models with selfdual vortices, Ecole Polytechnique preprint - A179.0692.
26.D.O'Sé, J. Burzlaff and D.H. Tchrakian, Lett. Math Phys. **13** (1987) 121.
27. Y. Yang, Lett. Math. Phys. **19** (1990) 257 ; Lett. Math. Phys. **20** (1990) 285.

NEW SOLITON SOLUTIONS TO THE YANG-MILLS GAUGE FIELD EQUATIONS

C. Au* P. C. W. Fung+ F. T. To*+

ABSTRACT

Within the differential geometry regime, we have arrived at two types of new solutions to the Yang-Mills gauge field equations under two different ansatz. First, starting from a diagonal metric, we have obtained four sets of analytical solutions of the O(4) gauge field. These solutions demonstrate the soliton-like properties as time evolves. The distribution of energy density of the soliton disturbance is nondispersive while the soliton propagates, and the total energy is found to be always positive definite. Secondly, starting from a spherical symmetric metric, we have also derived a spherically symmetric static solution of O(4) gauge field. Our numerical analysis show that the motion of a test particle in this gauge field will be confined within a finite range in space. This confinement property is found to be caused by the presence of non-linear term in the gauge field equations implying that there is an inherent confinement property of the gauge field.

I. Method of Construction of the Gauge Potential Using the Differential Geometrical Approach.

In order to solve the gauge field equations under certain specified conditions and to analyse properties of this gauge field, we introduce an O(4) gauge potential $A_\mu^{\hat\alpha\hat\beta}$, in the manner described in [1], where μ = 1-4 are the usual spacetime indices and $\hat\alpha,\hat\beta$ = 1-4 are the Lorentz indices. Based on the O(4) gauge potential, we can obtain the O(4) gauge field strength tensor

$$F_{\mu\nu}^{\hat\alpha\hat\beta} = \partial_\mu A_\nu^{\hat\alpha\hat\beta} - \partial_\nu A_\mu^{\hat\alpha\hat\beta} + A_\mu^{\hat\alpha\hat\sigma} A_\nu^{\hat\sigma\hat\beta} - A_\nu^{\hat\alpha\hat\sigma} A_\mu^{\hat\sigma\hat\beta}. \tag{1.1}$$

On the other hand, in flat spacetime the field equation is

$$\partial_\nu F^{\hat\alpha\hat\beta\mu\nu} + A_\nu^{\hat\alpha\hat\sigma} F^{\hat\sigma\hat\beta\mu\nu} + A_\nu^{\hat\beta\hat\sigma} F^{\hat\alpha\hat\sigma\mu\nu} + \Gamma^\mu_{\sigma\nu} F^{\hat\alpha\hat\beta\sigma\nu} + \Gamma^\nu_{\sigma\nu} F^{\hat\alpha\hat\beta\mu\sigma} = 0. \tag{1.2}$$

In order to integrate the gauge field equation (1.2), one usually takes an ansatz to simplify it. For example, the 't Hooft ansatz leads to the instanton solution [6]. Thus, it is fruitful to find suitable ansatz, arriving at certain solutions as accomplished in the past [2-9].

In this investigation, we propose a new ansatz based on a logical idea in the framework of differential geometry. First, we construct in R^4 space a spin connection whose geometrical properties are equivalent to that of gauge potential in the O(4) group. We can therefore assign this spin connection is simply our O(4) gauge potential. The first concrete process along this line is to express the line element ds in R^4 space as

$$ds^2 = - g_{\mu\nu}\, dx^\mu\, dx^\nu. \tag{1.3}$$

where μ, ν = 1-4. Within R^4 space, we can introduce the Euclidean metric

$$ds^2 = \delta_{\hat\alpha\hat\beta}\, dy^{\hat\alpha}\, dy^{\hat\beta}. \tag{1.4}$$

in which $dy^{\hat\alpha}$ is a unit orthogonal basis, with $\hat\alpha,\hat\beta$ = 1-4.

As pointed out by Wilczek [10], we can understand certain crucial properties of the gauge field by the introduction of the Vierbein field $L_\mu^{\hat\alpha}$, which is an orthogonal transformation relating the unit orthogonal basis $dy^{\hat\alpha}$ and another basis dx^μ:

$$dy^{\hat\alpha} = L_\mu^{\hat\alpha}\, dx^\mu. \tag{1.5}$$

* Department of Applied Mathematics, Hong Kong Polytechic, Hong Kong.
+ Department of Physics, University of Hong Kong, Hong Kong.

In view of (1.3)-(1.5), the metric $g_{\mu\nu}$ and the Vierbein field are related by

$$g_{\mu\nu} = L_{\mu}^{\hat{\alpha}} L_{\nu}^{\hat{\beta}} \delta_{\hat{\alpha}\hat{\beta}} . \qquad (1.6)$$

From the metric $g_{\mu\nu}$, we can write down the Riemann connection in R^4:

$$\Gamma^{\alpha}_{\mu\nu} = \frac{1}{2} g^{\alpha\beta} (\partial_{\mu} g_{\beta\nu} + \partial_{\nu} g_{\mu\beta} - \partial_{\beta} g_{\mu\nu}). \qquad (1.7)$$

Carrying out an orthogonal frame transformation on the Riemann connection by the Vierbein field, we derive the explicit form of our spin connection:

$$C_{\mu}^{\hat{\alpha}\hat{\beta}} = \frac{1}{2} L^{\lambda\hat{\beta}} (L^{\hat{\alpha}}_{\mu,\lambda} - L^{\hat{\alpha}}_{\lambda,\mu}) + \frac{1}{2} L^{\lambda\hat{\alpha}} (L^{\hat{\beta}}_{\lambda,\mu} - L^{\hat{\beta}}_{\mu,\lambda}) + \\ + \frac{1}{2} L^{\sigma\hat{\alpha}} L^{\lambda\hat{\beta}} (L^{\hat{\varepsilon}}_{\sigma,\lambda} L^{\hat{\varepsilon}}_{\mu} - L^{\hat{\varepsilon}}_{\lambda,\sigma} L^{\hat{\varepsilon}}_{\mu}) . \qquad (1.8)$$

It can proved that [10,11] the spin connection $C_{\mu}^{\hat{\alpha}\hat{\beta}}$ is invariant under simultaneous rotations of the frames x^{μ} and $y^{\hat{\alpha}}$. On the other hand, the O(4) gauge potential also has symmetric synchronization property: namely, in physics systems, $A_{\mu}^{\hat{\alpha}\hat{\beta}}$ is invariant under simultaneous rotations of spacetime and isotopic space. We can therefore assign

$$A_{\mu}^{\hat{\alpha}\hat{\beta}} \equiv C_{\mu}^{\hat{\alpha}\hat{\beta}}. \qquad (1.9)$$

Substituting (1.8) and (1.9) into (1.1) and (1.2) we can obtain the field strength and the gauge field equations, and arrive at our solutions.

II. Exact Soliton Solutions to the O(4) Gauge Field Equations.

Since the O(4) Y-M gauge field equation is nonlinear, we might expect soliton solutions to exist. Hence we apply this differential geometrical approach to look for exact solutions with propagating properties.

We limit ourselves to the diagonalized form of the metric:

$$g_{\mu\nu} = \begin{pmatrix} B^2(u) & & & \\ & C^2(u) & & \\ & & D^2(u) & \\ & & & A^2(u) \end{pmatrix} \qquad (2.1)$$

where $u \equiv ct - z$. First, from (1.8) we can write down the form of the Vierbein field:

$$L_{\mu}^{\hat{\alpha}} = \begin{pmatrix} B(u) & & & \\ & C(u) & & \\ & & D(u) & \\ & & & A(u) \end{pmatrix} \qquad (2.2)$$

Substituting (2.2) into (1.8), (1.1) and (1.2), we arrive at the relevant gauge potential, gauge field strength and derive our set of gauge fieldequations. There are four unknowns but six equations, implying that the solutions might be over-determined. In such a situation we first look for restriction(s) in which special solutions can be obtained. In particular, we observe that if $A'/D - D'/A = 0$, the set of gauge field equations are greatly simplified. based on some boundary conditions, we arrive at the four sets special solutions. For example, one of these four sets is as follows.

$$\begin{cases} A(u) = -\sqrt{u^2 + \frac{1}{2}}/(u^2 + 1), & D(u) = \sqrt{2u^4 + 2u^2 + \frac{3}{2}}/(u^2 + 1), \\ B(u) = \mathrm{sech}(u)\tanh(u) , & C(u) = 1 . \end{cases}$$

$$\begin{cases} A_1^{\hat{1}\hat{3}} = -\left[2\,\mathrm{sech}^3(u) - \mathrm{sech}(u)\right](u^2 + 1) \Big/ \sqrt{2u^4 + 2u^2 + \frac{3}{2}}, \\ A_1^{\hat{1}\hat{4}} = i\left[2\,\mathrm{sech}^3(u) - \mathrm{sech}(u)\right](u^2 + 1) \Big/ \sqrt{u^4 + \frac{1}{2}}, \end{cases}$$

$$\left\{\begin{array}{l} A_4^{\hat{3}\hat{4}} = -2u\left(u^2-\frac{1}{2}\right) \Big/ \left[(u^2+1)\sqrt{\left(u^4+\frac{1}{2}\right)\left(2u^4+2u^2+\frac{3}{2}\right)}\right] = iA_3^{\hat{3}\hat{4}}. \end{array}\right.$$

$$\left\{\begin{array}{l} F_{13}^{\hat{1}\hat{3}} = \dfrac{\mathrm{sech}(u)}{\sqrt{2u^4+2u^2+\frac{3}{2}}}\left\{\tanh(u)\ (u^2+1)(6\,\mathrm{sech}^2(u)-1)+\right. \\ \left. +[2\,\mathrm{sech}^2(u)-1](2u^3-3u)\left[\dfrac{1}{u^4+\frac{1}{2}}+\dfrac{1}{2u^4+2u^2+\frac{3}{2}}\right]\right\} = -iF_{14}^{\hat{1}\hat{3}}. \\ F_{14}^{\hat{1}\hat{4}} = \dfrac{\mathrm{sech}(u)}{\sqrt{u^4+\frac{1}{2}}}\left\{\tanh(u)\ (u^2+1)[6\,\mathrm{sech}^2(u)-1]+\right. \\ \left. +(2\,\mathrm{sech}^2(u)-1)(2u^3-u)\left(\dfrac{1}{u^4+\frac{1}{2}}+\dfrac{1}{2u^4+2u^2+\frac{3}{2}}\right)\right\} = iF_{13}^{\hat{1}\hat{4}}. \end{array}\right.$$

In Fig. 1 and 2 we plot the non-zero $A_\mu^{\hat{\alpha}\hat{\beta}}$ - z and $F_{\mu\nu}^{\hat{\alpha}\hat{\beta}}$ - z components for this solutions at t = 0. As the velocity of propagation is c, these "composite-pluse shaped" solitons travel to the right. The energy density T_{44} - z relations for this solutions is shown in Fig. 3. We see that an energy pulse travels along the z direction with velocity c. Here the total energy is positive definite.

III. A New Class of Solution With Confinement Properties in the Spherically Symmetric O(4) Gauge Field.

Consider a system with spherical symmetry, so that the line element ds can be expressed as

$$ds^2 = -e^{2\Phi}\,d\tau^2 - e^{2\Lambda}\,dr^2 - r^2\,(d\theta^2 + \sin^2\theta\,d\phi^2). \tag{3.1}$$

where $\tau = it$, $c = 1$. Λ and Φ being functions of r only. From (1.6) we can write down the form of Vierbein field:

$$L_\mu^{\hat{\alpha}} = \begin{pmatrix} e^{\Lambda} & & & \\ & r & & \\ & & r\sin\theta & \\ & & & e^{\Phi} \end{pmatrix} \tag{3.2}$$

Substituting (3.2) into (1.8), (1.1) and (1.2), we arrive at the expressions for the gauge potential, gauge field strength and a system of gauge field equations in the spherically symmetric case

$$(e^{\Phi-\Lambda}\,\Phi')'' + 2r^{-1}(e^{\Phi-\Lambda}\,\Phi')' - 2(re^{\Lambda})^{-2}(e^{\Phi-\Lambda}\,\Phi') = 0. \tag{3.3}$$

$$(e^{-\Lambda}\,\Lambda')' - r^{-2}e^{-\Lambda}(1 - e^{-2\Lambda}) + e^{-\Lambda}(e^{\Phi-\Lambda}\,\Phi')^2 = 0. \tag{3.4}$$

A simple ansatz for the system (3.3) and (3.4) is given by $\Phi = 0$. We can simplify this system to one equation

$$(e^{-\Lambda}\,\Lambda')' - r^{-2}e^{-\Lambda}\,(1 - e^{-2\Lambda}) = 0. \tag{3.5}$$

Note that under the condition $\Phi = 0$, the non-zero components of the gauge potential and field strength become simply

$$F^{\hat{r}\hat{\theta}r\theta} = f'/r^2, \quad F^{\hat{r}\hat{\phi}r\phi} = f'/(r^2\sin\theta), \quad F^{\hat{\theta}\hat{\phi}\theta\phi} = (1 - f^2)/(r^4\sin\theta). \tag{3.6}$$

where $f \equiv e^{-\Lambda(r)}$ and $f' \equiv df/dr$.

So far, the analytical solution to the field equation (3.5) cannot be obtained. However, crucial properties can be obtained by numerical solutions, and substituting them into (3.6) we can obtain the field strength components of $F^{\hat{\alpha}\hat{\beta}\mu\nu}$ in terms of r. For example, in Fig.4 we show the variation of $F^{\hat{r}\hat{\theta}r\theta}$ with respect to r. It is interesting to remark that clearly if a charged particle is influenced by the field of $F^{\hat{r}\hat{\theta}r\theta}$, the particle will be confined to a limited region in space.

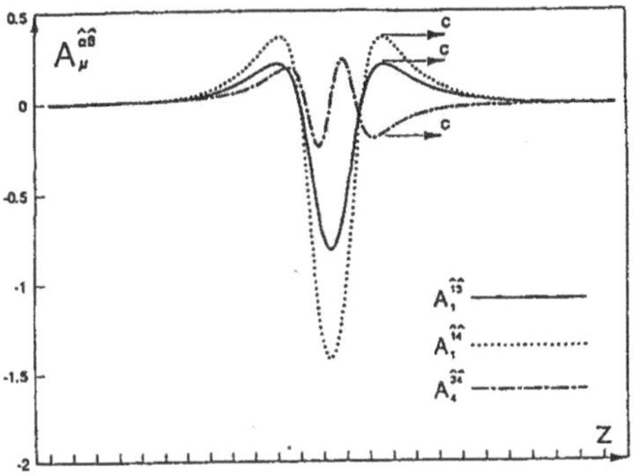

Fig.1

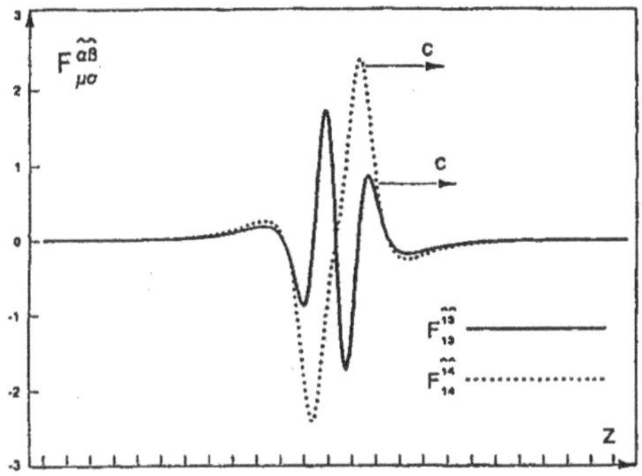

Fig.2

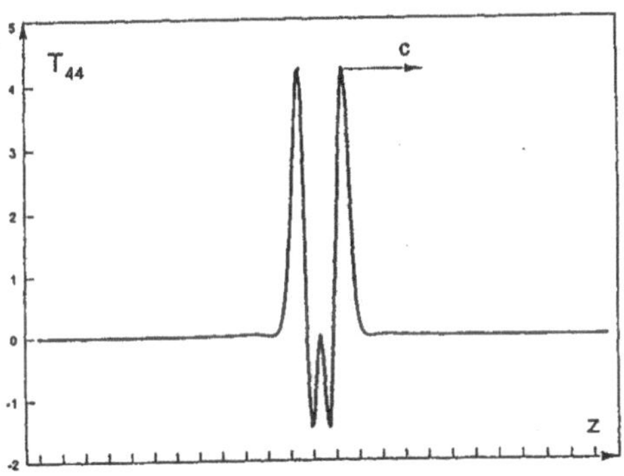

Fig.3

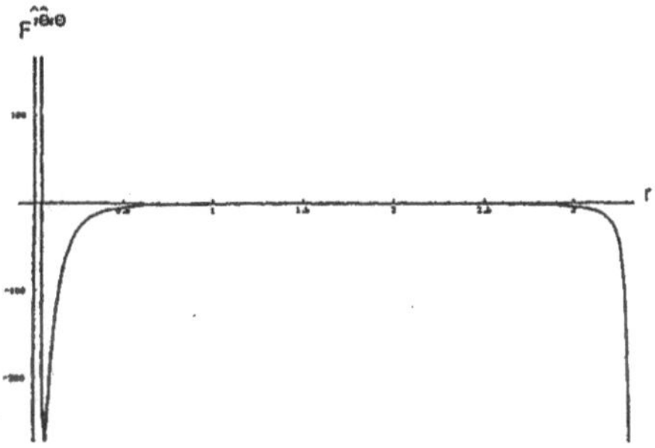

Fig.4

References

[1] Belavin A A, Polyakov A M, Schwartz A S and Tyupkin Y S 1975 Phys. Lett. 59B 85

[2] Wu T T and Yang C N 1968 Properties of Matter Under Unusual Conditions ed H Mark and S Fernbach (New York: Interscience)

[3] Julia B and Zee A 1975 Phys. Rev. D11 2227

[4] Hsu J P and Mac E 1977 J. Math. Phys. 18 100

[5] Rosen G 1977 J. Math.Phys. 13 595

[6] 't Hooft G 1976 Phys. Lett. 37 8

[7] Corrigan E and Fairlie D 1977 Phys. Lett. 67B 69

[8] 't Hooft G 1974 Nucl. Phys. B79 276

[9] Polyakov A M 1975 Sov. Phys. JETP 41 988

[10] Wilczek F 1977 Proc. Conf. on Quark Confinement and Field Theory (Rochester, NY)

[11] Choquet-Bruhat Y, DeWitt-Morette C and Dillard-Bleick M 1981 Analysis, Manifolds and Physics (Amsterdam: North-Holland)

Non-minimal Solutions to the Yang-Mills Equations

Hong-Yu Wang

Department of Mathematics
National University of Singapore
Singapore 0511

Abstract

By using Taubes' approach to self-dual connections [5], we establish the Morse theory for existence of non-minimal solutions to the Yang-Mills equations.

§1. Main results. The study of non-self-dual Yang-Mills connections [1] has developed rather slowly compared to the progress made in the understanding of self-dual connection [5]. Actually, the main difficulty in solving Yang-Mills equations lies in the failure of Palais-Smale conditions for the Yang-Mills functional [6]. Recently, L. & R. Sibner and K. Uhlenbeck [4], T. Parker [2] and L. Sadun and J. Segert [3] have found non-minimal solutions via reduction of dimensionality. In this paper, we present a Morse theory for existence of non-minimal solutions to the Yang-Mills equations. Our main results [7, 8, 9] are following:

Theorem 1. According to Sadun and Segert's construction of quadrupole bundles and $SO(3)$-equivariant non-minimal Yang-Mills connections, for any $C_2 \neq \pm 1$ there exists an infinite number of non-minimal $SU(2)$ solutions to the Yang-Mills equations on round 4-sphere S^4.

Theorem 2. Suppose that the base manifold is $S^1 \times S^3$ with standard product metric, there exists an infinite number of non-minimal $SU(2)$ Yang-Mills connections over $S^1 \times S^3$ with degree zero whose action $YM > 16\pi^2 k$, where k is a sufficiently large positive integer.

Theorem 3. Let M be a simply connected, oriented, compact 4-manifold with indefinite intersection matrix. Choose a generic Riemannian metric on TM. For any $\alpha \in H^2(M; Z)$, there exists an infinite number of irreducible $SU(2)$ non-minimal solutions to the Yang-Mills equations on M with degree $C_2 = -C_1^2$.

Remark here that none of the solutions found above is symmetric with respect to the Lie group actions on base manifolds. In fact, our solutions have a uniform background curvature, with concentrations near points, spaced evenly along a geodesic. Near half of these points, the solution looks self-dual, and near the other half, it looks anti-self-dual.

One more remark that the result of Sadun and Segert [3] is that on each quadrupole bundle $P_{(n_+,n_-)} \to S^4$ with $n_\pm \neq 1$, there exists a smooth non-self-dual Yang-Mills connection. In Theorem 1., we use their connections as background connections.

§2. Basic construction. The analytic techniques in this paper are entirely due to Taubes [5, 6], who has developed an extensive variational theory for existence of solutions to the Yang-Mills equations. His analytic argument and new implicit function theorem carry over exactly to our situation, and we depend on his methods and papers. Our approach as follows (Actually, this method was a joint work of Taubes and author.):

Let $P_0 \to M$ be a background principal $SU(2)$ bundle over M, let A_0 be a smooth background connection on P_0 (One requires that A_0 is an isolated solution to the Yang-Mills equations on M, this means that the Hessian of Yang-Mills functional at A_0 has no null space). Now, let C be a simple closed geodesic on M (Without loss of generality, assume that the length of C is equal to 2π exactly.). Fix a tubular neighborhood, V_0, of C. Let $\{s, y^\alpha\}_{\alpha=1}^3$ be a coordinate system on V_0 with $s: V_0 \to [0, 2\pi]$ restricting to C as arclength. Require $y^\alpha|_C = 0$ and $\{\frac{\partial}{\partial s}, \frac{\partial}{\partial y^\alpha}\}_{\alpha=1}^3$ orthonormal on C.

Choose $2k$ points $\{q_1, \cdots, q_k\}$ in V_0 such that

$$\frac{5\pi}{4k} > s(q_{i+1}) - s(q_i) > \frac{3\pi}{4k}.$$

For simplicity, set $d = \frac{\pi}{k}$. Over each q_i pick up $f_i \in F_M|_{q_i}$ (F_M is the frame bundle of TM). We have Gaussian coordinate system $\phi_{f_i}: U(q_i) \to B_\rho \subset \mathcal{R}^4$. Here, ρ be the injectivity radius of M, and require $d < \rho$. Choose $r \in (d, \rho)$ and for each i, choose $\lambda_i \in (0, \frac{r}{2})$. For each i, choose also $g_i \in SU(2)$. Together, the data C and $(\lambda_i, f_i, g_i)_{i=1}^{2k}$ gives our parameters for gluing. Indeed, by gluing operation one is able to graft instantons and auti-instantons over S^4 onto background connection, A_0, over M at distinct points $\{q_i\}_{i=1}^{2k}$, and so construct approximate solutions to the Yang-Mills equations.

However, the map from parameter space into "orbit space" of connections is not an injectivity map as it is invariant under the certain symmetry group. This invariance is due to the special structure group $SO(4)$ of the frame bundle F_M. It is well known that $SO(4) \cong SU(2) \times_{\{\pm\}} SU(2)$. Hence $AdP_0 \cong \bigwedge_\pm^2 T^*M$. Thus parameter space may be reduced. Let $\lambda_i = s_i d^2$. The parameters $y = \{(s_i, q_i, g_i)_{i=1}^{2k}\}$ are contained in

$$\prod_{i=1}^{2k} \{F_M^i \times_{SU(2)} SU(2)\}/\Gamma_{A_0} \times \Sigma_k \times \Sigma_k,$$

where F_M^i are frame bundles of $\bigwedge_\pm^2 T^*M$ whose signs, $\pm$, depend on the parity of i, Γ_{A_0} is the centralizer of the gauge group $Aut(A_0)$, and Σ_k is symmetry group on k-letters.

Hence we construct a parameter space for approximate solutions of Yang-Mills equations. Our strategy is to use the perturbation method. The obstruction to solving Yang-Mills equations is small eigenvalues of the Hessian of the Yang-Mills functional. Since the background connection, A_0, is an isolated solution, the parameters account for all small eigenvalues. Therefore we can solve Yang-Mills equations in the direction of othogonal complement of small eigenspace to obtain new approximate solutions. Then consider Taylor's expansion of Yang-Mills functional on these new approximate solutions in parameters $\{s_i\}_{i=1}^{2k}$. The first order term of the functional smoothly depends on parameters $\{(s_i, q_i, g_i)_{i=1}^{2k}\}$. Finally, by moving the positions of the parameters $\{(s_i, q_i, g_i)_{i=1}^{2k}\}$, we can find the unstable solutions of Yang-Mills equations:

Theorem 4. Suppose that M is a compact, oriented, Riemannian manifold. Let A_0 be an isolated non-minimal solution to the Yang-Mills equations with structure group $SU(2)$ on M, the norm of the curvature, F_{A_0}, a non-zero constant along a simple closed geodesic on M. By using global setting introduced above, one can construct an approximate solution space. The parameters $y = \{(s_i, q_i, g_i)_{i=1}^{2k}\}$ are contained in

$$\prod_{i=1}^{2k}\{F_m^i \times_{SU(2)} SU(2)\}/\Gamma_{A_0} \times \Sigma_k \times \Sigma_k.$$

There exists a positive integer $K_0 > 0$ such that for any positive even integer $k > K_0$, then exists an irreducible, non-minimal Yang-Mills $SU(2)$- connection with same degree as background connection A_0. In fact, this new non-minimal solution is a small perturbation of the approximate solution $A(\bar{y})$, here $\bar{y} = \{(\bar{s}_i, \bar{q}_i, \bar{g}_i)_{i=1}^{2k}\}$.

It is worth remarking here that our construction was inspired by Wente's [10] solution of the Hopf conjecture.

§3. Concluding comments. We now give remarks about gluing operation and expanding the Yang-Mills functional in parameters. It seems likely that there is physics intuition behind the expansion of the Yang-Mills functional. In physics, the connection is supposed to describe the force field due to certain pseudo-particles. For a physicist, the Yang-Mills functional measures the energy that is contained in the force field. The connect-sum construction (the gluing operation) can be interpreted in a heuristic sense as the addition of a specific, extra pseudo-particle to a system of pseudo-particles.

If the new pseudo-particle can be added to make the net force attractive (or repulsive), the normalized energy, the first order term of the functional, will decrease (in crease). Hence, the first order term of functional is viewed as the interaction between instantons and anti-instantons.

It is very interesting that how to find non-minimal solutions to the Yang-Mills equations for general cases. We believed that for general case, there are no non-minimal solutions of Yang-Mills equations.

Acknowledgments. Author is grateful to Nankai Institute of Mathematics and Professor M. L. Ge for their generosity, kindness and hospitality during this Conference.

References

[1] Atiyah, M. F., Jones, J. D. S., *Comm. Math. Phys. 61(1978), 97-118.*

[2] Parker, T., *Invent. Math. 107(1992), 397-420.*

[3] Sadun, L., Segert, J., *Preprint (1991).*

[4] Sibner, L., Sibner, R. and Uhlenbeck, K., *Proc. Natl. Acad. Sci. USA 86(1989), 8610-8613.*

[5] Taubes, C. H., *J. Diff. Geom. 19(1984), 517-560.*

[6] Taubes, C. H., *Invent. Math. 94(1988), 347-402.*

[7] Wang, H., *J. Diff. Geom. 34(1991), 701-767.*

[8] Wang, H., *Acta. Math. Sinica, New Series, 8(1992), 60-77.*

[9] Wang, H., *Preprint (1992).*

[10] Wente, H., *Pacific J. Math. 121 (1986), 193-243.*

Poster Sessions

SOME BOUNDS OF THE TOTAL CURVATURES OF IMMERSED MANIFOLDS[1]

Hengwu Cai

(*Department of Mathematics, Xiamen University P. R. C*)

Abstract In this paper, by using Morse theory and some wonderful methods, due to S. S. Chern, some bounds of the total absolute curvatures and the total mean curvatures of immersed submanifolds are considered.

Key words Total absolute curvature, Total Mean curvature, Generalized Gauss map, Tight immersion.

Main results

Theorem 1 Suppose M be a compact, non-negative curvature and orientable C^∞—manifold of dimension n. Let $X: M \to E^{n+N}$ be a tight immersion into the Euclidean space E^{n+N} with dimension $n+N$ $(N \geqslant 1)$, then there exists a constant $C=C(n)$, such that the total absolute curvature T of $x(M)$ satisfies $2 \leqslant T \leqslant C(n)$

Theorem 2 Let $X: M \to E^{n+N}$ be tight immersion into the Euclidean space given as theorem 1, and the sectional curvature $K \geqslant -n^2$, $n \geqslant 0$, then there exists a constant C, such that

$$2 \leqslant T \leqslant C^{1+nD}$$

where T is the total absolute curvature of x(M) and D is the diameter of M.

Theorem 3 Suppose M^2 be a compact, orientable smooth surface, and let $X: M^2 \to E^{N+2}$ be a c^∞—immersion into the Euclidean space E^{N+2} of dimension $N+2(N \geqslant 1)$. If the total absolute curvature $T<4$, then the genus of M^2 is zero.

Theorem 4 Let $X: M^2 \to E^{N+2}$ be the immersion given as theorem 3, if the genus of $X(M^2)$ is $g \geqslant k (k=0,1,2,\cdots,n)$ then the total absolute curvature of $X(M^2)$ satisfies $T \geqslant 2(1+k)(k=0,1,2,\cdots,n)$

Theorem 5 Let $X: M^n \to E^{n+N}$ be a compact orientable C^∞—immersion into the Euclidean space E^{n+N} of dimension $n+N(N \geqslant 1)$. If the principal

---

[1] This work was partially supported by the national Science Foundation and F. J. S. F. and XMUSF.

curvatures $k_i \geqslant 0 (i=1,2,\cdots,n)$ for any point $(p,v) \in B(p,v)$, then we have the following results;

(1) $L \geqslant 2$;

(2) If $L < 3$, then M^n is homeomorphic to a sphere of n—dimension.

(3) If $L = 2$, ,then M^n belongs to a linear subvariety E^{n+1} of dimension $n+1$, and is imbedded as a convex hypersphere in E^{n+1}. The converse of this is also true.

Reference

1. S. S. chern,R. K. Lashof"On the total curvature of immersed manifolds I". Amer. J. Math. 79 (1957). 306—313; II. Michigan Math. J. ,5(1958),5—12.
2. T. J. Willmore "Tight immersions and total absolute curvatures". Bull. London Math. Soc, 3 (1971), 129—151.
3. B. Y. Chen. "Mean curvature vector of a submanifold." Proceedings of symposia in Pure Math Vol. 27. 1975, 119—122.
4. N. H. Kuiper. "Immersions with minimal total absolute curvature". In Coll. de Geometrie Diff., Centre Belge de Recherches Math., Bruxelles,(1958) 75-88
5. N. H. Kuiper. "Tight Embeddings". The Chern symposium 1979 springer-verlag, Berlin, Heidelberg New York, 97-145
6. N. H. Kuiper "Minimal total absolute curvature for immersions". Invent. Math. 10, 209-238 (1970)
7. M. Gromov. Comment Math. Helv. 56(1981)179-195
8. T. E. Cecil & P. J. Ryan. "Tight and taut immersions of manifolds" Pitman Publishing INC 1985
9. J. Milnor"Morse theory" (Princeton Univ. Press, 1963).
10. Hengwu Cai"On the total curvature of immersed manifolds" Acta Sci. Natur. Unit. Amoien, Vol. 26(1987). No. 4 509—512.
11. Hengwu Cai "Immersed submanifolds and their tubular hypersurfaces" Acta. Sci. Natur, Univ. Amoien Vol. 27(1988).
12. Hengwu Cai " The splitting theorem for isoparametric submanifolds and their tubular hypersurfaces". Acta. Sci. Natur. Univ. Amoien Vol. 30(1991). No. 2 143-146.

A STATISTICAL MODEL FOR THERMODYNAMICS

SZE KUI NG
Department of Mathematics, Hong Kong Baptist College, 224 Waterloo Road, Kowloon, Hong Kong

ABSTRACT

In this paper a mathematical structure is set up which is considered as the basis of physics. From this mathematical structure a statistical model is set up which is considered as the micro basis of thermodynamics. A feature of this model is that the second law of thermodynamics can be derived from this model. Furthermore in this model the paradox between the second law of thermodynamics and the evolution phenomena of the world can be resolved. In this model a mathematical structure for the universe is set up and an explanation for the origin of matter and time is given.

1. Probability

Let us first consider the mathematical structure of probability and a difficulty of probability as an initial motivation of a mathematical structure of physics to be presented in this paper.

The basis of probability, proposed by Kolmogorov, is a mathematical structure $(\Omega,\mathfrak{F},P)$ which satisfies the following conditions as axioms of probability:

i) Ω is a set which is called the sample space.
ii) $\mathfrak{F}$ is a σ-algebra of subsets of Ω.
iii) P is a probability measure on $\mathfrak{F}$.

We call the triple $(\Omega,\mathfrak{F},P)$ as a probability space. Elements of $\mathfrak{F}$ are called events and $\mathfrak{F}$ is as a source of information.

The above structure of probability is mathematically well defined as a measure space. However, it is difficult to give a precise physical meaning to the number P(A). We may as usual call P(A) the probability that the event A occurs; however, it is just the meaning of probability that cannot be understood easily. This is the well known difficulty of the interpretation of probability.

We shall overcome the above difficulty by extending the above structure of probability. On the other hand we shall see that the most important achievement of this extension is that it gives us a mathematical structure of physics.

2. Physics

Considering the structure of probability, let us propose that a basic mathematical structure of physics is a triple $(\Omega,\mathfrak{F},M)$ which satisfies the following conditions as axioms of physics:

i) Ω is a set which is called a sample space.
ii) $\mathfrak{F}$ is a σ-algebra of events A in Ω.
iii) M is a complex measure on $\mathfrak{F}$.

In addition, we require that M satisfies the following condition

iv) M depends on two positive parameters h and β such that

$$M(h,\beta)(A) \to P(\beta)(A) \text{ as } h \to 0 \text{ for each A in } \mathfrak{F}$$

where $P(\beta)$ is a probability measure for each β and h denotes the Planck constant divided by 2π (Here for simple notation we use the symbol h instead of the usual symbol $h/2\pi$).

In the above structure the complex measure M is an extension of the probability measure P of the structure of probability. In this structure we can give M(A) a precise physical meaning as the wave amplitude of the event A which can physically be observed. $P(\beta)(A)$ then can have a physical meaning as the limit of the physical quantities $M(h,\beta)(A)$ as h tends to zero.

The parameters h and β will appear in the above structure in the form $-ih-\beta$. More specifically in the above structure we require that functional integrations with the measure M with the constant $-ih-\beta$ are of the same form of Feynman type path integrals with the constant -ih. i.e. we replace the constant -ih in Feynman integal with the constant $-ih-\beta$ to get integration with M. Since the parameter β is postive we have that integration with M is well defined while Feynman path integrals are ill defined.

We shall see that the parameter β will give the second and the third law of thermodynamics.

3. Thermodynamics

Let us call a mathematical structure $(\Omega,\mathfrak{F},M)$ as a physical system. We want to find out the thermodynamics of a physical system. In particular we want to clarify the meaning of entropy.

When a physical system consists of a very large amount of identical particles we may find ways to define the temperature of this system. In this paper because of limited pages we shall not specify these ways and they will appear elsewhere. Roughly these ways are analogous to the canonical ensemble of statistical mechanics.

With these ways we can find various distributions for the system such as the Boltzmann distribution, the Bose-Einstein distribution, the Fermi-Dirac distribution, e.t.c. depending on the property of the system. From these distributions we can define a function which is usually called entropy of the system. Here is a main point we want to clarify. In this paper let us call this function the information entropy of the system and denote it by S_1 as there will have another entropy.

Let us consider the basic structure $(\Omega,\mathfrak{F},M)$ where the measure M with the parameter $-ih-\beta$ is an extension of the Feynman type functional integration with the parameter ih. Analogous to that Schrodinger equation can be derived from Feynman integration we can derived from integration with M an equation which is a mixed type of Schrodinger and diffusion equation in the sense that the constants ih in the Schrodinger equation is replaced by $ih-\beta$ in this equation. Similar to the role of Schrodinger equation in quantum mechanics this equation will be the equation of states of the physical system. From this equation we can find the state of the system and then compute the energy of the system at this state. Let us denote this energy by E_1. Then we let E_2 be the energy of this state obtained by using the corresponding Schrodinger equation (with the constant ih instead of the constant $ih-\beta$). Then the dissipated energy D of the system at this state is defined to be $E_2 - E_1$ which is positive because of the parameter β.

Now let the absolute temperature T be determined from the above. Then we define an entropy S_2 by

$$E_1 = E_2 - TS_2 \quad (\text{or } D = TS_2)$$

where E_2 is regarded as total energy and the role of E_1 here is like free energy.

Then we define the total entropy S of the system at this state by

$$S = S_1 + S_2$$

The actual free energy F of the system at this state is given by

$$F = E_1 - TS_1$$

From this equation we obtain the first law of thermodynamics.

Now we have the dissipated energy D which is dissipated from the system into its external. The dissipated energy D can not be transformed into work by the system and this is the second law of thermodynamics.

From the equation defining S_2 we see that S_2 increases when E_2 is a constant and D increases and T is a constant or decreases. This may correspond to the usual statement that the entropy of an isolated system increases.

On the other hand the information entropy S_1 provides a measurement of information of the system and the heat form this entropy can be transformed into work by the system. The distinction of S_1 and S_2 clearly solves the paradox between the increasing of entropy which means more disordering and the evolution of the world which means more ordering.

We show elsewhere that the parameter β tends to infinite will imply that the absolute temperature T tends to 0 and that the parameter β never equals to infinite will imply that T never equals to 0. This is the third law of thermodynamics.

4. The Universe

Let us consider again the mathematical structure $(\Omega, \mathfrak{F}, M)$ which as a physical system will have all physical laws hold.

A physical system will dissipate energy due to the parameter β. Energy dissipated from a physical system is dissipated into the external of this system. Thus we see that due to the parameter β a physical system had to have external.

If a physical system is contained in a larger physical system, then the external of the smaller system restricted to the larger system is a part of the larger system and the larger system has external and will dissipate energy into its external.

Now let us suppose that a physical system plus its external can again be a physical system. Then this new system will again dissipate energy by the definition of physical system and energy is dissipated into its external and this is a contradiction since this new system has no external by the way it is formed. Thus a physical system plus its external can not be a physical system.

Now the universe has no external and can be defined as the union of a physical system and its external. From this we see that the universe is not a physical system. Thus physical laws may not hold for the whole universe. In particular, the second law of thermodynamics does not hold for the whole universe. The universe thus does not exist as a physical system and it exists only as a nonphysical object.

For a measure space it is usually not difficult to construct the sample space Ω and the σ-algebra $\mathfrak{F}$. The hard part is the construction of the measure M. Then we may consider the universe as a structure $(\Omega, \mathfrak{F})$ where we can not define a measure M having the properties of a physical system. Thus for the whole universe we can not define the wave amplitude of matter M(A) and thus we do not have the notion of matter for the whole universe. It is only when we consider a physical system $(\Omega, \mathfrak{F}', M)$ which is a part of the universe in the sense that $\mathfrak{F}'$ is a sub σ-algebra of $\mathfrak{F}$ which is for the whole universe that the notion of matter and time comes out.

ON THE CLASSICAL REALITY OF FOUR ELECTROMAGNETIC POTENTIALS

KIYOHISA TOKUNAGA

229, Shimonofu, Shingû-chô, Kasuya-gun, Fukuoka-ken 811-01, JAPAN

The rigorous application of the principle of least action and principal function of

$$d^4S=\left[G^{\mu\nu}dA_\mu\frac{1}{3!}\varepsilon_{\nu\beta\gamma\delta}-\left(\frac{-1}{2^4}G^{\lambda\rho}F_{\alpha\beta}\varepsilon_{\lambda\rho\gamma\delta}+\int\frac{1}{3!}\varepsilon_{\rho\beta\gamma\delta}J^\rho dA_\alpha\right)dx^\alpha\right]dx'^\beta dx''^\gamma dx'''^\delta \quad (1)$$

,

where $G^{\mu\nu}\equiv(-\mathbf{D},\ \mathbf{H})$, $F_{\mu\nu}\equiv(\mathbf{E},\ \mathbf{B})$, $A_\mu\equiv(\phi,\ -\mathbf{A})$ and $J^\mu\equiv(\rho,\ \mathbf{J})$ in MKSA unit system, simultaneously provide the two sets of Maxwell's equations. Regarding that four electromagnetic potentials describe charged matter, the conservation law of energy-momentum density tensor for electromagnetism exclusively involves the wave property as

$$\frac{\partial}{\partial x^\nu}\left(-G^{\mu\nu}F_{\mu\lambda}+\delta^\nu_\lambda\frac{1}{4}G^{\mu\rho}F_{\mu\rho}-\int J^\nu\, dA_\lambda+\delta^\nu_\lambda\int J^\mu\, dA_\mu\right)=0 \quad (2)$$

.

www.ingramcontent.com/pod-product-compliance
Lightning Source LLC
LaVergne TN
LVHW070906300926
849162LV00010B/1839
* 9 7 8 9 8 1 0 2 1 2 1 0 0 *